AF379291

Antioxidant Activity of Polyphenolic Plant Extracts

Antioxidant Activity of Polyphenolic Plant Extracts

Editor

Dimitrios Stagos

MDPI • Basel • Beijing • Wuhan • Barcelona • Belgrade • Manchester • Tokyo • Cluj • Tianjin

Editor
Dimitrios Stagos
University of Thessaly
Greece

Editorial Office
MDPI
St. Alban-Anlage 66
4052 Basel, Switzerland

This is a reprint of articles from the Special Issue published online in the open access journal *Antioxidants* (ISSN 2076-3921) (available at: https://www.mdpi.com/journal/antioxidants/special_issues/Polyphenolic_Plant_Extracts).

For citation purposes, cite each article independently as indicated on the article page online and as indicated below:

LastName, A.A.; LastName, B.B.; LastName, C.C. Article Title. *Journal Name* **Year**, *Volume Number*, Page Range.

ISBN 978-3-0365-0288-5 (Hbk)
ISBN 978-3-0365-0289-2 (PDF)

© 2021 by the authors. Articles in this book are Open Access and distributed under the Creative Commons Attribution (CC BY) license, which allows users to download, copy and build upon published articles, as long as the author and publisher are properly credited, which ensures maximum dissemination and a wider impact of our publications.
The book as a whole is distributed by MDPI under the terms and conditions of the Creative Commons license CC BY-NC-ND.

Contents

About the Editor

Dimitrios Stagos is an Assistant Professor (tenure) of "Animal Physiology—Toxicology" at the Department of Biochemistry and Biotechnology of the University of Thessaly, Greece. He obtained a B.Sc. in "Biology" from the Aristotle University, Greece, M.Sc. in "Human Genetics" from the University of Leeds, U.K. and Ph.D. in "Biochemistry and Biotechnology" from the University of Thessaly, Greece. During his work as a researcher in the School of Medicine at Penn State University, U.S.A., he discovered that plant polyphenols protected the SP-A protein of the human lung surfactant from environmental pollutant-induced oxidative stress. He also worked as a postdoctoral researcher at the School of Pharmacy at the University of Colorado, Denver, U.S.A., where he purified—for the first time—the human xenobiotic ALDH1B1 enzyme and demonstrated its role in ethanol metabolism, in addition to identifying a new transcript variant of the ALDH4A1 gene (submitted to the GENBANK database). He was appointed as a faculty member of the University of Thessaly in 2010. In 2017, he was then awarded the Certification of the "European Registered Toxicologist" by EUROTOX. Since 2002, his main research interests have been the antioxidant, chemopreventive and anticancer activity of plant polyphenols and other natural bioactive natural compounds as well as the adaptive physiological mechanisms of human organisms to oxidative stress. He has co-authored 1 textbook chapter, 74 articles in peer-reviewed journals and more than 70 abstracts in practicals of conferences. His publications have been cited 2216 times ("h-index": 27; Scopus). He has supervised more than 50 postgraduate theses and undergraduate dissertations. He is the owner of a patent regarding biofunctional animal feed. He has participated in several research grants as co-ordinator or partner. In 2019, he received a EUR 400,000 research grant from the Greek Secretariat of R&T for a 3-year investigation into the antioxidant and anticancer properties of bioactive compounds from algae species of the Aegean Sea. He is a member of the Editorial Boards of the *Antioxidants* and *Toxicology Reports* journals. He is a member of the Greek Association of Biosciences, the Greek Society of Toxicology, the Greek Geotechnical Chamber and the European Society for Free Radical Research.

Preface to "Antioxidant Activity of Polyphenolic Plant Extracts"

Over the last several years, plant polyphenols have received a great deal of research interest because of their potential beneficial effects on human health. Within the polyphenols' interesting biological properties, their antioxidant activity is considered as the most important. This book comprises twenty five peer-reviewed papers, including twenty four original research papers and one literature review. The scientific papers present studies on the isolation of individual polyphenols and/or polyphenolic extracts from different plants, evaluation of their antioxidant activity, prevention of oxidative stress-induced diseases and use as food additives. Specifically, some of the included studies provide evidence that polyphenols may be used for the prevention and treatment of diseases such as diabetes mellitus, Alzheimers' disease, cardiovascular and intestinal diseases. Importantly, in several of the studies "green extraction methods" for the isolation of polyphenols have been developed using modern technologies (e.g., ultrasound and microwave assisted extraction and far infrared irradiation), where few or no organic solvents were used, in order to minimize environmental and health impacts. Some of the plants used for the isolation of the polyphenols in these studies were *Prunus dulcis* (almond), *Salvia africana-lutea* (beach salvia), *Paeonia officinalis* (common peony), *Rubus idaeus* L. (raspberry), *Cornus stolonifera* Michx. (Red-osier dogwood), *Cornus mas* L. (Cornelian cherry), *Morus alba* and *Morus nigra* (mulberry), *Lavandula stoechas* (French lavender), *Opuntia ficus-indica* (prickly pear), *Dialium indum* (tamarind-plum), *Averrhoa bilimbi* (cucumber tree), *Rosmarinus officinalis* L. (rosemary), *Robiniae pseudoacaciae* (black locust), *Myrtus communis* (common myrtle), *Camellia sinensis* (tea), *Bixa orellana* (annatto) and *Angelica gigas* (Korean angelica). Special interest was also shown for polyphenols derived from olive oil or by-products of the olive oil production process.

As the Guest Editor, I would like to acknowledge the authors of all articles for their valuable contributions. I would also like to thank the publishing team of the *Antioxidants* journal for their professional help in the preparation of this book.

Dimitrios Stagos
Editor

 *antioxidants*

Editorial

Antioxidant Activity of Polyphenolic Plant Extracts

Dimitrios Stagos

Department of Biochemistry and Biotechnology, University of Thessaly, 41500 Biopolis, Greece;
stagkos@med.uth.gr

Received: 18 December 2019; Accepted: 19 December 2019; Published: 24 December 2019

Plant polyphenols are secondary metabolites characterized by one or more hydroxyl groups binding to one or more aromatic rings [1]. Several thousand polyphenolic molecules have been identified in higher plants, including edible ones. Plant polyphenols are divided into two major groups, flavonoids and non-flavonoids. Flavonoids can be divided into flavanols, flavonols, anthocyanidins, flavones, flavanones, and chalcones. Non-flavonoids include stilbene, phenolic acids, saponin, and tannins [1]. Among the important biological properties exhibited by plant polyphenols, their antioxidant activity has raised a great interest [1]. A number of studies have shown that plant polyphenols can be used as antioxidants against different oxidative stress-induced diseases [2,3].

This special issue consists of 25 articles related to isolation of polyphenols and/or polyphenolic extracts from different plants, assessment of their antioxidant activity, their prevention from oxidative stress-induced diseases, and their use as food additives.

Several studies have well established that oxidative stress plays a significant role in the manifestation and the health complication of diabetes mellitus [4]. Thus, antioxidant compounds such as plant polyphenols have been suggested that they may be used for the prevention and/or treatment of this disease afflicting millions of people worldwide [5]. Etsassala et al. [6] have reported for the first time that methanolic extract from *Salvia africana-lutea* is a rich source of terpenoids, especially abietane diterpenes (e.g., 19-acetoxy-12-methoxycarnosic acid, 3β-acetoxy-7α-methoxyrosmanol, 19-acetoxy-7α-methoxyrosmanol, 19-acetoxy-12-methoxy carnosol, safricana lactones A and B) and triterpenes (e.g., oleanolic and ursolic acids, 11,12-dehydroursolic acid lactone and β-amyrin). The methanolic extract exhibited in vitro strong antioxidant and antidiabetic properties [6]. Therefore, the authors suggested that extracts from *Salvia africana-lutea* may be used for both the prevention and/or amelioration of the symptoms of diabetes mellitus [6]. In another study, Dienaite et al. [7] showed that polyphenolic extracts from roots and leaves of *Paeonia officinalis* exhibited free radical scavenging activity and inhibited α-amylase suggesting their possible use as antidiabetic agents. The authors also performed UPLC-Q/TOF analysis supplemented with the on-line HPLC–DPPH•-scavenging method which revealed 20 gallic acid derivatives as radical scavenging compounds, which might account for the overall antioxidant potential of the extracts [7]. *Rubus idaeus* L. (raspberry) is another plant containing polyphenols combining antioxidant and antidiabetic properties [8]. Wu et al. [8] isolated extracts from leaves, fruit pulp, and seed of *R. idaeus* and identified their polyphenols by HR-HPLC-ESI-qTOF-MS/MS method. All the *R. idaeus'* extracts, especially those from leaves, exhibited strong antioxidant activity and inhibited the activity of the digestive enzymes α-glucosidase and α-amylase indicating antidiabetic properties [8]. The authors also used docking analysis in order to suggest possible mechanisms accounting for the digestive enzymes' inhibition by *R. idaeus'* polyphenols [8]. Moreover, in a review article, Vlavcheski et al. [9] presented studies investigating the antidiabetic activity of hydroxytyrosol, a polyphenol found in high concentration especially in olive oil. The main conclusions of their review literature were that hydroxytyrosol exhibits insulin-like effects in several cell types and antidiabetic activity in vivo [9]. Moreover, hydroxytyrosol has been shown to exert protection against diabetes-induced oxidative stress [9]. Thus, hydroxytyrosol given its low toxicity may be used for the prevention and/or treatment of diabetes mellitus, although further studies are needed to investigate its bioavailability and to fully elucidate its mechanism of antidiabetic action [9].

As mentioned hydroxytyrosol is present in high concentration in olive oil which also contains other potent polyphenolic antioxidants such as oleuropein, tyrosol, coumaric acid, caffeic acid, vanillic acid, ferulic acid, kaempferol, and quercetin [10]. Aprile et al. [11] assessed for the first time the antioxidant activity and the total polyphenolic content of olive fruits from the olive tree "Cellina di Nardò" (CdN), one of the most widespread cultivars in Southern Italy. The results showed that the fully maturation is the best harvest time to obtain a table olive with a high content in polyhpenols [11]. It was also shown that different treatments that are necessary to remove the bitterness of the raw olive and to stabilize them to obtain edible table olives, cause a loss in phenolic substances which also results in a loss of antioxidant activity [11]. In addition, Martinez et al. [12] examined if hydroxytyrosol extract from olive fruits could be used as preservative for fish patties. Their results showed that hydroxytyrosol extract had strong antioxidant activity in vitro and antimicrobial against *Staphylococcus aureus* [12]. Thus, hydroxytyrosol extract may be used in the food industry for extending shelf life of fish products. Apart from olive fruits and olive oil, olive mill wastewaters (OMWWs), the byproducts of the olive oil production process, are rich in polyphenolic compounds [13]. However, these polyphenols along with other organic compounds account for the environmental problems caused by the OMWW's discard [14]. Thus, the isolation of polyphenols from OMWW has been suggested as a way, on the one hand to reduce the environmental pollution around olive oil production industries, and on the other hand to produce polyphenolic extracts possessing strong antioxidant activity [13]. However, one of the main problems regarding the use of polyphenols as antioxidant supplements is their low availability [15]. In order to increase polyphenols' bioavailability, their encapsulation in different organic carriers has been suggested [16]. Thus, my research group conducted a study to find out the optimal conditions for the encapsulation of a polyphenolic extract from OMWW in maltodextrin and whey protein [17]. The results demonstrated that both tested carriers were effective for the production of antioxidant powder using encapsulation by spray drying, although in different conditions. Specifically, inlet/outlet temperature did not seem to affect maltodextrin samples' antioxidant activity, but whey protein samples showed better antioxidant activity at lower temperatures (within the temperature range used; 100–160 °C) [17]. In addition, the encapsulated OMWW extract exerted protection from DNA damage induced by ROS [17]. Furthermore, an OMWW extract (encapsulated or not) has been reported for the first time to increase antioxidant mechanisms such as glutathione levels in endothelial cells suggesting their possible use for protection from oxidative stress-induced pathologies associated with the cardiovascular system [17]. We also demonstrated in another study that polyphenolic extract from *Rosa canina* fruit possessed strong free radical scavenging activity, inhibited ROS-induced DNA damage and improved redox status in human endothelial cells by reducing ROS and increasing GSH levels [18].

In addition, in a clinical study, Chen et al. [19] also demonstrated that antioxidant polyphenols from almond skins may prevent oxidative stress induced cardiovascular disease. Specifically, they showed, in a placebo-controlled, three-way crossover trial with a 1-week washout period, that consumption of almond skin polyphenols increased the polyphenols catechin and naringenin, the ratio of reduced glutathione (GSH) to oxidized glutathione (GSSG) and the activity of the antioxidant enzyme glutathione peroxidase (GPx) in plasma [19]. Moreover, the consumption of almond skin polyphenols inhibited low density lipoprotein (LDL) oxidation which is an important etiological factor for cardiovascular diseases [19].

Apart from cardiovascular pathologies, another chronic disease associated with oxidative stress is Alzheimer's disease [20]. Diaz et al. [21] assessed in vivo the protective effect of epicatechin against oxidative stress-induced damage to neurons. In particular, they demonstrated that epicatechin administration to rats injected with $A\beta_{25-35}$ reduced neurotoxicity, oxidative stress, and inflammation in the hippocampus [21]. Moreover, epicatechin administration decreased the immunoreactivity to heat shock protein (HSP)-60, -70, and -90 and neuronal death in the Cornu Ammonis 1 (CA1) region of the hippocampus, which favors an improvement in the function of spatial memory [21].

Intestinal diseases have also been associated with oxidative stress [22]. Specifically, oxidative stress has been shown to cause a defective barrier function leading to intestinal pathologies [22]. Thus, Yang et al. [23] investigated the protective effects of Red-osier dogwood (*Cornus stolonifera* Michx.) polyphenolic extracts against hydrogen peroxide-induced damage in Caco-2 intestinal epithelial cells. The results showed that Red-osier dogwood extract's treatment increased cell viability and decreased ROS through increased expression of antioxidant enzymes such as hemeoxygenase-1 (HO-1), superoxide dismutase (SOD), and glutathione peroxidase (GPx) in Caco-2 cells [23]. The expression of all these enzymes was probably due to the enhanced protein expression of the nuclear factor (erythroid-derived 2)-like 2 (Nrf-2), the most important transcription factor regulating antioxidant genes' expression [23]. Red-osier dogwood extract was also shown to increase the transepithelial resistance (TEER) value through inhibition of disorganization of tight junction proteins such as zonula occludens-1 (ZO-1) and claudin-3 [23]. Finally, Red-osier dogwood extract decreased in Caco-2 cells markers (e.g., interleukin 8) of inflammation which plays important role in intestinal diseases [23]. In general, there is interdependence between oxidative stress and inflammation resulting in many chronic diseases [24]. Anti-inflammatory activity was also shown to be possessed by polyphenolic extracts from mulberry species [25]. In particular, Negro et al. [25] isolated polyphenolic extracts from Italian mulberry local varieties belonging to *Morus alba* and *Morus nigra* species. The *M. alba* and *M. nigra* extracts contained five main anthocyanin compounds as identified by HPLC/DAD/MS analysis [25]. The extracts from all the tested mulberry varieties exhibited in vitro strong free radical scavenging and inhibited cyclooxygenase (COX) activity (a marker of inflammation) [25]. It is not only the polyphenols that affect the gastrointestinal system, but also the gastrointestinal digestion may affect polyphenols activity [26]. For example, David et al. [26] used a simulated in vitro digestion model to investigate gastrointestinal digestion's effects on the antioxidant capacity of Cornelian (*Cornus mas* L.) cherry fruit extract. The results showed that presence of three anthocyanins (i.e., cyanidin-3-*O*-galactoside, pelargonidin-3-*O*-glucoside, and pelargonidin-3-*O*-rutinoside) found in Cornelian cherry fruits, was not significantly affected by the gastric digestion [26]. However, intestinal digestion decreased the anthocyanin content and antioxidant activity of the fruit extract indicating that its polyphenolic content stability during gastrointestinal digestion should be taken into consideration for estimating its bioavailability [26].

In addition, tea is one of the most known plants for its prevention from chronic diseases [27]. This tea's preventive activity is attributed mainly to the antioxidant properties of its polyphenolic content [27]. Although there have been several studies on tea's polyphenols, there is a need for ongoing research, since there are many tea varieties cultivated in various regions having different soil and climatic conditions which affect tea's chemical composition and consequently its bioactivities [27]. Tang et al. [28] isolated polyphenolic fractions from 30 Chinese teas belonging to six categories, namely green, black, oolong, dark, white, and yellow teas, identified their polyphenols by HPLC and assessed their antioxidant capacity. The results showed that yellow, and oolong teas exhibited greater antioxidant activity and had higher polyphenolic content (e.g., catechins like epicatechin, epigallocatechin, epicatechin gallate, and epigallocatechin gallate) compared to dark, black, and white teas, green, [28].

In general, the phytochemical and pharmacognostical studies of local plant varieties, especially those used as traditional medicinal plants, is considered currently important for drug and food supplement development [29]. Thus, Karabagias et al. [30] examined the antioxidant properties of *Lavandula stoechas* grown in Greece. *L. stoechas* has been used in the traditional medicine since ancient times [30]. Both methanolic and aqueous extracts of *L. stoechas* exhibited in vitro antioxidant activity, but the latter was the most potent [30]. The major polyphenols found in *L. stoechas* aqueous extract were caffeic acid, quercetin-*O*-glucoside, lutelin-*O*-glucuronide and rosmarinic acid [30]. Moreover, fifty volatile compounds belonging to alcohols, aldehydes, ketones, norisoprenoids, and numerous terpenoids were identified by using HS-SPME/GC-MS [30]. In another study, Blando et al. [31] assessed the antioxidant activity of polyphenolic extracts from cladodes of *Opuntia ficus-indica* (L.). In particular, *O. ficus-indica* cladodes extracts exhibited in vitro strong free radical scavenging activity and increased antioxidant activity in human erythrocytes [31]. The extracts also protected erythrocytes from ROS-induced hemolysis [31]. Moreover, the *O. ficus-indica* extracts exerted significant antimicrobial activity against *Staphylococcus aureus* biofilm formation [31]. The observed bioactivities of *O. ficus-indica* extracts were probably attributed to its polyphenolic content, especially to piscidic acid, eucomic acid, isorhamnetin derivatives, and rutin [31]. Moreover, Osman et al. [32] identified by GC-MS the polyphenols of *Dialium indum* L. fruit and assessed in vitro their antioxidant properties. *D. indum* is native to Southeast Asia and its fruit is edible [32]. The results showed that the *D. indum* fruit contains phenolics, amino acids, saccharides, fatty acids, sesquiterpene, polyols, and dicarboxylic acids [32]. Interestingly, it was also demonstrated for the first time that the exocarp of *D. indum* fruit contains thirteen phenolic antioxidants (i.e., vanillic acid, syringic acid, ferulic acid, isoferulic acid, sinapic acid, vanillin, syringic aldehyde, *p*-hydroxybenzaldehyde, coniferyl aldehyde, *p*-hydroxybenzoic acid, homovanillic acid, *p*-coumaric acid, and sinapic aldehyde) [32]. *Averrhoa bilimbi* is another plant that has been used in traditional medicine of Asian countries [33]. Ahmed et al. [33] investigated the antioxidant activity of polyphenolic extracts and fractions from *A. bilimbi* leaves. The results showed that *A. bilimbi* leaf extract had in vitro high free radical scavenging capacity, especially the n-butanol fraction [33]. Moreover, *A. bilimbi* leaf extracts inhibited xanthine oxidase, an enzyme involved in oxidative stress and metabolic disorders [33]. Finally, docking analysis indicated that 5,7,40-trihydroxy-6-(1-ethyl-4-hydroxyphenyl) flavone-8-glucoside (cucumerin A) and afzelechin 3-*O*-alpha-L-rhamnopyranoside are the possible compounds that are responsible for xanthine oxidase inhibition [33].

Currently, there is a great interest for natural preservatives of foods instead of synthetic, since the latter present often harmful effects on human health [34]. The dosage of antioxidant compounds used as food preservatives should be regulated and the functionality should be evaluated to ensure stability [34]. For example, polyphenolic extracts from *Rosmarinus officinalis* L. (rosemary) has been widely used in food industry as antioxidant additive [35]. In some countries, the addition of rosemary extract as food additive is permitted, while in other countries it is not allowed [35]. Thus, Choi et al. [35] developed a quantitative high-performance liquid chromatography-photodiode array (HPLC-PDA) method for the determination of the amount of rosemary extract in various food products (e.g., edible oils, processed meat products and dressings). Moreover, the method could evaluate the antioxidant activity of rosemary extract in foods, and consequently its functional stability [35]. Thus, they found that in terms of antioxidant activity carnosic acid of rosemary extract is more stable than carnosol [35]. Fruit vinegars containing polyphenolic antioxidants are also used widely in food industry as condiments [36]. Liu et al. [37] assessed in 23 fruit vinegars their phenolic components by HPLC coupled with photometric diode array detector (HPLC-PDA), and their antioxidant activity in vitro. The most common polyphenols in fruit vinegars were gallic acid, protocatechuic acid, chlorogenic acid, caffeic acid, and *p*-coumaric acid [37]. Among the 23 tested vinegars, the most potent were balsamic vinegar of Modena (Galletti), Aceto Balsamico di Modena (Monari Federzoni), red wine vinegar (Kühne), and red wine vinegar (Galletti) [37].

Since, polyphenols are used as food additives, it is also interesting to find out how cooking affects their antioxidant activity. Gunathilake et al. [38] examined the effect of cooking (boiling, steaming, and frying) on the antioxidant activity of polyphenols from edible leaves of six species, *Centella asiatica, Cassia auriculata, Gymnema lactiferum, Olax zeylanica, Sesbania granadiflora,* and *Passiflora edulis.* The findings demonstrated that frying decreased polyphenols, flavonoids, carotenoids, and their antioxidant activities in all leafy vegetables [38]. The effects of boiling and steaming on polyphenols, carotenoids, and their antioxidant properties, varied according to the leaf type [38]. Thus, the results of the study are useful for choosing the appropriate cooking method of leafy vegetables for their antioxidant properties to be maintained.

The use of polyphenols as food additives has raised concerns about the safety for human health of extraction methods using organic solvents [39]. Thus, "green extraction methods" are being developed using modern technology, where less or no organic solvents are used to minimize environmental and health impacts. The most common of such "green methods" are microwave assisted (MAE), ultrasound assisted (UAE), pulsed electric field assisted, and enzyme assisted extractions, infrared irradiation (IR), pressurized liquid and supercritical fluid extractions [39]. Gajic et al. [40] investigated the optimal conditions for isolating polyphenols from black locust (*Robiniae pseudoacaciae*) flowers using UAE. The results showed that extraction time had the greatest influence on the total polyphenolic content, followed by the extraction temperature and ethanol concentration [40]. The optimal extraction conditions were 60% (*v/v*) ethanol, 59 °C, and 30 min at the liquid-to-solid ratio of 10 cm^3 g^{-1} [40]. Moreover, the UAE gave a higher yield of polyphenols than the maceration and Soxhlet extraction [40]. The main polyphenols found in the extract were rutin, epigallocatechin, ferulic acid, and quercetin [40]. Bouaoudia-Madi et al. [41] have also used UAE to isolate polyphenolic extracts from *Myrtus communis* L. pericarp, a plant native of the Mediterranean basin. The results demonstrated that ethanol concentration, irradiation time, liquid solvent-to-solid ratio and amplitude affected significantly the yield of total polyphenolic content [41]. The optimal conditions for the isolation of *M. communis* polyphenolic extract using UAE were 70% *v/v* ethanol concentration, 7.5 min irradiation time and 30% liquid solvent-to-solid ratio [41]. In addition, isolation of *M. communis* polyphenolic extract using UAE was more efficient than MAE and conventional solvent extraction methods [41]. In another study, Quiroz et al. [42] used MAE to isolate polyphenolic extract from annatto (*Bixa orellana* L.) seeds. The results showed that the optimal extraction conditions were pH 7.0, solvent concentration 96% *v/v*, solvent-to-seed ratio 6:1 and microwave time 5 min [42]. The main polyphenols identified in the *B. orellana* seed extract were apigenin, hypolaetin, and caffeic acid derivatives [42]. Moreover, the *B. orellana* seed extract isolated by MAE had higher antioxidant activity and polyphenolic yield than that isolated by leaching [42]. Another "green extraction method," far IF, was used by Azad et al. [43] in order to isolate antioxidant polyphenolic compounds from *Angelica gigas* Nakai. The results indicated that the optimal conditions for the IF polyphenolic extraction from *A. gigas* Nakai were at 220 °C for 30 min [43]. In addition, HPLC analysis showed that the main polyphenols found in the *A. gigas* Nakai extract were decursin and decursinol angelate [43].

Funding: This research received no external funding.

Conflicts of Interest: The author declare no conflict of interest.

References

1. Zhou, Y.; Jiang, Z.; Lu, H.; Xu, Z.; Tong, R.; Shi, J.; Jia, G. Recent Advances of Natural Polyphenols Activators for Keap1-Nrf2 Signaling Pathway. *Chem. Biodivers.* **2019**, *16*, e1900400. [CrossRef] [PubMed]
2. Boo, Y.C. Can Plant Phenolic Compounds Protect the Skin from Airborne Particulate Matter? *Antioxidants* **2019**, *8*, 379. [CrossRef] [PubMed]
3. Pawlowska, E.; Szczepanska, J.; Koskela, A.; Kaarniranta, K.; Blasiak, J. Dietary Polyphenols in Age-Related Macular Degeneration: Protection against Oxidative Stress and Beyond. *Oxid. Med. Cell. Longev.* **2019**, *2019*, 9682318. [CrossRef]
4. Burgos-Morón, E.; Abad-Jiménez, Z.; Marañón, A.M.; Iannantuoni, F.; Escribano-López, I.; López-Domènech, S.; Salom, C.; Jover, A.; Mora, V.; Roldan, I.; et al. Relationship Between Oxidative Stress, ER Stress, and Inflammation in Type 2 Diabetes: The Battle Continues. *J. Clin. Med.* **2019**, *8*, 1385. [CrossRef] [PubMed]
5. Jin, T.; Song, Z.; Weng, J.; Fantus, I.G. Curcumin and other dietary polyphenols: Potential mechanisms of metabolic actions and therapy for diabetes and obesity. *Am. J. Physiol. Endocrinol. Metab.* **2018**, *314*, E201–E205. [CrossRef] [PubMed]
6. Etsassala, N.G.E.R.; Badmus, J.A.; Waryo, T.T.; Marnewick, J.L.; Cupido, C.N.; Hussein, A.A.; Iwuoha, E.I. Alpha-Glucosidase and Alpha-Amylase Inhibitory Activities of Novel Abietane Diterpenes from *Salvia africana-lutea*. *Antioxidants* **2019**, *8*, 421. [CrossRef] [PubMed]

7. Dienaitė, L.; Pukalskienė, M.; Pukalskas, A.; Pereira, C.V.; Matias, A.A.; Venskutonis, P.R. Isolation of Strong Antioxidants from Paeonia Officinalis Roots and Leaves and Evaluation of Their Bioactivities. *Antioxidants* **2019**, *8*, 249. [CrossRef]

8. Wu, L.; Liu, Y.; Qin, Y.; Wang, L.; Wu, Z. HPLC-ESI-qTOF-MS/MS Characterization, Antioxidant Activities and Inhibitory Ability of Digestive Enzymes with Molecular Docking Analysis of Various Parts of Raspberry (*Rubus ideaus* L.). *Antioxidants* **2019**, *8*, 274. [CrossRef]

9. Vlavcheski, F.; Young, M.; Tsiani, E. Antidiabetic Effects of Hydroxytyrosol: In Vitro and In Vivo Evidence. *Antioxidants* **2019**, *8*, 188. [CrossRef]

10. Serreli, G.; Deiana, M. Biological Relevance of Extra Virgin Olive Oil Polyphenols Metabolites. *Antioxidants* **2018**, *7*, 170. [CrossRef]

11. Aprile, A.; Negro, C.; Sabella, E.; Luvisi, A.; Nicolì, F.; Nutricati, E.; Vergine, M.; Miceli, A.; Blando, F.; De Bellis, L. Antioxidant Activity and Anthocyanin Contents in Olives (*cv* Cellina di Nardò) during Ripening and after Fermentation. *Antioxidants* **2019**, *8*, 138. [CrossRef] [PubMed]

12. Martínez, L.; Castillo, J.; Ros, G.; Nieto, G. Antioxidant and Antimicrobial Activity of Rosemary, Pomegranate and Olive Extracts in Fish Patties. *Antioxidants* **2019**, *8*, 86. [CrossRef] [PubMed]

13. Gerasopoulos, K.; Stagos, D.; Kokkas, S.; Petrotos, K.; Kantas, D.; Goulas, P.; Kouretas, D. Feed supplemented with byproducts from olive oil mill wastewater processing increases antioxidant capacity in broiler chickens. *Food Chem. Toxicol.* **2015**, *82*, 42–49. [CrossRef] [PubMed]

14. Kalogerakis, N.; Politi, M.; Foteinis, S.; Chatzisymeon, E.; Mantzavinos, D. Recovery of antioxidants from olive mill wastewaters: A viable solution that promotes their overall sustainable management. *J. Environ. Manag.* **2013**, *128*, 749–758. [CrossRef] [PubMed]

15. Marín, L.; Miguélez, E.M.; Villar, C.J.; Lombó, F. Bioavailability of dietary polyphenols and gut microbiota metabolism: Antimicrobial properties. *Biomed. Res. Int.* **2015**, *2015*, 905215. [CrossRef] [PubMed]

16. Fang, Z.; Bhandari, B. Encapsulation of polyphenols—A review. *Trends Food Sci. Technol.* **2010**, *21*, 510–553. [CrossRef]

17. Kreatsouli, K.; Fousteri, Z.; Zampakas, K.; Kerasioti, E.; Veskoukis, A.S.; Mantas, C.; Gkoutsidis, P.; Ladas, D.; Petrotos, K.; Kouretas, D.; et al. A Polyphenolic Extract from Olive Mill Wastewaters Encapsulated in Whey Protein and Maltodextrin Exerts Antioxidant Activity in Endothelial Cells. *Antioxidants* **2019**, *8*, 280. [CrossRef]

18. Kerasioti, E.; Apostolou, A.; Kafantaris, I.; Chronis, K.; Kokka, E.; Dimitriadou, C.; Tzanetou, E.N.; Priftis, A.; Koulocheri, S.D.; Haroutounian, S.A.; et al. Polyphenolic Composition of *Rosa canina*, *Rosa sempervivens* and *Pyrocantha coccinea* Extracts and Assessment of Their Antioxidant Activity in Human Endothelial Cells. *Antioxidants* **2019**, *8*, 92. [CrossRef]

19. Chen, C.O.; Milbury, P.E.; Blumberg, J.B. Polyphenols in Almond Skins after Blanching Modulate Plasma Biomarkers of Oxidative Stress in Healthy Humans. *Antioxidants* **2019**, *8*, 95. [CrossRef]

20. Butterfield, D.A.; Boyd-Kimball, D. Redox proteomics and amyloid β-peptide: Insights into Alzheimer disease. *J. Neurochem.* **2019**, *151*, 459–487. [CrossRef]

21. Diaz, A.; Treviño, S.; Pulido-Fernandez, G.; Martínez-Muñoz, E.; Cervantes, N.; Espinosa, B.; Rojas, K.; Pérez-Severiano, F.; Montes, S.; Rubio-Osornio, M.; et al. Epicatechin Reduces Spatial Memory Deficit Caused by Amyloid-β25–35 Toxicity Modifying the Heat Shock Proteins in the CA1 Region in the Hippocampus of Rats. *Antioxidants* **2019**, *8*, 113. [CrossRef] [PubMed]

22. Campbell, E.L.; Colgan, S.P. Control and dysregulation of redox signalling in the gastrointestinal tract. *Nat. Rev. Gastroenterol. Hepatol.* **2019**, *16*, 106–120. [CrossRef] [PubMed]

23. Yang, R.; Hui, Q.; Jiang, Q.; Liu, S.; Zhang, H.; Wu, J.; Lin, F.; Karmin, O.; Yang, C. Effect of Manitoba-Grown Red-Osier Dogwood Extracts on Recovering Caco-2 Cells from H_2O_2-Induced Oxidative Damage. *Antioxidants* **2019**, *8*, 250. [CrossRef] [PubMed]

24. Biswas, S.K. Does the Interdependence between Oxidative Stress and Inflammation Explain the Antioxidant Paradox? *Oxid. Med. Cell. Longev.* **2016**, *2016*, 5698931. [CrossRef]

25. Negro, C.; Aprile, A.; De Bellis, L.; Miceli, A. Nutraceutical Properties of Mulberries Grown in Southern Italy (Apulia). *Antioxidants* **2019**, *8*, 223. [CrossRef]

26. David, L.; Danciu, V.; Moldovan, B.; Filip, A. Effects of In Vitro Gastrointestinal Digestion on the Antioxidant Capacity and Anthocyanin Content of Cornelian Cherry Fruit Extract. *Antioxidants* **2019**, *8*, 114. [CrossRef]

27. Aboulwafa, M.M.; Youssef, F.S.; Gad, H.A.; Altyar, A.E.; Al-Azizi, M.M.; Ashour, M.L. A Comprehensive Insight on the Health Benefits and Phytoconstituents of *Camellia sinensis* and Recent Approaches for Its Quality Control. *Antioxidants* **2019**, *8*, 455. [CrossRef]
28. Tang, G.Y.; Zhao, C.N.; Xu, X.Y.; Gan, R.Y.; Cao, S.Y.; Liu, Q.; Shang, A.; Mao, Q.Q.; Li, H.B. Phytochemical Composition and Antioxidant Capacity of 30 Chinese Teas. *Antioxidants* **2019**, *8*, 180. [CrossRef]
29. Oyenihi, A.B.; Smith, C. Are polyphenol antioxidants at the root of medicinal plant anti-cancer success? *J. Ethnopharmacol.* **2019**, *229*, 54–72. [CrossRef]
30. Karabagias, I.K.; Karabagias, V.K.; Riganakos, K.A. Physico-Chemical Parameters, Phenolic Profile, In Vitro Antioxidant Activity and Volatile Compounds of Ladastacho (*Lavandula stoechas*) from the Region of Saidona. *Antioxidants* **2019**, *8*, 80. [CrossRef]
31. Blando, F.; Russo, R.; Negro, C.; De Bellis, L.; Frassinetti, S. Antimicrobial and Antibiofilm Activity against *Staphylococcus aureus* of *Opuntia ficus-indica* (L.) Mill. Cladode Polyphenolic Extracts. *Antioxidants* **2019**, *8*, 117. [CrossRef] [PubMed]
32. Osman, M.F.; Mohd Hassan, N.; Khatib, A.; Tolos, S.M. Antioxidant Activities of *Dialium indum* L. Fruit and Gas Chromatography-Mass Spectrometry (GC-MS) of the Active Fractions. *Antioxidants* **2018**, *7*, 154. [CrossRef] [PubMed]
33. Ahmed, Q.U.; Alhassan, A.M.; Khatib, A.; Shah, S.A.A.; Hasan, M.M.; Sarian, M.N. Antiradical and Xanthine Oxidase Inhibitory Activity Evaluations of *Averrhoa bilimbi* L. Leaves and Tentative Identification of Bioactive Constituents through LC-QTOF-MS/MS and Molecular Docking Approach. *Antioxidants* **2018**, *7*, 137. [CrossRef] [PubMed]
34. Franco, R.; Navarro, G.; Martínez-Pinilla, E. Antioxidants versus Food Antioxidant Additives and Food Preservatives. *Antioxidants* **2019**, *8*, 542. [CrossRef] [PubMed]
35. Choi, S.H.; Jang, G.W.; Choi, S.I.; Jung, T.D.; Cho, B.Y.; Sim, W.S.; Han, X.; Lee, J.S.; Kim, D.Y.; Kim, D.B.; et al. Development and Validation of an Analytical Method for Carnosol, Carnosic Acid and Rosmarinic Acid in Food Matrices and Evaluation of the Antioxidant Activity of Rosemary Extract as a Food Additive. *Antioxidants* **2019**, *8*, 76. [CrossRef] [PubMed]
36. Bakir, S.; Toydemir, G.; Boyacioglu, D.; Beekwilder, J.; Capanoglu, E. Fruit Antioxidants during Vinegar Processing: Changes in Content and in Vitro Bio-Accessibility. *Int. J. Mol. Sci.* **2016**, *17*, 1658. [CrossRef]
37. Liu, Q.; Tang, G.Y.; Zhao, C.N.; Gan, R.Y.; Li, H.B. Antioxidant Activities, Phenolic Profiles, and Organic Acid Contents of Fruit Vinegars. *Antioxidants* **2019**, *8*, 78. [CrossRef]
38. Gunathilake, K.D.P.P.; Ranaweera, K.K.D.S.; Rupasinghe, H.P.V. Effect of Different Cooking Methods on Polyphenols, Carotenoids and Antioxidant Activities of Selected Edible Leaves. *Antioxidants* **2018**, *7*, 117. [CrossRef]
39. Talmaciu, A.I.; Volf, I.; Popa, V.I. A Comparative Analysis of the 'Green' Techniques Applied for Polyphenols Extraction from Bioresources. *Chem. Biodivers.* **2015**, *12*, 1635–1651. [CrossRef]
40. Savic Gajic, I.; Savic, I.; Boskov, I.; Žerajić, S.; Markovic, I.; Gajic, D. Optimization of Ultrasound-Assisted Extraction of Phenolic Compounds from Black Locust (*Robiniae Pseudoacaciae*) Flowers and Comparison with Conventional Methods. *Antioxidants* **2019**, *8*, 248. [CrossRef]
41. Bouaoudia-Madi, N.; Boulekbache-Makhlouf, L.; Madani, K.; Silva, A.M.S.; Dairi, S.; Oukhmanou-Bensidhoum, S.; Cardoso, S.M. Optimization of Ultrasound-Assisted Extraction of Polyphenols from *Myrtus communis* L. Pericarp. *Antioxidants* **2019**, *8*, 205. [CrossRef] [PubMed]
42. Quiroz, J.Q.; Torres, A.C.; Ramirez, L.M.; Garcia, M.S.; Gomez, G.C.; Rojas, J. Optimization of the Microwave-Assisted Extraction Process of Bioactive Compounds from Annatto Seeds (*Bixa orellana* L.). *Antioxidants* **2019**, *8*, 37. [CrossRef] [PubMed]
43. Azad, M.O.K.; Piao, J.P.; Park, C.H.; Cho, D.H. Far Infrared Irradiation Enhances Nutraceutical Compounds and Antioxidant Properties in *Angelica gigas* Nakai Powder. *Antioxidants* **2018**, *7*, 189. [CrossRef] [PubMed]

 © 2019 by the author. Licensee MDPI, Basel, Switzerland. This article is an open access article distributed under the terms and conditions of the Creative Commons Attribution (CC BY) license (http://creativecommons.org/licenses/by/4.0/).

 *antioxidants*

Article

Alpha-Glucosidase and Alpha-Amylase Inhibitory Activities of Novel Abietane Diterpenes from *Salvia africana-lutea*

Ninon G.E.R. Etsassala [1], Jelili A. Badmus [2], Tesfaye T. Waryo [1], Jeanine L. Marnewick [2], Christopher N. Cupido [3], Ahmed A. Hussein [4,*] and Emmanuel I. Iwuoha [1]

[1]   Chemistry Department, University of the Western Cape, Private Bag X17, Bellville 7535, South Africa; 3415216@myuwc.ac.za (N.G.E.R.E.); twaryo@uwc.ac.za (T.T.W.); eiwuoha@uwc.ac.za (E.I.I.)
[2]   Oxidative Stress Research Unit, Cape Peninsula University of Technology, Symphony Rd. Bellville 7535, South Africa; jabadmus@lautech.edu.ng (J.A.B.); marnewickj@cput.ac.za (J.L.M.)
[3]   Department of Botany, University of Fort Hare, Private Bag X1314, Alice 5700, South Africa; ccupido@ufh.ac.za
[4]   Chemistry Department, Cape Peninsula University of Technology, Symphony Rd. Bellville 7535, South Africa
*   Correspondence: mohammedam@cput.ac.za; Tel.: +27-21-959-6193; Fax: +27-21-959-3055

Received: 26 June 2019; Accepted: 17 August 2019; Published: 20 September 2019

Abstract: The re-investigation of a methanolic extract of *Salvia africana-lutea* collected from the Cape Floristic Region, South Africa (SA), afforded four new abietane diterpenes, namely 19-acetoxy-12-methoxycarnosic acid (**1**), 3β-acetoxy-7α-methoxyrosmanol (**2**), 19-acetoxy-7α-methoxyrosmanol (**3**), 19-acetoxy-12-methoxy carnosol (**4**), and two known named clinopodiolides A (**5**), and B (**6**), in addition to four known triterpenes, oleanolic, and ursolic acids (**7, 8**), 11,12-dehydroursolic acid lactone (**9**) and β-amyrin (**10**). The chemical structural elucidation of the isolated compounds was determined on the basis of one and two dimensional nuclear magnetic resonance (1D and 2D NMR), high-resolution mass spectrometry (HRMS), ultra violet (UV), fourier transform infrared (IR), in comparison with literature data. The in vitro bio-evaluation against alpha-glucosidase showed strong inhibitory activities of **8, 10**, and **7**, with the half inhibitory concentration (IC_{50}) values of 11.3 ± 1.0, 17.1 ± 1.0 and 22.9 ± 2.0 µg/mL, respectively, while **7** demonstrated the strongest in vitro alpha-amylase inhibitory activity among the tested compounds with IC_{50} of 12.5 ± 0.7 µg/mL. Additionally, some of the compounds showed significant antioxidant capacities. In conclusion, the methanolic extract of *S. africana-lutea* is a rich source of terpenoids, especially abietane diterpenes, with strong antioxidant and anti-diabetic activities that can be helpful to modulate the redox status of the body and could therefore be an excellent candidate for the prevention of the development of diabetes, a disease where oxidase stress plays an important role.

Keywords: diabetes mellitus; Lamiaceae; alpha-glucosidase; alpha-amylase; *Salvia africana-lutea*; terpenoids; cape floristic region

---

1. Introduction

Diabetes mellitus (DM) is one of the most common metabolic disorders with significant morbidity and mortality rates around the world. It is caused either by deficiency in insulin secretion or degradation of secreted insulin [1], which is the result of cell alterations caused by many internal and external factors, such as obesity, sedentary lifestyle, and oxidative stress [2,3]. However, oxidative stress occurs when the production of free radicals, such as reactive oxygen species (ROS), overwhelms the detoxification capacity of the cellular antioxidant system, resulting in biological damages [4]. It plays a central role in the development of diabetes, such as microvascular and cardiovascular complications [5].

The metabolic abnormalities of diabetes cause mitochondrial superoxide overproduction, which is the main mediator of diabetes tissue damage, insulin resistance, β-cell dysfunction, and impaired glucose tolerance [6].

The prevalence of diabetes reported by the International Diabetes Federation (IFD) indicates that the number of adult diabetic patients globally was 366 million in November 2011, and it is projected to increase to 552 million by the 2030 [7]. In Africa, more than 14 million people have diabetes, accounting for about 4.3% of adults, and it caused about 401,276 deaths in 2012 [8]. Numerous synthetic anti-diabetic agents, such as acarbose, miglitol, sulfonylurea, metformin, and thiozolidinedione, are readily available on the market, but the effectiveness of these products is limited due to non-availability, cost, and high side effects, such as hypoglycaemia, damage to the liver, flatulence, diarrhea, abdominal pain, dropsy, drug-resistance, weight gain, and heart failure [9,10]. Therefore, due to the above-mentioned detrimental effects, there is a great need for developing alternative natural anti-diabetic products with a high safety margin.

Natural plants have been used as a source of medicine since time immemorial for the purpose of curing numerous human afflictions. Plants are known to be the main source of health-promoting substances because they are comprised of secondary metabolites, such as polyphenols, flavonoids, terpenoids, alkaloids, carotenoids, vitamins, and several other constituents, which are responsible for anti-diabetic, antioxidant, anti-hypertensive, and other health promoting effects [11].

Salvia africana-lutea, commonly known as *Bruinsalie* or beach sage, is distributed from Namaqualand to the Eastern Cape Province of South Africa (in the South-Western part of South Africa). It is traditionally used for the treatment of different kinds of ailments and/or diseases, such as coughs, sexual debility, mental and nervous conditions, throat inflammation, chronic bronchitis, tuberculosis, influenza, stomach ache, diarrhea, and urticarial [12]. A preliminary phytochemical and biological evaluation of *S. africana-lutea*, collected from Pretoria, South Africa, indicated a great potential of *S. africana-lutea* methanolic extract as a source of abietane diterpenes [13,14]. The ethno-medicinal report of the plant in controlling different symptoms related to ageing diseases justifies the present study to isolate and identify the chemical constituents and explore their in vitro bio-activities. Terpenoids isolated from Lamiaceae, especially abietane diterpenes and some classes of triterpenes, play a crucial role as eco-physiological mediators and are of interest for potential application as therapeutic agents in the treatment of diabetes [15,16]. Recent reports indicate the efficiency of carnosic acid derivatives in controlling the detrimental effects of diabetes by improving glucose and insulin secretion [17], improving glucose homeostasis, or stimulating glucose uptake through intensifying peripheral glucose clearance in tissues, thereby alleviating pathological effects related with the hyperglycaemic state [18]. Carnosol improved diabetes and its complications by modulation of oxidative stress and inflammatory responses [19] and stimulated glucose uptake [20]. Several triterpenes including ursolic acid, oleanolic acid, lupeol, and betulinic acid, have been reported to be potential anti-diabetic candidates with different mechanisms of action [21–23].

This work primarily examines the phytochemical isolation of different constituents present in the methanolic extract of *S. africana-lutea* as well as the anti-diabetic and the antioxidant activities of its isolated compounds.

2. Experimental Section

2.1. General Information

Epigallocatechingallate (EGCG), trolox (6-hydroxyl-2, 5, 7, 8- tetramethylchroman-2-carboxylic acid), and other reagents, including 2,2-azino-bis (3-ethylbenzothiazoline-6-sulfonic acid) (ABTS) diammonium salt, potassium peroxodisulphate, fluorescein sodium salt, 2,2-Azobis (2-methylpropionamidine) dihydrochloride (AAPH), perchloric acid, 2,4,6-tri[2-pyridyl]-s-triazine (TPTZ), iron (III) chloride hexahydrate, copper sulphate, hydrogen peroxide, α-glucosidase (*Saccharomyces cerevisiae*), α-amylase (procaine pancreas), 3,5-dinitro salicylic acid (DNS), p-nitro-phenyl-α-D-glucopyranoside (p-NPG), sodium carbonate (Na_2CO_3), sodium dihydrogen phosphate, and di-sodium hydrogen phosphate, secured and purchased from Sigma-Aldrich, Cape town, South Africa.

Organic solvents, methanol (HPLC grade), ethanol, ethyl acetate, and hexane, were supplied by Merck (Cape Town, South Africa). Thin layer chromatography (TLC) as conducted on normal-phase (Merck) Silica gel 60 PF254 pre-coated aluminum plates. Column chromatography was performed using silica gel 60 H (0.040–0.063 mm particle size, Merck, Cape Town, South Africa) and Sephadex LH-20 (Sigma-Aldrich, Cape Town, South Africa).

NMR spectra were recorded on an Avance 400 MHz NMR spectrometer (Bruker, Rheinstetten, Germany) in deuterated chloroform and acetone, using the solvent signals as the internal reference. HRMS analysis was conducted on an Ultimate 3000 LC (Dionex, Sunnyvale, CA, USA) coupled to a Bruker QTOF with an electrospray ionization (ESI) interface working in the positive ion mode. Preparative HPLC was used for further isolation of pure compounds using HPLC methanol and distilled water.

2.2. Plant Material

The plant material used in this study was collected from the Cape Flats Nature Reserve, University of the Western Cape, South Africa. A voucher specimen was identified at the Compton Herbarium, Kirstenbosch by Dr. Christopher Cupido (South African National Biodiversity Institute, Kirstenbosch), with herbarium number NBG1465544-0.

2.3. Extraction and Purification of Chemical Constituents

The aerial parts of the fresh plant material (2.5 kg) were blended and extracted with methanol (4.5 L) at room temperature (25 °C) for 24 h. The methanol extract was filtered and evaporated to dryness under reduced pressure at 40 °C to yield 97.77 g (3.9%). The total extract (97 g) was applied to a silica gel column (30 × 18 cm) and eluted using a gradient of hexane (Hex) and ethyl acetate (EtOAc) in order of increasing polarity: 94 fractions (500 mL each) were collected and combined according to their TLC profiles to yield 21 fractions, labeled I–XXI.

The main fraction XVII (500 mg) was subjected to a successive silica gel column using a Hex/EtOAc gradient (7:3; 100%) then Sephadex (using 95% aqueous ethanol) to produce 18-acetoxy-12-methoxy carnosic acid (compound **1**, 44.1 mg). The main fraction XIII (1.4 g) was chromatographed on silica gel using a Hex/EtOAc gradient (9:1; 7:3; 100%) then HPLC, using a gradient solvent system of MeOH and de-ionized water (70:30 to 100% MeOH in 45 min), producing compound **3** (R_t 23.62 min, 9 mg), compound **4** (R_t 25.19 min, 11.17 mg), compound **5** (R_t 32.17 min, 46.3 mg), and compound **6** (R_t 12.75 min, 40 mg). Main fraction XI was subjected to a successive silica gel column, under the same condition, then HPLC to produce compound **2** (R_t 19.67 min, 9.9 mg).

Main fraction X was subjected to a silica gel column under the same condition then chromatographed on a sephadex column (5 % aqueous ethanol) to produce compound **8** (34.6 mg), compound **7** (9.6 mg), compound **9** (12.5 mg), and compound **10** (54 4 mg).

2.4. Alpha-Glucosidase Inhibitory Activity

Alpha-glucosidase inhibitory activity of the isolated compounds was carried out according to the standard method, with a slight modification [24]. In a 96-well plate, the reaction mixture containing 50 µL of phosphate buffer (100 mM, pH = 6.8), 10 µL alpha-glucosidase (1 U/mL), and 20 µL of varying concentrations of isolated compounds was pre-incubated at 37 °C for 15 min. Next, 20 µL of p-NPG (5 mM) was added as a substrate and incubated further at 37 °C for 20 min. The reaction was stopped by adding 50 µL of sodium carbonate Na_2CO_3 (0.1 M). The absorbance of the released p-nitrophenol was measured at 405 nm using a Multiplate Reader (Multiskan thermo scientific, version 1.00.40, Vantaa, Finland). Acarbose at various concentrations was included as a standard. Each experiment was performed in triplicates. The results were expressed as a percentage inhibition, which was calculated using Formula (1).

$$\text{Inhibitory activity (\%)} = (1 - A/B) \times 100 \tag{1}$$

where A is the absorbance in the presence of the test substance and B is the absorbance of the control.

2.5. Alpha-Amylase Inhibitory Activity

The alpha-amylase inhibitory activity of the extract and fractions was carried out according to the standard method, with a slight modification [24]. In a 96-well plate, the reaction mixture containing 50 µL of phosphate buffer (100 mM, pH = 6.8), 10 µL alpha amylase (2 U/mL), and 20 µL of varying concentrations of isolated compounds was pre-incubated at 37 °C for 20 min. Next, 20 µL of 1% soluble starch (100 mM phosphate buffer pH 6.8) was added as a substrate and incubated further at 37 °C for 30 min. A volume of 100 µL of the color reagent (DNS) was added and then boiled for 10 min. The absorbance of the resulting mixture was measured at 540 nm using a Multiplate Reader (Multiskan thermo scientific, version 1.00.40). Acarbose at various concentrations was used as a standard. Each experiment was performed in triplicates. The results were expressed as the percentage inhibition, which was calculated using Formula (2).

$$\text{Inhibitory activity (\%)} = (1 - A/B) \times 100 \tag{2}$$

where A is the absorbance in the presence of the test substance and B is the absorbance of the control.

2.6. Antioxidant Assays

2.6.1. Ferric-Ion Reducing Antioxidant Power (FRAP) Assay

The FRAP assay was carried out in accordance with the method described previously [25]. Absorbance was measured at 593 nm. L-Ascorbic acid was used as a standard, and the results were expressed as µmole ascorbic acid equivalents per milligram of dry weight (µM AAE/g DW) of the test samples.

2.6.2. Trolox Equivalent Absorbance Capacity (TEAC) Assay

The total anti-oxidant activity of the test sample was measured using previously described methods [26]. Absorbance was read at 734 nm at 25 °C in a plate reader, and the results were expressed as µmole Trolox equivalents per milligram of dry weight (µM TE/g DW) of the test samples.

2.6.3. Oxygen Radical Absorbance Capacity (ORAC) Assay

ORAC assay was done according to the previous method [27]. ORAC values were expressed as micromoles of trolox equivalents (TE) per milligram of the test sample. Samples without a perfect curve were further diluted, and the dilution factors were used in the calculations of such samples.

2.7. Statistical Analysis

All the measurements were done in triplicate and IC_{50} values were calculated using GraphPad Prism 5 version 5.01 (Graph pad software, Inc., La Jolla, CA, USA.) statistical software.

3. Results and Discussion

3.1. Chemical Characterization of the Isolated Compounds

Chromatographic purification of a methanolic extract of *S. africana-lutea* was done using different techniques, including semi prep-HPLC yielded pure terpenoids (Figure 1), four of which were reported for the first time.

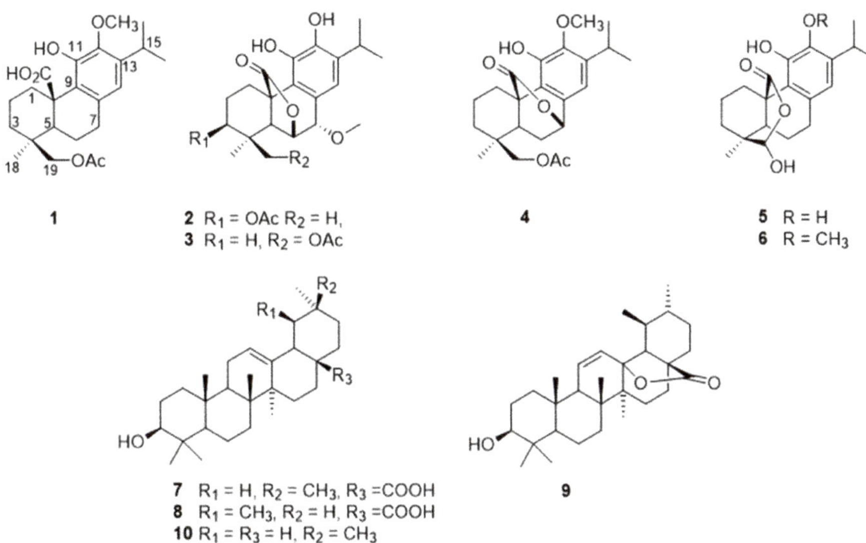

Figure 1. Chemical structures of the isolated compounds (**1–10**) from *S. africana-lutea*.

Compound **1** was isolated as amorphous yellowish-brown powder. The high-resolution mass spectrometry (HRMS) data indicated a molecular ion peak at [M]$^+$ 403.2115 *m/z*, suggesting a possible chemical formula of $C_{23}H_{32}O_6$. Its IR spectrum exhibited bands at 1725 and 3447 cm^{-1} forester and hydroxyl groups. The UV spectrum showed two absorptions at 210 and 280 nm. The ^{1}H NMR spectra (Table 1, Supplementary Materials) showed signals of three methyls at 1.19 ppm (*d*, *J* = 6.8 Hz), 1.17 (*d*, *J* = 6.8 Hz), and 1.06 (s); an aromatic proton at 6.46 (s); an oxygenated methylene proton at 3.97, 4.24 (*d*/each, *J* = 11.4 Hz); a multiplet signal at 3.17 (*sept*, *J* = 6.8 Hz); in addition to acetoxyl and a methoxyl groups at 1.86, 3.75. The ^{13}C NMR, distortionless enhancement by polarization transfer (DEPT-135), and heteronuclear multiple quantum coherence (HSQC) spectra indicated the presence of 23 carbons (Table 2) classified as four methyls, including a methoxy (61.1), two methylene, six methines, one of which was aromatic (117.8), and eight quaternary carbons, including a carbonyl (181.7) and five aromatic carbons (126.8, 133.8, 148.9, 143.7, and 139.6) in addition to the acetoxyl group (172.5, 20.7). The NMR data indicated an abietane diterpene similar to carnosic acid, previously isolated from the same source [14], with the only difference being the presence of extra acetoxyl and methoxyl groups. The methoxyl group was allocated at C-12 from the heteronuclear multiple bond correlation (HMBC) spectra, which showed correlations (among others) between the methoxyl protons and C-12 (143.7). The acetoxyl group was allocated at C-19 due to the presence of a methylene signal at 3.97 and 4.24, and both protons showed HMBC correlations with the acetoxyl's carbonyl group, C-4/C-5, and C-3. Other 2D data [heteronuclear multiple bond correlation (HMBC) and nuclear overhauser effect spectroscopy (NOESY)] as shown in Figure 2 confirmed the structure of compound **1** as 19-acetoxy-12-methoxycarnosic acid. The absolute configuration of the compound is proposed to belong to the normal abietane skeleton based on a biosynthetic basis, because normal abietane diterpene derivatives were previously isolated from the same source and directly related to the isolated compound [14,28].

The HRMS of compound **2** showed the molecular ion peak at [M]$^+$ 417.1907 *m/z*, corresponding to the molecular formula of $C_{23}H_{30}O_7$. Its IR spectrum exhibited bands at 1727 and 3447 cm^{-1} for ester and hydroxyl groups. The UV spectrum showed two peaks at 212 and 287 nm. Analysis of the ^{1}H NMR data (Table 1) showed the presence of an isopropyl group [a proton at δ_H = 3.04 (*sept*, *J* = 6.5 Hz); two methyls at 1.24 (*d*/both, *J* = 6.5 Hz)]; an aromatic proton (δ_H = 6.79, s); three separated protons attached to three oxygenated carbons at 4.66 (*dd*, *J* = 3.7, 12.1 Hz), 4.74 (*d*, *J* = 3.2 Hz), and 4.29 (*d*, *J* = 3.2 Hz); in addition to four methyl signals at 1.01 and 0.99 (s each), 2.07 (s, acetoxy), and 3.66 (s, methoxy). The ^{13}C NMR, DEPT-135 and HSQC showed 23 carbons classified as four methyls at (16.2, 26.7, 22.3, 22.4), an acetoxy (21.2), a methoxy (58.4), two methylene groups (25.3, 24.3), six methines, (one of which was aromatic), nine quaternary carbons, including two carbonyls (170.7, 178.1), and five aromatics (126.6, 123.4, 141, 141.8, and 134.7). According to HSQC, the proton resonating at δ_H 4.66 attached to carbon at 74.0 showed HMBC correlations with carbons at 174 (CO acetoxy), 77.4 (C-7), 51.5 (C-5), and 46.5 (C-10), which indicates the presence of a lactone ring at position six. Additionally, the proton at 4.29 (H-7) showed HMBC correlations with carbons at 74.0 (C-6), 126.6 (C-8), 123.4 (C-9), and 120.4 (C-14) as shown in Table 2. On the other hand, the spectroscopic data of compound **2** showed a close similarity with methoxyrosmanol, the only difference being the presence of an extra acetoxyl signal (2.07/170.7). The position of the acetate group was located at C-3 (from HMBC and NOESY correlations). Other 2D spectra confirmed the structure of compound **2** as 3β-acetoxy-7α-methoxyrosmanol [29–31].

Compound **3** showed NMR data similar to compound **2**. The ^{1}H NMR showed the absence of the C-3 methine signal at 4.7 and the appearance of the signal of methylene protons (4.05 s) attached to an acetoxyl group (from HMBC spectra). The acetoxyl group is attached to C19 because the NOESY and HMBC correlations showed cross-peaks between the acetoxy methyl/C-19, in addition to CH$_2$-19/C$_3$; C$_4$ and C$_5$. Other 2D data confirmed the structure of compound **3** as 19-acetoxy-7α-methoxyrosmanol [28].

The NMR of compound **4** (Tables 1 and 2) showed a close similarity with carnosol [32] and compounds **2** and **3**, except for the absence of the C-6α(OH)/C-7α(OH) system appearing at the C-20/C-7α lactone ring, in addition to a methoxy group, which was located at C-12, as shown in Figure 1. The HMBC spectra showed a correlation between H6/C7; C5; C4; C8 and H7/C6; C8; C13; C9. Other 2D data confirmed the structure of compound **4** as 19-acetoxy-12-methoxy carnosol.

Table 1. NMR spectroscopic data assignments (400 MHz) for compounds 1–6 (δ in ppm, m, J in Hz) in CDCl$_3$.

N°	1	2	3	4	5	6
	δ_H (*J* in Hz)	δ_H (*J* in Hz)	δ_H (*J* in Hz)	δ_H (*J* in Hz)	δ_H (*J* in Hz)	δ_H (*J* in Hz)
1	3.47 br d (10.7) 1.16 *	3.28 br d (15.5) 2.28 ddd (4.2, 14.3, 14.3)	3.22 br d (14.3) 2.06 dd (5.2, 14.1)	3.22 br d (13.4) 1.64 ddd (5.8, 11.0, 13.4)	3.34 br d (13.4) 1.47 ddd (5.4, 13.4, 13.4)	3.42 br d (13.0) 1.46 ddd (5.4, 13.4, 13.4)
2	2.42 m 1.48 br d (15.2)	1.81 m 1.62 m	1.53 ddddd (3.7, 14.0, 14.0, 14.0, 14.0) 1.69 *	1.40 ddd (4.7, 13.3, 13.3)	1.74 *	1.73 *
3	1.66 br d (13.7) 1.19 *	4.66 dd (3.7, 12.1)	1.83 br d (14.0) 1.23 d (6.8)	3.28 d (4.6) 1.64 br d (5.5)	2.3 br d (13.5) 1.23 *	2.3 br d (12.5) 1.21 *
5	1.58 d (12.0)	2.42 s	2.38 s	1.91 *	1.61 br d (12.4)	1.57 br d (13.4)
6	1.47 br d (14.3) 1.93 br d (16.7)	4.74 d (3.2)	4.26 d (3.2)	1.37 d (4.7) 1.92 d (2.2)	2.11 br d (13.5) 1.31 ddd (5.4, 12.4, 12.4)	2.11 br d (12.5) 1.43 ddd (5.4, 12.4, 12.4)
7	2.75 br d (5.3)	4.29 d (3.2)	4.9 d (3.2)	4.6 d (3.3)	2.77 m	2.8 m
14	6.46 s	6.79 s	6.82 s	6.54 s	6.55 s	6.56 s
15	3.17 sept (6.8)	3.04 sept (6.5)	3.06 sept (7.0)	3.23 sept (7.0)	3.24 sept (7.0)	3.28 sept (7.0)
16	1.19 d (6.8)	1.24 d (6.5)	1.23 d (7.0)	1.22 (7.0)	1.23 d (7.0)	1.2 d (7.0)
17	1.17 d (6.8)	1.24 d (6.5)	1.23 d (7.0)	1.22 (7.0)	1.23 d (7.0)	1.19 d (7.0)
18	1.06 s	1.01 s	1.05 s	1.19 s	1.15 s	1.13 s
19	3.97 d (11.4) 4.24 d (11.4)	0.99 s	4.05 s	4.16 d (11.8) 4.24 d (11.8)	5.61 s	5.61 s
OCOCH$_3$	1.86 s	2.07 s	2.09 s	2.08 s		
OCH$_3$	3.75 s	3.66 s	3.66s	3.76 s		3.86 s

Singlet (s); doublet (d); septuplet (sept); multiplet (m); broad doublet (br d); doublet doublet (dd); * not well defined; doublet doublet doublet (ddd).

Table 2. NMR spectroscopic data assignments (100 MHz) for compounds **1–6** (δ in ppm) in CDCl$_3$.

N°	1	2	3	4	5	6
	^{13}C	^{13}C	^{13}C	^{13}C	^{13}C	^{13}C
1	34.6	25.3	27.1	28.3	35.4	35.0
2	19.7	24.3	18.4	20.9	21.2	21.0
3	37.1	77.5	33.2	28.3	32.5	32.7
4	36.9	36.0	51.0	45.0	37.0	36.9
5	44.3	51.5	50.9	51.8	50.3	50.2
6	20.0	74.0	77.4	21.0	21.6	21.4
7	32.9	77.4	73.9	78.7	30.7	30.9
8	126.8	126.6	126.1	133.7	122.3	123.8
9	133.8	123.4	123.8	119.1	128.0	133.3
10	49.0	46.5	47.0	44.0	49.8	49.4
11	148.9	141.0	142.4	147.9	143.4	149.7
12	143.7	141.8	141.8	143.0	142.1	146.3
13	139.6	134.7	134.8	140.2	133.9	141.8
14	117.8	120.4	119.1	117.8	118.9	118.4
15	26.5	27.4	27.3	26.5	27.2	26.7
16	23.5	22.4	22.2	22.7	22.3	23.5
17	23.5	22.3	22.4	23.5	24.1	23.3
18	27.6	16.2	26.0	25.5	24.1	24.1
19	68.7	26.7	66.2	65.7	102.8	103.3
20	181.7	178.1	178.5	174.3	180.0	180.0
O<u>C</u>OCH$_3$	172.5	170.7	170.8	170.9		
OCO<u>C</u>H$_3$	20.7	21.2	20.9	20.9		
O<u>C</u>H$_3$	61.1	58.4	58.4	61.9		58.6

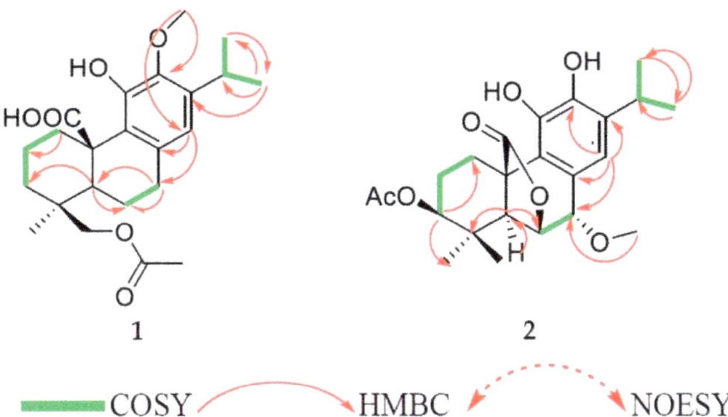

Figure 2. Key ^{1}H–^{1}H correlation spectroscopy (COSY), heteronuclear multiple bond correlation (HMBC), and nuclear Overhauser effect spectroscopy (NOESY) correlations of **1** and **2**.

Compound **5** was isolated as a white powder. The HRMS data indicated an ion peak at [M]$^+$ 345.1694 *m/z*, suggesting a possible chemical formula of C$_{20}$H$_{28}$O$_5$. Its IR spectrum exhibited bands at 1675 cm^{-1} for a lactone-carbonyl, as well as at 3500 and 3210 cm^{-1} attributed to hydroxyl groups. The UV spectrum showed two peaks at 208 and 284 nm. The NMR spectra of compound **5** (Tables 1 and 2) showed similar signals to compound **1**, the difference between them being the absence of the acetoxyl and methoxy1 groups and the appearance of the dioxygenated-carbon signal at 102.8 (C-19), attached to a singlet proton at 5.6, which forms a lactone ring with C-20 (180.1). In particular, the HMBC showed a correlation between a proton at 5.6 (H-19) and the C-20 (180.1)/C-4 (37.0)/C-5 (50.2). Other 2D spectra in comparison with literature data confirmed the structure of compound **5** as clinopodiolide A [33]. On the other hand, compound **6** showed typical NMR signals similar to the ones of compound **5** (Tables 1 and 2), except for the presence of an extra methoxyl signal, which was placed on C-12 from the HMBC correlations. Other 2D spectra in comparison with literature data confirmed the structure of compound **6** as clinopodiolide B [33]. The occurrence of the lactol moiety at C-19–C-20 is very unusual in nature and it has been recently isolated for the first time from *Salvia clinopodioide*. To the best of our knowledge, these compounds (**5** and **6**) have been isolated for the first time from *S. africana lutea*.

3.2. In Vitro Bioactivity

3.2.1. Alpha-Glucosidase and Alpha-Amylase Activities

Alpha-glucosidase is an enzyme located in the brush border of the small intestine epithelium, which catalyzes the breaking down of the reaction of disaccharides and starch to glucose. Glucosidase inhibitors reduce the rate of carbohydrate digestion and delay the carbohydrate absorption from the alimentary tract [29]. Alpha-amylase is one of the main enzymes in humans that is directly involved in the breakdown of starch to simpler sugars [30].

It hydrolyses complex polysaccharides to produce oligosaccharides and disaccharides, which are then hydrolyzed by alpha-glucosidase to monosaccharide, which are absorbed through the small intestines into the hepatic portal vein and increase postprandial glucose levels. The inhibitory mechanisms of these enzymes are characterized by delaying carbohydrate digestion and reducing the rate of glucose absorption [34]. The bio-evaluation of natural resources for the antidiabetic properties has been intensified, and a great deal of research is being carried out to identify plants with potent anti-diabetic activity with emphasis on the inhibition of the two enzymes, alpha-glucosidase and alpha-amylase. In this study, the inhibitory activity of the isolated compounds from *S. africana lutea* was investigated and the results showed that compound **8** demonstrated the highest alpha-glucosidase inhibitory activity with IC_{50} value of 11.3 ± 1.0 μg/mL, followed by compounds **10** and **7** with IC_{50} values of 17.1 ± 1.0 μg/mL and 22.9 ± 2.0 μg/mL, as indicated in Table 3. The IC_{50} value of compound **8** is consistent with the previously reported value of 12.1 ± 1.0 μM [35]. The higher inhibitory activity demonstrated by compound **8** (compared to compound **7**) could be explained by the shift of the C-29 methyl group from C-20 to C-19, which has enhanced the inhibition of the alpha-glucosidase enzyme [35]. In addition, the lowest alpha-glucosidase inhibitory activity demonstrated by compound **10** among the tested triterpenes might be due to the absence of the carboxylic group in its chemical skeleton. Among all the tested abietane diterpenes, only compound **6** demonstrated moderate alpha-glucosidase inhibitory activity, with an IC_{50} value of 81.7 ± 2.1 μg/mL.

Remarkably, compound **7** demonstrated the strongest alpha-amylase inhibitory activity among the tested compounds with an IC_{50} value of 12.5 ± 0.7 μg/mL, followed by compounds **8** and **10** with IC_{50} values of 66.1 ± 2.0 μg/mL and 76.6 ± 2.1 μg/mL, respectively. None of the tested abietane diterpenes showed alpha amylase inhibitory activity, as shown in Table 3.

Table 3. Inhibitory activities of the isolated compounds on alpha-glucosidase and alpha-amylase.

Items	Alpha-GlucosidaseIC_{50} (μg/mL)	Alpha-AmylaseIC_{50} (μg/mL)
1	NA	NA
2	NA	NA
3	NA	NA
4	NA	NA
5	NA	NA
6	81.7 ± 2.1	NA
7	22.9 ± 2.0	12.5 ± 0.7
8	11.3 ± 1.0	66.1 ± 2.0
9	85.8 ± 2.3	NA
10	17.1 ± 1.0	76.6 ± 2.1
Acarbose	610.4 ± 1.0	10.2 ± 0.6

Not active (NA) at the test concentrations. The results are expressed as mean ± SEM for $n = 3$.

3.2.2. Antioxidant Activity

The in vitro antioxidant activity of the isolated compounds of the methanolic extract of *S. africana-lutea* were investigated by evaluating their ferric-ion reducing antioxidant power (FRAP), trolox equivalent absorbance capacity (TEAC), and oxygen radical absorbance capacity (ORAC) activities. The TEAC and FRAP are assays based on a single electron transfer (SET) mechanism, in which the antioxidant transfers an electron to the corresponding cationic radical to neutralize it [36], while ORAC is based on a hydrogen atom transfer (HAT) mechanism, in which the antioxidant exhibits the potential health-beneficial roles via transferring hydrogen atom(s) to the reactive species, thereby deactivating them [37]. The results demonstrated that compounds **1** and **5** exhibited strong activity on ORAC (2588.2 ± 10.1; 2357.2 ± 0.1) µM TE/g, respectively. Compounds **5** and **6** showed moderate activities on TEAC (862.2 ± 1.4; 705.5 ± 2.0) µM TE/g, whereas compounds **5** and **2** demonstrated significant inhibitory activity on FRAP (2262.9 ± 11.0; 2200.9 ± 14.2) µM AAE/g when compared to the reference antioxidant epigallocatechingallate (EGCG), as shown in Table 4. Phenolic compounds have been reported to be responsible for the antioxidant activity of numerous plant species by stabilization of radicals by donating electrons or by metal ion complexation, among other mechanisms. Nevertheless, other aspects can be considered, for example, the presence of vicinal hydroxyl groups is essential in a pronounced antioxidant activity [38]. Abietane diterpenes are known for having strong antioxidant activity due to the presence of ortho-dihydroxyl groups/vicinal hydroxyls in the aromatic ring that serve as hydrogen and/or electron donating agents to the corresponding reactive species leading to the formation of the stable quinone derivatives [39]. It has been shown that the phenolic group at the 11 position would be more implicated in the antioxidant activity [40]. In general, phenolic compounds are expected to transfer electrons or donate protons to the reactive radicals because of the resonance stability of the phenoxy radical [41].

Table 4. Antioxidant activities of the isolated compounds.

Items	ORAC (µmole TE/g)	TEAC (µmole TE/g)	FRAP (µM AAE/g)
1	2588.2 ± 10.1	694.0 ±1.6	1217.4 ± 2.9
2	2233.9 ± 8.0	635.7 ± 0.8	2200.9 ± 14.2
3	735.4 ± 2.0	124.4 ± 0.6	1440.4 ± 9.1
4	559.7 ± 15.2	440.1 ± 1.5	1257.0 ± 6.7
5	2357.2 ± 0.1	862.2 ± 1.4	2262.9 ± 11.0
6	1502.5 ± 21.2	724.9 ± 1.3	1480.0 ± 3.9
7	NA	NA	347.8 ± 3.7
9	NA	NA	283.4 ± 3.9
8	NA	NA	778.9 ± 6.8
10	NA	NA	412.2 ± 13.0
EGCG	3976.8 ± 3.8	4146.4 ± 19.8	7525.0 ± 4.9

Not active (NA) at the test concentrations; epigallocatechingallate (EGCG). Trolox equivalent absorbance capacity (TEAC); oxygen radical absorbance capacity (ORAC); ferric-ion reducing antioxidant power (FRAP). The results are expressed as mean ± SEM for $n = 3$.

The structure-activity relationship (SAR) of compound **5** could be related to the presence of ortho-dihydroxyl groups/vicinal hydroxyls in the aromatic ring as well as the presence of the free hydroxyl groups at the 19 position in its chemical structure, which are responsible for its high activity, demonstrated when compared to compound **6**. However, the substitution of the free OH at the 12 position by a methoxyl group is directly related to the decrease of the activity observed. The presence of an acetoxy group in compound **1** could be responsible for its high capacity of hydrogen atom transfer, demonstrated in the ORAC assay. However, the antioxidant activity of compound **2** demonstrated in FRAP is high because of the high-stress lactone ring, which may open during the course of the chemical reaction leading to an extension of the conjugation and formation of the p-quininoidal structure.

Compound **1**: Yellow, amorphous solid; $[\alpha]^{28}_D$ +70.43 (0.1, MeOH); UV (MeOH) λ_{max} (log ε) 210 (4.31), 280 (3.79); nm; IR (KBr) ν_{max} 3447, 3320, 2954, 1725, 1570, 1381, 1250, 1039 cm^{-1}; ^{1}H; and ^{13}C

NMR data, see Tables 1 and 2; positive-ion high-resolution electrospray ionisation mass spectrometry (HRESIMS) [M−H]$^+$ *403.2115* (calcd for $C_{23}H_{32}O_6$, 404.2199).

Compound **2**: Red brownish powder; $[\alpha]^{28}_D$ +22.92 (0.2, MeOH); UV (MeOH) λ_{max} (log ε) 212 (4.32), 287 (4.09); nm; IR (KBr) ν_{max} 3447, 3300, 2958, 1727, 1575, 1246, 1034 cm^{-1}; ^{1}H; and ^{13}C NMR data, see Tables 1 and 2; positive-ion HRESIMS [M−H]$^+$ *417.1890* (calcd for $C_{23}H_{32}O_6$, 418.1932).

Compound **3**: Brown amorphous powder; $[\alpha]^{28}_D$ +49.04 (0.07, MeOH); UV (MeOH) λ_{max} (log ε) 210 (4.31), 295 (3.92); nm; IR (KBr) ν_{max} 3450, 2962, 1769, 1634, 1435, 1239, 1034 cm^{-1}; ^{1}H; and ^{13}C NMR data, see Tables 1 and 2; positive-ion HRESIMS [M−H]$^+$ *417.1907* (calcd for $C_{23}H_{32}O_6$, 418.20).

Compound **4**: Yellow amorphous powder; $[\alpha]^{28}_D$ -53.27 (0.03, MeOH); UV (MeOH) λ_{max} (log ε) 210 (4.31), 281 (4.13); nm; IR (KBr) ν_{max} 3330, 2950, 1750, 1617, 1446, 1243, 1050 cm^{-1}; ^{1}H; and ^{13}C NMR data, see Tables 1 and 2; positive-ion HRESIMS [M−H]$^+$ 401.1609 (calcd for $C_{23}H_{32}O_6$, 402.2042).

4. Conclusions

The phytochemical and in vitro bio-activity investigation of the *S. africana-lutea* methanolic extract revealed that this plant is a rich source of abietane diterpenes and triterpenes with significant alpha glucosidase and alpha amylase inhibitory activities, as well as significant antioxidant activity when considering the FRAP, TEAC, ORAC assays. The present work is the first scientific report on *S. africana-lutea*, and the results suggest that the methanolic extract of this plant and/or its individual constituents might become prominent natural therapeutic agents for the inhibition of alpha glucosidase and alpha amylase enzymes and oxidative stress, which both play an important role in the development of diabetic related diseases. Therefore, compounds with high antioxidant and anti-diabetic activities are the most logical choice for reducing diabetes-induced ROS.

Supplementary Materials: The supplementary materials are available online at http://www.mdpi.com/2076-3921/8/10/421/s1, Supplementary Figures 1–10: ^{1}H-NMR (400 MHz, CDCl3) Spectrum, ^{13}C-NMR (400 MHz, CDCl3) Spectrum, DEPT-NMR (400 MHz, CDCl3) Spectrum, COSY (400 MHz, CDCl3) Spectrum, HSQC (400 MHz, CDCl3) Spectrum, HMBC (400 MHz, CDCl3) Spectrum, NOESY (400 MHz, CDCl3) Spectrum, HR-MS Spectrum, UV Spectrum and FTIR Spectrum of Compound **1**; Supplementary Figures 11–20: ^{1}H-NMR (400 MHz, CDCl3) Spectrum, ^{13}C-NMR (400 MHz, CDCl3) Spectrum, DEPT-NMR (400 MHz, CDCl3) Spectrum, COSY (400 MHz, CDCl3) Spectrum, HSQC (400 MHz, CDCl3) Spectrum, HMBC (400 MHz, CDCl3) Spectrum, NOESY (400 MHz, CDCl3) Spectrum, HR-MS Spectrum, UV Spectrum and FTIR Spectrum of Compound **2**; Supplementary Figures 21–28: ^{1}H-NMR (400 MHz, CDCl3) Spectrum, ^{13}C-NMR (400 MHz, CDCl3) Spectrum, DEPT-NMR (400 MHz, CDCl3) Spectrum, HSQC (400 MHz, CDCl3) Spectrum, HMBC (400 MHz, CDCl3) Spectrum, HR-MS Spectrum, UV Spectrum, FTIR Spectrum of Compound **3**; Supplementary Figures 29–37: ^{1}H-NMR (400 MHz, CDCl3) Spectrum, ^{13}C-NMR (400 MHz, CDCl3) Spectrum, DEPT-NMR (400 MHz, CDCl3) Spectrum, COSY (400 MHz, CDCl3) Spectrum, HSQC (400 MHz, CDCl3) Spectrum, HMBC (400 MHz, CDCl3) Spectrum, HR-MS Spectrum, UV Spectrum, FTIR Spectrum of Compound **4**.

Author Contributions: Designing and performing of experiments and manuscript drafting, N.G.E.R.E.; biological activity experiments, data curation, and manuscript drafting, J.A.B.; characterization of compounds, T.T.W.; bioactivity experiments, data curation, and manuscript drafting, J.L.M.; bioactivity experiments and plant taxonomy, C.N.C.; conceptualization, project supervision, and manuscript review, A.A.H.; and project supervision and manuscript review, E.I.I.

Funding: The National Research Foundation (NRF) of South Africa provided the research grants (NRF SARChI Chair UID 85102 and NRF CPRR160506164193) for the project.

Acknowledgments: The authors would like to thank the National Research Foundation (NRF) of South Africa for providing the funding for this project.

Conflicts of Interest: The authors declare that there is no conflict of interest.

References

1. Cao, A.; Tang, Y.; Liu, Y. Novel Fluorescent Biosensor for α-Glucosidase Inhibitor ScreeningBased on Cationic Conjugated Polymers. *ACS Appl. Mater. Interfaces* **2012**, *4*, 3773–3778. [CrossRef] [PubMed]
2. Mamun-or-Rashid, A.N.M.; Hossain, M.S.; Hassan, N.; Dash, B.K.; Sapon, M.A.; Kumer, S.M.A. Review on medicinal plants with antidiabetic activity. *J. Pharmacogn. Phytochem.* **2014**, *3*, 149–159.

3. Ullah, A.; Khan, A.; Khan, I. Diabetes mellitus and oxidative stress—A concise review. *Pharm. J.* **2016**, *24*, 547–553.

4. Liguori, I.; Russo, G.; Curcio, F.; Bulli, G.; Aran, L.; Della-Morte, D.; Gargiulo, G.; Testa, G.; Cacciatore, F.; Bonaduce, D.; et al. Oxidative stress, aging, and diseases. *Clin. Interv. Aging* **2018**, *13*, 757–772. [CrossRef] [PubMed]

5. Wright, E.S.B. Oxidative stress in type 2 diabetes: The role of fasting and postprandial glycaemia. *Int. J. Clin. Pract.* **2006**, *60*, 308–314. [CrossRef] [PubMed]

6. Giacco, F.B. Oxidative stress and diabetic complications. *Circ. Res.* **2010**, *107*, 1058–1070. [CrossRef] [PubMed]

7. Mohiuddin, M.; Arbain, D.; Shafiqul Islam, A.K.M.; Ahmad, M.S.; Ahmad, M.N. Alpha-glucosidase enzyme Biosensor forthe Electrochemical Measurement ofAntidiabetic Potential of Medicinal Plants. *Nanoscale Res. Lett.* **2016**, *11*, 95. [CrossRef]

8. Mohammed, A.; Ibrahim, M.A.; Islam, M.S. African Medicinal Plants with Antidiabetic Potentials: A Review. *Planta Med.* **2014**, *80*, 354–377. [CrossRef]

9. Lorenzati, B.; Zucco, C.; Miglietta, S.; Lamberti, F.; Bruno, G. Oral Hypoglycemic Drugs: Pathophysiological Basis of Their Mechanism of Action. *Pharmaceuticals* **2010**, *3*, 3005–3020. [CrossRef]

10. Marín-Peñalver, J.J.; Martín-Timón, I.; Sevillano-Collantes, C.; del Cañizo-Gómez, F.J. Type 2 diabetes and cardiovascular disease: Have all risk factors the same strength. *World J. Diabetes* **2016**, *7*, 354–395. [CrossRef]

11. Sofowora, A.; Ogunbodede, E.; Onayade, A. The Role and Place of Medicinal Plants in the Strategies for Disease Prevention. *Afr. J. Tradit. Complement. Altern. Med.* **2013**, *10*, 210–229. [CrossRef] [PubMed]

12. Manning, J.; Goldblatt, P. *Plants of the Greater Cape Floristic Region 1: The Core Cape flora*; South African National Biodiversity Institute: Pretoria, South Africa, 2012.

13. Arief, M.M.H.; Hussein, A.A.F.; Mohammed, A.; Elmwafy, H.M. Chemical and Bioactivity Studies on Salvia sfricana-Lutea: Cytotoxicity and Apoptosis Induction by Abietane Diterpenes Isolated from Salvia Africana-Lutea. *J. Basic Environ. Sci.* **2018**, *5*, 72–79.

14. Hussein, A.; Meyer, J.M.; Jimeno, M.L.; Rodrıguez, B. Bioactive Diterpenes from Orthosiphon labiatus and Salvia africana-lutea. *J. Nat. Prod.* **2007**, *70*, 293–295. [CrossRef]

15. Gonzalez, M.A. Aromatic abietane diterpenoids: Their biological activity and synthesis. *Nat. Prod. Rep.* **2015**, *32*, 684–704. [CrossRef]

16. Nazaruk, J.; Borzym-Kluczyk, M. The role of triterpenes in the management of diabetes mellitus and its complications. *Phytochem. Rev.* **2015**, *14*, 675–690. [CrossRef] [PubMed]

17. Mao Song, H.; Li, X.; Liu, Y.Y.; Lu, W.P.; Cui, Z.H.; Zhou, L.; Yao, D.; Zhang, H.M. Carnosic acid protects mice from high-fat diet-induced NAFLD by regulating MARCKS. *Int. J. Mol. Med.* **2018**, *42*, 193–207.

18. Lipina, C.; Hundal, H.S. Carnosic acid stimulates glucose uptake in skeletal muscle cells via a PME-1/PP2A/PKB signalling axis. *Cell Signal* **2014**, *26*, 2343–2349. [CrossRef]

19. Samarghandian, S.; Borji, A.; Farkhondeh, T. Evaluation of Antidiabetic Activity of Carnosol (Phenolic Diterpene in Rosemary) in Streptozotocin-Induced Diabetic Rats. *Cardiovasc. Hematol. Disord. Drug Targets* **2017**, *17*, 11–17. [CrossRef]

20. Vlavcheski, F.; Baron, D.; Vlachogiannis, I.A.; MacPherson, R.E.K.; Tsiani, E. Carnosol Increases Skeletal Muscle Cell Glucose Uptake via AMPK-Dependent GLUT4 Glucose Transporter Translocation. *Int. J. Mol. Sci.* **2018**, *19*, 29. [CrossRef]

21. Castellano, J.M.; Guinda, A.; Delgado, T.; Rada, M.; Cayuela, J.A. Biochemical Basis of the Antidiabetic Activity of Oleanolic Acid and Related Pentacyclic Triterpenes. *Diabetes* **2013**, *62*, 1791–1799. [CrossRef]

22. Ramu, R.; Shirahatti, P.S.; Nanjunda, S.S.; Zameer, F.; Dhananjaya, B.L.; Nagendra, P.M.N. Assessment of In Vivo Antidiabetic Properties of Umbelliferone and Lupeol Constituents of Banana (Musa sp. var. Nanjangud Rasa Bale) Flower in Hyperglycaemic Rodent Model. *PLoS ONE* **2016**, *11*, e0151135.

23. Silva, F.S.G.; Oliveira, P.J.; Duarte, M.F. Oleanolic, Ursolic, and Betulinic Acids as Food Supplements orPharmaceutical Agents for Type 2 Diabetes: Promise or Illusion? *J. Agric. Food Chem.* **2016**, *64*, 2991–3008. [CrossRef] [PubMed]

24. Telagari, M.; Hullatti, K. In-vitro α-amylase and α-glucosidase inhibitory activity of Adiantum caudatum Linn. and Celosia argentea Linn. extracts and fractions. *Indian J. Pharmacol.* **2015**, *47*, 425–429.

25. Benzie, I.F.F.J.; Strain, J. Ferric reducing/antioxidant power assay: Direct measure of total antioxidant activity of biological fluids and modifies version for simultaneous measurement of total antioxidant power and ascorbic acid concentration. *Methods Enzymol.* **1999**, *299*, 15–27. [PubMed]

26. Fellegrini, N.; Re, R.; Yang, M.; Rice-Evans, C.A. Screening of dietary carotenoid-rich fruit extracts for antioxidant activities applying ABTS radical cation decolorisation assay. *Method Enzymol.* **1999**, *299*, 379–389.

27. Prior, R.L.; Hoang, H.; Gu, L.; Wu, X.; Bacchiocca, M.; Howard, L.; Hampschwoodill, M.; Huang, D.; Ou, B.; Jacob, R. Assays for hydrophilic antioxidant capacity (ORACFL). *J. Agric. Food Chem.* **2003**, *51*, 3273–3279. [CrossRef] [PubMed]

28. Gao, J.B.; Yang, S.J.; Yan, Z.R.; Zhang, X.J.; Pu, D.B.; Wang, L.X.; Li, X.L.; Zhang, R.H.; Xiao, W.L. Isolation, Characterization, and Structure–Activity Relationship Analysis of Abietane Diterpenoids from Callicarpa bodinieri as Spleen Tyrosine Kinase Inhibitors. *Nat. Prod.* **2018**, *81*, 998–1006. [CrossRef] [PubMed]

29. Kang, V.K.; BajpaiSun, C. Tyrosinase and α-Glucosidase Inhibitory Effects of an Abietane Type Diterpenoid Taxodone from Metasequoia glyptostroboides. *Natl. Acad. Sci. Lett.* **2015**, *38*, 399–402.

30. Nickavar, B.; Abolhasani, L.; Izadpanah, H. α-Amylase Inhibitory Activities of Six Salvia Species. *Iran. J. Pharm. Res.* **2008**, *7*, 297–303.

31. Batista, O.; Simoes, M.F.; Nascimento, J.; Riberio, S.; Duarte, A.; Rodriguez, B.; De La Torre, M.C. A rearranged abietane diterpenoid from plectranthus hereroensis. *Phytochemistry* **1996**, *41*, 571–573. [CrossRef]

32. Etsassala, N.G.E.R.; Adeloye, A.O.; El-Halawany, A.; Hussein, A.A.; Iwuoha, E.I. Investigation of In-Vitro Antioxidant and Electrochemical Activities of Isolated Compounds from Salvia chamelaeagnea P.J.Bergius Extract. *Antioxidants* **2019**, *8*, 98. [CrossRef] [PubMed]

33. Bustos-Brito, C.; Joseph-Nathan, P.; Burgueño-Tapia, E.; Martínez-Otero, D.; Nieto-Camacho, A.; Calzada, F.; Yépez-Mulia, L.; Esquivel, B.; Quijano, L. Structure and Absolute Configuration of Abietane Diterpenoids from Salvia clinopodioides: Antioxidant, Antiprotozoal, and Antipropulsive Activities. *J. Nat. Prod.* **2019**, *82*, 1207–1216. [CrossRef] [PubMed]

34. Thilagam, E.; Parimaladevi, B.; Kumarappan, C.; Mandal, S.C.J. α-Glucosidase and a-Amylase Inhibitory Activity of Senna surattensis. *Acupunct. Meridian Stud.* **2013**, *6*, 24–30. [CrossRef] [PubMed]

35. Zhang, B.W.; Xing, Y.; Wen, C.; Yu, X.X.; Sun, W.L.; Xiu, Z.L.; Dong, Y.S. Pentacyclic triterpenes as a-glucosidase and a-amylase inhibitors: Structure-activity relationships and the synergism with acarbose. *Bioorg. Med. Chem. Lett.* **2017**, *27*, 5065–5070. [CrossRef] [PubMed]

36. Pérez-Fons, L.; Garzón, M.T.; Micol, V. Relationship between the antioxidant capacity and effect of rosemary (*Rosmarinus officinalis* L.) polyphenols on membrane phospholipid order. *J. Agric. Food Chem.* **2010**, *58*, 161–171. [CrossRef]

37. Apak, R.; Ozyurek, M.; Güclu, K.; Çapanoglu, E. Antioxidant Activity/Capacity Measurement. 2. Hydrogen Atom Transfer (HAT)-Based, Mixed-Mode (Electron Transfer (ET)/HAT), and Lipid Peroxidation Assays. *J. Agric. Food Chem.* **2016**, *64*, 1028–1045. [CrossRef]

38. De Oliveira, F.D.; Aguiar, P.N.; Ribeiro, G.D.; de Amorim, M.L.; Guimaraes, P.S.; Mendonca Filho, C.V.; Sivieri, R.R.; Brandao, M.D.; dos Santos, W.T.; Grael, C.F. Antioxidant Activity and Phytochemical Screening of Extracts of Erythroxylum′ suberosum A.St.-Hil (Erythroxylaceae). *Res. J. Phytochem.* **2015**, *9*, 68–78.

39. Brewer, M.S. Natural Antioxidants: Sources, compounds, mechanisms of action, and potential applications. *Compr. Rev. Food Sci. Food Saf.* **2011**, *10*, 221–246. [CrossRef]

40. Masuda, T.; Kirikihira, T.; Takeda, Y. Recovery of antioxidant activity from carnosol quinone: Antioxidants obtained from a water-promoted conversion of carnosol quinone. *J. Agric. Food Chem.* **2005**, *53*, 6831–6834. [CrossRef]

41. Özgen, U.; Mavi, A.; Terzi, Z.; Kazaz, C.; Asci, A.; Kaya, Y.; Secen, H. Relationship between chemical structure and antioxidant activity of Luteolin and Its glycosides isolated from Thymus sipyleus subsp. sipyleus var. sipyleus. *Rec. Nat. Prod.* **2011**, *5*, 12–21. [CrossRef]

© 2019 by the authors. Licensee MDPI, Basel, Switzerland. This article is an open access article distributed under the terms and conditions of the Creative Commons Attribution (CC BY) license (http://creativecommons.org/licenses/by/4.0/).

 *antioxidants*

Article

Isolation of Strong Antioxidants from *Paeonia Officinalis* Roots and Leaves and Evaluation of Their Bioactivities

Lijana Dienaitė [1], Milda Pukalskienė [1], Audrius Pukalskas [1], Carolina V. Pereira [2], Ana A. Matias [2] and Petras Rimantas Venskutonis [1,*]

[1] Department of Food Science and Technology, Kaunas University of Technology, Radvilėnų. pl. 19, LT-50254 Kaunas, Lithuania
[2] IBET—Instituto de Biologia Experimental e Tecnológica, Food & Health Division Apartado 12, 2780-901 Oeiras, Portugal
* Correspondence: rimas.venskutonis@ktu.lt; Tel.: +370-699-40978 or +370-37-456647

Received: 27 June 2019; Accepted: 25 July 2019; Published: 27 July 2019

Abstract: *Paeonia officinalis* extracts from leaves and roots were tested for their antioxidant potential using in vitro chemical (Folin-Ciocalteu, 2,2-diphenyl-1-picrylhydrazyl radical (DPPH), 2,2′-azino-*bis*-3-ethylbenzothiazoline-6-sulfonic acid (ABTS), oxygen radical absorbance capacity (ORAC), hydroxyl radical antioxidant capacity (HORAC), hydroxyl radical scavenging capacity HOSC)) and cellular antioxidant activity (CAA) assays. Leaf extracts were stronger antioxidants than root extracts, while methanol was a more effective solvent than water in chemical assays. However, the selected water extract of leaves was a stronger antioxidant in CAA than the methanol extract (0.106 vs. 0.046 µmol quercetin equivalents/mg). Twenty compounds were identified by ultra performance liquid chromatography-quadrupole-time-of-flight (UPLC-Q-TOF) mass spectrometer, while on-line screening of their antioxidant capacity by high performance liquid chromatography (HPLC) with a DPPH$^\bullet$-scavenging detector revealed that gallic acid derivatives are the major peony antioxidants. Root water and leaf methanol extracts inhibited α-amylase in a dose dependent manner. The IC$_{50}$ value for the strongest inhibitor, the methanol extract of leaves, was 1.67 mg/mL. In addition, the cytotoxicity assessment of extracts using human Caco-2 cells demonstrated that none of them possessed cytotoxic effects.

Keywords: *P. officinalis*; phytochemicals; antioxidant capacity; cytotoxicity; α-amylase inhibition

---

1. Introduction

Plants biosynthesize a large number of various phytochemicals that have demonstrated antioxidant and health beneficial properties in numerous studies. Polyphenolics can be defined as a group of heterogeneous biologically active non-nutrients, belonging to the most important and widely investigated class of such phytochemicals [1]. In the last few decades, numerous publications have reported that high consumption of phytochemical-rich foods might reduce the risk of several diseases. Therefore, modifying one's diet by increasing the intake of fruits, vegetables, herbs and spices may be a promising strategy for cancer prevention.

Formation of free radicals and reactive oxygen species (ROS) is a normal process in human cells. However, excessive production of ROS, which may occur due to various exogenous factors, can disturb homeostatic conditions, resulting in oxidative stress, which may largely contribute to the development of chronic health problems, including cancer, inflammation, cardiovascular diseases and aging [2]. Therefore, it has been hypothesized that polyphenolic compounds and other dietary antioxidants, which are abundant in many fruits, vegetables, and botanicals, are essential nutrients protecting against harmful effects of the excessive free radicals [1]. Since ancient times, botanicals have been used in folk medicine to treat various diseases and health disorders, as well as in flavourings, fragrances, and colorants. However, only recently have the health benefits of many medicinal and spicy plants been explained by scientific evidence. For instance, a large number of polyphenolic antioxidants have been identified and characterized in aromatic herbs, spices, and other plant materials. Many of them have been proved as effective antioxidants [3], while their health benefits have been linked to various mechanisms, including the scavenging of harmful free radicals [1]. However, considering a vast number of species in the Plant Kingdom, there are still many poorly investigated plant species, which might be a good source of new natural bioactive substances including strong antioxidants.

Literature survey indicates that some species of the genus *Paeonia* (Paeoniaceae), which is divided into the three sections (*Moutan*, *Oneapia* and *Paeonia*) may be considered poorly studied plants [4]. The section of *Paeonia* consists of 25 species, which are widely distributed throughout temperate Eurasia. Many of them, including *P. officinalis*, remain under-investigated until now. For instance, among 262 compounds identified in different anatomical parts of *Paeonia* (terpenoids, tannins, flavonoids, stilbenes, steroids, paeonols, and phenols), only several anthocyanins were reported in *P. officinalis* flowers and one phenolic acid glycoside (1,2,3,6-*tetra*-O-galloyl-β-*d*-glucose) in its roots [4].

A large spectrum of bioactive substances found in different *Paeonia* spp. may be responsible for their biological and pharmacological activities. For instance, it has been used in folk medicine to treat epilepsy, liver diseases, diarrhea, and many other disorders [5]. Furthermore, antioxidant, antitumor, anti-inflammatory, anti-microbial, immune system modulation, central nervous, and cardiovascular system protective activities were also reported for *Paeonia* plants [6–10]. *P. lactiflora* roots are among the most traditional Chinese medicines and have been used for their anti-inflammatory, analgesic, blood tonifying, stringent, and menstruation regulation properties [11]. Besides the ornamental value of flowers, the seeds of some other *Paeonia* spp. (*P. lactiflora*, *P. suffruticosa*) have been considered as a rich source of polyunsaturated oil and proteins for foods [12]. *P. officinalis* has been used mainly for medicinal purposes, e.g., as an antiepileptic and antispasmodic drug, whereas its flowers were used to produce syrup [13].

Considering the rather scarce scientific knowledge reported for *P. officinalis*, the purpose of this study was to perform more systematic studies of its extracts, including its phytochemical composition and antioxidant activity, in order to provide more comprehensive data on the potential of *P. officinalis* as a source for functional ingredients for various applications.

2. Materials and Methods

2.1. Plant Material, Solvents and Chemicals

Paeonia officinalis roots and leaves of blooming plant were harvested at the Kaunas Botanical Garden of Vytautas Magnus University (Kaunas, Lithuania) and dried in a ventilated and protected from direct sunlight room. Dried leaves were separated from stems before further use.

Analytical grade methanol was purchased from StanLab (Lublin, Poland); formic acid (98%), liquid chromatography-mass spectrometry (LS-MS) and high performance liquid chromatography (HPLC) grade acetonitrile, 2,2′-azino-*bis*-3-ethylbenzothiazoline-6-sulfonic acid (ABTS, 98%), Folin-Ciocalteu reagent, monopotassium phosphate, gallic acid, 2,2-diphenyl-1-picrylhydrazyl radical (DPPH•, 98%), 2′,2′-azo-*bis*-(2-amidinopropane) dihydrochloride (AAPH), picolinic acid, 6-hydroxyl-2,5,7,8-*tetra*-methylchroman-2-carboxylic acid (Trolox), potassium chloride, H_2O_2, caffeic acid, cobalt (II) fluoride tetrahydrate, NaCl, and sodium phosphate monobasic monohydrate were from Sigma-Aldrich (Darmstadt, Germany); ethanol (99%) were from Scharlau (Barcelona, Spain); microcrystalline cellulose (20 μm) was from Sigma-Aldrich (Steinheim, Germany); sodium phosphate dibasic dihydrate, potassium iodine, and ferric chloride were from Riedel-de-Haen (Seelze, Germany); disodium fluorescein was from TCI Europe (Antwerpen, Belgium); α–amylase, type VI-B (from porcine pancreas), sodium carbonate, quercetin, and 2′,7′-dichlorofluorescin diacetate (DCFH-DA) were from

Sigma-Aldrich (St. Quentin Fallavier, France); potato starch was from Fluka (Buchs, Switzerland); acarbose was from Bayer Pharma AG (Leverkusen, Germany); epithelial colorectal adenocarcinoma cells (Caco-2) were from DSMZ (Braunschweig, Germany). Trypsin, Roswell Park Memorial Institute (RPMI) 1640, penicillin streptomycin, and heat inactivated fetal bovine serum were obtained from Invitrogen (Gibco, Paisley, UK); phosphate buffer saline solution (PBS) was from Sigma-Aldrich (St. Louis, MO, USA). Ultrapure water was produced in a Simplicity 185 system (Millipore, MA, USA).

2.2. Extraction Procedure

Dried leaves and roots were ground in a laboratory mill Vitek (An-Der, Austria) by using a 0.5 mm size sieve. Methanol extract was prepared in an accelerated solvent extractor ASE 350 (Dionex, Sunnyvale, CA, USA) from 10 g of material, which was mixed with 4 g of diatomaceous earth and placed in a 66 mL extraction cell. Extraction was carried out three times at 60 °C temperature and 10 MPa pressure with a 15 min static and a 90 s purge time for each extraction cycle (in total, 3 cycles). Peony leaves (10 g) were also extracted in a conical flask with 200 mL methanol by using a mechanical shaker (Sklo Union LT, Teplice, Czech Republic) at room temperature and 170 rpm. Water extracts were prepared from 5 g of leaves and 2 g of roots in a conical flask by suspending plant powder in 50 mL of distilled water at 80 °C and continuously stirring at 400 rpm by magnetic hotplate stirrer (IKA, Wilmington, DE, USA). The extraction procedures of methanol and water were repeated three times, each lasting 24 h and 15 min, respectively. After methanol (TR) and water extraction the solids were filtered through a 0.3 µm filter (Filtrac, Niederschlag, Germany) and combined. Methanol was removed in a Rotavapor R-114 (Büchi, Flawil, Switzerland), then additionally dried in a flow of nitrogen (20 min) and finally all methanol and water extracts were freeze dried in a Maxi Dry Lyo (Hetto-Holton AIS, Allerod, Denmark). Solid residues after each extraction were also dried and stored at −18 °C in a freezer until further analysis. Three replicate extractions were performed for each plant, material, solvent, and method. The extracts obtained are abbreviated by the following letters: P—peony, L—leaves, R—roots, M—methanol, W—water, ASE—accelerated solvent extraction, and TR—traditional extraction.

2.3. Determination of Antioxidant Potential by Single Electron Transfer Based Assays

Fast colorimetric methods were selected for the in vitro assessment of total phenolic content (TPC), DPPH$^\bullet$ scavenging, and ABTS$^{\bullet+}$ decolourization capacity. Detailed description of these methods are provided elsewhere [14]. Briefly, for TPC, 30 µL of extract at various concentrations were mixed in a 96-well microplate with 150 µL Folin-Ciocalteu reagent diluted in distilled water (1:10 *v/v*) and 120 µL 7% Na_2CO_3 solution. After shaking 10 s the absorbance was recorded at 765 nm in a FLUOstar Omega reader (BMG Labtech, Offenburg, Germany). The calibration curve was prepared using 10–250 µg/mL solutions of gallic acid in water. The TPC was expressed in mg gallic acid equivalents (GAE) per dry weight of plant (DWP) and dry weight of extract (DWE (GAE/g DWP and GAE/g DWE, respectively) from four replicate measurements.

For DPPH$^\bullet$ scavenging 8 µL of extract and 292 µL of DPPH$^\bullet$ (6×10^{-5} M) solutions in methanol were mixed in a 96-well microplate and the absorbance was recorded at 515 nm in a FLUOstar Omega reader every min during 60 min. Trolox solutions (299–699 µM/L) were used for the calibration curve, the results expressed as trolox equivalents (TE) in g of DWP and DWE from 4 replicate measurements.

For the ABTS$^{\bullet+}$ decolourization reaction, working solution of ABTS$^{\bullet+}$ was produced by mixing 50 mL of ABTS and 200 µL of potassium persulfate stock solutions and kept 15 h at room temperature in the dark. The absorbance was adjusted to 0.800 ± 0.020 at 734 nm with PBS (phosphate buffered saline) and 294 µL of ABTS$^{\bullet+}$ were mixed with 6 µL of methanolic extract solutions in a 96-well microplate. The absorbance was measured in a FLUOstar Omega reader at 734 nm during 30 min at 1 min intervals. Calibration curve was constructed by using Trolox solutions (399–1198 µM/L), and the results were expressed as µM TE/g DWP and DWE from 6 replicate measurements.

The QUick, Easy, New, CHEap and Reproducible (QUENCHER) method was applied to determine the antioxidant capacity of solid materials before and after extractions in order to evaluate the effectiveness of the recovery of antioxidants. Solid materials were mixed with microcrystalline cellulose at a ratio from 1:5 to 1:100 (TPC and ABTS$^{\bullet+}$), or from 1:1 to 1:100 (DPPH$^\bullet$). For TPC, 5 mg of sample/blank (cellulose) were mixed with 150 µL of MeOH:H$_2$O (1:4), 750 µL Folin-Ciocalteu's reagent, and 600 µL Na$_2$CO$_3$ solution, vortexed for 15 s, shaken at 250 rpm for 3 h in the dark, centrifuged (4500 rpm, 10 min), and the absorbance of the optically clear supernatant was measured at 765 nm. For the DPPH$^\bullet$ scavenging assay, 5 mg of sample/blank were mixed with 40 µL of MeOH:H$_2$O (1:4) and 1960 µL of DPPH$^\bullet$ methanolic solution, vortexed 60 s, shaken at 250 rpm for 25 min in the dark, centrifuged (for 4800 rpm, 3 min), and the absorbance of the optically clear supernatant was measured at 515 nm. For ABTS$^{\bullet+}$ decolorisation, 5 mg of sample/blank were mixed with 40 µL of PBS and 1960 µL of working ABTS$^{\bullet+}$ solution, vortexed for 60 s, shaken at 250 rpm for 30 min in the dark, centrifuged (4800 rpm, 3 min), and the absorbance of optically clear supernatant was measured at 734 nm. The values were calculated for g DWP from 6 replicate measurements.

2.4. Determination of Antioxidant Potential by Peroxyl-Radicals Inhibition Assays Based Hydrogen Atom Transfer

Oxygen radical absorbance capacity (ORAC), hydroxyl radical antioxidant capacity (HORAC), and hydroxyl radical scavenging capacity (HOSC) assays, which are based on peroxyl-radicals inhibition (HAT), and therefore are considered as more relevant to the processes in the biological systems [15], were selected for the characterisation of peony extracts. Detailed description of these methods are provided elsewhere [14]. Briefly, in ORAC, 25 µL of trolox standards, antioxidant, and 150 µL 2×10^{-7} mM of FL (fluorescein) solutions in PBS (75 mM, pH 7.4) were placed in a black 96-well microplate. Then, the mixture was preincubated in a FL800 microplate fluorescence reader (Bio-Tek Instruments, Winooski, VT, USA) for 10 min at 37 °C. To start the reaction 25 µL of AAPH (153 mM), which was used as a source of peroxyl radical, were added to each well automatically through the injector coupled with the FL800 microplate reader. The fluorescence was recorded every min during 1 h at 485 ± 20 nm excitation and 530 ± 25 nm emission. Trolox solutions (5–40 µM) were used for calibration, the results expressed in µM TE/g DWP and DWE.

For HORAC, 30 µL of sample/standard/blank, 170 µL of FL (9.28×10^{-8} M) and 40 µL of H$_2$O$_2$ (0.1990 M) solutions were pipetted into a black microplate, 60 µL of CoF$_2$ (3.43 mM) solution was added, and the microplate was placed on a FL800 microplate fluorescence reader for 60 min at 37 °C using 485 nm excitation and 530 nm emission filters. The calibration curve was constructed by using 50 to 250 µM caffeic acid containing solutions. SPB (sodium phosphate buffer) (75 mM, pH 7.4) was used for hydrogen peroxide and fluorescein preparation, while acetone:Milli-Q water (50:50 *v/v*) was used as a blank and for the preparation of sample, calibration, and CoF$_2$ solutions. Data was expressed as µM caffeic acid equivalents (CAE) per g DWP and DWE.

For the HOSC assay to 30 µL of blank/standard/sample prepared in acetone:MilliQ water (50:50 *v/v*), 40 µL of H$_2$O$_2$ (0.1990 M), and 170 µL of FL solution (9.28×10^{-8} M), 60 µL FeCl$_3$ solution (3.43 mM) were pipetted in a black microplate, which was immediately placed in the FL800 reader, and the fluorescence was recorded every min during 1 h at 37 °C using 485 ± 20 nm excitation and 530 ± 25 nm emission filters. FeCl$_3$ and H$_2$O$_2$ solutions were prepared in ultrapure water, while SPB (75 mM, pH = 7.4) was used to prepare the solution of FL. Trolox containing solutions (5–30 µL) were used for the calibration curve, and antioxidant capacity values were expressed as µM TE/g. Mean values in all these assays were calculated for g DWP and DWE from 4 replicate measurements.

2.5. Evaluation of HPLC-DPPH$^\bullet$ Scavenging On-Line

The HPLC (high performance liquid chromatography) system consisted of a Rheodyne 7125 manual injector (Rheodyne, Rohnert Park, CA, USA), a Waters 996 photodiode array detector (Milford, MA, USA), and a Waters 1525 binary pump (Milford, MA, USA). Separation of compounds was performed at 40 °C on a Hypersil C18 analytical column (5 µm, 250 × 0.46 cm, Thermo scientific, USA). The linear binary gradient was used at a constant flow rate 0.8 mL/min with solvent A (1% formic acid in ultrapure water) and B (acetonitrile) by using the following gradient elution order: 0 min 5% B, 0–2 min 10% B; 2–40 min 22% B; 40–55 min 100% B, 55–60 min 5%. 20 µL of the sample was injected and the spectra were recorded in the range from 210 to 450 nm. In order to detect radical scavengers, the HPLC system was coupled with an Agilent 1100 series pump (Agilent Technologies, USA) supplying freshly prepared methanolic DPPH$^\bullet$ (5×10^{-6} M) solution into a reaction coil (15 m, 0.25 mm ID) at a flow rate of 0.6 mL/min. A Shimadzu SPD-20A UV detector (Shimadzu Corporation, Kyoto, Japan) was used to record the negative peaks due to a decrease of absorbance at 515 nm after the reaction of radical scavengers with DPPH$^\bullet$. For the preliminary identification of compounds, a quadrupole mass detector Micromass ZQ (Waters, Milford, MA, USA) in negative ionization mode was used with the following parameters: cone voltage 30 V; cone gas flow 80 L/h; desolvation temperature 300 °C; desolvation gas flow 310 L/h; capillary voltage 3000 V; source temperature 120 °C; scanning range 100 to 1200 m/z. Chromatographic conditions were used as described above. Diluted extracts in methanol and water mixture (50:50, v/v) were analyzed.

2.6. Ultra Performance Liquid Chromatography/Electron Spray Ionisation Quadrupole Time-of-Flight Mass Spectrometry (UPLC/ESI-QTOF-MS) Analysis

UPLC/ESI-QTOF-MS analysis was carried out on a Waters Acquity UPLC system (Milford, MA) combined with a MaXis 4G Q-TOF (quadrupole time of flight) mass spectrometer, a sample manager, PDA detector, binary solvent manager, and controlled by HyStar software (Bruker Daltonic, Bremen, Germany). The Acquity C18 column (1.7 µm, 100 mm × 2.1 mm i.d. Waters, Milford, MA, USA) at a separation temperature 40 °C was used. The mobile phase consisting of solvent A (1% formic acid in ultrapure water) and solvent B (acetonitrile) was eluted in the following order: 95% A at 0.0 min; 0–5% B at 0 min; 5–20% B at 0–9 min; 20–50% B at 9–12 min, 50–100% B at 12–14 min, and held these condition for 1 min. Finally, the initial conditions were re-introduced over 1 min and held for 1 min. Before each run, the column was equilibrated for an additional 1 min. The following parameters were: a capillary voltage of 4 kV; an end plate offset of −0.5 kV; a flow rate of drying-gas of 10.0 L/min; a nebulizer pressure of 2.5 bar; a scanning range 79–2400 m/z; an injection volume of 1 µL; a flow rate of 0.45 mL/min. For fragmentation study, a data dependent scan was performed by deploying collision induced dissociation (CID) using nitrogen as a collision gas at 30 eV. Peaks were identified and analysed by comparing their retention times, accurate masses, and formulas by using external standards and commercial databases.

Quantitative analysis was performed by using external standards. Calibration curves were drawn using six concentrations of standard solutions and represented the dependence between the integrated chromatographic peak areas and the corresponding amounts of injected standards. According to the lowest point of the calibration curve, the LOQ (S/N = 10) and LOD (S/N = 3) were calculated.

2.7. Determination of α-Amylase Inhibitory Activity

An α-Amylase assay was carried out according to the procedure of Al-Dabbas et al. [16]. Briefly, 60 µL of blank (PBS)/sample/acarbose (0.02 mg/mL) solutions and 200 µL of a starch solution (400 µg/mL) were mixed in 6 Eppendorf tubes for incubation at 37 °C for 5 min. To start the reaction, twenty µL of α-amylase (5 mg/mL) was added to the three tubes and in the rest of three twenty µL of PBS (pH = 7.4) as a control for the sample. Afterwards, twenty µL of PBS was added to all tubes and incubated at 37 °C for 7.5 min. After incubation, in order to determine the degradation of the starch, 200 µL of I$_2$ solution (0.01 M) was transferred to the Eppendorf tubes. Finally, by adding 1 mL of a distilled H$_2$O reaction, the mixture was diluted and its absorbance was measured at 660 nm by using a GENESYS 10 spectrophotometer (Thermo Scientific, Waltham, MA, USA). For the calibration, a curve stock solution of acarbose (5 mg/mL) was used. The inhibition of enzyme activity (IC$_{50}$) was expressed as mg/mL. The experiments were performed in triplicate.

2.8. Cytotoxicity Assay in Caco-2 Cells

The cells were maintained as monolayers in 175 cm^2 culture flasks containing an RPMI 1640 medium supplemented with 10% fetal bovine serum and 1% penicillin-streptomycin, at 37 °C in humidified air with 5% CO_2. Caco-2 cells were seeded at a density of 2×10^4 cells/well in transparent 96-well plates and allowed to grow as a confluent and non-differentiated monolayer that can mimic the human intestinal epithelium. This cell model shares some characteristics with crypt enterocytes, and thus it has been considered to be an accepted intestinal model widely implemented to assess the effect of chemical, food compounds, and nano/microparticles on the intestinal function [17]. The medium was changed every 2 days. On the day of the experiment, cells were washed twice with pre-warmed PBS at ~37 °C. Water and ethanol extracts of peony were solubilized in H_2O and EtOH, respectively. All extracts were prepared with a final concentration of 16.7 mg/mL. Cell-based assays were performed using a maximum concentration of solvent—50% and 5% for H_2O and ethanol, respectively.

Cytotoxicity was assessed by the MTS (3-(4,5-dimethylthiazol-2-yl)-2,5-dyphenyltetrazolium bromide) method [17]. Prepared cells monolayers were incubated with various concentrations of peony extracts (100 µL) for 4, 24, and 48 h and washed with PBS. Their viability was determined by adding a 100 µL 10-fold diluted MTS reagent (according to the manufacturer's guidelines), incubating for 2.5 h at 37 °C 5% CO_2, measuring the absorbance at 490 nm in an Epoch Microplate Spectrophotometer (Bio-Tek, Instruments, Winooski, VT, USA). Data were expressed in a cellular viability percentage relative to control (%). The experiments were performed in triplicate.

2.9. Cellular Antioxidant Activity Assay (CAA)

The CAA assay was carried out by the procedure of Wolfe and Liu [18]. Cell monolayers were treated with 50 µL of PBS, sample and standard (quercetin, 2.5–20 µM) solution, and fifty µL of a DCFH-DA solution containing 50 µM of reagent and pre-incubated at 37 °C, 5% of CO_2 for 1 h. Afterwards, to each well containing standards/samples and tree wells containing PBS (control) was added 100 µL of AAPH (12 mM). In the rest of the six wells containing PBS (blank) 100 µL of PBS solution was added. The data were recorded every 5 min during 60 min by using a FL800 fluorescent reader (ex. 485 nm, em. 540 nm). CAA values were expressed as µM of quercetin equivalents (QE) per g of the extract from three independent experiments.

2.10. Statistical Data Handling

All results are presented as means ± standard deviation (SD), and all the experiments were completed at least three times. Significant differences among means were evaluated by one-way ANOVA, using the statistical package GraphPad Prism 6. Duncans' posthoc test was used to determine the significant difference among the treatments at $p < 0.05$.

3. Results

3.1. Total Yield, Total Phenolic Content and Radical Scavenging Capacity of Peony Extracts

The yields and antioxidant capacity values of extracts and plant materials are presented in Table 1. High polarity solvents, methanol, and water were compared in this study. In addition, methanol was applied in a pressurized liquid extraction (PLE). Methanol has been shown to be an effective solvent for polyphenolic antioxidants in numerous studies, while water is very attractive in terms of its favourable green chemistry principles and low cost. It may be observed that the total yields from leaves were remarkably higher than those from the roots in the case of both solvents, whereas PLE with methanol resulted in an approximately 10% higher yield.

Using several antioxidant activity assays is important for completing a more comprehensive evaluation of natural products. Thus, Huang et al. [15] recommended applying at least two single electron transfer (SET) and one hydrogen atom transfer (HAT) assays for this purpose. Following this recommendation, TPC (Folin-Ciocalteu), ABTS$^{\bullet+}$ and DPPH$^\bullet$ scavenging, as well as ORAC, HORAC, and HOSC assays, were used for measuring the antioxidant potential of the dried extracts (DWE) and recovery of antioxidants from the initial plant material (DWP). These values are useful for evaluating the antioxidant potential of extracts and the recovery of antioxidants from the raw material, respectively. For instance, a low yield extract may possess stronger antioxidant activity, while for high yields, a better recovery of antioxidants from the plant may be achieved, although the extracts may show lower antioxidant capacities due to the dilution of the active constituents with neutral ones.

Table 1. Yields, antioxidant capacity, and total phenolic content [1] of different *P. officinalis* extracts.

Samples	Yield, %	TPC, mg GAE/g		DPPH•, µM TE/g		ABTS•+, µM TE/g	
		DWE	DWP	DWE	DWP	DWE	DWP
PLASEM	47.6 ± 1.1 [a]	516.4 ± 14.53 [a]	245.6 ± 6.91 [a]	2424 ± 23.66 [a]	1153 ± 11.25 [a]	4524 ± 26.75 [a]	2151 ± 12.72 [a]
PRASEM	23.9 ± 0.6 [b]	247.3 ± 8.79 [b]	59.1 ± 2.1 [b]	567.7 ± 11.38 [b]	135.6 ± 2.72 [b]	935.7 ± 4.51 [b]	223.5 ± 1.08 [b]
PLTRM	43.5 ± 0.8 [c]	601.1 ± 10.51 [c]	285.9 ± 4.99 [c]	2553 ± 28.40 [c]	1110 ± 12.34 [c]	4610 ± 18.70 [c]	2004 ± 8.13 [c]
PLTRW	33.2 ± 0.2 [d]	430.6 ± 9.49 [d]	187.2 ± 4.13 [d]	2138 ± 13.41 [d]	710.7 ± 4.46 [d]	4231 ± 14.07 [d]	1406 ± 6.76 [d]
PRTRW	19.2 ± 0.5 [e]	215.7 ± 3.05 [e]	41.4 ± 0.6 [e]	343.4 ± 6.06 [e]	65.94 ± 1.16 [e]	886.0 ± 7.36 [e]	170.1 ± 1.41 [e]
		ORAC, µM TE/g		HOSC, µM TE/g		HORAC, µmol CAE/g	
PLASEM		1433 ± 6.93 [a]	681.4 ± 3.48 [a]	1957 ± 7.47 [a]	931.0 ± 9.53 [a]	1891 ± 13.06 [a]	899.4 ± 5.45 [a]
PLTRM		1257 ± 10.50 [b]	546.6 ± 2.59 [b]	2012 ± 10.22 [b]	874.6 ± 6.61 [b]	1758 ± 7.56 [b]	764.1 ± 8.69 [b]
PLTRW		1232 ± 7.61 [c]	409.6 ± 1.34 [c]	2010 ± 8.02 [b]	668.2 ± 4.15 [c]	1566 ± 8.45 [c]	520.3 ± 3.16 [b]

Values are represented as mean ± standard deviation ($n = 4$); different superscript letters for means down the vertical column that do not share common letters are significantly different ($p < 0.05$). The extracts isolated with methanol and water are further referred to by abbreviations composed of the first letter of the plant (P—peony), the botanical part (L—leaves, R—roots) and the solvent (M—methanol, W—water); ASE and TR mean extraction type: the accelerated solvent extraction and traditional extraction, respectively; DWE—dry weight of extract; DWP—dry weight of plant; ORAC—oxygen radical absorbance capacity; HOSC—hydroxyl radical scavenging capacity; HORAC—(hydroxyl radical antioxidant capacity).

Thus, TPC values measured for *P. officinalis* extracts (Table 1) were in the range of 215.7 (PRTRW)–601.1 (PLTRM) mg GAE/g DWE, while the recovery of polyphenols from the raw material was in the range of 41.42 (PRTRW)–285.9 (PLTRM) mg GAE/g DWP. Methanol extraction resulted in higher TPC values both for DWP and DWE, while the TPC in leaves was higher than that in roots. It may be observed that higher TPC values for leaves were found via conventional extraction (TR) than via PLE; most likely, this difference was due to the dilution of Folin-Ciocalteu reactive substances in the latter case (PLE gave higher yields than TR) and some other changes. Consequently, in terms of TPC, traditional methanol extraction was the most effective method.

DPPH• and ABTS•+ scavenging assays, which are also based on SET (single electron transfer) reaction, gave similar results for the antioxidant capacity of extracts and the recovery of radical scavengers (Table 1). However, ABTS•+ decolourisation values were remarkably higher than DPPH• scavenging values, which may be explained by the different reaction conditions [15]. Thus, DPPH• and ABTS•+ scavenging values were in the ranges of 343.4 ± 6.06–2553 ± 28.40 and 886.0 ± 7.36–4610 ± 18.70 µM TE/g DWE, respectively, while antioxidant recoveries were 65.94 ± 1.16–1153 ± 11.25 and 170.1 ± 1.41–2151 ± 12.72 µM TE/g DWP, respectively. It should be noted that the TPC and ABTS•+/DPPH• scavenging capacity was not reported for *P. officinalis* previously. Some antioxidants may be strongly bound to the insoluble plant matrix and are not available for any solvent without pre-treatment [19]. To evaluate the recovery of the active constituents, the antioxidant capacity of peony solids was monitored by using the so-called QUENCHER procedure [20]. The results presented in Figure 1 show antioxidant capacity values of peony solids before and after extraction. It may be observed that after extraction, they are remarkably reduced, indicating that the extraction processes were quite efficient for the recovery of antioxidants. Water was slightly more efficient than methanol.

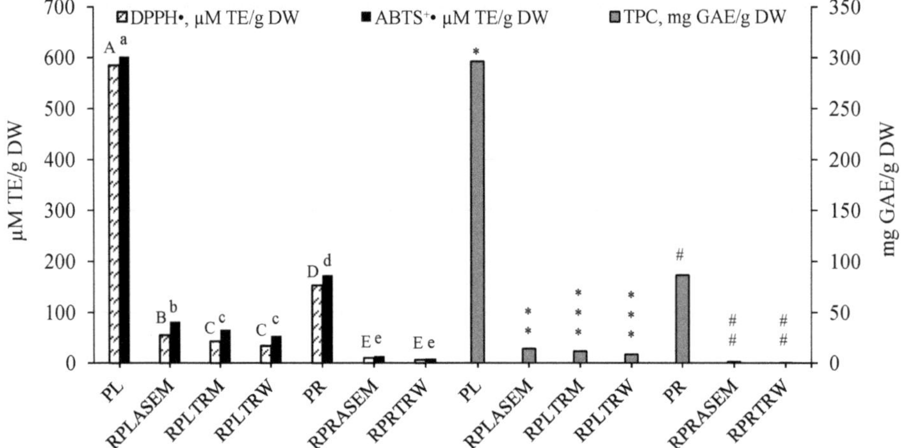

Figure 1. Antioxidant capacity indicators of solid substances determined by the QUENCHER (The QUick, Easy, New, CHEap and Reproducible) method. Values represented as mean ± standard deviation (n = 4); the mean values followed by different letters and symbols are significantly different ($p < 0.05$) (ABTS: small letters for leaves (a–c) and for roots (d–e); DPPH•: capital letters are used for leaves (A–C) and for roots (D–E) and TPC–symbols are used for leaves (*, **, ***) and for roots (#, ##). TPC is expressed in mg GAE/g DWP; DPPH•, ABTS+• in µM TE/g DWP. The residues after methanol and water extraction are further referred to by abbreviations composed of mean residue (R), peony (P), leaves or roots (L—leaves, R—roots), and solvent (M—methanol, W—water); ASE and TR mean extraction type: accelerated solvent extraction and traditional extraction, respectively.

3.2. Peroxyl and Hydroxyl Radicals Inhibition in ORAC, HORAC and HOSC Assays

Based on TPC and ABTS•+/DPPH•-scavenging values and chemical composition, methanol (PLASEM, PLTRM) and water (PLTRM) extracts were tested (Table 1). ORAC values varied in the ranges of 1232–1433 µmol TE/g DWE and 409.6–681.4 µmol TE/g DWP, HOSC in the ranges of 1957–2012 µmol TE/g DWE and 668.2–931.0 µmol TE/g DWP, and HORAC in the ranges of 1566–1891 µmol CAE/g DWE and 520.3–899.4 CAE/g DWP. These assays confirmed that methanol extracts were better antioxidants than water extracts. It may be observed that regardless of significant differences in antioxidant activity between some tested extracts, these differences were not remarkable. However, due to the differences in the extract yields, the recoveries of antioxidants from DWP were in a wider range. Thus, methanol extracted antioxidants more efficiently. For instance, for the ASE methanol extracted from 1 g of dried peony leaves, the quantity of antioxidants was equivalent to 0.17–0.23 g of Trolox.

3.3. Determination of Phytochemicals by UPLC-Q/TOF

In total, 23 compounds were detected and most of them were identified based on the measured accurate masses, suggested in various databases formulas, chromatographic retention times, MS/MS fragmentation, data obtained with authentic reference compounds, and various literature sources (Table 2, representative chromatogram in Figure 2A).

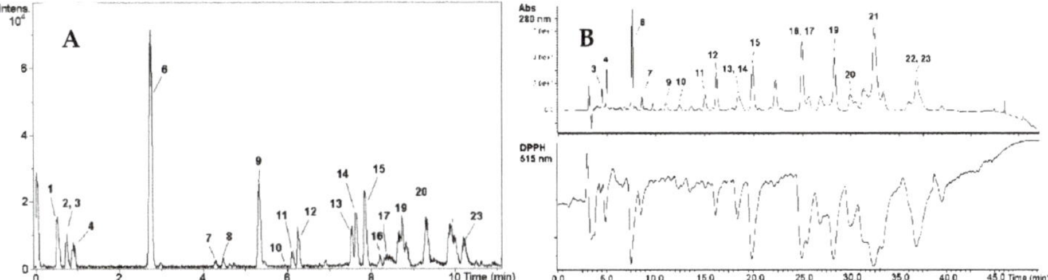

Figure 2. Representative chromatograms of methanol extract (PLASEM) of *P. officinalis*. (**A**) UPLC-Q-TOF chromatogram; (**B**) HPLC-UV-DPPH•-scavenging chromatogram showing 19 active compounds (negative peaks at 515 nm), which were detected by comparing their retention times with the UV chromatogram recorded at 280 nm: gallic acid derivatives (3, 4, 5, 6, 7, 10, 12, 14, 15, 16, 17, 19, 20, 21), quercetin derivatives (11, 13), paeoniflorin derivatives (9, 18), and unknown compounds (22, 23).

MS data and retention times of compounds **1** and **4** well matched the standards of quinic and gallic acids. The compound **2** with molecular ion $m/z = 341$ [M − H]⁻ and fragments of 191, 149, and 89 was assigned to dihexose. The molecular ion of compound **3** ($m/z = 331.0671$) corresponds to $C_{13}H_{15}O_{10}$, while the fragment ion ($m/z = 169.0140$) fits $C_7H_5O_5$, the residue of gallic acid. Based on the recorded m/z and literature data [21], the compound was tentatively identified as galloylhexose. The main ion of peak **5** ($m/z = 321.0253$) corresponds to $C_{14}H_9O_9$, while the MS/MS fragmentation, due to the loss of [M − H − 152]⁻ and [M − H–CO₂]⁻, gave the ions of $m/z = 169$ (gallic acid, $C_7H_5O_5$) and $m/z = 125.0240$ ($C_6H_5O_5$), respectively. This peak was assigned to digallic acid, while compound **6**, which gave $m/z = 183.0302$ ([M − H]⁻), was tentatively identified as methyl gallate [21]. The MS² spectrum of this compound produced characteristic ions at $m/z = 168.0060$, 140.0112, and 124.0166. Compound **7** gave a molecular ion [M − H]⁻ ($m/z = 635$), and fragments of 483, 465, and 169, fitting $C_{27}H_{23}O_{18}$, $C_{20}H_{19}O_{14}$, $C_{20}H_{17}O_{13}$, and $C_7H_5O_5$, respectively. The loss of 152, 170, and 466 amu was attributed to the loss of [M–H–galloyl]⁻, [M − H–galloyl–H₂O]⁻ and [M − H–2 galloyl–hexose]⁻ units, respectively. As a result, this compound was tentatively identified as *tri*-galloyl-hexoside. Compounds **8** (t$_R$ 4.5), **22** (t$_R$ 9.9), and **23** (t$_R$ 10.2) with $m/z = 491.1768$, 621.0637, and 697.0695, matching $C_{21}H_{31}O_{13}$, $C_{15}H_{25}O_{26}$, and $C_{31}H_{21}O_{19}$, respectively, have not been identified. Compounds **9** and **18** displayed a molecular ion [M − H]⁻ ($m/z = 525$, $C_{24}H_{29}O_{13}$) and several fragment ions in the MS/MS mode. The ion at m/z 479.1508/4791358 ($C_{23}H_{27}O_{11}$) was a basic fragment arising from the loss of [M − H–H₂O–CO]⁻ (46 mass units), which, by a further loss of CH₂O, $C_7H_6O_2$, and [M − H–paeonyl–CHOH–2H]⁻, produced the fragments at m/z 449.1448/449.1440 ($C_{22}H_{25}O_{10}$), 357.1191/357.1088 ($C_{16}H_{21}O_9$), and 283.0818/283.0714 ($C_{13}H_{15}O_7$), respectively. The ion at m/z 327.1086/327.1075 ($C_{15}H_{19}O_8$) can be derived by the loss of the $C_8H_8O_3$ from the basic ion, with $m/z = 449$, or from the ion with $m/z = 357$ (−30 mass unit; CH₂O). Finally, the fragments corresponded to the paeonyl ($m/z = 165.0556/165.0544$, $C_9H_9O_3$) and benzoyl ($m/z = 121.0294/121.0281$, $C_7H_5O_2$) units resulting from the cleavage of the paeoniflorin. The compounds shared the same m/z at 121, 165, 283, 327, 357, and 449 as previously reported for paeoniflorin [22]. Considering that molecular ion ($m/z = 479$) of **9** and **18** was higher by 46 Da, which can be attributed to CO (28 Da) and H₂O (18 Da), the peaks tentatively identified as carboxylated and hydratated derivatives of paeoniflorin.

Table 2. UPLC-Q/TOF identification data of phenolic compounds detected in *P. officinalis* extracts.

Peak No.	Compound	Molecular Formula	t_R (min)	m/z, $[M-H]^-$	PLASEM	PRASEM	PLTRM	PLTRW	PRTRW	MS Fragments
1	Quinic acid [a]	$C_7H_{12}O_6$	0.5	191.0563	+	−	+	+	+	-
2	Dihexose [c]	$C_{12}H_{21}O_{11}$	0.6	341.1097	+	+	+	+	+	191.0564; 149.0459; 89.0246
3	Galloyl-hexoside [b,d]	$C_{13}H_{15}O_{10}$	0.8	331.0671	+	−	+	+	+	169.0140
4	Gallic acid [a]	$C_7H_5O_5$	0.9	169.0144	+	+	+	+	+	-
5	Digallic acid [b,d]	$C_{14}H_9O_9$	1.9	321.0253	−	−	−	+	−	169.0138; 125.0240
6	Methyl galate [b,d]	$C_8H_7O_5$	2.8	183.0302	+	+	+	+	−	168.0060; 140.0112; 124.0166
7	Tri-galloyl-hexoside [d]	$C_{27}H_{23}O_{18}$	4.3	635.0889	+	−	−	−	−	483.0838; 465.0641; 169.0175
8	Unidentified	$C_{21}H_{31}O_{13}$	4.5	491.1768	+	−	−	−	+	-
9	Paeoniflorin derivative [b,d]	$C_{24}H_{29}O_{13}$	5.3	525.1617	+	+	+	+	+	479.1508; 449.1448; 357.1191; 327.1086; 283.0818; 165.0556; 121.0294
10	Tetra-galloyl-hexoside [b,d]	$C_{34}H_{27}O_{22}$	5.9	787.1006	+	−	+	+	−	617.0793; 456.0683; 169.0139
11	Quercetin dihexoside [c,d]	$C_{27}H_{29}O_{16}$	6.1	609.1460	+	−	+	+	−	463.0883; 301.0325;
12	Quercetin-galloyl-hexoside [b,d]	$C_{28}H_{23}O_{16}$	6.2	615.0993	+	−	+	+	−	463.0884; 301.0324; 169.0132
13	Quercetin pentoside [b,d]	$C_{20}H_{17}O_{11}$	7.5	433.0779	+	−	+	+	+	301.0340
14	Methyl digallate [b,d]	$C_{15}H_{11}O_9$	7.6	335.0410	+	+	+	−	+	183.0301; 124.0170
15	Penta-galloyl-hexoside [b,d]	$C_{41}H_{31}O_{26}$	7.8	939.1122	+	+	+	+	-	769.0901; 617.0795; 447.0569; 169.0132
16	Isorhamnetin-galloyl-hexosyde [b,d]	$C_{29}H_{25}O_{16}$	8.2	629.1149	+	+	+	+	+	477.1046; 315.0568; 169.0141
17	Dihydroxybenzoic acetate-digallate derivative [b,d]	$C_{24}H_{17}O_{15}$	8.4	545.0580	+	−	−	−	−	469.0489; 393.0466; 169.0135
18	Paeoniflorin derivative [b,d]	$C_{24}H_{29}O_{13}$	8.4	525.1615	−	−	+	+	−	479.1358; 449.1440; 357.1088; 327.1075; 283.0714; 165.0544; 121.0281
19	Dihydroxybenzoic acetate-digallate derivative [b,d]	$C_{24}H_{17}O_{15}$	8.7	545.0581	+	+	+	+	−	469.0407; 393.0461; 169.0139
20	Dihydroxybenzoic acetate-digallate derivative [b,d]	$C_{24}H_{17}O_{15}$	9.3	545.0582	+	-	+	+	−	469.0381; 393.0468; 169.0140
21	Hexa-galloyl-hexoside [b,d]	$C_{48}H_{35}O_{30}$	9.3	1091.1236	+	−	+	+	−	939.1101; 769.0895; 617.0811; 169.0143
22	Unidentified	$C_{15}H_{25}O_{26}$	9.9	621.0637	+	+	+	+	+	-
23	Unidentified	$C_{31}H_{21}O_{19}$	10.2	697.0695	+	−	−	+	−	-

[a] Confirmed by a standard; [b] Confirmed by a reference; [c] Confirmed by parent ion mass using free chemical database (Chemspider, PubChem); [d] Confirmed by MS/MS.

Compound **10** (t_R 5.9 min) gave m/z = 787.1006 ($C_{34}H_{27}O_{22}$), which was assigned to *tetra*-galloyl-hexoside because the main fragment at m/z = 617.0793 ($C_{27}H_{21}O_{17}$) may be obtained by the loss of gallic acid (m/z = 169.0139, $C_7H_5O_5$) residue from a deprotonated parent ion (m/z = 787.1006). The other characteristic fragment (m/z = 456.0683) may result due to the loss of the [M−H−hexosyl]⁻ moiety from m/z = 617.0793. Compound **11** was tentatively identified as quercetin-dihexoside: its molecular ion m/z = 609.1460 corresponds to $C_{27}H_{29}O_{16}$), [M − H − 463]⁻ indicates the loss of quercetin-hexose ($C_{21}H_{19}O_{12}$), [M − H − 146]⁻ is characteristic to hexose moiety, while m/z = 301.0325 represents quercetin ($C_{15}H_9O_7$). The compound **12** with a major ion of m/z = 615.0993 was assigned to quercetin-galloyl-hexoside; the loss of 463 Da indicates quercetin hexose moiety, while the fragments of m/z = 301.0324 ($C_{15}H_9O_7$) and m/z = 169.0132 ($C_7H_5O_5$) are characteristic of quercetin and gallic acid, respectively [23]. Compound **13** was assigned to quercetin pentoside (m/z = 433.0779, [M − H]⁻), which in MS/MS fragmentation lost pentose (132 Da) and gave a fragment ion of m/z = 301.0340 matching molecular formula ($C_{15}H_9O_7$) of quercetin [21]. The full scan mass spectra of compound **14** (methyl digallate) showed mainly an intense ion at m/z 335, which yielded an MS^2 ion at m/z 183 (methyl gallate) corresponding to the loss of a gallic acid moiety ([M − H − 152]⁻). Compound **15** (t_R = 7.8 min) gave an [M − H]⁻ ion with m/z = 939.1122, indicating a molecular formula of $C_{41}H_{31}O_{26}$. Its MS^2 spectrum contained four fragment ions: m/z = 769.0901 ($C_{34}H_{25}O_{21}$) showing the loss of [M−152 − H₂O]⁻ and matching *tetra*-galloyl-hexose moiety; m/z = 617.0795 ($C_{27}H_{21}O_{17}$), showing the loss of additional galloyl moiety [M − H − 152]⁻ and matching *tri*-galloyl-hexose moiety; m/z = 447.0569, indicating the split of [M − H − 152–H₂O]⁻ and indicating a *di*-galloyl-hexose unit; m/z = 169.0132 ($C_7H_5O_5$) matching gallic acid. All these data led to the identification of *penta*-galloyl-hexoside. Compound **16** (t_R 8.2 min) with the molecular ion of m/z = 629.1149 ($C_{29}H_{25}O_{16}$) was tentatively identified as isorhamnetin-galloyl-hexoside [21]. It produced a fragment ion of m/z = 477.1046 [M − H − 152]⁻ due to the loss of gallic acid; the fragment ion of m/z = 315.0568 indicates the loss of galloyl + hexose [M − H − 152 − 162]⁻, and finally the ion of m/z = 169.0141 indicates on the presence of gallic acid. Three compounds, **17**, **19**, and **20** (t_R 8.4, 8.7 and 9.3 min) with the precursor ion of m/z = 545 were assigned to dihydroxybenzoic acetate-digallate derivatives based on QTOF-MS and previously reported data [21]. In addition, their spectra showed the product ions of m/z = 469.0489/469.0407/469.0381 [M − H − 76]⁻ and 393.0466/393.0461/393.0468 [M − H − 152]⁻, corresponding to the neutral losses of acetyl + H₂O and galloyl moieties from the parent ion (m/z = 545), respectively. Finally, a fragment ion of m/z = 169.0135/169.0139/169.0140 indicates the presence of gallic acid. Compound **21** gave a [M − H]⁻ ion with m/z = 1091.1236 and similar to *penta*-galloyl-hexoside fragmentation pattern with a difference of 152 amu, suggesting an additional galloyl unit. Moreover, the presence of product ions of m/z = 939.1101, 769.0895, 617.0811, and 169.0143 suggests the presence of 6 galloyl groups. Consequently, compound **21** was tentatively identified as *hexa*-galloyl-hexoside.

3.4. Determination of Antioxidants by the On-Line HPLC-UV-DPPH•-Scavenging

HPLC-UV with the on-line DPPH•-scavenging detectors was used for the preliminary screening of antioxidant phytochemicals in paeony extracts. As it may be judged from the size of the negative peaks in the chromatogram (Figure 2B), gallic acid derivatives (**3, 4, 5, 6, 7, 10, 12, 14, 15, 16, 17, 19, 20, 21**), quercetin derivatives (**11, 13**), paeoniflorin derivatives (**9, 18**), and unknown compounds (**22, 23**) were the strongest radical scavengers in the investigated extracts. Compound **5** was detected only in the PLTRW extract (data not shown).

The constituents, for which reference compounds were available, were quantified and their amounts were expressed in mg/g DWE and mg/g DWP (Table 3). The concentration of quinic acid, gallic acid, and quercetin dihexoside in different peony fractions was in the ranges of 2.14–7.58, 0.44–12.33, and 1.04–1.64 mg/g DWE, respectively. Quinic acid was quantitatively a major constituent in *P. officinalis* extracts (except for PLTRW), while the amounts of quercetin *di*-hexoside and gallic acid varied depending on extract type. For instance, the PLTRW extract contained the highest amount of gallic acid (12.33 ± 0.87 mg/g DWE). Its recovery was 4.10 ± 0.25 mg/g DWP, while the best recovery of quercetin *di*-hexoside was found in case of PLTRM.

Table 3. Quinic acid, gallic acid, and quercetin dihexoside recovery (mg/g DWP) and their concentration in the extracts (mg/g DWE) obtained from *P. officinalis* by different solvents.

Sample	Quinic Acid		Gallic Acid		Quercetin Dihexoside *	
	DWP	DWE	DWP	DWE	DWP	DWE
PLASEM	3.61 ± 0.12 [a]	7.58 ± 0.99 [a]	2.59 ± 0.08 [a]	5.45 ± 0.09 [a]	0.50 ± 0.001 [a]	1.04 ± 0.05 [a]
PRASEM	nd	nd	0.11 ± 0.001 [b]	0.44 ± 0.001 [b]	nd	nd
PLTRM	1.52 ± 0.09 [b]	3.50 ± 0.31 [b]	0.22 ± 0.001 [b,c]	0.51 ± 0.002 [b]	0.66 ± 0.003 [b]	1.52 ± 0.02 [b]
PLTRW	1.83 ± 0.10 [c]	5.51 ± 0.32 [c]	4.10 ± 0.25 [d]	12.33 ± 0.87 [c]	0.55 ± 0.001 [c]	1.64 ± 0.01 [b]
PRTRW	0.41 ± 0.001 [d]	2.14 ± 0.12 [b]	0.34 ± 0.002 [b,c]	1.75 ± 0.44 [d]	nd	nd

nd—not detected; * Based on the calibration curve obtained by using rutin; values represented as mean ± standard deviation (n = 3); a–d: means down the vertical column not sharing common letters are significantly different ($p < 0.05$). The extracts isolated with methanol and water are further referred to by the abbreviation composed of the first letter of the plant (P—peony), the botanical part (L—leaves, R—roots), and the solvent (M—methanol, W—water); ASE and TR mean extraction type: accelerated solvent extraction and traditional extraction, respectively.

3.5. α-Amylase Inhibitory Properties of Selected Extracts

P. officinalis extracts applied in a concentration range of 1.25–5 mg/mL inhibited α-amylase in a dose dependent manner, whereas the differences for lower concentrations (0.83–1.25 mg/mL) in most cases were not significant (Figure 3). The PLTRM extract was a slightly, although significantly, stronger inhibitor of α-amylase than other leaf extracts in the concentration range of 0.83–2.5 mg/mL. The values calculated in mg/mL IC_{50} (inhibitory concentration) were in the following range: PLTRW (2.52 ± 0.32) > PLASEM (2.34 ± 0.18) > PLTRM (1.67 ± 0.17) > acarbose 0.3 ± 0.12 [d]. Consequently, the strongest α-amylase inhibitor PLTRM extract was 5.5 times less effective than acarbose.

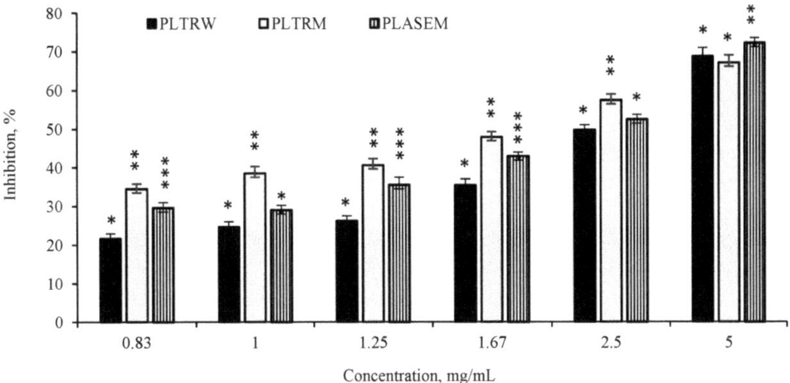

Figure 3. The inhibitory potential of *P. officinalis* methanol and water extracts against porcine α-amylase activity. The values represented as a mean ± standard deviation (n = 3). *–***: different symbols indicate significant differences ($p < 0.05$) between different extracts at the same concentration. The abbreviations are composed of the first letter of the plant (P—peony), the botanical part (L—leaves, R—roots), and the solvent (M—methanol, W—water); ASE and TR mean extraction type: accelerated solvent extraction and traditional extraction, respectively.

3.6. In Vitro Cytotoxic and Cellular Antioxidant Activity (CAA) Activity of Extracts

The cytotoxic effect was assessed on the Caco-2 cell line using the MTS method, where results showed no cytotoxic effect for each peony extract (PLTRW, PLASEM, PLTRM) in the range of the applied concentrations at 4 h, 24 h, and 48 h treatment (data not shown). Therefore, all extracts could be assessed in further research work. CAA-values for the peony extracts ranged from 0.046 to 0.106 μmol QE/mg extract (Figure 4).

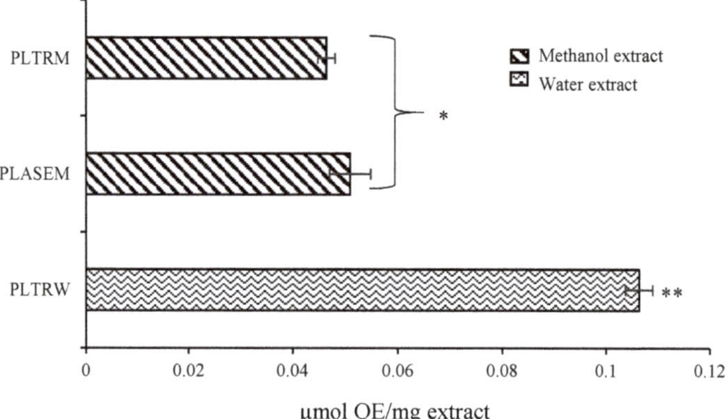

Figure 4. Antioxidant activity of *P. officinalis* extracts evaluated by the cellular antioxidant activity (CAA) method. *–**: the mean ± standard deviation ($n = 3$) values followed by different symbols are significantly different ($p < 0.05$). Other abbreviations are explained in the legend of Figure 3.

4. Discussion

The total extraction yield, the concentration of target compounds in the extracts, and their recovery from the plant materials are important process characteristics. Very high yields obtained with methanol (>43%) indicate the considerable content of soluble high polarity substances in *P. officinalis* leaves (Tale 1). The yields from plant roots were almost two-fold lower than those from leaves. For instance, Bae et al. [24] reported high yields of methanol extracts for the roots of other *Paeonia* species, namely *P. lactiflora* (30%) and *P. obovate* (32.5%), obtained during sonication, while in our study, the *P. officinalis* methanol extract obtained by PLE yielded 23.89%. Other studies [25,26] reported approximately two-fold lower yields from *P. officinalis* roots obtained at room temperature with ethanol and water than in our study, most likely, due to the differences in extraction temperature, which in our study for water extraction was 80 °C. It is well known that heat is one of the most important factors for extraction rate and yield.

The assessment of antioxidant capacity in vitro is widely used for preliminary evaluation of the beneficial properties of botanical extracts [27], and, although these values are not appropriate for direct assessment of physiological bioactivities, the experts in this area recently concluded that these assays, as low-cost and high throughput tools, cannot be ignored [28]. Plants are very complex biological structures containing a large diversity of antioxidants with different chemical structures and properties. Therefore, for a preliminary screening of antioxidant activity of plant preparations, several fast and simple in vitro assays have been developed and widely used. Basically, they may be classified into single electron (SET) or hydrogen atom (HAT) transfer methods [15]. The former is based on the ability of antioxidants to scavenge radicals (e.g., ABTS$^{•+}$ and DPPH$^{•}$) or reduce the compounds present in the reaction mixture (e.g., FRAP, Folin-Ciocalteu) by transferring one electron to them. The HAT method is based on inhibiting peroxyl-radicals (e.g., ORAC, HORAC, HOSC), and, therefore, they are more relevant to the processes in the biological systems [15]. Due to the different reaction conditions and mechanisms, the antioxidant capacity values determined by these assays may significantly differ; therefore, there is no perfect method for the comprehensive evaluation of antioxidant potential of complex biological systems [29].

Most recently, Camargo et al. (2019) [30] also recognized colorimetric methods as important screening tools by proving the links between TPC, ORAC, and FRAP values and the anti-inflammatory potential of grape by-products, which inhibited the activation of NF-κB in RAW 264.7 cells. Kleinrichert and Alappat [31] reported that the antioxidant efficacy of plant extracts positively correlated with the disruption of A-aggregation, which is accepted as an important biomarker in the pathogenesis of Alzheimer's Disease.

Li et al. [32] reported a remarkably lower antioxidant capacity for aqueous extracts of *P. lactiflora* and *P. suffruticosa* than the capacity determined in our study for *P. officinalis*. In the ABTS•+ scavenging assay, they were 365.72 ± 5.08, 85.25 ± 1.36, and 243.12 ± 4.65 μM TE/g DWE, respectively; in the TPC assay, they were 31.48 ± 1.52, 21.38 ± 0.26 and 22.37 ± 1.17 mg GAE/g DW, respectively. Cai et al. [33] reported even lower values for methanolic and aqueous root extracts of *P. lactiflora* and *P. suffruticosa*: 11.4/29.0 (methanol extracts) and 7.0/12.5 (aqueous extracts) mg GAE/g DW and 4.1/4.23 (methanol extracts) and 1.5/8.76 (water extracts μM TE/g DW, respectively. It may be assumed that these differences depend on the plant species, its anatomical parts, and the extraction procedure.

In general, in vitro SET based assays revealed the strong antioxidant potential of *P. officinalis* leaves and roots (Table 1). The highest values measured for 1 g of leaf extract were equivalent to approximately 0.64 and 1.15 g of trolox in DPPH and ABTS assays, respectively. In addition, the values obtained in the all three assays showed a good positive correlation, e.g., the extract with the highest TPC (601.1 mg GAE/g DWE) was also the strongest DPPH•/ABTS•+ scavenger. This correlation was reported in many previously published articles [34].

Dietary antioxidants are hypothesised to be possible exogenous protective agents in neutralising excessive reactive oxygen species (ROS), which may form in human organisms due to various factors. Among numerous in vitro antioxidant activity evaluation methods, peroxyl and hydroxyl radical inhibition assays (ORAC, HORAC, and HOSC) are recognised as better related to the antioxidant processes happening in the biological samples [35,36]. However, only a few studies performed ORAC, HORAC, and HOSC assays for peony extracts. For instance, Soare et al. [37] reported remarkably lower ORAC values for *P. officinalis* methanol and ethanol extracts (480.87 and 555.2 μmol TE/g, respectively). It has been shown in numerous articles that antioxidant activity may depend on various factors, such as harvesting time and climatic conditions.

The identification and characterization of individual bioactive constituents are among the most important tasks for valorising new plant materials. Bioactives in *P. officinalis* extracts were analysed by the UPLC-Q/TOF method (Table 2, Figure 2). It may be observed that gallic acid and its derivatives are the most abundant constituents in the *P. officinalis* plant (Table 3), as well as in other peony species [22,24,38]. For instance, Bae et al. [24] reported that the content of gallic acid in *P. lactiflora* and *P. obovate*, depending on plant collection site, varied from 2.4 ± 0.0 to 1740.4 ± 35.3 and from 1.6 ± 00 to 1426.2 ± 1.9 mg/100 g DW, respectively. Paeoniflorin and its derivatives were also among the main phytochemicals in peony seeds from various *Paeonia* species collected from different areas [39–42].

Type 2 diabetes mellitus is a progressive metabolic disorder of glucose metabolism, which could be treated by decreasing postprandial glucose levels. The enzyme α-amylase is responsible for the breakdown of complex polysaccharides into disaccharides. The inhibition of this enzyme could prolong overall carbohydrate digestion time, causing a reduction of glucose level in postprandial plasma. Consequently, α-amylase inhibitory activity is a useful indicator of the bioactivity of plant phytochemicals (Figure 3). To the best of our knowledge, α-amylase inhibitory activity was not previously reported for *P. officinalis* extracts.

New natural extracts or purified phytochemicals from poorly studied plants after comprehensive evaluation of their toxicity and safety may be promising disease risk reducing agents or even a platform for designing new nutraceuticals and medicines. For this purpose, cytotoxicity assays in human cells have been widely used. For instance, numerous studies have shown that many drugs used in medical treatments, such as chemotherapy, are cytotoxic to normal cells. Therefore, the search for effective compounds against cancer cells is necessary. Our results showed no cytotoxic effect for each tested peony extract. To the best of our knowledge, there are no reports on the cytotoxicity of *P. officinalis* extracts. In the literature regarding the cytotoxicity of the genus *Paeonia*, we found that

Lin et al. [43] determined that *Radix Paeoniae Rubra* extract, on normal urothelial SV-HUC-1 cells, was cytotoxic at 3.5 mg/mL at 48 h and at > 3.5 mg/mL at 24 h, while the cytotoxic effect on BFTC 905 and MB49 cancer cells was 1.4 and 2.8 mg/mL and 1.4 and 1.8 mg/mL at 48 and 24 h treatment, respectively. Furthermore, Lin et al. [44] observed cytotoxicity in normal SV-HUC-1 cells for *Cortex Moutan* extract (>3.5 mg/mL at 24 h and 1.6 mg/mL at 48 h). According to Almosnid et al. [45], *cis* and *trans*-suffruticosol D isolated from *P. suffruticosa* seeds was cytotoxic against A549 (lung), BT20 (breast), MCF-7 (breast), and U2OS (osteosarcoma) cancer cell lines at 9.93 to 46.79 µM concentrations and above, while on normal breast epithelial cells (HMEC) and normal lung epithelial cells (HPL1A; EC_{50} values were in range from 146.3 to 269.5 µM and from 78.3 to 177.5 µM, respectively), the cytotoxicity wasthe notably weaker. As can be seen, a cytotoxic effect on normal cells compared to cancer cells appears after applying considerably higher concentrations of various extracts derived from plants, demonstrating that natural phytochemicals derived from medical herbs are a promising source for the development of cancer drugs.

One of the disadvantages of in vitro colorimetric antioxidant assays is that they do not provide any information about the absorption and metabolism of antioxidants in a cellular environment. Therefore, the values measured by chemical assays often do not correlate with the antioxidant processes both in vitro and in vivo, meaning that strong radical scavenging capacity values are not always confirmed by biological activities [46]. To improve the biological relevance of antioxidant activity results, the cellular antioxidant activity (CAA) method was developed for evaluating phytochemicals that can penetrate the cell membrane and prevent oxidation [18]. This method gives additional information about the uptake, absorption, and metabolism in cell environments. For instance, the strong antioxidant activity by plant polyphenolics also demonstrated numerous protective effects against chronic diseases [47]. After a preliminary evaluation of cytotoxicity, the CAA method was used for a more comprehensive evaluation of the antioxidant potential of *P. officinalis* extracts (Figure 4). This technique is more biologically relevant than chemical antioxidant activity assays, because it accounts for the uptake, metabolism, and localization of antioxidant compounds within cells [18]. It is interesting to note that the water extract demonstrated higher CAA than the methanol extracts isolated both by conventional and PLE methods (PLTRM and PLASEM), while in SET and HAT based antioxidant capacity assays, methanol was a more effective solvent than water. Similarly, water extracts were remarkably stronger antioxidants than methanol extracts in the CAA assay in our previous study with *Nepeta* spp. [14]. This result is possibly linked with the higher affinity of water extracts to aqueous cellular environments, which might be favourable for crossing the cell membrane via water-soluble phytochemicals. Xiong et al. [48] and Huang et al. [49] reported CAA values of 26.7 ± 3.0 and 18.11 ± 1.84 µmol QE/100 g of flower, respectively, for *P. suffruticosa*, which was extracted with acetone.

5. Conclusions

In general, this study demonstrated that *P. officinalis* is a rich source of polyphenolic compounds with high antioxidant potential. Leaf extracts were stronger agents than root extracts, while methanol was a more effective solvent than water, as was demonstrated via its chemical antioxidant capacity and in α-amylase inhibitory assays. However, the selected water extract of leaves was a stronger antioxidant in the cellular antioxidant activity assay than the methanol extracts. These results confirm that plant extracts do not necessarily have the same response in a biological environment (in vitro) as chemical assays. Additionally, none of the extracts demonstrated associated cytotoxicity. UPLC-Q/TOF analysis supplemented with the on-line HPLC–DPPH•-scavenging method revealed 20 radical scavenging compounds, gallic acid derivatives, most likely, to be major contributors to the overall antioxidant potential of the extracts. Here, cellular antioxidant activity, cytotoxicity, and the inhibition of α-amylase *P. officinalis* extracts were reported for the first time. Therefore, *P. officinalis* may be a promising phytochemical to explore for its antidiabetic and antioxidant properties, with future applications in nutraceuticals, cosmeceuticals, and pharmaceuticals. For this purpose, further evaluation should focus on in vivo studies.

Author Contributions: Conceptualization, P.R.V., A.A.M.; methodology, L.D., M.P., A.P., C.V.P.; software: L.D.; validation, L.D., M.P., A.P., C.V.P.; formal analysis, L.D.; investigation, L.D., M.P., A.P., C.V.P.; resources, P.R.V., A.A.M..; data curation, P.R.V., A.A.M.; writing—original draft preparation, L.D.; writing—review and editing, P.R.V., A.A.M.; visualization, L.D.; supervision, P.R.V., A.A.M.; project administration, P.R.V., A.A.M.; funding acquisition, P.R.V., A.A.M.

Funding: The authors acknowledge the financial support received from the Portuguese Fundação para a Ciência e Tecnologia (FCT) through the PEst-OE/EQB/LA0004/2011 grant. The iNOVA4Health - UID/Multi/04462/2013 program is also acknowledged. A.A. Matias thanks for the IF Starting Grant – GRAPHYT (IF/00723/2014).

Acknowledgments: The authors would like to thank Kaunas Botanical Garden of Vytautas Magnus University (Kaunas, Lithuania) for supplying plant material.

Conflicts of Interest: The authors declare no conflict of interest.

References

1. Shahidi, F.; Ambigaipalan, P. Phenolics and polyphenolics in foods, beverages and spices: Antioxidant activity and health effects—A review. *J. Funct. Foods* **2015**, *18*, 820–897. [CrossRef]

2. Meng, D.; Zhang, P.; Zhang, L.L.; Wang, H.; Ho, C.T.; Li, S.M.; Shahidi, F.; Zhao, H. Detection of cellular redox reactions and antioxidant activity assays. *J. Funct. Foods* **2017**, *37*, 467–479. [CrossRef]

3. Venskutonis, P.R. Natural antioxidants in food systems. In *Food Oxidants and Antioxidants. Chemical, Biological and Functional Properties*; Grzegorz, B., Ed.; CRC Press: Boca Raton, FL, USA, 2013; pp. 235–301.

4. He, C.N.; Peng, Y.; Zhang, Y.C.; Xu, L.J.; Gu, J.; Xiao, P.G. Phytochemical and biological studies of Paeoniaceae. *Chem. Biodivers.* **2010**, *7*, 805–838. [CrossRef] [PubMed]

5. Ahmad, F.; Tabassun, N.; Rasool, S. Medicinal uses and phytoconstituents of Paeonia officinalis. *Int. Res. J. Pharm.* **2012**, *3*, 85–87.

6. Xu, S.P.; Sun, G.P.; Shen, Y.X.; Wei, W.; Peng, W.R.; Wang, H. Antiproliferation and apoptosis induction of paeonol in HepG2 cells. *World J. Gastroenterol.* **2007**, *14*, 250–256. [CrossRef] [PubMed]

7. Guo, J.P.; Pang, J.; Wang, X.W.; Shen, Z.Q.; Jin, M.; Li, J.W. In vitro screening of traditionally used medicinal plants in China against enteroviruses. *World J. Gastroenterol.* **2006**, *12*, 4078–4081. [CrossRef]

8. Zheng, Y.Q.; Wei, W.; Zhu, L.; Liu, J.X. Effects and mechanisms of paeoniflorin, a bioactive glucoside from peony root, on adjuvant arthritis in rats. *J. Inflamm. Res.* **2007**, *56*, 182–188. [CrossRef] [PubMed]

9. Yang, H.O.; Ko, W.K.; Kim, J.Y.; Ro, H.S. Paeoniflorin: An antihyperlipidemic agent from *Paeonia lactiflora*. *Fitoterapia* **2004**, *75*, 45–49. [CrossRef] [PubMed]

10. Zhang, X.J.; Li, Z.; Leung, W.M.; Liu, L.; Xu, H.X.; Bian, Z.X. The analgesic effect of paeoniflorin on neonatal maternal separation-induced visceral hyperalgesia in rats. *J. Pain* **2008**, *9*, 497–505. [CrossRef] [PubMed]

11. Liu, P.; Xu, Y.F.; Gao, X.D.; Zhu, X.Y.; Du, M.Z.; Wang, Y.X.; Deng, R.X.; Gao, J.Y. Optimization of ultrasonic-assisted extraction of oil from the seed kernels and isolation of monoterpene glycosides from the oil residue of *Paeonia lactiflora Pall. Ind. Crop Prod.* **2017**, *107*, 260–270. [CrossRef]

12. Ning, C.; Jiang, Y.; Meng, J.; Zhou, C.; Tao, J. Herbaceous peony seed oil: A rich source of unsaturated fatty acids and γ-tocopherol. *Eur. J. Lipid Sci. Technol.* **2015**, *117*, 532–542. [CrossRef]

13. Lieutaghi, P. *Badasson & Cie: Tradition Médicinale et Autres Usages des Plantes en Haute Provence*; Actes Sud: Arles, France, 2009; p. 715.

14. Dienaitė, L.; Pukalskienė, M.; Matias, A.A.; Pereira, C.V.; Pukalskas, A.; Venskutonis, P.R. Valorization of six *Nepeta* species by assessing the antioxidant potential, phytochemical composition and bioactivity of their extracts in cell cultures. *J. Funct. Foods* **2018**, *45*, 512–522. [CrossRef]

15. Huang, D.; Ou, B.; Prior, R.L. The chemistry behind antioxidant capacity assays. *J. Agric. Food Chem.* **2005**, *53*, 1841–1856. [CrossRef] [PubMed]

16. Al-Dabbas, M.M.; Kitahara, K.; Suganuma, T.; Hashimoto, F.; Tadera, K. Antioxidant and α-Amylase Inhibitory Compounds from Aerial Parts of *Varthemia iphionoides Boiss. Biosci. Biotechnol. Appl. Biochem.* **2006**, *70*, 2178–2184. [CrossRef] [PubMed]

17. Sambuy, Y.; De Angelis, I.; Ranaldi, G.; Scarino, M.L.; Stammati, A.; Zucco, F. The CaCo-2 cell line as a model of the intestinal barrier: Influence of cell and culture-related factors on CaCo-2 cell functional characteristics. *Cell Biol. Toxicol.* **2005**, *21*, 1–26. [CrossRef] [PubMed]

18. Wolfe, K.L.; Liu, R.H. Cellular antioxidant activity (CAA) assay for assessing antioxidants, foods, and dietary supplements. *J. Agric. Food Chem.* **2007**, *55*, 8896–8907. [CrossRef] [PubMed]

19. Saura-Calixto, F. Concept and health-related properties of nonextractable polyphenols: The missing dietary polyphenols. *J. Agric. Food Chem.* **2012**, *60*, 11195–11200. [CrossRef] [PubMed]

20. Serpen, A.; Capuano, E.; Fogliano, V.; Gökmen, V. A new procedure to measure the antioxidant activity of insoluble food components. *J. Agric. Food Chem.* **2007**, *55*, 7676–7681. [CrossRef] [PubMed]

21. Abu-Reidah, I.M.; Ali-Shtayeh, M.S.; Jamous, R.M.; Arríez-Romín, D.; Segura-Carretero, A. HPLC-DAD-ESI-MS/MS screening of bioactive components from *Rhus coriaria* L. (Sumac) fruits. *Food Chem.* **2015**, *166*, 179–191. [CrossRef]

22. He, C.; Peng, B.; Dan, Y.; Peng, Y.; Xiao, P. Chemical taxonomy of tree peony species from China based on root cortex metabolic fingerprinting. *Phytochemistry* **2014**, *107*, 69–79. [CrossRef] [PubMed]

23. Li, C.; Du, H.; Wang, L.; Shu, Q.; Zheng, Y.; Xu, Y.; Zhang, J.; Yang, R.; Ge, Y. Flavonoid composition and antioxidant activity of tree peony (*Paeonia* section *Moutan*) yellow flowers. *J. Agric. Food Chem.* **2009**, *57*, 8496–8503. [CrossRef] [PubMed]

24. Bae, J.Y.; Kim, C.Y.; Kim, H.J.; Park, J.H.; Ahn, M.J. Differences in the chemical profiles and biological activities of *Paeonia lactiflora* and *Paeonia obovate*. *J. Med. Food* **2015**, *18*, 224–232. [CrossRef] [PubMed]

25. Ahmad, F.; Tabassum, N. Effect of 70% ethanolic extract of roots of *Paeonia officinalis* Linn. on hepatotoxicity. *J. Acute Med.* **2013**, *3*, 45–49. [CrossRef]

26. Ahmad, F.; Tabassum, N. Preliminary phytochemical, acute oral toxicity and antihepatotoxic study of roots of *Paeonia officinalis* Linn. *Asian Pac. J. Trop. Biomed.* **2013**, *3*, 64–68. [CrossRef]

27. Shahidi, F.; Zhong, Y. Measurement of antioxidant activity. *J. Funct. Foods* **2015**, *18*, 757–781. [CrossRef]

28. Granato, D.; Shahidi, F.; Wrolstad, R.; Kilmartind, P.; Meltone, L.D.; Hidalgo, F.J.; Miyashita, K.; Camp, J.V.; Alasalvar, C.; Ismail, A.B.; et al. Antioxidant activity, total phenolics and flavonoids contents: Should we ban in vitro screening methods? *Food Chem.* **2018**, *264*, 471–475. [CrossRef] [PubMed]

29. Ou, B.; Huang, D.; Hampsch-Woodill, M.; Flanagan, J.A.; Deemer, E.K. Analysis of antioxidant activities of common vegetables employing oxygen radical absorbance capacity (ORAC) and ferric reducing antioxidant power (FRAP) assays: A comparative study. *J. Agric. Food Chem.* **2002**, *50*, 3122–3128. [CrossRef] [PubMed]

30. De Camargo, A.C.; Biasoto, A.C.T.; Schwember, A.R.; Granato, D.; Rasera, G.B.; Franchin, M.; Rosalen, P.L.; Alencar, S.M.; Shahidi, F. Should we ban total phenolics and antioxidant screening methods? The link between antioxidant potential and activation of NF-κB using phenolic compounds from grape by-products. *Food Chem.* **2019**, *290*, 229–238. [CrossRef]

31. Kleinrichert, K.; Alappat, B. Comparative analysis of antioxidant and anti-amyloidogenic properties of various polyphenol rich phytoceutical extracts. *Antioxidants* **2019**, *8*, 13. [CrossRef]

32. Li, S.; Li, S.K.; Gan, R.Y.; Song, F.L.; Kuang, L.; Li, H.B. Antioxidant capacities and total phenolic contents of infusions from 223 medicinal plants. *Ind. Crops Prod.* **2013**, *51*, 289–298. [CrossRef]

33. Cai, Y.; Luo, Q.; Sun, M.; Corke, H. Antioxidant activity and phenolic compounds of 112 traditional Chinese medicinal plants associated with anticancer. *Life Sci.* **2004**, *74*, 2157–2184. [CrossRef] [PubMed]

34. Minh, T.N.; Khang, D.T.; Tuyen, P.T.; Minh, L.T.; Anh, L.H.; Quan, N.V.; Ha, P.T.T.; Quan, N.T.; Toan, N.P.; Elzaawely, A.A.; et al. Phenolic compounds and antioxidant activity of Phalaenopsis orchid hybrids. *Antioxidants* **2016**, *5*, 31. [CrossRef] [PubMed]

35. Ou, B.; Hampsch-Woodill, M.; Flanagan, J.; Deemer, E.K.; Prior, R.L.; Huang, D. Novel fluorometric assay for hydroxyl radical prevention capacity using fluorescein as the Probe. *J. Agric. Food Chem.* **2002**, *50*, 2772–2777. [CrossRef]

36. Moore, J.; Yin, J.J.; Yu, L.L. Novel fluorometric assay for hydroxyl radical scavenging capacity (HOSC) estimation. *J. Agric. Food Chem.* **2006**, *54*, 617–626. [CrossRef] [PubMed]

37. Soare, L.C.; Ferde, M.; Stefanov, S.; Denkova, Z.; Nicolova, R.; Denev, P.; Ungureanu, C. Antioxidant and antimicrobial properties of some plant extracts. *Rev. Chim.* **2012**, *63*, 432–434.

38. Shi, Y.H.; Zhu, S.; Ge, Y.W.; Toume, K.; Wang, Z.; Batkhuu, J.; Komatsu, K. Characterization and quantification of monoterpenoids in different types of peony root and the related *Paeonia* species by liquid chromatography coupled with ion trap and time-of-flight mass spectrometry. *J. Pharm. Biomed. Anal.* **2016**, *129*, 581–592. [CrossRef]

39. He, C.; Peng, Y.; Xiao, W.; Liu, H.; Xiao, P.G. Determination of chemical variability of phenolic and monoterpene glycosides in the seeds of *Paeonia* species using HPLC and profiling analysis. *Food Chem.* **2013**, *38*, 2108–2114. [CrossRef]

40. Liu, P.; Zhang, Y.; Xu, Y.F.; Zhu, X.Y.; Xu, X.F.; Chang, S.; Deng, R.X. Three new monoterpene glycosides from oil peony seed cake. *Ind. Crop Prod.* **2018**, *111*, 371–378. [CrossRef]

41. Liu, P.; Zhang, Y.; Gao, J.Y.; Du, M.Z.; Zhang, K.; Zhang, J.L.; Xue, N.C.; Yan, M.; Qu, C.X.; Deng, R.X. HPLC-DAD analysis of 15 monoterpene glycosides in oil peony seed cakes sourced from different cultivation areas in China. *Ind. Crop Prod.* **2018**, *118*, 259–270. [CrossRef]

42. Zhang, Y.; Liu, P.; Gao, J.Y.; Wang, X.S.; Yan, M.; Xue, N.C.; Qu, C.X.; Deng, R.X. *Paeonia veitchii* seeds as a promising high potential by-product: Proximate composition, phytochemical components, bioactivity evaluation and potential applications. *Ind. Crop Prod.* **2018**, *125*, 248–260. [CrossRef]

43. Lin, M.Y.; Chiang, S.Y.; Li, Y.Z.; Chen, M.F.; Chen, Y.S.; Wu, J.Y.; Liu, Y.W. Anti-tumor effect of *Radix Paeoniae Rubra* extract on mice bladder tumors using intravesical therapy. *Oncol. Lett.* **2016**, *12*, 904–910. [CrossRef] [PubMed]

44. Lin, M.Y.; Lee, Y.R.; Chiang, S.Y.; Li, Y.Z.; Chen, Y.S.; Hsu, C.D.; Liu, Y.W. *Cortex Moutan* induces bladder cancer cell death *via* apoptosis and retards tumor growth in mouse bladders. *Evid. Based Complement. Alternat. Med.* **2013**, *2013*, 207279. [CrossRef] [PubMed]

45. Almosnid, N.M.; Gao, Y.; He, C.; Park, H.S.; Altman, E. In vitro antitumor effects of two novel oligostilbenes, *cis*- and *trans*-suffruticosol D, isolated from *Paeonia suffruticosa* seeds. *Int. J. Oncol.* **2015**, *48*, 646–656. [CrossRef] [PubMed]

46. López-Alarcón, C.; Denicola, A. Evaluating the antioxidant capacity of natural products: A review on chemical and cellular-based assays. *Anal. Chim. Acta* **2013**, *763*, 1–10.

47. Shahidi, F.; Yeo, J.D. Bioactivities of phenolics by focusing on suppression of chronic diseases: A Review. *Int. J. Mol. Sci.* **2018**, *19*, 1573. [CrossRef] [PubMed]

48. Xiong, L.; Yang, J.; Jiang, Y.; Lu, B.; Hu, Y.; Zhou, F.; Shen, C. Phenolic compounds and antioxidant capacities of 10 common edible flowers from China. *J. Food Sci.* **2014**, *79*, 517–525. [CrossRef]

49. Huang, W.; Mao, S.; Zhang, L.; Lu, B.; Zheng, L.; Zhou, F.; Li, M. Phenolic compounds, antioxidant potential and antiproliferative potential of 10 common edible flowers from China assessed using a simulated in vitro digestion–dialysis process combined with cellular assays. *J. Sci. Food Agric.* **2017**, *97*, 4760–4769. [CrossRef]

© 2019 by the authors. Licensee MDPI, Basel, Switzerland. This article is an open access article distributed under the terms and conditions of the Creative Commons Attribution (CC BY) license (http://creativecommons.org/licenses/by/4.0/).

 *antioxidants*

Article

HPLC-ESI-qTOF-MS/MS Characterization, Antioxidant Activities and Inhibitory Ability of Digestive Enzymes with Molecular Docking Analysis of Various Parts of Raspberry (*Rubus ideaus* L.)

Lingfeng Wu [1], Yufeng Liu [2], Yin Qin [3], Lu Wang [1,*] and Zhenqiang Wu [2,4,*]

[1] College of Food Science and Engineering, Hainan University, Haikou 570228, China
[2] School of Biology and Biological Engineering, South China University of Technology, Guangzhou 510006, China
[3] School of Pharmaceutical Engineering, Guizhou Institute of Technology, Guiyang 550003, China
[4] Pan Asia (Jiangmen) Institute of Biological Engineering and Health, Jiangmen 529080, China
* Correspondence: lwang@hainanu.edu.cn (L.W.); btzhqwu@scut.edu.cn (Z.W.)

Received: 30 June 2019; Accepted: 1 August 2019; Published: 3 August 2019

Abstract: The anti-oxidative phenolic compounds in plant extracts possess multiple pharmacological functions. However, the phenolic characterization and in vitro bio-activities in various parts of raspberry (*Rubus idaeus* L.) have not been investigated systematically. In the present study, the phenolic profiles of leaves (LE), fruit pulp (FPE), and seed extracts (SE) in raspberry were analyzed by HR-HPLC-ESI-qTOF-MS/MS method, and their antioxidant activities and digestive enzymes inhibitory abilities were also investigated. The molecular docking analysis was used to delineate their inhibition mechanisms toward type II diabetes related digestive enzymes. Regardless of LE, FPE, or SE, 50% methanol was the best solvent for extracting high contents of phenolic compounds, followed by 50% ethanol and 100% methanol. The LE of raspberry displayed the highest total phenolic content (TPC) and total flavonoid content (TFC). A total of nineteen phenolic compounds were identified. The quantitative results showed that gallic acid, ellagic acid, and procyanidin C3 were the major constituents in the three extracts. The various parts extracts of raspberry all exhibited the strong antioxidant activities, especially for LE. Moreover, the powerful inhibitory effects of the three extracts against digestive enzymes (α-glucosidase and α-amylase) were observed. The major phenolic compounds of the three extracts also showed good inhibitory activities of digestive enzyme in a dose-dependent manner. The underlying inhibitory mechanisms of the main phenolic compounds against digestive enzymes were clarified by molecular docking analysis. The present study demonstrated that the various parts of raspberry had strong antioxidant activities and inhibitory effects on digestive enzymes, and can potentially prevent oxidative damage or diabetes-related problems.

Keywords: *Rubus idaeus* L.; phenolic compounds; HPLC-ESI-HR-qTOF-MS/MS; antioxidant activities; digestive enzymes inhibitors; molecular docking analysis

1. Introduction

Nowadays, incidences of metabolic syndrome-related diseases have rapidly increased all over the world, such as oxidative damage, diabetes, obesity, hypertension, and cardiovascular disease [1]. Food nutritionists have revealed that oxidative stress and high-calorie diets was closely related to the development of chronic disease [2]. To some extent, chemical drugs have good curative effect for relieving these chronic diseases, but they may also bring seriously side effects and drug dependence to the human body. Polyphenolics, as one of the ubiquitous secondary metabolites in plants, have been proven to possess important physiological functions including anti-oxidant, anti-diabetes, anti-inflammatory, anti-carcinogenic, anti-obesity, and anti-proliferative activities [3,4]. Currently, many scientists have pointed out that phytochemicals (especially phenolics and flavonoids, etc.) from the plant-based foods may be used as natural antioxidants, which can inhibit oxidative damage. Moreover, long-term intake of plant-based foods enriched in polyphenolics compounds is conducive to prevent the occurrence of chronic diseases caused by oxidative stress [5,6].

Raspberry (*Rubus idaeus* L.) belongs to the *Rosaceae* family, which is widely cultivated in Asia, Europe, and North America [7]. Fruits of raspberry are widely consumed as fresh fruits, functional beverages, and fermented wine due to its attractive color, delicious taste, and excellent nutritional properties (enriched in anti-oxidative phenolics) [8]. The leaves of raspberry have also received considerable attention for its human health benefits including anti-oxidant, anti-diabetes, and anti-inflammatory activities, and have been made into tea product and its consumption has been also increasing [9–11]. Raspberry seeds are also good raw materials for food industries, but there is little research focused on the its phyto-constituents and bio-activities [12]. Qin et al. (2018) have reported that raspberry fruits and seeds include large amounts of anti-oxidative bound phenolics. During in vitro digestion, the released bound phenolics (non-extractable) possessed antioxidant activities and the inhibitory effect of α-glucosidase [13]. Moreover, different extraction methods for releasing the non-extractable phenolics in the leaves and seeds of raspberry have been investigated in our previous work [14]. To the best of our knowledge, the characterization and bio-activities (especially their antioxidant activities and the inhibitory properties on digestive enzymes) of phenolic extracts in different parts of raspberry have not been compared and investigated systematically.

The aim of the present work was to systematically investigate the phenolic profiles and in vitro antioxidant activities of leaves, fruit pulp, and seeds extracts of raspberry. Meanwhile, the inhibitory effects against digestive enzymes (α-glucosidase and α-amylase) were also evaluated. Importantly, the inhibition mechanisms of the main phenolic molecules toward digestive enzymes were investigated by molecular docking analysis. This work may supply important evidence for the comprehensive utilization of raspberry in food industries.

2. Materials and Methods

2.1. Chemicals and Reagents

The Folin–Ciocalteu phenol reagent and chemicals used for the antioxidant activity assay were purchased from Aladdin Industrial Corporation (Shanghai, China). Phenolic standards, trolox, *p*-Nitrophenyl-α-D-glucopyranoside (*p*-NPG, > 99.8%), porcine pancreatic α-amylase (13 U/mg), and α-glucosidase ($\geq$ 58 U/mg solid) from *Saccharomyces cerevisiae* were all purchased from Sigma-Aldrich (St. Louis, MO, USA). Analytical-grade materials were purchased from Guangzhou Reagent Co. (Guangzhou, China). Formic acid and acetonitrile used for the HPLC analysis were of chromatography grade. Water was purified by the Milli-Q (Millipore, MA, USA) system and filtered through membranes with a pore size of 0.22 μm before use.

2.2. Materials

Raspberry was provided from Guishanhong Agricultural Development Co. Ltd., (Guiyang, Guizhou, China). The leaves, fruit pulp, and seeds of raspberry were separated, vacuum freeze-dried to remove the water, and then ground in a mechanic micromill (BJFSJ-150G, Shanghai, China) to fine powder, respectively. All samples were stored at −20 °C until use.

2.3. Extraction of Phenolic Compounds

One gram of the above sample powder was soaked with 10 mL of different extractions solvents (50% ethanol (EtOH), 100% EtOH, 50% methanol, 100% methanol, ethyl acetate and acetone), and then sonicated for 30 min at 40°C, 320 W. The mixture was subsequently filtered through a Whatman No. 1 paper.

2.4. Determination of Total Phenolic Content (TPC) and Flavonoid Content (TFC)

TPC was measured according to the Folin–Ciocalteau method with gallic acid as the standard [15]. TPC were expressed as mg gallic acid equivalents (GAE)/g sample in dry weight (DW). TFC was determined based on the aluminium chloride colorimetric method with rutin as the standard [16]. TFC were expressed as mg rutin equivalents (RE)/g sample in DW.

2.5. Phenolic Compositions Analysis by HPLC-ESI-HR-qTOF-MS/MS

The phenolic compositions of the three parts extracts' (extracted by 50% methanol) of raspberry were separated by using an HPLC system (Agilent 1200, CA, USA) equipped with a Diode Array Detector (DAD, Aglient, CA, USA). The analytical column was 250 mm × 4.6 mm, Zorbax Eclipse C_{18} plus column (5 µm, Aligent, CA, USA). Acetonitrile including 0.1% formic acid (phase A) and water including 0.1% formic acid (phase B) were used as the mobile phases. The gradient elution program was 0–5 min, 15% A; 5–20 min, 15–25% A; 20–30 min, 25–35% A; 30–40 min, 35–50% A; 40–50 min, 80% A; 50–55 min, 15% A, with a flow rate of 0.8 mL/min. The injection volume was 10 µL, temperature of the column was set to 30°C, and the UV/DAD were monitored from 200 to 600 nm. The HR-qTOF-MS/MS analysis was performed with a high-resolution time-of-flight (HR-qTOF) mass detector (maXis, Bruker, Billerica, MA, USA) in the negative or positive mode (4.0 kV). Mass spectra were recorded over the mass range *m/z* 100 to 1000. The acquired MS data were processed by Bruker Daltonics DataAnalysis software. The contents of the analytes were expressed as mg per g DW (Table S1).

2.6. Antioxidant Activities Assays

2.6.1. DPPH Radical Scavenging Activity Assay

DPPH assay was measured based on the method described earlier [17]. The absorbance at 517 nm was recorded by a microplate reader (SpectraMax M5 Molecular Device, CA, USA). Trolox or Vc solution (5–100 µg/mL) were served as positive controls. The results were expressed as the percentage of inhibition in the following Equation (1).

$$\text{DPPH radical scavenging activity (\%)} = \left(1 - \frac{A_s - A_b}{A_c}\right) \times 100 \tag{1}$$

where A_s = the absorbance of the sample extracts with DPPH solution, A_b = the absorbance of the sample extracts without DPPH solution, A_c = the absorbance of DPPH solution.

2.6.2. ABTS Cation Radical Scavenging Activity Assay

ABTS assay was determined according to the method of Wang et al. (2017) [18]. The absorbance at 517 nm was recorded by the microplate reader. Trolox or Vc solution (5–100 µg/mL) was served as positive controls. The scavenging activity (%) was calculated by Equation (2).

$$\text{ABTS}^+ \text{ radical scavenging activity (\%)} = \left(1 - \frac{A_s - A_b}{A_c}\right) \times 100 \tag{2}$$

where A_s = the absorbance of the extracts with $ABTS^+$ solution, A_b = the absorbance of the extracts without $ABTS^+$ solutionl, A_c = the absorbance of $ABTS^+$ solution.

2.6.3. Hydroxyl (OH^-) Radical Scavenging Activity Assay

The scavenging activity of OH^- radicals was measured based on the method described by Liu et al., 2017 [19]. The reaction mixture included 100 µL of the extracts' dilutions, 100 µL of 6 mM Fe_2SO_4

solution, and 100 μL of 2.4 mM H_2O_2. After 10 min of incubation at 25°C, the mixture was incubated with 100 μL of 6 mM salicylic acid at 25 °C for 30 min, then the absorbance at 510 nm was measured. A Trolox or Vc solution (5–100 μg/mL) served as the positive control.

2.6.4. Ferric Reducing/Antioxidant Power (FRAP) Assay

The FRAP assay was measured according to the method reported by Wong, Li, Cheng, and Chen (2006) [20]. A standard curve was constructed using the $FeSO_4 \cdot 7H_2O$ (0–1000 μM) as the reference standard. The FRAP values were expressed in mM ferrous sulfate equivalents Fe(II)SE/g sample in DW (mM Fe(II)SE/g DW).

2.7. Type II Diabetes Related Enzyme Inhibition Properties

2.7.1. α-Glucosidase Inhibition Activity Assay

The inhibitory activity of α-glucosidase was performed as the previous reported method with modifications [21]. Briefly, 100 μL of the extracts' dilutions and 100 μL of 1 U/mL α-glucosidase in 0.1 M phosphate buffer solution (pH 6.8) was mixed and pre-incubated at 37 °C for 10 min. Then, 100 μL of 5 mM *p*-NPG solution was added, and the reaction solution was incubated for another 20 min. The reaction was terminated by adding 500 μL of 0.2 M Na_2CO_3 solution. The absorbance at 405 nm was recorded. Acarbose was used as a positive control. The inhibitory potency (%) was calculated by Equation (3).

$$\alpha - \text{Glucosidase inhibitory potency } (\%) = \left[1 - \frac{\Delta As}{\Delta Ac}\right] \times 100 \tag{3}$$

where $\Delta A_s = A_{extract+enzyme} - A_{extract}$, $\Delta A_c = A_{buffer+enzyme} - A_{buffer}$.

2.7.2. α-Amylase Inhibition Activity Assay

The inhibitory activity of α-amylase was carried out according to the literature with slight modification [22]. An amount of 200 μL of the extracts' dilutions were mixed with 200 μL of α-amylase (0.5 mg/mL in PBS, pH = 6.8), and the mixture was pre-incubated at 37 °C for 15 min. Then, 400 μL of 2 mg/mL soluble starch solution was added to start the hydrolysis at 37 °C for 10 min. Afterwards, 1 mL of DNS reagent was added to stop the hydrolysis reaction, and the mixture was placed in boiling water bath for 5 min. When being cooled to room temperature, the absorbance at 540 nm was measured. The solution without α-amylase was used as the blank. The inhibitory potency (%) was calculated by Equation (4).

$$\alpha - \text{Amylase inhibitory potency } (\%) = \left[1 - \frac{\Delta As}{\Delta Ac}\right] \times 100 \tag{4}$$

where $\Delta A_s = A_{extract+enzyme} - A_{extract}$, $\Delta A_c = A_{buffer+enzyme} - A_{buffer}$.

2.8. Molecular Docking Analysis

The 2D conformers of the main phenolic standards (gallic acid, ellagic acid, and procyanidin C3) and acarbose (an anti-diabetic drug) were drawn by Chem 3D software, and the PDB formats of the α-glucosidase (PDB ID:3A4A) and α-amylase (PDB ID:1PPI) were downloaded from RCSB PDB (http://www.rcsb.org/pdb/home/home.do) [23]. Because the structural information of α-glucosidase from *S. cerevisiae* was not available, the homology structural (isomaltase, PDB ID: 3A4A) with high similarity of α-glucosidase was used as the template to perform the α-glucosidase docking analysis [23]. The molecular docking analysis of the main phenolic standards and acarbose to digestive enzymes was performed by the Surflex-Dock Geom (SFXC) mode using SYBYL-X 2.0 software package (Tripos, Inc., St. Louis, MO, USA). The docking procedure of small molecules to digestive enzymes is referred to in the literature [24]. Subsequently, a docking score file was generated and saved as the SD format. A C-Score (≥ 4) was selected as the credible results for the next docking analysis. The relevant docking parameters (e.g., T-Score, PMF-Score, D-Score, CHEM-Score, amino acid residues with active site, and hydrogen bond distances, etc.) may be used to reveal the inhibition mechanism of small molecules to digestive enzymes. The Surflex-Dock scoring function is a weighted sum of non-linear functions involving van der Waals surface distances between the appropriate pairs of exposed enzyme and ligand atoms.

2.9. Data Analysis

All the experiments were carried out in triplicate. Values were presented as the mean values ± standard deviation (SD). Analyses of variance and significance differences were analyzed by SPSS Statistics version 17.0 (IBM SPSS, Chicago, USA).

3. Results and Discussion

3.1. Total Phenolic Content and Total Flavonoid Content

The choice of the extraction solvent is very important for the recovering of phenolic compounds in plant matrix [25]. Common solvents including ethanol, methanol, acetone, and ethyl acetate have been widely used for the extraction of nature antioxidants in plant matrix [26,27].

As seen from Figure 1, results of TPC and TFC showed that concentrations significantly varied between the different solvents in various parts of raspberry. Regardless of LE, FPE, or SE, 50% methanol/ethanol showed high efficiency in the extraction of total phenolic and flavonoid compounds, followed by 100% EtOH and 100% methanol. In contrast, water and ethyl acetate were the least efficient solvents for extracting the phenolics and flavonoids compounds. It is worth noting that 50% methanol was evidently better than 100% methanol for extracting phenolic compounds ($p < 0.01$), which may be due to that the extraction efficiency depends on the polarity of the compounds present in the samples [28,29]. P. López-Perea et al, (2019) also reported that 80% methanol showed higher efficiency in the extraction of phenolic compounds from wheat bran and barley husk than 100% methanol [30]. When extracted by 50% methanol solution, the highest TPC (63.79 ± 3.11 mg GAE/g DW) and TFC (38.68 ± 2.4 mg RE/g DW) were found in LE, followed by FPE (42.26 ± 3.11 mg GAE/g DW for TPC, 28.60 ± 2.12 mg RE/g DW for TFC), and comparatively low amounts in SE (20.25 ± 1.79 mg GAE/g DW for TPC, 15.03 ± 1.82 mg RE/g DW for TFC). Wanyo et al. (2014) reported that 64% ethanol was the most efficient phenolic compounds extraction solvent [31]. Djordjevic et al. (2011) showed that high concentrations of phenolic compounds in barley grain (*Hordeum vulgare*) were obtained with 70% ethanol [32]. Our results also confirmed the choice of solvent varieties and concentrations have very significant influences in the extraction efficiency of phenolic compounds. Therefore, in the next study, in order to investigate the phenolic compositions and their in vitro biological activities of various parts in raspberry, 50% methanol was chosen as the best extraction solvent.

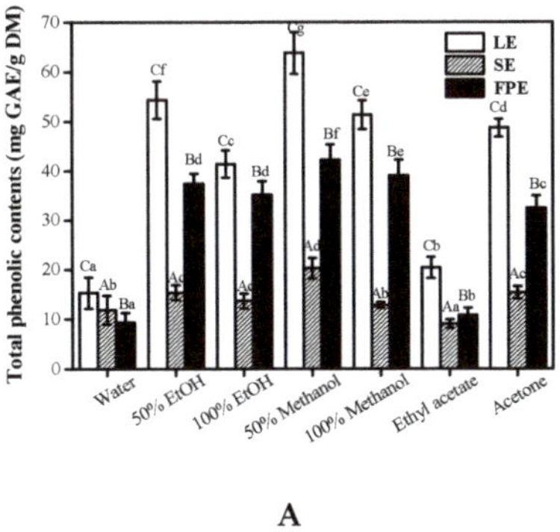
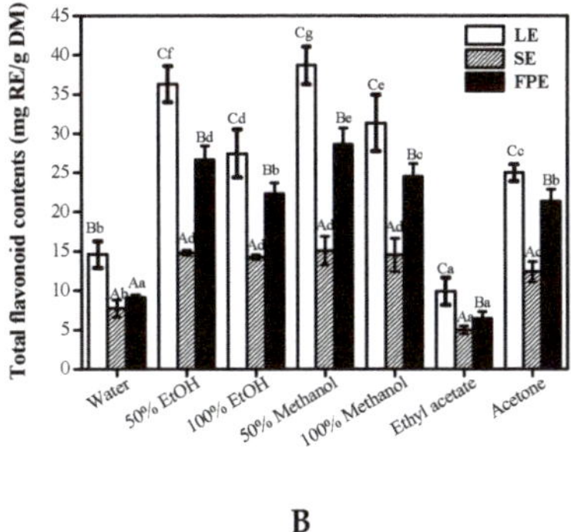

A   B

Figure 1. The effects of extraction solvents on total phenolic contents (**A**) and total flavonoid contents (**B**) in various parts of raspberry. LE, leaves extracts; SE, seeds extracts; FPE, fruit pulp extracts; EtOH, ethanol. Different lowercase letters (a–g) mean statistically significant differences following different extraction solvents. Different uppercase letters (A–C) mean statistically significant differences following different samples under same extraction solvents.

3.2. HPLC-ESI-HR-qTOF-MS/MS Characterization and Quantification of Phenolic Compositions

The phenolic compositions in different samples were identified by comparing their retention times and MS spectrum data with the authentic standards and reported data [14]. Table 1 and Figure 2 show the corresponding identification results of each peaks in HPLC chromatograms. Based on the MS/MS spectrum data, compound 1 displayed a $[M-H]^-$ in m/z 169.1221, which was easily identified as gallic acid. Chlorogenic acid (compound 2) and epicatechin (compound 3) were identified by their UV spectrum and MS spectral data, respectively. Compounds 4, 5, and 7 showed the parent ions at 579 $[C_{30}H_{26}O_{12}+H]^+$ and its fragments ions at m/z 291.0503 $[C_{15}H_{14}O_6+H]^+$, which can be tentatively identified as three isomers of procyanidin dimers. Among them, compound 7 can be identified as procyanidin B2 according to MS spectrum data and the retention time of the standard. Compounds 6 and 8 showed the parent ions $[M+H]^+$ at 867 and its MS/MS fragments ions at 579.1502 $[C_{30}H_{26}O_{12}+H]^+$ and 291.0155 $[C_{15}H_{14}O_6+H]^+$, respectively, which can be likely identified as two isomers of procyanidin trimer. By comparing the MS fragments information and the retention time of the standard, compound 8 can be identified as procyanidin C3. Ellagic acid pentoside was easily identified by the parent ion $[M+H]^+$ at m/z 435 and its ion fragment at m/z 303.0142 $[C_{14}H_6O_8+H]^+$. Compound 10 was identified as rutin by the parent ion $[M+H]^+$ at m/z 611 and its ions fragments m/z at 303.0563 $[M-glc+H]^+$ and 163.1221 $[M-C_{15}H_{10}O_7]^+$. Compound 11 was kaempferol-galactoside-glucoside by its ions $[M+H]^+$ at m/z 611 and two fragments ions at m/z 449.1338 $[M-glc+H]^+$ and 287.0716 $[C_{15}H_{10}O_6+H]^+$. Compound 13 with $[C_{14}H_6O_8+H]^+$ ion at m/z 303 was easily identified as ellagic acid. Compounds 14 and 15 with $[M+H]^+$ ions at m/z 465 giving MS/MS fragments ions m/z at 303 $[C_{15}H_{10}O_7+H]^+$ and 161 $[M-C_{15}H_{10}O_7+H]^+$ were very likely to be two isomers of quercetin glucosides. By comparing the retention time of two standard compounds, they were identified as quercetin-3-O-galactoside and quercetin-3-O-glucoside, respectively. Compound 16 with $[M+H]^+$ ion m/z at 435 giving fragment ion at 303.0500 $[C_{15}H_{10}O_7+H]^+$ was identified as avicularin. Kaempferol-7-O-glucuronide (Compound 17) was easily identified by the $[M+H]^+$ ion m/z at 463 and its MS/MS fragment ion at 287.0546 $[C_{15}H_{10}O_6+H]^+$. Quercetin-7-O-glucuronide (Compound 18) was identified by the ion $[M+H]^+$ at m/z 479 and its fragment ion at 303.0506 $[C_{15}H_{10}O_7+H]^+$. Compound 19 was easily identified as kaempferol-3-O-glucuronide by the $[M+H]^+$ ion at m/z 463 and its ion fragment at $[C_{15}H_{10}O_7+H]^+$.

The quantification results of fifteen phenolic compounds are shown in Table 2. The LE possessed the widest range of phenolic compositions, but also with the highest contents of individual phenolics. However, SE included the narrowest range of investigated compounds and the lowest contents of individual phenolics. Regardless of LE, FPE, or SE, gallic acid, ellagic acid, and procyanidin C3 were the most abundant of phenolic compounds. Meanwhile, the contents of gallic acid and ellagic acid in LE reached up to 539.42 ± 2.09 µg/g DW and 527.26 ± 3.27 µg/g DW. Their contents in FPE were 339.45 ± 2.17 µg/g DW and 95.42 ± 0.53 µg/g DW, respectively, which were significantly lower than those in LE. The contents of procyanidin B2 (10.72 ± 0.07 µg/g DW) and procyanidin C3 (252.37 ± 0.05 µg/g DW) in FPE were evidently higher than those in LE or SE. Particularly, some flavonols compounds (quercetin-3-glucoside, kaempferol-7-O-glucuronide, quercetin-7-O-glucuronide and kaempferol-3-O-glucuronide) except for rutin and avicularin were only found in LE and SPE of raspberry. Meanwhile, it can be found that there also existed high contents of some invididual phenolic compounds in SE, such as gallic acid (127.15 ± 3.21 µg/g DW), procyanidin C3 (29.12 ± 0.11 µg/g DW) and ellagic acid (48.32 ± 0.23 µg/g DW), which showed that SE may be used as a good food ingredient enriched in phenolic compounds. Qin et al. (2018) have confirmed that high levels of gallic acid and ellagic acid existing in the form of bound phenolics were found in raspberry fruit and seed extracts, which was consistent with the results of our study [13].

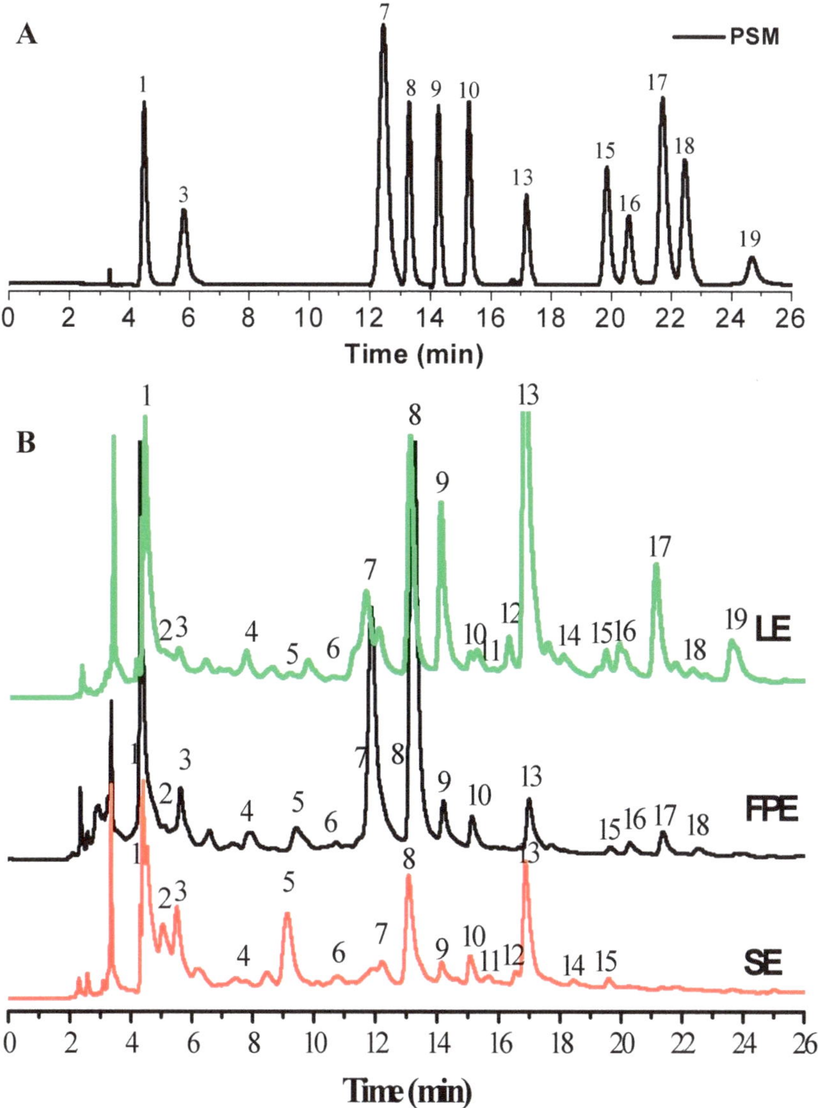

Figure 2. HPLC chromatograms (280 nm) of phenolic standards mixtures (**A**) and various parts extracts in raspberry (**B**). Peaks identification and their MS data are shown in Table 1. PSM, phenolic standard mixtures; 1, Gallic acid; 3, Epicatechin; 7, Procyanidin B2; 8, Procyanidin C3; 9, Ellagic acid pentoside; 10, Rutin; 13, Ellagic acid; 15, Quercetin 3-O-glucoside; 16, Avicularin; 17, Kaempferol-7-O-glucuronide; 18, Quercetin-7-O-glucuronide; 19, Kaempferol-3-O-glucuronide. LE, leaves extracts; SE, seeds extracts; FPE, fruit pulp extracts.

Table 1. Identification of phenolic compositions in various parts of raspberry by HPLC-ESI-HR-qTOF-MS/MS method.

Peak No.	Retention Time (min)	λ_{max} (nm)	Molecular ion (*m/z*)	MS (*m/z*)	Mw	Formula	Error (ppm)	Compounds (Abbreviation)	Reference	LE	FPE	SE
1	4.31	215, 271	169.2101[M-H]$^-$	169.1221	170	$C_7H_6O_5$	−0.7	Gallic acid (GA)	Standard	√	√	√
2	5.12	210, 268	353.2410[M+H]$^-$	191.0121, 98.9212	354	$C_{16}H_{17}O_9$	1.2	Chlorogenic acid	MS/MS			√
3	5.86	208, 279	291.0869 [M+H]$^+$	291.0869, 209.1545, 138.1423	290	$C_{15}H_{14}O_6$	−0.4	Epicatechin	Standard	√	√	√
4	7.87	215, 280	579.1503[M+H]$^+$	579.1503, 291.0708	578	$C_{30}H_{26}O_{12}$	2.7	Procyanidin dimer 1	MS/MS	√	√	
5	9.17	214, 280	579.1496[M+H]$^+$	579.1496, 291.0503	578	$C_{30}H_{26}O_{12}$	0.5	Procyanidin dimer 2	MS/MS		√	√
6	10.78	218, 284	867.2129 [M+H]$^+$	867.2131, 563.1548, 291.0294	866	$C_{45}H_{38}O_{18}$	2.2	Procyanidin trimer 1	MS/MS			√
7	11.98	220, 285	579.1514 [M+H]$^+$	579.1514, 447.0508, 303.0134, 291.032	578	$C_{30}H_{26}O_{12}$	0.3	Procyanidin B2	Standard	√	√	
8	13.19	225, 280	867.2140 [M+H]$^+$	867.2131, 579.1502, 291.0155,185.0085	866	$C_{45}H_{38}O_{18}$	0.2	Procyanidin C3 (PC)	Standard	√	√	√
9	14.27	253, 358	435.0564 [M+H]$^+$	435.0564, 303.0142, 185.0085	434	$C_{19}H_{14}O_{12}$	0.9	Ellagic acid pentoside	Standard	√	√	√
10	15.13	253, 354	611.1607 [M+H]$^+$	611.1607, 449.0557, 303.0563	610	$C_{27}H_{30}O_{16}$	−2.5	Rutin	Standard	√	√	√
11	15.87	254, 351	611.1607 [M+H]$^+$	611.1254, 449.1338, 435.0557, 287.0716	610	$C_{27}H_{30}O_{16}$	−2.5	kaempferol-galactoside-glucoside	MS/MS	√		
12	16.81	257, 353	867.2147 [M+H]$^+$	611.1607, 479.0557, 303.0515, 209.1547	610	$C_{27}H_{30}O_{16}$	1.7	Quercetin-3-O-glucuronide-arabinoside	MS/MS	√		√
13	16.95	256, 351	303.0501 [M+H]$^+$	303.0501, 193.1221	302	$C_{14}H_6O_8$	−0.3	Ellagic acid (EA)	Standard	√	√	√
14	18.21	254, 359	465.1032 [M+H]$^+$	465.1032, 303.0498	464	$C_{21}H_{20}O_{12}$	−0.5	Quercetin 3-O-galactoside	MS/MS	√		√
15	19.67	254, 356	465.1037 [M+H]$^+$	465.1037, 303.0506	464	$C_{21}H_{20}O_{12}$	−0.7	Quercetin 3-O-glucoside	Standard	√	√	√
16	21.57	262,391	435.0924 [M+H]$^+$	435.0924, 303.0500, 219.1754	434	$C_{20}H_{18}O_{11}$	−0.2	Avicularin	Standard	√	√	
17	20.42	253, 357	463.0807 [M+H]$^+$	463.0807, 287.0546, 133.1526	462	$C_{21}H_{18}O_{12}$	−0.6	Kaempferol-7-O-glucuronide	Standard	√	√	
18	22.72	254, 351	479.0827 [M+H]$^+$	479.0829, 303.0506	478	$C_{21}H_{18}O_{13}$	0.2	Quercetin-7-O-glucuronide	Standard	√	√	
19	23.51	257, 363	463.0878 [M+H]$^+$	463.0878, 317.0123, 287.0557	462	$C_{21}H_{18}O_{12}$	−1.4	Kaempferol-3-O-glucuronide	Standard	√		

Notes: LE, leaves extracts; FPE, fruit pulp extracts; SE, seed extracts; GA, gallic acid; EA, ellagic acid; PC, procyanidin C3.

Table 2. Contents of the main individual phenolics of various parts in raspberry.

Analytes	Contents (μg/g DW)		
	LE	FPE	SE
Gallic acid	539.42 ± 2.09c	339.45 ± 2.17b	127.15 ± 3.21a
Epicatechin	3.47 ± 0.02b	0.41 ± 0.07a	0.32 ± 0.12a
* Procyanidin dimer 1	6.71 ± 0.07c	4.39 ± 0.05b	2.13 ± 0.09a
* Procyanidin dimer 2	1.79 ± 0.05a	2.07 ± 0.03b	13.35 ± 1.12c
Procyanidin B2	6.82 ± 0.12b	21.72 ± 0.07c	0.17 ± 0.02a
# Procyanidin trimer 1	N.D.	2.34 ± 0.01a	3.25 ± 0.03a
Procyanidin C3	149.17 ± 0.01b	252.37 ± 0.05c	29.12 ± 0.11a
Ellagic acid pentoside	67.88 ± 0.12c	12.82 ± 0.09b	5.87 ± 0.11a
Rutin	2.53 ± 0.16a	5.07 ± 0.07b	4.89 ± 0.15b
Ellagic acid	527.26 ± 3.27c	95.42 ± 0.53b	48.32 ± 0.23a
Quercetin 3-glucoside	7.39 ± 0.03c	4.35 ± 0.02b	1.57 ± 0.23a
Avicularin	35.87 ± 0.12b	16.73 ± 0.09a	N.D.
Kaempferol-7-O-glucuronide	21.31 ± 0.01b	14.31 ± 0.02a	N.D.
Quercetin-7-O-glucuronide	2.32 ± 0.05a	2.14 ± 0.07a	N.D.
Kaempferol-3-O-glucuronide	5.47 ± 0.02	N.D.	N.D.

Different lowercase letters (a–c) mean statistically significant differences following different samples ($p < 0.05$). N.D., not detected; LE, leaves extracts; FPE, fruit pulp extracts; SE, seed extracts. * Procyanidin dimer and # procyanidin trimer were quantified by procyanidin B2 and procyanidin C3, respectively.

3.3. Antioxidant Activities

In order to fully reflect the antioxidant capacity of the samples extracts, four well-known chemical test methods including DPPH, ABTS$^+$, and OH$^-$ free radical scavenging activities and ferric reducing antioxidant activity (FRAP) were used to perform the antioxidant activity assays.

It was found that LE exhibited the strongest antioxidant activity, followed by FPE, and SE performed the weakest antioxidant activity. All of the samples' extracts showed the antioxidant activities in a concentration-dependent manner (Figure 3A–D). The insets of Figure 3A–C show the corresponding IC$_{50}$ values of the samples/controls. Notably, the lower the IC$_{50}$ values indicated, the stronger the antioxidant activity. For the DPPH assay, the IC$_{50}$ value of LE (5.60 ± 0.31 μg/mL) was lower than that of Vc (16.80 ± 1.31 μg/mL) and Trolox (39.90 ± 0.67 μg/mL), about one-third of that of FPE (18.71 ± 1.35 μg/mL), and one-sixth of that of SE (32.33 ± 1.42 μg/mL). LE also exhibited the strongest ABTS$^+$ free radical scavenging activity. The IC$_{50}$ value of LE (3.70 ± 0.17 μg/mL) was lower than that of Trolox (41.50 ± 1.97 μg/mL), FPE (29.43 ± 1.83 μg/mL), and SE (33.15 ± 2.19 μg/mL). Meanwhile, three various parts' extracts in raspberry (LE, FPE and SE) also exhibited the strong OH$^-$ free radical scavenging activity, and their corresponding IC$_{50}$ values were 3.01 ± 0.13 μg/mL, 6.40 ± 0.27 μg/mL, and 8.61 ± 0.52 μg/mL, respectively. The IC$_{50}$ value of LE was also lower than that of Vc (19.83 ± 0.37 μg/mL) and Trolox (33.91 ± 1.82 μg/mL). The insets of Figure 3A–C showed the results of FRAP assay. As is well known, higher FRAP values represent stronger anti-oxidant activity. The FRAP value of LE (198.32 ± 21.72 mM/g DW) was also higher than that of FPE (100.81 ± 12.31 mM/g DW) and SE (21.92 ± 3.72 mM/g DW).

The correlation co-efficient analysis results between the phenolic contents at different concentrations and antioxidant activities of the three extracts clearly verified that there was a good correlation between these two parameters (DPPH vs. TPC, $r = -0.807$, $p < 0.01$; ABTS vs. TPC, $r = -0.875$, $p < 0.01$; OH$^-$ vs. TPC, $r = -0.792$, $p < 0.01$), suggesting that the phenolic compounds significantly contributed to the antioxidant activities of three samples extracts. Qin et al. (2018) reported that the antioxidant activities of raspberry fruits and seeds extracts were strongly correlated with the released phenolic contents during in vitro digestion [13]. Wang et al. (2019) also confirmed that the released bound phenolics of raspberry leaves and seeds treated with different methods were responsible for their antioxidant activities [14]. In the present study, gallic acid, ellagic acid, and procyanidin C3 were the major phenolic compounds in different parts of raspberry. Many researches have verified that these three phenolics possess strong antioxidant activity [33]. Malinda et al. (2017) reported that gallic acid and ellagic acid showed strong DPPH radical scavenging activity with IC_{50} values of 2.24 μg/mL and 4.80 μg/mL, respectively [34]. Bialonska et al. (2009) have also verified that ellagic acid displayed no significant differences in the antioxidant activity with Vc by in vivo testing [35]. Because LE possessed higher contents of these main phenolic compounds (especially for gallic acid and ellagic acid) than FPE and SE, thus, LE showed the strongest antioxidant capacity, followed by FPE. The results of the above studies further revealed that the phenolic compounds of various parts in raspberry may contribute evidently to their antioxidant activities.

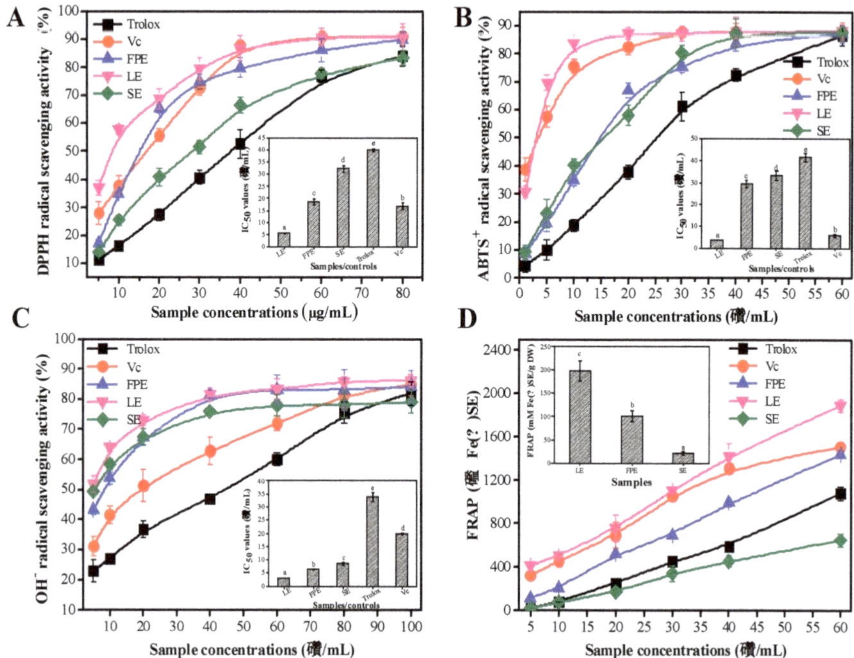

Figure 3. The antioxidant activities of various parts' extracts of raspberry and the positive controls (Vc and Trolox): DPPH (**A**), $ABTS^+$ (**B**), and OH^- radical scavenging activity (**C**) and Ferric reducing antioxidant power (FRAP) (**D**). The insets of Figure 3A–D represent the corresponding IC_{50} values of the samples/controls. LE, leaves extracts; FPE, fruit pulp extracts; SE, seed extracts. Vc, ascorbic acid. Different lowercase letters (a–e) mean statistically significant differences following different samples.

3.4. Type II Diabetes Related Enzymes Inhibitory Activities

Alpha-glucosidase or α-amylase, two key digestive enzymes in the digestive tract, can break down macromolecular carbohydrates into monosaccharide/glucose. As well known, excess glucose accumulates in the blood instead of being used for energy, which may cause type II diabetes. Many researches have confirmed that phenolic compounds may bind the amino acid residues with the active sites of digestive enzymes into complex formation by hydrogen bonding, and thereby inhibit the catalytic reaction of digestive enzymes on carbohydrates [36]. Therefore, the phenolic fractions in various parts of raspberry on the inhibition of digestive enzymes were used to evaluate their potential hypoglycemic effect.

Figure 4 exhibits that the various parts of raspberry extracts tended to be strong inhibitors of type II diabetes-related enzymes. It can be found that the samples' extracts or acarbose all showed the inhibition activity of digestive enzymes in a concentration-dependent manner (Figure 4A,B). Figure 4C,D show the IC_{50} values of the samples extracts or the main individual phenolic compounds on the inhibition effects on digestive enzymes. Similarly, the lower IC_{50} values indicated the stronger inhibition activity of digestive enzymes. The α-glucosidase inhibition activity of LE (IC_{50} = 96.50 ± 7.71 µg/mL) was evidently stronger than that of FPE (IC_{50} = 265.41 ± 20.7 µg/mL) and SE (IC_{50} = 218.5 ± 17.53 µg/mL) ($p < 0.01$). However, the α-glucosidase inhibition potency of acarbose (IC_{50} = 267.47 ± 19.72 µg/mL) was lower than that of LE ($p < 0.05$). FPE also possessed good inhibition activity against α-glucosidase, but there is no statistically significant difference with acarbose ($p > 0.05$). Figure 4C presents that the IC_{50} value for α-glucosidase inhibition activity of PC was 93.37 ± 5.79 µg/mL, which was significantly lower than that of GA (IC_{50} = 590.34 ± 15.71 µg/mL) and EA (IC_{50} = 976.32 ± 41.72 µg/mL) ($p < 0.01$). For α-amylase inhibition activity assay, LE have the lowest IC_{50} values (IC_{50} = 118.42 ± 2.79 µg/mL), followed by FPE (IC_{50} = 388.27 ± 2.47 µg/mL) and SE (IC_{50} = 891.12 ± 25.71 µg/mL). Moreover, the IC_{50} value of acarbose (IC_{50} = 442.23 ± 19.74 µg/mL) was higher than that of LE ($p < 0.05$), which indicates that LE may be used as a potential good anti-diabetic resource. Among the three main phenolic compounds, PC showed the strongest α-amylase inhibition activity (IC_{50} = 92.31 ± 3.51 µg/mL), followed by EA (IC_{50} = 516.73 ± 25.29 µg/mL), and GA (IC_{50} = 397.37 ± 12.37 µg/mL) ($p < 0.01$).

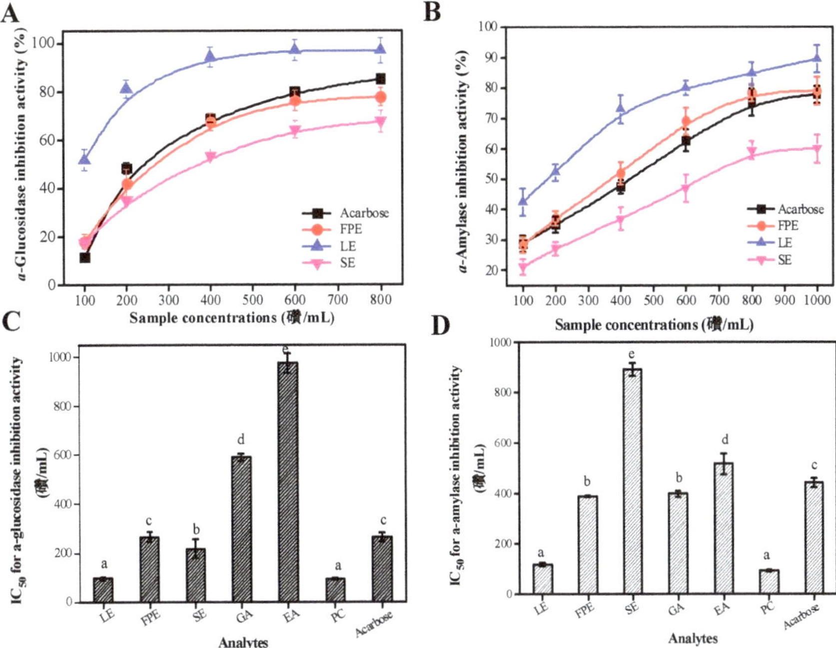

Figure 4. The digestive enzymes inhibitory abilities of various parts extracts in raspberry and acarbose: α-glucosidase inhibitory activity (**A**) and α-amylase inhibitory activity (**B**). Figure 4C,D represents the corresponding IC_{50} values of the samples/controls. LE, leaves extracts; FPE, fruit pulp extracts; SE, seed extracts; GA, gallic acid; EA, ellagic acid; PC, procyanidin C3. Different lowercase letters (a–e) mean statistically significant differences following different samples.

The correlation analysis results between the TPC and two digestive enzymes inhibition potency clearly verified that there was a good positive correlation between these two parameters (α-glucosidase inhibition potency vs. TPC, $r = 0.781$, $p < 0.05$; α-amylase inhibition potency vs. TPC, $r = 0.854$, $p < 0.01$). Many researches have confirmed that phenolic-rich extracts from leaf-tea, edible fruits, and natural products possess good inhibitory ability on digestive enzymes [23,37]. Wang et al. (2018) have confirmed that procyanidin C3 possessed strong α-glucosidase inhibitory capacity [38]. Zhang et al. (2010) reported that gallic acid, anthocyanins, and rutin in raspberry have strong α-glucosidase inhibitory capacity, but ellagic acid possessed the weakest inhibitory ability of α-glucosidase, which was consistent with the results of our study [39].

3.5. Molecular Docking Results

Molecular docking analysis was further done to analyze the digestive enzymes inhibitory mechanisms of the main phenolic compounds including GA, EA, and PC. The results clearly revealed that the structures of phenolic compounds significantly affect their inhibitory effects on α-glucosidase or α-amylase. Table 3 and Figure 5 show the molecular docking results with regard to interactions between α-glucosidase and several main phenolic molecules/acarbose binding. From Table 3, the molecular docking C-Scores values of those several molecules/acarbose were all ≥ 4, which indicates credible docking results. Figure 5 shows that GA interacted with the active sites of α-glucosidase and formed six H-bonds (yellow dotted line) with five amino acid residues (Asp 69, Arg 213, Asp 215, Asp 352, and His 351). The distances of H-bonds ranged from 1.896 Å to 2.600 Å. EA formed six H-bonds (the shortest distance was 1.700 Å and the longest distance was 2.433 Å) with five amino acid residues (Asp 215, Asp 352, Arg 213, Glu 411, and His 351). PC formed thirteen H-bonds within 4 Å (the distances ranged from 1.776 Å to 2.665 Å) with eleven amino acid residues (Asp 69, Asp 215, Asp 242, Glu 411, Gln 279, Gln 353, Leu 313, Lys 156, Tyr 158, The 314 and Pro 312). However, it was found that acarbose formed sixteen H-bonds within 4 Å (distances ranged from 1.776 Å to 2.732 Å) with ten amino acid residues (Asp 69, Asp 215, Asp 352, Arg 442, Glu 273, Glu 411, Gln 279, Tyr 158, Lys 156, and His 280). Some amino acid residues (Asp 69, Asp 215, and Asp 352) of α-glucosidase at least interacted with the above three investigated molecules, suggesting that these amino acid residues may play important roles in exerting the catalytic reaction of α-glucosidase.

Table 3. The analysis results of the main phenolic molecules and acarbose dockings into α-glucosidase or α-amylase ligands.

Digestive Enzymes	Main Phenolics	C-Score	T-Score	PMF-Score	CHEM-Score	G-Score	D-Score
α-Glucosidase	GA	4	3.86	−117.262	−15.181	−201.051	−108.348
	EA	5	1.66	−143.165	−25.320	−169.858	−134.180
	PC	4	5.58	−169.841	−20.602	−259.179	−186.583
	Acarbose	4	11.50	−267.829	−18.801	−399.410	−268.642
α-Amylase	GA	5	4.61	−100.463	−14.772	−137.871	−77.144
	EA	5	3.56	−108.516	−19.079	−156.583	−100.861
	PC	5	6.74	−157.306	−26.662	−326.221	−178.056
	Acarbose	5	7.07	−174.749	−7.459	−311.510	−211.197

Notes: GA, gallic acid; EA, ellagic acid; PC, procyanidin C3.

Table 3 and Figure 6 show the molecular docking results of α-amylase with the investigated phenolic molecules. The C-Scores of three main phenolic molecules and acarbose were all ≥ 4. Figure 6 shows that six H-bonds (yellow dotted line) were formed between the active site of α-amylase and gallic acid. The five amino acid residues with the active site were Asp 197, Arg 195, Glu 233, His 299, and His 305, respectively. The average distance of six H-bonds was 2.115 Å. Ellagic acid formed H-bonds with four amino acid residues with the active site, namely Asp 197, Asp 300, Arg 195, and His 299. The distances of H-bonds ranged from 1.937 Å to 2.735 Å. However, procyanidin C3 formed nine H-bonds with the active sites of six amino acid residues (Asp 197, Glu 233, Gly 306, Lys 200, Tyr 155, and His 305). The shortest distance was 1.864 Å and the longest distance was 2.724 Å. It was found that acarbose formed eleven H-bonds (the distances ranged from 1.776 Å to 2.732 Å) with seven amino acid residues, namely, Asp 300, Gln 63, Glu 240, Gly 306, Tyr 151, Lys 200, and His 305.

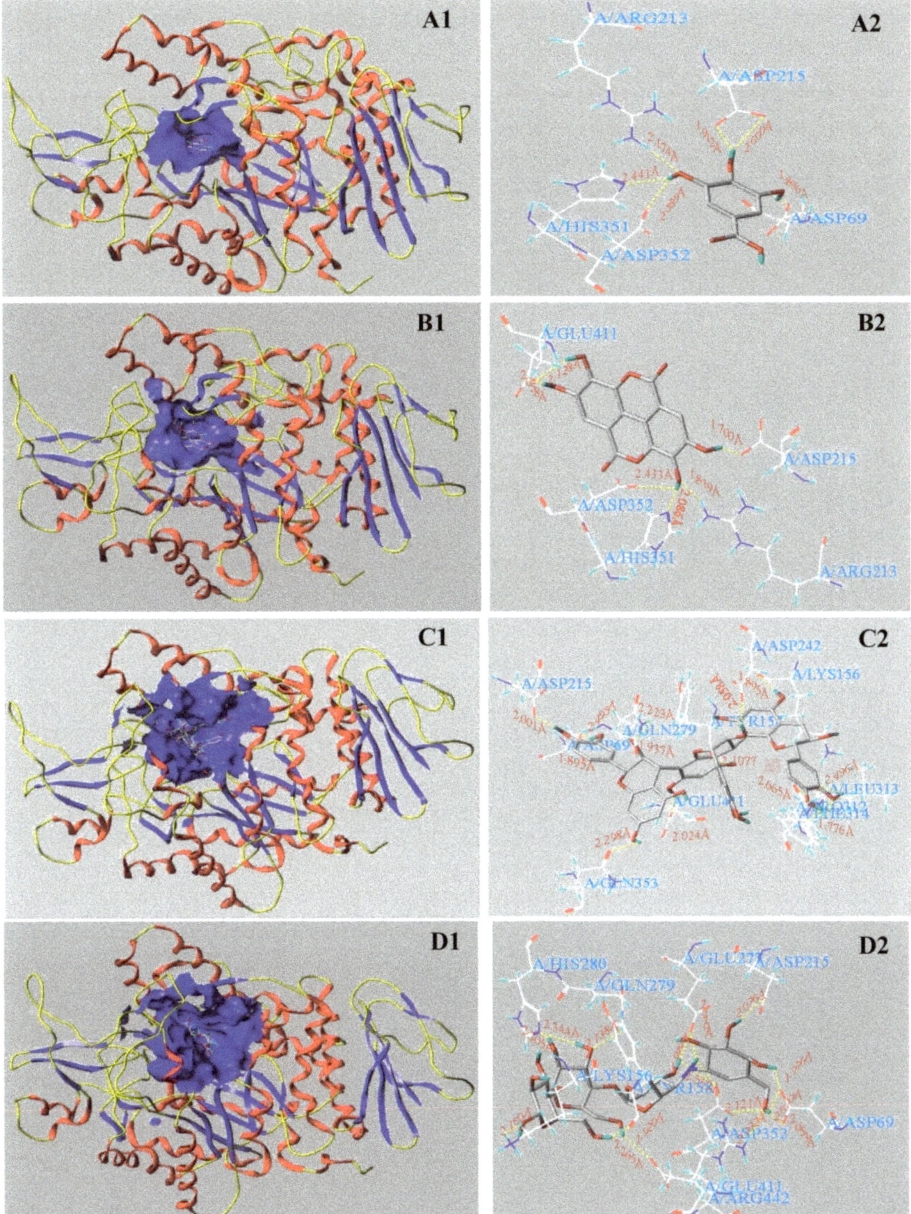

Figure 5. Molecular docking of the main three phenolic compounds and acarbose with the α-glucosidase. The 3D docking structures of three main phenolic compounds and acarbose were inserted into the hydrophobic cavity of the α-glucosidase (blue): gallic acid (**A1**); ellagic acid (**B1**); procyanidin C3 (**C1**); acarbose (**D1**). The conformation of active molecules interactions with amino acid residues in the active site of α-glucosidase: gallic acid (A2), ellagic acid (B2), procyanidin C3 (C2), and acarbose (D2) with residues in the active sites of the α-glucosidase, respectively. The dashed line stands for hydrogen bonds.

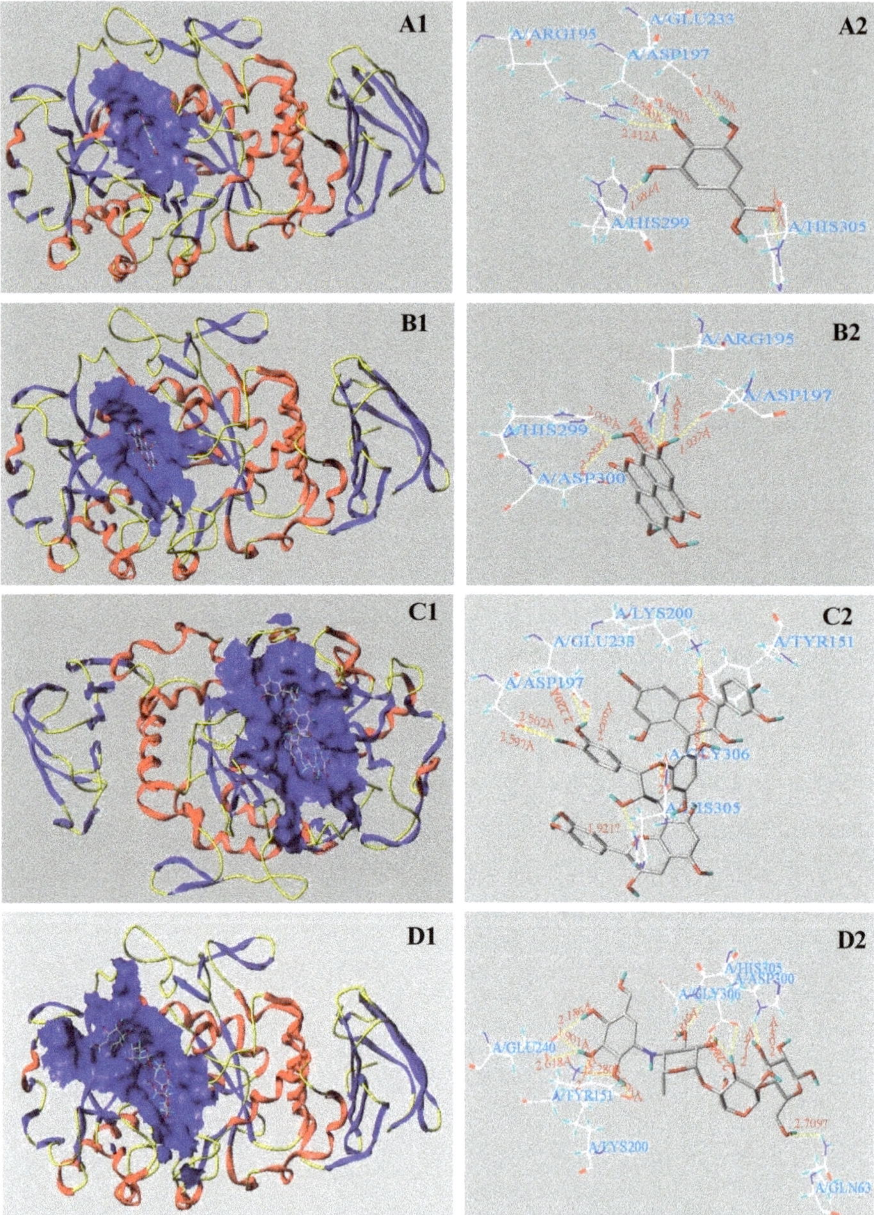

Figure 6. Molecular docking of the main three phenolic compounds and acarbose with α-amylase. The 3D docking structures of three phenolic compounds and acarbose were inserted into the hydrophobic cavity of α-amylase (blue): gallic acid (**A1**); ellagic acid (**B1**); procyanidin C3 (**C1**); acarbose (**D1**). The conformations of active molecules interactions with amino acid residues in the active site of α-amylase: gallic acid (**A2**), ellagic acid (**B2**), procyanidin C3 (**C2**), and acarbose (**D2**) with residues in the active sites of the α-amylase, respectively. The dashed line stands for hydrogen bonds.

It can be found that the numbers and distances of the H-bonds play important roles in exerting the catalytic reaction of the complex of digestive enzymes and ligands, and thereby cause the differences in inhibitory activity of digestive enzymes. Regardless of acarbose docked with α-glucosidase or α-amylase, the higher numbers of H-bonds and amino acid residues with active site were formed in the complex of digestive enzymes and acarbose. As a result, acarbose showed very good inhibitory effect of α-glucosidase and α-amylase. The numbers of H-bonds and amino acid binding sites formed by the two phenolic molecules (GA and EA) docked with α-glucosidase were equal, but there existed significant differences in inhibitory capacities of α-glucosidase. It may be due to that different interaction sites (amino acid residues) existed in between these molecules and α-glucosidase. Both of GA and EA all interacted with the amino acid residues His 351, Asp 215, Asp 352, and Arg 213 of α-glucosidase. EA also interacted with the amino acid residue Glu 411 of α-glucosidase. Some researchers have confirmed that some active sites (Glu 411) of α-glucosidase may inhibit the catalytic activity of this enzyme [37]. Consequently, ellagic acid showed the weakest α-glucosidase inhibitory effect. At least three investigated molecules formed H-bonds with the amino acid residues of Asp 69, Asp 215, and Asp 352, which may exert its α-glucosidase inhibitory effect. Hua et al. (2018) have also reported that Asp 69, Asp 215, and Arg 442 were the important residues involved in H-bond formation during the binding with α-glucosidase [37]. Zhang et al. (2018) also reported that the binding active sites (Asp 215 and Asp 352) in between the ligands and α-glucosidase played important roles in exert its α-glucosidase inhibitory effect [23]. Similarly, the order of amino acid residues numbers formed by four molecules (GA, EA, PC, and Acarbose) docked with α-amylase was: Acarbose (7) = PC (7) > GA (5) > EA (4). The order of H-bonds number was: Acarbose (11) > PC (9) > GA (6) > EA (5). Consequently, the order of the docking T-Score was: Acarbose > PC > GA > EA, which was consistent with the results of α-amylase inhibition activity. Moreover, at least three investigated molecules interacted with the amino acid residues Asp 197 and His 305 of α-amylase, indicating that these two amino acid residues may play critical roles in the catalytic reaction of α-amylase. Many reports have confirmed that the amino acid residues Asp 197 and His 305 played critical roles in the catalytic reaction of α-amylase [40,41]. The mechanisms of the digestive enzymes inhibitory activities of these compounds possibly involve the binding of compounds with the catalytic sites of digestive enzymes [37]. The results demonstrated that the hydrogen bonds and the binding residues with active sites have important effects on these digestive enzymes activities.

4. Conclusions

The solvents have significant impacts on the extraction of total phenolics and total flavonoids in various parts of raspberry. Fifty percent methanol was the best solvent for extracting high contents of phenolic and flavonoid compounds. The LE in raspberry displayed the highest TPC and TFC. A total of 19 phenolics compounds were identified. Gallic acid, ellagic acid, and procyanidin C3 were the main phenolic compositions existing in various parts of raspberry. Higher levels of phenolic compounds in raspberry showed the stronger anti-oxidant activities and inhibitory potency of digestive enzymes. The major phenolic compounds that were found in various parts of raspberry all showed good digestive enzyme inhibitory activities, especially for PC. Molecular docking analysis revealed the underlying inhibition mechanisms of these three main phenolic compounds against digestive enzymes, and the theoretical analysis data explained the experimental results very well.

Supplementary Materials: The following are available online at http://www.mdpi.com/2076-3921/8/8/274/s1, Table S1: Regression equation, R2, LOD, and LOQ results of the main individual phenolics of various parts in raspberry.

Author Contributions: L.W. (Lu Wang) and Z.W. designed the experiment; L.W. (Lingfeng Wu), Y.L., and Y.Q. carried out the experiments and the statistical analyses; L.W. (Lu Wang) analyzed the data and prepared the manuscript.

Funding: This work was supported by the Scientific Research Foundation of Hainan University (No. KYQD1901) and Guangdong special funds for science and technology innovation strategy, China (JK2018-352-0203).

Conflicts of Interest: The authors declared no conflict of interest.

Abbreviations

LE	leaves extracts
SE	seed extracts
FPE	fruit pulp extracts
DPPH	1,1-Diphenyl-2-picrylhydrazyl radical
ABTS	2-azino-bis (3-ethylbenzothiazoline-6-sulfonic acid) diammonium salt
TPTZ	2,4,6-tris(2-pyridyl)-s-triazine
FRAP	ferric reducing anti-oxidant power
p-NPG	p-Nitrophenyl-α-D-glucopyranoside
TPC	total phenolic content
TFC	flavonoid content
GA	gallic acid
EA	ellagic acid
PC	procyanidin C3
Vc	ascorbic acid
DW	dried weight
HPLC-ESI-HR-qTOF-MS/MS	High performance liquid chromatography-electrospray ionization-high resolution-quadrupole time of flight-tandem mass spectrometry
PBS	phosphate buffer solution
GAE	gallic acid equivalents
RE	rutin equivalents

References

1. Rodriguezmonforte, M.; Sanchez, E.; Barrio, F.; Costa, B.; Floresmateo, G. Metabolic syndrome and dietary patterns: a systematic review and meta-analysis of observational studies. *Eur. J. Nutr.* **2017**, *56*, 925–947. [CrossRef] [PubMed]

2. Furukawa, S.; Fujita, T.; Shimabukuro, M.; Iwaki, M.; Yamada, Y.; Nakajima, Y.; Shimomura, I. Increased oxidative stress in obesity and its impact on metabolic syndrome. *J. Clin. Investig.* **2004**, *114*, 1752–1761. [CrossRef] [PubMed]

3. Younger, K.M. Antioxidant, anti-inflammatory, anti-carcinogenic and anti-diabetic. *Nutr. Res. Rev.* **2010**, *23*, 181–193. [CrossRef] [PubMed]

4. Tsai, Y.D.; Hsu, H.F.; Chen, Z.H.; Wang, Y.T.; Huang, S.H.; Chen, H.J. Antioxidant, anti-inflammatory, and anti-proliferative activities of extracts from different parts of farmed and wild glossogyne tenuifolia. *Ind. Crop. Prod.* **2014**, *57*, 98–105. [CrossRef]

5. Zhan, J.; Liu, Y.J.; Cai, L.B.; Xu, F.R.; He, Q.Q. Fruit and vegetable consumption and risk of cardiovascular disease: a meta-analysis of prospective cohort studies. *Crit. Rev. Food. Sci.* **2015**, *57*, 1650–1663. [CrossRef] [PubMed]

6. Yamagata, K.; Tagami, M.; Yamori, Y. Dietary polyphenols regulate endothelial function and prevent cardiovascular disease. *Nutrition* **2015**, *31*, 28–37. [CrossRef] [PubMed]

7. Zafrilla, P.; Ferreres, F.; Tomás-Barberán, F.A. Effect of processing and storage on the antioxidant ellagic acid derivatives and flavonoids of red raspberry (*Rubus idaeus*) jams. *J. Agri. Food. Chem.* **2001**, *49*, 3651–3655. [CrossRef] [PubMed]

8. Figueira, M.E.; Câmara, M.B.; Direito, R.; Rocha, J.; Serra, A.T.; Duarte, C.M.; Sepodes, B. Chemical characterization of a red raspberry fruit extract and evaluation of its pharmacological effects in experimental models of acute inflammation and collagen-induced arthritis. *Food Funct.* **2014**, *5*, 3241–3251.

9. Han, N.; Gu, Y.; Ye, C.; Cao, Y.; Liu, Z.; Yin, J. Antithrombotic activity of fractions and components obtained from raspberry leaves (*Rubus chingii*). *Food Chem.* **2012**, *132*, 181–185. [CrossRef]

10. Buricova, L.; Andjelkovic, M.; Cermakova, A.; Reblova, Z.; Jurcek, O.; Kolehmainen, E.; Kvasnicka, F. Antioxidant capacity and antioxidants of strawberry, blackberry, and raspberry leaves. *Czech J. Food Sci.* **2018**, *29*, 181–189. [CrossRef]

11. Dvaranauskaite, A.; Venskutonis, P.R.; Labokas, J. Comparison of quercetin derivatives in ethanolic extracts of red raspberry (*Rubus idaeus* L.) leaves. *Acta Aliment.* **2008**, *37*, 449–461. [CrossRef]

12. Luther, M.; Parry, J.; Moore, J.; Meng, J.; Zhang, Y.; Cheng, Z.; Yu, L. Inhibitory effect of Chardonnay and black raspberry seed extracts on lipid oxidation in fish oil and their radical scavenging and antimicrobial properties. *Food Chem.* **2007**, *104*, 1065–1073. [CrossRef]

13. Qin, Y.; Wang, L.; Liu, Y.; Zhang, Q.; Li, Y.; Wu, Z.Q. Release of phenolics compounds from *Rubus idaeus* L. dried fruits and seeds during simulated in vitro digestion and their bio-activities. *J. Funct. Foods.* **2018**, *46*, 57–65. [CrossRef]

14. Wang, L.; Lin, X.; Zhang, J.; Zhang, W.; Hu, X.; Li, W.; Liu, S. Extraction methods for the releasing of bound phenolics from *Rubus idaeus* L. leaves and seeds. *Ind. Crop. Prod.* **2019**, *135*, 1–9. [CrossRef]

15. Wang, L.; Luo, Y.; Wu, Y.; Liu, Y.; Wu, Z. Fermentation and complex enzyme hydrolysis for improving the total soluble phenolic contents, flavonoid aglycones contents and bio-activities of guava leaves tea. *Food Chem.* **2018**, *264*, 189–198. [CrossRef] [PubMed]

16. Sheth, F.; De, S. Evaluation of comparative antioxidant potential of four cultivars of *Hibiscus rosasinensis* L. by HPLC-DPPH method. *Free Rad. Antioxid.* **2012**, *2*, 73–78. [CrossRef]

17. Rezaei-Sadabady, R.; Eidi, A.; Zarghami, N.; Barzegar, A. Intracellular ROS protection efficiency and free radical-scavenging activity of quercetin and quercetinencapsulated liposomes. *Artif. Cell. Nanomed. B.* **2016**, *44*, 128–134. [CrossRef]

18. Wang, L.; Bei, Q.; Wu, Y.; Liao, W.; Wu, Z.Q. Characterization of soluble and insoluble-bound polyphenols from *Psidium guajava* L. leaves co-fermented with *Monascus anka* and *Bacillus* sp. and their bio-activities. *J. Funct. Foods.* **2017**, *32*, 149–159. [CrossRef]

19. Liu, Q.; Cao, X.; Zhuang, X.; Han, W.; Guo, W.; Xiong, J.; Zhang, X. Rice bran polysaccharides and oligosaccharides modified by *Grifola frondosa* fermentation: Antioxidant activities and effects on the production of NO. *Food Chem.* **2017**, *223*, 49–53. [CrossRef]

20. Wong, C.C.; Li, H.B.; Cheng, K.W.; Chen, F. A systematic survey of antioxidant activity of 30 Chinese medicinal plants using the ferric reducing antioxidant power assay. *Food Chem.* **2006**, *97*, 705–711. [CrossRef]

21. Zhao, J.Q.; Wang, Y.M.; Yang, Y.L.; Zeng, Y.; Mei, L.J.; Shi, Y.P. Antioxidants and α-glucosidase inhibitors from "liucha" (young leaves and shoots of *Sibiraea laevigata*). *Food Chem.* **2017**, *230*, 117–124. [CrossRef] [PubMed]

22. Zaharudin, N.; Armando, A.S.; Dragsted, L.O. Inhibitory effects of edible seaweeds, polyphenolics and alginates on the activities of porcine pancreatic α-amylase. *Food Chem.* **2017**, *245*, 1196–1203. [CrossRef] [PubMed]

23. Zhang, X.; Jia, Y.; Ma, Y.; Cheng, G.; Cai, S. Phenolic composition, antioxidant properties, and inhibition toward digestive enzymes with molecular docking analysis of different fractions from *Prinsepia utilis* royle fruits. *Molecules* **2018**, *23*, 3373. [CrossRef] [PubMed]

24. Zhang, C.; Ma, Y.; Gao, F.; Zhao, Y.; Cai, S.; Pang, M. The free, esterified, and insoluble-bound phenolic profiles of *Rhus chinensis* mill. fruits and their pancreatic lipase inhibitory activities with molecular docking analysis. *J. Funct. Foods.* **2018**, *40*, 729–735. [CrossRef]

25. Ye, F.; Qiang, L.; Hang, L.; Zhao, G. Solvent effects on phenolic content, composition, and antioxidant activity of extracts from florets of sunflower (*Helianthus annuus* L.). *Ind. Crop. Prod.* **2015**, *76*, 574–581. [CrossRef]

26. A Comparative Study on Phytochemical Screening, Quantification of Phenolic Contents and Antioxidant Properties of Different Solvent Extracts from Various Parts of *Pistacia lentiscus* L. Available online: https://www.sciencedirect.com/science/article/pii/S1018364718303355 (accessed on 11 May 2018).

27. Prabakaran, S.; Ramu, L.; Veerappan, S.; Pemiah, B.; Kannappan, N. Effect of different solvents on volatile and non-volatile constituents of red bell pepper (*Capsicum annuum* L.) and their in vitro antioxidant activity. *J. Food. Meas. Charact.* **2017**, *11*, 1531–1541. [CrossRef]

28. Do, Q.D.; Angkawijaya, A.E.; Tran-Nguyen, P.L.; Huynh, L.H.; Soetaredjo, F.E.; Ismadji, S. Effect of extraction solvent on total phenol content, total flavonoid content, and antioxidant activity of *Limnophila aromatica*. *J. Food. Drug Anal.* **2014**, *22*, 296–302. [CrossRef] [PubMed]

29. Katarzyna, R.; Paweł, P.; Joanna, R.; Aneta, K.; Audrius, M.; Monika, N.; Boguslaw, B. Effect of solvent and extraction technique on composition and biological activity of *Lepidium sativum* extracts. *Food Chem.* **2019**, *289*, 16–25.

30. Rafińska, K.; Pomastowski, P.; Rudnicka, J.; Krakowska, A.; Maruśka, A.; Narkute, M.; Buszewski, B. Bioactive compounds and antioxidant activity of wheat bran and barley husk in the extracts with different polarity. *Int. J. Food Prop.* **2019**, *22*, 646–658.

31. Wanyo, P.; Meeso, N.; Siriamornpun, S. Effects of different treatments on the antioxidant properties and phenolic compounds of rice bran and rice husk. *Food Chem.* **2014**, *157*, 457–463. [CrossRef]

32. Djordjevic, T.M.; ller-Marinkovic, S.S.; Dimitrijevic-Brankovic, S.I. Antioxidant activity and total phenolic content in some cereals and legumes. *Int. J. Food Prop.* **2011**, *14*, 175–184. [CrossRef]

33. Hong, H.K.; Kwak, J.H.; Kang, S.C.; Lee, J.W.; Park, J.H.; Ahn, J.W. Antioxidative constituents from the whole plants of euphorbia supina. *Korean J. Pharmacogn.* **2008**, *39*, 260–264.

34. Characterization and Antioxidant Activity of Gallic Acid Derivative. Available online: https://aip.scitation.org/doi/abs/10.1063/1.5011887 (accessed on 27 November 2017).

35. Bialonska, D.; Kasimsetty, S.G.; Khan, S.I.; Ferreira, D. Urolithins, intestinal microbial metabolites of pomegranate ellagitannins, exhibit potent antioxidant activity in a cell-based assay. *J. Agr. Food. Chem.* **2009**, *57*, 10181–10186. [CrossRef] [PubMed]

36. Khaled, H.; Kais, M.; Zahra, A.; Ahmed, A.; Abdelfattah, E. Inhibition of key digestive enzymes related to diabetes and hyperlipidemia and protection of liver-kidney functions by trigonelline in diabetic rats. *Sci. Pharmaceut.* **2013**, *81*, 233–246.

37. Hua, F.; Zhou, P.; Wu, H.; Chu, G.; Xie, Z.; Bao, G. Inhibition of α-glucosidase and α-amylase by flavonoid glycosides from Lu'an GuaPian tea: Molecular docking and interaction mechanism. *Food Funct.* **2018**, *9*, 4173–4183.

38. Wang, L.; Liu, Y.; Luo, Y.; Huang, K.; Wu, Z. Quickly screening for potential α-glucosidase inhibitors from guava leaves tea by bio-affinity ultrafiltration coupled with HPLC-ESI-TOF/MS method. *J. Agri. Food. Chem.* **2018**, *66*, 1576–1582. [CrossRef]

39. Zhang, L.; Li, J.; Hogan, S.; Chung, H.; Welbaum, G.E.; Zhou, K. Inhibitory effect of raspberries on starch digestive enzyme and their antioxidant properties and phenolic composition. *Food Chem.* **2010**, *119*, 592–599. [CrossRef]

40. Rasouli, H.; Hosseinighazvini, S.M.; Adibi, H.; Khodarahmi, R. Differential α-amylase/α-glucosidase inhibitory activities of plant-derived phenolic compounds: a virtual screening perspective for the treatment of obesity and diabetes. *Food Funct.* **2017**, *8*, 1942–1954.

41. Sun, L.J.; Warren, F.J.; Gidley, M.J.; Guo, Y.L.; Miao, M. Mechanism of binding interactions between young apple polyphenols andporcine pancreatic α-amylase. *Food Chem.* **2019**, *283*, 468–474. [CrossRef]

© 2019 by the authors. Licensee MDPI, Basel, Switzerland. This article is an open access article distributed under the terms and conditions of the Creative Commons Attribution (CC BY) license (http://creativecommons.org/licenses/by/4.0/).

 *antioxidants*

Review

Antidiabetic Effects of Hydroxytyrosol: In Vitro and In Vivo Evidence

Filip Vlavcheski [1,2], Mariah Young [1] and Evangelia Tsiani [1,2,*]

[1] Department of Health Sciences, Brock University, St. Catharines, ON L2S 3A1, Canada;
    fvlavcheski@brocku.ca (F.V.); my14ac@brocku.ca (M.Y.)
[2] Centre for Bone and Muscle Health, Brock University, St. Catharines, ON L2S 3A1, Canada
* Correspondence: ltsiani@brocku.ca; Tel.: +1-905-688-5550 (ext. 3881)

Received: 17 May 2019; Accepted: 18 June 2019; Published: 21 June 2019

Abstract: Insulin resistance, a pathological condition characterized by defects in insulin action leads to the development of Type 2 diabetes mellitus (T2DM), a disease which is currently on the rise that pose an enormous economic burden to healthcare systems worldwide. The current treatment and prevention strategies are considerably lacking in number and efficacy and therefore new targeted therapies and preventative strategies are urgently needed. Plant-derived chemicals such as metformin, derived from the French lilac, have been used to treat/manage insulin resistance and T2DM. Other plant-derived chemicals which are not yet discovered, may have superior properties to prevent and manage T2DM and thus research into this area is highly justifiable. Hydroxytyrosol is a phenolic phytochemical found in olive leaves and olive oil reported to have antioxidant, anti-inflammatory, anticancer and antidiabetic properties. The present review summarizes the current in vitro and in vivo studies examining the antidiabetic properties of hydroxytyrosol and investigating the mechanisms of its action.

Keywords: insulin resistance; diabetes; hydroxytyrosol; olive oil; phenols; antioxidant; antidiabetic

---

1. Introduction

The composition of the culturally-based diet of the Greeks, Spanish, and Italians was first connected to cardiovascular health and lower cholesterol levels by American researcher Ancel Keys nearly fifty years ago [1–5]. This was termed the "Mediterranean diet" and has since been studied extensively, with research showing clear beneficial effects on health and lifespan. The impact of this type of diet is especially relevant to the studies of metabolic disease, cardiovascular well-being and cancer [5–9]. The core "Mediterranean triad" encompasses food derived from wheat, grapes, and olives [10]. As the principal source of lipids in the Mediterranean diet, olive oil has attracted the attention of many scientists attempting to determine how its chemical constituents affect the body. Fatty acids make up most of the mass of olive oil, but it also contains up to 1 gram per kilogram of phenolic compounds [11]. While thirty different phenols have been identified in olive oil, the ones with the highest relative abundances are hydroxytyrosol, tyrosol, and oleuropein (Figure 1). Hydroxytyrosol, often abbreviated as HT, and tyrosol are metabolic derivatives of oleuropein following hydrolysis and are structurally similar [11,12].

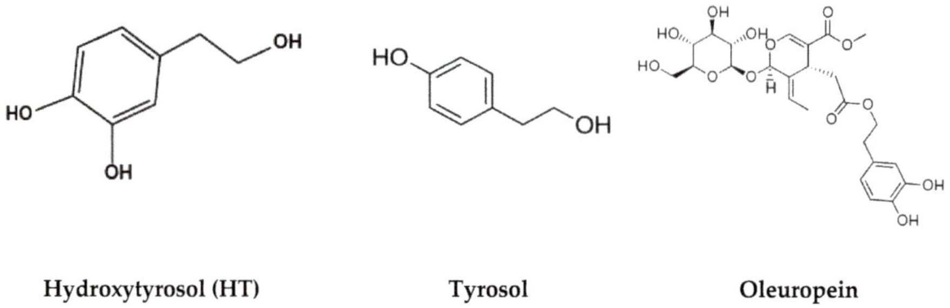

Hydroxytyrosol (HT)   Tyrosol   Oleuropein

Figure 1. Chemical structures of hydroxytyrosol, tyrosol, and oleuropein.

Hydroxytyrosol (HT) is the phenol present in the highest concentration in olive oil (Table 1) and is also a product of oleuropein metabolism in the body [12,13]. Much research has been conducted over the past twenty years attempting to determine the health effects of oleuropein [14], hydroxytyrosol, tyrosol, and other phenolic compounds found in olive oil [9,11,15–17]. Evidence shows that hydroxytyrosol has antioxidant properties, affects glucose and lipid homeostasis and may protect against diabetes [17–21]. In the present review, the relevant in vitro and in vivo studies examining the antidiabetic effects of HT are summarized.

Table 1. Most abundant phenolic compounds found in olive oil.

Polyphenol	Quantity	Olive Oil Type	Source
Hydroxytyrosol (3,4-dihydroxyphenyl ethanol) (HT)	0.93–14.64 mg/kg	Olive oil (various brands)	[21]
Tyrosol	0.25–14.97 mg/kg	Olive oil (various brands)	[21]
Oleuropein	0.0–4.7 mg/kg	Virgin olive oil	[22]

Currently, pharmacological agents used for the treatment of T2DM include drugs that increase peripheral (muscle and fat) glucose transport while decreasing hepatic glucose production by inhibiting gluconeogenesis (biguanides and thiazolidinediones (TZDs)) (Table 2). Metformin, a biguanide, is the first line of treatment for T2DM. However, its use is associated with an increased risk of lactic acidosis and gastrointestinal (GI) disturbances including cramping, nausea, vomiting, and diarrhea [23,24]. The use of TZDs (rosiglitazone and rioglitazone) is associated with bladder cancer, heart failure, hepatitis, and weight gain. Additionally, agents that stimulate insulin release from the pancreatic β-cells, sulfonylureas (glyburide and glipizide glimepiride) and meglitinides (repaglinide), are routinely prescribed but are associated with a high risk of hypoglycemia. Another class of medication, usually prescribed to prediabetics, is intestinal glucose absorption inhibitors that act by hindering α-glucosidase enzymes and, thereby, absorption of glucose (acarbose, voglibose, and miglitol). However, these drugs are not very potent and their use in North America is quite low. In the last decade, several new therapies for T2DM have become available. Gliptins and dipeptidyl peptide 4 (DDP-4) inhibitors (sitagliptin, saxagliptin, vidagliptin, linagliptin, and alogliptin) hinder glucagon release, thereby reducing blood glucose levels. However, significant side effects may occur including heart failure, pancreatitis, and pancreatic and prostate cancer [25]. Similarly, glucagon-like peptide 1 (GLP-1) receptor agonists (liraglutide, exenatide, and dulaglutide) inhibit glucagon release and stimulate insulin production, consequently lowering blood glucose levels. GLP-1 receptor agonists are associated with a lower risk of hypoglycemia in comparison to sulfonylureas and meglitinides. However, the mechanism of action is similar to the DPP-4 inhibitors. As a result, potential life-threatening conditions such as pancreatic cancer, pancreatitis and heart failure may occur [26]. Sodium–glucose cotransporter 2 (SGLT2) inhibitors (canagliflozin, capagliflozin) are the most recent antidiabetic drugs currently available and they exhibit glucose lowering properties by inhibiting renal glucose reabsorption and stimulating glucose excretion. Although initially promising, these drugs are also associated with adverse health effects including severe hypotension, urinary tract infections (UTI), and ketoacidosis [27,28].

Table 2. Current pharmacological treatments for T2DM.

Antidiabetic Agent	Target Tissues	Target Pathways	Effect	Side Effects
Biguanides *metformin*	Liver, fat, muscle	↑ AMPK activity ↓ Complex I of the respiratory chain	↑ glucose uptake (fat and muscle) ↓ hepatic glucose production ↑ glucose tolerance ↑ insulin sensitivity	lactic acidosis GI problems (camps, nausea, vomiting, diarrhea)
Thiazolidinediones (TZD) glitazones: *rosiglitazone* *rioglitazone*	Liver, fat, muscle	↑ PPARγ activity	↑ adipocyte lipid storage ↓ circulating FFA ↓ ectopic lipid accumulation ↑ glucose uptake (fat and muscle) ↑ insulin sensitivity ↑ β-cell function	bladder cancer heart failure hepatitis bone fractures weight gain edema
Sulfonylureas glinides: *glyburide, glipizide* *glinepiride* meglitinides: *repaglinide*	pancreas fat, muscle	↑ intracellular potassium concentration leading to depolarization of pancreatic β cells	↑ glucose-mediated insulin release ↓ blood glucose	hypoglycemia weight gain hunger skin reactions
α-glucosidase-inhibitors: *acarbose, voglibose, miglitol*	small intestine	Competitive inhibition of enzymes vital for carbohydrate digestion	↓ carbohydrate absorption ↓ blood glucose	abdominal pain bloating, nausea, vomiting diarrhea flatulence
Dipeptidyl peptide 4 (DDP-4) inhibitors) gliptins: *sitagliptin, saxagliptin* *vidagliptin, linagliptin* *alogliptin*	pancreas	↓ DDP-4 activity ↑ incretin levels (GLP-1 and GIP)	↓ glucagon secretion ↑ insulin release ↓ blood glucose	heart failure pancreatitis pancreatic cancer prostate cancer GI problems flu-like symptoms
Incretin mimetics (glucagon-like peptide 1 (GLP-1) receptor agonist *liraglutide, exenatide* *dulaglutide* Gastric inhibitory polypeptide (GIP)	direct effect on pancreas, stomach and brain indirect on liver and muscle	↑ activation of GLP-1 receptor	↓ glucagon secretion ↑ insulin release ↓ β-cell apoptosis ↓ gastric emptying ↓ appetite ↓ blood glucose	pancreatic cancer pancreatitis heart failure prostate cancer abdominal pain nausea, vomiting
Sodium–glucose cotransporter 2 (SGLT2) inhibitors gliflozins: *canagliflozin* *capagliflozin*	kidneys	↓ SGLT2 action in the proximal convoluted tubule	↓ reabsorption of glucose ↑ facilitate excretion in urine	Hypotension urinary tract infections ketoacidosis hyperkalemia kidney failure bone fractures

The risks associated with existing T2DM treatments emphasize a continued need to develop more effective T2DM therapeutics with fewer side effects. Despite the progress in the last decade in terms of available treatment/management approaches, T2DM is a disease currently on the rise and poses an enormous burden to our health care systems globally. In addition, approximately 50% of the T2DM-affected population are living in poverty-stricken areas in Africa and Asia. Therefore, there is an urgent need not only for more effective but also for affordable treatment options. Novel compounds, that exhibit insulin-like effects, improve insulin sensitivity, enhance the efficacy of already existing antidiabetic agents and have very few side effects, are greatly desired as they will broaden the spectrum of treatment options for insulin resistance and T2DM.

2. In Vitro Evidence: Antidiabetic Effects of Hydroxytyrosol (HT)

2.1. Effects of Hydroxytyrosol (HT) on Skeletal Muscle Cells

Treatment of C2C12 cells with HT (1–50 μM) increased creatine kinase activity and myosin heavy chain expression, which are indicators of muscle cell differentiation and strength of contraction, respectively, therefore demonstrating a possible improvement in muscle adaptation to exercise by HT (Table 3) [29]. In addition, treatment with HT attenuated the tumor necrosis factor-α (TNF-α)-induced downregulation of mitochondrial biogenesis by increasing peroxisome proliferator-activated receptor-gamma coactivator (PGC)-1α, mitochondrial complexes (I and II) and myogenin expression. This indicated that HT improves mitochondrial development and function in muscle cells under inflammatory stress [29]. In another study, treatment of C2C12 myoblasts with HT (5–20 μM) resulted in attenuation of H_2O_2-induced apoptosis and oxidative stress [30]. Additionally, HT completely prevented the H_2O_2-induced morphological changes, swollen mitochondria and presence of autophagic vacuoles [30]. Drira and Sakamoto (2013) found that exposure of C2C12 myotubes to hydroxytyrosol-acetate (12 h at 25–75 μM), which acts similarly to HT, resulted in significantly increased glucose uptake in a dose-dependent manner [31]. Another study found that treatment of C2C12 myotubes with hydroxytyrosol-acetate (1–50 μM for 24h) significantly attenuated the *tert*-butylhydroperoxide (t-BHP)-induced mitochondrial damage, optic atrophy 1 (OPA-1) cleavage and muscle degradation while increasing oxygen consumption capacity, ATP production, activities of mitochondrial complex I, II and V, and myosin heavy chain expression [32]. Additionally, the cell viability was markedly increased by HT treatment (Table 3) [32]. Altogether, the study indicates an effect of HT to protect against oxidative stress-induced skeletal muscle damage.

Table 3. Effects of hydroxytyrosol on skeletal muscle cells.

Cell Type	Hydroxytyrosol Concentration/Duration	Effect	Source
C2C12 myoblasts	1–50 μM for 30 min; TNFα for 4–5 days	↑ muscle cell differentiation (↑ creatine kinase & myosin heavy chain) ↑ PGC1α ↑ mitochondrial biogenesis	[29]
C2C12 myoblasts	5 or 20 μM HT for 3 h with 1 mM H_2O_2	↓ H_2O_2-induced apoptosis ↓ morphology changes ↓ oxidative stress	[30]
C2C12 myotubes	Hydroxytyrosol-acetate 0–75 μM for 12 h	↑ glucose uptake	[31]
C2C12 myotubes	1–50 μM for 24 h; 100 μM t-BHP	↑ cell viability ↓ mitochondrial dysfunction (↑ ATP production and activity of complex I, II and V) ↓ muscle cell degeneration (OPA cleavage) ↑ myosin heavy chain expression	[32]

The studies above show that HT has the potential to affect key signaling molecules involved in mitochondrial function and development, protect against oxidative stress-induced muscle damage [29,30,32] and significantly increase muscle cell glucose uptake [31]. These studies indicate that HT could have potential antidiabetic effects on muscle tissue directly and further studies are required to investigate this assumption.

2.2. Effects of Hydroxytyrosol (HT) on Adipocytes

In one study, exposure of 3T3-L1 adipocytes to HT (0.1-10 μM) resulted in enhanced mitochondrial biogenesis and oxygen uptake (Table 4) [33]. Markers of mitochondrial capacity including peroxisome proliferator-activated receptor γ (PPARγ), PPARγ coactivator (PGC)-1alpha (PGC1α), nuclear respiration factor 1 and 2 (NRF1 and NRF2), mitochondrial transcription factor A and mitochondrial complexes I, II, III and V were significantly upregulated. In addition, to the upregulation of mitochondria number and mtDNA, intracellular fatty acids were reduced and oxygen consumption was increased. More importantly, HT significantly increased the phosphorylation of the energy sensor AMP-activated protein kinase (AMPK) and its downstream target acetyl-CoA carboxylase (ACC). This finding indicates that activating AMPK may be the possible mechanism for HT-induced expression of PGC1α leading to mitochondrial biogenesis [33]. Exposure of mouse-derived C3H10 T1/T2 pre-adipocytes to HT (0.5–25 μM for 4 or 7 days) significantly inhibited adipocyte differentiation and lipid accumulation [34]. Fat droplet size and number were dose-dependently decreased and hormone-sensitive lipase (HSL) was downregulated by HT treatment. Furthermore, HT inhibited rosiglitazone-stimulated lipid synthesis shown by a decrease in carnitine palmitoyltransferase I (CPT1β) and downregulation of adipogenesis regulators such as PPARγ/α, CCAAT-enhancer-binding proteins (C/EBPα) and differentiation markers aP2 (adipocyte fatty acid binding protein) and adiponectin [34]. Interestingly, GLUT4 gene expression was also downregulated most likely as a consequence of HT inhibiting the two master regulators PPARγ and C/EBPα. Altogether, these studies present evidence showing HT suppressing adipogenesis and lipogenesis in adipocytes [34], however the effects on glucose uptake remain to be investigated. Treatment of 3T3-L1 preadipocytes with HT (100–150 μM) resulted in dose-dependent inhibition of cell division during mitotic clonal expansion thus causing cell cycle delay and increased lipid accumulation with no effect on cell viability [35]. Additionally, adipogenesis related genes were significantly downregulated including PPARγ, sterol regulatory element-binding transcription factor 1 (SREBF1), C/EBPα and their downstream target genes GLUT4, cluster of differentiation 36 (CD36) and fatty acid synthase (FAS) [35]. Similar effects were seen when 3T3-L1 adipocytes were treated with HT-acetate (25–75 μM) [31]. A significant inhibition of adipogenesis and lipid accumulation was seen with HT-acetate which was associated with downregulation of PPARγ, SREBP-1c, C/EBPα, GLUT4, CD36, and FAS [31]. Increased lipolysis was also indicated via increase of the quantity of glycerol being released and the activation of hormone-sensitive lipase (HSL) [31]. Exposure of 3T3-L1 adipocytes to HT (0-150 μM) resulted in a dose-dependent increase in lipolysis and glycerol release while decreasing adipocyte triglyceride accumulation [36]. The phosphorylation of proteins involved in lipolysis pathways, including hormone-sensitive lipase (HSL), ERK and perilipin, were significantly increased [36]. Overall expression of HSL, adipocyte triglyceride lipase (ATGL) and adipogenesis proteins PPARγ and C/EBPα were decreased. Despite the negative effect of HT on the expression levels of HSL and ATGL the phosphorylation and therefore activation of HSL and lipolysis was enhanced. The decrease in expression levels may be due to a compensatory mechanism or negative feedback loop resulting from the increased HSL activation. Furthermore, pretreatment with PKA and ERK1/2 inhibitor attenuated the HT-stimulated lipolysis, indicating that PKA and ERK may be involved in the HT-induced lipolysis [36]. Exposure of 3T3-L1 and human Simpson-Golabi-Behmel syndrome (SGBS) adipocytes to HT (0.1–20 μM) prevented the TNFα-induced suppression of total adiponectin secretion and protein expression [37]. Additionally, HT prevented the TNF-α-induced downregulation of PPARγ and JNK phosphorylation. This study showed that the deleterious effects of TNFα are attenuated by HT [37]. Treatment of 3T3-L1 preadipocytes with HT (10–100 μM) resulted in inhibition of cell differentiation, increased FAS and lipoprotein lipase (LPS) gene expression while simultaneously downregulating PPARγ and cannabinoid receptor type 1 (CB1) gene expression [38]. Recent studies on obesity emphasize the importance of discovering food components that have the ability to suppress adipocyte proliferation and differentiation and thus adipose tissue expansion. A new proposed approach includes molecules that may potentially modulate endocannabinoid receptor gene expression [39]. Another study showed that exposure of human primary visceral preadipocytes to

HT (5–70 µg/mL) resulted in decreased triglyceride accumulation and increased apoptosis, lipolysis, glycerol release and expression of adipogenesis inhibiting genes such as GATA- binding factor 2 and 3 (GATA2 and GATA3) [40], protein Wnt-3A (WNT3A), secreted frizzled-related protein 5 (SFRP5), hairy and enhancer of split-1 (HES1), and NAD-dependent deacetylase sirtuin-1 (SIRT1). Additionally, genes involved in promoting adipogenesis including leptin (LEP), sterol regulatory element-binding protein 1 (SREBF1), Cyclin D1 (CCND1) and fibroblast growth factor 1 (FGF1) were significantly downregulated [40]. On the other hand, Anter et al. 2016 showed that exposure of human bone marrow mesenchymal stem cells (MSCs) to HT (100 µM) as they differentiated into adipocytes, resulted inupregulated adipocyte differentiation genes such as PPARγ and fatty-acid binding protein 4 (FABP4) as well as fat vesicle formation in the MSC adipocytes. Additionally, lipid protein lipase (LPL) gene expression was modestly but non-significantly decreased (Table 4) [41].

The above studies suggest that HT decreases adipocyte differentiation and proliferation [31, 35,36,38,40] and reduces the number and size of adipocyte lipid droplets [34]. Additionally, HT significantly downregulates the genes involved in adipogenesis and obesity while exhibiting substantial anti-inflammatory properties [31,33–38]. Furthermore, these studies show significant increase in lipolysis and related enzymes such as HSL and ATGL [31,36,40]. It is important to note that the increase in adipocyte lipolysis in vitro may suggest antidiabetic properties by decreasing the size/expansion of adipocytes, however an upsurge in lipolysis in vivo may significantly increase circulating FFA levels leading to hyperlipidemia, lipid toxicity and exacerbation of insulin resistance and T2DM symptoms. Therefore, more in vivo studies investigating the effect of HT on adipose tissue and circulating lipid levels are needed.

Table 4. Effects of hydroxytyrosol on adipocytes.

Cell Type	Hydroxytyrosol Concentration/Duration	Effect	Source
3T3-L1 adipocytes	0.1–10 µM for 24–72 h	↑ mitochondrial biogenesis and O_2 consumption ↑ mitochondrial complexes I, II, III and V ↓ fatty acid content ↑ CPT-1, PPARα, PPARγ	[33]
C3H10 T1/T2 preadipocytes	25 µM for 4 or 7 days	↓ lipid differentiation and accumulation ↓ lipid droplet size and number ↓ adipogenesis-related genes (PPARγ and C/EBPα) ↓ differentiation markers (aP2 and adiponectin) ↓ GLUT4 gene expression	[34]
3T3-L1 preadipocytes	100 or 150 µM for 0–8 days	↓ cell division and lipid accumulation ↓ mitotic clonal expansion ↓ adipogenesis marker genes (PPARg, SREBP-1c, C/EBPα, GLUT4, CD36 and FAS)	[35]
3T3-L1 adipocytes	Hydroxytyrosol-acetate 25–75 µM for 12 h	↓ lipid accumulation ↓ adipogenesis (PPARγ, SREBP-1c, C/EBPα, GLUT4, CD36, and FAS) ↑lipolysis and glycerol release ↑ HSL	[31]
3T3-L1 adipocytes	0–150 µM for 24–72 h	↑ lipolysis and glycerol release ↓ triglyceride accumulation ↑ HSL, ERK, perilipin phosphorylation ↓ ATGL, HSL, C/EBPα	[36]
3T3-L1 adipocytes SGBS adipocytes	0.1–20 µM with 10 ng/mL TNFα	↓ adiponectin suppression ↓ PPARγ suppression ↓ JNK phosphorylation	[37]

Table 4. *Cont.*

Cell Type	Hydroxytyrosol Concentration/Duration	Effect	Source
3T3-L1 preadipocytes	10–100 μM for 24–48 h 24 h only	↓ cell proliferation ↓ CBI receptor ↑ FAS, lipoprotein lipase (LPS) ↓ PPARγ	[38]
Primary human visceral preadipocytes	5–70 μg/mL for 20 days	↓ triglyceride accumulation ↑ apoptosis, lipolysis, glycerol release ↑ adipogenesis inhibition markers (GATA2, GATA3, WNT3A, SFRP5, HES1, and SIRT1) ↓ adipogenesis promoting genes (LEP, FGF1, CCND1, and SREBF1)	[40]
Human bone marrow MSC adipocytes	1 or 100 μM for 7–14 days	↑ adipogenesis markers PPARγ, FABP4 ↑ fat vesicle formation ↓ LPS	[41]

2.3. Effects of Hydroxytyrosol (HT) on Hepatocytes

Treatment of primary cultured mouse hepatocytes with HT (100 μM) in conditions mimicking ischemia/reperfusion (I/R) resulted in a dose-dependent decrease in apoptosis and increased cell viability (Table 5) [42]. Additionally, HT significantly ameliorated the I/R-induced decrease in antioxidant enzymes such as SOD1, SOD2 and catalase (CAT). Overall, treatment with HT protected against I/R-induced injury and oxidative damage in mouse hepatocytes [42]. Treatment of rat hepatocytes with HT (25 μM) resulted in reduced de novo lipid synthesis including fatty acid, triglyceride and cholesterol without affecting cell viability [43]. Additionally, exposure to HT significantly decreased the activity of crucial enzymes involved in fatty acid synthesis (ACC), triglyceride synthesis (diacylglycerol acyltransferase) and cholesterogenesis (3-hydroxy-3-methyl-glutaryl-CoA reductase) [43]. Moreover, HT increased the phosphorylation of AMPK and its downstream target ACC suggesting that the mechanism of reduced lipid synthesis by HT is AMPK-mediated [43]. These findings indicate that HT decreases lipid synthesis in hepatic tissue. Another study showed that exposure of vitamin E-deficient rat liver microsomes to HT (0.05–2 mM) decreased the formation of thiobarbituric acid reactive substances (TBARS), a biomarker of lipid peroxidation, and thereby decreased lipid peroxidation [44]. TBARS are formed when malonaldehyde, a product of lipid oxidation reacts with thiobarbituric acid leading to cell damage [45]. In a more recent study, exposure of vitamin E-deficient rat liver microsomes to modified HT compounds (0.05–2 mM) resulted in a stronger, more potent inhibition of harmful lipid peroxidation and TBARS formation compared to pure HT (Table 5) [46].

Table 5. Effects of hydroxytyrosol on hepatocytes.

Cell Type	Hydroxytyrosol Concentration/Duration	Effect	Source
Mouse hepatocytes	100 μM for 4 h (hypoxia); followed by reoxygenation	↓ cell apoptosis ↑ hepatocyte viability ↑ SOD1, SOD2, CAT activity	[42]
Rat hepatocytes	25 μM for 2 h	↓ lipid synthesis (fatty acid, cholesterol and triglyceride) ↓ ACC, diacylglycerol acyltransferase, 3-hydroxy-3-methyl-glutaryl-CoA reductase ↑ AMPK and ACC phosphorylation	[43]
Vit. E-deficient rat liver microsomes	0.05–2 mM for 30 min	↓ lipid peroxidation, TBARS	[44]
Vit. E-deficient rat liver microsomes	0.05–0.25 mM for 20 min	↓ lipid peroxidation, TBARS	[46]

Overall, the studies above have shown that HT exhibits hepatoprotective properties by decreasing apoptosis, increasing antioxidant activity and hepatocyte viability [42,44,46]. More importantly, treatment with HT significantly reduced lipid synthesis [43]. Accumulation of lipids in hepatocytes is often associated with the development of hepatic steatosis which subsequently leads to lower glucose utilization, liver injury and fibrosis thus exacerbating the development of insulin resistance and T2DM. These studies indicate that HT may protect against hepatic steatosis and the development of hepatic insulin resistance.

2.4. Effects of Hydroxytyrosol (HT) on Pancreatic Cells

Treatment of rat pancreatic tissue with HT (50 µg/mL) attenuated the hyperglycemia-induced decline in insulin secretion (Table 6) [47]. This indicates that HT may enhance pancreatic insulin secretion in hyperglycemic conditions. In contrast to these finding, exposure of INS-1β cells to HT (0.1–30 µM) did not relieve the hyperglycemia-induced decline in insulin secretion [48]. However, 3-hydroxytyrosol (3-HT) significantly inhibited the formation and cytotoxicity of amylin aggregates in INS-1 β cells. The estimated HT concentration for half maximal inhibition (IC_{50}) of amylin aggregates formation was 100 µM [48]. These findings suggest that HT may have a dose-dependent remodeling and inhibitory effect on pancreatic amylin aggregates, which occur frequently in T2DM. Amyloid aggregates contribute to endoplasmic reticulum (ER) stress, mitochondrial damage and membrane disruption leading to β cell death and thereby T2DM [49–51]. One of the characteristics of T2DM is progressive deficit in β cell function and mass with increased apoptosis. Therefore preventing the progression of amyloid aggregates may protect the β cells from death (Table 6) [52]. Additionally, the formation of amyloids is often seen in several neurodegenerative diseases such as Huntington's, Alzheimer's and Parkinson's where similarly to T2DM, accumulation of locally expressed misfolded proteins share the tendency to produce amyloid aggregates [51].

Table 6. Effects of hydroxytyrosol on pancreatic cells.

Cell Type	Hydroxytyrosol Concentration/Duration	Effect	Source
Rat pancreatic tissue	50 µg/mL for 0–40 min with 4 g/L glucose	↓ decline in insulin secretion induced by hyperglycemia	[47]
Rat INS-1 β cells	0.1–30 µM 3-HT; 11 mM glucose for 1 h 3-HT:amylin ratio of 10:1 M for 0–40 h	↔ insulin secretion ↓ amylin amyloids	[48]

Experiments with hydroxytyrosol (HT) exposure of skeletal muscle cells, adipocytes, hepatocytes, and pancreatic cells indicate that HT may have effects that could be beneficial in the treatment of diabetes or metabolic syndrome. The few studies examining the effects of HT on skeletal muscle cells have found that HT may increase oxidative capacity and muscular health by supporting mitochondrial biogenesis and protecting myocytes from oxidative stress, in addition to increasing glucose uptake [29–32]. Multiple studies exposing adipocytes to HT report diminished adipogenesis, enhanced mitochondrial capacity, and decreased lipid accumulation with only one study reporting increased adipocyte differentiation [31,33–38,40,41]. Most cell culture studies of hepatocytes suggest that HT attenuates oxidative stress in the liver, with one reported instance of reduced lipid synthesis [42–44,46,53,54]. The impact of exposure of pancreatic cells to HT has not been established, but preliminary research indicates that it may enhance insulin secretion and inhibit amylin amyloid β cell damage [47,48].

3. In Vivo Evidence: Antidiabetic Effect of Hydroxytyrosol

3.1. Effect of Hydroxytyrosol (HT) on Alloxan-Induced Diabetes in Rodents

Hamden et al. induced diabetes in male Wistar rats via intraperitoneal injections of alloxan monohydrate (150 mg/kg), and animals with hyperglycemia (blood glucose levels of 2 g/L after 2 weeks) were retained for experimentation (Table 7) [47]. The treatment groups were given daily intraperitoneal injections of olive mill waste monomeric phenols (F1), olive mill waste polymeric phenols (F2), or purified hydroxytyrosol (F3) at 20 mg/kg for two months. All three treatments, especially purified HT (F3), resulted in significantly decreased blood glucose levels. The hepatic toxicity indicators TBARS, bilirubin, and fatty cysts were reduced in animals receiving HT treatment. Hepatic glycogen, circulating high-density lipoprotein (HDL), and antioxidant enzymes (SOD, CAT, and GPX) in the liver and kidney were increased by HT. Additionally, treatment with HT attenuated the deleterious effects of alloxan in pancreatic β cells [47]. Jemai et al. (2009) examined the possible antidiabetic and antioxidant benefits of oleuropein and HT administration in male Wistar rats. Diabetes was induced with an intraperitoneal injection of alloxan (180 mg/kg) preceding treatment for 4 weeks with HT dissolved in drinking water to reach concentrations of 8 or 16 mg/kg [55]. All treatment groups showed inhibition of hyperglycemia, hypercholesterolemia, and hepatic oxidative damage (TBARS) with simultaneous increases in hepatic glycogen and antioxidant enzymes (SOD, CAT). The 16 mg/kg dose had stronger effects than the 8 mg/kg dose, showing dose-dependence [55]. This study showed the ability of HT to retain healthier lipid profiles, glucose levels, and antioxidant activity in a diabetic rat model.

3.2. Effect of Hydroxytyrosol (HT) on Streptozotocin-Induced Diabetes in Rodents

Hamden et al. (2010) used intraperitoneal streptozotocin (STZ) or STZ (150 mg/kg) and nicotinamide (1000 mg/kg) injections to induce diabetes in male Wistar rats (Table 7) [56]. The rats that displayed a moderate diabetic phenotype with hyperglycemia after 2 weeks were treated with 20 mg/kg HT for two months before serum and tissues from the pancreas and small intestine were isolated. Treatment with HT significantly lowered blood glucose, low-density lipoprotein (LDL)-cholesterol and plasma triglycerides while HDL-cholesterol was increased. Additionally, HT reduced the STZ-induced increase of intestinal enzymes (maltase, lactase and sucrose) that are often elevated in diabetes. These enzymes are imperative for the digestion of disaccharides into simple glucose which is readily available for intestinal absorption [57] and therefore the increased number and activities of the enzymes may lead to hyperglycemia, a major characteristic of diabetes. Furthermore, HT treatment inhibited intestinal lipase and consequently, decreased lipid absorption [56]. In the pancreas, HT treatment increased the activities of antioxidant enzymes SOD, CAT, GSH, and GPX and decreased the formation of harmful advanced glycate end-products (AGE). Pancreatic cells showed lower levels of TBARS and LDH (lactate dehydrogenase) activity [56]. Overall, HT had a hypoglycemic and hypolipidemic effect on diabetic rats while providing protection from oxidative damage (Table 7). Ristagno et al. (2012) also induced diabetes with an intraperitoneal injection of STZ (60 mg/kg) in male Sprague-Dawley rats before administering HT daily by intragastric gavage in doses of either 10 or 100 mg/kg for 6 weeks [58]. HT was found to inhibit the hyperglycemia-induced increases in plasma TBARS and also prevented impairments in nerve conduction velocity (NCV), thermal nociception, and Na^+/K^+-ATPase activity [58]. This study showed that HT could mitigate peripheral neuropathy caused by diabetes. Male Wistar rats were given oral HT daily at doses of 0.5, 1, 2.5, 5, or 10 mg/kg for 7 days before diabetes was induced with 50 mg/kg injected STZ in the study by López-Villodres et al. (2016) [59]. HT treatment was continued for two months after the induction of diabetes. Treatment lowered oxidative and nitrosative stresses, inflammatory markers, platelet aggregation, and aortic wall area in comparison to the non-treated diabetic group [59]. It was concluded that HT may attenuate the vasculopathy or blood vessel inflammation induced by diabetes. Reyes et al. (2017) administered HT to male Wistar rats via gavage at dosages of 1, 5 or 10 mg/kg for 7 days before inducing diabetes with intravenous STZ (50 mg/kg) [60]. HT treatments continued for 2 months following the induction of diabetes. Analysis of brain tissue showed that all concentrations of HT attenuated the damage caused by an experimental model of hypoxia and reoxygenation. Reductions in lipid peroxidation, nitrosative stress, cell death, and markers of brain inflammation (IL-1β and prostaglandin E_2) were observed, showing a neuroprotective effect of HT against diabetes [60].

In another study, oral administration of HT (5 mg/kg) for 7 days before and 2 months after STZ injection resulted in reductions of diabetic retinopathy symptoms [61]. HT treatment attenuated the hyperglycemia-induced decrease in number of retinal ganglion cells, and increases in retinal thickness and retinal cell size [61]. Administration of HT (77 mg/kg/day) by intragastric gavage for 4 weeks decreased plasma glucose levels and markers of oxidative stress (nitric oxide (NO), malondialdehyde (MDA), while increasing levels of serum SOD and SIRT1 expression in the thoracic aorta and human umbilical vein endothelial cells (HUVEC) (Table 7) [62]. This study indicates that HT has some capacity to reduce oxidative stress and attenuate hyperglycemia and hyperlipidemia.

3.3. Effect of Hydroxytyrosol (HT) on Genetically-Induced Diabetes in Rodents

Administration of HT by oral gavage (10 mg/kg/day) for 8 weeks to male hyperglycemic db/db C57BL/6J mice, lacking functional leptin receptors significantly reduced fasting glucose and lipid serum levels, in addition to reducing liver and muscle oxidative stress compared to nontreated control mice (Table 7) [63]. Metformin treatment also attenuated the increased fasting glucose but was not as effective as HT in controlling serum triglycerides and cholesterol levels [63]. In another study, administration of HT (10 or 50 mg/kg) daily for 8 weeks in male C57BL/6J db/db mice resulted in increased expression of mitochondrial respiratory chain complexes I/II/IV and the activity of complex I in the brain [64]. Additionally, HT significantly enhanced the activities of the antioxidant transcription factor p62 (sequestosome-1), haeme oxygenase 1 (HO-1) SOD 1, and SOD2, while attenuating protein oxidation in the brain. HT activated AMPK, SIRT1, and PPARγ-1α in brain tissue, demonstrating a possible neuroprotective effect against damage caused by high glucose levels [64].

Table 7. Anti-diabetic Effects of Hydroxytyrosol: In vivo alloxan, streptozotocin- and genetic-induced diabetes animal studies.

Study Model	Hydroxytyrosol Concentration/Duration	Effect	Source
Alloxan-Induced Diabetes Model			
Alloxan-induced diabetic male Wistar rats	20 mg/kg for 2 months; intraperitoneal injection	↓ blood glucose levels ↓ liver TBARS, bilirubin, fatty cysts ↑ hepatic glycogen, HDL ↑ SOD, CAT, GPX in liver/kidney ↓ β cell damage	[47]
Alloxan-induced diabetic male Wistar rats	8 or 16 mg/kg orally for 4 weeks;	↓ blood glucose levels ↓ TC ↓ hepatic oxidative damage (TBARS) ↑ hepatic glycogen ↑ antioxidant enzymes (SOD, CAT)	[55]
Streptozotocin-Induced Diabetes Model			
STZ-induced diabetic male Wistar rats	20 mg/kg/day orally for 2 months	↓ blood glucose, HDL ↓ LDL cholesterol, TG ↓ intestinal enzymes (maltase, lactase and sucrose, lipase) ↑ pancreas SOD, CAT, GSH, GPX ↓ pancreas TBARS, AGE, LDH	[56]
STZ-induced male diabetic Sprague-Dawley rats	10 or 100 mg/kg/day for 6 weeks via gavage	↓ plasma TBARS ↑ NCV, thermal nociception Na+/K+-ATPase activity	[58]

Table 7. *Cont.*

Study Model	Hydroxytyrosol Concentration/Duration	Effect	Source
Streptozotocin-Induced Diabetes Model			
STZ-induced male diabetic Wistar rats	0.5-10 mg/kg/day orally for 7 days prior to STZ and 2 months thereafter	↓ nitrosative, oxidative stress ↓ inflammation, IL-1β ↓ platelet aggregation ↓ aortic wall area	[59]
STZ-induced diabetic male Wistar rats	1, 5, or 10 mg/kg/day orally for 7 days prior to STZ and 2 months thereafter	↓ brain lipid peroxides, ↓ nitrosative stress, cell death ↓ brain inflammation, IL-1β, prostaglandin E_2	[60]
STZ-induced diabetic male Wistar rats	5 mg/kg/day via endogastric cannula for 7 days prior to STZ and 2 months thereafter	↑ retinal ganglion cell number ↓ retinal thickness, cell size	[61]
STZ-induced diabetic KM mice	77 mg/kg/day HT or HT-NO via gavage for 4 weeks	↓ plasma glucose levels ↓ oxidative stress (↓ serum NO, MDA, ↑ SOD) ↑ SIRT1 expression in aorta and HUVEC)	[62]
Genetic Diabetes Model			
Male db/db C57BL/6J mice	10 mg/kg/day via gavage for 8 weeks	↓ fasting glucose levels ↓ serum lipids ↓ oxidative damage in liver and muscle	[63]
Male db/db C57BL/6J mice	10 or 50 mg/kg/day orally for 8 weeks	↑ mitochondrial complex I/II/IV expression (brain) ↑ activity complex I (brain) ↓ oxidative stress (↑ p62, HO-1) SOD1, SOD2) ↓ protein oxidation in brain ↑ AMPK, SIRT1, PPARγ -1α activation in brain	[64]

3.4. Effect of Hydroxytyrosol (HT) on Diet-Induced Diabetes in Rodents

Administration of hydroxytyrosol-rich extract (3 mg/kg, respectively) for 17 weeks in Wistar rats fed a cholesterol-rich diet attenuated the increase in serum triglycerides, and total and LDL cholesterol. Additionally, serum antioxidant capacity and CAT, SOD enzymes in liver were significantly increased while TBARS were reduced in the liver, heart, kidney and aorta (Table 8) [65]. Administration of HT (10 or 50 mg/kg/day via gavage) for 17 weeks in obese and insulin resistant C57BL/6J mice reduced fasting glucose and insulin levels, fasting leptin, serum inflammatory markers IL-6 and CRP, serum triglycerides, and the serum LDL/HDL ratio [63]. No significant effect on serum adiponectin was observed. Hepatic and muscle tissue lipid content was decreased by HT, possibly explained by decreases in SREBP-1c and FAS expression [63]. HT also showed antioxidative capacity by decreasing oxidative damage to proteins and lipids in the liver and increasing hepatic GST and SOD enzyme activity [63]. Tabernero et al. (2014) used a cholesterol-rich diet to induce metabolic disease and hypercholesterolemia in male Wistar rats [66]. Over the course of 8 weeks, high-cholesterol (2%) diets were supplemented (0.04%) with HT, hydroxytyrosol-acetate (HT-Ac), or ethyl-hydroxytyrosol-ether (HT-Et) in the treatment group. All three compounds decreased plasma levels of insulin, leptin, MDA, total cholesterol, LDL cholesterol, glucose, and inflammatory markers TNFα and IL-1β in the hypercholesterolemic rats while increasing antioxidant capacity [66]. HT-Ac and HT-Et also decreased MCP-1 in visceral adipose tissue (VAT), which is associated with inflammation. HT-Ac treatment alone decreased plasma cholesterol levels, but the decrease seen in plasma free fatty acid (FFA) levels for all of the treatment groups were non-significant [66]. In another study, treatment of high-fat diet-induced diabetic mice with HT (20 mg/kg/day, orally for 3 weeks) resulted in reduced weight gain, visceral fat deposits, inguinal white adipose tissue (WAT) mass, and blood glucose levels [67]. Serum insulin and non-esterified fatty acids (NEFA) were not significantly changed by HT administration. Plasma mRNA levels of adiposity marker *Mest* was decreased in the treatment group compared to the control [67]

(Table 8). Overall, HT demonstrated considerable potential in counteracting the effects of obesity and diabetes.

Pirozzi et al. (2016) administered a high-fat diet (58% fat) to male Sprague-Dawley rats for 6 weeks to induce non-alcoholic fatty liver disease (NAFLD), which starts with abnormal lipid accretion in hepatocytes (Table 8) [68]. The treatment group received 10 mg/kg of HT in addition to the high-fat diet via intragastric gavage. In the HT group, serum cholesterol and markers of liver damage (aspartate aminotransferase and alanine aminotransferase) were decreased while glucose tolerance and insulin sensitivity were improved [68]. Additionally, tissue assessments showed that HT inhibited hepatic inflammation (TNFα, IL-6, COX2), oxidative and nitrosative damage, and intestinal barrier damage. Carnitine palmitoyltransferase (CPT1a), PPARα, and acetyl CoA carboxylase (ACC) phosphorylation in the liver were all decreased by the high-fat diet, but this decrease was not observed in the treatment group, indicating the preservation of normal lipid metabolism in hepatic tissue by HT [68]. Administration of HT (20 mg/kg/day) for 8 weeks by oral gavage in high-fat, high-carbohydrate fed male Wistar rats showed attenuation of glucose intolerance, insulin resistance and weight gain [69]. Systolic blood pressure, heart inflammation, and liver damage were all reduced in the treatment group [69]. However, no effect was observed on plasma lipid levels. In this study, HT displayed protective effects on cardiac and hepatic tissue as well as improving glucose metabolism [69]. Administration of HT (20 mg/kg/day for 10 weeks) in high-fat diet-induced diabetic mice resulted in lower fasting glucose and insulin levels and increased GLUT4 expression in adipose and skeletal muscle tissue [70]. Modulation of insulin signaling in adipose tissue was shown by decreased serine phosphorylation of IRS-1 and increased Akt phosphorylation by HT [70]. Inflammation was decreased as indicated by lower levels of TNFα and IL-1β in the liver and adipose tissue as well as decreased serum CRP and IL-6 [70]. There were no significant changes in adiposity, serum adiponectin, serum lipids or liver (alanine aminotransferase (ALT), aspartate aminotransferase (AST) enzymes. However, HT was found to inhibit hepatic steatosis, lipid accumulation, and SREBP-1 mRNA expression in the liver; it also decreased indicators of endoplasmic reticulum stress in adipose tissue [70]. Similarly, oral administration of HT (5 mg/kg/day) to high-fat diet-induced diabetic mice for 12 weeks inhibited weight gain, hyperglycemia, insulin resistance, and the increases in hepatic and serum lipids caused by the high-fat diet. In the livers of the treated mice, steatosis scores were improved, anti-inflammatory markers eicosapentaenoic acid (EPA) and docosahexaenoic acid (DHA) were increased, and pro-inflammatory markers were decreased [71]. In this in vivo model of metabolic syndrome, HT demonstrated an ability to reduce hepatic inflammation and attenuate insulin resistance.

In STZ and high-fat diet (HFD)-induced male diabetic Institute of Cancer Research (ICR) mice hydroxytyrosol-fenofibrate (FF-HT) (36 μmol/kg via intragastric gavage) for 11 weeks resulted in decreased plasma glucose, glucose intolerance, lipid profile (total cholesterol, triglycerides, LDL-cholesterol), atherosclerotic index (AI) and hepatic lipid accumulation compared to the fenofibrate control [72]. Additionally, antioxidant enzyme activities (SOD and, GSH-PX) were increased, while MDA and inflammatory markers TNFα and CRP were decreased. Furthermore, FF-HT treatment exhibited protective effects on pancreatic and hepatic tissues. The study also examined the effect of HT on Triton WR-1339 induced-hyperlipidemic mice further confirming the improvement in lipid profile (decrease in total cholesterol, triglycerides, LDL-cholesterol) and antioxidant activity (MDA decrease) [72]. Another study administered HT-nicotinamide (HT-N) or pure HT (both 0.38 mmol/kg by intragastric gavage) for 4 weeks in streptozotocin and high-fat diet (60 mg/kg for 7 days after 6 weeks of high-fat diet) - induced diabetic Kun Ming (KM) mice [73]. Results showed that both HT-N and HT caused a reduction in plasma glucose levels, total cholesterol, and triglycerides, while increasing the antioxidant activities of SOD, CAT, and GSH-PX. Additionally, treatment with both HT-N and HT showed a protective effect on pancreatic tissue and inhibited the destruction of β cells. The study also investigated the effect of HT on hyperlipidemic mice (induced with 400 mg/kg Triton WR-1339) showing decreased plasma triglyceride, cholesterol, and MDA levels [73]. In another study, Xie et al. (2018) investigated the effects of hydroxytyrosol-clofibrate (CF-HT, 240 μmol/kg/day for 7 days) in hyperlipidemic mice (induced with Triton WR-1339) showing decreased plasma total cholesterol and triglyceride levels, a protective effect on hepatic tissue (decreased AST, ALT, total bilirubin (TBIL), alkaline phosphatase (ALP)), and increased serum SOD and CAT. The CF-HT treatment additionally decreased serum MDA and hepatic oxidized glutathione (GSSH) while increasing hepatic glutathione (GSH) (Table 8) [74].

Table 8. Anti-diabetic Effects of Hydroxytyrosol: In vivo high-fat diet (HFD)-induced diabetes animal studies.

Study Model	Hydroxytyrosol Concentration/Duration	Effect	Source
Diet-induced hypercholesterolemic male Wistar rats	Olive leaf hydrolysate extract for 3 weeks (3 mg/kg b. w. orally containing HT (1.4 g/100 g dry weight), oleuropein	↓ serum TC, TG, LDL ↑ serum HDL ↓ TBARS (heart, liver, kidney) ↑ serum antioxidant capacity ↑ liver CAT, SOD activity	[65]
Diet-induced diabetic/obese male C57BL/6 mice	10 or 50 mg/kg/day via gavage for 17 weeks	↓ serum glucose, insulin ↓ serum IL-6, CRP, TG, leptin ↓ LDL/HDL ratio ↓ lipid content (liver, muscle) ↓ liver SREBP-1c, FAS ↓ liver protein carbonyls, MDA ↑ liver GST, SOD activity	[63]
Diet-induced hypercholesterolemic male Wistar rats	0.04% of diet with added HT, HT-Ac, or HT-Et for 8 weeks	↓ serum glucose, insulin ↓ serum leptin, MDA ↓ serum IL-1β, TNFα ↑ serum antioxidant activity ↓ VAT MCP-1, IL-1β ↓ serum TC, LDL (HT-Ac only)	[66]
Male C57BL/6J mice with diet-induced metabolic syndrome	20 mg/kg/day orally for 3 weeks	↓ serum glucose, *Mest* expression ↓ weight gain, visceral fat ↓ inguinal WAT ↔ serum insulin	[67]
Male Sprague-Dawley rats with diet-induced NAFLD	10 mg/kg/day via gavage for 6 weeks	↑ glucose tolerance ↓ serum glucose, insulin ↓ serum AST, ALT, TC ↑ hepatic PPARα, CPT1a, ACC ↓ liver TNFα, IL-6, COX-2 ↓ intestinal barrier damage ↓ liver ROS, MDA, RNS damage	[68]
Male Wistar rats with diet-induced metabolic syndrome	20 mg/kg/day via gavage for 8 weeks	↑ glucose tolerance ↓ serum insulin ↓ weight gain and fat mass ↓ liver steatosis, ventricular fibrosis ↓ plasma ALT, AST activity ↔ serum lipids	[69]
Diet-induced obese male ICR mice	20 mg/kg/day via gavage for 10 weeks	↓ fasting glucose, insulin ↑ glucose and insulin tolerance ↑ GLUT4 (adipocytes, myocytes) ↓ phospho-IRS-1 (Ser307) ↑ phospho-Akt (Ser473) ↓ liver, adipose tissue TNFα, IL-1β ↓ serum CRP, IL-6 ↓ hepatic steatosis, TG, ER stress ↓ liver SREBP-1 ↔ adiposity, adiponectin ↔ serum lipids, liver enzymes	[70]
Diet-induced obese male C57BL/6J mice	5 mg/kg/day orally for 12 weeks	↓ weight gain, insulin resistance ↓ serum glucose, insulin ↓ serum FFA, TAG, TC, LDL ↓ hepatic steatosis, FFA, TC ↓ liver TNFα, IL-1β/IL-6 ↑ hepatic EPA, DHA ↔ serum AST	[71]

Table 8. *Cont.*

Study Model	Hydroxytyrosol Concentration/Duration	Effect	Source
STZ-induced diabetic male ICR miceTriton WR-1339 induced hyperlipidemic mice	36 µmol/kg/day FF-HT via gavage for 11 weeks 36 µmol/kg/day FF-HT via gavage for 7 days	↓ plasma glucose, lipids ↓ hepatic lipids ↓ TNFα, CRP ↑ glucose tolerance, antioxidants ↓ plasma TG, TC, MDA, atherosclerotic index (AI)	[72]
STZ-induced diabetic male KM mice Triton WR-1339 induced hyperlipidemic mice	0.38 mmol/kg/day via gavage for 4 weeks 0.38 mmol/kg/day via gavage for 7 days prior to Triton WR	↓ blood glucose, lipids ↑ plasma SOD, CAT, GSH-PX ↓ plasma TG, TC, MDA	[73]
Triton WR-1339 induced hyperlipidemic mice	240 µmol/kg/day for 7 days	↓ plasma TG, TC, MDA ↓ AST, ALT, TBIL, ALP, hepatic GSSH ↑ serum SOD, CAT ↑ hepatic GSH	[74]

Numerous studies using experimental rodent models of diabetes have shown that HT can have beneficial in vivo effects against diabetes, obesity, and metabolic diseases. Diabetes was induced chemically with alloxan or streptozotocin, both of which damage β cells and inhibit production of insulin by the pancreas [75]. The studies using rodents with diabetes induced by alloxan or streptozotocin showed that HT may decrease serum glucose and lipid levels, mitigate inflammation, and significantly reduce oxidative stress. The harmful impact of hyperglycemia and lack of insulin production on the liver, heart, brain, pancreas, and retinal cells of the eye were also alleviated by HT [47,55,56,58–62,72–74]. Secondary experiments using Triton WR-1339 to chemically induce hyperlipidemia also showed that HT decreased serum lipids and liver damage [72,74]. HT has also been studied in genetic db/db C57BL/6J mouse models of diabetes, which lack leptin receptors and develop obesity and hyperglycemia [75]. Administration of HT to these mice was effective in reducing serum glucose and lipids as well as controlling oxidative stress in liver, muscle, and brain tissue [63,64]. Many in vivo studies researching HT have also used high-fat diets to induce metabolic syndrome in rodents. Most of these studies have reported that HT attenuated the increase in serum lipids, glucose, and insulin caused by the high-fat diets. They have also observed reduced oxidative damage, enhanced antioxidant capacity and decreased inflammation, with HT treatment [63,65–71]. These data suggest that HT has the potential to protect organs and tissues from damage caused by diabetes.

4. Effects of Hydroxytyrosol (HT) on Cellular Signaling Cascades

At the molecular/cellular level, HT was shown to increase glucose uptake in both muscle [31] and fat [70] cells. HT phosphorylated/activated the energy sensor AMPK and its downstream effector ACC in fat [33] and liver [43] cells, and increased the expression of SIRT1 in adipocytes [40] (Figure 2). Furthermore, HT treatment increased the expression of PGC-1 in skeletal muscle [29] and adipocytes [33]. Mitochondrial biogenesis, oxygen consumption capacity, ATP production, and activity of complex I, II and V in muscle cells and adipocytes were all increased by HT treatment [29,32,33] (Figure 2). Furthermore, treatment with HT increased the levels of adipose tissue phosphorylated/activated Akt, a crucial protein involved in insulin signaling [70].

HT reduced oxidative stress in muscle [30,63], liver [42,44,46,47,55,63,68,74], pancreas [56], kidney [65], heart [65], and brain tissue [60,64], as shown by the decreased levels of markers of oxidative stress such as NO and MDA and the increased activity of antioxidant enzymes including SOD, CAT, GSH and GPX. Additionally, HT reduced the inflammatory effects of TNF-α, and inhibited downstream the activation of JNK in adipocytes [37].

HT had significant anti-inflammatory effects. Serum CRP, IL-1β, IL-6 and TNF-α [55,58,59,63, 66,70,71], liver TBARS, IL-1β, IL-6 and TNF-α [47,55,68], adipose IL-1β and TNF-α [70], pancreatic

TBARS, AGE, LDH, CRP and IL-1β [56], and heart and kidney IL-6, TBARS levels were all reduced by HT treatment [65].

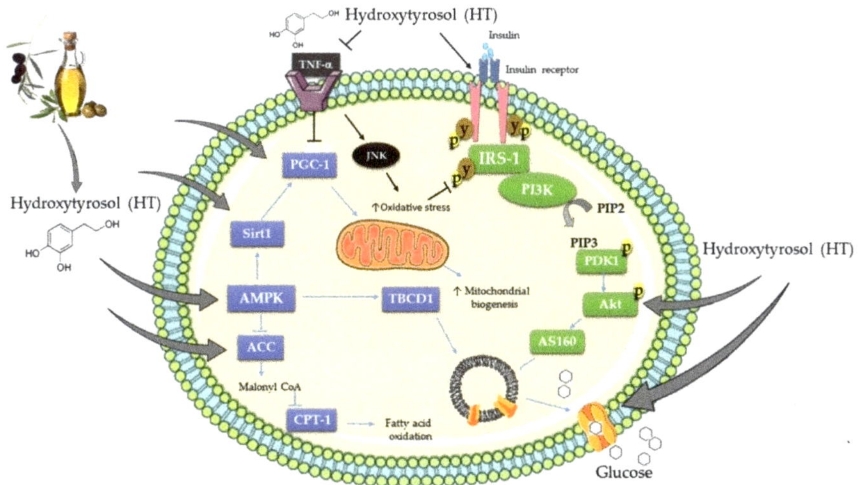

Figure 2. Effects of HT on cellular signaling molecules. The figure was created based on the evidence of the studies [33,37,43,64,70,76].

5. Summary, Conclusion and Future Directions

Existing studies indicate that hydroxytyrosol (HT) has insulin-like effects on insulin target cells including adipocytes, hepatocytes and muscle cells, and exerted significant anti-diabetic effects in animal models of T2DM. Moreover, HT exhibited protective effects against oxidative stress, inflammation, hyperglycemia and hyperlipidemia in chemically-, genetically- and dietary-induced animal models of T2DM.

Although the current literature regarding HT toxicity is scarce, the evidence suggests that HT is well-tolerated. A study showed that oral administration of pure HT at 5, 50 and 500 mg/kg/day for 13 weeks in Wistar rats resulted in no adverse effects, including micro or macro organ alterations, morbidity, or mortality [77]. More importantly, in 2011, the European Food Safety Authority (EFSA) approved the use of HT (5 mg/day) or its derivatives to provide protection against oxidative damage and inflammation and reduce the risk of cardiovascular disease and insulin resistance/diabetes [78]. However, more systematic long-term human studies are required to assess the optimal dose required for the health benefits while avoiding any potential toxicity/side effects. In addition, more clinical studies are required to examine the bioavailability and mechanism of action of HT and to fully understand its antioxidant, anti-inflammatory and antidiabetic effects.

Overall, the existing in vitro and in vivo studies of HT showed potent antioxidant, anti-inflammatory, insulin-like, and insulin sensitizing effects. This suggests a potential application for HT in the prevention and treatment of insulin resistance and T2DM.

Author Contributions: E.T. formulated the review topic. F.V. and E.T. wrote the majority of the manuscript. M.Y. contributed to the writing of the manuscript, table and figure preparation. All authors read and approved the final manuscript.

Acknowledgments: This work was supported by a Natural Sciences and Engineering Research Council of Canada (NSERC) to E.T.

Conflicts of Interest: The authors declare no conflict of interest.

References

1. Anderson, J.T.; Keys, A. Dietary protein and the serum cholesterol level in man. *Am. J. Clin. Nutr.* **1957**, *5*, 29–34. [PubMed]
2. Keys, A.; Anderson, J.T.; Grande, F. Serum cholesterol response to changes in the diet: II. The effect of cholesterol in the diet. *Metab. Clin. Exp.* **1965**, *14*, 759–765. [CrossRef]
3. Keys, A.; Anderson, J.T.; Grande, F. Serum cholesterol response to changes in the diet: III. Differences among individuals. *Metab. Clin. Exp.* **1965**, *14*, 766–775. [CrossRef]
4. Keys, A.; Anderson, J.T.; Grande, F. Serum cholesterol response to changes in the diet: IV. Particular saturated fatty acids in the diet. *Metab. Clin. Exp.* **1965**, *14*, 776–787. [CrossRef]
5. Altomare, R.; Cacciabaudo, F.; Damiano, G.; Palumbo, V.D.; Gioviale, M.C.; Bellavia, M.; Tomasello, G.; Lo Monte, A.I. The Mediterranean Diet: A History of Health. *Iran. J. Public Health* **2013**, *42*, 449–457. [PubMed]
6. Estruch, R.; Ros, E.; Salas-Salvadó, J.; Covas, M.-I.; Corella, D.; Arós, F.; Gómez-Gracia, E.; Ruiz-Gutiérrez, V.; Fiol, M.; Lapetra, J.; et al. Primary prevention of cardiovascular disease with a Mediterranean diet. *N. Engl. J. Med.* **2013**, *368*, 1279–1290. [CrossRef]
7. Barak, Y.; Fridman, D. Impact of Mediterranean Diet on Cancer: Focused Literature Review. *Cancer Genom. Proteom.* **2017**, *14*, 403–408.
8. Lăcătușu, C.-M.; Grigorescu, E.-D.; Floria, M.; Onofriescu, A.; Mihai, B.-M. The Mediterranean Diet: From an Environment-Driven Food Culture to an Emerging Medical Prescription. *Int. J. Environ. Res. Public Health* **2019**, *16*, 942. [CrossRef]
9. Finicelli, M.; Squillaro, T.; Di Cristo, F.; Di Salle, A.; Melone, M.A.B.; Galderisi, U.; Peluso, G. Metabolic syndrome, Mediterranean diet, and polyphenols: Evidence and perspectives. *J. Cell. Physiol.* **2019**, *234*, 5807–5826. [CrossRef]
10. Berry, E.M.; Arnoni, Y.; Aviram, M. The Middle Eastern and biblical origins of the Mediterranean diet. *Public Health Nutr.* **2011**, *14*, 2288–2295. [CrossRef]
11. Tuck, K.L.; Hayball, P.J. Major phenolic compounds in olive oil: Metabolism and health effects. *J. Nutr. Biochem.* **2002**, *13*, 636–644. [CrossRef]
12. Vissers, M.N.; Zock, P.L.; Roodenburg, A.J.C.; Leenen, R.; Katan, M.B. Olive Oil Phenols Are Absorbed in Humans. *J. Nutr.* **2002**, *132*, 409–417. [CrossRef] [PubMed]
13. Montedoro, G.; Servili, M.; Baldioli, M.; Miniati, E. Simple and hydrolyzable phenolic compounds in virgin olive oil. 1. Their extraction, separation, and quantitative and semiquantitative evaluation by HPLC. *J. Agric. Food Chem.* **1992**, *40*, 1571–1576. [CrossRef]
14. Shamshoum, H.; Vlavcheski, F.; Tsiani, E. Anticancer effects of oleuropein. *Biofactors* **2017**, *43*, 517–528. [CrossRef] [PubMed]
15. Tejada, S.; Pinya, S.; Del Mar Bibiloni, M.; Tur, J.A.; Pons, A.; Sureda, A. Cardioprotective Effects of the Polyphenol Hydroxytyrosol from Olive Oil. *Curr. Drug Targets* **2017**, *18*, 1477–1486. [CrossRef] [PubMed]
16. Imran, M.; Nadeem, M.; Gilani, S.A.; Khan, S.; Sajid, M.W.; Amir, R.M. Antitumor Perspectives of Oleuropein and Its Metabolite Hydroxytyrosol: Recent Updates. *J. Food Sci.* **2018**, *83*, 1781–1791. [CrossRef] [PubMed]
17. Alkhatib, A.; Tsang, C.; Tuomilehto, J. Olive Oil Nutraceuticals in the Prevention and Management of Diabetes: From Molecules to Lifestyle. *Int. J. Mol. Sci.* **2018**, *19*, 2024. [CrossRef]
18. Granados-Principal, S.; Quiles, J.L.; Ramirez-Tortosa, C.L.; Sanchez-Rovira, P.; Ramirez-Tortosa, M.C. Hydroxytyrosol: From laboratory investigations to future clinical trials. *Nutr. Rev.* **2010**, *68*, 191–206. [CrossRef]
19. Umeno, A.; Horie, M.; Murotomi, K.; Nakajima, Y.; Yoshida, Y. Antioxidative and Antidiabetic Effects of Natural Polyphenols and Isoflavones. *Molecules* **2016**, *21*, 708. [CrossRef]
20. Peyrol, J.; Riva, C.; Amiot, M.J. Hydroxytyrosol in the Prevention of the Metabolic Syndrome and Related Disorders. *Nutrients* **2017**, *9*, 306. [CrossRef]
21. Wani, T.A.; Masoodi, F.A.; Gani, A.; Baba, W.N.; Rahmanian, N.; Akhter, R.; Wani, I.A.; Ahmad, M. Olive oil and its principal bioactive compound: Hydroxytyrosol—A review of the recent literature. *Trends Food Sci. Technol.* **2018**, *77*, 77–90. [CrossRef]
22. Barbaro, B.; Toietta, G.; Maggio, R.; Arciello, M.; Tarocchi, M.; Galli, A.; Balsano, C. Effects of the Olive-Derived Polyphenol Oleuropein on Human Health. *Int. J. Mol. Sci.* **2014**, *15*, 18508–18524. [CrossRef] [PubMed]

23. Okayasu, S.; Kitaichi, K.; Hori, A.; Suwa, T.; Horikawa, Y.; Yamamoto, M.; Takeda, J.; Itoh, Y. The evaluation of risk factors associated with adverse drug reactions by metformin in type 2 diabetes mellitus. *Biol. Pharm. Bull.* **2012**, *35*, 933–937. [CrossRef]

24. DeFronzo, R.; Fleming, G.A.; Chen, K.; Bicsak, T.A. Metformin-associated lactic acidosis: Current perspectives on causes and risk. *Metab. Clin. Exp.* **2016**, *65*, 20–29. [CrossRef] [PubMed]

25. Wu, S.; Hopper, I.; Skiba, M.; Krum, H. Dipeptidyl peptidase-4 inhibitors and cardiovascular outcomes: meta-analysis of randomized clinical trials with 55,141 participants. *Cardiovasc. Ther.* **2014**, *32*, 147–158. [CrossRef] [PubMed]

26. Forsmark, C.E. Incretins, Diabetes, Pancreatitis and Pancreatic Cancer: What the GI specialist needs to know. *Pancreatology* **2016**, *16*, 10–13. [CrossRef]

27. Blau, J.E.; Tella, S.H.; Taylor, S.I.; Rother, K.I. Ketoacidosis associated with SGLT2 inhibitor treatment: Analysis of FAERS data. *Diabetes Metab. Res. Rev.* **2017**, *33*, 10–1002. [CrossRef]

28. Liu, J.; Li, L.; Li, S.; Jia, P.; Deng, K.; Chen, W.; Sun, X. Effects of SGLT2 inhibitors on UTIs and genital infections in type 2 diabetes mellitus: a systematic review and meta-analysis. *Sci. Rep.* **2017**, *7*, 2824. [CrossRef]

29. Friedel, A.; Raederstorff, D.; Roos, F.; Toepfer, C.; Wertz, K. Hydroxytyrosol Benefits Muscle Differentiation and Muscle Contraction and Relaxation. U.S. Patent Application No. 13/550,972, 7 March 2013.

30. Burattini, S.; Salucci, S.; Baldassarri, V.; Accorsi, A.; Piatti, E.; Madrona, A.; Espartero, J.L.; Candiracci, M.; Zappia, G.; Falcieri, E. Anti-apoptotic activity of hydroxytyrosol and hydroxytyrosyl laurate. *Food Chem. Toxicol.* **2013**, *55*, 248–256. [CrossRef]

31. Drira, R.; Sakamoto, K. Modulation of adipogenesis, lipolysis and glucose consumption in 3T3-L1 adipocytes and C2C12 myotubes by hydroxytyrosol acetate: A comparative study. *Biochem. Biophys. Res. Commun.* **2013**, *440*, 576–581. [CrossRef]

32. Wang, X.; Li, H.; Zheng, A.; Yang, L.; Liu, J.; Chen, C.; Tang, Y.; Zou, X.; Li, Y.; Long, J.; et al. Mitochondrial dysfunction-associated OPA1 cleavage contributes to muscle degeneration: Preventative effect of hydroxytyrosol acetate. *Cell. Death Dis.* **2014**, *5*, e1521. [CrossRef] [PubMed]

33. Hao, J.; Shen, W.; Yu, G.; Jia, H.; Li, X.; Feng, Z.; Wang, Y.; Weber, P.; Wertz, K.; Sharman, E.; et al. Hydroxytyrosol promotes mitochondrial biogenesis and mitochondrial function in 3T3-L1 adipocytes. *J. Nutr. Biochem.* **2010**, *21*, 634–644. [CrossRef] [PubMed]

34. Warnke, I.; Goralczyk, R.; Fuhrer, E.; Schwager, J. Dietary constituents reduce lipid accumulation in murine C3H10 T1/2 adipocytes: A novel fluorescent method to quantify fat droplets. *Nutr. Metab.* **2011**, *8*, 30. [CrossRef] [PubMed]

35. Drira, R.; Chen, S.; Sakamoto, K. Oleuropein and hydroxytyrosol inhibit adipocyte differentiation in 3 T3-L1 cells. *Life Sci.* **2011**, *89*, 708–716. [CrossRef] [PubMed]

36. Drira, R.; Sakamoto, K. Hydroxytyrosol stimulates lipolysis via A-kinase and extracellular signal-regulated kinase activation in 3T3-L1 adipocytes. *Eur. J. Nutr.* **2014**, *53*, 743–750. [CrossRef] [PubMed]

37. Scoditti, E.; Massaro, M.; Carluccio, M.A.; Pellegrino, M.; Wabitsch, M.; Calabriso, N.; Storelli, C.; De Caterina, R. Additive regulation of adiponectin expression by the mediterranean diet olive oil components oleic Acid and hydroxytyrosol in human adipocytes. *PLoS ONE* **2015**, *10*, e0128218. [CrossRef] [PubMed]

38. Tutino, V.; Orlando, A.; Russo, F.; Notarnicola, M. Hydroxytyrosol Inhibits Cannabinoid CB1 Receptor Gene Expression in 3T3-L1 Preadipocyte Cell Line. *J. Cell. Physiol.* **2016**, *231*, 483–489. [CrossRef]

39. Engeli, S.; Böhnke, J.; Feldpausch, M.; Gorzelniak, K.; Janke, J.; Bátkai, S.; Pacher, P.; Harvey-White, J.; Luft, F.C.; Sharma, A.M.; et al. Activation of the peripheral endocannabinoid system in human obesity. *Diabetes* **2005**, *54*, 2838–2843. [CrossRef]

40. Stefanon, B.; Colitti, M. Original Research: Hydroxytyrosol, an ingredient of olive oil, reduces triglyceride accumulation and promotes lipolysis in human primary visceral adipocytes during differentiation. *Exp. Biol. Med.* **2016**, *241*, 1796–1802. [CrossRef]

41. Anter, J.; Quesada-Gómez, J.M.; Dorado, G.; Casado-Díaz, A. Effect of Hydroxytyrosol on Human Mesenchymal Stromal/Stem Cell Differentiation into Adipocytes and Osteoblasts. *Arch. Med. Res.* **2016**, *47*, 162–171. [CrossRef]

42. Pan, S.; Liu, L.; Pan, H.; Ma, Y.; Wang, D.; Kang, K.; Wang, J.; Sun, B.; Sun, X.; Jiang, H. Protective effects of hydroxytyrosol on liver ischemia/reperfusion injury in mice. *Mol. Nutr. Food Res.* **2013**, *57*, 1218–1227. [CrossRef] [PubMed]

43. Priore, P.; Siculella, L.; Gnoni, G.V. Extra virgin olive oil phenols down-regulate lipid synthesis in primary-cultured rat-hepatocytes. *J. Nutr. Biochem.* **2014**, *25*, 683–691. [CrossRef] [PubMed]

44. Rubio-Senent, F.; de Roos, B.; Duthie, G.; Fernández-Bolaños, J.; Rodríguez-Gutiérrez, G. Inhibitory and synergistic effects of natural olive phenols on human platelet aggregation and lipid peroxidation of microsomes from vitamin E-deficient rats. *Eur. J. Nutr.* **2015**, *54*, 1287–1295. [CrossRef] [PubMed]

45. Jardine, D.; Antolovich, M.; Prenzler, P.D.; Robards, K. Liquid chromatography-mass spectrometry (LC-MS) investigation of the thiobarbituric acid reactive substances (TBARS) reaction. *J. Agric. Food Chem.* **2002**, *50*, 1720–1724. [CrossRef] [PubMed]

46. Rodríguez-Gutiérrez, G.; Rubio-Senent, F.; Gómez-Carretero, A.; Maya, I.; Fernández-Bolaños, J.; Duthie, G.G.; de Roos, B. Selenium and sulphur derivatives of hydroxytyrosol: Inhibition of lipid peroxidation in liver microsomes of vitamin E-deficient rats. *Eur. J. Nutr.* **2018**. [CrossRef] [PubMed]

47. Hamden, K.; Allouche, N.; Damak, M.; Elfeki, A. Hypoglycemic and antioxidant effects of phenolic extracts and purified hydroxytyrosol from olive mill waste in vitro and in rats. *Chem. Biol. Interact.* **2009**, *180*, 421–432. [CrossRef] [PubMed]

48. Wu, L.; Velander, P.; Liu, D.; Xu, B. Olive Component Oleuropein Promotes β-Cell Insulin Secretion and Protects β-Cells from Amylin Amyloid-Induced Cytotoxicity. *Biochemistry* **2017**, *56*, 5035–5039. [CrossRef] [PubMed]

49. Huang, C.; Lin, C.; Haataja, L.; Gurlo, T.; Butler, A.E.; Rizza, R.A.; Butler, P.C. High expression rates of human islet amyloid polypeptide induce endoplasmic reticulum stress mediated beta-cell apoptosis, a characteristic of humans with type 2 but not type 1 diabetes. *Diabetes* **2007**, *56*, 2016–2027. [CrossRef] [PubMed]

50. Gurlo, T.; Ryazantsev, S.; Huang, C.; Yeh, M.W.; Reber, H.A.; Hines, O.J.; O'Brien, T.D.; Glabe, C.G.; Butler, P.C. Evidence for proteotoxicity in beta cells in type 2 diabetes: Toxic islet amyloid polypeptide oligomers form intracellularly in the secretory pathway. *Am. J. Pathol.* **2010**, *176*, 861–869. [CrossRef] [PubMed]

51. Mulder, H. Transcribing β-cell mitochondria in health and disease. *Mol. Metab.* **2017**, *6*, 1040–1051. [CrossRef] [PubMed]

52. Butler, A.E.; Janson, J.; Bonner-Weir, S.; Ritzel, R.; Rizza, R.A.; Butler, P.C. Beta-cell deficit and increased beta-cell apoptosis in humans with type 2 diabetes. *Diabetes* **2003**, *52*, 102–110. [CrossRef] [PubMed]

53. Gutierrez, V.R.; de la Puerta, R.; Catalá, A. The effect of tyrosol, hydroxytyrosol and oleuropein on the non-enzymatic lipid peroxidation of rat liver microsomes. *Mol. Cell. Biochem.* **2001**, *217*, 35–41. [CrossRef] [PubMed]

54. Stupans, I.; Kirlich, A.; Tuck, K.L.; Hayball, P.J. Comparison of Radical Scavenging Effect, Inhibition of Microsomal Oxygen Free Radical Generation, and Serum Lipoprotein Oxidation of Several Natural Antioxidants. *J. Agric. Food Chem.* **2002**, *50*, 2464–2469. [CrossRef] [PubMed]

55. Jemai, H.; El Feki, A.; Sayadi, S. Antidiabetic and Antioxidant Effects of Hydroxytyrosol and Oleuropein from Olive Leaves in Alloxan-Diabetic Rats. *J. Agric. Food Chem.* **2009**, *57*, 8798–8804. [CrossRef] [PubMed]

56. Hamden, K.; Allouche, N.; Jouadi, B.; El-Fazaa, S.; Gharbi, N.; Carreau, S.; Damak, M.; Elfeki, A. Inhibitory action of purified hydroxytyrosol from stored olive mill waste on intestinal disaccharidases and lipase activities and pancreatic toxicity in diabetic rats. *Food Sci. Biotechnol.* **2010**, *19*, 439. [CrossRef]

57. Feher, J. 8.5-Digestion and Absorption of the Macronutrients. In *Quantitative Human Physiology*, 2nd ed.; Feher, J., Ed.; Academic Press: Boston, MA, USA, 2017; pp. 821–833.

58. Ristagno, G.; Fumagalli, F.; Porretta-Serapiglia, C.; Orrù, A.; Cassina, C.; Pesaresi, M.; Masson, S.; Villanova, L.; Merendino, A.; Villanova, A.; et al. Hydroxytyrosol attenuates peripheral neuropathy in streptozotocin-induced diabetes in rats. *J. Agric. Food Chem.* **2012**, *60*, 5859–5865. [CrossRef]

59. López-Villodres, J.A.; Abdel-Karim, M.; De La Cruz, J.P.; Rodríguez-Pérez, M.D.; Reyes, J.J.; Guzmán-Moscoso, R.; Rodriguez-Gutierrez, G.; Fernández-Bolaños, J.; González-Correa, J.A. Effects of hydroxytyrosol on cardiovascular biomarkers in experimental diabetes mellitus. *J. Nutr. Biochem.* **2016**, *37*, 94–100. [CrossRef]

60. Reyes, J.J.; Villanueva, B.; López-Villodres, J.A.; De La Cruz, J.P.; Romero, L.; Rodríguez-Pérez, M.D.; Rodriguez-Gutierrez, G.; Fernández-Bolaños, J.; González-Correa, J.A. Neuroprotective Effect of Hydroxytyrosol in Experimental Diabetes Mellitus. *J. Agric. Food Chem.* **2017**, *65*, 4378. [CrossRef]

61. González-Correa, J.A.; Rodríguez-Pérez, M.D.; Márquez-Estrada, L.; López-Villodres, J.A.; Reyes, J.J.; Rodriguez-Gutierrez, G.; Fernández-Bolaños, J.; De La Cruz, J.P. Neuroprotective Effect of Hydroxytyrosol in Experimental Diabetic Retinopathy: Relationship with Cardiovascular Biomarkers. *J. Agric. Food Chem.* **2018**, *66*, 637–644. [CrossRef]

62. Wang, W.; Shang, C.; Zhang, W.; Jin, Z.; Yao, F.; He, Y.; Wang, B.; Li, Y.; Zhang, J.; Lin, R. Hydroxytyrosol NO regulates oxidative stress and NO production through SIRT1 in diabetic mice and vascular endothelial cells. *Phytomedicine* **2019**, *52*, 206–215. [CrossRef]

63. Cao, K.; Xu, J.; Zou, X.; Li, Y.; Chen, C.; Zheng, A.; Li, H.; Li, H.; Szeto, I.M.-Y.; Shi, Y.; et al. Hydroxytyrosol prevents diet-induced metabolic syndrome and attenuates mitochondrial abnormalities in obese mice. *Free Radic. Biol. Med.* **2014**, *67*, 396–407. [CrossRef] [PubMed]

64. Zheng, A.; Li, H.; Xu, J.; Cao, K.; Li, H.; Pu, W.; Yang, Z.; Peng, Y.; Long, J.; Liu, J.; et al. Hydroxytyrosol improves mitochondrial function and reduces oxidative stress in the brain of db/db mice: Role of AMP-activated protein kinase activation. *Br. J. Nutr.* **2015**, *113*, 1667–1676. [CrossRef] [PubMed]

65. Jemai, H.; Bouaziz, M.; Fki, I.; El Feki, A.; Sayadi, S. Hypolipidimic and antioxidant activities of oleuropein and its hydrolysis derivative-rich extracts from Chemlali olive leaves. *Chem. Biol. Interact.* **2008**, *176*, 88–98. [CrossRef] [PubMed]

66. Tabernero, M.; Sarriá, B.; Largo, C.; Martínez-López, S.; Madrona, A.; Espartero, J.L.; Bravo, L.; Mateos, R. Comparative evaluation of the metabolic effects of hydroxytyrosol and its lipophilic derivatives (hydroxytyrosyl acetate and ethyl hydroxytyrosyl ether) in hypercholesterolemic rats. *Food Funct.* **2014**, *5*, 1556–1563. [CrossRef] [PubMed]

67. Voigt, A.; Ribot, J.; Sabater, A.G.; Palou, A.; Bonet, M.L.; Klaus, S. Identification of Mest/Peg1 gene expression as a predictive biomarker of adipose tissue expansion sensitive to dietary anti-obesity interventions. *Genes Nutr.* **2015**, *10*, 27. [CrossRef]

68. Pirozzi, C.; Lama, A.; Simeoli, R.; Paciello, O.; Pagano, T.B.; Mollica, M.P.; Di Guida, F.; Russo, R.; Magliocca, S.; Canani, R.B.; et al. Hydroxytyrosol prevents metabolic impairment reducing hepatic inflammation and restoring duodenal integrity in a rat model of NAFLD. *J. Nutr. Biochem.* **2016**, *30*, 108–115. [CrossRef] [PubMed]

69. Poudyal, H.; Lemonakis, N.; Efentakis, P.; Gikas, E.; Halabalaki, M.; Andreadou, I.; Skaltsounis, L.; Brown, L. Hydroxytyrosol ameliorates metabolic, cardiovascular and liver changes in a rat model of diet-induced metabolic syndrome: Pharmacological and metabolism-based investigation. *Pharmacol. Res.* **2017**, *117*, 32–45. [CrossRef]

70. Wang, N.; Liu, Y.; Ma, Y.; Wen, D. Hydroxytyrosol ameliorates insulin resistance by modulating endoplasmic reticulum stress and prevents hepatic steatosis in diet-induced obesity mice. *J. Nutr. Biochem.* **2018**, *57*, 180–188. [CrossRef]

71. Echeverría, F.; Valenzuela, R.; Espinosa, A.; Bustamante, A.; Álvarez, D.; Gonzalez-Mañan, D.; Ortiz, M.; Soto-Alarcon, S.A.; Videla, L.A. Reduction of high-fat diet-induced liver proinflammatory state by eicosapentaenoic acid plus hydroxytyrosol supplementation: Involvement of resolvins RvE1/2 and RvD1/2. *J. Nutr. Biochem.* **2018**, *63*, 35–43. [CrossRef]

72. Xie, Y.; Xu, Y.; Chen, Z.; Lu, W.; Li, N.; Wang, Q.; Shao, L.; Li, Y.; Yang, G.; Bian, X. A new multifunctional hydroxytyrosol-fenofibrate with antidiabetic, antihyperlipidemic, antioxidant and antiinflammatory action. *Biomed. Pharmacother.* **2017**, *95*, 1749–1758. [CrossRef]

73. Xie, Y.-D.; Chen, Z.-Z.; Li, N.; Lu, W.-F.; Xu, Y.-H.; Lin, Y.-Y.; Shao, L.-H.; Wang, Q.-T.; Guo, L.-Y.; Gao, Y.-Q.; et al. Hydroxytyrosol nicotinate, a new multifunctional hypolipidemic and hypoglycemic agent. *Biomed. Pharmacother.* **2018**, *99*, 715–724. [CrossRef] [PubMed]

74. Xie, Y.-D.; Chen, Z.-Z.; Shao, L.-H.; Wang, Q.-T.; Li, N.; Lu, W.-F.; Xu, Y.-H.; Gao, Y.-Q.; Guo, L.-Y.; Liu, H.-L.; et al. A new multifunctional hydroxytyrosol-clofibrate with hypolipidemic, antioxidant, and hepatoprotective effects. *Bioorg. Med. Chem. Lett.* **2018**, *28*, 3119–3122. [CrossRef] [PubMed]

75. King, A.J. The use of animal models in diabetes research. *Br. J. Pharmacol.* **2012**, *166*, 877–894. [CrossRef] [PubMed]

76. Petersen, M.C.; Shulman, G.I. Mechanisms of Insulin Action and Insulin Resistance. *Physiol. Rev.* **2018**, *98*, 2133–2223. [CrossRef] [PubMed]

77. Auñon-Calles, D.; Canut, L.; Visioli, F. Toxicological evaluation of pure hydroxytyrosol. *Food Chem. Toxicol.* **2013**, *55*, 498–504. [CrossRef] [PubMed]
78. Scientific Opinion on the substantiation of health claims related to polyphenols in olive and protection of LDL particles from oxidative damage (ID 1333, 1638, 1639, 1696, 2865), maintenance of normal blood HDL cholesterol concentrations (ID 1639), maintenance of normal blood pressure (ID 3781), "anti-inflammatory properties" (ID 1882), "contributes to the upper respiratory tract health" (ID 3468), "can help to maintain a normal function of gastrointestinal tract" (3779), and "contributes to body defences against external agents" (ID 3467) pursuant to Article 13(1) of Regulation (EC) No 1924/2006. *EFSA J.* **2011**, *9*, 2033.

© 2019 by the authors. Licensee MDPI, Basel, Switzerland. This article is an open access article distributed under the terms and conditions of the Creative Commons Attribution (CC BY) license (http://creativecommons.org/licenses/by/4.0/).

 *antioxidants*

Article

Antioxidant Activity and Anthocyanin Contents in Olives (*cv* Cellina di Nardò) during Ripening and after Fermentation

Alessio Aprile [1],[†], Carmine Negro [1],[†], Erika Sabella [1],[*], Andrea Luvisi [1], Francesca Nicolì [1], Eliana Nutricati [1], Marzia Vergine [1], Antonio Miceli [1], Federica Blando [2] and Luigi De Bellis [1]

[1] Department of Biological and Environmental Sciences and Technologies (DiSTeBA), Salento University, Via Prov. le Lecce-Monteroni, 73100 Lecce, Italy; alessio.aprile@unisalento.it (A.A.); carmine.negro@unisalento.it (C.N.); andrea.luvisi@unisalento.it (A.L.); francesca.nicoli@unisalento.it (F.N.); eliana.nutricati@unisalento.it (E.N.); marzia.vergine@unisalento.it (M.V.); antonio.miceli@unisalento.it (A.M.); luigi.debellis@unisalento.it (L.D.B.)
[2] Institute of Sciences of Food Production (ISPA), National Research Council (CNR), Research Unit of Lecce, Via Prov. le Lecce-Monteroni, 73100 Lecce, Italy; federica.blando@ispa.cnr.it
[*] Correspondence: erika.sabella@unisalento.it
[†] These authors contributed equally to this work.

Received: 30 April 2019; Accepted: 16 May 2019; Published: 18 May 2019

Abstract: The olive tree "Cellina di Nardò" (CdN) is one of the most widespread cultivars in Southern Italy, mainly grown in the Provinces of Lecce, Taranto, and Brindisi over a total of about 60,000 hectares. Although this cultivar is mainly used for oil production, the drupes are also suitable and potentially marketable as table olives. When used for this purpose, olives are harvested after complete maturation, which gives to them a naturally black color due to anthocyanin accumulation. This survey reports for the first time on the total phenolic content (TPC), anthocyanin characterization, and antioxidant activity of CdN olive fruits during ripening and after fermentation. The antioxidant activity (AA) was determined using three different methods. Data showed that TPC increased during maturation, reaching values two times higher in completely ripened olives. Anthocyanins were found only in mature olives and the concentrations reached up to 5.3 g/kg dry weight. AA was determined for the four ripening stages, and was particularly high in the totally black olive fruit, in accordance with TPC and anthocyanin amounts. Moreover, the CdN olives showed a higher TPC and a greater AA compared to other black table olives produced by cultivars commonly grown for this purpose. These data demonstrate the great potential of black table CdN olives, a product that combines exceptional organoleptic properties with a remarkable antioxidant capacity.

Keywords: olive; *Olea europaea*; anthocyanin; cyanidin 3-glucoside; cyanidin 3-rutinoside; oxygen radical absorbance capacity (ORAC); high performance liquid chromatography mass spectrometry (HPLC-MS).

---

1. Introduction

The olive cultivar "Cellina di Nardò" (CdN), also known as "Leccese", "Saracena", "Visciola", "Asciulo", or "Muredda", is an Italian variety that is widespread in Salento (Apulia, South Italy), especially in the Province of Lecce, although it is also grown in the territories of Taranto and Brindisi, covering a total of about 60,000 hectares. However, since 2013, Salento has been suffering from a devastating olive disease (the olive quick decline syndrome) caused by the bacteria *Xylella fastidiosa* [1,2] that has progressively destroyed hundreds of thousands of olive trees. Unfortunately, the CdN cultivar is sensitive to *Xylella fastidiosa*, and in the coming years is likely to disappear completely from the Salento area [3,4]. For this reason, a complete characterization of such traditional cultivars is required, as many plants are still cultivated in South Italy.

The CdN tree is vigorous; the branches at the top are erect, while the lateral ones almost pendulous, and the tree can reach a height of 20 m. The leaves have an elliptical elongated shape; the upper side of the leaf is dark green while the lower one is silvery gray. The blooming stage is quite early, and the inflorescence results in 15–20 flowers. The fruit is an elliptical drupe, slightly asymmetrical, with a color ranging from green to black; when ripening is completed, it has a reduced size, with a weight ranging from 1.5 to 2.0 g, and a low oil yield (15–17%).

The drupes have a high resistance to detachment and are not suitable for mechanical harvesting. On the contrary, this cultivar is valued by local farmers due to the slow vegetative growth, good fructification in adverse conditions, and good tolerance to cold and various pests. Despite the low oil yield and difficulties in harvesting, the CdN is cultivated mainly for oil production, which is characterized by an intense fruity flavor. The oil obtained from CdN and Ogliarola olives (another traditional cultivar widespread in Salento) is guaranteed by the brand "Terra d'Otranto" as a "Protected Designation of Origin", and presents some specific characteristics as a consequence of the geographical influence, pedoclimatic conditions, agronomic techniques, and oil processing.

The olive fruit consists of water (about 50%) and fats (20%), and the remaining part is made up of nitrogenous compounds, cellulose, sugars, and secondary metabolites [5]. The secondary metabolites of fresh fruit and fermented olives can vary greatly. In fact, the different processes of extraction and purification can modify the chemical structure of the molecules due to exposure to oxygen or solvents or even pH changes, situations that can commonly occur in phenolic metabolism [6]. The proportion of phenolic compounds within the edible part of the olive is considerable and can reach concentrations ranging from 1% to 3% of the fresh weight of the pulp [7]. There is a complex mixture of phenolic compounds in olives, some of which are present at very low concentrations and as a consequence are difficult to identify [8]. Moreover, phenolic and secondary metabolites are not uniformly present in diverse parts of the fruit: most of them are present in the pulp (about 85–90%), followed by the peel (about 8–12%), and then by the seed (1–2%).

Among phenolic compounds, oleuropein is generally the most represented among the various olive cultivars, reaching concentrations up to 140 mg/g fresh weight (FW) [9]. Oleuropein belongs to the secoiridoid family, as well as other compounds usually found in olives like dimethyl oleuropein, verbascoside, ligstroside, and nüzhenide [10]. In particular, oleuropein, dimethyl oleuropein, and verbascoside have been found in the all parts of the olive fruit (pulp, skin, and seed); conversely, the presence of nüzhenide was reported only in the seed [11]. In the olive fruit there are also flavones such as luteolin-7-glucoside, flavonols such as quercetin 3-rutinoside [12], anthocyanins like cyanidin-3-rutinoside and cyanidin-3-glucoside, and phenolic acids such as hydroxybenzoic, gallic, ferulic, caffeic, vanillic, and syringic acid [13].

All these secondary metabolites are of great interest for human health because of their antioxidant activity and properties with respect to cancer prevention, inflammatory disorders, and cardiovascular diseases [14,15].

The health properties of CdN when used as table olive have not yet been reported, and the *Xylella fastidiosa* threat suggests their urgent investigation due to the extinction risk caused by the pathogen. Therefore, in this paper a characterization of the secondary metabolites during ripening stages was carried out. Moreover, six different table olive cultivars were compared to CdN cultivars to establish the best food in terms of antioxidant effects and phenolic compounds.

2. Materials and Methods

2.1. Plant Material and Samples Preparation

To evaluate the antioxidant activity and phenolic content during ripening, the olives of CdN cultivar were collected from four of the eight Maturity Index (MI) classification groups (0–7) of olives described by Guzman et al. [16] (Figure 1).

- Green olives, characterized by green peel and light green pulp (MI group 0, indicated as Stage 0);
- Partially green olives with slightly pigmented peel and green pulp (MI group 2, indicated as Stage 2);
- Olives with purple peel and yellow pulp (MI group 4, indicated as Stage 4);
- Black coloration of the peel and pulp identified (MI Group 7, indicated as Stage 7)

Drupes were collected from three different orchards in Salento.

Figure 1. Cellina di Nardò olives. The images represent four of the eight Maturity Index (MI) classification groups (0–7) of olives described by Guzman et al. [16]. Stage 0: green olives; Stage 2: Olive peel partially pigmented and green pulp; Stage 4: dark/black peel and yellow pulp; Stage 7: dark/black peel and pulp.

To obtain CdN table olives, fully-ripened olives are traditionally first washed for two days (replacing the water several times) and then deposited in barrels of about 300 kg, covered with saturated brine, and left to ferment naturally (without the addition of chemical products for debittering) until they reach a pH below 4.0; the process takes an average of six months. CdN table olives were analyzed and compared with commercial table olives (Leccino, Blanqueta, Ogliarola, Empeltre, Hojiblanca, Kalamata) purchased on the market. Leccino and Ogliarola table olives were locally fermented essentially as described for CdN olives, whereas table olives of the other cultivars (much more common on the market) were harvested green and darkened by oxidation.

To extract the phenols from edible part of olives, the pulp of 60 olives was homogenized and 5 g were sampled. Then, 50 mL of a cold solution of methanol/water (80/20 *v/v*) acidified with HCl (pH 2.5) were added. Methanol and pulp were mixed by continuous agitation for 30 min. The homogenate was then filtered. The pulp was recovered from the filter and mixed again with 50 mL of the same methanol/water solution. A second vacuum filtration was then performed, and the filtrate was added to the previous one. Since the eluates could contain lipid traces, two treatments with an equal volume of hexane (≥97%, HPLC grade) were carried out to remove fatty acids. The extracts were finally concentrated by evaporation under low pressure and cold, avoiding any residues of methanol and obtaining, therefore, aqueous extracts (approximately 10 mL).

2.2. Chemicals

Water, methanol, acetonitrile, and hexane were HPLC/MS grade and were provided by Sigma Aldrich (Milan, Italy), as were 6-hydroxy-2,5,7,8-tetramethylchromane-2-carboxylic acid (Trolox), 2,2-diphenyl-1-picrylhydrazyl (DPPH), quinic acid, rutin, verbascoside, quercitrin, luteolin 7 glucoside, luteolin 7 rutinoside, oleuropein, luteolin, quercetin, apigenin 7 glucoside, cyanidin 3 glucoside, and cyanidin 3 rutinoside, which were of analytical standard grade.

2.3. Total Phenolic Content (TPC)

Phenols were determined using the Folin–Ciocalteau method described by Singleton and Rossi [17]. Here, 500 μL of olive extract were mixed with 2.5 mL of distilled water and 500 μL of Folin–Ciocalteau reagent. After 4 min, 2 mL of 10% $NaCO_3$ and 4.5 mL of distilled water were added. After one hour, the UV-VIS spectrophotometer absorbance was reported (λ = 756 nm).

A calibration curve was calculated using the gallic acid as reference. The calculated equation was:

$$\text{Optical Density (O.D.)} = 4.958X + 0.0437 \text{ with } R^2 = 0.9997$$

Results were reported as gallic acid equivalents (mg/g of dried pulp).

2.4. Antioxidant Activity

Antioxidant activity was evaluated using three different assays: the 2,2-diphenyl-1-picrylhydrazyl (DPPH) test, as reported by Goristen et al., [18]; the Oxygen Radical Absorbance Capacity (ORAC) test, as reported by Wang et al., [19]; and superoxide anion scavenging activity analysis, as described by Dasgupta et al. [20]. All the essays were performed in triplicate and the antioxidant activity were expressed as μmol of Trolox equivalent mg^{-1} of dry weight (DW).

2.5. HPLC ESI/MS-TOF Analysis of Olive Extracts

The phenolic compounds characterization was performed using Agilent 1200 High Pressure Liquid Chromatography (HPLC) System (Agilent Technologies, Palo Alto, CA, USA) equipped with a standard autosampler, as reported by Nicolì et al. [21]. The HPLC system was coupled to an Agilent diode-array detector (detection wavelength 280 nm) and an Agilent 6320 TOF mass spectrometer equipped with a dual ESI interface (Agilent Technologies) operating in negative ion mode. Detection was carried out within a mass range of 50–1700 *m/z*. Accurate mass measurements of each peak from the total ion chromatograms (TICs) were obtained by means of an ISO Pump (Agilent G1310B) using a dual nebulizer ESI source that introduces a low flow (20 μL·min^{-1}) of a calibration solution which contains the internal reference masses at *m/z* 112.9856, 301.9981, 601.9790, and 1033.9881, in negative ion mode. The identification of anthocyanins was carried out with the same method, but with positive ionization, using the internal reference masses at *m/z* 121.050873, 149.02332, 322.048121, and 922.009798.

The quantification of anthocyanins was achieved using calibration curves of authentic chemical standards cyanidin 3 glucoside and cyanidin 3 rutinoside [22], using an HPLC Agilent 1100 coupled with an Agilent DAD sensor (detection wavelength 280 nm and 520 nm). Separation was carried out at 30 °C with a gradient elution program at a flow rate of 0.8 mL/min using a Phenomenex Gemini C18 250 × 4.6 mm, 5-μm separation column. The mobile phases consisted of water plus 7.0% formic acid (A) and water:formic acid:acetonitrile 43:7:50 (B). The following multistep linear gradient was applied: 0 min, 6% B; 15 min, 30% B; 25 min, 50% B; 30 min, 60% B; The injection volume in the HPLC system was 5 μL.

2.6. Statistical Analysis

Data were reported as the mean ± SD and four biological replicates were carried out for each sample. Statistical evaluation was conducted by ANOVA, followed by multicomponent Duncan's test

($p < 0.05$) to discriminate among the mean values. The R^2 correlation coefficients between TPC and the antioxidant activities were also calculated.

3. Results

3.1. Total Phenolic Content during Maturation and after Fermentation Processes

To evaluate metabolic profiles and antioxidant activities, olives belonging to homogeneous classes of maturity were sampled according to Guzman et al. [16]. Their fresh and dry weight has been determined and the pulp percentage was calculated (data not shown). The TPC during maturation was quantified by the Folin–Ciocolteau method and the results are reported in Figure 2.

As reported in Figure 2, there is a progressive increase in the amount of total phenolic contents during maturation. Green and immature olives, corresponding to Stage 0, have the lowest number of polyphenols, equal to 14.0 mg of gallic acid equivalent (GAE)/g dry weight pulp. In Stage 2, the TPC value increased, reaching 26.18 mg (GAE)/g DW, whereas in Stage 4 TPC was 29.68 mg (GAE)/g DW. When the olives reached full maturity (stage 7), a further increase in total phenolic substances was recorded: 31.80 mg GAE/g DW.

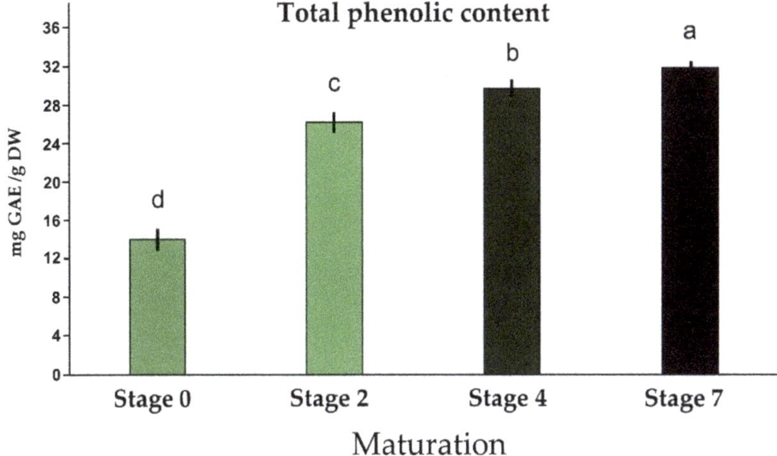

Figure 2. Total phenolic contents (mg GAE/g DW) in Cellina di Nardò olives at four different stages of the maturation process. Results are expressed as mg of GAE/g dried olive pulp. Same letters mean no statistical differences between averages (Duncan test, $n = 3$, $p = 0.05$). GAE: gallic acid equivalent; DW: dry weight.

Following fermentation/curing, the CdN olives were compared with other six commercial black table olives for the presence of phenolic compounds (Figure 3). It was found that among the seven black table olives, the CdN table olives were the richest in phenolic compounds, with a TPC equal to 13.08 mg/g DW. Since these table olives showed a TPC of 31.80 mg GAE/g DW at full maturity—Stage 7 (Figure 2)—the fermentation process drastically reduced the TPC. Kalamata olives also showed a high content of phenols (10.84 mg GAE/g DW). The lowest level of polyphenols was observed in Hojablanca cultivar (1.19 mg GAE/g DW) (Figure 3).

To identify the principal phenolic compounds in CdN olive extracts, a reverse-phase HPLC/MS-TOF was used. The identification was carried out by comparing the retention times, UV absorbance, and molecular masses with literature data and analytical standard when available.

Representative chromatograms of olive extracts during maturation are reported in Figure 4 (A–D) and the list of identified compounds are reported in Table 1.

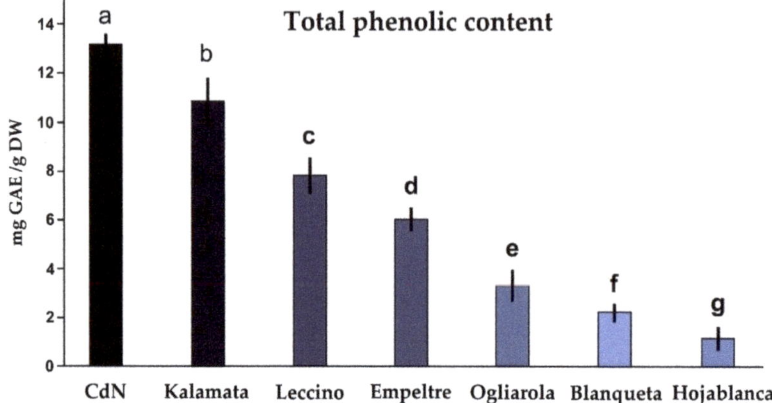

Figure 3. Total phenolic contents (mg GAE/g DW) in seven commercial black table olives. Results are expressed as mg of gallic acid equivalent (GAE)/g dried olive pulp. Same letters indicate no statistical differences between averages (Duncan test, $n = 3$, $p = 0.05$).

Table 1. List of chemicalss and anthocyanins putatively identified by High-Performance Liquid Chromatography coupled to Electrospray Ionization Time-of-Flight Mass Spectrometry (HPLC ESI/MS-TOF) following extraction from CdN olive pulp at different stages of maturation.

N.	Compound	RT [a] (min)	(M−H)⁻	m/z Exp [b]	m/z Clc [c]	Diff. (ppm) [d]	Score [e]	Ref.[f]
1	* Quinic acid	2.82	$C_7H_{11}O_6$	191.0510	191.0561	−5.89	90.44	[23–25]
2	Hydroxytyrosol glucoside	4.63	$C_{14}H_{19}O_8$	315.1095	315.1085	−1.26	96.62	[22,23,25]
3	Secologanoside is. 1	4.85	$C_{16}H_{21}O_{11}$	389.1095	389.1089	−1.11	88.91	[22,23,25]
4	Secologanoside is. 2	4.94	$C_{16}H_{21}O_{11}$	389.1101	389.1089	−2.62	96.13	[24,26]
5	* Rutin	5.83	$C_{27}H_{29}O_{16}$	609.1474	609.1461	−2.15	90.20	[24,26]
6	* Verbascoside	6.02	$C_{29}H_{35}O_{15}$	623.2013	623.1618	−0.05	93.73	[24,25]
7	Elenoic acid glucoside	6.31	$C_{17}H_{23}O_{11}$	403.1262	403.1246	−3.68	80.90	[24,26]
8	Oleuropein aglycon	7.02	$C_{16}H_{25}O_{10}$	377.1459	377.1453	−1.23	92.94	[24]
9	* Quercitrin	8.85	$C_{21}H_{19}O1_1$	447.0960	447.0933	−6.05	89.44	[27]
10	Hydroxyoleuropein	9.82	$C_{25}H_{31}O_{14}$	555.1773	556.1803	−2.04	97.55	[24,27]
11	* Luteolin 7 glucoside is. 1	10.03	$C_{21}H_{19}O_{11}$	447.0952	447.0933	−3.93	77.64	[24,25]
12	* Luteolin rutinoside	10.95	$C_{27}H_{29}O_{15}$	593.1517	593.1512	−0.87	97.79	[24]
13	* Luteolin 7 glucoside is. 2	11.87	$C_{21}H_{19}O_{11}$	447.0948	447.0933	−3.03	96.13	[24–26]
14	* Oleuropein	12.21	$C_{15}H_9O_{13}$	539.1772	539.1770	0.03	97.14	[23–25,27]
15	* Luteolin	12.53	$C_{15}H_9O_6$	285.0419	285.0405	−4.87	97.08	[23–25,27]
16	* Quercetin	13.07	$C_{15}H_9O_7$	301.0351	301.0354	1.10	96.04	[24,25]
17	Ligstroside	13.88	$C_{25}H_{31}O_{12}$	523.1823	523.1821	−0.03	97.55	[26]
18	* Apigenin 7 glucoside	14.31	$C_{15}H_9O_5$	269.0461	269.0455	−1.77	98.70	[23]
19	Diosmetin	14.72	$C_{16}H_{11}O_6$	299.0566	299.0561	−1.43	98.50	[23]
20	** Cyanidin 3 glucoside	15.03	$C_{21}H_{21}O_{11}$	449.1081	449.1078	0.66	92.21	[28]
21	** Cyanidin 3 rutinoside	15.82	$C_{27}H_{31}O_{15}$	595.1658	595.1657	0.16	95.23	[28]

[a] RT, Retention time; [b] *m/z* Exp, mass to charge experimental; [c] *m/z* Clc, mass to charge calculated; [d] Diff., difference between the observed mass and the theoretical mass of the compound (ppm); [e] Isotopic abundance distribution match: a measure of the probability that the distribution of isotope abundance ratios calculated for the formula matches the measured data; [f] Ref., References. * Confirmed by authentic chemical standard. ** These peaks were identified in positive ion mode (M−H)⁺.

The metabolic profile during olive maturation (Figure 4) highlighted a great variability among the four analyzed stages. The green olives (Stage 0) had a predominance of substances represented by the peaks 2, 6, 14, 15, and 17, corresponding to hydroxytyrosol glucoside, verbascoside, oleuropein, luteolin, and ligstroside, respectively. During drupe maturation, starting from Stage 4, the quantity of these compounds decreases. Conversely, the concentrations of the anthocyanins cyanidin 3 glucoside and cyanidin 3 rutinoside increase after Stage 4.

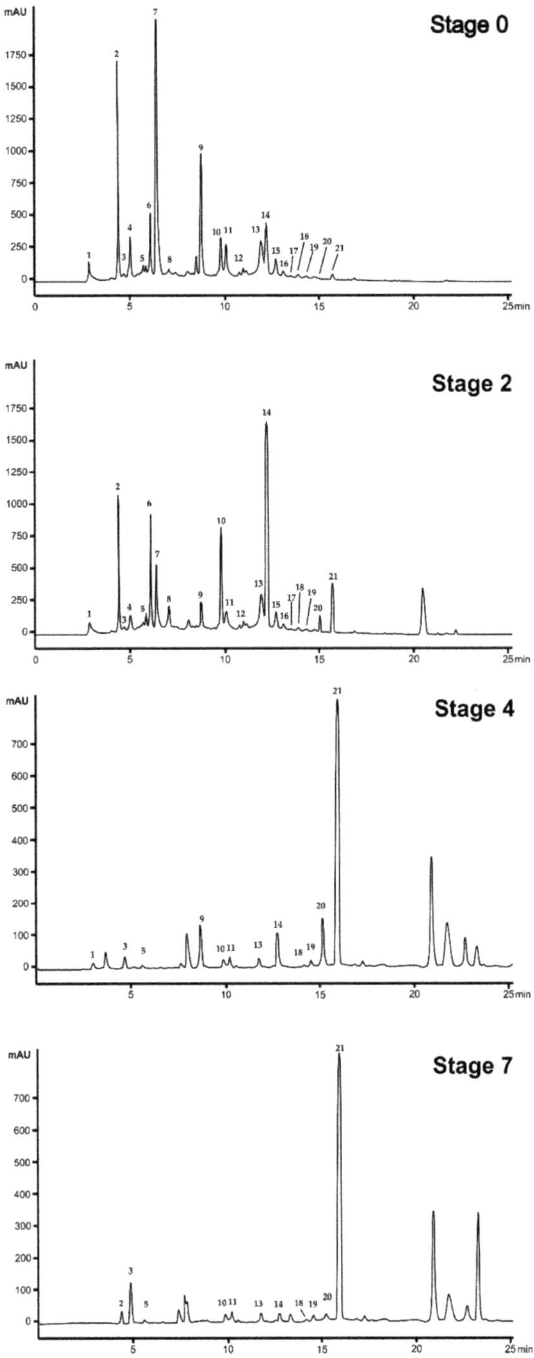

Figure 4. Representative chromatograms of Cellina di Nardò olive extracts during the maturation process. Detection at 280 nm. For the identification of the peaks and relative compounds, see Table 1.

3.2. Anthocyanin Quantification in CdN Olives

Anthocyanins were identified by 520 nm UV absorbance and molecular weight, and confirmed by authentic chemical standard spectra; the quantification was performed using the calibration curve obtained using the standard cyanidin 3-rutinoside (Table 2).

Table 2. Anthocyanin contents in Cellina di Nardò olive extracts during maturation stages and after fermentation. Data are reported as g/kg DW of cyanidin 3-rutinoside. Same letters indicate no statistical differences between averages (Duncan test, $n = 3$, $p = 0.05$).

Olive Extract	Cyanidin-3-Rutinoside (g/kg DW)
Stage 0	ND
Stage 2	Traces
Stage 4	3.22 [b] ± 0.22
Stage 7	4.62 [a] ± 0.06
Table olive (fermented)	1.16 [c] ± 0.16

With the maturation progresses, it is evident that there is a variation of anthocyanin contents: in green fruits (corresponding to Stage 0), there are no anthocyanins; some traces begin to appear in olives belonging to Stage 2; in the olives of Stage 4, anthocyanins are detectable and 3.22 g/kg DW were reported, while fully ripe olives showed 4.62 g/kg DW. The quantity of anthocyanin in fermented CdN olives was lower: 1.16 g/kg DW.

3.3. Antioxidant Activity of Olive Extracts

To evaluate the antioxidant properties of the olive fruit extracts during ripening stages, three different antioxidant assays were carried out (ORAC, DPPH, and superoxide anion scavenging activity). Data are reported in Table 3.

Table 3. Antioxidant activity detected in extracts of Cellina di Nardò olives at four different maturation stages. Results are expressed as μmol Trolox Equivalents/100 g FW (ORAC and DPPH tests) and as Inibitory Concentration (IC$_{50}$, μg of FW olive pulp). Same letters indicate no statistical differences between averages (Duncan test, $n = 3$, $p = 0.05$).

Olive Extract	ORAC Test μmol TE/100 g FW	DPPH μmol TE/100 g FW	Superoxide Anion Test IC$_{50}$ (μg FW)
Stage 0	11,412 [b] ± 1722	2888 [d] ± 234	3.15 [a] ± 0.13
Stage 2	13,565 [b] ± 2173	4212 [c] ± 351	2.15 [b] ± 0.35
Stage 4	15,990 [a,b] ± 486	6285 [b] ± 312	1.45 [b,c] ± 0.49
Stage 7	18,788 [a] ± 3298	9062 [a] ± 302	1.05 [c] ± 0.07

As reported in Table 3, the three antioxidant in vitro assays provided similar results among stages. The olive extracts with the highest antioxidant activities were the olives of the Stage 7 (complete maturation). Conversely, the antioxidant activity was lower at earlier maturation stages. However, the ratio and the increase in the antioxidant activity between the olives of Stage 0 and those of Stage 7 were different among the antioxidant assays. The ratio between CdN Stage 0 and CdN Stage 7 in ORAC assay was less than two-fold, whereas in DPPH and superoxide anion assays the ratio was about three-fold. The same analyses were conducted on commercial black table olives, and the results are reported in Table 4.

Table 4. Antioxidant activity detected in commercial black table olives of seven different cultivars. Results are expressed as μmol of Trolox Equivalents (TE)/100 g FW (ORAC and DPPH tests) and as IC_{50} per μg FW (superoxide anion assay) of olive pulp. Same letters indicate no statistical differences between averages (Duncan test, $n = 3$, $p = 0.05$).

Black Table Olive Extract	ORAC Test μmol TE/100 g FW	DPPH μmol TE/100 g FW	Superoxide Anion Test (IC_{50} μg FW)
Cellina di Nardò	7415 [a] ± 353	2920 [a] ± 51	4.25 [c] ± 0.21
Kalamata	4717 [b] ± 96	2533 [a] ± 135	6.10 [c] ± 0.42
Leccino	3964 [c] ± 213	2612 [a] ± 686	9.55 [c] ± 1.34
Empeltre	2355 [d] ± 224	1209 [b] ± 247	22.00 [b] ± 2.83
Ogliarola	1676 [e] ± 87	1186 [b] ± 398	22.95 [b] ± 2.62
Blanqueta	1537 [e] ± 164	1299 [b] ± 416	42.21 [a] ± 11.60
Hojiblanca	747 [f] ± 10	1356 [b] ± 246	47.55 [a] ± 2.05

Whereas the fully ripened CdN olive extracts showed an antioxidant activity of 18,788 ± 3298, 9062 ± 302, and 1.05 ± 0.07, respectively in ORAC, DPPH, and superoxide anion assay, the antioxidant activities of table CdN olives dropped dramatically (an about three-fold reduction) after the fermentation process. However, despite this drastic decrease, the antioxidant activity of CdN olives was the highest among the analyzed table olives. Kalamata and Leccino had similar antioxidant activities of CdN using the DPPH and superoxide anion assays, but the values reported by ORAC assay were statistically different among CdN, Kalamata, and Leccino. Blanqueta and Hojiblanca black table olives showed the lowest antioxidant activities.

4. Discussion

Cellina di Nardò is an olive cultivar closely connected to the Salento region by a long cultural tradition. Many plants are over 100 years old, and some of them (the monumental ones) are more than 1000 years old. People of this region have developed an empathic sentiment towards these plants, especially since the *Xylella* epidemic in 2013 [1].

In addition to oil production, CdN is used, in different local food preparations and after appropriate tanning, to produce table olives with a unique and recognizable flavor. Moreover, this olive is harvested when it is completely black (fully ripened), whereas most of the commercial black table olives are darkened by oxidation. The intense black color of the CdN natural black table olive is due to the accumulation in the skin and pulp tissues of a considerable quantity of anthocyanins, pigments that belong to the large family of polyphenols. The anthocyanins, as with polyphenols in general, have considerable beneficial properties as a consequence of their ability to scavenge oxygen radicals. They have good antioxidant, anti-inflammatory, and preventive effects against various pathological states, such as cardiovascular diseases and tumors [29].

The results reported in this work strongly support the possibility to use this table olive as a functional food. We demonstrated that the fully maturation is the best harvest time to obtain a table olive with a high content in phenolic compounds, anthocyanins, and other health valuable compounds. In fact, to assess whether and how the phenolic components vary during maturation, the phenolic content was studied during different ripening stages. The results showed an increase in the total quantity of polyphenols in relation to the stage of maturation; from a value of total phenolic substances of 14.00 mg GAE/g DW (Stage 0) to a value of 31.80 mg of GAE/g DW (Stage 7). Therefore, a two-fold intensification of total phenolic compounds was observed. These data are in accordance with the values reported by other authors [5]. The same authors observed that phenolic fraction in fresh pulp is about the 2% of the total weight of the pulp. We detected a similar value in fully ripened CdN olives where the phenolic fraction represents the 1.6% of the pulp total weight.

The qualitative analysis of phenolic compounds during maturation (Figure 3 and Table 1) demonstrated a great variation of the phenolic composition during ripening. In particular, from Stage 4, olives start to accumulate anthocyanins (cyanidin-3-rutinoside and cyanidin-3-glucoside), so that at Stage 7, the anthocyanin level reached 4.62 g/kg DW. Amiot et al. [9] reported an increase in anthocyanins matching to the progress of fruit ripening. In fact, the anthocyanin biosynthesis starts when the oleuropein decreases as direct result of an increase of the enzymatic activity of phenylalanine-ammonium lyase during drupe maturation [9]. The anthocyanin content found in CdN olives is higher than the quantity of anthocyanins present in most of other commercial olive fruits [28], as reported by a study conducted on several Italian cultivars. The black olives of Frantoio, a cultivar widespread in Tuscany, has a quantity of anthocyanins equal to 1.25 g/kg of fresh pulp; Ciliegino olives have an even lower quantity of phenolic compounds: 0.50 g/kg of fresh pulp [30].

After fermentation process, the level of anthocyanins in CdN decreased to 1.16 g/kg DW, a reduction of about 75% from the freshly harvested olives.

The same trend was observed for the total phenolic content. In fact, after the fermentation process of CdN, we noticed a drop in phenolic content (about 60%) to 13.08 GAE mg/g DW as reported also by Zou and colleagues [31], who observed how fermentation processes can reduce the total phenolic content by 50%. In the Spanish method fermentation for example, as well as in the Greek method, the debittering phase with NaOH causes the higher loss in phenolic compounds, probably due to the polyphenol oxidation under alkaline conditions [32]. During this process the polyphenols undergo to chemical transformations (hydrolysis of oleuropein into hydroxytyrosol) and their concentrations usually decrease. The following processing steps such as lactic fermentation, on the contrary, do not seem to affect total phenolic compound reduction [32].

To produce table olives from the CdN cultivar, completely ripened drupes (Stage 7) are usually employed, determining a high phenolic content after fermentation which is certainly higher than that of the black table olives on the market (Figure 3). The value is even higher than others found in literature concerning black table olive varieties. For example, in the Chondrolia table olives a total amount of phenolic substances of about 5.4 mg GAE/g FW was reported, while in the Amfiss olives the value is even lower and is equal to 2.3 mg GAE/g of fresh pulp [33]. Ben Othman et al. [34] estimated that the total phenolic content in black table olives is about 4–8 mg GAE/g DW. These data suggest the great potential of CdN cultivar to be used as a food reach in phenolic compounds.

The antioxidant capacities of fresh CdN olives were also investigated at four different maturation stages and after the fermentation process. CdN olives showed antioxidant properties (7415 μmol TE/100 g FW) higher than the average of other vegetables. About other fruits, an ORAC value of 4275 μmol TE/100 g fresh weight was reported for the Red Delicious apple [35], and a value of 1837 μmol TE/100 g fresh weight was reported for grapes [36], although the latter fruit is considered rich in antioxidant substances. Moreover, we observed an increase of the antioxidant activity positively correlated to the increase in phenolic substances ($R^2 = 0.832$, $R^2 = 0.756$ and $R^2 = 0.957$, respectively, for ORAC, DPPH, and the superoxide anion test) as already reported in Tunisian olives [37].

5. Conclusions

Different treatments (or curing methods) that are necessary to remove the bitterness of the raw olive and to stabilize them to obtain edible table olives, causing a loss in phenolic substances which also results in a loss of anthocyanins and antioxidant activity. However, CdN black table olives were the richest in polyphenols, consequently possessing the best antioxidant activity among the analyzed black table olives and among other black table olives reported in literature [37]. Moreover, it is plausible that regular consumption of CdN table olives can give real returns in terms of prevention of oxidative stress.

Author Contributions: Conceptualization, A.A., E.S., C.N., F.B.; methodology, C.N., A.A., A.M., E.N.; formal analysis, A.A., E.S., M.V.; investigation, F.N.; resources, A.L.; data curation, A.A., C.N; writing—original draft preparation, A.A., C.N., E.S.; writing—review and editing, F.B., L.D.B; supervision, A.M.; project administration, L.D.B.; funding acquisition, A.L. and L.D.B.

Funding: This research received no external funding.

Conflicts of Interest: The authors declare no conflict of interest.

References

1. Saponari, M.; Boscia, D.; Nigro, F.; Martelli, G.P. Identification of DNA sequences related to Xylella fastidiosa in oleander, almond and olive trees exhibiting leaf scorch symptoms in Apulia (Southern Italy). *J. Plant. Pathol.* **2013**, *95*, 668.
2. Luvisi, A.; Aprile, A.; Sabella, E.; Vergine, M.; Nicolì, F.; Nutricati, E.; Miceli, A.; Negro, C.; De Bellis, L. *Xylella fastidiosa* subsp. pauca (CoDiRO strain) infection in four olive (*Olea europaea* L.) cultivars: Profile of phenolic compounds in leaves and progression of leaf scorch symptoms. *Phytopathol. Mediterr.* **2017**, *56*, 259–273.
3. Martelli, G.P. The current status of the quick decline syndrome of olive in Southern Italy. *Phytoparasitica* **2016**, *44*, 1–10. [CrossRef]
4. Sabella, E.; Luvisi, A.; Aprile, A.; Negro, C.; Vergine, M.; Nicolì, F.; Miceli, A.; De Bellis, L. *Xylella fastidiosa* induces differential expression of lignification related-genes and lignin accumulation in tolerant olive trees cv. Leccino. *J. Plant Physiol.* **2018**, *220*, 60–68. [CrossRef] [PubMed]
5. Ryan, D.; Robards, K. Phenolic compounds in olives. *Analyst* **1998**, *123*, 31–44. [CrossRef]
6. Baiano, A.; Gambacorta, G.; Terracone, C.; Previtali, M.A.; Lamacchia, C.; La Notte, E. Changes in Phenolic Content and Antioxidant Activity of Italian Extra-Virgin Olive Oils during Storage. *J. Food Sci.* **2009**, *74*, 177–183. [CrossRef]
7. Garrido Fernandez, A.; Adams, M.R.; Fernandez-Diez, M.J. *Table Olives*, 1st ed.; Springer: New York, NY, USA, 1997; pp. 67–109.
8. Cioffi, G.; Pesca, M.S.; De Caprariis, P.; Braca, A.; Severino, L.; De Tommasi, N. Phenolic compounds in olive oil and olive pomace from Cilento (Campania, Italy) and their antioxidant activity. *Food Chem.* **2010**, *121*, 105–111. [CrossRef]
9. Amiot, M.J.; Fleuriet, A.; Macheix, J.J. Importance and evolution of phenolic compounds in olive during growth and maturation. *J. Agr. Food Chem.* **1986**, *34*, 823–826. [CrossRef]
10. Servili, M.; Baldioli, M.; Selvaggini, R.; Macchioni, A.; Montedoro, G.F. Phenolic Compounds of Olive Fruit: One- and Two-Dimensional Nuclear Magnetic Resonance Characterization of Nüzhenide and Its Distribution in the Constitutive Parts of Fruit. *J. Agr. Food Chem.* **1999**, *47*, 12–18. [CrossRef]
11. Ryan, D.; Antolovich, M.; Prenzler, P.; Robards, K.; Lavee, S. Biotransformations of phenolic compounds in *Olea europaea* L. *Scie. Horticult.* **2002**, *92*, 147–176. [CrossRef]
12. Bouaziz, M.; Grayer, R.J.; Simmonds, M.S.J.; Damak, M.; Sayadi, S. Identification and Antioxidant Potential of Flavonoids and Low Molecular Weight Phenols in Olive Cultivar Chemlali Growing in Tunisia. *J. Agr. Food Chem.* **2005**, *53*, 236–241. [CrossRef] [PubMed]
13. Vlahov, G. Flavonoids in three olive (*Olea europaea*) fruit varieties during maturation. *J. Sci. Food Agricul.* **1992**, *58*, 157–159. [CrossRef]
14. Cicerale, S.; Lucas, L.; Keast, R. Biological activities of phenolic compounds present in virgin olive oil. *Int. J. Mol. Sci.* **2010**, *11*, 458–479. [CrossRef]
15. Omar, S.H. Oleuropein in olive and its pharmacological effects. *Scie. Pharm.* **2010**, *78*, 133–154. [CrossRef]
16. Guzman, E.; Baeten, V.; Pierna, J.A.F.; Garcia-Mesa, J.A. Determination of the olive maturity index of intact fruits using image analysis. *J. Food Sci. Technol.* **2015**, *52*, 1462–1470. [CrossRef] [PubMed]
17. Singleton, V.L.; Rossi, J.A. Colorimetry of total phenolics with phosphomolybdic-phosphotungstic acid reagents. *Am. J. Enol. Viticult.* **1965**, *16*, 144–158.
18. Goristen, S.; Martin-Belloso, O.; Katrich, E.; Lojek, A.; Ciz, M.; Gligelmo-Miguel, N. Comparison of the contents of the main biochemical compounds and the antioxidant activity of some Spanish olive oils as determined by four different radical scavenging tests. *J. Nutr. Biochem.* **2003**, *14*, 154–159. [CrossRef]
19. Wang, H.; Cao, G.; Prior, R.L. Total Antioxidant Capacity of Fruits. *J. Agr. Food Chem.* **1996**, *1996*. *44*, 701–705. [CrossRef]
20. Dasgupta, N.; De, B. Antioxidant Activity of *Piper Betle* L. Leaf Extract in Vitro. *Food Chem.* **2004**, *88*, 219–224. [CrossRef]
21. Nicolì, F.; Vergine, M.; Negro, C.; Luvisi, A.; Nutricati, E.; Aprile, A.; Rampino, P.; Sabella, E.; De Bellis, L.; Miceli, A. *Salvia clandestina* L.: Unexploited source of danshensu. *Nat. Prod. Res.* **2019**, *19*, 1–4. [CrossRef]

22. Lee, M.J.; Park, J.S.; Choi, D.S.; Jung, M.Y. Characterization and quantification of anthocyanins in Purple-fleshed sweet potatoes cultivated in Korea by HPLC-DAD and HPLC-ESI-QTOF-MS/MS. *J. Agric. Food Chem.* **2013**, *61*, 3148–3158. [CrossRef]

23. Fu, S.; Arráez-Roman, D.; Segura-Carretero, A.; Menéndez, J.A.; Menéndez-Gutiérrez, M.P.; Micol, V.; Fernández-Gutiérrez, A. Qualitative screening of phenolic compounds in olive leaf extracts by hyphenated liquid chromatography and preliminary evaluation of cytotoxic activity against human breast cancer cells. *Anal. Bioanal. Chem.* **2010**, *397*, 643–654. [CrossRef] [PubMed]

24. Talhaoui, N.; Gómez-Caravaca, A.M.; León, L.; De la Rosa, R.; Segura-Carretero, A.; Fernández-Gutiérrez, A. Determination of phenolic compounds of 'Sikitita' olive leaves by HPLC-DAD-TOF-MS. Comparison with its parents 'Arbequina' and 'Picual' olive leaves. *LWT - Food Sci. Technol.* **2014**, *58*, 28–34. [CrossRef]

25. Talhaoui, N.; Gomez-Caravaca, A.M.; Rolda, C.; Leo, L.; De La Rosa, R.; Fernandez-Gutierrez, A.; Segura-Carretero, A. Chemo-metric analysis for the evaluation of phenolic patterns in olive leaves from six cultivars at different growth stages. *J. Agric. Food Chem.* **2015**, *63*, 1722–1729. [CrossRef] [PubMed]

26. Lozano-Sánchez, J.; Segura-Carretero, A.; Menéndez, J.; Oliveras-Ferraros, C.; Cerretani, L.; Fernández-Gutiérrez, A. Prediction of extra virgin olive oil varieties through their phenolic profile. Potential cytotoxic activity against human breast cancer cells. *J. Agric. Food Chem.* **2010**, *58*, 42–55. [CrossRef]

27. Quirantes-Pine, R.; Zurek, G.; Barrajon-Catalan, E.; Bassmann, C.; Micol, V.; Segura-Carretero, A.; Fernandez-Gutierrez, A. A metabolite-profiling approach to assess the uptake and metabolism of phenolic compounds from olive leaves in SKBR3 cells by HPLC-ESI-QTOF-MS. *J. Pharm. Biomed. Anal.* **2013**, *72*, 121–126. [CrossRef]

28. Yorulmaz, A.; Poyrazoglu, E.S.; Ozcan, M.M.; Tekin, A. Phenolic profiles of Turkish olives and olive oils. *Eur. J. Lipid Sci. Technol.* **2012**, *114*, 1083–1093. [CrossRef]

29. Li, D.; Wang, P.; Luo, Y.; Zhao, M.; Chen, F. Health benefits of anthocyanins and molecular mechanisms: Update from recent decade. *Crit Rev Food Sci Nutr.* **2017**, *57*, 1729–1741. [CrossRef] [PubMed]

30. Romani, A.; Mulinacci, N.; Pinelli, P.; Vincieri, F.F.; Cimato, A. Polyphenolic content in five tuscany cultivars of *Olea europaea* L. *J. Agric. Food Chem.* **1999**, *47*, 964–967. [CrossRef]

31. Zou, B.; Wu, J.; Yu, Y.; Xiao, G.; Xu, Y. Evolution of the antioxidant capacity and phenolic contents of persimmon during fermentation. *J. Agr. Food Chem.* **2017**, *26*, 563–571. [CrossRef]

32. Brenes, M.; Rejano, L.; Garcia, P.; Sanchez, A.H.; Garrido, A. Biochemical changes in phenolic compounds during Spanish-style green olive processing. *J. Agr. Food Chem.* **1995**, *43*, 2702–2706. [CrossRef]

33. Ziogas, V.; Tanou, G.; Molassiotis, A.; Diamantidis, G.; Vasilakakis, M. Antioxidant and free radical-scavenging activities of phenolic extracts of olive fruits. *Food Chem* **2010**, *120*, 1097–1103. [CrossRef]

34. Ben Othman, N.; Roblain, D.; Chammen, N.; Thonart, P.; Hamdi, M. Antioxidant phenolic compounds loss during the fermentation of Chetoui olives. *Food Chem.* **2009**, *116*, 662–669. [CrossRef]

35. Wolf, K.L.; Kang, X.; He, X.; Dong, M.; Zhang, Q.; Liu, R.H. Cellular antioxidant activity of common fruits. *J. Agr. Food Chem.* **2008**, *56*, 8418–8426. [CrossRef] [PubMed]

36. Wu, X.; Beecher, G.R.; Holden, J.M.; Haytowitz, D.B.; Gebhardt, S.E.; Prior, R.L. Lipophilic and hydrophilic antioxidant capacities of common foods in the United States. *J. Agr. Food Chem.* **2004**, *52*, 4026–4037. [CrossRef] [PubMed]

37. Bouaziz, M.; Chamkha, M.; Sayadi, S. Comparative study on phenolic content and antioxidant activity during maturation of the olive cultivar Chemlali from Tunisia. *J. Agr. Food Chem.* **2004**, *52*, 5476–5481. [CrossRef] [PubMed]

 © 2019 by the authors. Licensee MDPI, Basel, Switzerland. This article is an open access article distributed under the terms and conditions of the Creative Commons Attribution (CC BY) license (http://creativecommons.org/licenses/by/4.0/).

 *antioxidants*

Article

Antioxidant and Antimicrobial Activity of Rosemary, Pomegranate and Olive Extracts in Fish Patties

Lorena Martínez [1], Julián Castillo [2], Gaspar Ros [1] and Gema Nieto [1,*]

[1] Department of Food Technology, Nutrition and Food Science, Veterinary Faculty University of Murcia, Campus de Espinardo, 30100 Espinardo, Murcia, Spain; lorena.martinez23@um.es (L.M.); gros@um.es (G.R.)
[2] Research and Development Department of Nutrafur-Frutarom Group, Camino Viejo de Pliego s/n, 80320 Alcantarilla, Murcia, Spain; j.castillo@Nutrafur.com
* Correspondence: gnieto@um.es; Tel.: +34-868-889624

Received: 6 March 2019; Accepted: 28 March 2019; Published: 3 April 2019

Abstract: Natural extracts (rich in bioactive compounds) that can be obtained from the leaves, peels and seeds, such as the studied extracts of Pomegranate (P), Rosemary (RA, Nutrox OS (NOS) and Nutrox OVS (NOVS)), and olive (*Olea europaea*) extracts rich in hydroxytyrosol (HYT-F from olive fruit and HYT-L from olive leaf) can act as antioxidant and antimicrobial agents in food products to replace synthetic additives. The total phenolic compounds, antioxidant capacity (measured by 2,2-diphenyl-1-picrylhydrazyl (DPPH), 2,2-Azinobis (3-ethylbenzothiazolin) -6-sulphonic acid (ABTS), Ferric Reducing Antioxidant Power (FRAP), and Oxygen Radical Absorbance Capacity (ORAC$_H$)) and their antimicrobial power (using the diffusion disk method with the *Escherichia Coli*, *Lysteria monocytogenes*, and *Staphilococcus Aureus* strains) were measured. The results showed that all the extracts were good antioxidant and antimicrobial compounds in vitro. On the other hand, their antioxidant and antimicrobial capacity was also measured in fish products acting as preservative agents. For that, volatile fatty acid compounds were analysed by GS-MS at day 0 and 11 from elaboration, together with total vial count (TVC), total coliform count (TCC), *E. Coli*, and *L. monocytogenes* content at day 0, 4, 7 and 11 under refrigerated storage. The fish patties suffered rapid lipid oxidation and odour and flavour spoilage associated with slight rancidity. Natural extracts from pomegranate, rosemary, and hydroxytyrosol delayed the lipid oxidation, measured as volatile compounds, and the microbiological spoilage in fish patties. Addition of natural extracts to fish products contributed to extend the shelf life of fish under retail display conditions.

Keywords: antioxidant; antimicrobial; hydroxytyrosol; rosemary; pomegranate; fish; volatile compounds

---

1. Introduction

The food industry generates an enormous amount of waste in form of skins, seeds and leaves, whose disposal is a problem for the environment and expensive for companies concerned. Many residues from fruits, which are rich source of phenolic compounds, can be extracted and used by food industries as antioxidant and antimicrobial preservatives.

In this work, the antimicrobial and antioxidant capacity of several extracts from rosemary, pomegranate, and hydroxytyrosol were measured. These extracts were chosen because of their beneficial effects for human health and their notable antioxidant properties. *Rosmarinus officinalis* L. is a natural woody perennial green herb from the Mediterranean region rich in vitamins A, C, B1, B6 and B9, minerals such as Mg, Ca, Cu, Fe and Mn, as well as phenolic compounds (diterpenes and rosmarinic acid). The regular consumption of this herb has been seen to have many beneficial effects for human health [1,2] acting as a powerful antioxidant and antibacterial agent [3]. Furthermore, pomegranate extract, from *Punica granatum* fruit, contains large quantities of phenolic compounds (ellagitannins, flavonoids, punicalagin, ellagic acid, vitamin C and minerals). For this reason, pomegranate has been used in medical applications for more than 2000 years. Among its beneficial effects for the human body [4,5], pomegranate peel extracts and other subproducts obtained from this fruit, such as juice or seeds, have shown high antioxidant and antimicrobial capacities, with a great scavenging power, preventing microbiological growth of different bacteria [6,7]. On the other hand, oleuropein is the main phenolic compound in olive (*Olea europaea*) tree and the precursor of hydroxytyrosol, which is also extracted from leaves and vegetation waters from the manufacture of olive oil. This phenolic phytochemical is considered the most antioxidant compound after gallic acid, whose consumption has many beneficial effects for the skin, eyes and immune system, and acting as an antimicrobial, anti-infammatory and anticancer agent [8–10].

As these extracts act as antioxidants, antimicrobials and hence preservatives of shelf life food products, it is interesting to investigate about their action in a food matrix. In this case, fish has been chosen due to it is an important source or omega-3, particularly eicosapentaenoic acid (EPA), and docosahexaenoic acid (DHA), essential fatty acids for the protection against autoimmune, inflammatory and cardiovascular diseases [11]. However, fish consumption has decreased in populations groups under 14 years old, thus manufacture of fish products as "burgers" could be a good option to stimulate fish consumption among young people [12]. In the same way, synthetic additives are widely used to avoid the microbiological, enzymatic and oxidative degradation of these kinds of products. As substitution of synthetic additives by natural extracts is gaining in importance in the food industry [13], the use of natural extracts as preservatives in fish products could be an excellent way to produce a saludable product with high nutritional level, Clean label, and without synthetic additives. Alternatively, pomegranate has already demonstrated its antioxidant capacity after addition to chub mackerel minced muscle during frozen storage [14]. In this case, 2% pomegranate seed extract inhibited the formation of substances related with lipid oxidation regard to the control sample. Similarly, rosemary oil at 0.2%, 1% and 3% added to minced rainbow trout had a positive effect on the freshness indicators, oxidative stability, fatty acid and biogenic amine contents during refrigerated storage [15]. In addition, hydroxytyrosol has also showed an important inhibition of the formation of lipid oxidation products in foodstuffs rich in fish lipids (bulk cod liver oil, cod liver oil-in water emulsions and frozen minced horse mackerel) [16], while in a general view, hydroxytyrosol has also demonstrated to have excellent antioxidant properties in a nanostructured starch developed as active food packaging [17].

Briefly, the objective of this work was to make a comparative study of pomegranate, rosemary and hydroxytyrosol extracts measuring their antioxidant and antimicrobial capacities, as in vitro, following several methods, as antioxidants and antimicrobials in a food matrix: fish patties.

2. Materials and Methods

2.1. Plant Extracts

The plant extracts used were: Pomegranate (P), Rosemary: Rosmarinic acid extract (RA), Nutrox OS (NOS), Nutrox OVS (NOVS), Hydroxytyrosol from fruit (HYT-F), and from leaf (HYT-L). All extracts were supplied by Nutrafur-Frutarom Group (Alcantarilla, Murcia, Spain) and obtained from the corresponding dry vegetal materials using a maceration process including, as an initial step, solid-liquid extraction with different ethanol-water mixtures. Pomegranate extract (P) was obtained from dehydrated lignocellulosic materials of fruits by means of a hydro-alcoholic extraction, filtration of the vegetal material, evaporation of the ethanol and crystallisation in aqueous medium, followed by concentration and drying. De-oiled rosemary leaf was used as raw material to obtain the rosemary extracts used in this study. The water-soluble rosemary extract (RA) was obtained by extraction in aqueous medium and simple filtration of the plant material, concentration and drying. Rosemary extracts not soluble in water were obtained by extraction with acetone-water, filtration of the plant material, concentration and drying. Subsequently, the solid obtained was dissolved in different excipients to obtain the two extracts used in this study, sunflower oil (NOS) and sunflower oil plus lecithin (NOVS). HYT-L was made by extraction in ethanol-water medium, filtration of the plant material, evaporation of the solvent and subsequent thermal treatment. Finally, the water insoluble materials were filtered and dried. While HYT-F was obtained by extraction from the dry plant material with ethanol (vegetation waters), decantation, evaporation of the solvent, drying, recrystallisation of aqueous medium, elimination of insoluble materials in water and drying. All the drying processes were carried out in a vacuum and at a reduced temperature (50–70 °C).

2.1.1. Determination of Extract Composition (HPLC)

For the quantification of phenolics in the P extract, it was dissolved in dimethylsulfoxide (DMSO) in the ratio of 4 mg/mL; this solution was filtered through a 0.45 μm nylon membrane. The HPLC equipment used for all the extracts was a Hewlett-Packard Series HP 1100 equipped with a diode array detector. The stationary phase was a C_{18} LiChrospher 100 analytical column (250 × 4 mm i.d.) with a particle size of 5 μm (Merck, Darmstadt, Germany) thermostated at 30 °C. The flow rate was 1 mL/min and the absorbance changes were monitored at 280 nm. The mobile phases for chromatographic analysis were: (A) acetic acid/water (0.1:99.5) and (B) methanol. An initial isocratic period for 15 min with 100% (A) was run, after, a linear gradient was run from 100% (A) to 90% (A) and 10% (B) for 15 min (30 min total time), and it was maintained for 10 min (40 min total time), before reequilibrating in 10 min (50 min, total time) to initial composition. Punicalagin in P was identified and quantified by comparation of their retention times with the corresponding standard and by their UV spectra obtained with the diode array detector.

For the quantification of diterpenes in rosemary extracts (mainly carnosic acid and carnosol), for NOS and NOVS extracts, each extract was dissolved in methanol in variable ratios between 0.2 and 4 mg/mL, depending on the diterpene concentration expected for each extract; the solution was filtered through a 0.45 μm nylon membrane. The flow rate was 0.75 mL/min, and the elution was monitored at 230 nm. HPLC was used for the separation of the different diterpenes present in the rosemary extracts. The mobile phase for chromatographic analysis was an isocratic single step of acetonitrile (65%), water (35%) and phosphoric acid (0.2%) for 25 min. Diterpenes were quantified by comparing the chromatographic areas of the corresponding standards. On the other hand, for the quantification of rosmarinic acid in rosemary extracts (RA), the solid was dissolved in water at 5 mg/mL for analytical chromatography; this solution was filtered through a 0.45 μm nylon membrane. The flow rate was 1 mL/min. The absorbance changes were monitored at 340 nm.

For the quantification of phenolics in the olive (*Olea europaea*) extracts (HYT-F and HYT-L), the extract was dissolved in dimethylsulfoxide (DMSO) in the ratio of 5 mg/mL; this solution was filtered through a 0.45 μm nylon membrane. The flow rate was 1 mL/min and the absorbance changes were monitored at 280 nm. The mobile phases for chromatographic analysis were: (A) acetic acid/water (2.5:97.5) and (B) acetonitrile. A linear gradient was run from 95% (A) and 5% (B) to 75% (A) and 25% (B) for 20 min; it was then changed to 50% (A) and (B) in 20 min (40 min, total time); in 10 min it was changed to 20% (A) and 80% (B) (50 min, total time), before reequilibrating in 10 min (60 min, total time) to initial composition. Phenolic compounds in olive extracts were identified and quantified by comparison of their retention times with the corresponding standard and by their UV spectra obtained with the diode array detector [18].

Pure standards for HPLC quantification: hydroxytorosol (Code 4999S, Extrasynthése, Genay, France); carnosic acid (Sigma, Code C-0609, Madrid, Spain); carnosol (Sigma, Code C-9617, Darmstadt, Germany); rosmarinic acid (Extrasynthése, Code 4957S, Genay, France); punicalagin (Sigma, Code P-0023, Darmstadt, Germany).

2.1.2. Total phenolic content (TPC)

The total phenolic content (TPC) was determined quantitatively using the Folin–Ciocalteu reagent and gallic acid as the standard [19,20]. Each extract was diluted with water or ethanol, according to its polarity, in a 1000 ppm solution. Then, 2 mL of 2% Na_2CO_3 was added to 100 μL of sample and a standard solution of gallic acid. After, 100 µL of 10:1 Folin–Ciocalteau phenol reagent was added. Following incubation for 30 min, the absorbance was measured at 750 nm. The TPC was expressed as mg gallic acid equivalents (GAE) per g extract.

2.1.3. Antioxidant Activity

The antioxidant activity was also measured using the 2,2-diphenyl-1-picrylhydrazyl (DPPH) free radical scavening method, described by Brand-Williams et al. [21] and Sánchez-Moreno et al. [22]. Each extract was diluted with water or ethanol in a 1000 ppm solution. A DPPH radical solution was prepared with 0.0063 g DPPH in 250 mL ethanol. Then, 3.9 mL of DPPH radical solution was added to 100 µL of sample and a standard solution of 500 µM Trolox. After incubation in the dark at room temperature for 30 min, the absorbance was measured at 515 nm. The % chelating activity was calculated using the following formula: [(Abs DPPH − Abs Sample)/Abs DPPH] × 100 [23].

Radical cation scavenging capacity against ABTS+ radical (2,2-Azinobis (3-ethylbenzothiazolin) -6-sulphonic acid) was measured as by Re et al. [24]. Each extract was diluted with water or ethanol in a 1000 ppm solution. ABTS radical cations were prepared by reacting 7 mM ABTS (2,2-Azinobis (3-ethylbenzothiazolin) -6-sulphonic acid) with 2.45 mM potassium persulphate (1:1, v/v) pH = 7.4. This solution was diluted with distilled water to an absorbance of 0.7000 at 734 nm. Then, 1 mL of ABTS solution was added to 100 µL of sample and standard solution of 500 µM Trolox. After incubation for 2 min, the absorbance was measured at 734 nm. The % chelating activity was calculated using this formula: [(Abs ABTS − Abs Sample)/Abs ABTS] × 100.

The hydrophilic antoxidant capacity was measured using the ORAC (Oxygen Radical Absorbance Capacity) method described by Prior et al. [25]. The reaction was carried out in a phosphate buffer (0.075 M, pH 7.0). For this, 20 µL of sample, at different concentrations, and Trolox standard solutions (6.25, 12.5, 25, 50 µM), were pipetted into the wells of a 96-well black microplate. When the microplate was ready, 200 µL of 0.04 µM fluorescein were dispensed into each well. Samples were incubated for 15 min at 37 °C in the dark and the reaction was started by adding 20 µL of 40 mM AAPH (2,2′-Azobis(2-amidinopropane)dihydrochloride) to each well. The flourescence decay was measured every 2 min at 37 °C using the microplate reader Biotek Synergy HT at 485 nm excitation and 538 nm emission until zero fluorescence was detected. All samples were prepared in triplicate and at a minimum of three different concentrations. The antioxidant activity of the sample was expressed as µM of Trolox Equivalents (TE) per g of extract, with the following formula: *(C × DF)/a*; where C is obtained from the area under the fluorescence decay curve of fluorescein in the presence of sample (AUC net = AUC sample − AUC blank); *DF* is the dilution factor; and *a* is the weight of the sample. A Biotek Synergy HT fluorescent microplate reader (Biotek Instruments, Winooski, VT, USA) was used with an excitation wavelength of 485 nm and an emission wavelength of 528 nm; and a UV2 spectrophotometer (Pye Unicam Ltd., Cambridge, UK) at different wavelengths depending on the method to be performed.

The total antioxidant activity was also determined using the method described by Benzie and Strain [26] with some modifications. Each extract was diluted with water or ethanol in a 1000 ppm solution. Then, the FRAP reagent was prepared with 20 mL 300 mmol/L acetate buffer pH = 3.6, 2 mL 20 mmol/L FeCl$_3$·6H$_2$O and 2 mL 10 mmol/L TPTZ (2,4,6-tripyridyl-s-triazine) in 40 mmol/L HCl. Then, 1 mL of the FRAP reagent was added to 100 µL of sample and a standard solution of 500 µM Trolox. After incubation for 4 min, the absorbance was measured at 593 nm. The antioxidant power was expressed as µM Trolox Equivalents (TE) per g extract.

2.1.4. Antimicrobial Activity

All extracts were tested against Gram-negative bacteria: *Escherichia Coli* O157:H7 ATCC 25922 CECT 434; and Gram-positive bacteria: *Lysteria Monocytogenes* KCTC 3569 CECT 7467 and *Staphilococcus Aureus* ATCC 25923 CECT 435. The antimicrobial activity of the different extracts was measured following the diffusion disk method described by Chanwitheesuk et al. [27] with some modifications, using a bacterial cell suspension with an equilibrated concentration to a 0.5 McFarland standard or 10^5–10^6 cfu/mL. Each bacterial suspension was spread on a Mueller-Hinton agar plate. Sterile paper discs of 6 mm diameter were impregnated with 30, 60 and 90 µL of each test sample. Chloramphenicol (30 µg), a standard antibiotic was used as a positive control, while discs with the solvents used for extraction were used as negative controls, water or MeOH. The plates were incubated at 37 °C for 24 h. After incubation, the inhibition zones appearing around the discs were measured and recorded. The values, expressed in mm, were averages of six measurements per disc, taken at six different points in order to minimise errors.

2.2. Fish Patties

Antioxidant and antimicrobial extracts were also tested in fish patties that incorporated them in each sample formula.

2.2.1. Samples Elaboration

Ultrafrozen skinless hake fillets (*Merluccius capensis, Merluccius paradoxus*) (Pescanova España S.L.U.) were bought in a local supermarket and thawed in refrigeration for 24 h before making into fish patties. Fish was minced in an electric mincer (Bosch, Germany) for 2 min, mixing all the ingredients of each one of seven samples, represented in Table 1. Afterwards, fish patties were formed (50 g) and packed in aerobic conditions. Samples were stored at 4 °C until analysis, at day 0, 4, 7 and 11. 20 fish patties were formed for each batch and analysis were carried out by triplicated.

Table 1. Fish patties formulation.

Ingredients	Control	P	RA	NOS	NOSV	HYT-L	HYT-F
Hake (g)	852	875	875	875	875	875	875
Water (ml)	100	100	100	100	100	100	100
Commercial mix (g/kg)	48						
Fibers (g/kg)		25	25	25	25	25	25
Natural extracts (ppm)		200	200	200	200	200	200

Commercial mix®: supplied by Catalina Food Solutions, S.L. (El Palmar, Murcia, Spain) and composed by: vegetables fibers, salt, potato starch, stabiliser (Pocessed euchema seaweed (PES) E-407-a), acidity correctors (sodium citrate E-331, and sodium acetate E-262), spices, spice extracts and antioxidant (sodium ascorbate E-301). P: Pomegranate extract, RA: Rosemary extract rich in Rosmarinic Acid; NOS: Rosemary extract rich in diterpenes; NOVS: Rosemary extract rich in diterpenes and with lecitin as emulsifier; HYT-L: Hydroxytyrosol extract obtained from olive leaf; HYT-F: Hydroxytyrosol extract obtained from olive fruit.

2.2.2. Volatile Organic Compounds Analysis

Lipid oxidation was related with the concentration of volatile organic compounds (VOCs). For that, 5 g of fish burger samples were placed in glass vials. Volatile compounds were extracted using solid-phase microextraction (SPME). A SPME device (Supelco, Bellefonte, PA, USA) containing a fused-silica fibre (10 mm length) was used. Extraction was performed at 35 °C for 30 min. Once sampling was finished, the fibre was withdrawn into the needle and transferred to the injection port of the gas chromatograph-mass spectrometer (GC-MS) system. Volatiles were analysed in duplicate in all samples at days 0 and 11 from elaboration. Analyses were performed on a Hewlett-Packard 6890 N Series GC gas chromatograph fitted with a HP 5973 mass spectrometer and an MSD Chemstation (Hewlett-Packard, Palo Alto, CA, USA). A split injection port was used to thermally desorb the volatiles from the SPME fibre onto the front of the DB-624 capillary column (J&W scientific: 30 m × 0.25 mm id, 1.4 µm film thickness). Helium was used as a carrier gas with a linear velocity of 36 cm/s. The temperature programme was: 40 °C for 2 min and then raised to 100 °C at 3 °C/min; then from 100 to 180 °C at 5 °C/min; total run time 50.8 min. The mass spectra were obtained using a mass selective detector working in electronic impact at 70 eV, with a multiplier voltage of 1953 V and collecting data at a rate of 6.34 scans/s over the range m/z 40–300. Compounds were identified by comparing their mass spectra with those contained in the NIST05 (National Institute of Standards and Technology, Gaithersburg) library.

2.2.3. Microbiological Analysis

Microbiological growth of total vial count (TVC), total coliform count (TCC), *E. Coli*, *S. Aureus*, and *L. monocytogenes* was determined on days 0, 4, 7 and 11 from elaboration. Mass seeding was performed on Rapid E. Coli (to determine *E. Coli* and TCC), PCA (TVC), and Rapid L. mono (*L. monocytogenes*). All media were sterilised at 121 °C for 20 min according to product indications. Peptone water (OXOID, Ltd. CM0087 Basingstoke, Hampshire, United Kingdom) was used to make the dilutions. A laminar flow hood (Telstar, BIO-II-A, Spain) was used to carry out the analysis. Finally, plates were incubated for 24 h at 45 °C for *E. Coli*, 24 h at 37 °C for TCC, 48 h at 37 °C for TVC, and 48 h at 37 °C for *L. monocytogenes*. Results were obtained by triplicated and expressed in cfu/g.

2.3. Statistic Analysis

Data were analysed with the statistical package SPSS 15.0 (Statistical Package for the Social Science for Window (IBM, Armonk, New York, USA). The antioxidant and antimicrobial capacity were analysed using ANOVA. A value of $p < 0.05$ was considered statistically significant. Pearson's correlation was applied out to test differences between groups.

3. Results

3.1. Characterisation of Natural Extracts (HPLC)

The antioxidant and antimicrobial capacities of these extracts depend on the concentration of the phenolic compounds. Extracts obtained from *Rosmarinus officinalis* L. contained 8.10% of rosmarinic acid (RA), and 5.76% of diterpenes (NOS and NOVS), more specifically, carnosic acid and carnosol. P had 41.38% of total punicalagins, as principal bioactive compounds. Otherwise, hydroxytyrosol extracts, obtained from different parts of the olive tree (*Olea europaea*) during the manufacture of olive oil, had different concentrations of the bioactive compound: HYT-F had 11.25% hydroxytyrosol, while HYT-L presented 7.26%. HPLC chromatograms for each extract are presented in Figure 1.

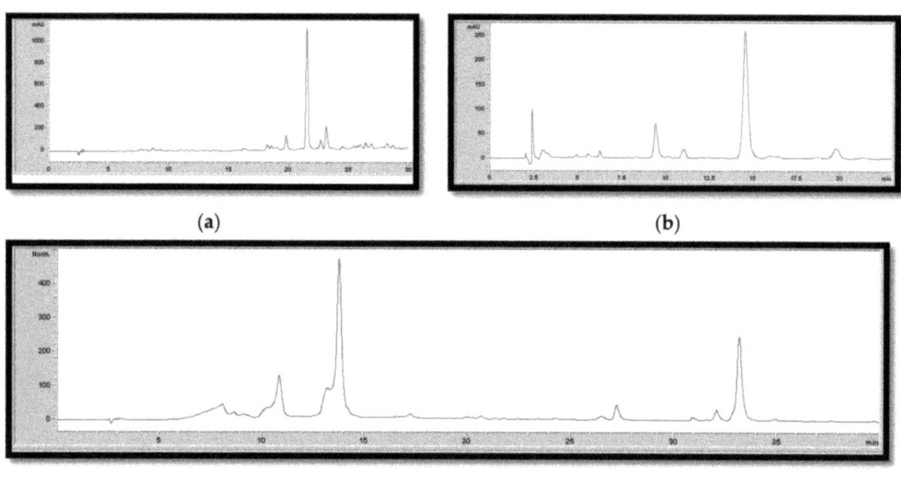

(a)   (b)

(c)

Figure 1. *Cont.*

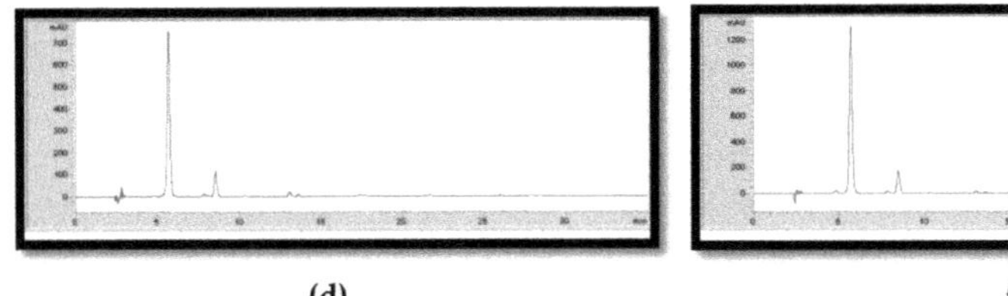

(d)   (e)

Figure 1. HPLC chromatograms for natural extract. (**a**) RA: Rosemary extract rich in Rosmarinic Acid, (**b**) NOS: Rosemary extract rich in diterpenes and NOVS: Rosemary extract rich in diterpenes and with lecitin as emulsifier, (**c**) P: Pomegranate extract, (**d**) HYT-F: Hydroxytyrosol extract obtained from olive fruit, (**e**) HYT-L: Hydroxytyrosol extract obtained from olive leaf.

3.1.1. Total Phenolic Content (TPC)

Determination of the total phenolic content by means of the Folin–Ciocalteau method allows a comparative evaluation of the content of this kind of compounds, considering, at molecular level, the significant structural difference between the various polyphenols present in the extracts being analysed. The results obtained (mg GAE/g) are shown in Table 2.

HYT-F showed the highest quantity of phenolic compounds, with 41.44 mg GAE/g, followed by P, NOS, RA, HYT-L, and NOVS. This last one with 35.95 mg GAE/g, 13.2% less than the first extract. All the extracts showed more than 35 mg GAE/g. In that way, hydroxytyrosol (HYT-F and HYT-L), rosmarinic (RA, NOS and NOVS), and pomegranate extracts had similar total phenolic amounts, between 36 and 41 mg GAE/g.

3.1.2. Antioxidant Activity

The chelating activity percentages obtained using two different methods, are also shown in Table 2. ABTS + radical cation assay and the DPPH free radical scavenging method were used. The capacity of P, HYT-L and RA to scavenge the ABTS + radical was higher than 80%, while the chelating activity of HYT-F, NOVS, and NOS was of 79.62%, 70.61% and 70.17%, respectively. On the other hand, scavenging ability of P was also the highest by measuring the stability of the DPPH radical, 92.55%. The chelating activity of HYT-L, NOVS, NOS, RA, and HYT-F ranged from 89% to 78%. Thus, the extracts with greatest chelating capacity were those obtained from pomegranate, hydroxytyrosol and rosemary, which is related with their high content of phenols, such as, punicalagin, hydroxytyrosol and rosmarinic compounds (carnosic acid, carnosol, and rosmarinic acid), respectively.

Table 2. Total phenolic content (TPC) of natural extracts (mg GAE/g) and their antioxidant activity by measuring their ABTS, and DPPH radical scavenging activity, together to their ORAC$_{HP}$, and FRAP (μM TE/g).

Samples	TPC	ABTS	DPPH	ORAC$_{HP}$	FRAP
	mg GAE/g $\pm$ SD	% Chelating Activity	% Chelating Activity	μM TE/g $\pm$ SD	μM TE/g $\pm$ SD
RA	36.37 $\pm$ 0.007 [b]	81.1 [a]	81.29 [b]	123.2 $\pm$ 0.24 [b]	73.81 $\pm$ 1.52 [a]
NOS	36.49 $\pm$ 0.014 [b]	70.17 [b]	87.98 [ab]	45.18 $\pm$ 0.23 [c]	64.17 $\pm$ 1.47 [b]
NOVS	35.95 $\pm$ 0.009 [b]	70.61 [b]	88.76 [a]	49.16 $\pm$ 0.54 [c]	73.43 $\pm$ 2.01 [a]
P	40.74 $\pm$ 0.023 [a]	83.12 [a]	92.55 [a]	146.39 $\pm$ 0.85 [a]	61.28 $\pm$ 0.98 [b]
HYT-F	41.44 $\pm$ 0.056 [a]	79.62 [ab]	77.96 [b]	140.54 $\pm$ 0.98 [a]	71.16 $\pm$ 1.94 [a]
HYT-L	36.34 $\pm$ 0.021 [b]	82.11 [a]	88.95 [a]	147.46 $\pm$ 1.36 [a]	64.96 $\pm$ 2.88 [b]

GAE: Gallic acid equivalents; SD: Standard deviation; Superscript letters indicate significant differences ($p < 0.05$) between natural extracts. P: Pomegranate extract, RA: Rosemary extract rich in Rosmarinic Acid; NOS: Rosemary extract rich in diterpenes; NOVS: Rosemary extract rich in diterpenes and with lecitin as emulsifier; HYT-L: Hydroxytyrosol extract obtained from olive leaf; HYT-F: Hydroxytyrosol extract obtained from olive fruit.

The hydrophilic antioxidant capacity of the natural extracts obtained by measuring the oxygen radical absorbance is shown in Table 2. However, in this case, the extract with the greatest antioxidant activity was HYT-L (147.46 µM TE/g), followed by P (146.39 µM TE/g), HYT-F (140.54 µM TE/g), RA (123.2 µM TE/g), NOVS (49.16 µM TE/g) and NOS (45.18 µM TE/g). As it can be appreciated, a great significant difference ($p < 0.05$) exists among rosmarinic extracts (NOS and NOVS) rich in diterpenes (carnosic acid and carnosol) and RA rich in rosmarinic acid, together with P, HYT-L and HYT-F. This fact could be explained by lipophilic activity of NOS and NOVS, because of this measurement is carried out in a hydrophilic system. In this way, the rest of extracts are watersoluble, as P, as well as HYT-L, HYT-F and RA present dual affinities, both to polar and non-polar solvents.

The efficiency of the natural plant extracts to reduce Fe^{+++} to Fe^{++} as a antioxidant power measured is also presented in Table 2 expressed in µM Trolox Equivalents (TE)/g. Obtained data showed as that all the extracts have a good and similar ferric reducing antioxidant power, ranging from 73.8 µM TE/g (RA) to 61.3 µM TE/g (P). The order of antioxidant activity using this method was RA > NOVS > HYT-F > HYT-L > NOS > P. All the extracts had high levels of reducing power, which indicated the presence of some compounds that are electron donors and could react with free radicals to convert them into more stable products.

3.1.3. Antimicrobial Activity

Data obtained from measuring the antimicrobial capacity of different extracts are presented in Table 3. The results differed according to the bacterial strain used, *L. monocytogenes* KCTC 3569 CECT 7467 (Gram-positive), *S. Aureus* ATCC 25923 CECT 435 (Gram-positive) or *E. Coli* O157:H7 ATCC 25922 CECT 434 (Gram-negative). All the extracts showed a lower growth inhibition capacity than cloramphenicol (positive control), a broad spectrum antibiotic.

Table 3. Antimicrobial activity of natural extracts measured by the disc difussion method (mm ± SD).

Samples	Concentration (ppm)	Dilution (µL)	L. monocytogenes KCTC 3569 CECT 7467 (Gram-positive)	S. Aureus ATCC 25923 CECT 435 (Gram-positive)	E. Coli O157:H7 ATCC 25922 CECT 434 (Gram-negative)
RA	1000	30	8.0 ± 0.0	7.8 ± 0.7	6.9 ± 1.0
		60	9.6 ± 0.6	10.5 ± 0.5	12.3 ± 0.6
		90	12.3 ± 0.6	16.8 ± 1.8	15.2 ± 0.3
NOS	1000	30	8.0 ± 0.5	7.5 ± 0.0	8.0 ± 0.5
		60	10.8 ± 0.7	9.0 ± 0.5	10.5 ± 1.0
		90	14.0 ± 0.7	13.1 ± 0.5	20.0 ± 1.0
NOVS	1000	30	-	-	-
		60	7.7 ± 1.0	6.7 ± 0.5	7.2 ± 0.7
		90	14.7 ± 1.0	11.9 ± 0.3	18.5 ± 1.0
P	1000	30	10.3 ± 0.6	9.0 ± 0.0	8.0 ± 0.0
		60	12.5 ± 0.5	11.2 ± 0.3	10.2 ± 0.3
		90	15.3 ± 0.6	13.8 ± 0.8	12.6 ± 0.6
HYT-F	1000	30	6.0 ± 0.6	7.8 ± 0.5	-
		60	10.0 ± 0.6	14.5 ± 0.3	-
		90	13.0 ± 1.5	25.2 ± 0.5	11.3 ± 0.8
HYT-L	1000	30	10.0 ± 0.5	9.0 ± 0.5	-
		60	12.5 ± 0.6	18.5 ± 0.6	8.0 ± 0.5
		90	15.0 ± 0.5	28.1 ± 0.5	12.0 ± 0.5
Chloramphenicol	30 µM		34.7 ± 0.6	32.7 ± 0.5	33.7 ± 0.7

Chloramphenicol: positive control; SD: Standard Deviation. P: Pomegranate extract, RA: Rosemary extract rich in Rosmarinic Acid; NOS: Rosemary extract rich in diterpenes; NOVS: Rosemary extract rich in diterpenes and with lecitin as emulsifier; HYT-L: Hydroxytyrosol extract obtained from olive leaf; HYT-F: Hydroxytyrosol extract obtained from olive fruit.

HYT-L, followed by HYT-F had the highest antimicrobial capacity values against *S. Aureus* ATCC 25923 CECT 435 (gram-positive) growth, because of they prestented 28.1 and 25.2 mm of growth inhibition, respectively. These extracts were followed by RA, P, NOS and NOVS. In this case, clearly, the most active compound was the hydroxytyrosol, as obtained from olive leaves as from olive fruits. In second place was RA, which is the most similar in terms of its molecular structure. Both of them provide an opportunity to study the action mechanism of these compounds in future research. However, the inhibitory capacity of the rest of the compounds was lower and does not allow any structure-activity hypothesis to be proposed.

On the other hand, the gram-positive bacterium *L. monocytogenes* KCTC 3569 CECT 7467 is the most resistant strain to phenolic compounds. In this case, P was the most antimicrobial with 15.3 mm growth inhibition, which corresponds with 44.1% of the positive control, chloramphenicol. This was followed by HYT-L, NOVS, NOS, HYT-F and RA. It can be considered, then, that all the studied compounds showed similar inhibitory activities against this microorganism.

Finally, NOS, was the most antimicrobial extract against the gram-negative bacteria *E. Coli* O157:H7 ATCC 25922 CECT 434, with 20 mm of growth inhibition, 40.6% less than chloramphenicol. This phenolic extact was followed by NOVS, RA, P, HYT-L and HYT-F. In this case, of great interest and significance is the fact that fat-soluble compounds such as terpenoids had a higher inhibitory capacity against the growth of gram-negative bacteria. This was followed by the rest of extracts, all of which showed similar values of antimicrobial activity, making it difficult to offer any considerations on their structure-activity.

3.2. Influence of Natural Extracts in Oxidative and Microbiological Damage in Fish Products

3.2.1. Volatile Organic Compounds

Table 4 shows the results obtained from the of GS-MS analysis of the volatile organic compounds: 1-Penten-3-ol, hexanal, 2-nonanone, 1,6-octadien-3-ol, octanal, pentadecane in fish patties. In general, all the volatile compounds analysed increased ($p < 0.05$) from the beginning of storage for all the treatments. These results point to the degradation that is shown in fish patties, that is due to oxidation phenomena, as most straight chain aldehydes are derived from the oxidation of unsaturated fatty acids.

Table 4. Average values and standard deviations of organic compounds (mg/g) in fish patties stored for 11 days, under retail conditions.

	Sample	Day 0 (M ± SD)	Day 11 (M ± SD)
	Control	0.05 ± 0.01	5.14 ± 0.01 [a]
	P	0.06 ± 0.05	3.86 ± 0.05 [b]
	RA	0.04 ± 0.02	3.54 ± 0.02 [b]
1-Penten-3-ol	NOS	0.05 ± 0.04	1.95 ± 0.04 [c]
	NOVS	0.01 ± 0.03	2.01 ± 0.03 [c]
	HYT-F	0.04 ± 0.05	4.24 ± 0.05 [b]
	HYT-L	0.07 ± 0.04	4.27 ± 0.04 [b]
	Control	0.03 ± 0.00	4.25 ± 0.02 [a]
	P	0.09 ± 0.01	2.01 ± 0.03 [b]
	RA	0.04 ± 0.02	2.96 ± 0.04 [b]
Hexanal	NOS	0.07 ± 0.01	1.53 ± 0.02 [b]
	NOVS	0.02 ± 0.05	1.47 ± 0.02 [b]
	HYT-F	0.03 ± 0.01	2.88 ± 0.05 [b]
	HYT-L	0.03 ± 0.01	2.96 ± 0.03 [b]
	Control	0.16 ± 0.01	0.39 ± 0.02
	P	0.51 ± 0.01	0.46 ± 0.04
	RA	0.37 ± 0.01	0.22 ± 0.03
2-nonanone	NOS	0.33 ± 0.01	0.16 ± 0.01
	NOVS	0.29 ± 0.01	0.25 ± 0.03
	HYT-F	0.47 ± 0.01	0.27 ± 0.05
	HYT-L	0.39 ± 0.01	0.32 ± 0.01

Table 4. *Cont.*

	Sample	Day 0 (M ± SD)	Day 11 (M ± SD)
1,6-octadien-3-ol	Control	16.79 ± 0.04 [a]	22.32 ± 0.21 [a]
	P	2.05 ± 0.04 [b]	2.32 ± 0.05 [b]
	RA	2.05 ± 0.03 [b]	2.55 ± 0.03 [b]
	NOS	0.65 ± 0.01 [b]	0.44 ± 0.01 [b]
	NOVS	0.77 ± 0.02 [b]	0.77 ± 0.02 [b]
	HYT-F	0.57 ± 0.02 [b]	0.57 ± 0.02 [b]
	HYT-L	0.87 ± 0.02 [b]	0.87 ± 0.02 [b]
Nonanal	Control	0.16 ± 0.01	1.81 ± 0.02
	P	0.51 ± 0.01	1.30 ± 0.02
	RA	0.37 ± 0.01	1.75 ± 0.02
	NOS	0.33 ± 0.01	0.66 ± 0.02
	NOVS	0.29 ± 0.01	0.66 ± 0.01
	HYT-F	0.47 ± 0.01	1.20 ± 0.03
	HYT-L	0.39 ± 0.01	1.57 ± 0.02
Pentadecane	Control	1.13 ± 0.02	1.60 ± 0.02
	P	1.17 ± 0.03	1.44 ± 0.02
	RA	0.10 ± 0.0	1.04 ± 0.03
	NOS	0.62 ± 0.01	1.08 ± 0.01
	NOVS	0.57 ± 0.01	1.05 ± 0.03
	HYT-F	0.82 ± 0.03	1.54 ± 0.01
	HYT-L	0.75 ± 0.03	1.27 ± 0.01

Results are expressed as mean ± standard deviation in arbitrary area units ($\times 10^6$). P: Pomegranate extract, RA: Rosemary extract rich in Rosmarinic Acid; NOS: Rosemary extract rich in diterpenes; NOVS: Rosemary extract rich in diterpenes and with lecitin as emulsifier; HYT-L: Hydroxytyrosol extract obtained from olive leaf; HYT-F: Hydroxytyrosol extract obtained from olive fruit.

Hexanal and 1,6-octadien-3-ol were the dominant aldehyde in the fish patties meat in all the groups. Hexanal values ranged from 0.1 mg/kg, in day 0, to 5.14 mg/kg after 11 days, in control samples. These values are in the same line than those reported by Brunton et al. [28], who found hexanal values of 4.01 l/g in cooked turkey stored for 6 days at 4 °C.

Differences in the mean hexanal levels between C and patties with natural extracts were significant ($p < 0.05$) on day 11. On day 11, NOVS showed lower (39%) hexanal values than C, meaning that Rosemary extracts improved lipid stability of the fish patties. In this sense, Shahidi, Yun, and Rubin [29] reported that such an increase in hexanal is a good indicator of lipid oxidation. Indeed, these authors suggested hexanal as a valid indicator of oxidative stability and flavour acceptability in cooked ground meat.

The behaviour of nonanal and 1-penten-3-ol was similar to hexanal, both increasing ($p < 0.05$) during storage and showing significant differences between C and natural extracts samples on day 11. Nonanal is a waxy flavor and descriptors, while 1-penten-3-ol is amongst the compounds responsible for the rancid odour in mayonnaise. Note the absence of significant differences in 2-nonanone and pentadecane on day 11.

3.2.2. Microbiological Content

Antimicrobial capacity of different extract was also proved in fish products elaborated from thawed hake. Consequently, microbiological results of fish patties at days 0, 4, 7 and 11 from elaboration, are shown in Table 5.

Table 5. Microbiological results (cfu/g) of fish patties analysis at days 0, 4, 7 and 11 under refrigerated storage.

Analysis	Sample	Storage day			
		0	**4**	**7**	**11**
TVC	Control	1.98×10^3	6.20×10^3	2.77×10^4	5.55×10^7
	P	1.52×10^3	5.12×10^3	1.28×10^4	5.35×10^7
	RA	9.10×10^2	4.25×10^3	2.01×10^4	7.29×10^7
	NOS	7.25×10^2	3.62×10^3	1.10×10^4	6.50×10^7
	NOVS	1.01×10^3	4.05×10^3	1.56×10^4	6.90×10^7
	HYT-L	1.88×10^3	6.22×10^3	1.79×10^4	3.48×10^7
	HYT-F	2.01×10^3	5.98×10^3	2.10×10^4	3.15×10^7
TCC	Control	<10	8.40×10^2	4.75×10^3	5.45×10^4
	P	<10	1.40×10^3	7.80×10^3	6.75×10^3
	RA	<10	1.95×10^3	6.10×10^3	3.78×10^4
	NOS	<10	1.32×10^3	5.30×10^3	2.63×10^4
	NOVS	<10	1.69×10^3	6.00×10^3	3.12×10^4
	HYT-L	<10	4.4×10^2	3.30×10^3	7.15×10^4
	HYT-F	<10	5.9×10^2	3.90×10^3	7.81×10^4
E. Coli	Control	<10	<10	<10	<10
	P	10	<10	<10	<10
	RA	<10	<10	<10	<10
	NOS	<10	<10	<10	<10
	NOVS	<10	<10	<10	<10
	HYT-L	<10	<10	20	<10
	HYT-F	<10	<10	<10	<10
L. monocytogenes	Control	Absence in 25 g	Absence in 25 g	Absence in 25 g	Absence in 25 g
	P	Absence in 25 g	Absence in 25 g	Absence in 25 g	Absence in 25 g
	RA	Absence in 25 g	Absence in 25 g	Absence in 25 g	Absence in 25 g
	NOS	Absence in 25 g	Absence in 25 g	Absence in 25 g	Absence in 25 g
	NOVS	Absence in 25 g	Absence in 25 g	Absence in 25 g	Absence in 25 g
	HYT-L	Absence in 25 g	Absence in 25 g	Absence in 25 g	Absence in 25 g
	HYT-F	Absence in 25 g	Absence in 25 g	Absence in 25 g	Absence in 25 g

TVC: Total Viable Count; TCC: Total Coliform Count P: Pomegranate extract, RA: Rosemary extract rich in Rosmarinic Acid; NOS: Rosemary extract rich in diterpenes; NOVS: Rosemary extract rich in diterpenes and with lecitin as emulsifier; HYT-L: Hydroxytyrosol extract obtained from olive leaf; HYT-F: Hydroxytyrosol extract obtained from olive fruit.

As it can be appreciated, TVC results after 11 days of refrigerated storage were lower in samples that incorporated HYT-F and HYT-L in their formula, followed by P, control sample, NOVS, NOS and RA. While, TCC results showed as both pomegranate and rosemary extracs, diterpens rich extracts (NOS and NOVS) and rosmarinic acid rich extract (RA) obtained the lowest results in comparison with the control sample, or samples that incorporated HYT to their formulas. These last results can be related with antimicrobial activity of all the extracts against *E. Coli* O157:H7ATCC 25922 CECT 434. On the other hand, results obtained from analysis of *E. Coli* and *L. monocytogenes* did not present significant differences ($p < 0.05$) among incorporation of natural extracts. In addition, it must be taken into account that only samples enriched with HYT, both from fruit or leaf, did not exceed the limit stablished by European legislation regarding to TVC in fish products (5,000,000 cfu/g). In parallell, only P and rosemary extracts (RA, NOS, and NOVS) kept fish product samples below legal microbiological safety limits of TCC (50,000 cfu/g). However, all the natural extracts, including the control sample prevented against microbiological growth of *E. Coli* and *L. monocytogenes*.

4. Discussion

Firstly, the obtained results of total phenolic content agrees with previous findings by other authors using the Folin-Ciocalteau method or by HPLC [30–32]. Results obtained in the present spectrophotometric determination were not strictly correlated with the data obtained by HPLC analysis, which is due to the different response factors of each of the polyphenol structures present in the extracts (punicalagins, rosmarinic acid, carnosic acid, carnosol and hydroxytyrosol) regarding the pattern used, as the gallic acid in this case. It is difficult to make a structural interpretation of the results obtained for the antioxidant capacity measurements using the studied methods, although, clearly, some factors are related with the molecular structure of the active substances: the presence of phenols, their conjugation and polymerisation, cathecol and/or gallate groups presence, etc. In both methods, P shows the best results, probably, due to the presence of some conjugated polyphenol structures and a significant amount of gallic acid groups (tri-hydroxy phenol structures). Regarding the olive extracts, HYT-L, with its lower level of hydroxytyrosol than HYT-F, as principal active compound, showed a higher antioxidant capacity in both models, making it one of the most powerful extracts. This fact could originate from the presence of flavonoid compounds in combination with the hydroxytyrosol, providing a synergistic effect in terms of antioxidant activity. RA is the most structurally similar extract to olive extracts, due to the presence of rosmarinic acid as an active compound. This substance could be termed "double-hydroxytyrosol," only for their structural similarity, perhaps for this reason both extracts showed proximate chelating activity values. The difference among rosemary extracts was significant. The water-soluble extract (RA) was more active in the ABTS method, while liposoluble extracts (NOS, and NOVS) showed a higher activity in the DPPH model. This behaviour could be explained by the different chemical structure of the radical used in each technique and by the different properties of the molecular structures of phenylpropanoids (rosmarinic acid) and diterpens (carnosic acid and carnosol). However, both structures have a cathecol group and a carborxylic acid group. These results can be compared with previous research. For example, Hmid et al. [33] and Elfalleh et al. [34] obtained similar values for pomegranate extracts using the same methods, as well as hydroxytyrosol [35] and rosemary extracts [36,37].

Hydrophilic ORAC is one of the most widely used methods for evaluating antioxidant capacity, but it is clear that the results may be conditioned not only by the antioxidant capacity of each compound, but also by the physical and chemical properties, particularly its water solubility. Pomegranate and olive extracts obtained similar values for their antioxidant activity unrelated to their origin (leaves, fruit or vegetation water). In this case, the different of cathecol and gallate groups did not seem to be significant. Despite this, HYT-L again showed a higher activity. The antioxidant capacity of RA was lower than that of the above (−15%), although it followed the same order. It seems obvious that the structural similarity goes on establishing a parallellism in the antioxidant activity, also in this model. If not, the lower ORAC activity of the fat-soluble rosemary extracts (diterpens) compared with the hydrosoluble extracts that were already described. Previous researchers, such as Azaizeh et al. [38], obtained similar results analysing hydroxytyrosol in olive (*Olea europaea*) vegetation waters, while Sueishi et al. [39] obtained results that were 50% higher when measuring the seasonal variations of oxygen radical scavenging ability in rosemary leaf extract using the same method. In research carried out by Durante et al. [40], the authors measured the antioxidant activity of diferent extracts from tomato, grape and pomegranate seeds, obtaining similar results as the last. In the same way, previous research obtained similar results to that obtained results by the FRAP method regarding to rosemary [36], pomegranate [33] and hydroxytyrosol [35,41].

Regarding antimicrobial activity, the terpenoid structure did not provide good results, although this does not mean that this compound has a lower antioxidant capacity. While not significant, it is interesting to point out that NOVS, which contains lecitin as an emulsionant, shows higher antioxidant activity than NOS, which does not contain this excipient. This method has been used in much research to test the antimicrobial capacity of many drugs and natural extracts. For example, Laincer et al. [39] measured the antimicrobial activity of several olive phenolics, including HYT, against *E. Coli* and *S. Aureus*, obtaining similar results. Weckesser et al. [42] analysed the antimicrobial activity of plant extracts, such as *Rosmarinus officinalis* L. against bacteria of dermatological relevance, among them *E. Coli* and *S. Aureus*, obtaining similar results using the diffusion disk method. Regarding the P extract, Kharchoufi et al. [7] obtained similar results for pomegranate peel extracts against *Pseudomonas putida*,

Penicillium digitatum and *Saccharomyces cerevisae,* but not against any strains used in the present study. Finally, applying rosemary extracts, Santomauro et al. [43] obtained similar results (more than 10 mm of inhibition) in different strains.

Regarding the oxidative and antimicrobial damage of fish products under refrigerated storage for 11 days, all the natural extracts showed an antioxidative effect against formation of volatile compounds related to lipid oxidation.

Table 4 shows that all the volatile compounds analysed are the main components that contribute the most to the emergence of unpleasant notes of flavour, due to the low flavour threshold [44]. In general, the presence of natural extract (especially NOVS) in the fish patties delayed the formation of all the volatile lipid-derived compounds. In the same line, Nieto et al. [45,46] reported lower hexanal values, rancid odour and rancid flavour scores in lamb meat from ewes fed thyme leaves or rosemary by-products, respectively.

As it can be appreciated in Table 5, natural extracts also acted as antimicrobial agents against TVC and TCC proliferation. This behaviour has been previously observed by other researchers using other natural extracts. For example, Del Nobile et al. [47] studied the combined effect of different gas mix compositions (MAP 30:40:30 O_2:CO_2:N_2, 50:50 O_2:CO_2, and 5:95 O_2:CO_2) and three essential oils (thymol, lemon and grapefruit seed extracts) on fresh blue fish burgers. Results obtained showed as the combination of 110 ppm of thymol, 100 ppm of grapefruit seed extract, or 120 ppm of lemon extract with MAP 5:95 O_2:CO_2 was able to maintain the microbial quality of fish burgers for 28 days under refrigerated storage. In the same way, the combined effect of antimicrobial mixtures of chitosan, nisin and sodium lactate with MAP 55:45 CO_2:N_2 was able to guarantee the microbial acceptance of hake burgers for 30 days of refrigerated storage [48]. On the other hand, Smaldone et al. [49] have observed that only MAP 5:60:35 O_2:CO_2:N_2 application can extend the microbiological shelf-life of hake burgers for 15 days after elaboration. However, in the present study, modified atmosphere package treatment was not assessed, neither in previous research on fish products using natural extracts from pomegranate, rosemary or olive tree (*Olea europaea*). Likewise, with obtained results it can be concluded that bioactive compounds from studied extracts (P, RA, NOS, NOVS, HYT-L and HYT-F) act as antimicrobial agents, which has also demonstrated in vitro and it is due to their high concentration of phenolic compounds (punicalagin, carnosic acid, carnosol, rosmarinic acid and hydroxytyrosol) with known antimicrobial activity, as it has been exposed in the introduction of this work. For this reason, it is not surprising that their application avoided *L. monocytogenes* or *E. Coli* growth. Nevertheless, is important to know that samples that incorporated rosemary extracts presented TVC growth higher than the control sample at day 11, similar to hydroxytyrosol extracts that showed higher TCC growth than the control. This fact can be explained by the great amount of antioxidant compounds in combination with spices and spice extracts that the commercial mix contained, and which can produce a synergism between them, increasing the shelf-life of fish products.

5. Conclusions

It is necessary to use several methods to know the antioxidant capacity of the studied extracts since they are rich in phenolic compounds with different molecular structures and react in different ways when reducing Fe^{+++} or acting against hydroxyl, ABTS and DPPH radicals. Consequently, natural extracts obtained from olive (*Olea europaea*), pomegranate (*Punica granatum*) and rosemary (*Rosmarinus officinalis* L.) are excellent antioxidant agents. The fish patties made with natural extracts showed lower volatile compounds related with lipid oxidation throughout the 11 days of storage under retail display conditions. On the other hand, the most antimicrobial compounds were P against *L. monocytogenes*, HYT against *S. Aureus*, and NOS against *E. Coli*. After applying these extracts in fish products, all of them acted as preservatives, extending the shelf-life for 11 days with a lower microbiological growth in regard to the control sample. Then, it can be concluded that all the extracts are good antioxidant and antimicrobial agents and could be applied in the food industry to extend shelf-life of Clean label fish products.

Author Contributions: Conceptualization, L.M. and G.N.; methodology, L.M. and G.N.; software, L.M.; validation, L.M., J.C., G.R. and G.N.; formal analysis, L.M.; investigation, L.M. and G.N.; resources, J.C., G.R. and G.N.; data curation, L.M. and G.N.; writing—original draft preparation, L.M., J.C. and G.N.; writing—review and editing, G.N.; visualization, L.M. and G.N.; supervision, G.N.; project administration, G.R. and G.N.; funding acquisition, G.R. and G.N.

Funding: This research received no external funding.

Conflicts of Interest: The authors declare no conflict of interest.

References

1. Sánchez-Camargo, A.P.; Herrero, M. Rosemary (Rosmarinus officinalis) as a functional ingredient: Recent scientific evidence. *Curr. Opin. Food Sci.* **2017**, *14*, 13–19. [CrossRef]
2. Alu'datt, M.H.; Rababah, T.; Alhamad, M.N.; Gammoh, S.; Al-Mahasneh, M.A.; Tranchant, C.C.; Rawshdeh, M. Chapter 15—Pharmaceutical Nutraceutical and Therapeutic properties of selected wild medicinal plants: Thyme, spearmint and Rosemary. In *Therapeutic, Probiotic and Unventional Food*; Grumezescu, A.M., Holban, A.M., Eds.; Academic Press: Cambridge, MA, USA, 2018; pp. 275–290.
3. Andrade, M.A.; Ribeiro-Santos, R.; Costa-Bonito, M.C.; Saraiva, M.; Sanches-Silva, A. Characterization of rosemary and thyme extracts for incorporation into a whey protein based film. *LWT-Food Sci. Technol.* **2018**, *92*, 497–508. [CrossRef]
4. Karimi, M.; Sadegui, R.; Kokini, J. Pomegranate as a promising opportunity in medicine and nanotechnology. *Trends Food Sci. Technol.* **2017**, *69*, 59–73. [CrossRef]
5. Khwairakpam, A.D.; Bordoloi, D.; Thakur, K.K.; Monisha, J.; Arfuso, F.; Sethi, G.; Mishra, S.; Kumar, A.P.; Kunnumakkara, A.B. Possible use of Punica gratum (Pomegranate) in cáncer therapy. *Pharmacol. Res.* **2018**, *133*, 53–64. [CrossRef]
6. Derakhshan, Z.; Ferrante, M.; Tadi, M.; Ansari, F.; Heydari, A.; Hosseini, M.S.; Conti, G.O.; Sabradad, E.K. Antioxidant activity and total phenolic content of ethanolic extract of pomegranate peels, juice and seeds. *Food Chem. Toxicol.* **2018**, *114*, 108–111. [CrossRef] [PubMed]
7. Kharchoufi, S.; Licciardello, F.; Siracusa, L.; Muratore, G.; Hamdi, M.; Restuccia, C. Antimicrobial and antioxidant features of 'Gabsi' pomegranate peel extracts. *Ind. Crops Prod.* **2018**, *111*, 345–352. [CrossRef]
8. Martínez, L.; Ros, G.; Nieto, G. Hydroxytyrosol: Health benefits and use as functional ingredient in meat. *Medicines* **2018**, *5*, 13. [CrossRef]
9. Bernini, R.; Gilardini-Montani, M.S.; Merendino, N.; Romani, A.; Velotti, F. Hydroxytyrosol-derived compounds: A basis for the creation of new pharmacological agents for cancer prevention and therapy. *J. Med. Chem.* **2015**, *58*, 9089–9107. [CrossRef]
10. Bernini, R.; Carastro, I.; Palmini, G.; Tanini, A.; Zonefrati, R.; Pinelli, P.; Brandi, M.L.; Romani, A. Lipophilization of Hydroxytyrosol-enriched fractions from *Olea europaea* L. byproducts and evaluation of the in vitro effects on a model of colorectal cancer cells. *J. Agric. Food Chem.* **2017**, *65*, 6506–6512. [CrossRef]
11. Albat, V. Estudio de Nuevas Formulaciones Para la Obtención de Patés a Base de Pescado y Algas. Final Degree Proyect, Universidad Politécnica de Valencia, Valencia, Spain, 2015.
12. Martín-Cerdeño, V.J. Consumo de pescados y mariscos en España. Un análisis de los perfiles de la demanda. *Distrib. Consumo* **2017**, *27*, 5–18.
13. Ghaly, A.E.; Dave, D.; Budge, S.; Brooks, M.S. Fish spoilage mechanisms and preservation techniques: Review. *Am. J. Appl. Sci.* **2010**, *7*, 859–877. [CrossRef]
14. Özalp Özen, B.; Eren, M.; Pala, A.; Özmen, İ.; Soyer, A. Effect of plant extracts on lipid oxidation during frozen storage of minced fish muscle. *Int. J. Food Sci. Technol.* **2011**, *46*, 724–731. [CrossRef]
15. Peiretti, P.G.; Gai, F.; Ortoffi, M.; Aigotti, R.; Medana, C. Effects of rosemary oil (Rosmarinus officinalis) on the shelf-life of minced rainbow trout (Oncorhynchus mykiss) during refrigerated storage. *Foods* **2012**, *1*, 28–39. [CrossRef]
16. Pazos, M.; Alonso, A.; Sánchez, I.; Medina, I. Hydroxytyrosol prevents oxidative deterioration in foodstuffs rich in fish lipids. *J. Agric. Food Chem.* **2008**, *56*, 3334–3340. [CrossRef]

17. Luzi, F.; Fortunati, E.; Di Michele, A.; Pannucci, E.; Botticella, E.; Santi, L.; Kenny, J.M.; Torre, L.; Bernini, R. Nanostructured starch combined with Hydroxytyrosol in poly(vinyl alcohol) based ternary films as active packaging system. *Carbohydr. Polym.* **2018**, *193*, 239–248. [CrossRef]

18. Benavente, O.; Castillo, O.; Lorente, J.; Ortuño, A.; Del Rio, J.A. Antioxidant activity of phenolic extracted from Olea europaea L. leaves. *Food Chem.* **2000**, *68*, 457–462. [CrossRef]

19. Singleton, V.L.; Rossi, J.A., Jr. Colorimetry of total phenolics with phosphomolybdic-phosphotungstic acid reagents. *Am. J. Enol. Vitic.* **1965**, *16*, 144–158.

20. Zheng, W.; Wang, S.Y. Antioxidant activity and phenolic compounds in selected herbs. *J. Agric. Food Chem.* **2001**, *49*, 5165–5170. [CrossRef]

21. Brand-Williams, W.; Cuvelier, M.E.; Berset, C. Use of a free radical method to evaluate antioxidant avtivity. *LWT Food Sci. Technol.* **1995**, *28*, 25–30. [CrossRef]

22. Sánchez-Moreno, C.; Larrauri, J.A.; Saura-Calixto, F. A procedure to measure the antiradical efficiency of polyphenols. *J. Sci. Food Agric.* **1998**, *76*, 270–276. [CrossRef]

23. Barontini, M.; Bernini, R.; Carastro, I.; Gentili, P.; Romani, A. Synthesis and DPPH radical scavenging activity of novel compounds obtained from tyrosol and cinnamic acid derivatives. *New J. Chem.* **2014**, *38*, 809–816. [CrossRef]

24. Re, R.; Pellegrini, N.; Proteggente, A.; Pannala, A.; Yang, M.; Rice-Evans, S. Antioxidant activity applying and omproved ABTS radical catión decolorization assay. *Free Radic. Biol. Med.* **1999**, *26*, 1231–1237. [CrossRef]

25. Prior, R.L.; Hoang, H.; Gu, L.; Wu, X.; Bacchiocca, M.; Howard, L.; Hampsch-Woodil, M.; Huang, D.; Ou, B.; Jacob, R. Assays for hydrophilic and lipophilic antioxidant capacity (oxygen radical absorbance capacity (ORACFL)) of plasma and other biological and food samples. *J. Agric. Food Chem.* **2003**, *51*, 3273–3279. [CrossRef] [PubMed]

26. Benzie, I.F.F.; Strain, J.J. The Ferric Reducing Ability of Plasma (FRAP) as a Measure of "Antioxidant Power": The FRAP Assay. *Anal. Biochem.* **1996**, *239*, 70–76. [CrossRef]

27. Brunton, N.P.; Cronin, D.A.; Monahan, F.J.; Durcan, R. A comparison of solid-phase microextraction (SPME) fibres for measurement of hexanal and pentanal in cooked turkey. *Food Chem.* **2000**, *68*, 339–345. [CrossRef]

28. Shahidi, F.; Yun, J.; Rubin, L.J. The hexanal content as an indicator of oxidative stability and flavour acceptability in cooked ground meat. *Can. Inst. Food Sci. Technol. J.* **1987**, *20*, 104–106. [CrossRef]

29. Chanwitheesuk, A.; Teerawutgulrag, A.; Kilburn, J.D.; Rakariyatham, N. Antimicrobial gallic acid from Caesalpinia mimosoides Lamk. *Food Chem.* **2007**, *100*, 1044–1048. [CrossRef]

30. Fuentes, E.; Paucar, F.; Tapia, F.; Ortiz, J.; Jimenez, P.; Romero, N. Effect of the composition of extra virgin olive oils on the differentiation and antioxidant capacities of twelve monovarietals. *Food Chem.* **2018**, *243*, 285–294. [CrossRef] [PubMed]

31. Balasundram, N.; Sundram, K.; Samman, S. Phenolic compounds in plants and agri-industrial by-products: Antioxidant activity, occurrence and potential uses. *Food Chem.* **2006**, *99*, 191–203. [CrossRef]

32. Presti, G.; Guarrasi, V.; Gulotta, E.; Provenzano, F.; Provenzano, A.; Giulano, S.; Monfreda, M.; Mangione, M.R.; Passantino, R.; San Biagio, P.L.; et al. Bioactive compounds from extra virgin olive oils: Correlation between phenolic content and oxidative stress cell protection. *Biophys. Chem.* **2017**, *230*, 109–116. [CrossRef]

33. Hmid, I.; Elothmani, D.; Hanine, H.; Oukabli, A.; Mehinagic, E. Comparative study of phenolic compounds and their antioxidant attributes of eighteen pomegranate (Punica granatum L.) cultivars grown in Morocco. *Arab. J. Chem.* **2017**, *10*, S2675–S2684. [CrossRef]

34. Elfalleh, W.; Nasri, N.; MArzougui, N.; Thabti, I.; M'Rabet, A.; Yahya, Y.; Lachiheb, B.; Guasmi, F.; Ferchichi, A. Physico-chemical properties and DPPH-ABTS scavenging activity of some local pomegranate (Punica granatum) ecotypes. *Int. J. Food Sci. Nutr.* **2009**, *60*, 197–210. [CrossRef]

35. Kouka, P.; Priftis, A.; Stagos, D.; Angelis, A.; Stathopoulos, P.; Xinos, N.; Skaltsounis, A.L.; Mamoulakis, C.; Tsatsakis, A.M.; Spandidos, D.A.; et al. Assessment of the antioxidant activity of an olive oil total polyphenolic fraction and hydroxytyrosol from a Greek Olea europea variety in endothelial cells and myoblasts. *Int. J. Mol. Med.* **2009**, *70*, 703–712.

36. Pereira, D.; Pinheiro, R.S.; Heldt, L.F.S.; Mour, C.; Bianchin, M.; Almedia, J.F.; Reis, A.S.; Ribeiro, I.S.; Haminiuk, C.W.I.; Carpes, S.T. Rosemary as natural antioxidant to prevent oxidation in chicken burgers. *Food Sci. Technol. Camp.* **2017**, *37*, 17–23. [CrossRef]

37. Erkan, N.; Ayranci, G.; Ayranci, E. Antioxidant activities of rosemary (Rosmarinus Officinalis L.) extract, blackseed (Nigella sativa L.) essential oil, carnosic acid, rosmarinic acid and sesamol. *Food Chem.* **2008**, *110*, 76–82. [CrossRef] [PubMed]

38. Azaizeh, H.; Halahlih, F.; Najami, N.; Brunner, D.; Faulstich, M.; Tafesh, A. Antioxidant activity of phenolic fractions in olive mill wastewater. *Food Chem.* **2012**, *134*, 2226–2234. [CrossRef]

39. Sueishi, Y.; Sue, M.; Masamoto, H. Seasonal variations of oxygen radical scavenging ability in rosemary leaf extract. *Food Chem.* **2018**, *245*, 270–274. [CrossRef]

40. Durante, M.; Montefusco, A.; Marrese, P.P.; Soccio, M.; Pastore, D.; Piro, G.; Mita, G.; Lenucci, M.S. Seeds of pomegranate, tomato and grapes: An underestimated source of natural bioactive molecules and antioxidants from agri-food by-products. *J. Food Compos. Anal.* **2017**, *63*, 65–72. [CrossRef]

41. Laincer, A.; Laribi, R.; Tamendjari, A.; Arrar, L.; Rovellini, P.; Venturini, S. Olive oils from Algeria: Phenolic compounds, antioxidant and antibacterial activities. *Grasas y Aceites* **2014**, *65*, 1.

42. Weckesser, S.; Engel, K.; Simon-Haarhaus, B.; Wittmer, A.; Pelz, K.; Schempp, C.M. Screening of plant extract for antimicrobial activity against bacteria and yeast with dermatological relevance. *Phytomedicine* **2007**, *14*, 508–516. [CrossRef]

43. Santomauro, F.; Sacco, C.; Donato, R.; Bellumori, M.; Innocenti, M.; Mulinacci, N. The antimicrobial effects of three phenolic extracts from Rosmarinus officinalis L.; Vitis vinifera L. and Polygonum cuspidatum L. on food pathogens. *Nat. Prod. Res.* **2017**. [CrossRef]

44. Kerler, J.; Grosch, W. Character impact odorants of boiled chicken: Changes during refrigerated storage and reheating. *Z. Lebensm. Unters. und-Forsch.* **1997**, *205*, 232–238. [CrossRef]

45. Nieto, G.; Estrada, M.; Jordán, M.J.; Garrido, M.D.; Bañón, S. Effects in ewe diet of rosemary by-product on lipid oxidation and the eating quality of cooked lamb under retail display conditions. *Food Chem.* **2011**, *124*, 1423–1429. [CrossRef]

46. Nieto, G.; Bañon, S.; Garrido, M.D. Effectofsupplementingewes'diet with thyme (Thymus zygis ssp. Gracilis) leaves on the lipid oxidation of cooked lamb meat. *Food Chem.* **2011**, *125*, 1147–1152. [CrossRef]

47. Del Nobile, M.A.; Corbo, M.R.; Speranza, B.; Sinigaglia, M.; Conte, A.; Caroprese, M. Combined effect of MAP and active compounds on fresh blue fish burger. *Int. J. Food Microbiol.* **2009**, *135*, 281–287. [CrossRef]

48. Schelegueda, L.I.; Delcarlo, S.B.; Gliemmon, M.F.; Campos, C.A. Effect of antimicrobial mixtures and modified atmosphere packaging on the quality of Argentine hake (*Merluccius hubbsi*) burgers. *LWT-Food Sci. Technol.* **2016**, *68*, 258–264. [CrossRef]

49. Smaldone, G.; Marrone, R.; Zottola, T.; Vollano, L.; Grossi, G.; Cortesi, M.L. Formulation and shelf-life of fish burgers served to preschool children. *Italian J. Food Saf.* **2017**, *6*, 6373. [CrossRef] [PubMed]

© 2019 by the authors. Licensee MDPI, Basel, Switzerland. This article is an open access article distributed under the terms and conditions of the Creative Commons Attribution (CC BY) license (http://creativecommons.org/licenses/by/4.0/).

 *antioxidants* 

Article

A Polyphenolic Extract from Olive Mill Wastewaters Encapsulated in Whey Protein and Maltodextrin Exerts Antioxidant Activity in Endothelial Cells

Konstantina Kreatsouli [1], Zinovia Fousteri [1], Konstantinos Zampakas [1], Efthalia Kerasioti [1], Aristidis S. Veskoukis [1], Christos Mantas [2], Paschalis Gkoutsidis [2], Dimitrios Ladas [1], Konstantinos Petrotos [2], Demetrios Kouretas [1] and Dimitrios Stagos [1,*]

[1] Department of Biochemistry and Biotechnology, University of Thessaly, Viopolis, 41500 Larissa, Greece
[2] Department of Biosystem Engineering, Technical Education Institute of Thessaly, 41110 Larissa, Greece
* Correspondence: stagkos@med.uth.gr; Tel.: +30-2410-565229

Received: 15 June 2019; Accepted: 2 August 2019; Published: 5 August 2019

Abstract: The aim of the present study was to compare maltodextrin and whey protein as encapsulation carriers for olive mill wastewater (OMWW) phenolic extract for producing antioxidant powder, by using spray drying under 17 different conditions. In some samples, gelatin was also added in the encapsulation mixture. The antioxidant activity was assessed in vitro by using the DPPH$^\bullet$, ABTS$^{\bullet+}$, reducing power and DNA plasmid strand breakage assays. The results showed that both materials were equally effective for producing antioxidant powder, although by using different conditions. For example, inlet/outlet temperature of the spray drying did not seem to affect the maltodextrin samples' antioxidant activity, but whey protein samples showed better antioxidant activity at lower temperatures. Gelatin use decreased antioxidant activity, especially in whey protein samples. The two most potent samples, one encapsulated in maltodextrin and the other in whey protein, were examined for their antioxidant effects in human endothelial cells by assessing glutathione (GSH) and reactive oxygen species (ROS) levels. Both samples significantly enhanced the antioxidant molecule of GSH, while maltodextrin sample also decreased ROS. The present findings suggested both materials for encapsulation of OMWW extract for producing antioxidant powder which may be used in food products, especially for the protection from ROS-induced endothelium pathologies.

Keywords: olive mill wastewater; encapsulation; maltodextrin; whey protein; gelatin; spray drying; antioxidants; glutathione; endothelial cells

---

1. Introduction

Olive mill wastewaters (OMWWs) are byproducts of the olive oil production process, causing significant problems such as soil contamination and eutrophication, when they are discarded in the environment. Although OMWWs have been extensively considered as a byproduct, several studies have shown that they are rich in polyphenolic compounds (e.g., oleuropein, tyrosol, hydroxytyrosol, coumaric acid, caffeic acid, vanillic acid, ferulic acid, kaempferol, and quercetin) with important biological activities [1]. For example, phenolic compounds are widely known for their antioxidant properties [2]. Our research group has previously demonstrated that administration of feed supplemented with polyphenols from OMWW improves the redox status in farm animals [3–5].

Polyphenolic extracts from OMWW may be used as food supplements or preservatives, but their unpleasant bitter taste is a significant problem. Furthermore, the polyphenolic compounds can be chemically unstable under the influence of temperature, light, and oxygen [6]. Therefore, the encapsulation of polyphenolic extracts is an effective way to preserve their stability and bioactivity, while overcoming the problem of bad taste and simultaneously enhancing their bioavailability [7].

Spray drying encapsulation is a common and popular technique in terms of production cost, short process time, and low thermal stresses [8].

The entire vascular system consists of a monolayer of endothelial cells, the integrity of which is necessary to maintain a harmonic circulatory function. Apart from that, the endothelium regulates the homeostasis as well as the immune and inflammatory responses of the body [9,10]. Oxidative stress can severely damage the endothelium; thus, it constitutes one of the most important factors of pathological conditions of the vessels, along with atherosclerosis and thrombosis [10].

Thus, the aim of the present study was initially to evaluate the antioxidant potency of a polyphenolic extract from OMWW, encapsulated in three different carrier agents (i.e., maltodextrin, whey protein and gelatin) under different conditions, thus producing 17 samples. Then, the two samples exhibiting the greatest antioxidant activity (i.e., 3 and 9) were chosen in order to assess their antioxidant effects on the redox status of human endothelial cells.

2. Materials and Methods

2.1. Preparation of Encapsulation Powders

The liquid raw material that was used for the production of the OMWW originating antioxidant powders was supplied by Polyhealth S.A. (Larissa, Greece). This liquid raw material was a standard product of Polyhealth S.A., commercialized under the trademark 'MEDOLIVA® Liquid'. The polyphenolic composition of MEDOLIVA® liquid as assessed by HPLC analysis has been previously reported [11]. Thus, a mixed liquid material was prepared by mixing MEDOLIVA® liquid (an aqueous solution) with Twin 80 (liquid surfactant) and with either plain or mixed encapsulation agents (i.e., whey protein C80, maize maltodextrin DE18, and gelatin) according to the recipes given in detail in Table 1. Consequently, the mixture was thoroughly homogenized by ultrasonic energy. Finally, the homogenized aqueous solution was spray dried for water removal and production of 17 dry encapsulated OMWW polyphenol powders. A BUCHI Spray Dryer model B-290 connected with a B296 Dehumidifier was used to carry out the spray drying. The operating conditions used for each recipe are presented in Table 1. The conditions and the ingredient ratios given in Table 1 were selected after more than 100 trials (data not shown) and only 17 were found to be acceptable on the basis of giving good quality and stable powder at a yield of more than 75% of the initial aqueous solution dry matter. The initial temperature range tested was from 100 to 220 °C, while the gelatin percentage tested was from 5 to 30 *w/w*% of the maltodextrin or whey protein. In many cases, the tested conditions led to a failure to obtain powder and the finished product was slurry type or very sticky due to unfavorable glass transition conditions.

Table 1. The tested powders and their encapsulation conditions.

Sample	Inlet Temperature (°C)	Outlet Temperature (°C)	Aspirator (%)	Pump (%)	MEDOLIVA (mL)	Tween80 (mL)	Whey Protein (gr)	Maltodextrin (gr)	Gelatin (gr)	H$_2$O (mL)	Total Volume (mL)
1	100	66	100	5	37.5	12.5	25	-	-	175	250
2	100	72	100	5	37.5	12.5	25	-	-	175	250
3	120	82	100	5	37.5	12.5	25	-	-	175	250
4	140	90	100	5	37.5	12.5	25	-	-	175	250
5	160	98	100	5	37.5	12.5	25	-	-	175	250
6	120	71	100	5	37.5	-	25	-	5	175	242.5
7	140	79	100	5	37.5	-	25	-	5	175	242.5
8	160	96	100	5	37.5	-	25	-	5	175	242.5
9	100	57	100	10	37.5	12.5	-	25	-	175	250
10	100	67	100	5	37.5	12.5	-	25	-	175	250
11	120	67	100	10	37.5	12.5	-	25	-	175	250
12	120	76	100	5	37.5	12.5	-	25	-	175	250
13	140	84	100	5	37.5	12.5	-	25	-	175	250
14	160	86	100	10	37.5	12.5	-	25	-	175	250
15	160	96	100	5	37.5	12.5	-	25	-	175	250
16	100	61	100	5	37.5	-	-	25	5	175	242.5
17	140	80	100	5	37.5	-	-	25	5	175	242.5

2.2. Free-Radical Scavenging Activity

The free-radical scavenging activity of the powders was evaluated using the 2,2'-azino-bis (3-ethylbenzthiazoline-6-sulfonic acid) ($ABTS^{\bullet+}$) and 2,2-diphenyl-picrylhydrazyl ($DPPH^{\bullet}$) radical scavenging assays as previously described [12].

In the DPPH assay, 950 µL of 100 µM methanolic solution of $DPPH^{\bullet}$ was mixed with 50 µL of the tested powder at different concentrations. Regarding the $DPPH^{\bullet}$ assay, 1.0 mL of freshly made methanolic solution of $DPPH^{\bullet}$ radical (100 µM) was mixed with the powder solution at different concentrations. The contents were vigorously mixed, incubated at room temperature in the dark for 20 min and the absorbance was measured at 517 nm. The measurement was conducted on a Hitachi U-1900 ratio beam spectrophotometer (Tokyo, Japan). In each experiment, the tested powder alone in methanol was used as blank and $DPPH^{\bullet}$ alone in methanol was used as control.

In the $ABTS^{\bullet+}$ assay, $ABTS^{\bullet+}$ radical was produced by mixing 2 mM ABTS with 30 µM H_2O_2 and 6 µM horseradish peroxidase (HRP) enzyme in 1 mL of distilled water. The solution was vigorously mixed and incubated at room temperature in the dark for 45 min until $ABTS^{\bullet+}$ radical formation. Then, 10 µL of different powder concentrations were added in the reaction mixture and the absorbance at 730 nm was read. The measurement was conducted on a Hitachi U-1900 ratio beam spectrophotometer (Tokyo, Japan). In each experiment, the tested powder in distilled water containing ABTS and H_2O_2 was used as blank, and the $ABTS^{\bullet+}$ radical solution with 10 µL H_2O was used as control.

The percentage of the radical scavenging capacity (RSC) of the tested powders for both assays was calculated according to the following equation:

$$\text{Radical scavenging capacity (\%)} = [(A_{control} - A_{sample})/A_{control}] \times 100 \tag{1}$$

where, $A_{control}$ and As_{ample} are the absorbance values of the control and the tested samples, respectively. Moreover, in order to compare the radical scavenging capacity of the samples, the IC_{50} value showing the concentration that induced 50% scavenging of $DPPH^{\bullet}$ and $ABTS^{\bullet+}$ was calculated. In both assays vitamin C was used as a positive control. All experiments were carried out in triplicate and at least on two separate occasions.

2.3. Peroxyl-Radical-Induced Plasmid DNA Strand Cleavage

The peroxyl-radical-induced DNA plasmid strand cleavage assay was performed as described previously [12]. In brief, peroxyl radicals (ROO) were produced from thermal decomposition of 2,2'-azobis(2-amidinopropane hydrochloride) (AAPH). The reaction mixture (10 µL) containing 1 µg Bluescript-SK+ plasmid DNA, 2.5 mM AAPH in phosphate-buffered saline (PBS) and the tested powder at different concentrations was incubated in the dark for 45 min at 37 °C. Then, the reaction was stopped by the addition of 3 µL loading buffer (0.25% bromophenol blue and 30% glycerol). After analyzing the DNA samples by agarose gel electrophoresis, they were photographed and analyzed using the Alpha Innotech Multi Image (ProteinSimple, San Jose, CA, USA). In addition, plasmid DNA was treated with each powder alone at the highest concentration used in the assay in order to test their effects on plasmid DNA conformation. The percentage of the protective activity of the tested powders from ROO-induced DNA strand breakage was calculated using the following equation:

$$\% \text{ Inhibition} = [(S - S_o)/(S_{control} - S_o)] \times 100 \tag{2}$$

where, $S_{control}$ is the percentage of supercoiled DNA in the negative control (plasmid DNA alone), S_o is the percentage of supercoiled plasmid DNA in the positive control (without the tested powder but in the presence of the radical initiating factor), and S is the percentage of supercoiled plasmid DNA in the tested powder along with the radical initiating factor. Moreover, the IC_{50} values showing the concentration that inhibited the AAPH-induced DNA relaxation by 50% were calculated. Vitamin C was used as a positive control. At least three independent experiments were performed for each sample.

2.4. Reducing Power

Reducing power was determined spectrophotometrically as described previously [13]. _The $RP_{0.5AU}$ value showing the powder concentration that caused an absorbance of 0.5 at 700 nm was calculated from the graph plotting absorbance against powder concentration. Vitamin C was used as a positive control. At least two independent experiments in triplicate were performed for each tested sample.

2.5. Evaluation of Relative Antioxidant Capacity Index (RACI)

In order to find out which samples exhibited the highest antioxidant activity in all antioxidant assays, the RACI was evaluated for each sample as described in Sun and Tanumihardjo [13]. RACI is the mean value of standard scores evaluated by initial data generated with different methods for a sample. A standard score was calculated according to the following equation:

$$\text{Standard score} = (x - \mu)/\sigma \tag{3}$$

where x was the raw data, μ was the mean of all samples' values of each method, and σ was the standard deviation.

Since in all antioxidant assays the raw data were IC_{50} values, the lower the RACI value the higher the antioxidant capacity.

2.6. Cell Culture Conditions

As previously described [14], human endothelial EA.hy926 cells gifted from Prof. Koukoulis (University of Thessaly, Greece) were cultured in normal Dulbecco's modified Eagle's medium (DMEM) in plastic disposable tissue culture flasks at 37 °C in 5% carbon dioxide.

2.7. XTT Cell Viability Assay

The antioxidant activity of the powders in EA.hy926 cells was examined using non-cytotoxic concentrations. In order to select these concentrations, the cytotoxicity of the powders was checked using the XTT cell viability assay kit (Sigma) as previously described [14]. Briefly, EA.hy926 cells were seeded into a 96-well plate (1×10^4 cells per well) in DMEM containing 10% fetal bovine serum FBS. After 24 h incubation at 37 °C in 5% CO_2, the cells were treated with different concentrations of the powders in FBS-free DMEM and incubated for another 24 h. Then, 50 µL of XTT test solution was added to each well. After 4 h of incubation, absorbance was measured at 450 nm and also at 630 nm as a reference wavelength in a Bio-Tek ELx800 microplate reader (Winooski, VT, USA). Negative control was DMEM serum-free medium. The absorbance values of the control and powders were used for calculating the percentage inhibition of cell growth caused by the powder treatment. All experiments were carried out in triplicate and on two separate occasions.

2.8. Treatment of EA.hy926 Cells with the Powders

The cells were cultured until 75%–80% confluence of the flask. Afterwards the medium was replaced with serum-free medium containing the tested powders at non-cytotoxic concentrations. The cells were treated with the powders for 24 h, and then they were trypsinized, collected, and centrifuged twice at 300× *g* for 10 min at 5 °C. At the end of the first centrifugation, the supernatant fluid was discarded and the cellular pellet was resuspended in PBS. After the second centrifugation, the cell pellet was collected and used for measuring the glutathione (GSH) and reactive oxygen species (ROS) levels.

2.9. Assessment of GSH and ROS Levels in EA.hy926 Cells by Flow Cytometry

The GSH and ROS levels in EA.hy926 cells were assessed using mercury orange and 2,7-dichlorofluorescein diacetate (DCF-DA), respectively, as described previously [9]. In brief, the cells were re-suspended in PBS (10^6 cells/mL) and incubated in the presence of mercury orange (10 μM) or DCF-DA (40 μM) respectively, in the dark at 37° C for 30 min. Then, the cells were washed, re-suspended in PBS, and subjected to flow cytometric analysis using a FACSCalibur flow cytometer (Becton Dickinson, Franklin Lakes, NJ, USA) with excitation and emission wavelengths at 488 and 530 nm for ROS, and at 488 and 580 nm for GSH. Data were analyzed using 'BD Cell Quest' software (Becton Dickinson). Each experiment was repeated at least three times.

2.10. Statistical Analysis

All results were expressed as mean ± standard deviation (SD). Differences were considered significant at $p < 0.05$. One-way ANOVA was performed followed by Tukey's test for multiple pair-wise comparisons using the SPSS 20.0 software (SPSS, Inc., Chicago, IL, USA).

3. Results

3.1. Free-Radical Scavenging Activity of the Powders

In the present study, 17 powders produced by encapsulation of an OMWW polyphenolic extract by spray drying under different conditions in whey protein, maltodextrin, and gelatin were tested for their free-radical scavenging activity against DPPH• and ABTS•+ radicals. All of them were able to scavenge DPPH• with IC_{50} values ranging from 295 ± 18 to 660 ± 73 μg/mL (Table 2). Moreover, all the extracts demonstrated a potent ABTS•+ scavenging capacity with IC_{50} values ranging from 290 ± 9 to 710 ± 68 μg/mL (Table 2).

Table 2. Free-radical scavenging activity against DPPH• and ABTS•+ radicals, protective activity against peroxyl radical (ROO)-induced DNA damage, and reducing power of the powders.

Sample	ABTS [a] IC_{50} (μg/mL)	DPPH [a] IC_{50} (μg/mL)	ROO [b] IC_{50} (μg/mL)	Reducing power [a] 0.5 AU	RACI [c]
1	428 ± 26 *	295 ± 24 *	800 ± 88 *	660 ± 33 *	−0.76
2	355 ± 14 *	340 ± 41 *	1580 ± 63 *	707 ± 28 *	−0.41
3	450 ± 36 *	352 ± 46 *	520 ± 41 *	434 ± 30 *	−1.13
4	500 ± 45 *	320 ± 10 *	935 ± 84 *	890 ± 18 *	−0.06
5	547 ± 16 *	330 ± 26 *	960 ± 57 *	700 ± 21 *	−0.27
6	300 ± 22 *	390 ± 16 *	725 ± 67 *	ND	−1.03
7	510 ± 44 *	385 ± 19 *	1190 ± 111 *	ND	−0.13
8	710 ± 68 *	409 ± 37 *	2300 ± 227 *	ND	1.19
9	290 ± 9 *	295 ± 18 *	590 ± 65 *	689 ± 55 *	−1.11
10	390 ± 39 *	526 ± 42 *	2075 ± 103 *	866 ± 30 *	0.55
11	590 ± 51 *	465 ± 47 *	1600 ± 149 *	ND	0.56
12	530 ± 46 *	638 ± 51 *	1250 ± 113 *	ND	0.64
13	540 ± 38 *	540 ± 16 *	2250 ± 180 *	747 ± 67 *	0.78
14	520 ± 48 *	650 ± 59 *	1200 ± 113 *	ND	0.61
15	430 ± 17 *	660 ± 73 *	1515 ± 136 *	825 ± 8 *	0.59
16	460 ± 42 *	465 ± 65 *	820 ± 77 *	ND	−0.29
17	590 ± 48 *	510 ± 25 *	1900 ± 86 *	ND	0.86
Vitamin C	4 ± 0.3 *	5 ± 0.4 *	69 ± 5 *	3 ± 0.4 *	-

[a] Values are the mean ± SD of at least two separate triplicate experiments. [b] Values are the mean ± SD from three independent experiments. * $p < 0.05$, indicates significant difference from the control values. [c] RACI: Relative antioxidant capacity index. ND: Non-detectable at the tested concentrations.

3.2. Protection from ROO•-Induced Plasmid DNA Strand Cleavage

All the powders protected from ROO-induced DNA damage with IC_{50} values ranging from 520 ± 41 to 2300 ± 227 µg/mL (Table 2).

3.3. Reducing Capacity of the Extracts

The estimation of the extracts' reducing capacity was based on the reducing power assay by determining the $RP_{0.5AU}$ values, which ranged from 434 ± 30 to 890 ± 18 µg/mL (Table 2). Samples 9, 12, 13, 15, 16, 17, 18, and 19 did not show absorbance of 0.5 at 700 nm at the tested concentrations (Table 2).

3.4. Determination of Extracts' Non-Cytotoxic Concentrations in EA.hy926 Cells

Based on the results from DPPH•, ABTS•+, ROO-induced DNA damage, and reducing power, the two most potent of the powders were selected in order to examine their antioxidants effects in endothelial EAhy.926 cells. Specifically, the selection was made according to the RACI value that each powder had in all the above assays (Table 2). Since the RACI values were based on the IC_{50} values, the lower the RACI value the higher the antioxidant capacity. Thus, the two most potent samples were 3 and 9. However, before examining the powders' antioxidant effects in EAhy.926 cells, their cytotoxicity was assessed using the XTT assay in order to select non-cytotoxic concentrations. The results showed that both powders had no significant cytotoxicity at concentrations up to 1600 µg/mL (Figure 1).

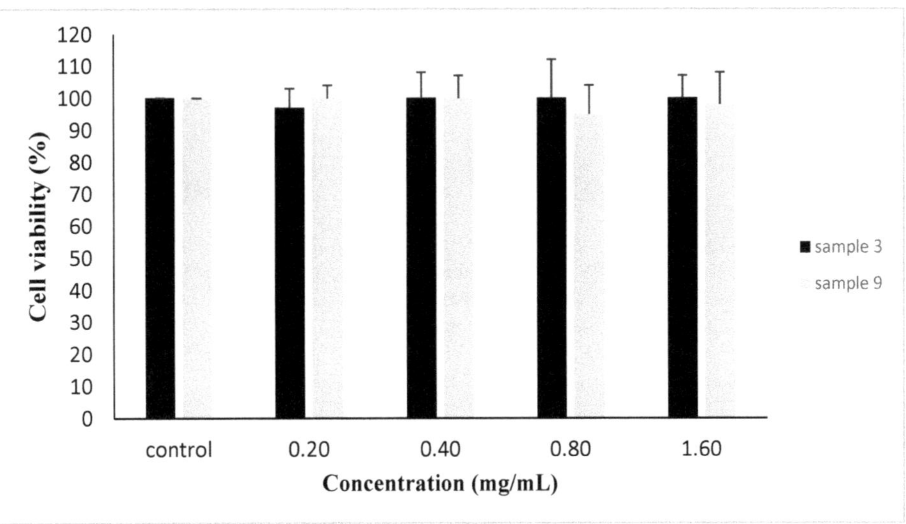

Figure 1. Cell viability following the treatment of EAhy.926 cells with samples 3 and 9. The results are presented as the means ± SEM of three independent experiments carried out in triplicate. * $p < 0.05$ indicates significant difference from the control value.

3.5. Effects of Powders on GSH Levels in EA.hy926 Cells

For assessing the effects of samples 3 and 9 on GSH levels in EA.hy926 cells, flow cytometry analysis was used. The results showed that treatment of EA.hy926 cells with sample 3 significantly increased GSH levels by 22%, 59%, and 82% at 400, 800, and 1600 µg/mL, respectively compared to control (Figure 2). When cells were treated with sample 9, GSH levels were significantly increased by 51% and 88% at concentrations of 800 and 1600 µg/mL, respectively, compared to the control (Figure 3).

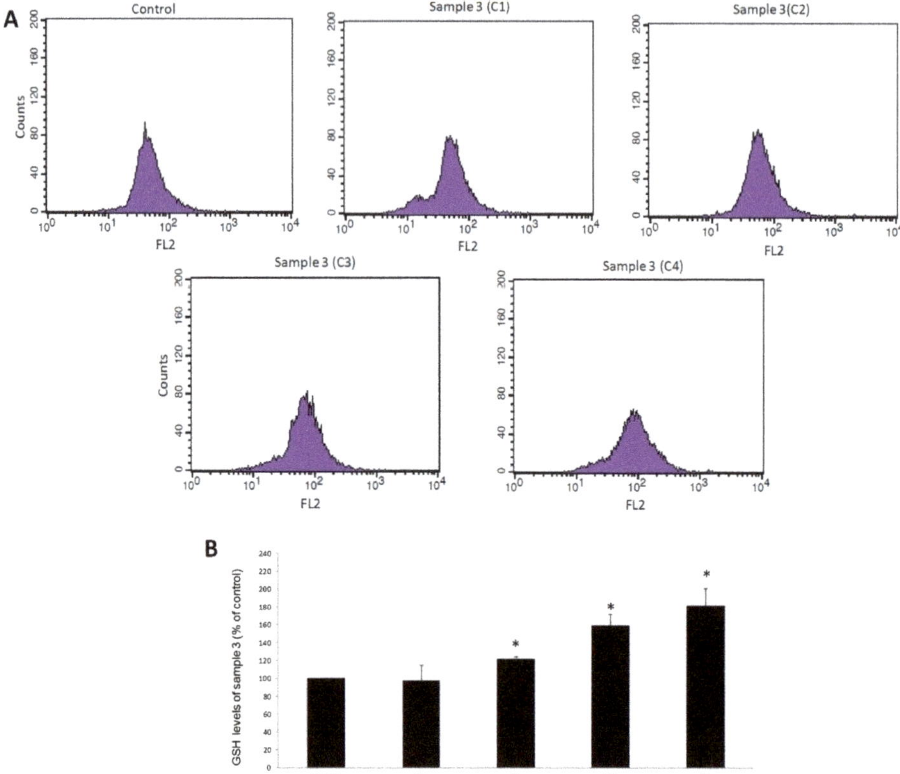

Figure 2. The effects of sample 3 on glutathione (GSH) levels in EA.hy926 cells after treatment for 24 h, as assessed by flow cytometry. (**A**) The histograms of cell counts versus fluorescence of 10,000 cells after treatment with sample 3. (**B**) Bar charts demonstrate the GSH levels as % of control as estimated by the histograms in EA.hy926 cells after treatment. C1, C2, C3, C4: 200, 400, 800, and 1600 µg/mL, respectively. * Statistically significant compared to the control cells. FL2: The detection of fluorescence using 488 and 580 nm as the excitation and emission wavelength, respectively.

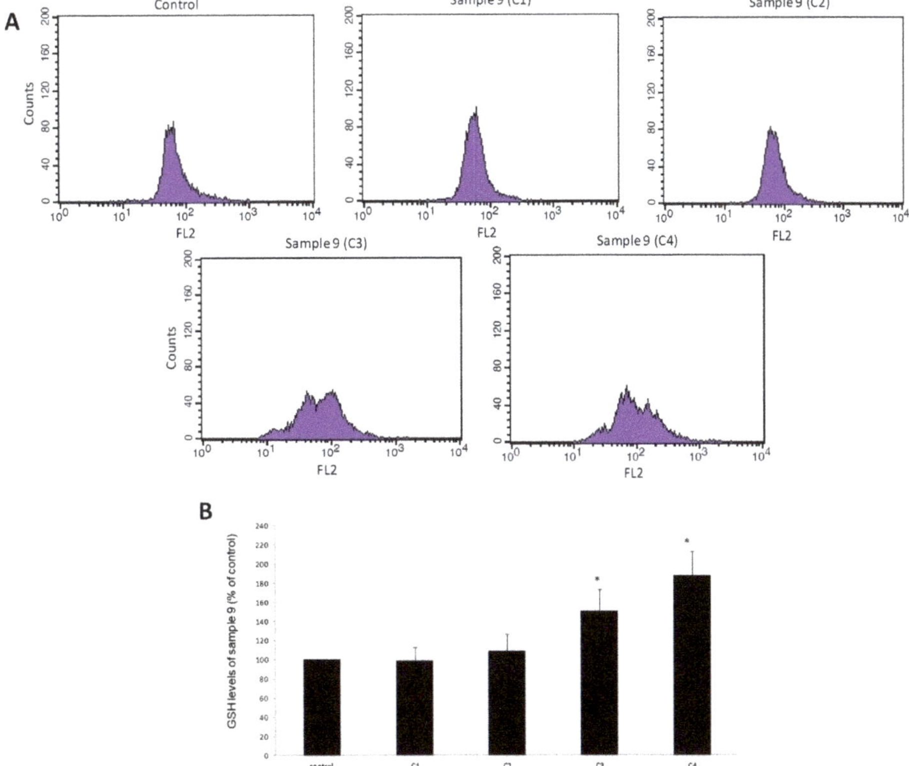

Figure 3. The effects of sample 9 on GSH levels in EA.hy926 cells after treatment for 24 h, as assessed by flow cytometry. (**A**) The histograms of cell counts versus fluorescence of 10,000 cells after treatment with sample 9. (**B**) Bar charts demonstrate the GSH levels as % of control as estimated by the histograms in EA.hy926 cells after treatment. C1, C2, C3, C4: 200, 400, 800, and 1600 µg/mL, respectively. * Statistically significant compared to the control cells. FL2: The detection of fluorescence using 488 and 580 nm as the excitation and emission wavelength, respectively.

3.6. Effects of Powders on ROS Levels in EA.hy926 Cells

Like GSH, the levels of ROS in EAhy.926 cells after treatment with samples 3 and 9 were evaluated by flow cytometry analysis. The results demonstrated that after treatment of cells with sample 3, ROS levels did not change at any concentration compared to control (Figure 4). Moreover, after cell treatment with sample 9, ROS were decreased by 25% at 1600 µg/mL, compared to control (Figure 5).

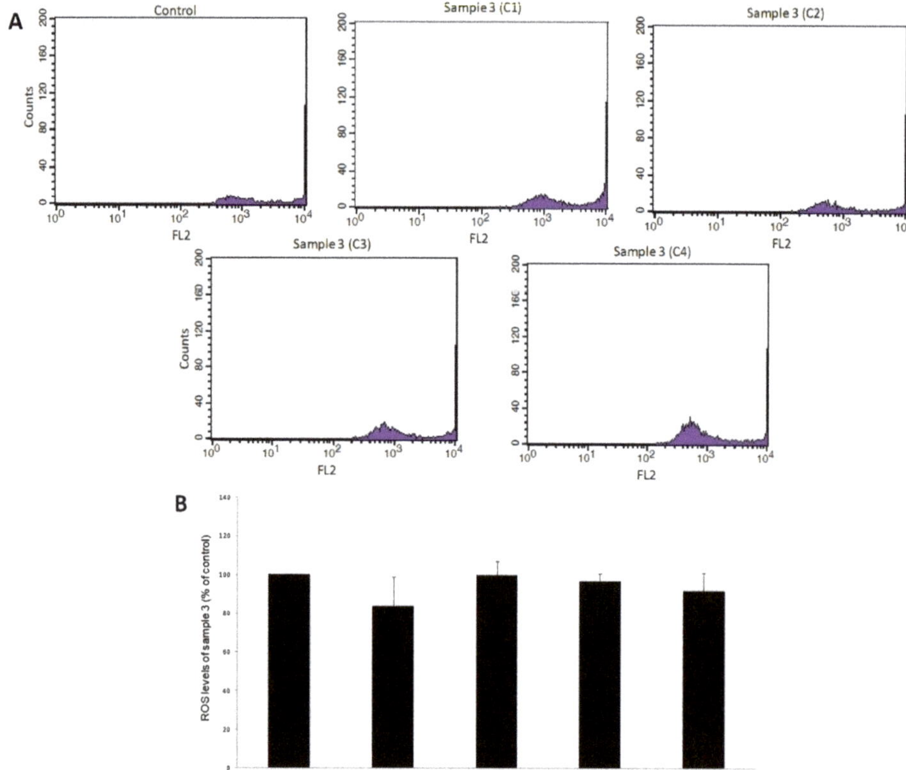

Figure 4. The effects of sample 3 on reactive oxygen species (ROS) levels in EA.hy926 cells after treatment for 24 h, as assessed by flow cytometry. (**A**) The histograms of cell counts versus fluorescence of 10,000 cells after treatment with sample 3. (**B**) Bar charts demonstrate the ROS levels as % of control as estimated by the histograms in EA.hy926 cells after treatment. C1, C2, C3, C4: 200, 400, 800, and 1600 µg/mL, respectively. * Statistically significant compared to the control cells. FL2: The detection of fluorescence using 488 and 580 nm as the excitation and emission wavelength, respectively.

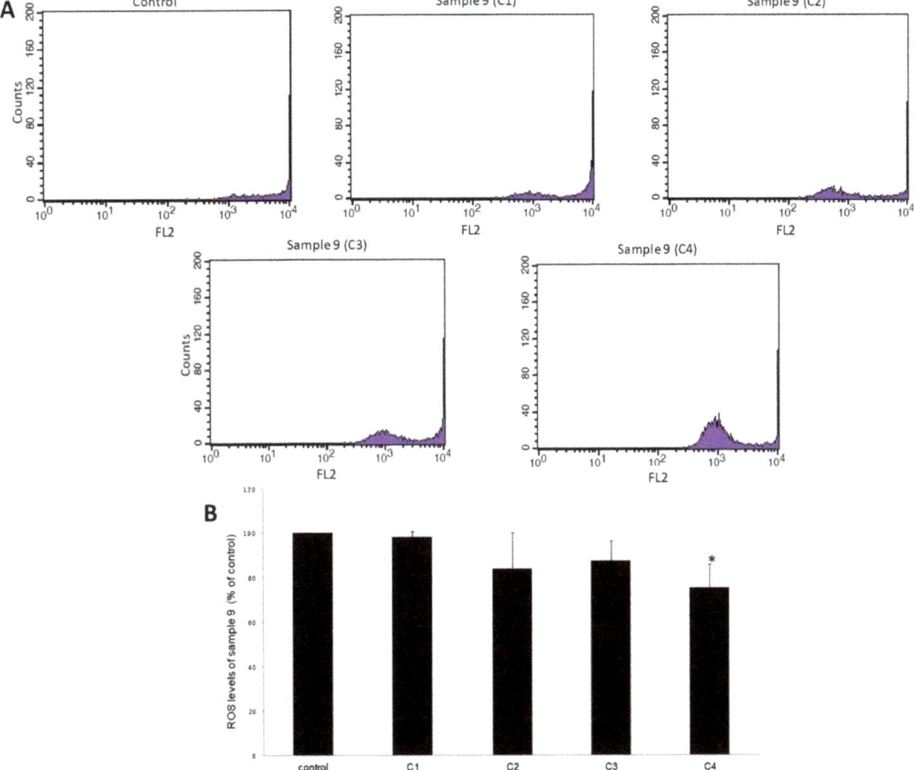

Figure 5. The effects of sample 9 on ROS levels in EA.hy926 cells after treatment for 24 h, as assessed by flow cytometry. (**A**) The histograms of cell counts versus fluorescence of 10,000 cells after treatment with sample 9. (**B**) Bar charts demonstrate the ROS levels as % of control as estimated by the histograms in EA.hy926 cells after treatment. C1, C2, C3, C4: 200, 400, 800, and 1600 μg/mL, respectively. * Statistically significant compared to the control cells. FL2: The detection of fluorescence using 488 and 580 nm as the excitation and emission wavelength, respectively.

4. Discussion

The composition and the amounts of OMWW cause serious environmental problems in areas of olive oil production, especially when they are discharged without any previous treatment [15]. On the other hand, OMWWs contain high quantities of polyphenols with important bioactivities such as antioxidant property [1,11]. Thus, polyphenolic extracts from OMWW can be used as natural alternatives to commercial synthetic antioxidants with applications in the food industry and in the development of nutraceutical products [3,4,11,16].

However, when polyphenols are added in foods, there are problems regarding bad or bitter taste and discoloration. The encapsulation of polyphenolic extracts has been suggested as a method to overcome these problems as well as to improve polyphenols' stability, half-life, and bioavailability [10,17]. Thus, the aim of this study was to evaluate the effects on the redox status of endothelial cells of an OMWW extract encapsulated under different conditions and encapsulation carriers using spray drying. Apart from the encapsulation carrier, the spray drying conditions of the different samples differed in the inlet and outlet temperature, percentage of pump function, and use of Tween 80. Spray drying is a process widely used for encapsulation of oils and flavors [18]. Some of the advantages of this procedure are the generation of sample powders with good quality, low water activity, and easier handling and storage, while it also protects the active material against undesirable reactions [19]. The encapsulation carriers used were maltodextrin, maltodextrin/gelatin (5:1), whey protein, and whey protein/gelatin (5:1). Maltodextrin, a hydrolyzed starch, offers advantages for encapsulation such as relatively low cost, neutral aroma and taste, low viscosity at high solid concentrations, and protection against oxidation [20]. However, the most serious drawback of this material is its low emulsifying capacity; thus, it is desirable to use it in combination with other surface-active biopolymers such as gelatin [21]. Whey proteins, like other milk proteins, are hydrocolloids exhibiting good solubility and behavior and are used over the last years in food industry due to the increasingly need for natural products [22]. There have so far been only a few studies on the encapsulation of polyphenols from OMWW. For example, it has been reported that OMWW polyphenols encapsulated in maltodextrin and maltodextrin/acacia fiber (1:1) using spray drying exhibited antioxidant and antiglycative activities as well as inhibited Maillard reaction in milk [23–25]. Moreover, Caporaso et al. [22] have produced antioxidant powder by encapsulating OMWW polyphenols in whey protein/xanthan gum.

At first, in vitro antioxidant assays were applied in order to select the most potent samples that were used for the examination in endothelial cells. The findings showed that although all the 17 samples exhibited free-radical scavenging activity in DPPH$^{\bullet}$ and ABTS$^{\bullet+}$ assays, there was a great variation in their effect up to about two- and threefold, respectively. In ABTS$^{\bullet+}$ assay, the samples encapsulated in whey protein were more potent at lower inlet/outlet temperature than at higher ones. However, inlet/outlet temperature did not seem to affect the potency of samples encapsulated in maltodextrin. Moreover, the scavenging activity against ABTS$^{\bullet+}$ radical was not dependent on the encapsulation carrier. For example, in the ABTS$^{\bullet+}$ assay, the first (No. 9) and third (No. 2) most potent samples were encapsulated in maltodextrin and whey protein, respectively. Finally, in the ABTS$^{\bullet+}$ assay, the use of gelatin did not significantly affect the antioxidant potency of samples encapsulated either in whey protein or maltodextrin. Unlike the ABTS$^{\bullet+}$ assay, in the DPPH$^{\bullet}$ assay, inlet/outlet temperatures had no effect on the potency of samples encapsulated in whey protein. Moreover, like the ABTS$^{\bullet+}$ assay, these temperatures did not seem to affect scavenging activity against DPPH$^{\bullet}$ of samples encapsulated in maltodextrin. Furthermore, in the DPPH$^{\bullet}$ assay, all the samples encapsulated in whey protein exhibited higher antioxidant activity than the samples encapsulated in maltodextrin (except No. 9). In the DPPH$^{\bullet}$ assay, the addition of gelatin decreased antioxidant activity in samples encapsulated in whey protein, while it did not affect the activity of samples encapsulated in maltodextrin. Moreover, some samples (e.g., No. 9) exhibited high potency in both DPPH$^{\bullet}$ and ABTS$^{\bullet+}$ assay, while other samples (e.g., No. 5) had high potency in one assay and low in the other. The observed differences between DPPH$^{\bullet}$ and ABTS$^{\bullet+}$ may be explained by the different solvents used in these assays, that is, methanol and water, respectively. Thus, lipophilic compounds are more active in the DPPH$^{\bullet}$ assay, while hydrophilic compounds are more active in the ABTS$^{\bullet+}$ assay.

Apart from the free-radical scavenging activity, the samples' reducing capacity was assessed using the reducing power assay, since many antioxidants act as hydrogen donors. Again, the reducing capacity varied up to twofold between the samples exhibiting the higher and the lower activity. The type of encapsulation carrier did not affect reducing activity, since samples of both whey protein (e.g., No. 3) and maltodextrin (e.g., No. 9) demonstrated high potency. Moreover, reducing activity did not seem to depend on inlet/outlet temperatures in samples encapsulated either in whey protein or maltodextrin. It was also remarkable that the addition of gelatin in the encapsulation mixture decreased reducing activity in both whey protein and maltodextrin samples.

Furthermore, all the samples protected from ROS-induced DNA damage, but, like the other assays, the IC$_{50}$ varied greatly up to about 4.4-fold. Samples encapsulated in both whey protein (e.g., No. 3) and maltodextrin (e.g., No. 9) exhibited high protection. The samples' protective activity was independent of inlet/outlet temperatures. As in the DPPH$^{\bullet}$ assay, the use of gelatin decreased the potency of samples encapsulated in whey protein but not in maltodextrin. Previous studies have shown that OMWW extracts inhibited ROS-induced DNA damage using pure DNA or cells [26,27].

However, to the best of our knowledge, this is the first study reporting the protective effect of an encapsulated OMMW polyphenolic extract against DNA damage caused by free radicals.

Based on the RACI value that each sample had in all the above assays (i.e., DPPH$^\bullet$, ABTS$^{\bullet+}$, ROO-induced DNA damage, and reducing power), the two most potent samples were selected in order for their antioxidant activity to be examined in endothelial cells at non-cytotoxic concentrations. These two samples were No. 3 encapsulated in whey protein and No. 9 encapsulated in maltodextrin. For assessing the samples' antioxidant activity, GSH and ROS levels were evaluated by flow cytometry in endothelial cells. Cell treatment with both encapsulated samples showed significantly increased GSH levels, one of the most important antioxidant molecules [28]. The OMWW sample-induced increase in GSH may be due to the rescue of GSH from reaction with ROS by their direct scavenging, since OMWW encapsulated samples were shown to possess free-radical scavenging activity. Moreover, the observed increase in GSH may be attributed to increase in activity of enzymes involved in GSH synthesis and metabolism. For example, polyphenols found in olive oil have been reported to increase the expression and/or activity of glutathione peroxidase (GPx) and glutathione reductase (GR) [29]. The expression of such enzymes is mainly regulated by the transcription factor nuclear factor (erythroid-derived 2)-like2 (Nrf2) [30]. Hydroxytyrosol, one of the polyphenols identified in our OMWW extract used for the encapsulation [11], has been reported to activate Nrf2 in mouse heart [31]. Interestingly, we have also demonstrated that administration of feed containing polyphenolic extract from OMWW increased GSH in different tissues including the heart of farm animals [3,4]. Moreover, ROS levels were decreased in endothelial cells after treatment with encapsulated OMWW samples. This decrease in ROS was in agreement with the OMWW sample-induced increase in antioxidant mechanisms such as GSH in the endothelial cells as well as with OMWW samples' free-radical scavenging activity. However, this ROS decrease was observed only in cells treated with sample No. 9 encapsulated in maltodextrin. The absence of ROS decrease in cells treated with whey protein sample No. 3 may be explained by the fact that samples' antioxidant effects were examined in naïve cells, that is, cells that were not treated with an oxidative agent. Thus, in such cells, the baseline ROS levels are low. Moreover, sample No. 9 exhibited higher potency in free-radical scavenging assays than sample No. 3. Overall, the above findings suggested the use of OMWW polyphenolic extract encapsulated either in maltodextrin or whey protein for the development of food supplements or biofunctional foods possessing antioxidant activity, especially protection from oxidative stress-induced pathologies associated with the cardiovascular system. Interestingly, OMWW polyphenolic extract has been reported to decrease cholesterol levels in rats [32].

5. Conclusions

In few previous studies, maltodextrin and whey protein have been used for the encapsulation of OMWW polyphenolic extracts [21–24]. However, in the present study, there was for the first time a direct comparison between the two materials for their efficiency for the production of antioxidant powder. The results showed that both maltodextrin and whey protein were almost equally effective for the production of antioxidant powder, although in slightly different encapsulation conditions. For example, the findings indicated that, in general, inlet/outlet temperature did not affect maltodextrin samples' antioxidant activity, but whey protein samples exhibited better antioxidant activity at lower temperatures (within the temperature range used; 100–160 °C). Gelatin was also first used in the mixture for OMWW extract's encapsulation, but the results showed that it decreased antioxidant activity, especially in whey protein samples. Furthermore, the present study is the first demonstrating that encapsulated OMWW extract protected from ROS-induced DNA damage. Finally, it is the first time, that an encapsulated OMWW extract (as well as in general an OMWW extract) has been shown to enhance antioxidant mechanisms (i.e., GSH) in endothelial cells. Of course, further studies are needed for the elucidation of mechanisms accounting for these possible beneficial effects.

Author Contributions: Conceptualization, D.S., K.P., and D.K.; methodology, D.S., K.P., and D.K.; supervision, D.S., K.P., and D.K.; carried out the experiments, K.K, Z.F., K.Z., E.K., C.M., P.G., and D.L.; writing, K.K., D.S., and A.S.V.; funding acquisition, D.S., K.P., and D.K.; resources, D.S., K.P., and D.K.

Funding: The work was funded in part by the "Biotechnology—Nutrition and Environment" and "Application of Molecular Biology—Genetics" MSc programs in the Department of Biochemistry and Biotechnology at the University of Thessaly.

Conflicts of Interest: The authors declare no conflict of interest.

References

1. Frankel, E.; Bakhouche, A.; Lozano-Sánchez, J.; Segura-Carretero, A.; Fernández-Gutiérrez, A. Literature review on production process to obtain extra virgin olive oil enriched in bioactive compounds. Potential use of byproducts as alternative sources of polyphenols. *J. Agric. Food Chem.* **2013**, *61*, 5179–5188. [CrossRef] [PubMed]

2. Stagos, D.; Amoutzias, G.D.; Matakos, A.; Spyrou, A.; Tsatsakis, A.M.; Kouretas, D. Chemoprevention of liver cancer by plant polyphenols. *Food Chem. Toxicol.* **2012**, *50*, 2155–2170. [CrossRef] [PubMed]

3. Gerasopoulos, K.; Stagos, D.; Kokkas, S.; Petrotos, K.; Kantas, D.; Goulas, P.; Kouretas, D. Feed supplemented with byproducts from olive oil mill wastewater processing increases antioxidant capacity in broiler chickens. *Food Chem. Toxicol.* **2015**, *82*, 42–49. [CrossRef] [PubMed]

4. Gerasopoulos, K.; Stagos, D.; Petrotos, K.; Kokkas, S.; Kantas, D.; Goulas, P.; Kouretas, D. Feed supplemented with polyphenolic byproduct from olive mill wastewater processing improves the redox status in blood and tissues of piglets. *Food Chem. Toxicol.* **2015**, *86*, 319–327. [CrossRef] [PubMed]

5. Makri, S.; Kafantaris, I.; Savva, S.; Ntanou, P.; Stagos, D.; Argyroulis, I.; Kotsampasi, B.; Christodoulou, V.; Gerasopoulos, K.; Petrotos, K.; et al. Novel Feed Including Olive Oil Mill Wastewater Bioactive Compounds Enhanced the Redox Status of Lambs. *In Vivo* **2018**, *32*, 291–302. [PubMed]

6. Mohan, A.; Rajendran, S.R.C.K.; He, Q.S.; Bazinet, L.; Udenigwe, C.C. Encapsulation of food protein hydrolysates and peptides: A review. *RSC Adv.* **2015**, *97*, 79270–79278. [CrossRef]

7. Fang, Z.; Bhandari, B. Encapsulation of polyphenols—A review. *Trends Food Sci. Technol.* **2010**, *21*, 510–523. [CrossRef]

8. Jafari, S.M.; Assadpoor, E.; He, Y.; Bhandari, B. Encapsulation efficiency of food flavors and oils during spray drying. *Dry Technol.* **2008**, *26*, 816–835. [CrossRef]

9. Kerasioti, E.; Stagos, D.; Georgatzi, V.; Bregou, E.; Priftis, A.; Kafantaris, I.; Kouretas, D. Antioxidant Effects of Sheep Whey Protein on Endothelial Cells. *Oxid. Med. Cell Longev.* **2016**, *2016*, e6585737. [CrossRef]

10. Deanfield, J.E.; Halcox, J.P.; Rabelink, T.J. Endothelial function and dysfunction: Testing and clinical relevance. *Circulation* **2007**, *115*, 1285–1295. [CrossRef]

11. Papadopoulou, A.; Petrotos, K.; Stagos, D.; Gerasopoulos, K.; Maimaris, A.; Makris, H.; Kafantaris, I.; Makri, S.; Kerasioti, E.; Halabalaki, M.; et al. Enhancement of Antioxidant Mechanisms and Reduction of Oxidative Stress in Chickens after the Administration of Drinking Water Enriched with Polyphenolic Powder from Olive Mill Waste Waters. *Oxid. Med. Cell. Longev.* **2017**, *2017*, 8273160. [CrossRef] [PubMed]

12. Stagos, D.; Portesis, N.; Spanou, C.; Mossialos, D.; Aligiannis, N.; Chaita, E.; Panagoulis, C.; Reri, E.; Skaltsounis, L.; Tsatsakis, A.M.; et al. Correlation of total polyphenolic content with antioxidant and antibacterial activity of 24 extracts from Greek domestic Lamiaceae species. *Food Chem. Toxicol.* **2012**, *50*, 4115–4124. [CrossRef] [PubMed]

13. Stagos, D.; Balabanos, D.; Savva, S.; Skaperda, Z.; Priftis, A.; Kerasioti, E.; Mikropoulou, E.V.; Vougogiannopoulou, K.; Mitakou, S.; Halabalaki, M.; et al. Extracts from the Mediterranean Food Plants Carthamus lanatus, Cichorium intybus, and Cichorium spinosum Enhanced GSH Levels and Increased Nrf2 Expression in Human Endothelial Cells. *Oxid. Med. Cell. Longev.* **2018**, *2018*, 6594101. [CrossRef] [PubMed]

14. Sun, T.; Tanumihardjo, S.A. An integrated approach to evaluate food antioxidant capacity. *J. Food Sci.* **2007**, *72*, R159–R165. [CrossRef] [PubMed]

15. Rinaldi, M.; Rana, G.; Introna, M. Olive-mill wastewater spreading in southern Italy: Effects on a durum wheat crop. *Field Crops Res.* **2003**, *84*, 319–326. [CrossRef]

16. Caporaso, N.; Formisano, D.; Genovese, A. Use of phenolic compounds from olive mill wastewater as valuable ingredients for functional foods. *Crit. Rev. Food Sci. Nutr.* **2018**, *58*, 2829–2841. [CrossRef] [PubMed]

17. Caporaso, N.; Genovese, A.; Burke, R.; Barry-Ryan, C.; Sacchi, R. Effect of olive mill wastewater phenolic extract, whey protein isolate and xanthan gum on the behaviour of olive O/W emulsions using response surface methodology. *Food Hydrocoll.* **2016**, *61*, 66–76. [CrossRef]

18. Ahn, J.H.; Kim, Y.P.; Lee, Y.M.; Seo, E.M.; Lee, K.W.; Kim, H.S. Optimization of microencapsulation of seed oil by response surface methodology. *Food Chem.* **2008**, *107*, 98–105. [CrossRef]

19. Carneiro, H.C.F.; Tonon, R.V.; Grosso, C.R.F.; Hubinger, M.D. Encapsulation efficiency and oxidative stability of flaxseed oil microencapsulated by spray drying using different combinations of wall materials. *J. Food Eng.* **2013**, *115*, 443–451. [CrossRef]

20. Gharsallaoui, A.; Roudaut, G.; Chambin, O.; Voilley, A.; Saurel, R. Applications of spray drying in microencapsulation of food ingredients: An overview. *Food Res. Int.* **2007**, *40*, 1107–1121. [CrossRef]

21. Bae, E.K.; Lee, S.J. Microencapsulation of avocado oil by spray drying using whey protein and maltodextrin. *J. Microencapsul.* **2008**, *25*, 549–560. [CrossRef] [PubMed]

22. Caporaso, N.; Genovese, A.; Burke, R.; Barry-Ryan, C.; Sacchi, R. Physical and oxidative stability of functional olive oil-in-water emulsions formulated using olive mill wastewater biophenols and whey proteins. *Food Funct.* **2016**, *7*, 227–238. [CrossRef] [PubMed]

23. Navarro, M.; Fiore, A.; Fogliano, V.; Morales, F.J. Carbonyl trapping and antiglycative activities of olive oil mill wastewater. *Food Funct.* **2015**, *6*, 574–583. [CrossRef] [PubMed]

24. Troise, A.D.; Fiore, A.; Colantuono, A.; Kokkinidou, S.; Peterson, D.G.; Fogliano, V. Effect of olive mill wastewater phenol compounds on reactive carbonyl species and Maillard reaction end-products in ultrahigh-temperature-treated milk. *J. Agric. Food Chem.* **2014**, *62*, 10092–10100. [CrossRef] [PubMed]

25. Sedej, I.; Milczarek, R.; Wang, S.C.; Sheng, R.; de Jesús Avena-Bustillos, R.; Dao, L.; Takeoka, G. Membrane-Filtered Olive Mill Wastewater: Quality Assessment of the Dried Phenolic-Rich Fraction. *J. Food Sci.* **2016**, *81*, E889–E896. [CrossRef] [PubMed]

26. Nousis, L.; Doulias, P.T.; Aligiannis, N.; Bazios, D.; Agalias, A.; Galaris, D.; Mitakou, S. DNA protecting and genotoxic effects of olive oil related components in cells exposed to hydrogen peroxide. *Free Radic. Res.* **2005**, *39*, 787–795. [CrossRef] [PubMed]

27. Obied, H.K.; Prenzler, P.D.; Konczak, I.; Rehman, A.U.; Robards, K. Chemistry and bioactivity of olive biophenols in some antioxidant and antiproliferative in vitro bioassays. *Chem. Res. Toxicol.* **2009**, *22*, 227–234. [CrossRef]

28. Gould, R.L.; Pazdro, R. Impact of Supplementary Amino Acids, Micronutrients, and Overall Diet on Glutathione Homeostasis. *Nutrients* **2019**, *11*, 1056. [CrossRef]

29. Masella, R.; Varì, R.; D'Archivio, M.; Di Benedetto, R.; Matarrese, P.; Malorni, W.; Scazzocchio, B.; Giovannini, C. Extra virgin olive oil biophenols inhibit cell-mediated oxidation of LDL by increasing the mRNA transcription of glutathione-related enzymes. *J. Nutr.* **2004**, *134*, 785–791. [CrossRef]

30. Kumar, H.; Kim, I.S.; More, S.V.; Kim, B.W.; Choi, D.K. Natural product-derived pharmacological modulators of Nrf2/ARE pathway for chronic diseases. *Nat. Prod. Rep.* **2014**, *31*, 109–139. [CrossRef]

31. Bayram, B.; Ozcelik, B.; Grimm, S. A diet rich in olive oil phenolics reduces oxidative stress in the heart of SAMP8 mice by induction of Nrf2-dependent gene expression. *Rejuvenation Res.* **2012**, *15*, 71–81. [CrossRef] [PubMed]

32. Fki, I.; Sahnoun, Z.; Sayadi, S. Hypocholesterolemic effects of phenolic extracts and purified hydroxytyrosol recovered from olive mill wastewater in rats fed a cholesterol-rich diet. *J. Agric. Food Chem.* **2007**, *55*, 624–631. [CrossRef] [PubMed]

© 2019 by the authors. Licensee MDPI, Basel, Switzerland. This article is an open access article distributed under the terms and conditions of the Creative Commons Attribution (CC BY) license (http://creativecommons.org/licenses/by/4.0/).

 *antioxidants*

Article

Polyphenolic Composition of *Rosa canina*, *Rosa sempervivens* and *Pyrocantha coccinea* Extracts and Assessment of Their Antioxidant Activity in Human Endothelial Cells

Efthalia Kerasioti [1], Anna Apostolou [2], Ioannis Kafantaris [1], Konstantinos Chronis [1], Eleana Kokka [1], Christina Dimitriadou [1], Evangelia N. Tzanetou [2,3], Alexandros Priftis [1], Sofia D. Koulocheri [2], Serkos A. Haroutounian [2], Demetrios Kouretas [1] and Dimitrios Stagos [1,*]

[1] Department of Biochemistry and Biotechnology, University of Thessaly, Viopolis, 41500 Larissa, Greece; e-f-thalia@hotmail.com (E.K.); kafantarisioannis@gmail.com (I.K.); kochronis@uth.gr (K.C.); elkokka@uth.gr (E.K.); xristina_211@hotmail.com (C.D.); priftis@uth.gr (A.P.); dkouret@uth.gr (D.K.)

[2] Laboratory of Nutritional Physiology and Feeding, Agricultural University of Athens, Iera Odos 75, 11855 Athens, Greece; annapost21@msn.com (A.A.); eyatza@gmail.com (E.N.T.); skoul@aua.gr (S.D.K.); sehar@aua.gr (S.A.H.)

[3] Department of Pesticides Control and Phytopharmacy, Benaki Phytopathological Institute, 8 St. Delta Street, Kifissia, 14561 Athens, Greece

* Correspondence: stagkos@med.uth.gr; Tel.: +30-241-056-5229

Received: 6 March 2019; Accepted: 1 April 2019; Published: 6 April 2019

Abstract: The aim of the present study was the investigation of the antioxidant activity of plant extracts from *Rosa canina*, *Rosa sempervivens* and *Pyrocantha coccinea*. The results showed that the bioactive compounds found at higher concentrations were in the *R. canina* extract: hyperoside, astragalin, rutin, (+)-catechin and (-)-epicatechin; in the *R. sempervirens* extract: quinic acid, (+)-catechin, (−)-epicatechin, astragalin and hyperoside; and in the *P. coccinea* extract: hyperoside, rutin, (−)-epicatechin, (+)-catechin, astragalin, vanillin, syringic acid and chlorogenic acid. The total polyphenolic content was 290.00, 267.67 and 226.93 mg Gallic Acid Equivalent (GAE)/g dw, and the total flavonoid content 118.56, 65.78 and 99.16 mg Catechin Equivalent (CE)/g dw for *R. caninna*, *R. sempervirens* and *P. coccinea* extracts, respectively. The extracts exhibited radical scavenging activity in DPPH and 2,2′-azino-bis(3-ethylbenzothiazoline-6-sulphonic acid) $(ABTS)\bullet^{+}$ assays and protection from $ROO\bullet$-induced DNA damage in the following potency order: *R. canina* > *R. sempervirens* > *P. coccinea*. Finally, treatment with *R. canina* and *P. coccinea* extract significantly increased the levels of the antioxidant molecule glutathione, while *R. canina* extract significantly decreased Reactive Oxygen Species (ROS) in endothelial cells. The results herein indicated that the *R. canina* extract in particular may be used for developing food supplements or biofunctional foods for the prevention of oxidative stress-induced pathological conditions of endothelium.

Keywords: Rosa canina; Rosa sempervivens; Pyrocantha coccinea; polyphenols; antioxidants; endothelial cells; glutathione

---

1. Introduction

It is well known that within living organisms, the reactive oxygen species (ROS) are produced endogenously under physiological processes such as metabolism and inflammation [1,2]. Low ROS levels are needed for the progression of several basic biological processes including cellular proliferation and differentiation [2,3]. However, excessive intracellular concentration of ROS can lead to oxidative stress, a pathological condition associated with the development of various of diseases including cancer, diabetes, rheumatoid arthritis and other degenerative diseases in humans [4,5]. Oxidative stress causes healthy cells of the body to lose their function and structure as a consequence of ROS interaction with proteins, lipids and DNA [6].

One of the tissues that is especially susceptible to oxidative stress is vascular endothelium, a critical regulator of vascular homeostasis. Endothelial cell injury or dysfunction by ROS can be both a cause and consequence of many vascular complications including atherosclerosis, thrombosis and cardiovascular diseases [7,8]. In particular, vascular endothelium cells are susceptible to ROS damage, since ROS derived from different tissues circulate in the bloodstream and can interact directly with endothelial cells in the inner wall of blood vessels [9,10]. In addition, oxidative stress may cause damage to the endothelium through leukocyte adhesion [11].

However, human organism possesses several antioxidant mechanisms to counteract the overproduction and harmful effects of ROS [12]. The antioxidant protective mechanisms act in order to keep a balance between free radical production and scavenging. These antioxidants are either produced endogenously or obtained through diet, especially from plant foods [13]. The antioxidant proprieties of plant foods are mainly attributed to polyphenols, a large group of secondary metabolites, that possess antiradical activities due to their phenolic hydroxyls acting as reducing agents, metal ion chelators, antioxidant enzymes activators, and oxidases inhibitors [14,15]. Thus, there is currently a great interest in natural sources of antioxidants in order to improve the redox status and protect the organism from the detrimental effects of oxidative stress.

Many plant species of the *Rosaceae* family are considered to be of high importance because of their use in various food preparations as jam, tea and beverages. The *Rosa* genus of *Rosaceae* family consists of approximately 200 species located mainly in the Northern hemisphere in rainy areas and deserts [16]. *Rosa* species produce rose hips, a pseudocarp or a fruit. A number of studies have shown that rose hips demonstrate a great variety of bioactivities such as antioxidant, anticancer, anti-inflammatory and anti-obesity activities [17–20]. Wild fruits of *Rosa canina* are rich in vitamin C and are used for the prevention and/or treatment of many diseases including diabetes, flu, arthritis, inflammations, pain and diarrhea [17,18,21]. *Rosa sempervirens*, the evergreen rose, is a representative member of the *Rosa* species. It is a thorny, climbing rose characterized by long branches with few or no prickles. Moreover, it is an important source of vitamin C, carotenoids, polyphenols, organic acids and tocopherols [22]. *Pyracantha coccinea* of the genus *Pyracantha* of the *Rosaceae* family is a plant growing from South Europe to South-East Asia. Its fruits are known for their rich content in fatty acids, polyphenolic compounds, phytosterols and vitamins. Thus, they have been used in traditional medicines for cardiac and diuretic properties [23].

The aim of the present study was to investigate the antioxidant properties of polyphenolic extracts derived from the following three wild *Rosaceae* species of Greece; *Pyracantha coccinea* (fruit extract), *Rosa sempervirens* (fruit extract) and *Rosa canina* (fruit extract). The extracts were initially examined for their free radical scavenging activity against DPPH• and 2,2′-azino-bis(3-ethylbenzothiazoline-6-sulphonic acid) (ABTS)•$^+$ radicals and for their protective effects against peroxyl (ROO•) radical-induced DNA damage. Moreover, for the first time, the ability of these extracts to enhance the antioxidant defense in human endothelial cells was assessed. As mentioned above, the oxidative stress-induced endothelium damage is a crucial etiological factor for cardiovascular diseases, and so the identification of compounds that could protect from such damage is important.

2. Materials and Methods

2.1. Plant Material and Preparation of the Extracts

Fresh plant material of the investigated *Rosaceae* family plants *Rosa sempervirens* (fruit), *Rosa canina* (fruit) and *Pyracantha coccinea* (fruit) were collected during early summer of 2014, from respective wild plants colonies that grow on Parnitha mountain (Attica Region, Greece). A voucher specimen for each sampling material has been deposited in the herbarium of the Agricultural University of Athens (Athens, Greece).

After air-drying, 500 g of each plant material were crushed, homogenized with a blender and lyophilized to provide a powder, which was stirred in darkness with 1 L of methanol (HPLC-analytical grade) for 48 h. Then, the solution was filtered and the solid was re-extracted twice following the same procedure. The combined methanolic extracts (3 L in total) were evaporated to dryness under reduced pressure to provide crude extracts, which were further elaborated for the assessment of their chemical content and bioactivities.

2.2. Determination of Extracts' Polyphenolic Content Using HPLC and Ultra Performance Liquid Chromatography–Tandem Mass Spectrometer (UPLC-MS-MS) Analysis

The chemical composition of the extracts was determined using HPLC analysis performed on a Hewlett Packard HP1100 (Hewlett Packard, Palo Alto, CA, USA) equipped with an Agilent 1100 diode-array detector (Agilent Technologies, Santa Clara, CA, USA) (measuring absorbance over the full wavelength range during the entire run), a quaternary pump, degasser and coupled to HP ChemStation utilizing the manufacturer's 5.01 software package system (Hewlett Packard, Palo Alto, CA, USA).

The column used was a Zorbax Eclipse Plus C18, 5 μm, 150 × 4.6 mm i.d. chromatographic column (Agilent Technologies, Santa Clara, CA, USA), connected with a guard column of the same material (8 × 4 mm). Injection was by means of a Rheodyne injection valve (model 7725I) with a 20 μL fixed loop. For the chromatographic analyses HPLC-grade water was prepared using a Milli-Q system (Merck Millipore, Burlington, MA, USA), whereas all HPLC solvents (except acetonitrile) were filtered prior to use through cellulose acetate membranes of 0.45 μm pore size.

The mobile phase was composed of a gradient system of 0.3% acetic acid in water (A) and acetonitrile (B). The flow rate was maintained at 1 mL/min and the column gradient elution program consisted of: 25% B (0 min), 25% B (5 min), 30% B (10 min), 40% B (15 min), 50% B (20 min) and 70% B (30 min) where it remained for additional 5 min, and returned during 2 min to initial conditions, where it stayed for additional 2 min. This routine was followed by a 15 min equilibration period with the zero-time mixture prior to injection of the next sample. Prolonged runtimes (extended until 100 min) were also applied to determine constituents that elute after the 35 min (betulinic and ursolic acids). Peaks were identified by comparing their retention times and UV–vis spectra with the reference compounds, and data were quantitated using the corresponding curves of the reference compounds as standards (Extrasynthese, Genay Cedec, France; Sigma-Aldrich, St. Louis, MO, USA; Alfa Aesar, Haverhill, MA, USA; Fluka, Buchs, Switzerland). Confirmatory UPLC-MS-MS analysis was carried out on a Thermo Scientific Ultra High Performance Liquid Chromatography system (Waltham, MA, USA) coupled to a TSQ Quantum Vantage (Thermo Fischer Scientific, San Jose, CA, USA) triple quadrupole mass spectrometer. Mass spectrometric analysis was conducted using a heated electrospray ionization (HESI) operating in two complementary modes (positive and negative mode). Selected ion monitoring (SIM) mode was primarily used to confirm the presence of analytes. In selected cases of compounds, tandem mass spectrometry (MS/MS) utilizing the multiple reaction monitoring mode (MRM) was employed for additional confirmation [quinic acid: parent ion, m/z 191 (m/z 85, 93 product ions), rutin: parent ion, m/z 609 (m/z 300, 271 product ions), quercetin: parent ion, m/z 300.9 (m/z 179, 151 product ions), quercitrin: parent ion, m/z 447.1 (m/z 301, 300, 271 product ions)]. The working conditions were the following: spray voltage 4.2 kV; vaporizer and capillary temperatures 280 and 260 °C respectively, while sheath and auxiliary gas at 60 and 40 arbitrary units respectively. The LC separation was achieved on a Hypersil Gold. 3 μm. 150 × 3 mm i.d. chromatographic column (Thermo Fischer Scientific, San Jose, CA, USA). The mobile phase and the gradient system were identical to the above mentioned for the HPLC-UV analysis, using a flow rate of 0.3 mL/min. Water, acetonitrile, and acetic acid were purchased from Merck (Darmstadt, Germany) and all were LC-MS grade. PTFE filters (0.45 μm) were obtained from Macherey-Nagel, Duren, Germany. All measurements were repeated three times.

2.3. Analytical Method Validation

With respect to the analytical method validation part, the linearity for all analytes was determined within the ranges of 10–1000 ng/mL (using matrix matched calibration standards), demonstrating acceptable correlation coefficient values ($r^2 \geq 0.99$). Recovery of the investigated compounds (as a criterion of the trueness of the method) was evaluated at two concentration levels (40 and 200 ng/mL) by the addition of mixed solutions of the standards into the respective extract and fell within the acceptable range of 70%−120%. Precision values were always acceptable with percent Relative Standard Deviation (RSD%) < 15%.

2.4. Assessment of the Total Polyphenolic Content of the Extracts

The total polyphenolic content (TPC) of the extracts was evaluated by the Folin-Ciocalteu method as described previously [24]. Briefly, 20 µL of the extract were added to a tube containing 1 mL of deionized water. 100 µL of Folin-Ciocalteu reagent was added to the reaction mixture, followed by incubation for 3 min at room temperature. Afterwards, 280 µL of 25% *w/v* sodium carbonate solution and 600 µL of deionized water were added to the mixture. Following 1 h incubation at room temperature in the dark, the absorbance was measured at 765 nm versus a blank containing Folin-Ciocalteu reagent and distilled water without the extract. The measurement of absorbance was conducted on a Hitachi U-1900 ratio beam spectrophotometer (Tokyo, Japan). The optical density of the sample (20 µL) in 25% *w/v* solution of sodium carbonate (280 µL) and distilled water (1.7 mL) at 765 nm was also measured. TPC was determined by a standard curve of absorbance values in correlation with standard concentrations (50–1500 µg/mL) of gallic acid. The total polyphenolic content was expressed as mg of gallic acid equivalents (GAE) per gram of dried weight (dw) of extract.

2.5. Total Flavonoid Content of the Extracts

The total flavonoid content (TFC) of the extracts was evaluated as described previously with minor changes [25]. In particular, 1 mL of the methanolic extract was added into a 10 mL flask containing 4 mL of deionized water. Then, 0.3 mL of sodium nitrite (5%) were added to this mixture and allowed to stand for 5 min at room temperature. Then, 0.3 mL of $AlCl_3 \cdot 6H_2O$ (10% ethanolic) was added, the mixture was allowed to stand for 1 min at room temperature and 2 mL of potassium hydroxide (1 M) was added. The solution was diluted to 10 mL with the addition of deionized water and the absorbance of the solution versus a blank at 510 nm was measured immediately. Flavonoid content was expressed as mg of catechin equivalents (CE) per gram of dry weight of extract by using a standard curve (absorbance versus concentration) prepared from authentic catechin sample.

2.6. Free Radical Scavenging Activity

Free radical scavenging activity of the extracts was evaluated using the 2,2-diphenyl-picrylhydrazyl (DPPH•) and 2,2′-azino-bis(3-ethylbenzthiazoline-6-sulfonic acid) (ABTS•$^+$) radical scavenging assays [26,27]. Regarding the DPPH• assay, 1.0 mL of freshly-made methanolic solution of DPPH• radical (100 µM) was mixed with the tested extract solution at different concentrations. The contents were vigorously mixed, incubated at room temperature in the dark for 20 min and the absorbance was measured at 517 nm. The measurement was conducted on a Hitachi U-1900 ratio beam spectrophotometer (Tokyo, Japan). In each experiment, the tested sample alone in methanol was used as blank and DPPH• alone in methanol was used as control. ABTS•$^+$ radical scavenging activity of the extracts was determined as described previously [27] with slight modifications. Briefly, ABTS•$^+$ radical was produced by mixing 2 mM ABTS with 30 µM H_2O_2 and 6 µM horseradish peroxidase (HRP) enzyme in 1 mL of distilled water. The solution was vigorously mixed, incubated at room temperature in the dark for 45 min until ABTS•$^+$ radical formation. Then, 10 µL of different extract concentrations were added in the reaction mixture and the absorbance at 730 nm was read. The measurement was conducted on a Hitachi U-1900 ratio beam spectrophotometer (Tokyo, Japan). In each experiment, the tested sample in distilled water containing ABTS and H_2O_2 was used as blank, and the $ABTS\bullet^+$ radical solution with 10 µL H_2O was used as control.

The percentage of radical scavenging capacity (RSC) of the tested extracts, for both assays was calculated according to the following equation:

$$\text{Radical scavenging capacity (\%)} = [(A_{control} - A_{sample})/\ A_{control}] \times 100$$

where $A_{control}$ and A_{sample} are the absorbance values of the control and the tested sample respectively. Moreover, in order to compare the radical scavenging efficiency of the extracts, the IC_{50} value showing the concentration that caused 50% scavenging of $DPPH\bullet$ and $ABTS\bullet^+$ radical was calculated from the graph plotted RSC percentage against extract concentration. All experiments were carried out in triplicate and at least on two separate occasions.

2.7. Peroxyl Radical-Induced DNA Plasmid Strand Cleavage

The peroxyl radical-induced DNA plasmid strand cleavage assay was performed as described previously [28]. In brief, peroxyl radicals ($ROO\bullet$) were produced from thermal decomposition of 2,2′-azobis(2-amidinopropane hydrochloride) (AAPH). The reaction mixture (10 µL) containing 1 µg Bluescript-SK+ plasmid DNA, 2.5 mM AAPH in phosphate-buffered saline (PBS) and the tested extract at different concentrations was incubated in the dark for 45 min at 37 °C. Then, the reaction was stopped by the addition of 3 µL loading buffer (0.25% bromophenol blue and 30% glycerol). After analyzing the DNA samples by agarose gel electrophoresis, they were photographed and analyzed using the Alpha Innotech Multi Image (ProteinSimple, San Jose CA, USA). In addition, plasmid DNA was treated with each extract alone at the highest concentration used in the assay in order to test their effects on plasmid DNA conformation. The percentage of the protective activity of the tested extracts from $ROO\bullet$-induced DNA strand breakage was calculated using the following formula:

$$\% \text{ Inhibition} = [(S - S_o)\ /\ (S_{control} - S_o)] \times 100$$

where $S_{control}$ is the percentage of supercoiled DNA in the negative control sample (plasmid DNA alone), S_o is the percentage of supercoiled plasmid DNA in the positive control sample (without tested extracts but in the presence of the radical initiating factor) and S is the percentage of supercoiled plasmid DNA in the sample with the tested extracts and the radical initiating factor. Moreover, IC_{50} values showing the concentration that inhibited the AAPH-induced relaxation by 50% were calculated from the graph plotted percentage inhibition against extract concentration. At least two independent experiments in triplicate were performed for each tested compound.

2.8. Cell Culture Conditions

As previously described [29], human endothelial EA.hy926 cells were cultured in normal Dulbecco's modified Eagle's medium (DMEM) in plastic disposable tissue culture flasks at 37 °C in 5% carbon dioxide.

2.9. XTT Cell Viability Assay

For examining the extracts' antioxidant activity in endothelial cells, non-cytotoxic concentrations were used. For selecting these concentrations, extracts' cytotoxicity in endothelial cells was checked using the XTT cell viability assay kit (Roche, Switzerland) as previously described [30]. Briefly, EA.hy926 cells were seeded into a 96-well plate with 1×10^4 cells per well in DMEM medium. After 24 h incubation, the cells were treated with different concentrations of the extracts in serum-free DMEM medium for 24 h. Then, 50 µL of XTT test solution was added to each well. After 4 h of incubation, absorbance was measured at 450 nm and also at 630 nm as a reference wavelength in a Bio-Tek ELx800 microplate reader (Winooski, VT, USA). The negative control was DMEM serum-free medium. The absorbance values of the control and samples were used for calculating the percentage inhibition of cell growth caused by the extract treatment. All experiments were carried out in triplicate and on two separate occasions.

2.10. Treatment of EA.hy926 Cells with the Extracts

The extracts from *R. sempervirens*, *R. canina* and *P. coccinea* were examined for their antioxidant capacity in endothelial EA.hy926 cells. The cells were cultured in flasks for 24 h. Afterwards the medium was replaced with a serum-free medium containing the tested extracts at non-cytotoxic concentrations. The cells were treated with the extracts for 24 h, and then they were trypsinized, collected and centrifuged twice at 300× g for 10 min at 5 °C. At the end of the first centrifugation, the supernatant fluid was discarded and the cellular pellet was resuspended in PBS. After the second centrifugation, the cell pellet was collected and used for measuring the glutathione (GSH) and ROS levels.

2.11. Assessment of GSH and ROS Levels by Flow Cytometry Analysis in Endothelial Cells

The GSH and ROS levels in EA.hy926 cells were assessed using mercury orange and DCF-DA, respectively, as described previously [31,32]. In brief, the cells were resuspended in PBS at 1×10^6 cells/mL and incubated in the presence of mercury orange (10 µM) or 2′,7′-Dichlorofluorescin diacetate (DCF-DA) (40 µM) respectively, in the dark at 37 °C for 30 min. Then, the cells were washed, resuspended in PBS, and submitted to flow cytometric analysis using a FACSCalibur flow cytometer (Becton Dickinson, Franklin Lakes, NJ, USA) with excitation and emission wavelengths at 488 and 530 nm for ROS and at 488 and 580 nm for GSH. Data were analyzed using 'BD Cell Quest' software (Becton Dickinson, Franklin Lakes, NJ, USA). Each experiment was repeated at least three times.

2.12. Statistical Analysis

All results were expressed as mean ± SD. Differences were considered significant at $p < 0.05$. One-way ANOVA was performed followed by Tukey's test for multiple pair-wise comparisons using the SPSS 20.0 software (IBM, Armonk, NY, USA).

3. Results

3.1. Extract Composition in Bioactive Compounds

The TPC values of the extracts were 267.67, 290.00 and 226.93 mg GAE/g dw, while the TFC values were 65.78, 118.56 and 99.16 mg CE/gr dw for *R. sempervivens*, *R. canina* and *P. coccinea*, respectively (Table 1).

Table 1. Bioactive compounds of extracts from dried fruits *R. canina*, *R. semprevirens* and *P. coccinea*.

Compound	*R. canina* [a]	*R. semprevirens* [a]	*P. coccinea* [a]
Quinic acid	1102.59 ± 38.91	389.95 ± 10.29	ND
(+)-Catechin	134.75 ± 1.02	25.48 ± 0.68	7.93 ± 0.31
Gallic acid	2.21 ± 0.09	0.44 ± 0.03	ND
Protocatechuic acid	2.09 ± 0.06	ND	ND
Syringic acid	ND	ND	6.23 ± 0.17
Caffeic acid	ND	ND	1.49 ± 0.08
(−)-Epicatechin	120.99 ± 1.18	34.01 ± 0.51	10.23 ± 1.10
Hyperoside	308.11 ± 7.10	8.31 ± 0.19	170.72 ± 3.49
Rutin	25.64 ± 0.48	2.62 ± 0.14	25.82 ± 0.98
Chlorogenic acid	ND	ND	4.82 ± 0.06
Taxifolin	ND	ND	0.09 ± 0.02
p-Coumaric acid	2.44 ± 0.07	ND	ND

Table 1. *Cont.*

Compound	*R. canina* [a]	*R. semprevirens* [a]	*P. coccinea* [a]
Vanillin	ND	ND	7.89 ± 0.06
Astragalin	172.48 ± 7.48	9.16 ± 0.10	9.13 ± 0.05
Phloridzin	3.41 ± 0.10	ND	ND
Eriodictyol	ND	0.05 ± 0.01	0.06 ± 0.01
Quercitrin	ND	0.44 ± 0.07	ND
Quercetin	0.67 ± 0.05	0.19 ± 0.02	0.06 ± 0.01
Genistein	ND	0.03 ± 0.01	ND
Kaempferol	0.46 ± 0.02	ND	0.05 ± 0.01
Isorhamnetin	ND	ND	ND
Isosakuranetin	ND	ND	0.03 ± 0.01
Betulinic acid	0.47 ± 0.03	ND	ND
Ursolic acid	138.23 ± 4.44	ND	ND
TPC [b]	290.00 ± 2.10 [d]	267.67 ± 1.78 [e]	226.93 ± 1.11 [f]
TFC [c]	118.56 ± 1.69 [g]	65.78 ± 0.93 [h]	99.16 ± 1.22 [i]

ND: not detected. [a] Values are expressed as µg/g of dried weight of extract and are the mean ± SD from three measurements. [b] TPC: Total Polyphenolic Content, expressed as mg of gallic acid equivalent/g dried weight extract. [c] TFC: Total Flavonoid Content, expressed as mg of catechin equivalent/g dried weight extract. [d,e,f] Values with different superscript letters are significantly different between them ($p < 0.05$). [g,h,i] Values with different superscript letters are significantly different between them ($p < 0.05$).

The results from the qualitative and quantitative assessment of the chemical composition of the extracts as assessed by using HPLC with Diode-Array Detection complemented by UPLC-MS-MS (especially for the quantitative analysis of non-UV sensistive compounds) analysis are depicted in Table 1. The HPLC analysis of the extracts identified polyphenols belonging to different subclasses of flavonoids such as flavanols (e.g., (+)-catechin and (−)-epicatechin), flavonols (e.g., hyperoside, rutin, astragalin, quercitrin, quercetin and kaempferol), flavanonols (e.g., taxifolin), flavanones (e.g., eriodictyol and isosakuranetin), isoflavones (e.g., genistein). In this respect, various polyphenolic acids were also detected such as hydroxybenzoic acids (e.g., gallic acid, syringic acid and protocatechuic acid), hydroxycinnamic acids (e.g., caffeic acid and *p*-coumaric acid) and the chlorogenic acid. Finally, a series of additional molecules were identified such as phloridzin and vanillin polyphenols, quinic acid and the terpenoids betulinic acid and ursolic acid. The latter compounds were detected only in *R. canina* extract.

Specifically, the chemical analysis showed that the *R. canina* extract is particularly rich in polyphenols hyperoside (308.11 µg/g dw), astragalin (172.48 µg/g dw), (+)-catechin (134.75 µg/g dw) and (−)-epicatechin (120.99 µg/g dw) (Table 1). Moreover, *R. canina* extract contained significant concentration of quinic acid (1102.59 µg/g dw) and the terpenoid, ursolic acid (138.23 µg/g dw) (Table 1). In *R. sempervirens* extract, the compounds identified at higher concentrations were quinic acid (389.95 µg/g dw), and the polyphenols (−)-epicatechin (34.01 µg/g dw), (+)-catechin (25.48 µg/g dw), astragalin (9.16 µg/g dw), and hyperoside (8.31 µg/g dw) (Table 1). Finally, *P. coccinea* extract exhibited higher concentrations of hyperoside (170.72 µg/g dw), rutin (25.82 µg/g dw), (−)-epicatechin (10.23 µg/g dw), astragalin (9.13 µg/g dw), (+)-catechin (7.93 µg/g dw), vanillin (7.89 µg/g dw) and syringic acid (6.23 µg/g dw) (Table 1).

3.2. Free Radical Scavenging Activity of the Extracts

All three of the extracts exhibited strong free radical scavenging activity. As known, the lower the IC_{50} value, the higher the antioxidant activity. Thus, in the DPPH assay the potency order and the IC_{50} values were: *R. canina* (100 µg/mL) > *R. sempervivens* (130 µg/mL) > *P. coccinea* (500 µg/mL) (Table 2). Similar order of portency was observed in ABTS$^{\bullet+}$ assay; *R. canina* (60 µg/mL) > *R. sempervivens* (85 µg/mL) > *P. coccinea* (140 µg/mL) (Table 2).

Table 2. Free radical scavenging activity against DPPH and ABTS radicals, protective activity against peroxyl radical (ROO•)-induced DNA damage of the extracts.

Plant Extracts	IC$_{50}$ (µg/mL)		
	DPPH [a]	ABTS [a]	ROO [b]
Rosa sempervivens (fruit)	130 ± 7.8 [a,*]	85 ± 10.0 [d,*]	570 ± 51.3 [g,*]
Rosa canina (fruit)	100 ± 7.0 [b,*]	60 ± 6.6 [e,*]	530 ± 68.9 [g,*]
Pyracantha coccinea (fruit)	500 ± 40.0 [c,*]	140 ± 4.2 [f,*]	950 ± 47.5 [h,*]

[a] Values are the mean ± SD of at least two separate triplicate experiments. [b] Values are the mean ± SD from three independent experiments. * $p < 0.05$, indicates significant difference from the control values. [a,b,c] Values with different superscript letters are significantly different between them ($p < 0.05$). [d,e,f] Values with different superscript letters are significantly different between them ($p < 0.05$). [g,h] Values with different superscript letters are significantly different between them ($p < 0.05$).

Finally, all three extracts exhibited protective activity against ROO•-induced DNA plasmid breakage with IC$_{50}$ values and potency order *R. canina* (530 µg/mL) > *R. sempervivens* (570 µg/mL) > *P. coccinea* (950 µg/mL) (Table 2 and Figure 1).

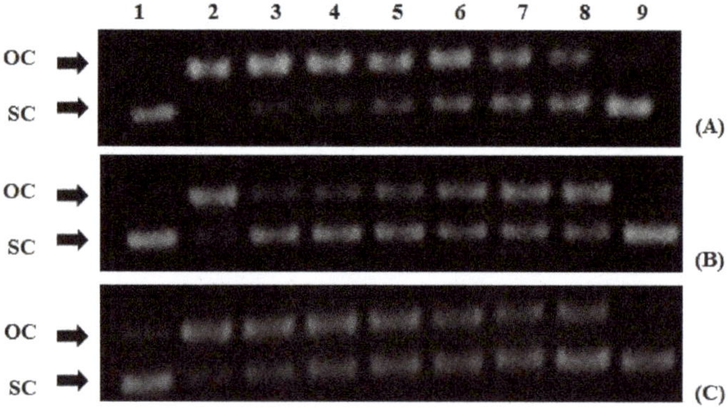

Figure 1. Protective activity of polyphenolic extracts from (**A**) *Pyracantha coccinea*, (**B**) *Rosa sempervivens* and (**C**) *Rosa canina* species against ROO• radical: Lane 1, pBluescript-SK+ plasmid DNA without any treatment; lane 2, plasmid DNA exposed to ROO• radical alone; lanes 3–8 plasmid DNA exposed to ROO• radical in the presence of different concentrations of extracts (*P. coccinea*: 0.063, 0.125, 0.250, 0.500, 1.0 and 1.5 mg/mL; *R. sempervivens*: 2.0, 1.5, 1.0, 0.500, 0.250, 0.125 mg/mL; *R. canina*: 0.063, 0.125, 0.250, 0.500, 1.0 and 1.5 mg/mL); lane 8, plasmid DNA exposed to the maximum tested concentration of each extract alone. OC: open circular; SC: supercoiled.

3.3. Effects of Extracts on the Antioxidant Status of Endothelial Cells

To examine the extracts' antioxidant activity in endothelial cells, flow cytometry analysis was performed. At first, the extract's effect on cell viability was assessed using the XTT assay, in order to use non-cytotoxic concentrations. The cell viability assay showed that significant cytotoxicity was exhibited at concentrations above 2.5 mg/mL for *R. sempervivens* and 2.0 mg/mL for *R. canina* (Figure 2B,C). None of the concentrations used for *P. coccinea* had cytotoxicity (Figure 2A). Thus, the selected non-cytotoxic concentrations of the extracts in the following assays were up to 1.00 mg/mL.

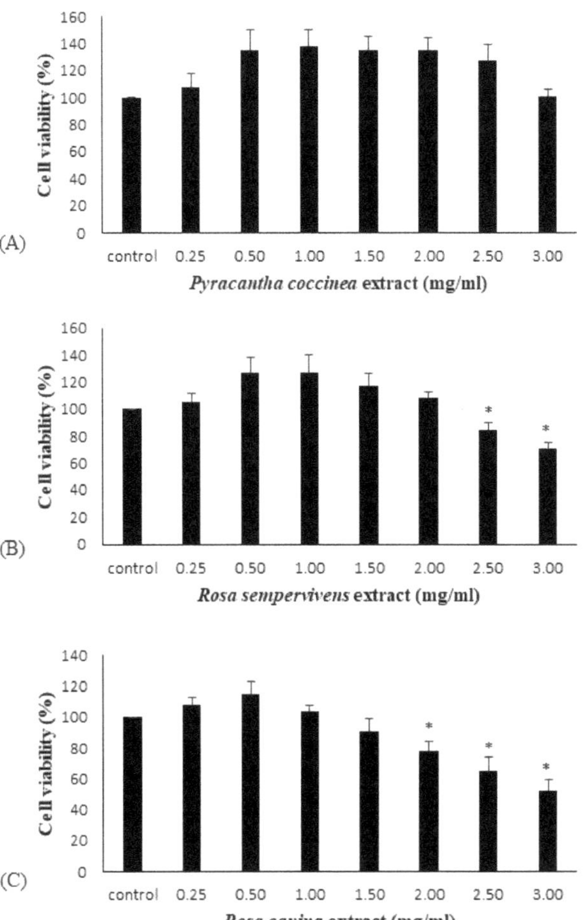

Figure 2. Cell viability following the treatment with polyphenolic extracts from (**A**) *Pyracantha coccinea*, (**B**) *Rosa sempervivens* and (**C**) *Rosa canina* species. The results are presented as the means ± SD of three independent experiments carried out in triplicate. * $p < 0.05$ indicates significant difference from the control value.

The assessment of the extracts' effects on the antioxidant capacity of endothelial cells was based on the measurement of GSH and ROS levels by flow cytometry analysis. The results demonstrated that *R. canina extract* significantly increased GSH levels by 15.0, 10.4, 28.4 and 43.1% at 0.13, 0.25, 0.50 and 1.00 mg/mL, respectively compared to control (Figure 3C). *P. coccinea* extract also significantly increased GSH levels by 29.2 and 32.3% at 0.50 and 1.00 mg/mL, respectively, compared to control (Figure 3A). However, *R. sempervirens* extract did not affect GSH levels at any of the examined concentrations (Figure 3B).

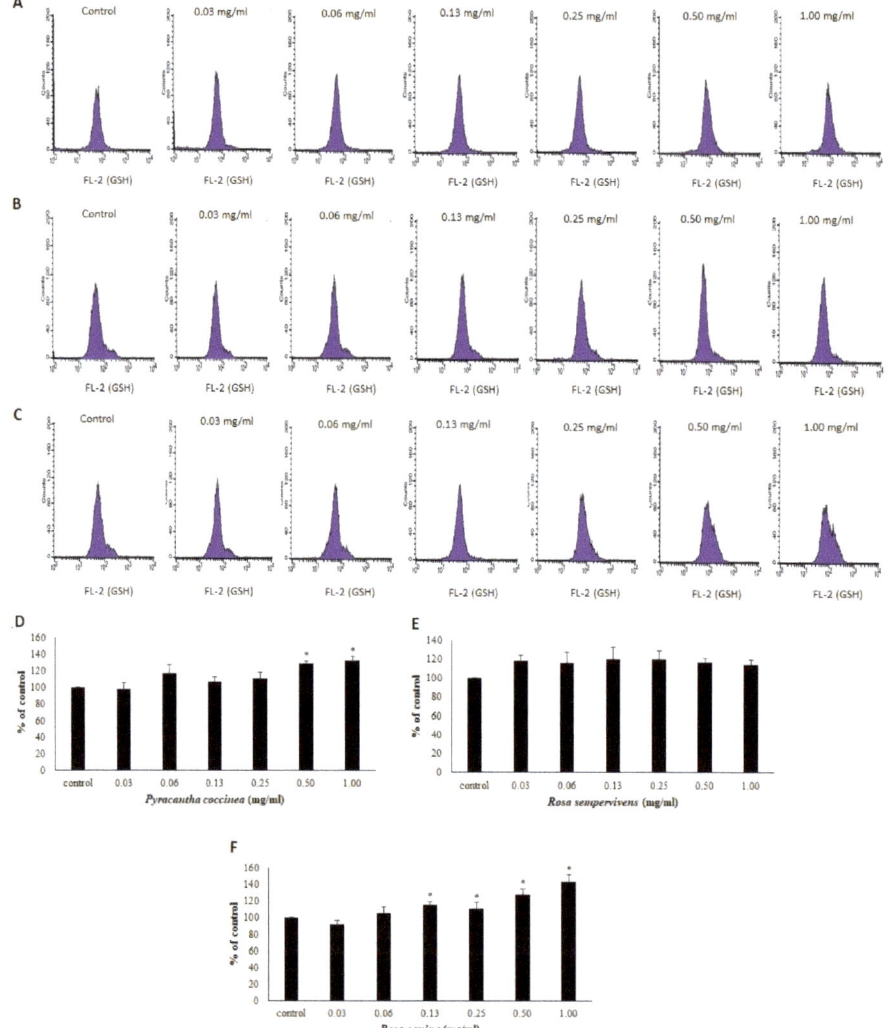

Figure 3. Effects on GSH levels after treatment with *P. coccinea*, *R. sempervivens* and *R. canina* extracts at different concentrations for 24 h in EA.hy926 cells. The histograms of cell counts versus fluorescence of 10,000 cells analyzed by flow cytometry for the detection of GSH levels after treatment with (**A**) *P. coccinea*, (**B**) *R. sempervivens* and (**C**) *R. canina*. FL-2 represents the detection of fluorescence in the FL-2 channel using 488 and 580 nm as the excitation and emission wavelength, respectively. Bar charts indicate the GSH levels as % of control as estimated by the histograms in EA.hy926 cells after treatment with (**D**) *P. coccinea*, (**E**) *R. sempervivens* and (**F**) *R. canina* extracts. All values of bar charts are presented as the mean ± SD of three independent experiments. * $p < 0.05$ indicates significant difference from the control.

The results from the assessment of extracts' effects on ROS levels are shown in Figure 4. According to the results, only one of the three extracts affected ROS levels. In particular, *R. canina extract* significantly reduced ROS by 9.73 and 13.37% at 0.50 and 1.00 mg/mL, respectively, compared to control

(Figure 4C). However, *P. coccinea* and *R. sempervivens* extracts did not significantly affect ROS levels, compared to control (Figure 4A,B).

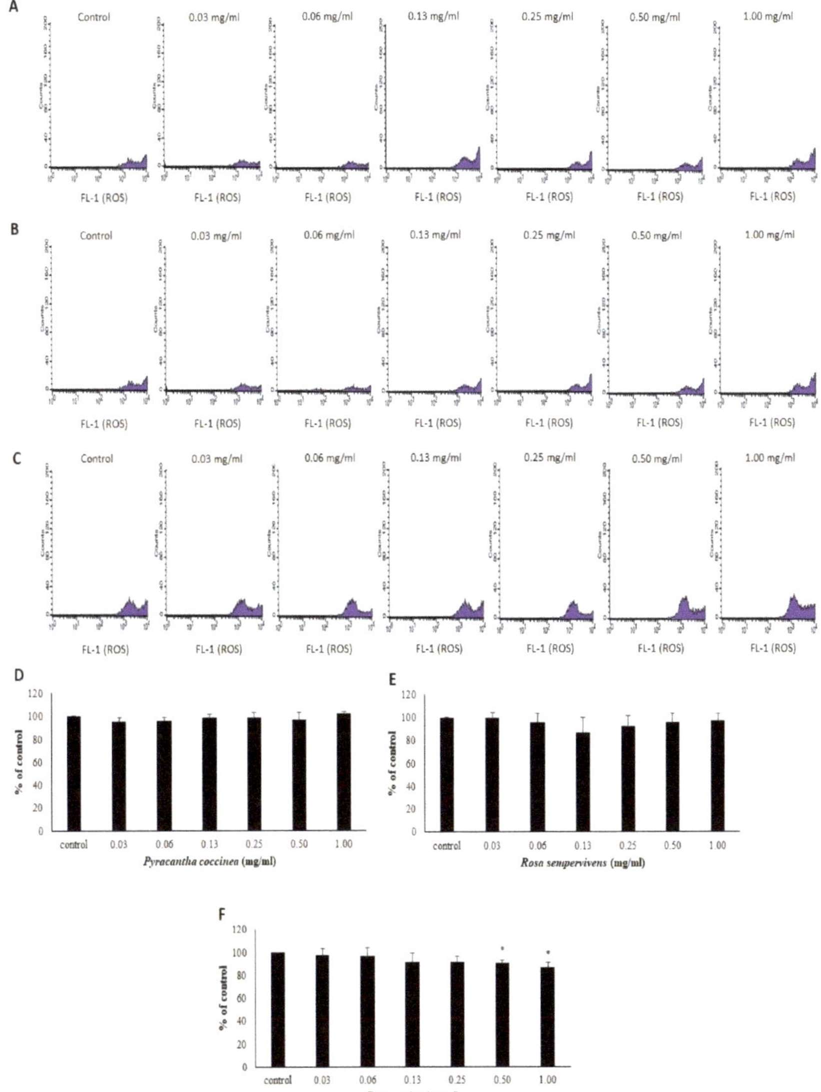

Figure 4. The diagrams show the changes in ROS levels after treatment with *P. coccinea*, *R. sempervivens* and *R. canina* extracts in EA.hy926 cells. The histograms demonstrate the cell counts versus fluorescence of 10,000 cells analyzed by flow cytometry for the detection of ROS levels after treatment with (**A**) *P. coccinea*, (**B**) *R. sempervivens* and (**C**) *R. canina*. FL-1 represents the detection of fluorescence in the FL-1 channel using 488 and 530 nm as the excitation and emission wavelength, respectively. Bar charts indicate the ROS levels as % of control as estimated by the histograms in EA.hy926 cells after treatment with (**D**) *P. coccinea*, (**E**) *R. sempervivens* and (**F**) *R. canina* extracts. All values of bar charts are presented as the mean ± SD of three independent experiments. * $p < 0.05$ indicates significant difference from the control.

4. Discussion

The aim of the present study was to assess for the first time the antioxidant effects in endothelial cells of the polyphenolic extracts obtained from the fruits of three wild growing plant species, *R. canina*, *R. sempervirens* and *P. coccinea*. It should be noted that *R. canina* is a well-studied plant species [33], but there is only one study for the polyphenolic composition and the antioxidant activity of *R. semprevirens* fruit extracts [34] and there are only few reports for *P. coccinea* fruit extracts [23].

Initially, the polyphenolic composition of the extracts was analyzed. *R. canina* extract exhibited the highest TPC among the three tested extracts. So, as expected *R. canina* had also the highest TFC. However, although the *R. sempervirens* extract had higher TPC than *P. coccinea*, the TFC of the former was lower than that of the latter. This was probably due to the high content of the *R. sempervirens* extract in polyphenols other than flavonoids. Other studies showed that the TPC values of *R. canina* extracts were from 50 to 500 mg/g dw, and so our results were within this range [35,36]. There has been so far only one study on the polyphenols of *R. sempervirens*, which reported 57.9 mg GAE/g dw for TPC and 0.47 mg CE/g dw for TFC, while our values were 267.67 mg GAE/g dw and 65.78 mg CE/g dw, respectively [34]. Nadpal et al., [34] have used different solvents (i.e., methanol and water) for isolating polyphenolic ectracts from *R. sempervirens*. Their results showed that water extraction yielded higher TPC and TFC values compared to methanol extraction [34]. Moreover, another study demonstrated that the TPC values were different between *R. canina* extracts isolated from plants grown in different locations [36]. In a previous study, it was also shown that ethanol extracts of *R. canina* had higher TPC values than water extracts [36]. Bhave et al., [37] have also shown that the content of biologically active compounds in Rosa species depended on specific genotypes. In addition, extracts from more ripened fruits of *R. canina* have been reported to have higher TPC compared to less ripened fruits [38]. Thus, the differences between our results and those from other studies could be atrributed to different factors such as differences in the analytical methods, different extraction methods, genetic and environmental factors and the maturity stage of fruits and harvesting time [37–39].

The HPLC analysis showed that the *R. canina* extract was especially rich in the polyphenols hyperoside, astragalin, rutin, (+)-catechin and (−)-epicatechin as well as in the terpenoid ursolic acid and a poly-hydroxylated organic acid organic acid, which is the quinic acid. The presence of these compounds have also been reported in previous studies [33,40]. Moreover, several other polyphenols have been identified in trace amounts in *R. canina* extracts such as ellagic acid, salicylic acid, vanillic acid, ferulic acid and caffeic acid (not found by us) [33]. The *R. sempervirens* extract had higher amounts of quinic acid, (+)-catechin, (−)-epicatechin, astragalin, and hyperoside. Like our findings, Nadpal et al. (2018) [34] have also identified gallic acid, quercitrin, quercetin, hyperoside and (+)-catechin in *R. sempervirens* extract, although not at the same concentrations as our samples. However, Nadpal et al. (2018) [34] have also reported ellagic acid, protocatechuic acid (not found in our extract), ferulic acid and kaempferol-3-*O*-glucoside. In general, our study is the first that identified quinic acid, (−)-epicatechin, rutin, astragalin, eriodictyol, and genistein in *R. sempervirens* extract. The *P. coccinea* extract contained higher concentrations of hyperoside, rutin, (−)-epicatechin, (+)-catechin, astragalin, vanillin, syringic acid and chlorogenic acid. As expected, the polyphenolic profiles of the two extracts from the *Rosa* genus (i.e., *R. canina* and *R. sempervirens*) had more similaritities with each other than with the extract from *P. coccinea* of *Pyracantha* genus.

All three extracts exhibited strong free radical scavenging activity in DPPH and ABTS•$^+$ assays. The highest potency of the *R. canina* extract in both assays was in accordance with its highest TPC value. Some studies have reported IC$_{50}$ values in DPPH assay for *R. canina* extracts, which were similar to ours, but other studies have found values that were significantly different from ours [18,40]. Previous studies have also shown that polyphenolics were the most important compounds of *R. canina* extracts for the DPPH radical scavenging activity [40]. *R. sempervirens* extract having higher TPC than *P. coccinea* extract also exhibited greater free radical scavenging activity. Nadpal et al. (2018) [34] have reported 28 µg/mL as IC$_{50}$ value in DPPH assay for a *R. sempervirens* extract, while our value was 130 µg/mL.

The differences between our values and those from other studies are probably, as mentioned above, due to different factors which lead to differences in the polyphenolic composition of the extracts.

Moreover, all three extracts exerted protective activity against ROO•-induced DNA damage. Like in DPPH and ABTS•$^+$ assays, the potency order in this assay followed the order of polyphenolic concentration, that is, *R. canina* > *R. sempervirens* > *P. coccinea*. Therefore, once again, the polyphenolic concentration seemed to play a crucial role in the protective activity from ROO•-induced DNA damage. Among the polyphenols identified in the extracts, (+)-catechin, (−)-epicatechin, rutin, vanillin, astragalin, phloridzin and gallic acid have been reported to scavenge ROO• radical [41–47]. As far as we know, protective activity of *R. canina*, *R. sempervirens* and *P. coccinea* extracts from free radical-induced DNA damage was examined for the first time in this study. Thus, based on these results, extracts from the tested plant species may be used for protection against diseases caused by ROS-induced DNA damage.

Since the tested extracts showed strong free radical scavenging activity, their antioxidant effects were also investigated at noncytotoxic concentrations in endothelial cells. Thus, the extracts' ability to increase GSH, one of the most important antioxidant molecules within cells, was assessed by flow cytometry [48]. In agreement with the free radical scavenging assays, *R. canina* extract exhibited the greatest capacity to increase GSH in the cells. However, it should be noted that when humans consumed rose hip powder from *R. canina*, there were no effects on the activity of enzymes related to GSH metabolism in erythrocytes [49]. Treatment of cells with *P. coccinea* extract also increased significantly GSH levels in EA.hy926 cells. However, *R. sempervirens* extract treatment had no effect on GSH levels. This finding was intriguing, since *R. sempervirens* extract contained more polyphenols than *P. coccinea* extract, while *R. sempervirens* extract's polyphenolic profile was quite similar to that of *R. canina*. This contradiction could be explained by examining the polyphenols that were found at a greater concentration in *P. coccinea* extract than in *R. sempervirens*. For example, hyperoside and rutin found at higher concentration in *R. canina* and *P. coccinea* extracts compared to *R. sempervirens* have been reported to increase GSH levels in endothelial cells [50,51]. The observed increase in GSH levels induced by *R. canina* and *P. coccinea* extract treatment is important, since GSH apart from its antioxidant role is a crucial regulator of cell signaling in endothelial cells [52,53].

Regarding extracts' effects on ROS levels in endothelial cells, only *R. canina* extract treatment exerted a significant decrease. This result was in accordance with *R. canina* extract's highest antioxidant potency exhibited in all the other assays. Moreover, *R. canina* extract-induced decrease in ROS levels was attributed, at least in part, to extract's capacity to increase antioxidant defense mechanisms such as GSH. Moreover, a *R. canina* extract has been shown to inhibit an H_2O_2-induced increase in ROS in colon cancer cells [54]. Furthermore, some polyphenols such as (+)-catechin, (−)-epicatechin and ursolic acid identified at higher concentrations in *R. canina* than in the other two tested extracts have been demonstrated to decrease ROS in endothelial EA.hy926 and human umbilical vein endothelial cells (HUVEC) cells [55–57].

In conclusion, the results of the present study provided new information concerning the polyphenolic composition of the tested extracts, especially those of *R. sempervirens* and *P. coccinea* fruit extracts. Moreover, all three tested extracts were demonstrated for the first time to protect against ROS-induced DNA damage, which thus suggests their possible use for prevention of relative diseases. In addition, the extracts from *R. canina* and *P. coccinea* were shown to increase GSH levels, the most important antioxidant molecule in endothelial cells. This finding suggests that these extracts may be used for the development of food supplements or biofunctional foods preventing diseases caused by oxidative damage to endothelium such as cardiovascular. Interestingly, previous *in vivo* studies have reported that administration of rose hip powder to humans or experimental animals could reduce the risk for cardiovascular diseases [22,58]. Of course, further studies are needed in order to investigate further the molecular mechanisms through which these protective activities are exerted. Moreover, since the environmental variability between different years may affect the chemical composition of the plant extracts and consequently their bioactivities, it should also be examined how the tested activities are changing from one year to the next.

Author Contributions: Conceptualization, D.S., S.A.H., D.K.; methodology, D.S., S.A.H., D.K.; supervision, D.S., S.A.H., D.K.; carried out the experiments, E.K. (Eleana Kokka), A.A., K.C., E.K. (Efthalia Kerasioti), C.D., E.N.T., A.P., S.D.K., I.K.; data curation, E.K. (Efthalia Kerasioti), A.P., A.A., D.S., S.A.H.; writing, D.S., S.A.H., K.C., E.K. (Eleana Kokka), C.D., I.K.; funding acquisition, D.S., S.A.H., D.K.; resources, D.S., S.A.H., D.K.; K.C., E.K. (Eleana Kokka), C.D., have contributed equally to the study.

Funding: The work was funded in part by the "Toxicology" MSc program (grant no.: 5835) in the Department of Biochemistry and Biotechnology at the University of Thessaly.

Conflicts of Interest: The authors declare no conflict of interest.

References

1. Dupre-Crochet, S.; Erard, M.; Nüβe, O. ROS production in phagocytes: Why, when, and where? *J. Leukoc. Biol.* **2013**, *94*, 657–670. [CrossRef]

2. Halliwell, B. Free radicals and other reactive species in disease. In *Encyclopedia of Life Sciences*; Wiley, John & Sons: London, UK, 2001; pp. 1–7.

3. Zhang, J.; Wang, X.; Vikash, V.; Ye, Q.; Wu, D.; Liu, Y.; Dong, W. ROS and ROS-Mediated Cellular Signaling. *Oxid. Med. Cell Longev.* **2016**, *2016*, e4350965. [CrossRef] [PubMed]

4. Rochette, L.; Zeller, M.; Cottin, Y.; Vergely, C. Diabetes, oxidative stress and therapeutic strategies. *Biochim. Biophys. Acta* **2014**, *1840*, 2709–2729. [CrossRef] [PubMed]

5. Lipinski, B. Pathophysiology of oxidative stress in diabetes mellitus. *J. Diabetes Complic.* **2001**, *15*, 203–210. [CrossRef]

6. Halliwell, B. Free radicals and antioxidants: Updating a personal view. *Nutr. Rev.* **2012**, *70*, 257–265. [CrossRef] [PubMed]

7. Deanfield, J.E.; Halcox, J.P.; Rabelink, T.J. Endothelial function and dysfunction: Testing and clinical relevance. *Circulation* **2007**, *115*, 1285–1295. [CrossRef] [PubMed]

8. Victor, V.M.; Rocha, M.; Solá, E.; Bañuls, C.; Garcia-Malpartida, K.; Hernández-Mijares, A. Oxidative stress, endothelial dysfunction and atherosclerosis. *Curr. Pharm. Des.* **2009**, *15*, 2988–3002. [CrossRef]

9. Closa, D.; Folch-Puy, E. Oxygen free radicals and the systemic inflammatory response. *IUBMB Life* **2004**, *56*, 185–191. [CrossRef]

10. Muller, M.M.; Griesmacher, A. Markers of endothelial dysfunction. *Clin. Chem. Lab. Med.* **2000**, *38*, 77–85. [CrossRef] [PubMed]

11. Zou, Y.; Yoon, S.; Jung, K.J.; Kim, C.H.; Son, T.G.; Kim, M.S.; Kim, Y.J.; Lee, J.; Yu, B.P.; Chung, H.Y. Upregulation of aortic adhesion molecules during aging. *J. Gerontol. A Biol. Sci. Med. Sci.* **2006**, *61*, 232–244. [CrossRef]

12. Karpinska, A.; Gromadzka, G. Oxidative stress and natural antioxidant mechanisms: The role in neurodegeneration. From molecular mechanisms to therapeutic strategies. *Postepy Hig. Med. Dosw.* **2013**, *67*, 43–53. [CrossRef]

13. Landete, J.M. Dietary intake of natural antioxidants: Vitamins and polyphenols. *Crit. Rev. Food Sci. Nutr.* **2013**, *53*, 706–721. [CrossRef] [PubMed]

14. Procházková, D.; Boušová, I.; Wilhelmová, N. Antioxidant and prooxidant properties of flavonoids. *Fitoterapia* **2011**, *82*, 513–523. [CrossRef] [PubMed]

15. Landete, J.M. Updated knowledge about polyphenols: Functions, bioavailability, metabolism, and health. *Crit. Rev. Food Sci. Nutr.* **2012**, *52*, 936–948. [CrossRef]

16. Jiménez, S.; Jiménez-Moreno, N.; Luquin, A.; Laguna, M.; Rodríguez-Yoldi, M.J.; Ancín-Azpilicueta, C. Chemical composition of rosehips from different Rosa species: An alternative source of antioxidants for food industry. *Food Addict. Contam. Part A* **2017**, *34*, 1121–1130. [CrossRef] [PubMed]

17. Roman, I.; Stanila, A.; Stanila, S. Bioactive compounds and antioxidant activity of *Rosa canina* L. Biotypes from spontaneous flora of Transylvania. *Chem. Cent. J.* **2013**, *7*, 73. [CrossRef] [PubMed]

18. Guimarães, R.; Barros, L.; Calhelha, R.C.; Carvalho, A.M.; Queiroz, M.J.R.; Ferreira, I.C. Bioactivity of different enriched phenolic extracts of wild fruits from northeastern portugal: A comparative study. *Plant Foods Hum. Nutr.* **2014**, *69*, 37–42. [CrossRef] [PubMed]

19. Nadpal, J.D.; Lesjak, M.M.; Šibul, F.S.; Anačkov, G.T.; Četojević-Simin, D.D.; Mimica-Dukić, N.M.; Beara, I.N. Comparative study of biological activities and phytochemical composition of two rose hips and their preserves: *Rosa canina* L. and Rosa arvensis Huds. *Food Chem.* **2016**, *192*, 907–914. [CrossRef] [PubMed]
20. Serteser, A.; Kargioğlu, M.; Gök, V.; Bağci, Y.; Ozcan, M.M.; Arslan, D. Determination of antioxidant effects of some plant species wild growing in Turkey. *Int. J. Food Sci. Nutr.* **2008**, *59*, 643–651. [CrossRef]
21. Tumbas, V.T.; Canadanovic-Brunet, J.M.; Cetojevic-Simin, D.D.; Cetkovic, G.S.; Ethilas, S.M.; Gille, L. Effect of rosehip (*Rosa canina* L.) phytochemicals on stable free radicals and human cancer cells. *J. Sci. Food Agric.* **2012**, *92*, 1273–1281. [CrossRef]
22. Andersson, S.C.; Olsson, M.E.; Gustavsson, K.E.; Johansson, E.; Rumpunen, K. Tocopherols in rose hips (*Rosa* spp.) during ripening. *J. Sci. Food Agric.* **2012**, *92*, 2116–2121. [CrossRef] [PubMed]
23. Keser, S. Antiradical activities and phytochemical compounds of firethorn (*Pyracantha coccinea*) fruit extracts. *Nat. Prod. Res.* **2014**, *28*, 1789–1794. [CrossRef] [PubMed]
24. Singleton, V.L.; Orthofer, R.; Lamuela-Raventos, R. Analysis of Total Phenols and Other Oxidation Substrates and Antioxidants by Means of Folin-Ciocalteu Reagent. *Methods Enzymol.* **1999**, *299*, 152–178.
25. Jia, Z.; Tang, M.C.; Wu, J.M. The determination of flavonoid contents in mulberry and their scavenging effects on superoxide radicals. *Food Chem.* **1999**, *64*, 555–559.
26. Brand-Williams, W.; Cuvelier, M.E.; Berset, C. Use of a free radical method to evaluate antioxidant activity. *LWT Food Sci. Technol.* **1995**, *28*, 25–30. [CrossRef]
27. Cano, A.; Hernández-Ruíz, J.; García-Cánovas, F.; Acosta, M.; Arnao, M.B. An end-point method for estimation of the total antioxidant activity in plant material. *Phytochem. Anal.* **1998**, *9*, 196–202. [CrossRef]
28. Chang, S.T.; Wu, J.H.; Wang, S.Y.; Kang, P.L.; Yang, N.S.; Shyur, L.F. Antioxidant activity of extracts from Acacia confusa bark and heartwood. *J. Agric. Food Chem.* **2001**, *49*, 3420–3424. [CrossRef] [PubMed]
29. Priftis, A.; Panagiotou, E.M.; Lakis, K.; Plika, C.; Halabalaki, M.; Ntasi, G.; Veskoukis, A.S.; Stagos, D.; Skaltsounis, L.A.; Kouretas, D. Roasted and green coffee extracts show antioxidant and cytotoxic activity in myoblast and endothelial cell lines in a cell specific manner. *Food Chem. Toxicol.* **2018**, *114*, 119–127. [CrossRef] [PubMed]
30. Kerasioti, E.; Stagos, D.; Priftis, A.; Aivazidis, S.; Tsatsakis, A.M.; Hayes, A.W.; Kouretas, D. Antioxidant effects of whey protein on muscle C_2C_{12} cells. *Food Chem.* **2014**, *155*, 271–278. [CrossRef] [PubMed]
31. Goutzourelas, N.; Stagos, D.; Demertzis, N.; Mavridou, P.; Karterolioti, H.; Georgadakis, S.; Kerasioti, E.; Aligiannis, N.; Skaltsounis, L.; Statiri, A.; et al. Effects of polyphenolic grape extract on the oxidative status of muscle and endothelial cells. *Hum. Exp. Toxicol.* **2014**, *33*, 1099–1112. [CrossRef]
32. Kerasioti, E.; Stagos, D.; Georgatzi, V.; Bregou, E.; Priftis, A.; Kafantaris, I.; Kouretas, D. Antioxidant Effects of Sheep Whey Protein on Endothelial Cells. *Oxid. Med. Cell Longev.* **2016**, *2016*, e6585737. [CrossRef]
33. Ayati, Z.; Amiri, M.S.; Ramezani, M.; Delshad, E.; Sahebkar, A.; Emami, S.A. Phytochemistry, traditional uses and pharmacological profile of rose hip: A review. *Curr. Pharm. Des.* **2018**, *24*, 4101–4124. [CrossRef]
34. Nadpal, J.D.; Lesjak, M.M.; Mrkonjić, Z.O.; Majkić, T.M.; Četojević-Simin, D.D.; Mimica-Dukić, N.M.; Beara, I.N. Phytochemical composition and in vitro functional properties of three wild rose hips and their traditional preserves. *Food Chem.* **2018**, *241*, 290–300. [CrossRef]
35. Gao, X.; Bjo, L.; Ttrajkovski, V.; Uggla, M. Evaluation of antioxidant activities of rosehip ethanol extracts in different test systems. *J. Sci. Food Agric.* **2000**, *80*, 2021–2027. [CrossRef]
36. Koczka, N.; Stefanovits-Banyai, E.; Ombodi, A. Total Polyphenol Content and Antioxidant Capacity of Rosehips of Some Rosa Species. *Medicines* **2018**, *5*, 84. [CrossRef]
37. Bhave, A.; Schulzova, V.; Chmelarova, H.; Mrnka, L.; Hajslova, J. Assessment of rosehips based on the content of their biologically active compounds. *J. Food Drug Anal.* **2017**, *25*, 681–690. [CrossRef]
38. Elmastas, M.; Demir, A.; Genç, N.; Dölek, Ü.; Günes, M. Changes in flavonoid and phenolic acid contents in some Rosa species during ripening. *Food Chem.* **2017**, *25*, 154–159. [CrossRef]
39. Cunja, V.; Mikulic-Petkovsek, M.; Zupan, A.; Stampar, F.; Schmitzer, V. Frost decreases content of sugars, ascorbic acid and some quercetin glycosides but stimulates selected carotenes in *Rosa canina* hips. *J. Plant Physiol.* **2015**, *178*, 55–63. [CrossRef]
40. Wenzig, E.M.; Widowitz, U.; Kunert, O.; Chrubasik, S.; Bucar, F.; Knauder, E.; Bauer, R. Phytochemical composition and in vitro pharmacological activity of two rose hip (*Rosa canina* L.) preparations. *Phytomedicine* **2008**, *15*, 826–835. [CrossRef] [PubMed]

41. Fusi, J.; Bianchi, S.; Daniele, S.; Pellegrini, S.; Martini, C.; Galetta, F.; Giovannini, L.; Franzoni, F. An in vitro comparative study of the antioxidant activity and SIRT1 modulation of natural compounds. *Biomed. Pharmacother.* **2018**, *101*, 805–819. [CrossRef]

42. Yan, X.T.; Li, W.; Sun, Y.N.; Yang, S.Y.; Lee, S.H.; Chen, J.B.; Jang, H.D.; Kim, Y.H. Identification and biological evaluation of flavonoids from the fruits of *Prunus mume*. *Bioorg. Med. Chem. Lett.* **2014**, *24*, 1397–1402. [CrossRef] [PubMed]

43. Kim, G.N.; Jang, H.D. Flavonol content in the water extract of the mulberry (*Morus alba* L.) leaf and their antioxidant capacities. *J. Food Sci.* **2011**, *76*, C869–C873. [CrossRef] [PubMed]

44. Galano, A.; León-Carmona, J.R.; Alvarez-Idaboy, J.R. Influence of the environment on the protective effects of guaiacol derivatives against oxidative stress: Mechanisms, kinetics, and relative antioxidant activity. *J. Phys. Chem. B* **2012**, *116*, 7129–7137. [CrossRef] [PubMed]

45. Choi, J.; Kang, H.J.; Kim, S.Z.; Kwon, T.O.; Jeong, S.I.; Jang, S.I. Antioxidant effect of astragalin isolated from the leaves of *Morus alba* L. against free radical-induced oxidative hemolysis of human red blood cells. *Arch. Pharm. Res.* **2013**, *36*, 912–917. [CrossRef] [PubMed]

46. Vasantha Rupasinghe, H.P.; Yasmin, A. Inhibition of oxidation of aqueous emulsions of omega-3 fatty acids and fish oil by phloretin and phloridzin. *Molecules* **2010**, *15*, 251–257. [CrossRef] [PubMed]

47. Marino, T.; Galano, A.; Russo, N. Radical scavenging ability of gallic acid toward OH and OOH radicals. Reaction mechanism and rate constants from the density functional theory. *J. Phys. Chem. B* **2014**, *118*, 10380–10389. [CrossRef] [PubMed]

48. Aquilano, K.; Baldelli, S.; Ciriolo, M.R. Glutathione: New roles in redox signaling for an old antioxidant. *Front. Pharmacol.* **2014**, *5*, 196. [CrossRef]

49. Kirkeskov, B.; Christensen, R.; Bügel, S.; Bliddal, H.; Danneskiold-Samsøe, B.; Christensen, L.P.; Andersen, J.R. The effects of rose hip (*Rosa canina*) on plasma antioxidative activity and C-reactive protein in patients with rheumatoid arthritis and normal controls: A prospective cohort study. *Phytomedicine* **2011**, *18*, 953–958. [CrossRef]

50. Li, H.B.; Yi, X.; Gao, J.M.; Ying, X.X.; Guan, H.Q.; Li, J.C. The mechanism of hyperoside protection of ECV-304 cells against tert-butyl hydroperoxide-induced injury. *Pharmacology* **2008**, *82*, 105–113. [CrossRef]

51. Gong, G.; Qin, Y.; Huang, W.; Zhou, S.; Yang, X.; Li, D. Rutin inhibits hydrogen peroxide-induced apoptosis through regulating reactive oxygen species mediated mitochondrial dysfunction pathway in human umbilical vein endothelial cells. *Eur. J. Pharmacol.* **2010**, *628*, 27–35. [CrossRef]

52. Elliott, S.J.; Koliwad, S.K. Redox control of ion channel activity in vascular endothelial cells by glutathione. *Microcirculation* **1997**, *4*, 341–347. [CrossRef]

53. Espinosa-Díez, C.; Miguel, V.; Vallejo, S.; Sánchez, F.J.; Sandoval, E.; Blanco, E.; Cannata, P.; Peiró, C.; Sánchez-Ferrer, C.F.; Lamas, S. Role of glutathione biosynthesis in endothelial dysfunction and fibrosis. *Redox Biol.* **2018**, *14*, 88–99. [CrossRef]

54. Jiménez, S.; Gascón, S.; Luquin, A.; Laguna, M.; Ancin-Azpilicueta, C.; Rodríguez-Yoldi, M.J. *Rosa canina* Extracts Have Antiproliferative and Antioxidant Effects on Caco-2 Human Colon Cancer. *PLoS ONE* **2016**, *11*, e0159136.

55. Zhang, T.; Mu, Y.; Yang, M.; Al Maruf, A.; Li, P.; Li, C.; Dai, S.; Lu, J.; Dong, Q. (+)-Catechin prevents methylglyoxal-induced mitochondrial dysfunction and apoptosis in EA.hy926 cells. *Arch. Physiol. Biochem.* **2017**, *123*, 121–127. [CrossRef]

56. Ruijters, E.J.; Weseler, A.R.; Kicken, C.; Haenen, G.R.; Bast, A. The flavanol (−)-epicatechin and its metabolites protect against oxidative stress in primary endothelial cells via a direct antioxidant effect. *Eur. J. Pharmacol.* **2013**, *715*, 147–153. [CrossRef]

57. Li, Q.; Zhao, W.; Zeng, X.; Hao, Z. Ursolic Acid Attenuates Atherosclerosis in ApoE(−/−) Mice: Role of LOX-1 Mediated by ROS/NF-κB Pathway. *Molecules* **2018**, *23*, 1101. [CrossRef]

58. Cavalera, M.; Axling, U.; Rippe, C.; Swärd, K.; Holm, C. Dietary rose hip exerts antiatherosclerotic effects and increases nitric oxide-mediated dilation in ApoE-null mice. *J. Nutr. Biochem.* **2017**, *44*, 52–59. [CrossRef]

© 2019 by the authors. Licensee MDPI, Basel, Switzerland. This article is an open access article distributed under the terms and conditions of the Creative Commons Attribution (CC BY) license (http://creativecommons.org/licenses/by/4.0/).

 *antioxidants*

Article

Polyphenols in Almond Skins after Blanching Modulate Plasma Biomarkers of Oxidative Stress in Healthy Humans

C.-Y. Oliver Chen [1,2], **Paul E. Milbury** [2] **and Jeffrey B. Blumberg** [1,2,*]

[1] Antioxidants Research Laboratory, Jean Mayer U.S. Department of Agriculture Human Nutrition Research Center on Aging at Tufts University, Boston, MA 02111, USA; oliver.chen@tufts.edu
[2] Friedman School of Nutrition Science and Policy, Tufts University, Boston, MA 02111, USA; tollers@verizon.net
* Correspondence: jeffrey.blumberg@tufts.edu

Received: 23 March 2019; Accepted: 8 April 2019; Published: 10 April 2019

Abstract: Almond skins are a waste byproduct of blanched almond production. Polyphenols extracted from almond skins possess antioxidant activities in vitro and in vivo. Thus, we examined the pharmacokinetic profile of almond skin polyphenols (ASP) and their effect on measures of oxidative stress. In a randomized crossover trial, seven adults consumed two acute ASP doses (225 mg (low, L) or 450 mg (high, H) total phenols) in skim milk or milk alone. Plasma flavonoids, glutathione peroxidase (GPx), glutathione (GSH), oxidized GSH (GSSG), and resistance of low-density lipoprotein (LDL) to oxidation were measured over 10 h. The H dose increased catechin and naringenin in plasma, with maximum concentrations of 44.3 and 19.3 ng/mL, respectively. The GSH/GSSG ratio at 3 h after the H doses was 212% of the baseline value, as compared to 82% after milk ($p = 0.003$). Both ASP doses upregulated GPx activity by 26–35% from the baseline at 15, 30, 45, and 120 min after consumption. The in vitro addition of α-tocopherol extended the lag time of LDL oxidation at 3 h after L and H consumption by 144.7% and 165.2% of that at 0 h compared to no change after milk ($p \leq 0.05$). In conclusion, ASP are bioavailable and modulate GSH status, GPx activity, and the resistance of LDL to oxidation.

Keywords: almond skins; bioavailability; waste by-products; flavonoids; oxidative stress; human

1. Introduction

Nuts, including almonds, are a nutrient-dense food containing minerals, vitamins, unsaturated fats, protein, and fiber [1]. An association between nut consumption and a reduced risk of cardiovascular disease (CVD), certain types of cancer, type 2 diabetes mellitus, and all-cause mortality has been reported in the ever-growing literature [2–5]. Due in part to the recognition of these health benefits, nut consumption has been gradually increasing around the globe since 2006 [6]. Among tree nuts, almonds are the most consumed. Commercially, almonds can be found in the shell, shelled, and peeled, with peeled almonds (devoid of the seed coat or skin) being commonly used as a raw material for confectionary and bakery applications. In the industrial setting, peeled almonds are produced via hot water blanching, a process which generates blanched water and almond skins, byproducts that are sometimes characterized as environmental pollutants. Almond skins are generally utilized in livestock feed and composting [7,8]. To promote sustainable agricultural and food systems, almond skins, which are rich in fiber and phenols, have the potential for valorization in functional foods, nutraceuticals, and/or food additives [9–12].

Almond skins have been recognized as a potential functional ingredient in foods due to their antioxidant polyphenols and prebiotic fiber [13–15]. Mandalari et al. [14] reported that almond skins have a favorable probiotic index (a comparative relationship between the growth of beneficial bacteria and less desirable ones), comparable to that of fructo-oligosaccharides. The profile of polyphenolics in almond skins includes flavonoids, phenolic acids, and proanthocyanins [16,17]. As ubiquitous secondary metabolites, polyphenols are found in relatively high concentrations in the seed coats or skins as phytoalexins where they act to protect nutrients in the seed kernel from oxidation and microbial action until germination [18]. Meta-analytic studies are largely consistent in showing that foods rich in flavonoids ameliorate CVD risk factors and reduce all-cause mortality outcomes [19–21]. The health benefits of flavonoids have been ascribed to their multifaceted bioactions, such as antioxidation, anti-inflammation, vasodilation, anti-hypertension, platelet function, insulin sensitization, and cholesterol reduction [18,22]. In addition, polyphenols present in almond skins have been reported to display antimicrobial effects against a range of food-borne pathogens, such as *Listeria monocytogenes*, *Staphylococcus aureus*, and *Salmonella enterica* [23].

We have previously reported that almond skin polyphenols (ASP) are bioavailable and protect low-density lipoprotein (LDL) against oxidation in hamsters [24]. We have also demonstrated that ASP act as antioxidants to scavenge superoxide, peroxynitrite, and hypochlorite and also to induce quinine reductase and stabilize LDL conformation during oxidation in vitro [25,26]. However, to date, there is no clinical evidence on the antioxidant action of ASP. Thus, the objective of this study was to examine the effect of ASP on plasma oxidative stress and redox status in a randomized crossover trial with seven healthy older adults consuming ASP in skim milk. Furthermore, the bioavailability of flavonoids in ASP was examined.

2. Materials and Methods

2.1. Chemicals

The following reagents were obtained from Sigma Co. (St. Louis, MO, USA): catechin, kaempferol, isorhamnetin, naringenin, quercetin, butylated hydroxytoluene (BHT), Folin–Ciocalteu's phenol reagent, copper sulfate, ethylenediaminetetraacetic acid (EDTA), β-glucuronidase type H-2 (containing sulfatase from *Helix pomatia*), glutathione reductase (type III), nicotinamide adenine dinucleotide phosphate (NADP), sodium 1-octanesulfonate monohydrate, reduced glutathione, oxidized glutathione, sodium azide, sodium chloride, sodium phosphate monobasic, sodium phosphate dibasic, 1,1,3,3-tetraethoxypropane, 2′,3′,4′-trihydroxyacetophenone, Tris buffer, magnesium chloride, triphenylphosine, and α-tocopherol. Diisopropylethylamine (DIPEA), O-bis (trimethylsilyl) trifluoroacetamide (BSTFA), and pentafluorobenzyl bromide (PFBBr) were purchased from Thermo (Boston, MA, USA). Deuterated prostaglandin (PG) $F_{2\alpha}$ (internal standard), $PGF_{2\beta}$, and 8-iso-$PGF_{2\alpha}$ were purchased from Cayman (Ann Arbor, MI, USA). All organic solvents, glacial acetic acid, and potassium bromide were purchased from Fisher Co. (Fair Lawn, NJ, USA).

2.2. Preparation of ASP

Almond skin powder was generously provided by the Almond Board of California. Polyphenols in 150 and 300 g almond skin powder were extracted sequentially twice using a ratio of 0.5:10 (w/v) with acidified ethanol solution (white vinegar/H_2O/200 proof ethanol at 1/19/80) over 16 h at 4 °C. The extraction solution was selected with consideration of the extract for human use. The resulting solution was centrifuged at 1000× g for 15 min at 4 °C, followed by ethanol removal using a Buchi R215 Rotary Evaporator equipped with a Vacuum Controller V-850 (Flawil, Switzerland). The total phenol content in the resulting aqueous syrup was determined by Folin–Ciocalteu's assay [27] and expressed as gallic acid equivalents (GAE). The GAE content produced from 150 and 300 g almond skin powder was 225 and 450 mg, respectively. ASP were stored overnight at 4 °C before administration to the volunteers the following morning.

2.3. Subjects

Seven generally healthy older subjects (3 male, 4 female) aged 63.3 ± 9.1 years (mean $\pm$ SEM) were recruited from Boston metropolitan area by the Metabolic Research Unit (MRU) of the Jean Mayer USDA Human Nutrition Center on Aging (HNRCA) at Tufts University. All study participants completed the trial. The mean baseline lipid profile values of the seven subjects were as follows: total cholesterol (TC), 206 ± 18; triglycerides (TG), 94 ± 13; low-density lipoprotein-cholesterol (LDL-C]), 135 ± 14; high-density lipoprotein-cholesterol (HDL-C) 52 ± 7 mg/dL. Mean plasma vitamin A, C, and E concentrations were 57 ± 14 µg/dL, 0.9 ± 0.1 mg/dL, and 978 ± 123 µg/dL, respectively. Mean body weight of participants was 71 ± 6 kg; body mass index (BMI) was 25.4 ± 1.3 kg/m^2; and systolic and diastolic blood pressure values were 128 ± 8 and 74 ± 6 mmHg, respectively. All participants were in good health and had no evidence of chronic disease on the basis of a medical history questionnaire, physical examination, electrocardiogram test, and results within normal limits of standard clinical laboratory tests, as well as fulfilling the following eligibility criteria: (1) no history of CVD, hepatic, gastrointestinal, or renal disease; (2) no alcoholism; (3) no use of antibiotics; (4) no use of supplemental multivitamins or minerals for ≥ 4 weeks (3 months for 60 mg vitamin C, 30 IU vitamin E, and/or 70 µg selenium) before the start of the study; and (5) no recent history of smoking. The study was approved by and performed under the guidelines of the Institutional Review Board at Tufts University Health Sciences Campus (project code: #5822). Informed consent was obtained from each subject before any study procedures were performed. Subjects were instructed not to consume alcohol or take other medications for 1 week prior to all study visits. They were also asked to consume a low-flavonoid diet for 1 week before study visits, according to the low-flavonoid diet guideline designed by the study dietitian, in which all berries, apples, pears, citrus fruits, fruit juices, onions, chocolate, wine, coffee, any kind of tea, beans, nuts, soy products, and most spices were excluded from their daily diets.

2.4. Study Design

The study design was a placebo-controlled, three-way crossover trial with a 1-week washout period between study visits. Subjects were randomly assigned to consume 360 mL skim milk (C), or 225 mg (L) or 450 mg (H) GAE ASP in skim milk. Although ASP were not completely soluble in skim milk, residual ASP in the glass were rinsed with water and then completely consumed. Each subject was admitted to the MRU in the morning after a 12-h overnight fast. Following a check of vital signs, an intravenous catheter was inserted into an antecubital vein of one forearm and a baseline blood sample was obtained. After subjects drank the test beverage in ~5 min, blood samples were collected at 15, 30, and 45 min and 1, 2, 3, 5, and 10 h. Lunch and dinner meals designed to contain low flavonoids and meet the recommended dietary allowances for protein and energy [28] were prepared under the supervision of the study dietitian. The same meals were served during each visit. These meals were provided at 4 and 9 h after the ASP consumption. The consumption of water, salt, sugar, and ginger ale was not limited but food and other beverages were not allowed during the intervention.

2.5. Sample Collection and Storage

Blood was collected into EDTA vacutainers and processed within 10 min. Whole blood was centrifuged at $1000 \times g$ at 4 °C for 15 min using a Sorvall RT6000 (Du Pont Co. Newtown, CT, USA). Plasma aliquots of 1.5 mL were flushed with N_2, stored at 4 °C, and used for the Cu^{2+}-induced LDL oxidation assay within 3 days. An aliquot of plasma was treated with an equivalent volume of 10% trichloroacetic acid, vortexed, and then centrifuged at $14,000 \times g$ for 10 min at 4 °C; then the resulting supernatant was snap-frozen for determination of reduced glutathione (GSH) and oxidized GSH (GSSG). Aliquots of plasma were snap frozen and stored at -80 °C until analysis of flavonoids, malondialdehyde (MDA), glutathione peroxidase (GPx), and $F_{2\alpha}$-isoprostanes.

2.6. Analysis of Plasma Flavonoids

Flavonoids in plasma were determined using the HPLC method of Chen et al. [24]. Briefly, plasma was first mixed with vitamin C-EDTA solution, internal standard (2′,3′,4′-trihydroxyacetophenone), and β-glucuronidase/sulfatase. After incubation at 37 °C for 45 min, flavonoids were extracted with acetonitrile. After centrifugation, supernatant was transferred, dried under purified N_2, and reconstituted in aqueous HPLC mobile phase for HPLC analysis. Flavonoids were determined by an HPLC system equipped with a Zorbax ODS C18 column (4.6 × 150 mm, 3.5 μm) and the Coularray 5600 A detector (ESA, Inc. Chelmsford, MA, USA). The quantification of plasma catechin, quercetin, naringenin, kaempferol, and isorhamnetin was calculated according to calibration curves constructed with authentic standards, with linear relationships of $R^2 > 0.999$. The limit of detection on column for flavonoids was 0.5 pmol. The coefficient of variation (CV) values of intra- and inter-day assays were 3.0% and 9.0%, respectively. The recovery rate for the internal standard was 97.0 ± 0.1%.

2.7. Biomarkers of Antioxidant Capacity and Lipid Peroxidation

Blood for plasma glutathione analysis was collected using a drip approach from the catheter to avoid potential hemolysis which increases plasma GSH through contamination from red blood cell GSH. Blood from the catheter collected into an EDTA vacutainer. GSH and GSSG in the supernatant collected from acidified plasma were determined using the high-performance liquid chromatography with electrochemical detection (HPLC-ECD) method of Chen et al. [28]. The concentrations of plasma GSH and GSSG were calculated based on calibration curves of authentic GSH and GSSG. The intra- and inter-day assay CV values for GSH were 2.7% and 2.9% and for GSSG the values were was 6.7% and 8.1 %, respectively.

Oxygen radical absorbance capacity (ORAC) in heparinized plasma-treated 0.5 M perchloric acid (PCA) (1:1 v/v) was determined according to the method of Ou et al. [29]. The assay provides an integrated and quantitative determination of "total antioxidant capacity" by employing the area under the curve (AUC) of the magnitude and time to the oxidation of fluorescein due to peroxyl radicals generated by the addition of 2,2′-azobis (2-amidinopropane) dihydrochloride (AAPH). ORAC values were calculated according to the method described by Cao et al. [30] and are expressed as μmol/L Trolox equivalents (TE).

Plasma GPx activity was determined using the spectrophotometric method of Pleban et al. [31] with a Cobas Fara II centrifugal analyzer (Roche Diagnostics, Nutley, NJ, USA). This assay measures GPx activity on the basis of the oxidation of GSH to GSSG and the reduction of H_2O_2 to H_2O, which is coupled to the oxidation of NADPH by glutathione reductase. The intra- and inter-assay CV values were 3.4% and 3.2%, respectively.

The LDL resistance against Cu^{2+}-induced oxidation was determined according to the method of Chen et al. [25]. Briefly, LDL particles were collected using an ultracentrifugation protocol. LDL (182 nmol/L) was oxidized by 10 μmol/L $CuSO_4$ with (or without) in vitro addition of a final concentration of 6 μmol/L α-tocopherol in a total volume of 1.0 mL phosphate buffer. The addition of α-tocopherol was intended to amplify any potential antioxidant effect of absorbed ASP on LDL resistance against oxidation [24]. Formation of conjugated dienes was monitored by absorbance at 234 nm at 37 °C using a UV1601 spectrophotometer (Shimadzu Corp, Kyoto, Japan). The results of the LDL oxidation are expressed as lag time (defined as the intercept at the abscissa in the diene-time plot). The intra- and inter-assay CV was 1.8% and 7.5%, respectively.

Plasma MDA was determined by the HPLC method of Volpi and Tarugi [32], in which a thiobarbituric acid-MDA conjugate product was separated by a C18 column and fluorometrically quantified at an excitation of 515 nm and emission of 553 nm. Plasma MDA concentration was calculated based on calibration curves of authentic standard 1,1,3,3-tetraethoxypropane, with a linear relationship of $R^2 > 0.995$. The intra- and inter-assay CV values were 3.9% and 12.3%, respectively.

Plasma $F_{2\alpha}$-isoprostanes were measured by gas chromatograph-mass spectrometry (GC-MS) with negative chemical ionization, as described by Walter et al. [33]. Briefly, plasma lipids were isolated with Folch extraction, followed by gentle alkaline saponification to release isoprostanes. Isoprostanes were then converted to pentafluorobenzyl esters using PFBBr and DIPEA. The resulting PFB esters of $F_{2\alpha}$-isoprostanes were isolated using an HPLC system equipped with an amino column, dried under N_2, and silylated with BSTFA and DIPEA. The silylated product was dried, resuspended in undecane, and analyzed by GC-MS. The final data are expressed as ng/mL.

2.8. Statistical Analysis

All results are reported as mean ± SEM. Percent change from the respective baseline (0 h) value of each visit was calculated to construct the area under the curve (AUC) using the linear trapezoidal integration [34]. All percent change and AUC data were normalized by a $\log_{10}$ conversion before statistical analysis. The mixed model procedure (PROC GLM) was used to test the effects of ASP dose, time point, and their interaction on study outcomes, followed by the Tukey–Kramer honestly significant difference (Tukey's HSD) test. Either paired *t*-test or one-way ANOVA followed by Tukey's HSD test was employed to test the difference between the high ASP dose and control in the AUC data of MDA, $F_{2\alpha}$-isoprostanes, GSH, GSSG, GSH/GSSG ratio, lag time of LDL oxidation, GPx activity, and ORAC$_{pca}$. Differences with $p \leq 0.05$ were considered significant. The SAS 9.2 statistical software package (SAS Institute Inc. Cary, NC, USA) was used to perform all statistical analyses.

3. Results

3.1. Bioavailability of Flavonoids

Five flavonoids were quantified in the plasma of the subjects consuming H-ASP (450 mg GAE) or milk vehicle only. The baseline plasma concentration of catechin, quercetin, naringenin, kaempferol, isorhamnetin, and total flavonoid were 17.0 ± 4.0, 7.3 ± 1.2, 11.5 ± 5.7, 10.1 ± 3.2, 2.3 ± 0.4, and 48.1 ± 9.3 ng/mL, respectively. Milk alone did not significantly affect their values. The H dose led to significant increases in plasma catechin, naringenin, and sum of five flavonoids, as compared to the corresponding baseline value ($p \leq 0.05$). Their maximum concentrations (C_{max}) were 44.3 ± 15.6, 19.3 ± 8.2, and 82.3 ± 17.6 ng/mL and times to reach C_{max} (T_{max}) were 1.4 ± 0.2, 3.3 ± 0.5, and 1.7 ± 0.3 h, respectively (Figure 1). At 10 h, flavonoid concentrations returned to their respective baseline values. The other three measured flavonoids, quercetin, kaempferol, and isorhamnetin, were not significantly increased by the H-ASP.

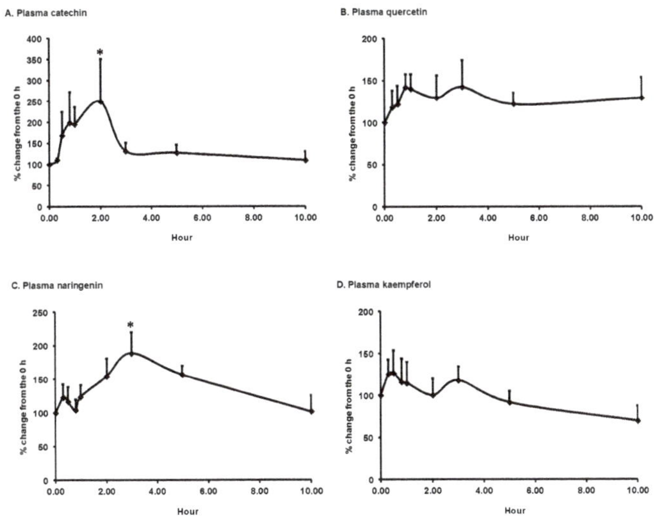

Figure 1. *Cont.*

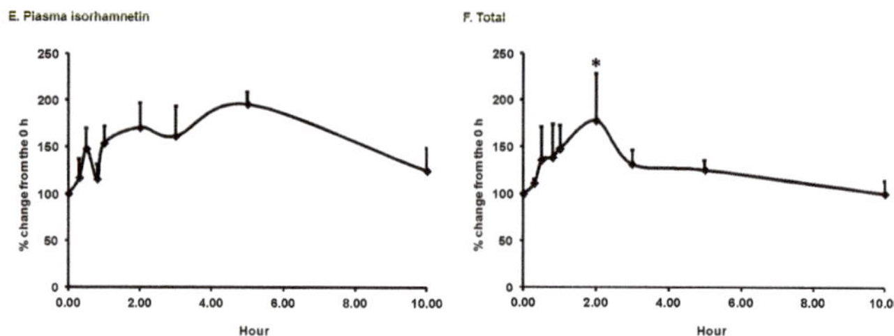

Figure 1. Time course of plasma flavonoids, catechin (**A**), quercetin (**B**), naringenin (**C**), kaempferol (**D**), isorhamnetin (**E**), and total (**F**) in older adults after acute intake of skim milk vehicle (C) and 450 (H) mg GAE ASP. The data are presented as the percent change from the respective baseline (0 h). The baseline concentrations of catechin, quercetin, naringenin, kaempferol, isorhamnetin, and total were 17.0 ± 4.0, 7.3 ± 1.2, 11.5 ± 5.7, 10.1 ± 3.2, 2.3 ± 0.4, and 48.1 ± 9.3 ng/mL, respectively. Values are expressed as mean ± SEM, $n = 7$. Means with a mark are significantly different from that of the baseline, $p \leq 0.05$, tested using PROC GLM, followed by Tukey's honestly significant difference (HSD) multi-comparison test. ASP: almond skin polyphenols; GAE: gallic acid equivalents.

3.2. Changes in Plasma Biomarkers of Antioxidant Defense

Mean baseline plasma values of GSH, GSSG, and their ratio were 1.16 ± 0.14, 0.11 ± 0.01 μmol/L, and 12.1 ± 2.4, respectively. A marked inter-individual variation in GSH, GSSG, and GSH/GSSG was noted. After consumption of skim milk, GSH values tended to decrease, GSSG values tended to increase, and the ratio remained unaltered (Figure 2). H-ASP tended to increase GSH by 25% at 3 h as compared to the respective baseline value and to decrease GSSG by 31% at 2 h. A favorable effect of H-ASP on GSH status was noticeable at 15 min post-consumption but retreated at 45 min and 1 h before the next favorable changes occurred at 2 and 3 h. At 3 h, the GSH/GSSG ratio of H-ASP was 212% of the baseline, which was significantly different from 82% of C at the same time point ($p = 0.0033$). The increased ratio at 3 h after H-ASP consumption was driven by both the increased GSH and the decreased GSSG. The AUC of GSH and GSSG did not differ between C and H-ASPs; the AUC of the GSG/GSSG ratio was significantly increased by 70% by H-ASP as compared to C (Table 1).

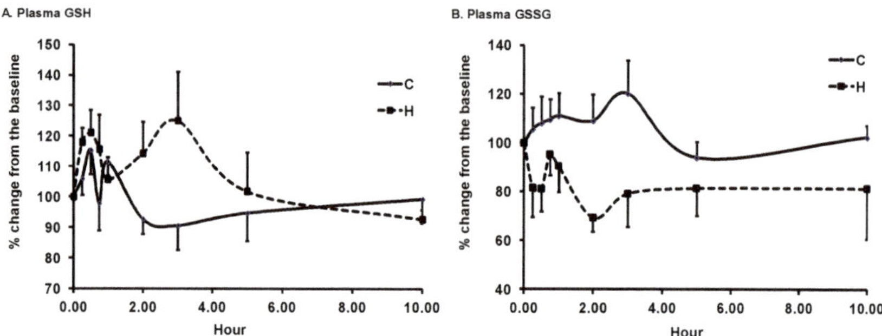

Figure 2. *Cont.*

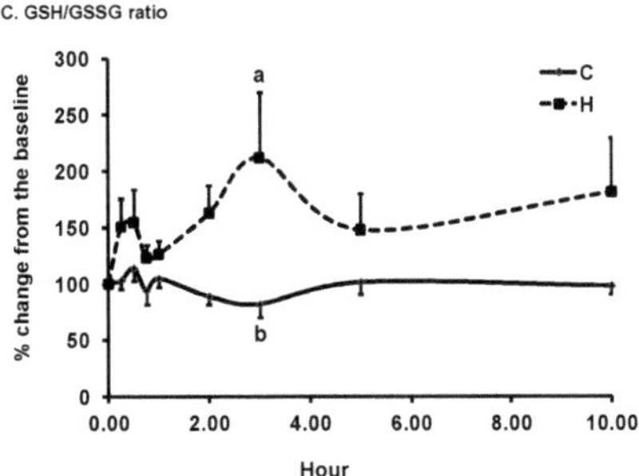

Figure 2. Time course of plasma glutathione (GSH) (**A**) and oxidized GSH (GSSG) (**B**) and their ratio (**C**) in older adults after acute intake of skim milk vehicle (C) and 450 (H) mg GAE ASP. The data are presented as the percent change from the respective baseline (0 h). Baseline values of GSH, GSSH, and their ratio were 1.16 ± 0.14, 0.11 ± 0.01 µmol/L, and 12.1 ± 2.4, respectively. Values are expressed as mean $\pm$ SEM, $n = 7$. Means not sharing the same letter at the same time point are significantly different, $p \leq 0.05$, tested using PROC GLM, followed by Tukey's HSD multi-comparison test. *p*-values of the dose effect for GSH, GSSG, and their ratio were 0.029, 0.013, and 0.014, respectively.

Table 1. The percent change of area under curve (AUC) of plasma glutathione, glutathione peroxidase activity, and oxygen radical absorbance capacity ($ORAC_{pca}$) in older adults after acute intake of 250 mg (L) or 450 mg (H) almond skin polyphenols (ASP) or skim milk (C) [1]. GPx: glutathione peroxidase.

ASP	GSH	GSSG	GSH/GSSG	GPx	$ORAC_{pca}$ [2]
			AUC (%·h) [3]		
C	971 ± 64	1037 ± 55	972 ± 63 [a]	987 ± 43	1114 ± 50
L	-	-	-	1142 ± 101	1026 ± 51
H	1057 ± 67	810 ± 106	1649 ± 300 [b]	1116 ± 47	1108 ± 33

[1] Values are mean $\pm$ SEM, $n = 7$. [a,b] Means in the same column not sharing the same letter differ, $p \leq 0.05$, using an ANOVA *t*-test, followed by Tukey's HSD test. [2] Plasma was treated with perchloric acid (PCA) first before testing by ORAC assay. [3] Percent change from the respective baseline value (0 h) of each visit is calculated to construct the area under the curve (AUC) using the linear trapezoidal integration.

Mean baseline plasma GPx activity was 197 ± 11 U/L. Skim milk did not affect GPx activity. L- and H-ASP up-regulated GPx activity in a two-phase mode with an initial increase occurring between 15 and 45 min, followed by the second one at 2 h (Figure 3). The effect of ASP on GPx activity was independent of the dose. The magnitude of the ASP-induced increase in GPx activity ranged between 26% and 35% from the baseline at 15, 30, and 45 min and 2 h, while skim milk slightly increased the activity during the same period. The AUC of GPx activity did not differ among three treatments (Table 1). Plasma ORAC value was not affected by ASP and milk up to 10 h. Mean baseline $ORAC_{pca}$ value was 896 ± 28 µmol/L TE (Table 1).

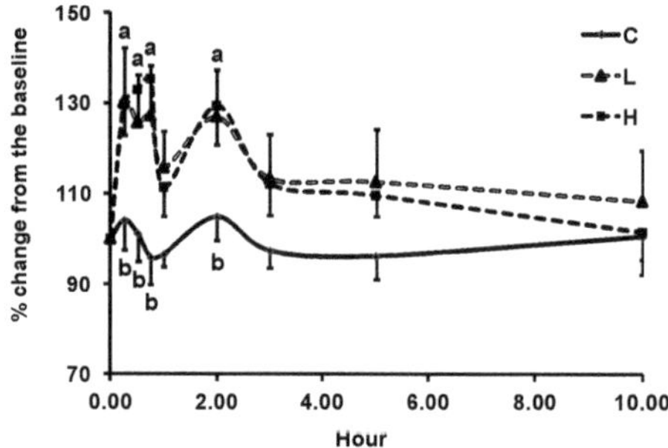

Figure 3. Time course of the percent change of plasma GPx activity in older adults after acute intake of skim milk vehicle (C), 225 mg (L), or 450 mg (H) GAE ASP. The data are presented as the percent change from the respective baseline (0 h). The mean baseline activity was 197.0 ± 10.6 U/L. Values are expressed as mean $\pm$ SEM, $n = 7$. Means not sharing the same letter at the same time point are significantly different, $p \leq 0.05$, using PROC GLM followed by Tukey's HSD multi-comparison test. *p*-Values for dose and time effect and their interaction were 0.039, $\leq$0.001, and 0.001, respectively.

3.3. Changes in LDL Resistance to Oxidation

The mean baseline lag time of LDL oxidation was 45.1 ± 1.7 min, and its value was not extended by either ASP dose. The AUC of lag time was comparable between treatments. The addition of 6 μmol/L α-tocopherol prior to the initiation of Cu^{2+}-induced LDL oxidation increased lag time to 95.7 ± 2.6 min at baseline (Figure 4). At 3 h, the lag time with added α-tocopherol after intake of L- and H-ASP was 144.7 ± 13.1 and $165.2 \pm 25.0\%$ of that at baseline, respectively, as compared to the $102.2 \pm 2.4\%$ observed after the skim milk ($p \leq 0.05$). There was no difference in the lag time between the two ASP doses. To reduce the influence of the variation of the baseline values, the percentage change from the baseline was calculated to assess the change obtained with in vitro addition of α-tocopherol.

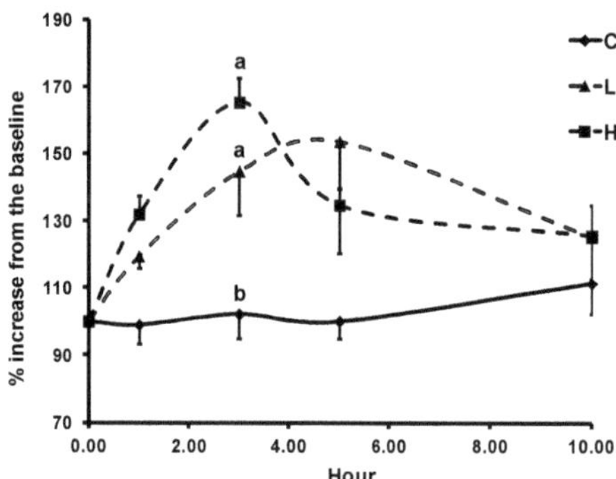

Figure 4. Time course of lag time of LDL oxidation with in vitro addition of 6 μmol/L α-tocopherol in older adults after acute intake of skim milk vehicle (C), 225 mg (L), or 450 mg (H) GAE ASP. The data are presented as the percent change from the respective baseline (0 h). The mean baseline lag time was 95.7 ± 2.6 min. Values are expressed as mean $\pm$ SEM, $n = 7$. Means not sharing the same letter at the same time point are significantly different, $p \leq 0.05$, tested using PROC GLM followed by Tukey's HSD multi-comparison test. *p*-Values of the dose and time effect and their interaction were 0.001, 0.014, and 0.049, respectively.

3.4. Changes in Plasma Biomarkers of Oxidative Stress

The mean baseline plasma MDA value was 2.4 $\pm$ 0.3 μmol/L. Skim milk and H-ASP did not affect MDA up to 10 h (data not shown) or its AUC (Table 2). Mean baseline plasma $F_{2\alpha}$-isoprostanes were 5.0 $\pm$ 0.2 ng/mL. Given that there were marked inter-individual variations, $F_{2\alpha}$-isoprostanes in plasma (data not shown) and their AUC (Table 2) were not affected by skim milk and H-ASP.

Table 2. The percent change of area under curve (AUC) of oxidative stress biomarkers in older adults after acute intake of 250 mg (L) or 450 mg (H) almond skin polyphenols (ASP) or skim milk (C) [1].

ASP	Plasma MDA	Plasma ISOPROSTANES	Lag Time of LDL Oxidation	Lag Time of LDL Oxidation with α-Tocopherol [2]
			AUC (%·h) [3]	
C	954 $\pm$ 56	948 $\pm$ 125	973 $\pm$ 18	1031 $\pm$ 44 [a]
L	-	-	1010 $\pm$ 26	1369 $\pm$ 152 [b]
H	993 $\pm$ 70	1154 $\pm$ 198	990 $\pm$ 18	1364 $\pm$ 108 [b]

[1] Values are mean $\pm$ SEM, $n = 7$. [a,b] Means in the same column not sharing the same letter differ, $p \leq 0.05$, using an ANOVA t-test, followed by Tukey's HSD test. [2] α-Tocopherol (6 μmol/L) was added to LDL before the initiation of oxidation; [3] Percent change from the respective baseline value (0 h) of each visit is calculated to construct the area under the curve (AUC) using the linear trapezoidal integration.

4. Discussion

From production and processing to storage, retailing, and consumption, waste is produced in all the phases of food life cycle [10]. According to the Food and Agriculture Organization of the United Nations [35], approximately one-third of the food produced in the world for human consumption is lost or wasted annually. For example, ~30% of nonedible products of vegetables and some fruits, mainly skins and seeds, are commonly wasted and discarded [9]. Similarly, almonds, the most commonly consumed tree nut, can generate $\geq$4% waste as skin during the production of blanched almonds. This type of food waste has historically been used as low-value livestock feed or compost. More recently, the agri-food industry has made substantial progress in utilizing waste by-products for development of novel ingredients or products [36,37]. Successful examples include the recovery of oil from olive kernel, the production of essential oils, flavonoids, and pectin from citrus peel, and the recapture of protein concentrates from cheese whey [12]. Some of these plant-based wastes contain a variety of phytochemicals, particularly polyphenols, which are often abundant in skins and seeds [9]. For example, almond skins contain an array of flavonoids and phenolic acids, with isorhamentin-3-rutinoside being the main polyphenol [13,17]. The result of several experimental studies suggests the potential for utilizing this by-product of almond processing as a value-added ingredient useful in the development of functional foods or nutraceuticals because of its antioxidant, anti-microbial, anti-viral, neuroprotective, photoprotective, and/or prebiotic activities [13–15,23,24,38–42]. Within the last decade and with availability of new equipment, the almond blanching process has evolved to use substantially less water and more steam, reducing the loss of polyphenols into the blanch water and allowing a higher polyphenol content of the almond skin after processing. We demonstrate here the acute bioavailability and antioxidant actions of polyphenols derived from almond skins in humans without apparent untoward side effects.

Flavonoid bioavailability is dependent on a wide array of factors, e.g., type of flavonoid, food matrix, co-consumed food components, polymorphism of detoxification mechanisms, and aging. Using a simulated human digestion system, Mandalari et al. [43] demonstrated that ASP were bioaccessible for the absorption in the upper gastrointestinal tract. In this study, the subjects consumed ASP at the dose of either 225 mg or 450 mg GAE, and plasma flavonoids were monitored over 10 h. As previously characterized in almonds, isorhamnetin is one of its principal flavonoids [13,17]; however, we found only a modest trend toward an increase in its plasma concentration. This result is in contrast with the marked increase we found in our hamster study [24]. Nonetheless, our inability to detect isorhamnetin in plasma is consistent with the report by Bartolomé et al. [44] who found it undetectable at 2.5 h post consumption of 884 mg GAE total phenols of almond skin extract. The dose employed in that study was about twice our highest dose. Bartolomé et al. study [44] estimated this dose is ~8-times higher than the dietary intake (102–121 mg/person/d) of nut polyphenols in the Spanish diet. While reports on the clinical pharmacokinetics of isorhamnetin following the consumption of isorhamnetin-rich foods or supplements are limited, Schulz et al. [45] observed a significant increase in plasma isorhamnetin with C_{max} in the 10-ng/mL range in 18 healthy men consuming one dose of 900 mg dry extract of St. John's wort. Besides isorhamnetin, we monitored catechin, naringenin, kaempferol, and quercetin values and found significant increases in plasma catechin and naringenin with the T_{max} at 2 and 3 h, respectively. Similarly, Garrido et al. [46] reported that consumption of almond skin extract containing 884 mg GAE total phenols increased urinary excretion of epicatechin and naringenin conjugates derived from phase II metabolism. We also noted a marked inter-individual variation in the concentration of these flavonoids, which is consistent with a range of other reports that ascribe this phenomenon to the wide range of endogenous and exogenous factors mentioned above [47,48]. We did not examine urinary flavonoids or phenolic acids derived from bacteria-catalyzed flavonoids in this study. However, Llorach et al. [49] noted a plethora of phenolic acids in urine of 24 people consuming an ASP extract. This study underscores the significant role of gut microbiota on the catabolism of flavonoids in the formation of phenolic acids from flavonols, e.g., hydroxyphenylvaleric, hydroxyphenylpropionic, and hydroxyphenylacetic acids. Together with low concentrations of detected flavonoids in plasma, future research is warranted to examine the effect of phenolic acids derived from polyphenols via catabolism of colonic microflora on human health.

Flavonoids are regarded as strong antioxidants, acting via scavenging (reducing) reactive oxidants, chelating transient metals, and/or modulating endogenous antioxidant defense mechanisms. However, the efficacy of such mechanisms of action post absorption has been questioned because of the low concentrations of flavonoids in blood and tissues as compared to other abundant endogenous antioxidants [50]. Previously, we demonstrated in hamsters that absorbed ASP enhance LDL resistance against ex vivo Cu^{2+}-induced oxidation [24] and worked with the in vitro addition of vitamin E to further bolster the LDL resistance. Instead, we found the effect of the absorbed ASP on the protection of LDL against oxidation was only unmasked when α-tocopherol (in a dose-dependent manner) was added in a physiologically relevant concentration. Thus, absorbed ASP may incorporate into LDL particles and then exert antioxidative actions or/and stabilize LDL structure to enhance LDL resistance. This speculation is also based on our in vitro study showing sand α-tocopherol work in a synergistic manner to stabilize LDL conformation during oxidation [25]. While the magnitude of polyphenol bioavailability and circulating flavonoid metabolites vary between species, this study extends the putative benefit of ASP on LDL resistance to oxidation from hamsters to humans.

Flavonoids are regarded as a class of beneficial phytonutrients. However, they also display characteristics as xenobiotics and are subject to phase I, II, and III detoxification metabolism known to be involved in drug clearance from the body. In addition, they may modulate the expression of cytochrome P450 monooxygenases, phase II conjugation enzymes, and/or on membrane transporters [51]. These actions are likely attributed to the effect of flavonoids on activating xenobiotic response elements and/or antioxidant/electrophil response elements (AREs/EpREs) [51,52]. For example, via these signal transduction pathways, quercetin enhanced the expression and activities of GSH reductase, GPx, and catalase [53,54]. We determined plasma GSH and GPx activity in support of the effect of ASP on endogenous antioxidant defense systems, a result that could implicate the activation of ARE by the absorbed ASP. We found that the high-dose ASP tended to increase plasma GSH between 1 and 5 h after consumption as compared to the skim milk vehicle, suggesting the absorbed ASP either increased its production in the liver via up-regulating γ-glutamylcysteine synthetase [51] or decreased its utilization or excursion. Further, the GSH/GSSG ratio, regarded as an index of redox status, was elevated at 3 h post-consumption of the high-dose ASP as compared to the skim milk vehicle, as well as the significantly larger AUC of the ratio, substantiating the absorbed

ASP could boost antioxidant defense capacity. Similarly, there was an early, transient effect of the absorbed ASP on plasma GPx activity. Although GSH is the substrate for the GPx reaction in the reduction of H_2O_2 and lipid peroxides to water and alcohols, we did not find a correlation between plasma GPx activity and GSH, GSSG, or their ratio (data not shown). In a previous human study, we also found that polyphenol avenanthramides from oats enabled a similar effect on GSH and GPx in humans [28]. However, neither blackberry nor cranberry juice, both rich in anthocyanins, affected erythrocyte GPx activity [55,56]. Thus, the effect of polyphenols on GPx activity and GSH status appears to depend on sample type (plasma vs. erythrocytes), flavonoid class (anthocyanins vs. flavonols), and study design (chronic vs. acute). Establishing an efficient extraction protocol to produce a commercial, high-quality polyphenol-rich product from almond skins for human use remains a challenge, though use of almond skin powder can already be found in the marketplace [57].

There are some limitations to this study, including its small sample size and absence measures of the oxidative modification of DNA and proteins. The inclusion of only healthy older adults also limits the generalizability of the results. There were significant inter-individual temporal differences in the pattern changes of plasma flavonoids and biomarkers of oxidative stress. This common phenomenon suggests marked individual genetic variation in enzymes that influence polyphenol absorption, metabolism, disposition, and excretion. Future studies are warranted to better understand interactions between genomics and flavonoid metabolism and utilization and their impact on biomarkers of oxidative stress and health outcomes.

5. Conclusions

During the production of blanched almonds, almond skins are generated as a waste by-product and commonly utilized as livestock feed or fertilizer. Like other waste produced during food processing, almonds skins (as well as skins from peanuts, grapes, and oranges) contain a variety of phytochemicals with a potential for development of nutraceuticals and functional foods [9]. A growing body of the literature has revealed that phytochemicals, including polyphenols and fiber in almond skins, display properties associated with health benefits, including antioxidant, anti-microbial, anti-viral, neuroprotective, photoprotective, and prebiotic activities [14,23–25,38–42]. Despite large inter-individual differences in metabolic changes, acute intake of almond skin extract increased plasma catechin, naringenin, GSH/GSSG ratio, and GPx activity at 2–3 h post-consumption. Further, the interaction of absorbed almond skin polyphenols with the in vitro addition of vitamin E unmasked the protection of the absorbed almond polyphenols on the ex vivo resistance of LDL to oxidation. In conclusion, polyphenolic constituents in blanched almond skins are capable of up-regulating antioxidant defense mechanisms and enhancing the resistance of LDL to oxidation. Thus, almond skins generated during the production of blanched almonds have the potential for use in the development of value-added ingredients and food products.

Author Contributions: C.-Y.O.C., P.E.M., and J.B.B. designed research and conducted research; C.-Y.O.C. analyzed the data and prepared the draft report. J.B.B. acquired funding. All authors reviewed, approved, and accept responsibility for the final manuscript.

Funding: Support was provided by the U.S. Department of Agriculture (USDA) Agricultural Research Service under Cooperative Agreement No. 58-1950-7-014 and The Almond Board of California. The contents of this publication do not necessarily reflect the views or policies of the USDA, nor does mention of trade names, commercial products or organizations imply endorsement by the U.S. government.

Acknowledgments: We would like to express our gratitude to the Almond Board of California for providing financial support and blanched almond skin powder for the study and to the volunteers for participating in the clinical trial.

Conflicts of Interest: The authors declare no conflict of interest.

References

1. Kamil, A.; Chen, C.Y. Health benefits of almonds beyond cholesterol reduction. *J. Agric. Food Chem.* **2012**, *60*, 6694–6702. [CrossRef] [PubMed]
2. Morgillo, S.; Hill, A.M.; Coates, A.M. The effects of nut consumption on vascular function. *Nutrients* **2019**, *11*, 116. [CrossRef] [PubMed]
3. Neale, E.P.; Tapsell, L.C.; Guan, V.; Batterham, M.J. The effect of nut consumption on markers of inflammation and endothelial function: A systematic review and meta-analysis of randomised controlled trials. *BMJ Open* **2017**, *7*, e016863. [CrossRef]
4. Chen, G.C.; Zhang, R.; Martinez-Gonzalez, M.A.; Zhang, Z.L.; Bonaccio, M.; van Dam, R.M.; Qin, L.Q. Nut consumption in relation to all-cause and cause-specific mortality: A meta-analysis 18 prospective studies. *Food Funct.* **2017**, *8*, 3893–3905. [CrossRef] [PubMed]
5. Aune, D.; Keum, N.; Giovannucci, E.; Fadnes, L.T.; Boffetta, P.; Greenwood, D.C.; Tonstad, S.; Vatten, L.J.; Riboli, E.; Norat, T. Nut consumption and risk of cardiovascular disease, total cancer, all-cause and cause-specific mortality: A systematic review and dose-response meta-analysis of prospective studies. *BMC Med.* **2016**, *14*, 207. [CrossRef] [PubMed]
6. International Nuts & Dried Fruits Council. Nuts & Dried Fruits. Statistical Yearbook 2017/2018. Available online: https://www.nutfruit.org/consumers/news/detail/inc-2017-2018-statistical-yearbook (accessed on 7 May 2018).
7. Grasser, L.A.; Fadel, J.G.; Garnett, I.; DePeters, E.J. Quantity and economic importance of nine selected by-products used in California dairy rations. *J. Dairy Sci.* **1995**, *78*, 962–971. [CrossRef]
8. González, J.F.; Gañán, J.; Ramiro, A.; González-García, C.M.; Encinar, J.M.; Sabio, E.; Román, S. Almond residues gasification plant for generation of electric power. Preliminary study. *Fuel Process. Technol.* **2006**, *87*, 149–155. [CrossRef]
9. Varzakas, T.; Zakynthinos, G.; Verpoort, F. Plant food residues as a source of nutraceuticals and functional foods. *Foods* **2016**, *5*. [CrossRef]
10. Kumar, K.; Yadav, A.N.; Kumar, V.; Vyas, P.; Dhaliwal, H.S. Food waste: A potential bioresource for extraction of nutraceuticals and bioactive compounds. *Bioresour. Bioprocess.* **2017**, *4*, 18. [CrossRef]
11. Babbar, N.; Oberoi, H.S.; Sandhu, S.K. Therapeutic and nutraceutical potential of bioactive compounds extracted from fruit residues. *Crit. Rev. Food Sci. Nutr.* **2015**, *55*, 319–337. [CrossRef]
12. Esfahlan, A.J.; Jamei, R.; Esfahlan, R.J. The importance of almond (*Prunus amygdalus* L.) and its by-products. *Food Chem.* **2010**, *120*, 349–360. [CrossRef]
13. Monagas, M.; Garrido, I.; Lebron-Aguilar, R.; Bartolome, B.; Gomez-Cordoves, C. Almond (*Prunus dulcis* (Mill.) D.A. Webb) skins as a potential source of bioactive polyphenols. *J. Agric. Food. Chem.* **2007**, *55*, 8498–8507. [CrossRef]
14. Mandalari, G.; Faulks, R.M.; Bisignano, C.; Waldron, K.W.; Narbad, A.; Wickham, M.S. In vitro evaluation of the prebiotic properties of almond skins (*Amygdalus communis* L.). *FEMS Microbiol. Lett.* **2010**, *304*, 116–122. [CrossRef]
15. Mandalari, G. Potential health benefits of almond skin. *J. Bioprocess. Biotech.* **2012**, *2*, e110. [CrossRef]
16. Bolling, B.W.; Dolnikowski, G.; Blumberg, J.B.; Oliver Chen, C.Y. Quantification of almond skin polyphenols by liquid chromatography-mass spectrometry. *J. Food Sci.* **2009**, *74*, C326–C332. [CrossRef] [PubMed]
17. Milbury, P.E.; Chen, C.Y.; Dolnikowski, G.G.; Blumberg, J.B. Determination of flavonoids and phenolics and their distribution in almonds. *J. Agric. Food Chem.* **2006**, *54*, 5027–5033. [CrossRef]
18. Ann Lila, M. The nature-versus-nurture debate on bioactive phytochemicals: The genome versus terroir. *J. Sci. Food Agric.* **2006**, *86*, 2510–2515. [CrossRef]
19. Hooper, L.; Kroon, P.A.; Rimm, E.B.; Cohn, J.S.; Harvey, I.; Le Cornu, K.A.; Ryder, J.J.; Hall, W.L.; Cassidy, A. Flavonoids, flavonoid-rich foods, and cardiovascular risk: A meta-analysis of randomized controlled trials. *Am. J. Clin. Nutr.* **2008**, *88*, 38–50. [CrossRef]
20. Kim, Y.; Je, Y. Flavonoid intake and mortality from cardiovascular disease and all causes: A meta-analysis of prospective cohort studies. *Clin. Nutr. ESPEN* **2017**, *20*, 68–77. [CrossRef]
21. Grosso, G.; Micek, A.; Godos, J.; Pajak, A.; Sciacca, S.; Galvano, F.; Giovannucci, E.L. Dietary flavonoid and lignan intake and mortality in prospective cohort studies: Systematic review and dose-response meta-analysis. *Am. J. Epidemiol.* **2017**, *185*, 1304–1316. [CrossRef]

22. Bolling, B.W.; Chen, C.Y.; McKay, D.L.; Blumberg, J.B. Tree nut phytochemicals: Composition, antioxidant capacity, bioactivity, impact factors. A systematic review of almonds, Brazils, cashews, hazelnuts, macadamias, pecans, pine nuts, pistachios and walnuts. *Nutr. Res. Rev.* **2011**, *24*, 244–275. [CrossRef]

23. Mandalari, G.; Bisignano, C.; D'Arrigo, M.; Ginestra, G.; Arena, A.; Tomaino, A.; Wickham, M.S. Antimicrobial potential of polyphenols extracted from almond skins. *Lett. Appl. Microbiol.* **2010**, *51*, 83–89. [CrossRef] [PubMed]

24. Chen, C.Y.; Milbury, P.E.; Lapsley, K.; Blumberg, J.B. Flavonoids from almond skins are bioavailable and act synergistically with vitamins C and E to enhance hamster and human LDL resistance to oxidation. *J. Nutr.* **2005**, *135*, 1366–1373. [CrossRef] [PubMed]

25. Chen, C.Y.; Milbury, P.E.; Chung, S.K.; Blumberg, J.B. Effect of almond skin polyphenolics and quercetin on human LDL and apolipoprotein B-100 oxidation and conformation. *J. Nutr. Biochem.* **2007**, *18*, 785–794. [CrossRef] [PubMed]

26. Chen, C.Y.; Blumberg, J.B. In vitro activity of almond skin polyphenols for scavenging free radicals and inducing quinone reductase. *J. Agric. Food Chem.* **2008**, *56*, 4427–4434. [CrossRef] [PubMed]

27. Singleton, V.L.; Orthofer, R.; Lamuela-Ravent, R.M. Analysis of total phenols and other oxidation substrates and antioxidants by means of Folin–Ciocalteu reagent. *Methods Enzymol.* **1999**, *299*, 152–178.

28. Chen, C.Y.; Milbury, P.E.; Collins, F.W.; Blumberg, J.B. Avenanthramides are bioavailable and have antioxidant activity in humans after acute consumption of an enriched mixture from oats. *J. Nutr.* **2007**, *137*, 1375–1382. [CrossRef]

29. Ou, B.; Hampsch-Woodill, M.; Prior, R.L. Development and validation of an improved oxygen radical absorbance capacity assay using fluorescein as the fluorescent probe. *J. Agric. Food Chem.* **2001**, *49*, 4619–4626. [CrossRef]

30. Cao, G.; Alessio, H.M.; Cutler, R.G. Oxygen-radical absorbance capacity assay for antioxidants. *Free Radic. Biol. Med.* **1993**, *14*, 303–311. [CrossRef]

31. Pleban, P.A.; Munyani, A.; Beachum, J. Determination of selenium concentration and glutathione peroxidase activity in plasma and erythrocytes. *Clin. Chem.* **1982**, *28*, 311–316.

32. Volpi, N.; Tarugi, P. Improvement in the high-performance liquid chromatography malondialdehyde level determination in normal human plasma. *J. Chromatogr. Biomed. Sci. Appl.* **1998**, *713*, 433–437. [CrossRef]

33. Walter, M.F.; Blumberg, J.B.; Dolnikowski, G.G.; Handelman, G.J. Streamlined F2-isoprostane analysis in plasma and urine with high-performance liquid chromatography and gas chromatography/mass spectroscopy. *Anal. Biochem.* **2000**, *280*, 73–79. [CrossRef] [PubMed]

34. Nielsen, I.L.; Chee, W.S.; Poulsen, L.; Offord-Cavin, E.; Rasmussen, S.E.; Frederiksen, H.; Enslen, M.; Barron, D.; Horcajada, M.N.; Williamson, G. Bioavailability is improved by enzymatic modification of the citrus flavonoid hesperidin in humans: A randomized, double-blind, crossover trial. *J. Nutr.* **2006**, *136*, 404–408. [CrossRef] [PubMed]

35. FAO. *Global Food Losses and Food Waste—Extent, Causes and Prevention*; FAO: Rome, Italy, 2011.

36. Baiano, A. Recovery of biomolecules from food wastes—A review. *Molecules* **2014**, *19*, 14821–14842. [CrossRef] [PubMed]

37. Rudra, S.G.; Nishad, J.; Jakhar, N.; Kaur, C. Food industry waste: Mine of nutraceuticals. *Int. J. Sci. Environ. Technol.* **2015**, *4*, 205–229.

38. Bisignano, C.; Mandalari, G.; Smeriglio, A.; Trombetta, D.; Pizzo, M.M.; Pennisi, R.; Sciortino, M.T. Almond skin extracts abrogate HSV-1 replication by blocking virus binding to the cell. *Viruses* **2017**, *9*, 178. [CrossRef] [PubMed]

39. Bisignano, C.; Filocamo, A.; La Camera, E.; Zummo, S.; Fera, M.T.; Mandalari, G. Antibacterial activities of almond skins on cagA-positive and -negative clinical isolates of *Helicobacter pylori*. *BMC Microbiol.* **2013**, *13*, 103. [CrossRef]

40. Mandalari, G.; Bisignano, C.; Genovese, T.; Mazzon, E.; Wickham, M.S.; Paterniti, I.; Cuzzocrea, S. Natural almond skin reduced oxidative stress and inflammation in an experimental model of inflammatory bowel disease. *Int. J. Immunopharmacol.* **2011**, *11*, 915–924. [CrossRef]

41. Mandalari, G.; Genovese, T.; Bisignano, C.; Mazzon, E.; Wickham, M.S.; Di Paola, R.; Bisignano, G.; Cuzzocrea, S. Neuroprotective effects of almond skins in experimental spinal cord injury. *Clin. Nutr.* **2011**, *30*, 221–233. [CrossRef]

42. Arena, A.; Bisignano, C.; Stassi, G.; Mandalari, G.; Wickham, M.S.; Bisignano, G. Immunomodulatory and antiviral activity of almond skins. *Immunol. Lett.* **2010**, *132*, 18–23. [CrossRef]

43. Mandalari, G.; Tomaino, A.; Rich, G.T.; Lo Curto, R.B.; Arcoraci, T.; Martorana, M.; Bisignano, C.; Saija, A.; Parker, M.L.; Waldron, K.; et al. Polyphenol and nutrient release from skin of almonds during simulated human digestion. *Food Chem.* **2010**, *122*, 1083–1088. [CrossRef]

44. Bartolome, B.; Monagas, M.; Garrido, I.; Gomez-Cordoves, C.; Martin-Alvarez, P.J.; Lebron-Aguilar, R.; Urpi-Sarda, M.; Llorach, R.; Andres-Lacueva, C. Almond (*Prunus dulcis* (Mill.) D.A. Webb) polyphenols: From chemical characterization to targeted analysis of phenolic metabolites in humans. *Arch. Biochem. Biophys.* **2010**, *501*, 124–133. [CrossRef] [PubMed]

45. Schulz, H.U.; Schurer, M.; Bassler, D.; Weiser, D. Investigation of pharmacokinetic data of hypericin, pseudohypericin, hyperforin and the flavonoids quercetin and isorhamnetin revealed from single and multiple oral dose studies with a hypericum extract containing tablet in healthy male volunteers. *Arzneimittelforschung* **2005**, *55*, 561–568. [CrossRef] [PubMed]

46. Garrido, I.; Urpi-Sarda, M.; Monagas, M.; Gomez-Cordoves, C.; Martin-Alvarez, P.J.; Llorach, R.; Bartolome, B.; Andres-Lacueva, C. Targeted analysis of conjugated and microbial-derived phenolic metabolites in human urine after consumption of an almond skin phenolic extract. *J. Nutr.* **2010**, *140*, 1799–1807. [CrossRef] [PubMed]

47. Manach, C.; Williamson, G.; Morand, C.; Scalbert, A.; Remesy, C. Bioavailability and bioefficacy of polyphenols in humans. I. Review of 97 bioavailability studies. *Am. J. Clin. Nutr.* **2005**, *81*, 230s–242s. [CrossRef]

48. Williamson, G.; Manach, C. Bioavailability and bioefficacy of polyphenols in humans. II. Review of 93 intervention studies. *Am. J. Clin. Nutr.* **2005**, *81*, 243s–255s. [CrossRef] [PubMed]

49. Llorach, R.; Garrido, I.; Monagas, M.; Urpi-Sarda, M.; Tulipani, S.; Bartolome, B.; Andres-Lacueva, C. Metabolomics study of human urinary metabolome modifications after intake of almond (*Prunus dulcis* (Mill.) D.A. Webb) skin polyphenols. *J. Proteome Res.* **2010**, *9*, 5859–5867. [CrossRef]

50. Hollman, P.C.; Cassidy, A.; Comte, B.; Heinonen, M.; Richelle, M.; Richling, E.; Serafini, M.; Scalbert, A.; Sies, H.; Vidry, S. The biological relevance of direct antioxidant effects of polyphenols for cardiovascular health in humans is not established. *J. Nutr.* **2011**, *141*, 989s–1009s. [CrossRef]

51. Moon, Y.J.; Wang, X.; Morris, M.E. Dietary flavonoids: Effects on xenobiotic and carcinogen metabolism. *Toxicol. In Vitro* **2006**, *20*, 187–210. [CrossRef]

52. Myhrstad, M.C.; Carlsen, H.; Nordstrom, O.; Blomhoff, R.; Moskaug, J.O. Flavonoids increase the intracellular glutathione level by transactivation of the gamma-glutamylcysteine synthetase catalytical subunit promoter. *Free Radic. Biol. Med.* **2002**, *32*, 386–393. [CrossRef]

53. Milner, J.A.; McDonald, S.S.; Anderson, D.E.; Greenwald, P. Molecular targets for nutrients involved with cancer prevention. *Nutr. Cancer* **2001**, *41*, 1–16. [PubMed]

54. Moskaug, J.O.; Carlsen, H.; Myhrstad, M.C.; Blomhoff, R. Polyphenols and glutathione synthesis regulation. *Am. J. Clin. Nutr.* **2005**, *81*, 277s–283s. [CrossRef] [PubMed]

55. Hassimotto, N.M.; Pinto Mda, S.; Lajolo, F.M. Antioxidant status in humans after consumption of blackberry (*Rubus fruticosus* L.) juices with and without defatted milk. *J. Agric. Food Chem.* **2008**, *56*, 11727–11733. [CrossRef] [PubMed]

56. Duthie, S.J.; Jenkinson, A.M.; Crozier, A.; Mullen, W.; Pirie, L.; Kyle, J.; Yap, L.S.; Christen, P.; Duthie, G.G. The effects of cranberry juice consumption on antioxidant status and biomarkers relating to heart disease and cancer in healthy human volunteers. *Eur. J. Nutr.* **2006**, *45*, 113–122. [CrossRef] [PubMed]

57. Garrido, I.; Monagas, M.; Gomez-Cordoves, C.; Bartolome, B. Polyphenols and antioxidant properties of almond skins: Influence of industrial processing. *J. Food Sci.* **2008**, *73*, C106–C115. [CrossRef] [PubMed]

© 2019 by the authors. Licensee MDPI, Basel, Switzerland. This article is an open access article distributed under the terms and conditions of the Creative Commons Attribution (CC BY) license (http://creativecommons.org/licenses/by/4.0/).

 *antioxidants* 

Article

Epicatechin Reduces Spatial Memory Deficit Caused by Amyloid-β25–35 Toxicity Modifying the Heat Shock Proteins in the CA1 Region in the Hippocampus of Rats

Alfonso Diaz [1], Samuel Treviño [1], Guadalupe Pulido-Fernandez [1], Estefanía Martínez-Muñoz [2], Nallely Cervantes [2], Blanca Espinosa [3], Karla Rojas [4], Francisca Pérez-Severiano [5], Sergio Montes [6], Moises Rubio-Osornio [7] and Jorge Guevara [2,*]

[1] Facultad de Ciencias Químicas, Benemérita Universidad Autónoma de Puebla, Puebla, Pue. PC. 72540, Mexico; alfonso.diaz@correo.buap.mx (A.D.); samuel_trevino@hotmail.com (S.T.); lupita.pulido.99@gmail.com (G.P.-F.)

[2] Departamento de Bioquímica, Facultad de Medicina, Universidad Nacional Autónoma de México, Ciudad de México PC. 04510, Mexico; estefania.mtz17@gmail.com (E.M.-M.); nallely.co@gmail.com (N.C.)

[3] Departamento de Bioquímica, Instituto Nacional de Enfermedades Respiratorias, SSA, Ciudad de Mexico, PC. 14269, Mexico; bespinosa1118@yahoo.com.mx

[4] Departamento de Ciencias de la Salud, Psicologia. Universidad del Valle de México, sede Sur., Ciudad de Mexico, PC. 04910, Mexico; karlacarmina_rojas@my.uvm.edu.mx

[5] Laboratorio de Neurofarmacología Molecular y Nanotecnología, Instituto Nacional de Neurología, SSA, Ciudad de Mexico, PC. 14269, Mexico; fperez@innn.edu.mx

[6] Departamento de Neuroquímica, Instituto Nacional de Neurología, SSA, Ciudad de Mexico, PC. 14269, Mexico; smontes@innn.edu.mx

[7] Laboratorio Experimental de Enfermedades Neurodegenerarivas, SSA, Ciudad de Mexico, PC. 14269, Mexico; ruomon@gmail.com

* Correspondence: jorge.guevara@comunidad.unam.mx; Tel.: +52-562-32510

Received: 8 February 2019; Accepted: 1 April 2019; Published: 30 April 2019

Abstract: Alzheimer's disease (AD) is a neurodegenerative disorder characterized by dementia and the aggregation of the amyloid beta peptide (Aβ). Aβ$_{25-35}$ is the most neurotoxic sequence, whose mechanism is associated with the neuronal death in the Cornu Ammonis 1 (CA1) region of the hippocampus (Hp) and cognitive damage. Likewise, there are mechanisms of neuronal survival regulated by heat shock proteins (HSPs). Studies indicate that pharmacological treatment with flavonoids reduces the prevalence of AD, particularly epicatechin (EC), which shows better antioxidant activity. The aim of this work was to evaluate the effect of EC on neurotoxicity that causes Aβ$_{25-35}$ at the level of spatial memory as well as the relationship with immunoreactivity of HSPs in the CA1 region of the Hp of rats. Our results show that EC treatment reduces the deterioration of spatial memory induced by the Aβ$_{25-35}$, in addition to reducing oxidative stress and inflammation in the Hp of the animals treated with EC + Aβ$_{25-35}$. Likewise, the immunoreactivity to HSP-60, -70, and -90 is lower in the EC + Aβ$_{25-35}$ group compared to the Aβ$_{25-35}$ group, which coincides with a decrease of dead neurons in the CA1 region of the Hp. Our results suggest that EC reduces the neurotoxicity induced by Aβ$_{25-35}$, as well as the HSP-60, -70, and -90 immunoreactivity and neuronal death in the CA1 region of the Hp of rats injected with Aβ$_{25-35}$, which favors an improvement in the function of spatial memory.

Keywords: reactive oxygen species; proinflammatory cytokines; Alzheimer's disease

---

1. Introduction

Alzheimer's disease (AD) is a neurodegenerative disorder that affects the senile population and is characterized clinically by loss of memory and difficulty in reasoning. At the histopathological level, neurofibrillary tangles (MNF) are formed by the hyperphosphorylated tau protein. Neuritic plaques (PNs) are formed by the amyloid-β peptide, which coexists with reactive astrogliosis and neuronal death in brain regions such as the cerebral cortex and the hippocampus (Hp) [1]. The amyloid-β (Aβ), the main component of the PNs, comes from the alternative hydrolysis of the amyloid precursor protein (PPA) [2]. There are several functional domains, characterized by a neurotoxic domain (25–35), of the amino acid sequence of Aβ that stand out. Several research groups propose the use of this neurotoxic domain as an experimental model to study AD [3,4].

The intrahippocampal injection of $A\beta_{25-35}$ promotes the neurodegeneration accompanied by a deterioration in spatial memory [5,6]. Our previous results show that the undecapeptide generates an inflammatory response. This is evidenced by reactive astrogliosis and the release of proinflammatory cytokines such as interleukin-1 beta (IL-1β) and Tumor necrosis factor-alpha (TNF-α) [7], besides exacerbation of the production of reactive oxygen species (ROS) and lipid peroxidation, implying the generation of a chronic state of oxidative stress and cell death [4]. At the cellular level, the expression of heat shock proteins (HSPs), chaperones that participate in the assembly, transport, and degradation of proteins under both normal and stress conditions, has been demonstrated to be a cell survival mechanism [8]. Several reports indicate that the Aβ increases the expression of HSP-60, HSP-70, and HSP-90 as a protection mechanism against the toxicity of this peptide [9,10]. HSPs participate in cellular proteostasis and a reduction in the death of hippocampal neurons [11]. However, these reports indicate that the increase in these HSPs is not enough to reverse the neurotoxicity caused by the injection of Aβ into the rat Hp. Consequently, this causes neuronal and cognitive impairment [9].

In this sense, it is necessary to evaluate new molecules that, together with the HSP activity, can help to prevent or inhibit oxidative stress, the inflammatory response, and thus dementia. This could reverse the neurodegeneration induced by $A\beta_{25-35}$ and be considered as a therapeutic alternative for AD. Recent reports indicate that epicatechin (EC), a flavonoid present in fruit and vegetables, has aroused great interest because of its beneficial antioxidant properties, also being used in the treatment and prevention of cancer as well as delaying aging [12,13]. The main biological activity of EC is the formation of protein complexes, the inhibition of free radicals, and the reduction of lipid peroxidation, making it an excellent antioxidant [14]. In addition, it can act as an anti-inflammatory by inhibiting cyclo-oxygenases and improving endothelial function [15,16].

Previously, it has been observed that EC treatment promotes the neuronal plasticity in the Hp and cortex, moreover improving spatial memory processes [16]. Also, EC prevents the toxicity induced by $A\beta_{25-35}$ by reducing ROS and LP levels and reactive astrogliosis, and by not causing a deterioration of spatial memory [17,18]. However, it is not known how the anti-inflammatory and antioxidative activity of EC modifies the immunoreactivity of HSP-60, HSP-70, and HSP-90 in response to $A\beta_{25-35}$ toxicity. The objective of this work is to evaluate the effect of EC administration on spatial memory, the oxidative-inflammatory response, and its relationship with HSPs after an intrahippocampal injection of $A\beta_{25-35}$.

2. Materials and Methods

2.1. Animals

Adult male Wistar rats (200–250 g, $n = 60$) were obtained from the vivarium of the faculty of medicine of the Universidad Nacional Autonoma de Mexico (UNAM). Animals were individually housed in a temperature- and humidity-controlled environment in a 12:12 h light-dark cycle with free access to food and water. The procedures described in this study were carried out in accordance with current national and international regulations for the use and care of laboratory animals of the Mexican Council of Animal Care (NOM-062-ZOO-1999), Guide of the National Institute of Health for the Care and Use of Laboratories, as well as Animals and Ethics Committee of the UNAM (UNAM-FMED-104-2017), in addition to minimizing the number of animals used for experimental purposes, which ensured generating the minimum possible pain.

2.2. Epicatechin Administration Protocol

Four experimental groups were considered for the study (15 rats per group): (1) Control (animals were given only sterile water), (2) $A\beta_{25-35}$ [100 µM], (3) Epicatechin (EC) (200 mg/kg/day × 4 days), and (4) EC + $A\beta_{25-35}$. The oral administration of EC started 1 day before the surgical injection and continued for 3 days more (4 days total). The treatment doses employed were chosen or calculated based on previous reports [18]. The $A\beta_{25-35}$ peptide [100 µM] was dissolved in sterile water and the solution incubated at 37 °C for 24 h. The animals were then anesthetized with ketamine-xylazine (0.1 mL/100 g, i.p.) and placed in a stereotaxic frame (Stoelting Co. Wood Dale, Illinois, USA). The stereotaxic coordinates used to produce a bilateral lesion in the Hp were A: −4.2 mm from bregma, L: −2.0 mm from the midline, V: −2.2 below dura) [17]. Injections of $A\beta_{25-35}$ or Control (1 µL) per side were administered for 5 min with a Hamilton syringe. After surgery, the animals were returned to their cages to recover.

2.3. Water Maze Spatial Task

After the treatment, the spatial memory was evaluated in the water maze (WM) at the Laboratory of Excitatory Amino Acids, Instituto Nacional de Neurología, INNN (Mexico City; Mexico), which consisted of a circular water pool (140 cm diameter and 80 cm high) that was filled with water to a height of 42 cm at 23 ± 24 °C and dyed with 0.01% white titanium oxide (TiO2). Four quadrants (N, S, E, or W) were traced into the pool. A platform of 20 cm diameter and 40 cm high, located at a constant position in the middle of one quadrant, was submerged 2 cm below the water's surface. This procedure was repeated during 4 days of training. The animals (n = 15/group) were allowed to search for the platform for 90 s and were gently guided to it if they did not reach the target on their own. The time of 90 s was assigned as the maximum score. The parameter to be evaluated was the time to find the platform. The memory test was performed 2 days after training; in this test, the platform was removed from the pool. The test consisted of each of the animals performing a single try to find the place where the platform was. The parameter to be evaluated was the latency time at the first crossing, with the aim of evaluating if the animal memorized where the platform was located. The trials were recorded by a video camera mounted above the center of the pool (Sharp VL-WD450 U, SHARP Corporation, Osaka, Japan).

2.4. Quantifying IL-1β and TNF-α Cytokines

After the spatial memory test, the animals were decapitated (n = 8/group). The brain was removed from the cranial cavity of the animals. After, the Hp was extracted according to the protocol described by Diaz et al. [6] The hippocampal tissue was placed in a microcentrifuge tube with 1.5 mL of a phosphate buffer solution (PBS) (0.1 M; pH 7.4) at a temperature of 4 °C, to be liquefied for one minute at 100 rpm, with the help of a homogenizer. Afterward, homogenate hippocampal tissue was centrifuged at 12,500 rpm at 4 °C [19,20]. The supernatant was extracted from the centrifuged tissue and subsequently aliquoted in microtubes (200 mL) and stored at −70 °C. The supernatants were used for protein and proinflammatory cytokine measurements. The concentrations of IL-1β and TNF-α in the supernatants were quantified by an immunoassay procedure, as specified in the kit protocols (R&D Systems, Minneapolis, MN, USA). Samples were treated with a monoclonal antibody in precoated wells where the immobilized antibody bound to the protein. After washing away any unbound substances, an enzyme-linked specific antibody was added to the wells. The enzyme reaction yielded a blue product that turned yellow when the stop solution was added. Samples were read in a microplate reader at a wavelength of 450 nm where it was found that the intensity of the measured color was in proportion to the amount of each cytokine.

2.5. Assay of Lipid Peroxidation

The lipid-soluble fluorescent compounds were determined by means of the method established by Cuevas et al., and Perez Severiano et al. [17,21]. The supernatant obtained from the hippocampal tissue was mixed with chloroform-methanol (2:1) placed on ice and in a dark room. Subsequently, the chloroform phase was taken to quantify the fluorescence at an excitation of 370 nm and emission wavelengths of 430 nm on a Perkin Elmer LS50-B luminescence spectrometer (Waltham, MASS, USA). The fluorescent signal from the kit was adjusted to 140 fluorescence units (FU) with a standard quinine solution (0.001 mg/mL quinine in 0.05 M H_2SO_4). The results were expressed as relative FU (RFU) per mg of protein. [19].

2.6. Assay of Reactive Oxygen Species

We used 5 μL of the supernatant of the hippocampal tissue previously centrifuged, which was diluted in 9 volumes of TRIS and HEPES (40 mM). Subsequently, samples were incubated with 2'7'-dichlorodihydrofluorescein diacetate (DCFH-DA) (5 μM) [22] for one hour at 37 °C under constant agitation. The fluorescence signals were determined at 488 nm excitation and emission wavelengths of 525 nm (Perkin Elmer LS50-B luminescence spectrometer, Waltham, MASS, USA). The results were plotted as the mean of the DFC nm formed by mg of protein per minute. [20].

2.7. Determination of Superoxide Dismutase Activity

To evaluate the activity of superoxide dismutase (SOD), 20 μL of the supernatant of the hippocampal tissue previously centrifuged was used, which was incubated with reduced cytochrome-C solution (10 μM), sodium azide (NaN3, 10 μM), disodium ethylenediaminetetraacetic acid (EDTA, 10 mM), sodium bicarbonate (NaHCO3, 20 mM), xanthine (100 μM), and Triton X-100 with a pH of 10.2. To the reaction was added xanthine oxidase (0.1 mM EDTA) and the absorbance was determined at 550 nm. The activity of Zn-SOD was calculated as the total activity minus the activity measured in the presence of potassium cyanide (KCN, 1 mM) (Mn-SOD).

2.8. Histological Examination

After the spatial memory test, the animals ($n = 7$ per group) were anesthetized with pentobarbital (40 mg/kg). Once the animal was in a state of hypnosis, the rib cage was opened and the pericardium was separated. Next, the vertex of the left ventricle was sectioned (with fine-tipped scissors) and a rigid cannula that was connected to the peristaltic pump was inserted, at a pressure of 80 mmHg that generated a continuous flow of 200 mL of PBS in the ascending aorta. The cannula was adjusted to the ventricle with the help of flat forceps. When the perfusion was started, the evacuation of the perfusion blood was facilitated by a cut in the right atrium. Once the perfusion with PBS was finished, perfusion was continued with 300 mL of 4% paraformaldehyde in a continuous flow and to the same pressure to ensure a better fixation of brain tissue. The brains were removed and postfixed in the same fixative solution for 48 h. The brain tissue was dehydrated and embedded in paraffin with the help of a histoquinet (Leica). Subsequently, the tissues were included in paraffin blocks and cut coronally (5 mm) with the help of a microtome (Leica), at the anterior temporal region level, in approximately 3.8 to 6.8 mm of bregma.

2.9. Immunofluorescence

The slides were deparaffinized and rehydrated using conventional histological techniques. They were then rinsed with a PBS (0.1 M; pH 7.4). Nonspecific binding sites were blocked by incubating in IgG-free 2% bovine serum albumin (Sigma) for 30 min at room temperature (RT). Afterward, specimens were permeabilized with 0.2% Triton X-100 (Sigma, St. Louis, MO, USA) for 10 min at RT. Sections were then incubated overnight at 4 °C to 8 °C with primary antibodies HSP-60, -70, -90 (1:200 dilution), and caspase-3 (1:100 dilution) (all primary antibodies were obtained from

Santa Cruz Biotechnology Inc., CA, USA), followed by anti-rabbit Fluorescein Isothiocyanate (FITC) conjugated secondary antibodies and counterstained (1:100, Jackson Immuno Research Laboratories Inc., West Grove, PA, USA) with VectaShield-DAPI (Vector Labs., CA, USA) for nuclei staining. The photomicrographs were taken near the site of the injection using a fluorescence microscope (Leica Microsystems, Wetzlar, GmBH, Germany) and the number of immunoreactive cells to HSPs and caspase-3 were quantified in the CA1 subfield of the Hp. All counting procedures were made blindly by an expert in morphology.

2.10. Hematoxylin and Eosin (H&E)

We evaluated the effect of EC on the neuronal loss caused by $A\beta_{25-35}$ injection in the Hp by mean hematoxylin and eosin staining (H&E). The neurons were observed at 40× (Leica DM-LS, Leica Microsystems, Wetzlar, GmBH, Germany). Undamaged neurons were recognized as cells with round, blue nuclei and a clear perinuclear cytoplasm [19–24]. Damaged neurons were cells with changed nuclei (pyknosis, karyorrhexis, and karyolysis) and cytoplasmatic eosinophilia or loss of hematoxylin affinity.

2.11. Statistical Analysis

The results were expressed as the mean ± standard error (SE) for all experiments. Statistical analyses were done using variance analysis and multiple comparisons were made using Bonferroni post-test, considering $p < 0.05$ significant.

3. Results

3.1. Effect of Epicatechin Treatment on Spatial Memory in Rats Injected with $A\beta_{25-35}$ in the Hippocampus of Rats

The animals of each group performed the spatial training test in the WM, where the latency time to find the escape platform was quantified. The behavior of the animals of all the experimental groups to find the platform with respect to the days of training showed a progressive decrease in the latency time. The comparative analysis indicates that the group administered with $A\beta_{25-35}$ showed a significantly longer latency time compared to the control group and the group administered only with EC on the different days that the training test was performed (one-way Analysis of Variance (ANOVA), $p < 0.05$). On the other hand, the group administered with EC + $A\beta_{25-35}$ showed a latency time to find the platform lower in comparison with the group administered only with $A\beta_{25-35}$ with a statistically significant difference that was observed from day 2 of training until the end of the test (One-way ANOVA, $p < 0.05$) (Figure 1A).

For the memory test, carried out two days after the learning test (Figure 1B), it is shown that the latency time at the first crossing of the target quadrant was significantly higher in the group treated with $A\beta_{25-35}$ with respect to the other three experimental groups, respectively. Particularly, when comparing the group of $A\beta_{25-35}$ with respect to the group EC + $A\beta_{25-35}$, it is shown that the administration of EC significantly lowered the latency time at the first crossing of the white quadrant, suggesting a diminishing of cognitive damage caused by the neurotoxicity of the peptide into the Hp (one-way ANOVA, $p < 0.05$) (Figure 1A).

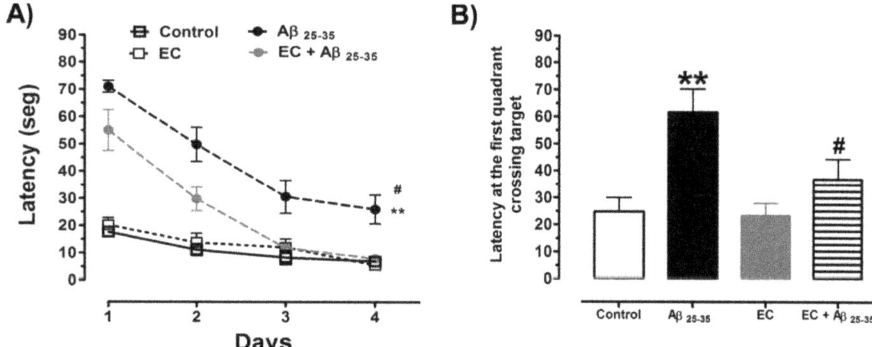

Figure 1. The administration of EC prevents the deterioration in spatial memory that induces the injection of $A\beta_{25-35}$ in the rat Hp. The spatial training test and spatial memory test of rats administered with saline solution, epicatechin, $A\beta_{25-35}$, and epicatechin (EC) + $A\beta_{25-35}$ were performed in the water maze. The parameters to be quantified were the latency time to find the escape platform (**A**) and latency of the first crossing in the objective quadrant (**B**). The values shown represent the standard error (SE) mean one-way ANOVA (** $p < 0.01$) comparing all groups with respect to the vehicle group. (# $p < 0.05$) compared the $A\beta_{25-35}$ group versus EC + $A\beta_{25-35}$ group.

3.2. Effect of the Administration of Epicatechin on the Oxidative Response Induced by $A\beta_{25-35}$ in the Hippocampus of Rats

The lipid lipoperoxidation data obtained from the Hp are shown in Figure 2A. The group injected with $A\beta_{25-35}$ showed a significant fourfold increase compared to the basal levels of peroxidation in the control group. Treatment with EC + $A\beta_{25-35}$ demonstrated its antioxidant effect by significantly decreasing (65%) the lipoperoxidation levels with respect to the group injected intrahippocampally with $A\beta_{25-35}$ (one-way ANOVA with significance of $p < 0.05$), while the group treated only with EC did not show any considerable changes in lipid peroxidation when compared to the control group. The amount of ROS by 2,7-dichlorodihydrofluorescein found in the Hp is shown in Figure 2B. The statistical analysis reveals that the group injected with $A\beta_{25-35}$ presented a significant increase of 59% regarding the control group. Likewise, when analyzing the data obtained from the hippocampi of the group treated with EC + $A\beta_{25-35}$, there was a significant decrease of 28% in the levels of ROS in relation to the group injected only with $A\beta_{25-35}$ (one-way ANOVA with a significance of $p < 0.05$). Regarding the group only injected with EC, the data obtained did not show a significant difference when compared with the control group. Figure 2C shows the enzymatic activity of Zn-SOD in the Hp. The $A\beta_{25-35}$ group presented a decrement of 38% in SOD activity in relation to the vehicle group. In the same way, the SOD activity in the EC + $A\beta_{25-35}$ group decreased by 25% with respect to the $A\beta_{25-35}$ group (one-way ANOVA with a significance of $p < 0.05$), while between the control group and EC-treated group there were no differences. The Mn-SOD activity is shown in Figure 2D. The $A\beta_{25-35}$ group and control group differed significantly by 35%. Likewise, when comparing the Mn-SOD activity in the EC + $A\beta_{25-35}$ group in relation to the $A\beta_{25-35}$ group, there was a significant change of 21% (one-way ANOVA with a significance of $p < 0.05$), while the EC group showed no differences regarding the control group.

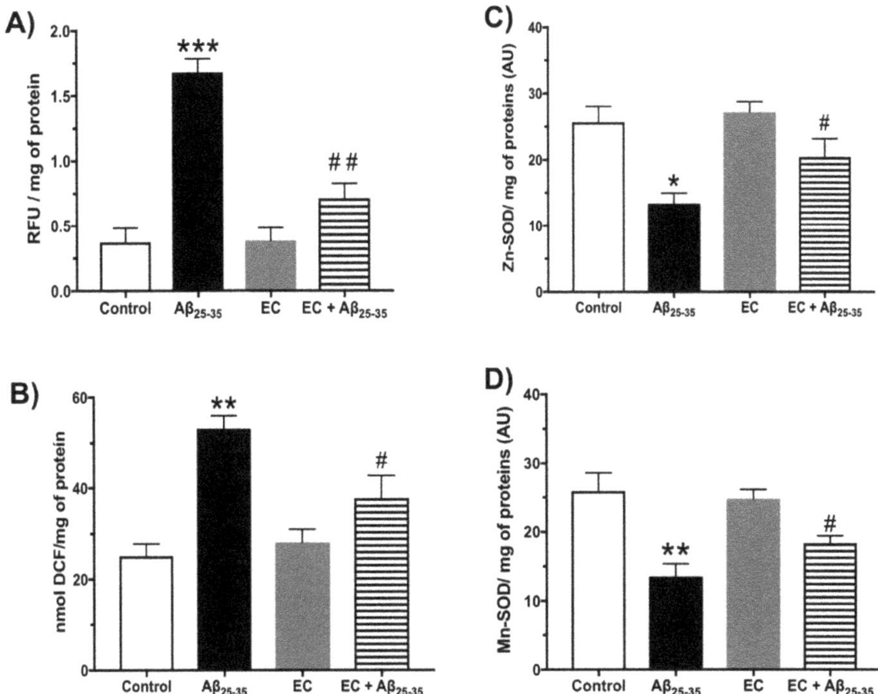

Figure 2. Effect of epicatechin on oxidative stress in the Hp of rats injected with $A\beta_{25-35}$. (**A**) Reactive Oxygen Species (ROS) assay; (**B**) lipid peroxidation assay; (**C**) Zn-Superoxide activity assay; (**D**) Mn-Superoxide activity assay. The mean ± SE is plotted. Data were analyzed with one-way ANOVA and post-test Bonferroni test. (* $p < 0.05$; ** $p < 0.01$; and *** $p < 0.001$) comparing all groups with respect to the vehicle group. (# $p < 0.05$ and ## $p < 0.01$) compared the $A\beta_{25-35}$ group versus EC + $A\beta_{25-35}$ group.

3.3. *Effect of Epicatechin on the Production of IL-1ß and TNF-α in the Hippocampus of Rats Injected with Aß$_{25-35}$*

The concentrations of IL-1β and TNF-α were determined from a supernatant of the hippocampal tissues of each of the experimental groups (Figure 3). The concentration of IL-1β clearly shows that the intrahippocampal injection of $A\beta_{25-35}$ was significantly higher compared to the control group. The EC treatment prevented the increase of the IL-1β in the EC + $A\beta_{25-35}$ group (one-way ANOVA with a significance of $p < 0.05$), but the EC-treatment-only group did not show changes in relation to the control group. Similar behavior was observed in the TNF-α level in the Hp of the $A\beta_{25-35}$ group, which was significantly higher compared to the rest of the experimental groups, and EC treatment prevented the exacerbation of the cytokine level after $A\beta_{25-35}$ injection; however, TNF-α did not have changes in the EC group (one-way ANOVA with a significance of $p < 0.05$).

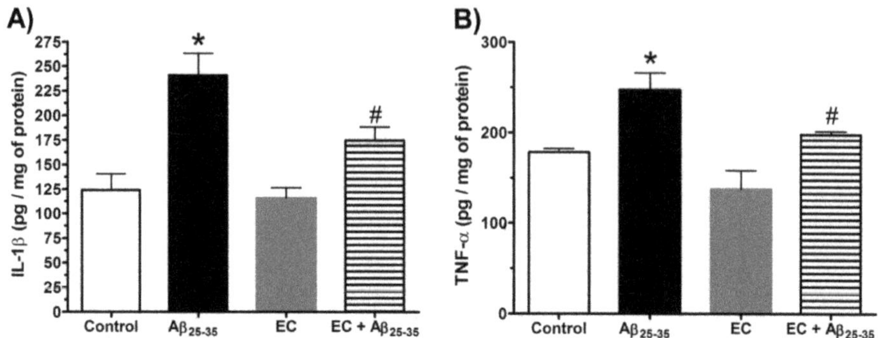

Figure 3. The epicatechin treatment decreases the concentration of proinflammatory cytokines in Hp of rats with $A\beta_{25-35}$. (**A**) IL-1β in Hp and (**B**) TNF-α in Hp. The mean $\pm$ SE is plotted. Data were analyzed with one-way ANOVA and post-test Bonferroni test (* $p < 0.05$) comparing all groups with respect to the vehicle group. # $p < 0.05$ compared the $A\beta_{25-35}$ group versus EC + $A\beta_{25-35}$ group.

3.4. The Administration of Epicatechin Changes the Immunoreactivity of HSP-60, HSP-70, and HSP-90 in the Hippocampus of Rats Injected with $A\beta_{25-35}$

To understand the reactivity of HSP in response to the toxicity of $A\beta_{25-35}$ and the EC treatment in rat hippocampi, immunofluorescence was performed to identify the HSP-60, HSP-70, and HSP-90 in the Hp-CA1 region in each experimental group. The photomicrographs are shown in Figure 4A. Qualitative analysis indicates that the group injected with $A\beta_{25-35}$ generated a greater immunoreactivity in the HSP (green color) in the CA1 region of the Hp compared to the control group. Particularly, HSP-60 showed a more intense label compared to HSP-90 and HSP-70. The treatment with EC in animals injected with $A\beta_{25-35}$ caused a decrease in the immunoreactivity of the three chaperone proteins (green color). Specifically, the HSP-70 immunoreactivity decreased in greater proportion with respect to HSP-90 and HSP-60 in the CA1 region of the Hp.

The statistical analysis of the number of cells reactive to HSP-60, HSP-70, and HSP-90 (Figure 4B–D) indicates that the group with $A\beta_{25-35}$ shows a significant increase in the number of immunoreactivity cells of 400%, 83%, and 200% for HSP-60, HSP-70, and HSP-90, respectively, in the CA1 region of the Hp, whereas the treatment with EC plus $A\beta_{25-35}$ caused a significant reduction of 17%, 72%, and 58% for HSP-60, HSP-70, and HSP-90, respectively, in the CA1 region of the Hp (one-way ANOVA with significance of $p < 0.05$).

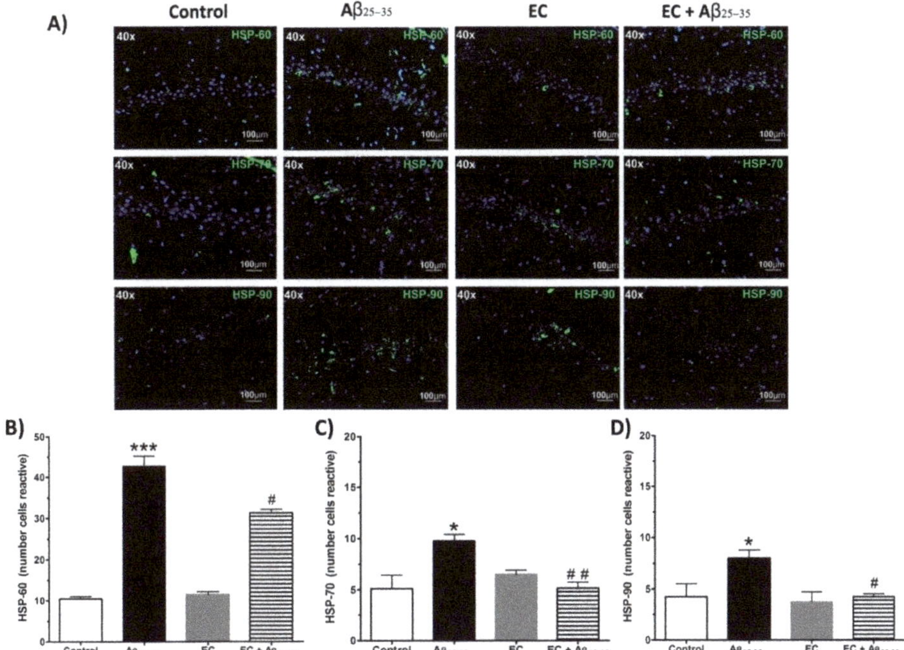

Figure 4. Effect of epicatechin on immunoreactivity of HSPs in CA1 subfields of the Hp of rats injected with Aβ$_{25-35}$. In (**A**) we observed the immunoreactivity (green color) for HSP-60, HSP-70, and HSP-90 in the CA1 subfield of the Hp of rats with different treatments: control, Aβ$_{25-35}$, EC, and EC + Aβ$_{25-35}$. The epicatechin treatment decreases the number of immunopositive cells to HSP-60 (**B**) HSP-70 (**C**) and HSP-90 (**D**) in the Hp of rats injected with Aβ$_{25-35}$ (**A**). The mean ± SE is plotted. Data were analyzed with one-way ANOVA and post-test Bonferroni test (* $p < 0.05$ and *** $p < 0.001$) comparing all groups with respect to the vehicle group. (# $p < 0.05$ and ## $p < 0.01$) compared the Aβ$_{25-35}$ versus EC + Aβ$_{25-35}$ groups.

3.5. The Administration of Epicatechin Reduces the Immunoreactivity of Caspase-3 in the Hippocampus of Rats Injected with Aβ$_{25-35}$

We performed an immunohistochemistry for caspase-3 and hematoxylin and eosin staining to investigate whether EC reduces neuronal death. The photomicrographs are shown in Figure 5A. The images reveal that Aβ$_{25-35}$ induces a greater immunoreactivity to caspase-3 (green color) in the cells of the CA1 region of the Hp, compared to the control group, while for the EC + Aβ$_{25-35}$ group, there exists a lower marker to caspase-3, with respect to the group injected only with Aβ$_{25-35}$. The number of immunoreactive cells to caspase-3 indicates that Aβ$_{25-35}$ increases the reactivity to caspase-3 by 40% with respect to the control group. However, treatment with EC in animals with Aβ$_{25-35}$ reduces the immunoreactivity by 45% in the cells of the CA1 region of the Hp (Figure 5B) (one-way ANOVA with a significance of $p < 0.05$). Hematoxylin and eosin staining show damage to the neurons through cytoplasmic changes of eosinophilia, in addition to the observed pyknosis, karyorrhexis, and karyolysis. This was more severe in the CA1 region of the Hp of the group treated with Aβ$_{25-35}$ than in the EC + Aβ$_{25-35}$ group. These indicators were not observed in the control and EC group (Figure 5B). Figure 5A shows the viable cell number in the CA1 subfield of the Hp. The number of viable neurons in the Aβ$_{25-35}$ group was less compared with the control group (54%), whereas the number of viable cells in the EC + Aβ$_{25-35}$ group was greater with respect to the Aβ$_{25-35}$ group (65%) (one-way ANOVA with a significance of $p < 0.05$). The group treated only with EC showed no changes with respect to the control group. This indicates that EC decreases the neuronal death that $A\beta_{25-35}$ induces in rat hippocampi.

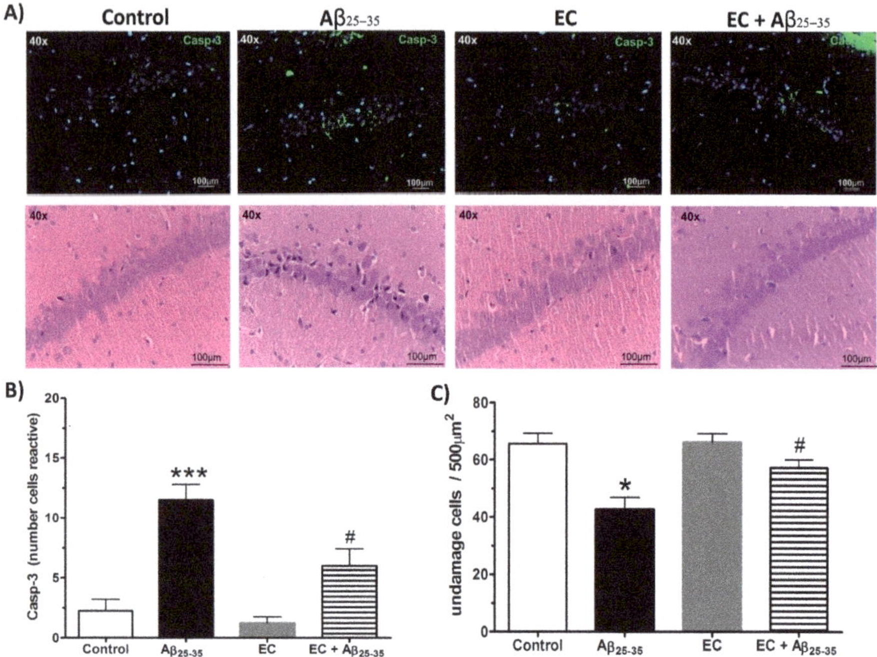

Figure 5. The epicatechin decreased the immunoreactivity of caspase-3 and the number of damaged cells in CA1 subfields of the Hp rats of injected with $A\beta_{25-35}$. In (**A**) we observed the immunoreactivity for caspase-3 (green color) and the cells stained with H & E of the CA1 subfield of the Hp of rats control, $A\beta_{25-35}$, EC, and EC + $A\beta_{25-35}$ groups. The epicatechin treatment decreases the number of immunoreactive cells to caspase-3 (**B**) and a number of damaged cells in the CA1 Hp of rats injected with $A\beta_{25-35}$ (**C**). The mean ± SE is plotted. Data were analyzed with one-way ANOVA and post-test Bonferroni test (* $p < 0.05$ and *** $p < 0.001$) comparing all groups with respect to the vehicle group. # $p < 0.05$ compared the $A\beta_{25-35}$ and EC + $A\beta_{25-35}$ groups.

4. Discussion

In the present work, we investigated whether treatment with EC is beneficial in combating the toxic effects of $A\beta_{25-35}$ in the Hp. The results show an improvement in the spatial memory task, as well as a reduction in markers of oxidative stress and inflammation. Additionally, a decrease in the immunoreactivity of HSP-60, HSP-70, and HSP-90 was detected, strongly suggesting a positive regulating of the processes of protein aggregation and neurodegeneration. Therefore, the treatment with EC reduces the hippocampal death events induced by the injection of $A\beta_{25-35}$.

Catechins are a group of polyphenolic compounds belonging to the flavonoid class, present in high concentrations in a variety of plant-based fruits, vegetables, and beverages [25–27]. Particularly, EC shows protection against dementia, as well as diseases in which oxidative stress seems to be highly prevalent [13]. Oxidative stress is considered a key factor in the AD pathogenesis [28]. $A\beta$ is responsible for promoting the generation of ROS, the oxidation of lipids to destabilize the neuronal membrane, and decreasing the activity of antioxidant enzymes such as SOD and catalase, which encourage the necrosis and apoptosis states [29]. Reports indicate that $A\beta_{25-35}$ interacts with the cell membrane, altering the structure and function of the lipid bilayer and augmenting the concentration of ROS and lipid peroxidation, as shown in our results. In addition, the injection of $A\beta_{25-35}$ into the Hp of rats produces a decrease in activity of SOD and catalase, indicating that the Hp is a brain region susceptible to oxidative and neuronal damage [30].

The interactions that $A\beta_{25-35}$ has with the cell membrane induces ion channel dysfunction, which is associated with dysregulation of neuronal Ca^{2+}. The intracellular increase of Ca^{2+} in neurons provokes vulnerability to the apoptotic processes [31]. In other reports, neuronal death is mainly associated with the generation of peroxidation products induced by $A\beta_{25-35}$, such as 4-hydroxy-nonenal (4-HNE), which alters the glutamate transporter (GLT1). This causes the excessive entry of Ca^{2+} to activate the nitric oxide neuronal synthase (nNOS) responsible for catalyzing the formation of nitric oxide (NO) in neurons. The interaction of NO with the superoxide ion (O^{2-}) promotes the formation of peroxynitrite ($ONOO^-$) [7]. This reactive species alters the structure and function of proteins of the cytoskeleton and the mitochondria by means of a nitration or nitrosylation reaction, structures considered as cell damage markers which have been reported on several occasions by our research team [6]. The inflammatory response is a complementary mechanism of $A\beta_{25-35}$ toxicity. The inflammation is mediated by the production and excessive release of proinflammatory cytokines, such as IL-1β and TNF-α. These cytokines are considered regulators of the intensity and duration of the inflammatory processes. TNF-induced cell death signaling is carried out by the TNF receptor (TNFR) [32]. Essentially, TNFR permits the recruitment of intracellular "death signaling inducing signaling complex" (DISC) proteins [33,34]. These proteins generate a scaffolding, so that recruitment is favored and caspase-8 is activated [35]. The freed, active caspase-8 then enzymatically processes procaspase-3, converting this executioner procaspase into active enzyme [35]. The activation of caspase-3 is essential for TNF-induced cell death, as it targets a latent DNase that degrades genomic DNA, thus causing apoptotic cell death, as was observed in the $A\beta_{25-35}$ group [6,36]. On the other hand, IL-1 is a pivotal mediator of the inflammation that is expressed rapidly in response to neuronal injury, predominantly by microglia, and elevated levels markedly exacerbate the injury [37,38]. The associated mechanism to IL-1β release is the increase of proinflammatory molecules that results in rapidly recruited neutrophils, which cause tissue damage through the generation of free radicals, proteolytic enzymes, and more proinflammatory cytokines such as IL-1β and TNF-α [39]. The activation of both Nuclear Factor kappa B (NF-κB) and TNF-α triggers the activation of Mitogen-Activated Protein Kinase (MAPK)-related stress-activated Jun kinase (JNK) that induces cell death [40,41]. Studies have shown that increases in both IL-1β and TNF-α produce a ROS accumulation and cell death as we have observed in our results, where the intrahippocampal injection of $A\beta_{25-35}$ showed a significant increase in the concentration of these cytokines compared to the control group [42]. Therefore, it is suggested that $A\beta_{25-35}$ promotes an inflammatory response mediated by the production of proinflammatory cytokines, which could trigger the generation of ROS and the proliferation of glial cells, leading to reactive gliosis [43]. All these mechanisms of neurodegeneration cause damage to the structure and function of the limbic system, particularly in the Hp, a brain region that has a key role in the development of spatial memory, in such a way that a characteristic of the neurotoxicity of $A\beta_{25-35}$ is the deterioration in learning and spatial memory, which is supported by a large number of publications [17,18,24,28,29].

On the other hand, unregulated ROS levels can activate oxidant stress sensor proteins, particularly the heat shock factor (HSF), a transcription factor responsible for increasing heat shock proteins (HSPs) because they react at the cellular level in response to environmental changes [44,45]. Under physiological conditions, HSPs act as molecular chaperones, preventing changes in the functional folding of proteins, while under stress conditions, they prevent protein aggregation [46]. Likewise, HSPs promote the regulation of specific signaling pathways that respond to inhibit apoptosis [47]. The neurotoxic changes caused by $A\beta_{25-35}$ provoke conformational changes in HSF and induce the expression and synthesis of HSPs [48]. It is proposed that these proteins are activated as a neuroprotective mechanism. In this work, it is demonstrated that the increase in ROS and lipid peroxidation in response to the toxicity of $A\beta_{25-35}$ promotes the increase in the immunoreactivity of HSP-60, HSP-70, and HSP-90 at 30 days postinjection with respect to the control group but does not prevent neuronal death.

Studies indicate that HSPs can be activated directly or indirectly. The ROS oxidize the thiol groups of proteins and cause the conformational change that activates the HSPs [49,50]. The in vitro and in vivo models indicate that each HSP performs specific functions in cells under conditions of oxidative stress. The HSP-90, being a folding chaperone, prevents the aggregation of the Aβ peptide (a necessary condition to cause neurotoxicity) and promotes its degradation. The increase in the expression of HSP-70 favors interaction with the intracellular Aβ, suggesting the sequestering of the peptide to avoid its interaction with other proteins and thus inhibiting its toxicity [9]. HSP-70 also activates microglia to synthesize and release cytokines and stimulate phagocytosis. There is evidence that HSP-70 promotes cell survival and reduces neuronal death [51]. It is also proposed that the overexpression of HSP-70 participates in inflammatory processes by reducing the levels of Cyclooxygenase (COX)-2 and the production of NO. In the same way, HSP-60, a mitochondrial chaperonin that is typically held responsible for the transportation and refolding of proteins from the cytoplasm into the mitochondrial matrix, also prevents the release of cytochrome-C in the mitochondria, the formation of apoptosome, and the activation of caspases, and therefore inhibits neuronal death. Reports indicate that Aβ impairs mitochondrial activity, however, HSP-60 prevents this damage and thus reduces the formation of ROS and promotes the native and functional conformation of SOD [52]. Our results indicate that all these chaperones act as damage biomarkers or neurosensors, suggesting their neuroprotective activity when increasing their immunoreactivity. These are active in different cellular compartments and interfere with oxidative, inflammatory processes and in the apoptotic signaling cascade that induces the toxicity of Aβ$_{25-35}$ in the rat Hp.

The administration of EC increased spatial memory, decreased ROS concentrations and peroxidation levels, increased the enzymatic activity of SOD and catalase, and decreased IL-1β and TNF-α levels. It is proposed that EC captures oxidizing species and prevents lipid peroxidation, which results in blocking the neurotoxicity caused by Aβ$_{25-35}$, therefore, the treatment of EC shows an improvement in spatial memory. It is also suggested that the catechol group in EC confers the ability to donate a hydrogen atom and therefore stabilizes the activity of free radicals, favoring antioxidant enzymes to help maintain redox balance [53]. Catechin treatments have demonstrated that they inactivate the Signal Transducer and Activator of Transcription 3 (STAT3) pathway, which plays a critical role in inflammation promotion, as well as being able to bind the p65 subunit of the transcription factor NF-κB and to inhibit cytokine and chemokines transcription stimulated by IL-1β and TNF-α. EC has also been shown to inhibit the aggregation of amyloidogenic proteins, including Aβ, present in AD [54]. Interestingly, in the presence of EC, it has been observed that the Aβ monomers adopt a new conformational form with an increased inter-center-of-mass distance with reduced β-sheet content [55,56]. This thereby affects its inclination to form fibril-prone states that otherwise increase the severity of Alzheimer's disease [56]. At a peripheral level, EC at concentrations of 0.1–1 mM was found to inhibit nitrite formation and the IL-1β-induced expression of iNOS by blocking the nuclear localization of the p65 subunit of NF-κB [57]. Therefore, it is reasonable to hypothesize that at the central level and, particularly, in the CA1 hippocampal region, EC can use similar pathways.

In addition, EC possesses the ability to directly or indirectly scavenge ROS by chemically reacting with ROS or by modulating pathways that regulate ROS scavenging compounds and enzymes, respectively [18]. EC can directly reduce ROS levels because the inhibitory strength of this compound is in the number of hydroxyl groups present [17]. In support of this idea, a [51] recent study confirmed that the o-catechol moiety of EC is essential for the direct detoxifying effects of the reaction with superoxide and hydrogen peroxide, or by blocking ROS generation [57–59]. EC also modulates the signaling of nuclear factor erythroid 2p45-related factor-2 (Nrf2), an important factor in cellular detoxification, proliferation, survival, and differentiation [60] that increases the synthesis and activity of antioxidant enzymes, such as SOD y catalase, correlating with our results. Likewise, chronic degenerative diseases have shown affectation on the oxidative phosphorylation machinery. Cytochrome c (Cyt-c) and cytochrome c oxidase (COX) catalyze the terminal and proposed rate-limiting step in the mitochondrial electron transport chain [60], thus their correct function is decisive in the dysregulations in mitochondrial respiration and cell signaling [51]. Since the oxidative phosphorylation machinery is suppressed in most of the degenerative diseases, it can be speculated that reactivation of mitochondrial function might be a strategy that interferes with the progress of the degenerative process. In this sense, EC stimulates mitochondrial respiration [51], significantly stimulating the expression of oxidative phosphorylation protein complexes. Also, EC can inhibit the decrease in mitochondrial succinate dehydrogenase activity, cytochrome c release, mitochondrial fragmentation, and cytochrome c oxidase protein levels [61], and thus inhibit the activation of ERK, caspase-3, and JNK, and the release of cytochrome c [51], diminishing cell death, as was observed. In the same way, oxidative and inflammation changes caused by the Aβ peptide can induce the unfolded, mitochondrial protein stress response. However, the expression of the nuclear-encoded HSP-60 mitochondrial chaperones prevents mitochondrial damage and reduces cell death [61]. It is suggested that EC modulates the cellular signaling cascade through the activation of different pathways, such as serine/threonine kinase 1 (AKT), PKC (protein kinase-C), and MAP kinases. All these pathways participate in the processes that mediate the activation of transcription factors and antiapoptotic signaling, promoting the regulation of the heat shock factor and consequently the downregulation of the HSPs. As far as we know, there are no reports that indicate the potential role of EC on the activity of HSP under conditions of cell stress.

5. Conclusions

This work provides information on the activity that this flavonoid induces on the reactivity of HSP and its consequences on the oxidative and inflammatory response to favor protection in the structure and function of hippocampal neurons and consequently an improvement in spatial memory. In this sense, it provides some data about flavonoid activity on the molecular mechanism for regulation of these chaperones, which assures cell survival in the Hp of animals intoxicated with $A\beta_{25-35}$, and which could be soon applied to AD therapy.

Author Contributions: A.D. and S.T. designed the study and wrote the protocol. A.D., G.P.-F., S.T., B.E., E.M.-M., and K.R. performed the experiments. A.D., S.T., F.P.-S., S.M., M.R.O., and J.G. managed the literature searches and analysis, F.P.-S. and S.M. undertook the statistical analysis. A.D., S.T., and J.G. wrote the first draft of the manuscript. All contributing authors have approved the final manuscript.

Funding: Funding for this study was provided by grants from VIEP-BUAP grant (No. DIFA-NAT18-G) to A.D., (No. TEMS-NAT18-I) to S.T. and PAPIIT-UNAM (IN214117) to J.G. None of the funding institutions had any further role in the study design, the collection of data, analyses and interpretation of data, and writing of the report, or in the decision to submit the paper for publication and interpretation of data.

Acknowledgments: We want to thank Enrique Moreno for his help with animal care. We give special thanks to Estephania Cortéz for technical support in the realization of this research project and Abel Santamaria for the water maze facilities. A.D., S.T., B.E., F.P.-S., S.M., M.R.O., and J.G. acknowledge the "Sistema Nacional de Investigadores" of Mexico for membership. Thanks to Thomas Edwards PhD., for editing the English language text.

Conflicts of Interest: Authors have no conflict of interest to declare. Also, all authors declare the availability of this manuscript.

References

1. Hampel, H.; Vergallo, A.; Aguilar, L.F.; Benda, N.; Broich, K.; Cuello, A.C.; Cummings, J.; Dubois, B.; Federoff, H.J.; Fiandaca, M.; et al. Precision pharmacology for Alzheimer's disease. *Pharmacol. Res.* **2018**, *130*, 331–365. [CrossRef]
2. De Oliveira, W.F.; Dos Santos Silva, P.M.; Coelho, L.; Dos Santos Correia, M.T. Biomarkers, Biosensors and Biomedicine. *Curr. Med. Chem.* **2019**. [CrossRef] [PubMed]
3. Millucci, L.; Ghezzi, L.; Bernardini, G.; Santucci, A. Conformations and biological activities of amyloid beta peptide 25-35. *Curr. Protein Pept. Sci.* **2010**, *11*, 54–67. [CrossRef] [PubMed]
4. Zussy, C.; Brureau, A.; Keller, E.; Marchal, S.; Blayo, C.; Delair, B.; Ixart, G.; Maurice, T.; Givalois, L. Alzheimer's disease related markers, cellular toxicity and behavioral deficits induced six weeks after oligomeric amyloid-beta peptide injection in rats. *PLoS ONE* **2013**, *8*, e53117. [CrossRef] [PubMed]

5.   Maurice, T.; Lockhart, B.P.; Privat, A. Amnesia induced in mice by centrally administered beta-amyloid peptides involves cholinergic dysfunction. *Brain Res.* **1996**, *706*, 181–193. [CrossRef]

6.   Diaz, A.; Rojas, K.; Espinosa, B.; Chavez, R.; Zenteno, E.; Limon, D.; Guevara, J. Aminoguanidine treatment ameliorates inflammatory responses and memory impairment induced by amyloid-beta 25-35 injection in rats. *Neuropeptides* **2014**, *48*, 153–159. [CrossRef]

7.   Diaz, A.; Mendieta, L.; Zenteno, E.; Guevara, J.; Limon, I.D. The role of NOS in the impairment of spatial memory and damaged neurons in rats injected with amyloid beta 25-35 into the temporal cortex. *Pharmacol. Biochem. Behav.* **2011**, *98*, 67–75. [CrossRef] [PubMed]

8.   Sottile, M.L.; Nadin, S.B. Heat shock proteins and DNA repair mechanisms: An updated overview. *Cell Stress Chaperones* **2018**, *23*, 303–315. [CrossRef]

9.   Ortega, L.; Calvillo, M.; Luna, F.; Perez-Severiano, F.; Rubio-Osornio, M.; Guevara, J.; Limon, I.D. 17-AAG improves cognitive process and increases heat shock protein response in a model lesion with Abeta25-35. *Neuropeptides* **2014**, *48*, 221–232. [CrossRef]

10.   Lackie, R.E.; Maciejewski, A.; Ostapchenko, V.G.; Marques-Lopes, J.; Choy, W.Y.; Duennwald, M.L.; Prado, V.F.; Prado, M.A.M. The Hsp70/Hsp90 Chaperone Machinery in Neurodegenerative Diseases. *Front. Neurosci.* **2017**, *11*, 254. [CrossRef]

11.   Liu, Y.; Zhang, X. Heat Shock Protein Reports on Proteome Stress. *Biotechnol. J.* **2018**, *13*. [CrossRef] [PubMed]

12.   Natsume, M. Polyphenols: Inflammation. *Curr. Pharm. Des.* **2018**, *24*, 191–202. [CrossRef]

13.   Bernatova, I. Biological activities of (-)-epicatechin and (-)-epicatechin-containing foods: Focus on cardiovascular and neuropsychological health. *Biotechnol. Adv.* **2018**, *36*, 666–681. [CrossRef]

14.   Patel, A.K.; Rogers, J.T.; Huang, X. Flavanols, mild cognitive impairment, and Alzheimer's dementia. *Int. J. Clin. Exp. Med.* **2008**, *1*, 181–191.

15.   Jiang, Z.; Zhang, J.; Cai, Y.; Huang, J.; You, L. Catechin attenuates traumatic brain injury-induced blood-brain barrier damage and improves longer-term neurological outcomes in rats. *Exp. Physiol.* **2017**, *102*, 1269–1277. [CrossRef]

16.   Li, Q.; Zhao, H.F.; Zhang, Z.F.; Liu, Z.G.; Pei, X.R.; Wang, J.B.; Li, Y. Long-term green tea catechin administration prevents spatial learning and memory impairment in senescence-accelerated mouse prone-8 mice by decreasing Abeta1-42 oligomers and upregulating synaptic plasticity-related proteins in the hippocampus. *Neuroscience* **2009**, *163*, 741–749. [CrossRef]

17.   Cuevas, E.; Limon, D.; Perez-Severiano, F.; Diaz, A.; Ortega, L.; Zenteno, E.; Guevara, J. Antioxidant effects of epicatechin on the hippocampal toxicity caused by amyloid-beta 25-35 in rats. *Eur. J. Pharmacol.* **2009**, *616*, 122–127. [CrossRef]

18.   Cruz-Gonzalez, T.; Cortez-Torres, E.; Perez-Severiano, F.; Espinosa, B.; Guevara, J.; Perez-Benitez, A.; Melendez, F.J.; Diaz, A.; Ramirez, R.E. Antioxidative stress effect of epicatechin and catechin induced by Abeta25-35 in rats and use of the electrostatic potential and the Fukui function as a tool to elucidate specific sites of interaction. *Neuropeptides* **2016**, *59*, 89–95. [CrossRef]

19.   Trevino, S.; Aguilar-Alonso, P.; Flores Hernandez, J.A.; Brambila, E.; Guevara, J.; Flores, G.; Lopez-Lopez, G.; Munoz-Arenas, G.; Morales-Medina, J.C.; Toxqui, V.; et al. A high calorie diet causes memory loss, metabolic syndrome and oxidative stress into hippocampus and temporal cortex of rats. *Synapse* **2015**, *69*, 421–433. [CrossRef] [PubMed]

20.   Diaz, A.; Trevino, S.; Guevara, J.; Munoz-Arenas, G.; Brambila, E.; Espinosa, B.; Moreno-Rodriguez, A.; Lopez-Lopez, G.; Pena-Rosas, U.; Venegas, B.; et al. Energy Drink Administration in Combination with Alcohol Causes an Inflammatory Response and Oxidative Stress in the Hippocampus and Temporal Cortex of Rats. *Oxid. Med. Cell. Longev.* **2016**, *2016*, 8725354. [CrossRef] [PubMed]

21.   Perez-Severiano, F.; Rodriguez-Perez, M.; Pedraza-Chaverri, J.; Maldonado, P.D.; Medina-Campos, O.N.; Ortiz-Plata, A.; Sanchez-Garcia, A.; Villeda-Hernandez, J.; Galvan-Arzate, S.; Aguilera, P.; et al. S-Allylcysteine, a garlic-derived antioxidant, ameliorates quinolinic acid-induced neurotoxicity and oxidative damage in rats. *Neurochem. Int.* **2004**, *45*, 1175–1183. [CrossRef]

22.   Ali, S.F.; David, S.N.; Newport, G.D. Age-related susceptibility to MPTP-induced neurotoxicity in mice. *Neurotoxicology* **1993**, *14*, 29–34.

23.   Ramos-Martinez, I.; Martinez-Loustalot, P.; Lozano, L.; Issad, T.; Limon, D.; Diaz, A.; Perez-Torres, A.; Guevara, J.; Zenteno, E. Neuroinflammation induced by amyloid beta25-35 modifies mucin-type O-glycosylation in the rat's hippocampus. *Neuropeptides* **2018**, *67*, 56–62. [CrossRef]

24. Diaz, A.; De Jesus, L.; Mendieta, L.; Calvillo, M.; Espinosa, B.; Zenteno, E.; Guevara, J.; Limon, I.D. The amyloid-beta25-35 injection into the CA1 region of the neonatal rat hippocampus impairs the long-term memory because of an increase of nitric oxide. *Neurosci. Lett.* **2010**, *468*, 151–155. [CrossRef]

25. Kim, H.; Ramirez, C.N.; Su, Z.Y.; Kong, A.N. Epigenetic modifications of triterpenoid ursolic acid in activating Nrf2 and blocking cellular transformation of mouse epidermal cells. *J. Nutr. Biochem.* **2016**, *33*, 54–62. [CrossRef]

26. Adler, G.; Hey, W.; Achenbach, C.; Kinzer, A. Pretreatment interhemispheric EEG coherence is related to seizure duration in right unilateral electroconvulsive therapy. *Neuropsychobiology* **2003**, *48*, 143–145. [CrossRef]

27. Wang, Z.F.; Liu, J.; Yang, Y.A.; Zhu, H.L. A Review: The Anti-inflammatory, Anticancer, Antibacterial Properties of Four Kinds of Licorice Flavonoids Isolated from Licorice. *Curr. Med. Chem.* **2018**. [CrossRef] [PubMed]

28. Nalivaeva, N.N.; Turner, A.J. Role of Ageing and Oxidative Stress in Regulation of Amyloid-Degrading Enzymes and Development of Neurodegeneration. *Curr. Aging Sci.* **2017**, *10*, 32–40. [CrossRef]

29. Gulyaeva, N.V.; Bobkova, N.V.; Kolosova, N.G.; Samokhin, A.N.; Stepanichev, M.Y.; Stefanova, N.A. Molecular and Cellular Mechanisms of Sporadic Alzheimer's Disease: Studies on Rodent Models in vivo. *Biochem. Biokhimiia* **2017**, *82*, 1088–1102. [CrossRef]

30. Kaminsky, Y.G.; Marlatt, M.W.; Smith, M.A.; Kosenko, E.A. Subcellular and metabolic examination of amyloid-beta peptides in Alzheimer disease pathogenesis: Evidence for Abeta(25-35). *Exp. Neurol.* **2010**, *221*, 26–37. [CrossRef]

31. Maurice, T.; Strehaiano, M.; Duhr, F.; Chevallier, N. Amyloid toxicity is enhanced after pharmacological or genetic invalidation of the sigma1 receptor. *Behav. Brain Res.* **2018**, *339*, 1–10. [CrossRef]

32. Tartaglia, L.A.; Rothe, M.; Hu, Y.F.; Goeddel, D.V. Tumor necrosis factor's cytotoxic activity is signaled by the p55 TNF receptor. *Cell* **1993**, *73*, 213–216. [CrossRef]

33. Zhang, M.; Wang, J.; Jia, L.; Huang, J.; He, C.; Hu, F.; Yuan, L.; Wang, G.; Yu, M.; Li, Z. Transmembrane TNF-alpha promotes activation-induced cell death by forward and reverse signaling. *Oncotarget* **2017**, *8*, 63799–63812. [CrossRef] [PubMed]

34. Boldin, M.P.; Goncharov, T.M.; Goltsev, Y.V.; Wallach, D. Involvement of MACH, a novel MORT1/FADD-interacting protease, in Fas/APO-1- and TNF receptor-induced cell death. *Cell* **1996**, *85*, 803–815. [CrossRef]

35. Lim, M.C.C.; Maubach, G.; Sokolova, O.; Feige, M.H.; Diezko, R.; Buchbinder, J.; Backert, S.; Schluter, D.; Lavrik, I.N.; Naumann, M. Pathogen-induced ubiquitin-editing enzyme A20 bifunctionally shuts off NF-kappaB and caspase-8-dependent apoptotic cell death. *Cell Death Differ.* **2017**, *24*, 1621–1631. [CrossRef] [PubMed]

36. Enari, M.; Sakahira, H.; Yokoyama, H.; Okawa, K.; Iwamatsu, A.; Nagata, S. A caspase-activated DNase that degrades DNA during apoptosis, and its inhibitor ICAD. *Nature* **1998**, *391*, 43–50. [CrossRef]

37. Simi, A.; Tsakiri, N.; Wang, P.; Rothwell, N.J. Interleukin-1 and inflammatory neurodegeneration. *Biochem. Soc. Trans.* **2007**, *35*, 1122–1126. [CrossRef]

38. Morganti-Kossmann, M.C.; Rancan, M.; Stahel, P.F.; Kossmann, T. Inflammatory response in acute traumatic brain injury: A double-edged sword. *Curr. Opin. Crit. Care* **2002**, *8*, 101–105. [CrossRef]

39. Ferrara-Bowens, T.M.; Chandler, J.K.; Guignet, M.A.; Irwin, J.F.; Laitipaya, K.; Palmer, D.D.; Shumway, L.J.; Tucker, L.B.; McCabe, J.T.; Wegner, M.D.; et al. Neuropathological and behavioral sequelae in IL-1R1 and IL-1Ra gene knockout mice after soman (GD) exposure. *Neurotoxicology* **2017**, *63*, 43–56. [CrossRef]

40. De Smaele, E.; Zazzeroni, F.; Papa, S.; Nguyen, D.U.; Jin, R.; Jones, J.; Cong, R.; Franzoso, G. Induction of gadd45beta by NF-kappaB downregulates pro-apoptotic JNK signalling. *Nature* **2001**, *414*, 308–313. [CrossRef] [PubMed]

41. Tang, Y.; Lopez, I.; Baloh, R.W. Age-related change of the neuronal number in the human medial vestibular nucleus: A stereological investigation. *J. Vestib. Res. Equilib. Orientat.* **2001**, *11*, 357–363.

42. Diaz, A.; Limon, D.; Chavez, R.; Zenteno, E.; Guevara, J. Abeta25-35 injection into the temporal cortex induces chronic inflammation that contributes to neurodegeneration and spatial memory impairment in rats. *JAD* **2012**, *30*, 505–522. [CrossRef]

43. Benjamin, I.J.; McMillan, D.R. Stress (heat shock) proteins: Molecular chaperones in cardiovascular biology and disease. *Circ. Res.* **1998**, *83*, 117–132. [CrossRef] [PubMed]

44. Driedonks, N.; Xu, J.; Peters, J.L.; Park, S.; Rieu, I. Multi-Level Interactions Between Heat Shock Factors, Heat Shock Proteins, and the Redox System Regulate Acclimation to Heat. *Front. Plant Sci.* **2015**, *6*, 999. [CrossRef] [PubMed]

45. Gething, M.J.; Sambrook, J. Protein folding in the cell. *Nature* **1992**, *355*, 33–45. [CrossRef] [PubMed]

46. Calvillo, M.; Diaz, A.; Limon, D.I.; Mayoral, M.A.; Chanez-Cardenas, M.E.; Zenteno, E.; Montano, L.F.; Guevara, J.; Espinosa, B. Amyloid-beta(25-35) induces a permanent phosphorylation of HSF-1, but a transitory and inflammation-independent overexpression of Hsp-70 in C6 astrocytoma cells. *Neuropeptides* **2013**, *47*, 339–346. [CrossRef] [PubMed]

47. Choi, J.; Malakowsky, C.A.; Talent, J.M.; Conrad, C.C.; Carroll, C.A.; Weintraub, S.T.; Gracy, R.W. Anti-apoptotic proteins are oxidized by Abeta25-35 in Alzheimer's fibroblasts. *Biochim. Biophys. Acta* **2003**, *1637*, 135–141. [CrossRef]

48. Choi, Y.J.; Kim, N.H.; Lim, M.S.; Lee, H.J.; Kim, S.S.; Chun, W. Geldanamycin attenuates 3nitropropionic acidinduced apoptosis and JNK activation through the expression of HSP 70 in striatal cells. *Int. J. Mol. Med.* **2014**, *34*, 24–34. [CrossRef]

49. Shabbir, A.; Bianchetti, E.; Cargonja, R.; Petrovic, A.; Mladinic, M.; Pilipovic, K.; Nistri, A. Role of HSP70 in motoneuron survival after excitotoxic stress in a rat spinal cord injury model in vitro. *Eur. J. Neurosci.* **2015**, *42*, 3054–3065. [CrossRef]

50. Wiesneth, S.; Jurgenliemk, G. Total phenolic and tannins determination: A modification of Ph. Eur. 2.8.14 for higher throughput. *Pharmazie* **2017**, *72*, 195–196. [CrossRef]

51. Bieschke, J.; Russ, J.; Friedrich, R.P.; Ehrnhoefer, D.E.; Wobst, H.; Neugebauer, K.; Wanker, E.E. EGCG remodels mature alpha-synuclein and amyloid-beta fibrils and reduces cellular toxicity. *Proc. Natl. Acad. Sci. USA* **2010**, *107*, 7710–7715. [CrossRef]

52. Meng, F.; Abedini, A.; Plesner, A.; Verchere, C.B.; Raleigh, D.P. The flavanol (-)-epigallocatechin 3-gallate inhibits amyloid formation by islet amyloid polypeptide, disaggregates amyloid fibrils, and protects cultured cells against IAPP-induced toxicity. *Biochemistry* **2010**, *49*, 8127–8133. [CrossRef]

53. Porat, Y.; Abramowitz, A.; Gazit, E. Inhibition of amyloid fibril formation by polyphenols: Structural similarity and aromatic interactions as a common inhibition mechanism. *Chem. Biol. Drug Des.* **2006**, *67*, 27–37. [CrossRef]

54. Jung, H.A.; Jung, M.J.; Kim, J.Y.; Chung, H.Y.; Choi, J.S. Inhibitory activity of flavonoids from Prunus davidiana and other flavonoids on total ROS and hydroxyl radical generation. *Arch. Pharmacal Res.* **2003**, *26*, 809–815. [CrossRef]

55. Ruijters, E.J.; Weseler, A.R.; Kicken, C.; Haenen, G.R.; Bast, A. The flavanol (-)-epicatechin and its metabolites protect against oxidative stress in primary endothelial cells via a direct antioxidant effect. *Eur. J. Pharmacol.* **2013**, *715*, 147–153. [CrossRef]

56. Shin, H.A.; Shin, Y.S.; Kang, S.U.; Kim, J.H.; Oh, Y.T.; Park, K.H.; Lee, B.H.; Kim, C.H. Radioprotective effect of epicatechin in cultured human fibroblasts and zebrafish. *J. Radiat. Res.* **2014**, *55*, 32–40. [CrossRef]

57. Granado-Serrano, A.B.; Martin, M.A.; Bravo, L.; Goya, L.; Ramos, S. Quercetin modulates NF-kappa B and AP-1/JNK pathways to induce cell death in human hepatoma cells. *Nutr. Cancer* **2010**, *62*, 390–401. [CrossRef]

58. Kudin, A.P.; Malinska, D.; Kunz, W.S. Sites of generation of reactive oxygen species in homogenates of brain tissue determined with the use of respiratory substrates and inhibitors. *Biochim. Biophys. Acta* **2008**, *1777*, 689–695. [CrossRef]

59. Huttemann, M.; Lee, I.; Malek, M.H. (-)-Epicatechin maintains endurance training adaptation in mice after 14 days of detraining. *FASEB J. Off. Publ. Fed. Am. Soc. Exp. Biol.* **2012**, *26*, 1413–1422. [CrossRef]

60. Elbaz, H.A.; Lee, I.; Antwih, D.A.; Liu, J.; Huttemann, M.; Zielske, S.P. Epicatechin stimulates mitochondrial activity and selectively sensitizes cancer cells to radiation. *PLoS ONE* **2014**, *9*, e88322. [CrossRef]

61. Jovaisaite, V.; Auwerx, J. The mitochondrial unfolded protein response-synchronizing genomes. *Curr. Opin. Cell Biol.* **2015**, *33*, 74–81. [CrossRef] [PubMed]

© 2019 by the authors. Licensee MDPI, Basel, Switzerland. This article is an open access article distributed under the terms and conditions of the Creative Commons Attribution (CC BY) license (http://creativecommons.org/licenses/by/4.0/).

 *antioxidants*

Article

Effect of Manitoba-Grown Red-Osier Dogwood Extracts on Recovering Caco-2 Cells from H₂O₂-Induced Oxidative Damage

Runqiang Yang [1,2], Qianru Hui [2], Qian Jiang [2], Shangxi Liu [2], Hua Zhang [3], Jiandong Wu [4], Francis Lin [4], Karmin O [2,5] and Chengbo Yang [2,*]

[1] College of Food Science and Technology, Nanjing Agricultural University, Nanjing 210095, China
[2] Department of Animal Science, University of Manitoba, Winnipeg, MB R3T 2N2, Canada
[3] Guelph Research & Development Centre, Agriculture and Agri-Food Canada, 93 Stone Road West, Guelph, ON N1G 5C9, Canada
[4] Department of Physics, University of Manitoba, Winnipeg, MB R3T 2N2, Canada
[5] St. Boniface Hospital Research Centre, Winnipeg, MB R2H 2A6, Canada
* Correspondence: Chengbo.Yang@umanitoba.ca; Tel.: +1-204-474-8188; Fax: +1-204-474-7628

Received: 27 June 2019; Accepted: 26 July 2019; Published: 28 July 2019

Abstract: Red-osier dogwood, a native species of flowering plant in North America, has been reported to have anti-oxidative properties because of abundant phenolic compounds; this could be promising as a functional food or a feed additive. In the present study, an oxidative damage model using 1.0 mM hydrogen peroxide (H₂O₂) in Caco-2 cells was established to evaluate the antioxidative effects of red-osier dogwood extracts (RDE). The results showed that 1.0 mM H₂O₂ pre-exposure for 3 h significantly decreased cell viability, and increased interleukin 8 (IL-8) secretion and the intracellular reactive oxygen species (ROS) level. Caco-2 cells were treated with 100 μg/mL RDE for 24 h after pre-exposure to H₂O₂. It was found that the decreased cell viability caused by H₂O₂ was significantly restored by a subsequent 100 μg/mL RDE treatment. Furthermore, the IL-8 secretion and ROS level were significantly blocked by RDE, accompanied by the enhanced gene expression of hemeoxygenase-1 (HO-1), superoxide dismutase (SOD), and glutathione peroxidase (GSH-Px), and the enhanced protein expression of the nuclear factor (erythroid-derived 2)-like 2 (Nrf-2). Moreover, RDE improved barrier functions in Caco-2 cells. Using RDE reduced the diffusion of fluorescein isothiocyanate (FITC)-dextran and increased the transepithelial resistance (TEER) value. The relative mRNA level of tight junction claudin-1, claudin-3, and occludin was elevated by RDE. These extracts also repaired the integrity of zonula occludens-1 (ZO-1) damaged by H₂O₂ and increased the protein expressions of ZO-1 and claudin-3 in the H₂O₂-pretreated cells. These results illustrated that RDE reduced the ROS level and enhanced the barrier function in oxidative-damaged epithelial cells.

Keywords: red-osier dogwood; antioxidative effect; H₂O₂; transepithelial resistance (TEER); Caco-2 cells

---

1. Introduction

Phytochemicals, also referred to as phytobiotics or phytogenics, are natural bioactive products derived from plants. They include mainly essential oils [1], flavonoids [2], terpenes [3], polysaccharides [4], and phenolic acid [5]. Currently, phytochemicals were studied widely because of their multiple functions in human and animal health [6,7]. Phytochemicals have broad spectrum functions that are antioxidative [8], anti-microbial [9], anti-inflammatory [10], and anticancer in nature [11]. Research has investigated the profile of phenolic compounds and antioxidant activity of finger millet varieties [12,13]. Twenty phenolic compounds were identified including 18 flavonoids with the predominant flavonoids of catechin and epicatechin. Chen et al. [14] revealed that the dominant phytochemicals in wheat were phenolic acids with high antioxidative capacity in vitro. In addition, the beneficial functions of soy isoflavone [15,16] and lutein [8] have drawn extensive attention recently. Burgeoning evidence indicates that phytochemicals extracted from the plants could be a good source of functional foods or potential additives for functional foods and animal feeds.

Red-osier dogwood (*Cornus stolonifera* Michx.) is native to North America and is commonly found in marshes, low meadows, and along river banks from Labrador to Alaska [17]. In Canada, it is a precious plant resource and rich in phenolic components, including anthocyanins with strong antioxidative activity [18]. At present, ground red-osier dogwood has been investigated in animal feeds to improve growth performance, feed intake, disease prevention, and meat quality [19,20]. Researchers [19,20] investigated the effects of ground red-osier dogwood supplementation on the digestibility, blood metabolites, and acute phase response in beef heifers. The results showed that ground red-osier dogwood could increase the feed digestibility and plasma concentration of haptoglobin and serum amyloid A in a ground red-osier dogwood dose-dependence. Dietary 4% ground red-osier dogwood supplementation attenuated oxidative stress and improved the growth performance of the *E. coli* infected weaned piglets, as revealed by Koo et al. [21,22] and Jayaraman et al. [21,22]. The beneficial effects of ground red-osier dogwood on animal growth and feed digestion might be associated with the oxidative homeostasis of the intestinal cells and the integrity of the intestinal barrier that are maintained by the antioxidative function of phytochemicals in the ground red-osier dogwood.

Cellular oxidative and inflammatory models have been widely used to investigate the potential mechanisms of phytochemicals involved in human and animal health. The intestinal porcine epithelial cell line J2 (IPEC-J2) and human colon adenocarcinoma cell line-2 (Caco-2) provide useful cell models for investigating the antioxidant mechanism of phytochemicals in the intestine [8,21,22]. Omonijo et al. [10] revealed the anti-inflammatory mechanism of thymol in the inflammatory models of lipopolysaccharide (LPS)-induced IPEC-J2 cells. Yang et al. [23] investigated the effects of resveratrol on the oxidative model of IPEC-J2 cells, and found that resveratrol protects IPEC-J2 cells from oxidative damage by stimulating the Nrf-2 pathway. Wu et al. [24] revealed that phytochemicals increase the expressions of tight junction proteins including occludin, zona occludens (ZO-1), and induce the upregulation of antioxidant enzymes including superoxidedismutase (SOD), hemeoxygenase-1 (HO-1), catalase (CAT), and glutathione peroxidase (GSH-Px) in the oxidative model of Caco-2 cells. A variety of adverse factors could cause the production of reactive oxygen species (ROS), resulting in oxidative damage of the intestinal barrier. Moreover, persistent oxidative damage could trigger the development of the intestinal diseases, including colon cancer [25]. Therefore, the intake of natural phytochemicals could maintain the health of the gut and be considered an effective way to reduce the occurrence of intestinal diseases. Although ground red-osier dogwood has been investigated as potential animal feed additives to improve animal performance and meat quality, the mechanisms of ground red-osier dogwood in maintaining the oxidative homeostasis and the integrity of the intestinal barrier are still not fully understood.

In the present study, it is hypothesized that red-osier dogwood extracts (RDE) act as antioxidant agents capable of exerting protective effects against chemical-induced damage in intestinal epithelial cells. Therefore, an oxidative damage model of Caco-2 cells induced by hydrogen peroxide (H_2O_2) was established, and the mechanisms of RDE in recovering the intestinal cells from oxidative damage were investigated by examining the antioxidant system and the tight junction protein expression.

2. Materials and Methods

2.1. Materials and Reagents

RDE powder (moisture 7.3%; total phenolic content 26.51 gallic acid equivalent (GAE)/100 g including gallic acid 1670.09 mg/100 g and rutin 2745.49 mg/100 g) from leaves was provided by Red Dog Enterprises Ltd. (Swan River, MB, Canada). Hydrogen peroxide (H_2O_2), 2′,7′dichlorofluorescein diacetate (DCFH-DA), 4 kDa fluorescein isothiocyanate-dextran (FD4) and paraformaldehyde (PFA) were purchased from Sigma-Aldrich (Oakville, ON, Canada). Hanks' balanced salt solution (HBSS), DMEM/F12 medium, trypsin (0.25%), anti-anti (penicillin (100 IU/mL), streptomycin (100 µg/mL), and 0.25 µg/mL of amphotericin B) and fetal bovine serum (FBS) were purchased from Invitrogen (Fisher Scientific, Ottawa, ON, Canada). T-25 or T-75 culture flasks, Transwell permeable supports (0.4 µm pore size), and cell culture plates were purchased from Corning (Fisher Scientific, Ottawa, ON, Canada).

2.2. Cell Culture

Caco-2 human colon cancer cell was purchased from the American Type Culture Collection (ATCC, Manassas, VA, USA). Cells were grown in DMEM-Ham's F-12 (1:1) supplemented with 10% FBS and 1% anti-anti, and maintained in an atmosphere of 5% CO_2 at 37 °C. Cells were cultured in a T-25 or T-75 culture flask with a medium change every two days until 90% confluence. Then, they were digested with 0.25% trypsin (at 37 °C for 1–2 min) and seeded in a cell culture plate (96-, 24-, 12-, or 6-well) to culture for further experiments.

2.3. Experiment Design

Cytotoxicity of H_2O_2 (ranges from 0.2 to 2.0 mM) on Caco-2 cells was tested by a viability assay, and the optimal concentration of H_2O_2 was chosen for inducing oxidative damage of Caco-2 cells. Cells were treated with 1 mM of H_2O_2 for 3 h firstly, and then washed with PBS (a pre-warmed at 37 °C water bath) one time. Afterward, the cells treated with 100 µg/mL RDE for 24 h were considered the treatment group (T) according to the results. The cells treated with H_2O_2 and followed with a standard medium [DMEM-Ham's F-12 (1:1) supplemented with 10% FBS and 1% anti-anti] for 24 h were defined as treatment control (TC). The cells without H_2O_2 and RDE treatment were negative control (NC). In brief, the H_2O_2 work solution was directly diluted using a medium from 10 M H_2O_2. The RDE solution was prepared according to the required concentrations using a medium.

2.4. Cell Viability Assay

After treatments, cell viability was measured using the water-soluble tetrazolium salts (WST-1) Cell Proliferation Reagent (Sigma Aldrich, Roche, Indianapolis, IN, USA) according to the manufacturer's instructions. Briefly, Caco-2 cells were seeded into 96-well plates at a density of 2×10^4 cells/well and cultured in a complete medium for about 5 days (90% confluent). After the different treatments (NC, TC, and T), cells were washed one time with PBS and then treated with 100 µL fresh medium containing 10% WTS-1 and incubated for 1.5 h at 37 °C in an atmosphere of 5% CO_2. The absorbance at 450 nm was measured using a Synergy™ H4 Hybrid Multi-Mode Microplate Reader (BioTek, Winooski, VT, USA). Cell viability was presented as a percentage of control cells.

2.5. IL-8 Determination

IL-8 was determined using IL-8 Human Uncoated ELISA Kit (cat. #88-8086, Invitrogen, Carlsbad, CA, USA). Cells were cultured in a 12-well plate with the inoculum density of 2×10^5 cells/well and treated with H_2O_2 and RDE as described above. After, 100 µL of the medium was used to determine IL-8 content according to the kit instructions.

2.6. Reactive Oxygen Species (ROS) Assay

Cellular ROS was measured using a BD FACSCalibur flow cytometer (BD Biosciences, San Jose, CA, USA) [26]. Cells were cultured in a T-25 flask with a inoculum density of 1×10^6 cells/well until 90% confluence reached. They were treated with H_2O_2 and RDE as described above. After treatments, cells were washed with PBS two times and treated with 1 mL DCFH-DA solution (10 µM in PBS) for 30 min at 37 °C. Then, cells were washed with PBS two times and fixed with 4% PFA for 15 min at room temperature. After that, they were digested with 0.25% trypsin (at 37 °C) to monoplast, washed with PBS, and then centrifuged to collect cells for flow cytometry analysis. A total of 10,000 cells was calculated, and fluorescence intensity was measured.

2.7. Glutathione Determination

Total glutathione (GSH) and oxidized glutathione (GSSG) were determined using a Glutathione Colorimetric Detection Kit (cat. #EIAGSHC, Invitrogen). Cells were cultured in 12-well plate with a inoculum density of 2×10^5 cells/well and treated with H_2O_2 and RDE as described above. Afterward, cells were washed with ice-cold PBS and resuspended in 200 µL of ice-cold 5% aqueous 5-sulfo-salicylic acid dehydrate (SSA) and followed by the freeze–thaw cycling to lyse cells. After, lysed cells were centrifuged at 14,800× g at 4 °C for 15 min. The dilution and assay were conducted by following the kit instructions. The reduced GSH was calculated by an equation: reduced GSH = total GSH − GSSG.

2.8. Transepethial Electrical Resistant (TEER) Measurement

The TEER of cell monolayers was measured using a Millicell Electrical Resistance System (ESR-2) (Millipore-Sigma) according to our previously published procedure [10]. Caco-2 cells were seeded onto Transwells (Polyester membrane, 12 wells, 12 mm diameter inserts, Corning Costar, Fisher Scientific) at a density of 1×10^5 cells/insert. The TEER value was monitored every other day. When a monolayer of cells was completely differentiated (about 10 days), cells were treated with H_2O_2 and RDE, and TEER values were measured after H_2O_2 treatment for 6 h and after RED treatment for 24 h, respectively. The data were presented as a percentage of initial value before treatments.

2.9. Measurement of Cell Permeability

To quantify the paracellular permeability of cell monolayers, 1 mg/mL of FD4 (46944-500MG-F, Sigma-Aldrich) was added to the apical side of the inserts (Polyester membrane, 12 wells, 12 mm diameter inserts, Corning Costar, Fisher Scientific, Ottawa, ON, Canada). The cell culture and the treatments were the same as that of TEER measurement. The basolateral medium aliquots were taken after 2 h of incubation at 37 °C in an atmosphere of 5% CO_2. Then, 100 µL of the medium was transferred into 96-well plates and the diffused fluorescent tracer was then measured by fluorometry (excitation, 485 nm; emission, 528 nm) using a Synergy™ H4 Hybrid Multi-Mode Microplate Reader (BioTek, Winooski, VT, USA) [10].

2.10. RNA Extraction and Real-Time PCR

Total RNA isolation and real-time PCR were conducted according to Omonijo et al. [10]. The reference gene was GAPDH (glyceraldehyde-3-phosphate dehydrogenase). The primers for real-time PCR analysis were designed with the Primer-Blast based on the published cDNA sequence in the Gene Bank (https://www.ncbi.nlm.nih.gov/tools/primer-blast/). The information of genes detected and the primers are shown in Table 1.

2.11. Immunofluorescent Staining

Cells with a inoculum density of 1×10^5 cells/well were cultured onto coverslips (24-well plate, Fisher Scientific) pre-coated with collagen and fixed with 4% PFA after treatments as described above. The cells were washed with PBS 2 times and incubated with 0.3% Triton X-100 in PBS (PBS/Triton) at room temperature for 10 min. For β-actin staining, the treated cells were incubated with Phalloidin, CF488A (1:100 dilution in PBS, Biotium, Inc., Fremont, CA, USA) at room temperature for 1 h. The cells were then washed three times with PBS and mounted with Vectashield mounting medium with DAPI (Vector Laboratories, Inc., Burlingame, CA, USA). For ZO-1 staining, cells were blocked with 5% goat serum (Jackson ImmunoResearch Laboratories, West Grove, PA, USA) for 1 h and then incubated with an anti-rabbit ZO-1 polyclonal antibody (cat. # 61-7300, 1:100 dilution, Thermal Scientific, Ottawa, Canada) at 4 °C overnight. The cells were then washed three times with PBS and incubated with an

Alexa fluor 488 goat anti-rabbit antibody (Thermal Scientific, cat. # A-11034) for 1 h at room temperature. Rinsed cells were mounted with Vectashield Mounting Medium with DAPI (Vector Laboratories, Inc.). The images were taken by a Zeiss Fluorescence Microscope (Carl Zeiss Ltd., Toronto, ON, Canada) [10].

Table 1. Primers used in this study

Gene	Primer Sequences (5′ →3′)	Length (bp)	Access No.
ZO-1	CAACATACAGTGACGCTTCACA CACTATTGACGTTTCCCCACTC	105	NM_001355013.1
Occludin	GACTATGTGGAAAGAGTTGAC ACCGCTGCTGTAACGAG	174	XM_017008914.2
Claudin-1	TGGTGGTTGGCATCCTCCTG AATTCGTACCTGGCATTGACTGG	232	NM_021101.4
Claudin-3	GCCACCAAGGTCGTCTACTC CCTGCGTCTGTCCCTTAGAC	101	NM_001306.3
SOD	ATCCTCTATCCAGAAAACACG ACACCACAAGCCAAACGAC	250	NM_000454.4
HO-1	TTTGAGGAGTTGCAGGAGC GTAAGGACCCATCGGAGAA	179	NM_002133.2
CAT	TCCAAGGCAAAGGTATTTGAGCA CAACGAGATCCCAGTTACCATCTTC	157	NM_001752.3
GSH-Px	CCTCTAAACCTACGAGGGAGGAA GGGAAACTCGCCTTGGTCT	107	NM_001329455.1
GAPDH	GCACCGTCAAGGCTGAGAAC ATGGTGGTGAAGACGCCAGT	142	NM_001289745.2

Note. ZO-1: zonula occludens-1, SOD: superoxidedismutase, HO-1: hemeoxygenase-1, CAT: catalase, GSH-Px: glutathione peroxidase, GAPDH: glyceraldehyde-3-phosphate dehydrogenase, bp: base pair.

2.12. Western Blotting

Cells were cultured in a 6-well plate with a inoculum density of 4×10^5 cells/well and total protein was extracted after the treatments described above. Protein extraction was conducted using a total protein extraction kit according to instructions provided by the manufacturer (Thermo Scientific, Ottawa, ON, Canada). A portion of protein was quantified using a BCA protein assay kit (Thermo Scientific). Then proteins were heat-denatured in an SDS-PAGE loading buffer. Proteins were electrophoresed on polyacrylamide gels and electro-transferred to mini-size nitrocellulose membranes (Bio-Rad, Laboratories Ltd., Montreal, QC, Canada). The immunoreaction was achieved by incubation of the membranes, previously blocked with 5% non-fat dry milk in Tris-buffered saline with 0.1% of Tween 20 (TBST), followed by incubation with rabbit anti-Nrf-2 (Thermal Scientific, cat. #PA5-68817), rabbit anti-ZO-1 (Thermal Scientific, cat. #61-7300), rabbit anti-claudin-1(Thermal Scientific, cat. #34-1700), rabbit anti-claudin-3 (Thermal Scientific, cat. #51-9000), and rabbit anti-occludin (Thermal Scientific, cat. #71-1500) proteins after being diluted (1:1000, Abcam Inc., Toronto, ON, Canada) overnight at 4°C. Afterward, it was washed five times with TBST. Detection of the immune complexes was performed with a horseradish peroxidase-conjugated secondary antibody (1:5000, goat anti-rabbit, Jackson ImmunoResearch Laboratories), then the membrane was washed 5 times, 5 min per time. The Clarity™ Western ECL Substrate was applied to the blot according to the manufacturer's recommendations (Bio-Rad Laboratories Ltd.). The chemiluminescent signals were captured using a ChemiDoc MP imaging system (Bio-Rad Laboratories Ltd.), and Image Lab 6.0 was used to quantify the intensity of the bands (Bio-Rad Laboratories Ltd.). β-actin (from mouse, Thermal Scientific, cat. #AM4302) was set as the internal reference.

2.13. Statistical Analysis

Data were presented as means ± standard deviations (SD). Statistical analysis was performed using GraphPad Prism 7 (GraphPad Software, La Jolla, CA, USA). Differences between the means were evaluated by one-way ANOVA. Duncan's multiple range test was used. Level of significance was set at $P < 0.05$.

3. Results

As shown in Figure 1A, the cell viability significantly decreased when the concentrations of H_2O_2 were more than 0.8 mM ($P < 0.05$). The cell viability was reduced at 35.15% when 1 mM H_2O_2 was added in the medium compared with the control (0 mM H_2O_2). Therefore, 1 mM H_2O_2 was selected to induce oxidative damage in this study. As shown in Figure 1B, the cell viability was not changed when cells were treated with the RDE at the concentrations of 0–200 μg/mL ($P > 0.05$). As shown in Figure 1C, when cells were pre-treated with 1 mM H_2O_2 for 3 h and then incubated with different concentrations of RDE for 24 h, RDE showed a dose-dependent recovery of the cell viability that was compromised by 1 mM H_2O_2 and the cells treated with 100 μg/mL of RDE had similar viability with the control cells ($P > 0.05$). Therefore, 100 μg/mL of RDE was used in the subsequent experiments.

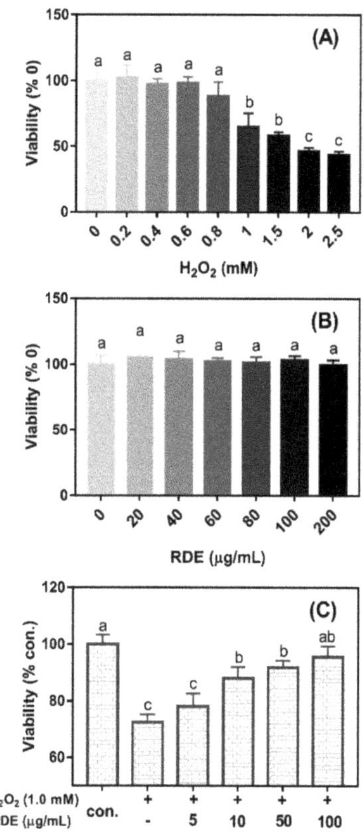

Figure 1. Effect of H_2O_2 and red-osier dogwood extracts (RDE) on the cell viability of Caco-2 cells. (**A**) Cells were treated with different concentrations of H_2O_2 for 24 h, and then cell viability was detected. (**B**) Cells were treated with RDE for 24 h, and then cell viability was detected. (**C**) Cells were treated with 1.0 mM H_2O_2 for 3 h, and then treated with a medium containing different RDE concentrations for 24 h. Cells were seeded in a 96 well plate at a density of 2×10^4/well and cultured for 5 d. Cell viability was expressed as a percentage of control (without H_2O_2 and RDE). The data were presented as mean ± SD, $n = 5$. Different lower case letters indicate a significant difference at $P < 0.05$.

H$_2$O$_2$ treatment induced a significant amount of IL-8 secretion and they were increased almost two-fold when compared with the control cells (NC) ($P < 0.05$) (Figure 2A). The 100 µg/mL of RDE reduced IL-8 secretion in the cells when compared with the treatment control cells (TC) ($P < 0.05$). However, IL-8 secretion was still higher in the cells treated with RDE (T) than in the control cells (NC) ($P < 0.05$). As shown in Figure 2B, the intracellular fluorescent intensity of cells treated with H$_2$O$_2$ (TC) was higher than those of the control cells (NC) and the incubation of H$_2$O$_2$-pretreated cells with RDE for 24 h reduced the ROS to the level of control. The H$_2$O$_2$ treatment increased total GSH content ($P < 0.05$) while the reduced GSH/ GSSS ratio remained the same ($P > 0.05$) when compared with the control cells (NC). The RDE treatment reduced both total GSH and GSSG ($P < 0.05$, Figure 2C) and had a higher reduced GSH/GSSG ratio when compared with both the NC and TC groups ($P < 0.05$, Figure 2D).

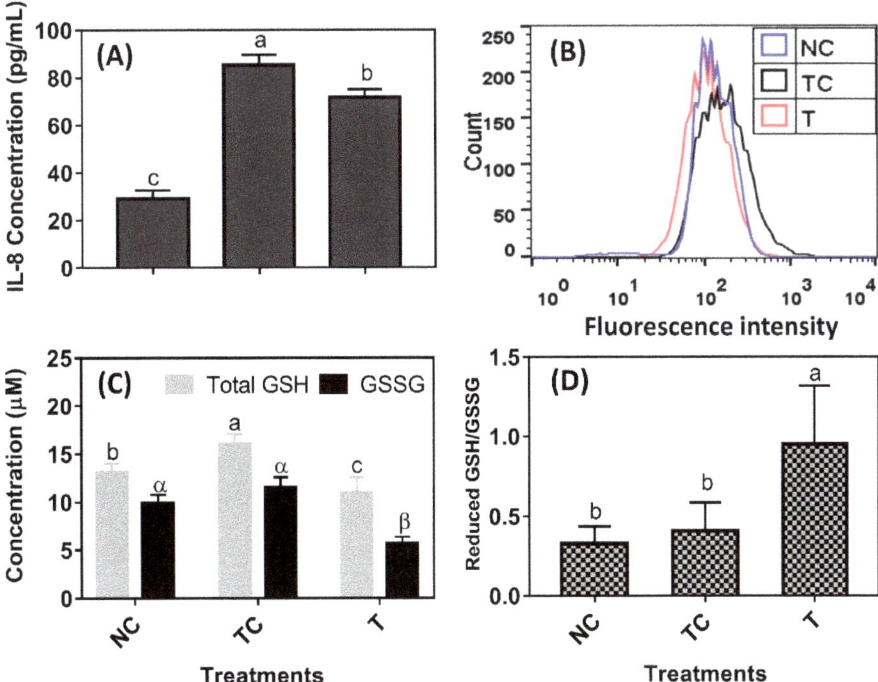

Figure 2. The effect of red-osier dogwood extracts (RDE) on interleukin (IL)-8 secretion (**A**), reactive oxygen species (ROS) level (**B**), glutathione (GSH) content (**C**), and reduced GSH/oxidized GSH (GSSG) (**D**) of Caco-2 cells after H$_2$O$_2$ treatment. NC indicates negative control: cells were cultured with a medium without H$_2$O$_2$ and RDE for 27 h (3 h + 24 h). TC indicates treatment control: cells were cultured with a medium containing 1 mM of H$_2$O$_2$ for 3 h, and then treated with a medium for 24 h. T indicates RDE treatment: cells were cultured with a medium containing 1 mM of H$_2$O$_2$ for 3 h, and then treated with a medium containing 100 µg/mL of RDE for 24 h. The data were presented as mean ± SD, $n = 3$. Different lower case letters and Greek letters indicate a significant difference at $P < 0.05$.

As shown in Figure 3A, the Nrf-2 protein abundance was lower ($P < 0.05$) in the cells treated with H$_2$O$_2$ (TC) than in the control cells (NC). The Nrf-2 protein abundance was higher ($P < 0.05$) in the cells treated with RDE (T) than in the cells treated with H$_2$O$_2$ (TC) but still lower ($P < 0.05$) than in the control cells (NC). HO-1 mRNA abundance was increased ($P < 0.05$) by 5.29 folds under H$_2$O$_2$ treatment compared with that of the control cells (NC) and HO-1 mRNA abundance was the highest ($P < 0.05$) in the cells treated with RDE (Figure 3B). H$_2$O$_2$ treatment significantly decreased the mRNA levels of SOD and GSH-Px, and the mRNA levels of the two enzymes were increased by followed RDE treatment ($P < 0.05$). However, there was no significant difference ($P > 0.05$) observed in the mRNA level of CAT among the three treatment groups (Figure 3C).

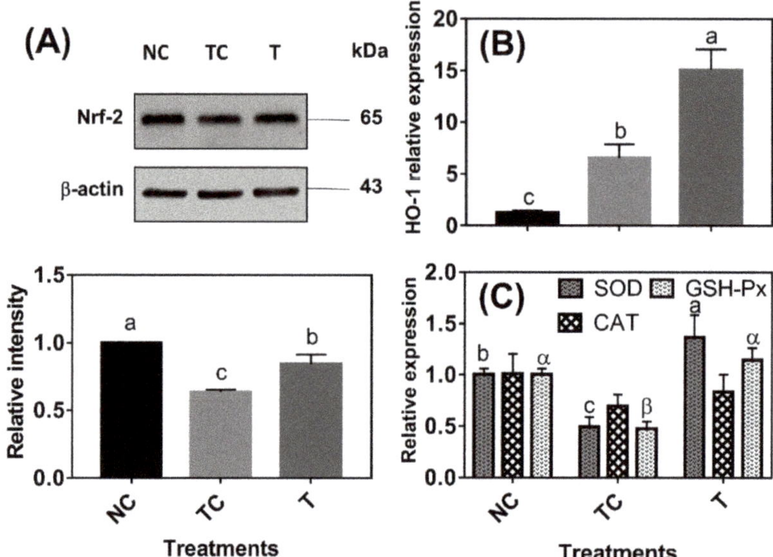

Figure 3. The effect of red-osier dogwood extracts (RDE) on the Nrf-2 protein level (**A**), HO-1 mRNA expression (**B**), and antioxidative enzymes mRNA levels (**C**) of Caco-2 cells after H_2O_2 treatment. NC indicates negative control: cells were cultured with a medium without H_2O_2 and RDE for 27 h (3 h + 24 h). TC indicates treatment control: cells were cultured with a medium containing 1 mM of H_2O_2 for 3 h, and then treated with a normal medium for 24 h. T indicates RDE treatment: cells were cultured with medium containing 1 mM of H_2O_2 for 3 h, and then treated with a medium containing100 µg/mL of RDE for 24 h. The data were presented as mean ± SD, $n = 4$. Different lower case letters and Greek letters indicate a significant difference at $P < 0.05$. Nrf-2: nuclear factor erythroid 2-related factor 2, SOD: superoxide dismutase; HO-1: hemeoxygenase-1; CAT: catalase; GSH-Px: glutathione peroxidase.

In Figure 4A, after a 3 h H_2O_2 treatment, the TEER values were significantly decreased (set as 0 h from RDE treatment) ($P < 0.05$). There was no significant difference ($P > 0.05$) observed in the TEER values between the TC and NC groups and between the NC and T groups when cells were further cultured for 24 h. However, the TEER value was higher ($P < 0.05$) in the cells treated with RDE (T) than in the treatment control cells (TC). As shown in Figure 4B, the leakage of FD4 was higher ($P < 0.05$) in the TC group than in the NC and T groups. However, there was no significant difference ($P > 0.05$) in the leakage of FD4 observed between the NC and T groups. As shown in Figure 5, the cytoskeletal structure of the β-actin fiber was partially disorganized in the TC group by H_2O_2 treatment, while incubation with RDE after H_2O_2 treatment could alleviate the damage induced by H_2O_2. The morphology of tight junction had similar changes as cytoskeleton after different treatments.

As shown in Figure 6A, ZO-1 mRNA abundance was significantly increased by H_2O_2 treatment ($P < 0.05$) while there was no difference ($P > 0.05$) observed in the mRNA abundance of claudin-1, claudin-3, and occludin when compared with the control cells (NC). However, the cells treated with RDE had a higher level of claudin-1, claudin-3, and occludin mRNA abundance and a lower level of ZO-1 mRNA abundance than those in the other two treatments ($P < 0.05$). As shown in Figure 6B, H_2O_2 treatment (TC) significantly decreased the protein abundance of ZO-1, claudin-1, claudin-3, and occludin ($P < 0.05$) when compared with the control cells (NC). RDE treatment (T) significantly increased the protein abundance of ZO-1 and claudin-3 ($P < 0.05$) when compared with the TC group.

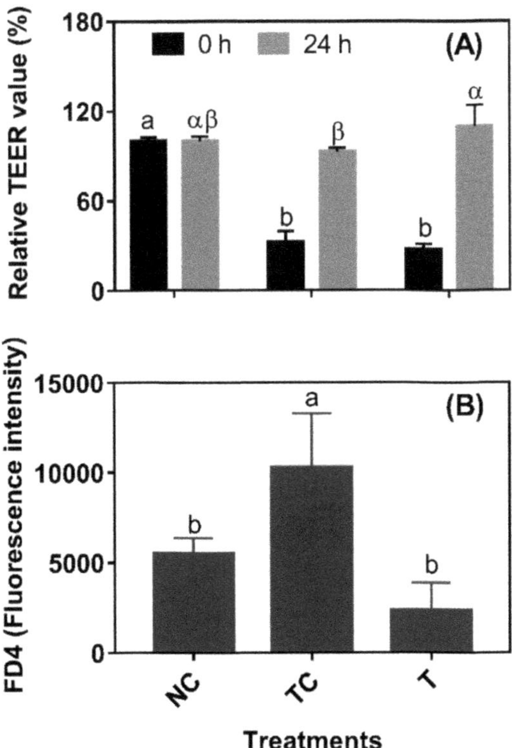

Figure 4. The effect of red-osier dogwood extracts (RDE) on the transepithelial resistance (TEER) (**A**) and permeability (**B**) of Caco-2 cells after H_2O_2 treatment. NC indicates negative control: cells were cultured with DMEM solution without H_2O_2 and RDE for 27 h (3 h + 24 h). TC indicates treatment control: cells were cultured with a medium containing 1 mM of H_2O_2 for 3 h, and then treated with a medium for 24 h. T indicates RDE treatment: cells were cultured with a medium containing 1 mM of H_2O_2 for 3 h, and then treated with medium containing 100 µg/mL of RDE for 24 h. TEER was measured after H_2O_2 treatment for 3 h and after RDE treatment for 24 h, respectively. The data were presented as mean ± SD, n = 4. Different lower case letters and Greek letters indicate a significant difference at $P < 0.05$.

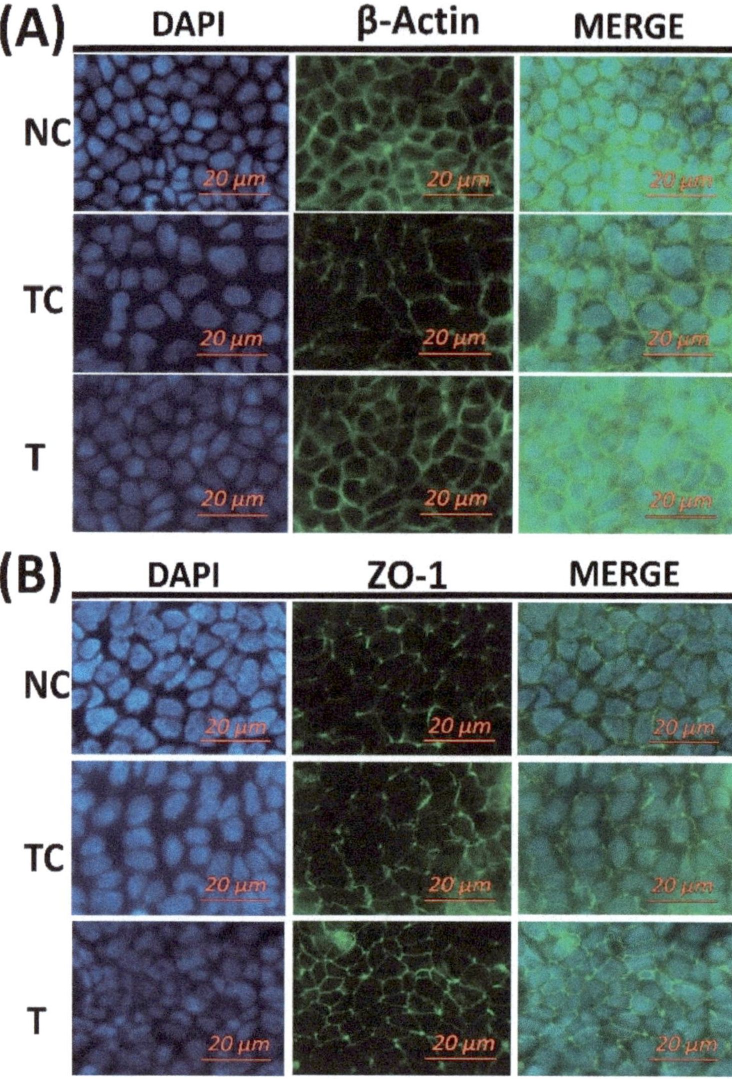

Figure 5. The effect of red-osier dogwood extracts (RDE) on the morphological changes of β-actin fiber and tight junction in Caco-2 cells after H_2O_2 treatment. (**A**) The morphological changes of β-actin and (**B**) The morphological changes of ZO-1. NC indicates negative control: cells were cultured with a medium without H_2O_2 and RDE for 27 h (3 h + 24 h). TC indicates treatment control: cells were cultured with a medium containing 1 mM of H_2O_2 for 3 h, and then treated with a medium for 24 h. T indicates RDE treatment: cells were cultured with a medium containing 1 mM of H_2O_2 for 3 h, and then treated with medium containing 100 μg/mL of RDE for 24 h. ZO-1: zonula occludens-1.

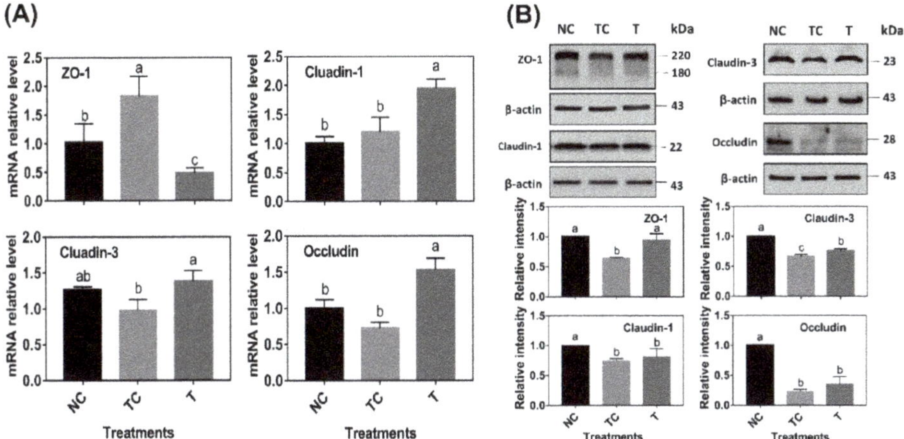

Figure 6. The effect of red-osier dogwood extracts (RDE) on the tight junction mRNA abundance (**A**) and protein abundance (**B**) of Caco-2 cells after H_2O_2 treatment. NC indicates negative control: cells were cultured with a medium without H_2O_2 and RDE for 27 h (3 h + 24 h). TC means treatment control: cells were cultured with a medium containing 1 mM of H_2O_2 for 3 h, and then treated with a medium for 24 h. T indicates RDE treatment: cells were cultured with a medium containing 1 mM of H_2O_2 for 3 h, and then treated with a medium containing 100 μg/mL of RDE for 24 h. The data were presented as mean ± SD, n = 3. Different lower case letters indicate a significant difference at $P < 0.05$. ZO-1: zonula occludens-1. The black bar, gray bar, and light black bar indicated the treatments NC, TC, and T, respectively.

4. Discussion

Numerous studies have shown that the occurrences of the intestinal disease are associated with a defective barrier function that was caused by oxidative damage [27]. Therefore, controlling oxidative damage and repairing the intestinal barrier could be effective ways to prevent many intestinal diseases. Phytochemicals have been reported to reduce the production of cellular ROS [28] and the incidence of intestinal inflammation [10]. The supplementation of ground red-osier dogwood in feeds decreased the serum malondialdehyde content and increased the SOD levels in the *Escherichia coli* F4 K88$^+$ infected animals, as reported by a previous study [21]. In the present study, an oxidative damage model was established by using H_2O_2 in Caco-2 cells, aiming to evaluate the therapeutic effects of RDE on intestinal oxidative damage.

To establish an H_2O_2-induced oxidative injury model with Caco-2 cells, the dosage effects of H_2O_2 on cell viability was investigated firstly. The results showed that the H_2O_2 treatment significantly reduced cell viability at more than 0.8 mM, indicating that a high dosage of H_2O_2 inhibited the growth of the cells or was toxic to the cells. These findings are consistent with previous studies of H_2O_2 and Caco-2 cells [29]. However, a wide concentration range of RDE (up to 200 μg/mL, equal to 53 μg/mL total phenolic compounds) did not show a negative effect on cell viability, suggesting that higher RDE containing higher phenolic compounds might have better pharmaceutical effectiveness. Moreover, our results demonstrated that 100 μg/mL of RDE completely recovered the cell viability that was compromised by 1 mM H_2O_2. Therefore, our results suggest that RDE exerts protective effects against H_2O_2-induced oxidative damage in intestinal epithelial cells in vitro.

H_2O_2-induced oxidative stress triggered an imbalanced redox state and excessive ROS accumulation in Caco-2 cells, which in turn induced inflammatory responses evidenced by an increased IL-8 level. IL-8, a chemokine that can be produced by epithelial cells, has been recognized as an indicator of the inflammatory response [30]. The mediators can activate signal transduction cascades as well as inducing variations in transcription factors, such as Nrf-2, which mediate immediate cellular stress responses [31]. The oxidative stress and inflammatory responses together exacerbated the damage of both cellular structure and biomacromolecules. After a 3 h treatment with H_2O_2 and following a 24 h treatment with 100 μg/mL of RDE, the viability of Caco-2 cells was significantly restored, accompanied by decreased levels of IL-8 and ROS, indicating the inflammatory responses and oxidative damage were alleviated. The anti-oxidative components in RDE including rutin and phenolic acids are highly effective at scavenging oxygen radicals [32].

A recent study shows that rutin attenuates inflammatory responses induced by lipopolysaccharide in an in vitro mouse muscle cell (C2C12) model and supplementation of rutin or rutin-containing plant extracts may hold promising potential for attenuating oxidative stress and inflammation [33]. In addition, RDE increased the ratio of reduced GSH/GSSG, and reduced the consumption of reduced GSH. The results indicate that RDE can alleviate the occurrence of inflammation and oxidative damage by scavenging ROS production directly.

To further investigate the therapeutic mechanism of RDE on H_2O_2-induced oxidative damage, the protein expression of Nrf-2 and gene expression of responsive gene HO-1 and antioxidative enzyme SOD, CAT, and GSH-Px were measured. The protein expression of Nrf-2 was induced (Figure 3A) and the mRNA expression of HO-1 was significantly upregulated after the addition of RDE (Figure 3B). In addition, the mRNA abundance of SOD and GSH-Px also showed the same trend as HO-1 (Figure 3C). This study revealed that RDE activated Nrf-2, mediated the antioxidant gene expression, and enhanced the antioxidant capacity of cells, thereby preventing the inflammation response and oxidative damage.

The intact gut barrier is important to prevent cell damage and restore cellular function caused by intestinal inflammation [10] and oxidative stress [34]. The TEER and permeability, as measured by FD4 diffusion, were often used to evaluate the integrity and tightness of the barrier formed by epithelial cells. Caco-2 cells were measured until the stable TEER formed by the tight junction. The permeability of the barrier after the 24 h treatment was also investigated by measuring the fluorescence intensity of FITC-Dextran (FD4) in the basolateral side diffused from the apical side. The H_2O_2 treatment lowered the TEER value and increased the permeability, suggesting that H_2O_2 treatment led to the increase in cell permeability [35], and resulted in a defective barrier function of cells. However, TEER values were increased and permeability was decreased by RDE treatment for 24 h, indicating that the barrier function of Caco-2 cells was improved significantly. A recent study showed that intact epithelial barrier functions are necessary in order to maintain the homeostatic state in response to physiological host-gut microbiome cross-talks, and therefore the maintenance of epithelial barrier function is crucial for alleviating gut inflammation [36]. It has been reported that phytochemicals extracted from apples can increase the TEER value of Caco-2 cells [37], and this might be attributed to the antioxidative effects of phytochemicals.

The morphology of the cytoskeleton and tight junction was visualized by β-actin and ZO-1 immunofluorescent staining. H_2O_2 treatment disorganized the β-actin fibrin and diffused cell tight junction protein ZO-1. This indicates that H_2O_2 caused oxidative damage to the cell membrane monolayer. The addition of RDE obviously restored cell structure and stabilized morphological characteristics of ZO-1.

In this study, the mRNA and protein abundance of tight junction proteins, including ZO-1, claudin-1, claudin-3, and occluding, were determined. When compared with the sole H_2O_2 treatment group, RDE supplemented to cells enhanced the mRNA abundance of tight junction protein claudin-1, claudin-3, and occludin, but decreased the ZO-1 mRNA level in the Caco-2 cells. Firstly, this might be attributed to the fact that the sensitivity of gene expression of the tight junction proteins was different when responding to H_2O_2 [38]. The transcription of ZO-1 was more sensitive to H_2O_2, and it could rapidly respond to oxidative stress. Secondly, it may also be related to the ordinal expression of different tight junction proteins. Within 24 h after H_2O_2 treatment, intracellular H_2O_2 was still playing a role in cells, and the expression of ZO-1, which was sensitive to oxidative stress, was induced. However, when cells were treated with H_2O_2 for 3 h, followed by RDE incubation of cells for 24 h, ROS was partially removed by RDE due to its strong antioxidant effect. Therefore, the expression of ZO-1 was not upregulated.

Interestingly, claudin-3 and occludin could be upregulated by RDE treatment, indicating that there were some unknown components in RDE that could regulate claudin-3 and occludin gene expression. In terms of tight junction proteins expression, H_2O_2 treatment reduced the contents of ZO-1, claudin-1, claudin-3, and occludin significantly. This was not completely consistent with their mRNA expression results. Firstly, it can be attributed to protein expression lagging behind mRNA expression [24]. In addition, H_2O_2 treatment significantly disrupted the structure of tight junction proteins, resulting in a decrease in the protein content. After RDE treatment, the protein content of ZO-1 and claudin-3 increased significantly, but the protein content of claudin-1 and occludin did not increase. It was also evident from the staining results that the ZO-1 structure has also been significantly repaired to enhance the cellular barrier function and viability. Based on our observations in this study, there could be two reasons to explain why RDE enhances cellular barrier function. Firstly, RDE is rich in antioxidants [19,32] that have the ability to scavenge ROS directly and repair the structure of tight junction proteins. Moreover, RDE is rich in unknown components that activate the Nrf-2 pathway to enhance the antioxidant capacity of cells and induce the expression of tight junction proteins.

5. Conclusions

In conclusion, RDE enhances cell activity by directly scavenging ROS and strengthening the cellular antioxidant system, as well as by repairing tight junction proteins and inducing their expression.

Author Contributions: Conceptualization, C.Y. and R.Y.; Data analysis, R.Y., S.L., Q.H., H.Z. and Q.J.; Financial funding, C.Y.; Methodology, F.L., S.L., J.W. and R.Y.; Supervision, C.Y. and K.O; Writing, R.Y., C.Y. and K.O.

Funding: This project was funded by the Natural Sciences and Engineering Research Council (NSERC) Engage grant (C. Yang, EGP 523472-2018) and the University of Manitoba Start-Up Grant (C. Yang, 46561). Runqiang Yang was a Visiting Scholar at the Department of Animal Science in the University of Manitoba and supported by the Chinese Scholarship Council (CSC).

Acknowledgments: The authors are grateful to Red Dog Enterprises Ltd. for providing red-osier dogwood extracts.

Conflicts of Interest: The authors declare no conflict of interest.

References

1. María José, A.; Luis Miguel, B.; Luis, A.; Paulina, B. The *Artemisia* L. genus: A review of bioactive essential oils. *Molecules* **2012**, *17*, 2542–2566.
2. Han, B.; Xin, Z.; Ma, S.; Liu, W.; Zhang, B.; Ran, L.; Yi, L.; Ren, D. Comprehensive characterization and identification of antioxidants in *Folium Artemisiae Argyi* using high-resolution tandem mass spectrometry. *J. Chromatogr. B* **2017**, *1063*, 84–92. [CrossRef]
3. Patil, S.P.; Kumbhar, S.T. Evaluation of terpene-rich extract of *Lantana camara* L. leaves for antimicrobial activity against mycobacteria using Resazurin Microtiter Assay (REMA). *Beni-Suef Univ. J. Basic Appl. Sci.* **2018**, *7*, 511–515. [CrossRef]
4. Zhang, P.; Ran, R.K.; Abdullahi, A.Y.; Shi, X.L.; Huang, Y.; Sun, Y.X.; Liu, Y.Q.; Yan, X.X.; Hang, J.X.; Fu, Y.Q.; et al. The mitochondrial genome of Dipetalonema gracile from a squirrel monkey in China. *J. Helminthol.* **2018**, 1–8. [CrossRef]
5. Xiang, J.; Zhang, M.; Apea-Bah, F.B.; Beta, T. Hydroxycinnamic acid amide (HCAA) derivatives, flavonoid C-glycosides, phenolic acids and antioxidant properties of foxtail millet. *Food Chem.* **2019**, *295*, 214–223. [CrossRef]
6. Lillehoj, H.; Liu, Y.; Calsamiglia, S.; Fernandez-Miyakawa, M.E.; Chi, F.; Cravens, R.L.; Oh, S.; Gay, C.G. Phytochemicals as antibiotic alternatives to promote growth and enhance host health. *Vet. Res.* **2018**, *49*, 76. [CrossRef]
7. Hassan, Y.I.; Lahaye, L.; Gong, M.M.; Peng, J.; Gong, J.; Liu, S.; Gay, C.G.; Yang, C. Innovative drugs, chemicals, and enzymes within the animal production chain. *Vet. Res.* **2018**, *49*, 71. [CrossRef]
8. Yang, C.; Fischer, M.; Kirby, C.; Liu, R.; Zhu, H.; Zhang, H.; Chen, Y.; Sun, Y.; Zhang, L.; Tsao, R. Bioaccessibility, cellular uptake and transport of luteins and assessment of their antioxidant activities. *Food Chem.* **2018**, *249*, 66–76. [CrossRef]

9.   Kim, J.-S.; Kwon, Y.-S.; Chun, W.-J.; Kim, T.-Y.; Sun, J.; Yu, C.-Y.; Kim, M.-J. *Rhus verniciflua* Stokes flavonoid extracts have anti-oxidant, anti-microbial and α-glucosidase inhibitory effect. *Food Chem.* **2010**, *120*, 539–543. [CrossRef]

10. Omonijo, F.A.; Liu, S.; Hui, Q.; Zhang, H.; Lahaye, L.; Bodin, J.-C.; Gong, J.; Nyachoti, M.; Yang, C. Thymol improves barrier function and attenuates inflammatory responses in porcine intestinal epithelial cells during lipopolysaccharide (LPS)-induced inflammation. *J. Agric. Food Chem.* **2019**, *67*, 615–624. [CrossRef]

11. Akhtar, M.S.; Swamy, M.K. *Anticancer Plants: Natural Products and Biotechnological Implements*; Springer: Singapore, 2018.

12. Xiang, J.; Apea-Bah, F.B.; Ndolo, V.U.; Katundu, M.C.; Beta, T. Profile of phenolic compounds and antioxidant activity of finger millet varieties. *Food Chem.* **2019**, *275*, 361–368. [CrossRef]

13. Xiang, J.; Li, W.; Ndolo, V.U.; Beta, T. A comparative study of the phenolic compounds and in vitro antioxidant capacity of finger millets from different growing regions in Malawi. *J. Cereal Sci.* **2019**, *87*, 143–149. [CrossRef]

14. Chen, Z.; Ma, Y.; Yang, R.; Gu, Z.; Wang, P. Effects of exogenous Ca^{2+} on phenolic accumulation and physiological changes in germinated wheat (*Triticum aestivum* L.) under UV-B radiation. *Food Chem.* **2019**, *288*, 368–376. [CrossRef]

15. Jeong, Y.J.; An, C.H.; Park, S.-C.; Pyun, J.W.; Lee, J.; Kim, S.W.; Kim, H.-S.; Kim, H.; Jeong, J.C.; Kim, C.Y. Methyl jasmonate increases isoflavone production in soybean cell cultures by activating structural genes involved in isoflavonoid biosynthesis. *J. Agric. Food Chem.* **2018**, *66*, 4099–4105. [CrossRef]

16. Hsu, C.; Wu, B.-Y.; Chang, Y.-C.; Chang, C.-F.; Chiou, T.-Y.; Su, N.-W. Phosphorylation of isoflavones by *Bacillus subtilis* BCRC 80517 may represent xenobiotic metabolism. *J. Agric. Food Chem.* **2018**, *66*, 127–137. [CrossRef]

17. Smithberg, M.H.; Weiser, C.J. Patterns of variation among climatic races of Red-Osier Dogwood. *Ecology* **1968**, *49*, 495–505. [CrossRef]

18. Feild, T.S.; Lee, D.W.; Holbrook, N.M. Why leaves turn red in autumn. The role of anthocyanins in senescing leaves of red-osier dogwood. *Plant Physiol.* **2001**, *127*, 566–574. [CrossRef]

19. Wei, L.Y.; Gomaa, W.M.S.; Ametaj, B.N.; Alexander, T.W.; Yang, W.Z. Feeding red osier dogwood (*Corns sericea*) to beef heifers fed a high-grain diet affected feed intake and total tract digestibility. *Anim. Feed Sci. Tech.* **2019**, *247*, 83–91. [CrossRef]

20. Wei, L.Y.Y.; Jiao, P.X.X.; Alexander, T.W.; Yang, W.Z. Inclusion of Red osier dogwood in high-forage and high-grain diets affected in vitro rumen fermentation. *Ann. Anim. Sci.* **2018**, *18*, 453–467. [CrossRef]

21. Koo, B.; Amarakoon, S.; Jayaraman, B.; Siow, Y.; Prashar, S.; Shang, Y.; Nyachoti, C. Effects of dietary supplementation with ground red-osier dogwood (*Cornus stolonifera*) on oxidative status in weaned pigs challenged with *Escherichia coli* K88⁺. *J. Anim. Sci.* **2018**, *96*, 308. [CrossRef]

22. Jayaraman, B.; Amarakoon, S.; Koo, B.; O, K.; Nyachoti, C. Effects of dietary supplementation with ground red-osier dogwood (*Cornus stolonifera*) on growth performance, blood profile, and ileal histomorphology in weaned pigs challenged with *Escherichia coli* K88⁺. *J. Anim. Sci.* **2018**, *96*, 322. [CrossRef]

23. Yang, J.; Zhu, C.U.I.; Ye, J.l.; Lv, Y.; Wang, L.; Chen, Z.; Jiang, Z.y. Resveratrol protects porcine intestinal epithelial cells from deoxynivalenol induced damage via the Nrf2 signaling pathway. *J. Agric. Food Chem.* **2019**, *67*, 1726–1735. [CrossRef]

24. Wu, H.; Luo, T.; Li, Y.M.; Gao, Z.P.; Zhang, K.Q.; Song, J.Y.; Xiao, J.S.; Cao, Y.P. Granny Smith apple procyanidin extract upregulates tight junction protein expression and modulates oxidative stress and inflammation in lipopolysaccharide-induced Caco-2 cells. *Food Funct.* **2018**, *9*, 3321–3329. [CrossRef]

25. Tian, T.; Zhang, J.; Wang, Z. Pathomechanisms of oxidative stress in inflammatory bowel disease and potential antioxidant therapies. *Oxid. Med. Cell. Longev.* **2017**, *2017*, 1–18. [CrossRef]

26. Wu, J.; Hillier, C.; Komenda, P.; Faria, R.L.D.; Santos, S.; Levin, D.; Zhang, M.; Lin, F. An all-on-chip method for testing neutrophil chemotaxis induced by fMLP and COPD patient's sputum. *Technology* **2016**, *4*, 104–109. [CrossRef]

27. Serra, G.; Incani, A.; Serreli, G.; Porru, L.; Melis, M.P.; Tuberoso, C.I.G.; Rossin, D.; Biasi, F.; Deiana, M. Olive oil polyphenols reduce oxysterols -induced redox imbalance and pro-inflammatory response in intestinal cells. *Redox Biol.* **2018**, *17*, 348–354. [CrossRef]

28. Tao, L.; Forester, S.C.; Lambert, J.D. The role of the mitochondrial oxidative stress in the cytotoxic effects of the green tea catechin, (-)-epigallocatechin-3-gallate, in oral cells. *Mol. Nutr. Food Res.* **2014**, *58*, 665–676. [CrossRef]

29. Sakuma, S.; Abe, M.; Kohda, T.; Fujimoto, Y. Hydrogen peroxide generated by xanthine/xanthine oxidase system represses the proliferation of colorectal cancer cell line Caco-2. *J. Clin. Biochem. Nutr.* **2015**, *56*, 15–19. [CrossRef]

30. Wang, X.; Zhao, Y.; Yao, Y.; Xu, M.; Du, H.; Zhang, M.; Tu, Y. Anti-inflammatory activity of di-peptides derived from ovotransferrin by simulated peptide-cut in TNF-α-induced Caco-2 cells. *J. Funct. Food.* **2017**, *37*, 424–432. [CrossRef]

31. Mahmoud, A.M.; Germoush, M.O.; Al-Anazi, K.M.; Mahmoud, A.H.; Farah, M.A.; Allam, A.A. Commiphora molmol protects against methotrexate-induced nephrotoxicity by up-regulating Nrf2/ARE/HO-1 signaling. *Biomed. Pharmacother.* **2018**, *106*, 499–509. [CrossRef]

32. Isaak, C.K.; Petkau, J.C.; Karmin, O.; Ominski, K.; Rodriguez-Lecompte, J.C.; Siow, Y.L. Seasonal variations in phenolic compounds and antioxidant capacity of *Cornus stolonifera* plant material: Applications in agriculture. *Can. J. Plant Sci.* **2013**, *93*, 725–734. [CrossRef]

33. Yu, L.; Liu, S.; Yang, C.; O, K.; Adewole, D.; Sid, V.; Wang, B. Rutin attenuates inflammatory responses induced by lipopolysaccharide in an in vitro mouse muscle cell (C2C12) model. *Poultry Sci.* **2019**, *98*, 2756–2764. [CrossRef]

34. Aggarwal, S.; Suzuki, T.; Taylor, W.L.; Bhargava, A.; Rao, R.K. Contrasting effects of ERK on tight junction integrity in differentiated and under-differentiated Caco-2 cell monolayers. *Biochem. J.* **2011**, *433*, 51–63. [CrossRef]

35. Wijeratne, S.S.K.; Cuppett, S.L.; Schlegel, V. Hydrogen peroxide induced oxidative stress damage and antioxidant enzyme response in Caco-2 human colon cells. *J. Agric. Food Chem.* **2005**, *53*, 8768–8774. [CrossRef]

36. Shin, W.; Kim, H.J. Intestinal barrier dysfunction orchestrates the onset of inflammatory host–microbiome cross-talk in a human gut inflammation-on-a-chip. *Proc. Natl. Acad. Sci. USA* **2018**, *115*, E10539–E10547. [CrossRef]

37. Vreeburg, R.A.; van Wezel, E.E.; Ocaña-Calahorro, F.; Mes, J.J. Apple extract induces increased epithelcial resistance and claudin 4 expression in Caco-2 cells. *J. Sci. Food Agric.* **2012**, *92*, 439–444. [CrossRef]

38. Suzuki, T.; Hara, H. Quercetin enhances intestinal barrier function through the assembly of zonnula occludens-2, occludin, and claudin-1 and the expression of claudin-4 in Caco-2 cells. *J. Nutr.* **2009**, *139*, 965–974. [CrossRef]

© 2019 by the authors. Licensee MDPI, Basel, Switzerland. This article is an open access article distributed under the terms and conditions of the Creative Commons Attribution (CC BY) license (http://creativecommons.org/licenses/by/4.0/).

 *antioxidants* 

Article

Nutraceutical Properties of Mulberries Grown in Southern Italy (Apulia)

Carmine Negro *, Alessio Aprile, Luigi De Bellis and Antonio Miceli

Department of Biological and Environmental Sciences and Technologies (DiSTeBA), Salento University, Via Prov. le Lecce-Monteroni, 73100 Lecce, Italy

* Correspondence: carmine.negro@unisalento.it

Received: 11 June 2019; Accepted: 12 July 2019; Published: 16 July 2019

Abstract: In this work, for the first time, were analyzed mulberry genotypes grown in Apulia (Southern Italy, Salento region) were analyzed. Two local varieties of *Morus alba* (*cv. Legittimo nero* and *cv. Nello*) and one of *Morus nigra* were characterized for content in simple sugars, organic acids, phenols, anthocyanins; fruit antioxidant activity (AA) was also evaluated by three different methods (2,2-Diphenyl-1-picrylhydrazyl, DPPH; 2,2'-Azino-bis(3-ethylbenzothiazoline-6-sulfonic acid), ABTS; and Ferric reducing antioxidant potential, FRAP test). The results showed that the sugars amount ranged between 6.29 and 7.66 g/100 g fresh weight (FW) while the malic and citric acids content was low, at about 0.1–1 g/100 g FW. Mulberries are a good source of phenols which are present in higher values in *M. nigra* and *M. alba cv. Legittimo nero* (485 and 424 mg Gallic Acid Equivalent (GAE)/ 100 g FW, respectively). The high performance liquid chromatography/diode array detector/mass spectrometry (HPLC/DAD/MS) analysis identified 5 main anthocyanin compounds present in different concentrations in each variety of mulberry: cyanidin 3-sophoroside, cyanidin 3-glucoside, cyanidin 3-rutinoside, pelargonidin 3-glucoside, pelargonidin 3-rutinoside. The highest concentration of anthocyanins was determined in *Morus alba Legittimo* (about 300 mg/100 g FW) while the lowest content (about 25 mg/100 g FW) was measured in *M. alba cv. Nello*. *Morus nigra* showed a good AA in comparison with the different *M. alba* genotypes with all the used methods; its AA was equal to 33, 26 and 21 μmols Trolox/g FW when using DPPH, ABTS and FRAP tests, respectively. All genotypes showed an anti-inflammatory activity (measured by cyclooxygenase (COX) inhibitory assay) which was also compared with two commercial anti-inflammatory drugs. The data obtained support the high biological qualities of mulberry fruits and their diffusion in human nutrition.

Keywords: mulberry (*Morus nigra*; *Morus alba*); simple sugars; organic acids; phenol compounds; high performance liquid chromatography/diode array detector/mass spectrometry anthocyanins; antioxidant activity; anti-inflammatory activity

---

1. Introduction

The mulberry is a dicotyledon plant belonging to the genus *Morus* of the Moraceae family and generally it is well known in sericulture and silk industry, since it is the only food plant for the domesticated silkworm *Bombyx mori* [1]. There are 24 species of *Morus* with at least 100 known varieties; it is widely distributed in Asia, Europe, North and South America, and Africa [2]. The most important species are *Morus alba*, with fruit colors ranging from white to dark red, and *Morus nigra* with dark red fruits mainly; both species have excellent productions in Mediterranean climate areas [3].

Almost all the parts of the mulberry-tree are used for their pharmacological actions. The leaves have been shown to have diuretic, hypoglycemic, and hypotensive activities [4,5], whereas the root bark has long been used for anti-inflammatory, antitussive, and antipyretic purposes [5]. Different studies underlined that these properties are due to hydroxylated alkaloids, flavonoids and Diels-Alder type adducts which seem to be responsible for a high α-glucosidase inhibition [6–8], improving the glucose levels in patients affected by type II diabetes [9,10].

Moreover, the mulberry fruit has numerous biologically active compounds like phenols, flavonoids, anthocyanins, carotenoids, essential fatty acids, ascorbic acid, and various organic acids [11,12], resulting in high antioxidant activities, especially in fully ripened fruits [10]. Mulberries can be consumed both fresh and processed such as syrup, jam, pulp, ice-cream, etc. Traditionally they are used as a warming agent, as a preparation against dysentery and as a tonic, sedative, laxative, odontalgic, anthelmintic, expectorant, and emetic. Moreover, a great deal of evidence suggests their potential role in the prevention of cancers, cardiovascular diseases, apoptosis, neurotoxicity and neurodegeneration [10,13] and oxidative stress. Several studies suggest that the reactive oxygen species (ROS) and reactive nitrogen species (RNS) play important negative roles by oxidizing DNA and other molecules, leading to age-related diseases [14–16]. In particular the brain, which has a fundamental role in cognitive dysfunction usually associated with neuro-degenerative problems, is the most sensitive organ to the oxidative stress because of its high oxygen need, high metabolic rate and relative low antioxidative defense mechanisms [17–19]. In addition, recently, higher antinociceptive properties have been reported for black mulberry fruits, which inhibited the expression of inflammation-related proteins [20].

These biological activities are correlated to their polyphenol components, in particular anthocyanins, a large group of water-soluble pigments responsible for colors (orange, red and blue) of flowers, fruits and vegetables and principally known for the high antioxidant activity, anti-inflammatory potential and reduction of liver injury (chemo-and hepatic protective role) [21,22]. Cyanidin 3-*O*-glucoside (C3G), cyanidin 3-*O*-rutinoside (C3R), pelargonidin 3-*O*-glucoside (P3G), pelargonidin 3-*O*-rutinoside (P3R) are the main anthocyanins identified in mulberries and their antioxidant power is very high [21,23].

Mulberry grows in a wide range of climatic, topographical, and soil conditions which can affect the chemical composition and nutritional status of the fruits. Although mulberry fruits have been characterized in different parts of the world, information on fruits from plants grown in Southern Italy (Salento, South Apulia) has not been obtained. In the present study, therefore, fruits of local variety of *Morus alba* and *Morus nigra* have been characterized for their nutritional components by evaluating the content of simple sugars, organic acids, total phenolic compounds, *o*-diphenolic compounds, anthocyanins, and antioxidant and anti-inflammatory properties.

2. Materials and Methods

2.1. Chemicals

2,2-Diphenyl-1-picrylhydrazyl (DPPH), 2,2′-Azino-bis(3-ethylbenzothiazoline-6-sulfonic acid) diammonium salt (ABTS), Gallic acid, Folin-Ciocalteu reagent, Trolox (6-hydroxy-2,5,7,8-tetramethylchroman-2-carboxylic acid), acetonitrile, acetone, methanol and water HPLC grade were purchased from Sigma-Aldrich Chemical Co. (St. Louis, MO) and they are of analytical grade. Cyanidin-3-O-Glucoside analytical grade was purchased from Extrasyntese (Genay, France), and anti-inflammatory Kit was purchased from Cayman Chemicals cat. Number 560131 (Ann Abor, MI, USA). Sugar kit and organic acids kit were purchased from R-Biopharm Italia Srl, cat. Numbers 10716260035, 10139076035, 10139068035 (Cerro al Lambro, Milano, Italy).

2.2. Extraction and Purification Juice

The analytical determinations were carried out on one local variety of *Morus nigra* and on two varieties of *Morus alba* known as *Morus alba cv Legittimo nero* and *Morus alba cv Nello*, which were harvested at full maturity in the Corigliano town fields (40°9′38″,16 N, 18°15′36″,72 E) (Province of Lecce) during June–July 2015 and stored at −20°C prior to the analysis.

Juice was obtained by centrifugation and used to quantify the sugars (sucrose, fructose and glucose) and citric and malic acid amounts by an enzymatic spectrophotometric kit provided by

R-Biopharm Italia (see "Chemicals" paragraph 2.1) and expressing the results as g/100 g of fresh weight (FW) and, after filtration with Watmann n. 1, for determination of antioxidant activity. The moisture was determined in accordance to the AOAC method [24].

For the determination of phenolic compounds, a raw extract of each mulberry sample was obtained homogenizing 25 g of fresh plant material (3 min. at 10,000 rpm with Ultraturrax) using 100 mL of cold acetone 70% (*v/v*) acidified with 0.1% HCl. The homogenate was stirred for two hours in the dark at 4 °C, then centrifuged at 5000× *g* for 10 min. On the pellet, two further extractions were carried out in the same way. The supernatants were dried at reduced pressure and re-suspended with ultrapure water (HPLC grade) acidified with 0.1% HCl, thus obtaining a final volume of 100 mL (1:4 *w/v*).

2.3. Total Phenolics and Anthocyanins Determination

Total phenolics compounds (TPC) were measured by the Folin Ciocalteau spectrophotometric method using gallic acid as a standard and expressing the results as mg Gallic Acid Equivalent (GAE)/100 g FW [25]. Moreover, *o*-diphenolics compounds (ODC) were determined by the Arnow spectrophotometric method and expressing the results as mg chlorogenic acid/100 g FW [26].

In order to evaluate the anthocyanin amount using high performance liquid chromatography/diode array detector/mass spectrometry (HPLC/DAD/MS), the mulberry raw extracts were purified by solid phase extraction (SPE) using cartridge Strata X (Phenomenex Italia, Castel Maggiore, Bologna, Italy). After activation of the SPE cartridge with 2 mL pure methanol and 5 mL bi-distilled water, 20 mL raw extract were loaded and washed with 10 mL acidified bi-distilled water and 5 mL ethyl acetate to remove sugars and less polar flavons respectively; finally, 20 mL acid methanol were used to recover the anthocyanins. The purified extract was dried under vacuum and re-suspended with HPLC water acidified with 0.1 % HCl.

The identification was completed using an HPLC Agilent 1100 system (Agilent Tecnnologies, palo Alto, CA, USA) coupled with an Agilent DAD sensor (detection wavelength 520 nm) and Agilent ESI/MS spectrometer 6100 in positive ionization mode as reported by Negro et al. [27]. The separation was carried out at 30 °C with a gradient elution program at a flow rate of 0.8 mL/min using a Phenomenex Gemini C18 250 × 4.6 mm, 5 μm separation column. The mobile phases consisted of water plus 2% formic acid (A) and water:formic acid:acetonitrile 48:2:50 (B). The following multistep linear gradient was applied: 0 min, 6% B; 15 min, 30% B; 25 min, 50% B; 30 min, 60% B. The injection volume in the HPLC system was 20 μL. The quantification of anthocyanin was achieved using calibration curves of the authentic chemical standards cyanidin 3-glucoside (C3G) and moreover, total anthocyanins (TA) was determined as being the sum of the area of the single compounds.

2.4. Antioxidant Tests

The antioxidant activity tests were performed by spectrophotometric assays using three different methods: DPPH, ABTS and FRAP (Ferric Reducing Antioxidant Power) test using a 96-well microplate according to Oki et al. [10], Re et al. [28] and Benzie and Strain [29], respectively. In all the assays, Trolox was used as a standard and results are expressed in terms of microgram of Trolox Equivalent Antioxidant Capacity (TEAC) per g of FW of sample.

2.5. Anti-Inflammatory Test

Anti-inflammatory activity (AI) was evaluated determining the cyclooxygenase activity (COX). The COX-1 and COX-2 inhibitory assay was carried out using COX Inhibitor Screening Assay Kit (Catalogue N° 560131, Cayman Chemicals, Ann Abor, MI, USA) according to the instructions provided by the manufacturer and as previously reported [27]. The AI was determined using the phenolic extract and results are expressed as inhibitory activity (IC_{50}, μg/mL).

2.6. Statistical Analysis

Data are reported as the mean ± SD and three biological replicates are carried out for each sample. Statistical evaluation was conducted by ANOVA, followed by multicomponent Duncan's test ($p < 0.05$) to discriminate among the mean values.

3. Results and Discussion

3.1. Simple Sugars, Organic Acids and Total Phenolics Content

The moisture and quantity of simple sugars, malic and citric acids, TPC and ODC were reported in Table 1. The moisture content in the fruits is very close, about 78%, according to the literature [1,11]. Sucrose was completely absent in all the samples analyzed; *M. nigra* had 7.66 g/100 g FW of sugars whereas the glucose and fructose contents were similar and equal to 3.94 and 3.72 g/100 g FW, respectively.

Table 1. Amounts of moisture (%), glucose, fructose, malic and citric acid (expressed as g/100 g FW), total phenolic compounds (TPC) (mg Gallic Acid Equivalent (GAE)/100 g FW) and *o*-diphenolic compounds (ODC) (mg of chlorogenic acid /100 g FW) in different mulberry fruit varieties. Values represent the results of three determinations ± SD; means with different letters in the same column are significantly different from each other ($p < 0.05$) according to the multicomponent Duncan's test.

Genotype	Moisture	Glucose	Fructose	Malic Acid	Citric Acid	TPC	ODC
M. nigra	78.2 ± 1.1 [a]	3.9 ± 0.1 [a]	3.7 ± 0.1 [a]	0.1 ± 0.1 [a]	0.9 ± 0.1 [a]	485.5 ± 7.1 [a]	101.2 ± 6.2 [a]
M. alba Legittimo nero	77.7 ± 1.3 [a]	3.3 ± 0.1 [a]	3.0 ± 0.1 [b]	0.1 ± 0.1 [a]	0.2 ± 0.1 [b]	423.6 ± 4.2 [b]	60.4 ± 3.1 [b]
M. alba Nello	77.6 ± 1.2 [a]	3.2 ± 0.1 [a]	3.2 ± 0.1 [b]	0.1 ± 0.1 [a]	0.1 ± 0.1 [b]	141.2 ± 6.1 [c]	26.2 ± 2.1 [c]

The fruits of other varieties showed a similar trend and although the glucose and fructose were equivalent, the total amount of sugars was lower corresponding to about 6 g/100 g FW. These values are in agreement as reported for the species of mulberry (*M. nigra*, *M. alba* and *M. laevigata*) cultivated in Pakistan [1], but they are notably inferior when compared to *M. nigra*, *M. alba* and *M. rubra* grown in Turkey for which the content of glucose and fructose was corresponding to about 6–7 g/100 g FW, respectively [30]. The organic acid content (malic and citric) was low and ranged between 0.13 and 1.02 g/100 g FW for *M. alba cv Nello* and *M. nigra*; *M. nigra* showed a greater quantity of citric acid of 0.92 g/100 g FW and this was in accordance with what was found in different varieties of Turkish mulberries where the amount of citric acid ranged between 0.39 and 1.08 g/100 g FW [30].

The differences between species in terms of citric and malic acid content might be caused by genetic factors as well as agronomic practices and ecological factors (temperature, soil conditions, humidity, ect.). In fact, Koyuncu [31] reported a variable amount of citric and malic acid that ranged between 0.5 to 2.3 g/100 g FW and 3.5 to 19.8 g/100 g FW for different genotypes of mulberry fruits cultivated in Turkey. These differences are also evident in fruits grown in different localities of the same country; Koyuncu [31] reported different amounts in citric and malic acid for black mulberry fruits from two locations ranged between 0.8 to 1.3 g/100g FW and 5.7 to 9.9 g/100g FW, respectively. In other Turkish mulberry genotypes, the content of malic and citric acid was very different, varying from 12.9 to 21.8 g/100 g FW and from 2.1 to 4.1 g/100 g FW, respectively [11].

Organic acids are water soluble and together with the sugars contribute to the taste of vegetables and fruits. The ratio of the total acid amount to the content of sugars in fruits is a criterion for the maturity; moreover, organic acids have a high impact on taste because of their conditioning (a reduction) on sweetness and their favoring effect on sourness. In addition, the type and the amount of acidity could be used for food decay; if the fruit is molded during the wait, there is an increment of some organic acids which seems also to have a significant impact on the purity control [30].

The amount of TPC and ODC, evaluated as mg of gallic acid equivalent (GAE) and mg of chlorogenic acid, respectively, is also reported in Table 1. The data showed that in *M. nigra* and

M. alba Legittimo the content of TPC was equal to about 485 and 423 mg GAE/100 g FW, respectively and corresponding to more than three times the TPC present in *M. alba* cv *Nello*. *M. nigra* showed an ODC content of 102.21 mg/100g FW, an amount about 1.8 and 4 times higher than that present in *M. alba* cv *Legittimo nero* and *M. alba* cv. *Nello*, respectively.

Mulberry fruits are a good source of phenolic compounds and the results clearly showed that fruit analyzed had high total phenolic content, nevertheless they showed wide differences in comparison to the literature data. In fact, TPC in mulberry fruits was reported which ranged from 104.8 to 213.5 GAE mg/100 g FW for eight Thai genotypes [32], from 158.4 to 249 mg/GAE 100g FW and from 100.5 to 348.8 mg GAE/100g FW, respectively for mulberries harvested in different Turkish sites [33,34], and from 76.7 to 180 mg GAE/100g FW in fruits grown in Spain [35]. However, data reported by Imran et al. [1] and by Ercisli et al. [12] showed that the TPC content in fruits collected in the northern region of Pakistan and in North-East Anatolia (Turkey) was very high, ranging from 880 to 1650 and from 1943 to 2237 mg GAE/100g FW, respectively. This great variability in the content of total phenols is related to the genotype, the conditions of growth and cultivation [22] and could be influenced by fruit moisture, too. It is known, in fact, that the plant can accumulate phenolic compounds under various stress conditions (heat, UV light, pathogen attack, etc.); in particular, climatic changes like low or high-temperature stress can promote the production of the phenolic compounds [36–38].

3.2. Anthocyanins Analysis

Figure 1 reports a typical HPLC/DAD/MS separation of the anthocyanins present in mulberry fruits. Five components were identified on the basis of the UV and MS spectra and retention time, corresponding to cyanidin-3-O-sophoroside (C3S, m/z 611, peak 1), cyanidin-3-O-glucoside (C3G, m/z 449, peak 2), cyanidin-3-O-rutinoside (C3R, m/z 595, peak 3), pelargonidin-3-O-glucoside (P3G, m/z 433, peak 4) and pelargonidin-3-O-rutinoside (P3R, m/z 579, peak 5). Our results are in agreement with the data reported some years ago by Dugo et al. [39] that identified a mixture of five different anthocyanidin glycosides in Italian mulberries cultivars but are discordant with Pawloska et al. [13], which revealed four anthocyanins in *M. nigra* fruits harvested in Benevento (South Italy).

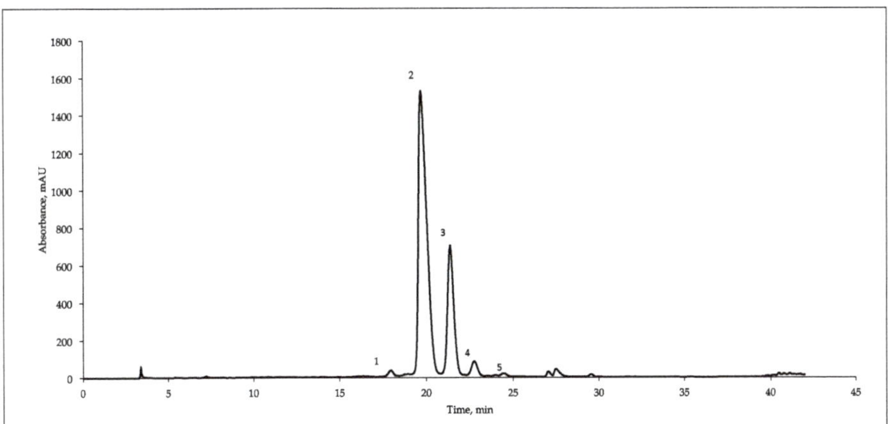

Figure 1. Typical HPLC/DAD/MS analysis (recorded at 520 nm) of the anthocyanins present in mulberry fruit extracts (*M. nigra*). (1) cyanidin-3-*O*-sophoroside, (2) cyanidin-3-*O*-glucoside, (3) cyanidin-3-*O*-rutinoside, (4) pelargonidin -3-*O*-glucoside and (5) pelargonidin -3-*O*-rutinoside.

The anthocyanins identified and the total anthocyanin content (TA), calculated as the sum of the single compounds, were shown in Table 2. The TA in *M. alba Nello* was 24.7 mg/100 g FW, was about 8 and 12 times lower than that determined in *M. nigra* and *M. alba Legittimo nero*, corresponding to 206 and 289 mg/100 g FW, respectively. According to earlier reports, TA content was 2–30, 0.3–83,

1.5–615 mg/100g FW respectively for genotypes grown in Turkey, Korea and India [34,40,41]. C3G and C3R are the anthocyanins most represented with amounts ranging from 18.2 to 212 and from 5.2 to 71.8 mg/100 g FW, respectively. P3G ranged from 1.6% to 5% of the TA content while C3S was sligtly higher in *M. nigra*, representing about 1.5% of the TA in comparison to 0.17% and 0.56% for the other genotypes.

Table 2. Amounts of cyanidin-3-*O*-sophoroside (C3S), cyanidin-3-*O*-glucoside (C3G), cyanidin-3-*O*-rutinoside (C3R), pelargonidin-3-*O*-glucoside (P3G) and pelargonidin-3-*O*-rutinoside (P3R) and Total anthocyanins (TA) in mulberry fruits. Values are expressed as C3G mg/100 g FW and represent the results of three determinations ± SD. Means with different letters in the same column are significantly different from each other ($p < 0.05$) according to the multicomponent Duncan's test.

Genotype	C3S	C3G	C3R	P3G	P3R	TA
M. nigra	3.1 ± 0.3 [a]	138.6 ± 1.5 [b]	52.3 ± 2.1 [b]	10.4 ± 1.1 [a]	1.7 ± 0.1 [a]	206.1 ± 1.8 [b]
M. alba Legittimo nero	0.5 ± 0.1 [b]	212.2 ± 1.1 [a]	71.8 ± 0.3 [a]	4.7 ± 0.2 [b]	0.1 ± 0.1 [c]	289.2 ± 0.9 [a]
M. alba Nello	0.1 ± 0.1 [c]	18.2 ± 0.4 [c]	5.2 ± 0.2 [c]	1.0 ± 0.1 [c]	0.2 ± 0.1 [b]	24.7 ± 0.3 [c]

3.3. Antioxidant Activity

Table 3 reports the mulberry AA measured by three different tests. Data indicate clearly that *M. nigra* showed the highest antioxidant activity corresponding to about 33, 26 and 21 μmol Trolox/g FW, respectively by the DPPH, ABTS and FRAP tests.

Table 3. Antioxidant activity of different mulberry genotypes evaluated by DPPH, ABTS and FRAP test. Results are expressed as μmol Trolox Equivalents/g FW. Values represent the results of three determinations ± SD; means with different letters in the same column are significantly different from each other ($p < 0.05$) according to the multicomponent Duncan's test.

Genotype	DPPH Test	ABTS Test	FRAP Test
M. nigra	32.9 ± 0.7 [a]	26.1 ± 1.5 [a]	21.3 ± 1.1 [a]
M. alba Legittimo nero	22.7 ± 1.4 [b]	11.6 ± 2.3 [b]	10.6 ± 1.5 [b]
M. alba Nello	18.5 ± 2.3 [c]	7.3 ± 2.1 [b]	1.7 ± 0.9 [c]

In the DPPH scavenging activity test, all the mulberry genotypes showed an AA significantly higher than ABTS and FRAP test. In fact, *M. alba Legittimo nero* exhibited an antioxidant capacity that was about two times greater than that measured with the other methods and *M. alba Nello* showed a capacity that was about 10 times lower when assessed by the FRAP test.

These data could be the result of a different qualitative and quantitative composition of phenolic compounds of the three mulberries. Further studies are needed to prove this. Moreover, we cannot exclude the possibility that the quenching mechanisms of the different compounds are more efficient against DPPH• than ABTS•⁺.

However, the observed antioxidant activity is probably due to the set of phenolic compounds present in the juice; in addition to the anthocyanin effect, it is known that there are significant quantities of other phenolic compounds such as gallic and cinnamic acid, procyanidin B1, catechin and quercetin [32].

Moreover, these results are in agreement with the literature data, in particular with the work of Ercisli et al. [33] that reports for *M. nigra* and *M. rubra* an AA ranging between 16 and 21.2 μmol Trolox/g FW and between 9.2 and 12.1 μmol Trolox/g FW when it was estimated by DPPH; instead, the range was between 12.3 to 14.1 μmol Trolox/g FW and from 4.9 to 8.1 μmol Trolox/g FW, respectively for *M. nigra* and *M. rubra* when measured by the FRAP test [33]. Also, the radical scavenging activity measured using ABTS system was in agreement with the values reported for different mulberry genotypes (black and/or red mulberries) grown in various Turkish regions, ranging from 6.8 to 14.4 μmol Trolox/g FW and from 5.1 to 7.3 μmol Trolox/g FW, respectively, in *M. nigra* and *M. rubra* [34].

3.4. Anti-Inflammatory Activity (AI)

The anti-inflammatory activity is related to the ability of some compounds to inhibit the two isoforms of the cyclooxygenase enzyme, COX-1 and COX-2; this activity was demonstrated for the anthocyanins isolated from raspberries and sweet cherries (cyanidin-3-glucosyl-rutinoside and cyanidin-3-rutinoside [42] and also for the anthocyanidins as cyanidin and malvidin [43].

The results of the AI in vitro assay are reported in Table 4; data indicate for all the extracts a relevant biological activity. The IC_{50} values relative to COX-1 ranged between 125 and 185 μg/mL TA and from 64 and 97 μg/mL TA relative to COX-2 for *M. nigra* and *M. alba Nello* genotypes, respectively. Employing identical analysis conditions, the AI of ibuprofen and nimesulide (positive controls, two synthetic anti-inflammatories compounds), were equal to 6 and 5 μg/mL, respectively, for COX-1 assay, and to 5 and 2 μg/mL for COX-2 assay. To our knowledge, it is the first time that mulberry extracts AI was determined in vitro using the ELISA methodology. Several authors have evaluated pure anthocyanins or different fruit extracts (such as cherry, blueberry, blackberry, etc) employing different assay methods showing that among aglycones, cyanidin was particularly active in the COX-2 assay [39]. Comparing IC_{50} values for different fruits from literature data, for pomegranate values were reported relative to COX-1 between 249 and 145 μg/mL and from 175 and 75 relative to COX-2, respectively for different genotypes grown in Salento area (Southern Apulia) [27]; the cherries showed values up to 130 μg/mL whereas blueberries, which are known to have high AI, had a range of 400–800 μg/mL depending to the ecotypes [42].

Table 4. Anti-inflammatory activity (AI) of mulberry genotypes measured as COX-1 and COX-2 inhibitory activity expressed as IC_{50}, μg/mL extract. Values represent the results of three determinations ± SD; means with different letters in the same column are significantly different from each other ($p < 0.05$) according to the multicomponent Duncan's test.

Genotype/Component	AI (IC_{50}, μg/mL)	
	COX1	COX2
M. nigra	125 ± 5 [b]	64 ± 7 [b]
M. alba Legittimo nero	140 ± 7 [c]	81 ± 8 [c]
M. alba Nello	185 ± 9 [d]	97 ± 5 [c]
Nimesulide	5 ± 1 [a]	2 ± 1 [a]
Ibuprofen	6 ± 1 [a]	5 ± 1 [a]

4. Conclusions

To our knowledge, this is the first time that mulberry genotypes growing in the Salento area have been characterized, particularly in terms of phenols, anthocyanin analysis and biological properties. The biochemical analysis of different mulberry genotypes confirmed that this fruit is rich in biologically active substances such as total phenols, o-diphenols, and anthocyanins, which are also responsible for strong antioxidant and anti-inflammatory properties. Considering the high concentration of total phenols, particularly anthocyanins, and the free anti-scavenging properties, this "minor" fruit can be considered an excellent source of antioxidant compounds and *M. nigra* and *M. alba Legittimo nero* seem to be the most interesting genotypes. Moreover, in relation to the sugars and organic acid contents, it is important to observe that their contents positively influence consumer opinions.

Mulberry should be classified as a "functional food" and its use as fruit, juice, jams, or as a muesli component may be recommended in human nutrition because of its biological qualities. In fact, the antioxidant and anti-inflammatory activities of the mulberry suggest that its diffusion and consumption would be beneficial to human health; hopefully, this fruit, if included in a regular diet, might be able to alleviate and/or control different pathologies.

Moreover, the mulberry may represent a good plant species for more extensive re-implantation in marginal zones and/or degraded lands. Its cultivation in a territory such as the Salento, a site strongly suffering from a devastating *Xylella fastidiosa* infection, might give a positive boost to agriculture and, at the same time, protect the environment from further degradation.

Author Contributions: Conceptualization, C.N. and A.M.; methodology C.N. and A.M.; formal analysis, C.N.; investigation, C.N. and A.M.; resources, A.M.; data curation, C.N., A.A.; writing—original draft preparation, A.M.; writing—review and editing, A.M., A.A.; supervision, L.D.B.; project administration, A.M. and L.D.B.; funding acquisition, A.M. and L.D.B.

Funding: This research reived no external funding.

Conflicts of Interest: The authours declare no conflict of interest.

References

1. Imran, M.; Khan, H.; Shah, M.; Khan, R.; Khan, F. Chemical composition of certain *Morus* species. *J. Zhejiang Univ.Sci. B (Biomed. Biotechnol.)* **2010**, *11*, 973–980. [CrossRef] [PubMed]

2. Thabti, I.; Elfalleh, W.; Hannachi, H.; Ferchichi, A.; Da Graça Campos, M. Identification and quantification of phenolic acids and flavonol glycosides in Tunisian *Morus* species by HPLC-DAD and HPLC-MS. *J. Funct. Foods* **2012**, *4*, 367–374. [CrossRef]

3. Gerasopouls, D.; Stravroulakis, G. Quality characteristics of four mulberry (*Morus* sp) cultivars in the area of Chania, Greece. *J. Sci. Food Agric.* **1997**, *73*, 261–264. [CrossRef]

4. Kelkar, S.M.; Bapat, V.A.; Ganapathi, T.R.; Kalig, G.S.; Rao, P.S.; Heble, M.R. Determination of hypoglycemic activity in *Morus indica* L. (mulberry) shoot cultures. *Curr. Sci.* **1996**, *71*, 71–72.

5. Asano, N.; Yamashita, T.; Ikeda, K.; Kizu, H.; Kameda, Y.; Kato, A.; Nash, R.J.; Lee, H.; Ryu, K.S. Polyhydroxylated alkaloids isolated from mulberry trees (*Morus alba* L.) and silkworms (*Bombyx mori* L.). *J. Sci. Food Agric.* **2001**, *49*, 4208–4213. [CrossRef]

6. Asano, N.; Oseky, K.; Tomioka, E.; Kizu, H.; Matsui, K. N-containing sugars from *Morus alba* and their glycosidase inhibitory activities. *Carbohydr. Res.* **1994**, *259*, 243–255. [CrossRef]

7. Du, J.; He, Z.D.; Jiang, R.W.; Ye, W.C.; Xu, H.X.; But, P.P.H. Antiviral flavonoids from the root bark of *Morus alba* L. *Phytochemistry* **2003**, *62*, 1235–1238. [CrossRef]

8. Daj, S.J.; Ma, Z.B.; Wu, Y.; Yu, D.Q. Guangsangons F-J, anti-oxidant and anti-inflammatory Diels-Alder type adducts, from *Morus macroura* Miq. *Phytochemistry* **2004**, *65*, 3135–3141.

9. Andallu, B.; Suryakantham, V.; Srikanthi, B.L.; Reddy, G.K. Effect of mulberry (*Morus indica* L.) therpy on plasma and erythrocyte membrane lipids in patients with type 2 diabetes. *Clin. Chim. Acta* **2001**, *314*, 47–53. [CrossRef]

10. Oki, T.; Kobayashi, M.; Nakamura, T.; Okuyama, A.; Masuda, M.; Shiratusuchi, H.; Suda, I. Changes in radical-scavenging activity and components of mulberry fruit during maturation. *J. Food Sci.* **2006**, *71*, C18–C22. [CrossRef]

11. Ercisli, S.; Orhan, E. Chemical composition of white (*Morus alba*), red (*Morus rubra*) and black (*Morus nigra*) fruits. *Food Chem.* **2007**, *103*, 1380–1384. [CrossRef]

12. Ercisli, S.; Orhan, E. Some physico-chemical characteristics of black mulberry (*Morus nigra* L.) genotypes from Northeast Anatolia region of Turkey. *Sci. Hortic.* **2008**, *116*, 41–46. [CrossRef]

13. Pawloska, A.M.; Oleszek, W.; Braca, A. Quali-quantitative analyses of flavonoids of *Morus nigra* L and *Morus alba* L. fruits. *J. Agric. Food Chem.* **2008**, *56*, 3377–3380. [CrossRef] [PubMed]

14. Calabrese, V.; Cornelius, C.; Stella, A.M.; Calabrese, S.J. Cellular stress responses, mitostress and carnitine insufficiencies as critical determinants in aging and neurodegenerative disordes: role of hormosis and vitagenes. *Neurochem. Res.* **2010**, *35*, 1880–1915. [CrossRef] [PubMed]

15. Obrenovich, M.E.; Li, Y.; Parvathaneni, K.; Yendluri, B.B.; palacios, H.H.; Leszek, J.; Aliev, G. Antioxidants in healths, disease and aging. *CNS Neurol. Disord. Drug Targest* **2011**, *10*, 192–207. [CrossRef]

16. Fransen, M.; Norgren, M.; Wang, B.; Apanasets, O.; Van Veldhoven, P.P. Aging, age-related diseases and peroxisomes. *Subcell. Biochem.* **2013**, *69*, 45–65. [PubMed]

17. Hélie, S.; Paul, E.J.; Ashby, F.G. A neurocomputational account of cognitive deficits in Parkinson's disease. *Neuropsychologia* **2011**, *50*, 2290–2302. [CrossRef]

18. O'Neill, C. PI3-kinase/Akt/mTOR signaling: impaired on/off switches in aging, cognitive decline and Alzheimer's disease. *Exp. Gerontol.* **2013**, *48*, 647–653. [CrossRef]

19. Turgut, N.H.; Mert, D.G.; Kara, H.; Egilmez, H.R.; Arslanbas, E.; Tepe, B.; Gungor, H.; Yilmaz, N.; Tuncel, N.B. Effect of black mulberry (Morus nigra) extract treatment on cognitive impairment and oxidative stress status of D-galactosee-induced aging mice. *Pharm. Biol.* **2016**, *54*, 1052–1064. [CrossRef]

20. Chen, H.; Yu, W.; Chen, G.; Meng, S.; Xiang, Z.; He, N. Antinociceptive and antibacterial properties of anthocyanins and flavonols from fruits of black and non-black mulberries. *Molecules* **2018**, *23*, 1004. [CrossRef]

21. Liu, X.; Xiao, G.; Chen, W.; Xu, Y.; Wu, J. Quantification and purification of mulberry anthocyanins with macroporus resins. *J. Biomed. Biotechnol.* **2004**, *5*, 326–331. [CrossRef] [PubMed]

22. Sývacý, A.; Sökmen, M. Seasonal changes in antioxidant activity, total phenolic and anthocyanin constituent of the steams of two *Morus* species (*Morus alba* L. and *Morus nigra* L.). *Plant Growth Regul.* **2004**, *44*, 251–256. [CrossRef]

23. Du, Q.; Zheng, J.; Xu, Y. Composition of anthocyanins in mulberry and their antioxidant activity. *J. Food Comp. Anal.* **2008**, *21*, 390–395. [CrossRef]

24. *Official Methods of Analysis of AOAC International*, 15th ed.; Association of Official Analytical Chemist: Washington DC, USA, 1990.

25. Singleton, V.L.; Orthofer, R. Analysis of Total Phenols and Other Oxidation Substrates and Antioxidants by Means of Folin-Ciocalteu Reagent. *Methods Enzymol.* **1999**, *299*, 152–178.

26. Bendini, A.; Bonoli, M.; Cerretani, L.; Biguzzi, B.; Lercker, G.; Gallina Toschi, T. Liquid–liquid and solid-phase extractions of phenols from virgin olive oil and their separation by chromatographic and electrophoretic methods. *J. Chrom. A* **2003**, *985*, 425–433. [CrossRef]

27. Negro, C.; Longo, L.; Vasapollo, G.; De Bellis, L.; Miceli, A. Biochemical, antioxidant and anti-inflammatory properties of promegranate fruits growing in Southern Italy (Salento, Apulia). *Acta Aliment.* **2012**, *41*, 190–199. [CrossRef]

28. Re, R.; Pellegrini, N.; Proteggente, A.; Pannala, A.; Yang, M.; Rice-Evans, C.A. Antioxidant activity applyng an improved ABTS radical cation decolorization assay. *Free Radic. Bio. Med.* **1999**, *26*, 1231–1237. [CrossRef]

29. Benzie, I.F.F.; Strain, J.J. The ferric reducing ability of plasma as measure of "antioxidant power": The FRAP assay. *Anal. Biochem.* **1996**, *239*, 70–76. [CrossRef]

30. Gundogdu, M.; Muradoglu, F.; Gazioglu Sensoy, R.I.; Ylmaz, H. Determination of fruit chemical properties of *Morus nigra* L., *Morus alba* L. and *Morus rubra* L. by HPLC. *Sci. Hortic.* **2011**, *132*, 37–41. [CrossRef]

31. Koyuncu, F. Organic acid composition of native black mulberry fruit. *Chem. Nat. Compd.* **2004**, *40*, 367–369. [CrossRef]

32. Butkhup, L.; Samappito, W.; Samappito, S. Phenolic composition and antioxidant activity of whyte mulberry (*Morus alba* L.) fruits. *Int. J. Food Sci. Technol.* **2013**, *48*, 934–940. [CrossRef]

33. Ercisli, S.; Tosun, M.; Duralija, B.; Voca, S.; Sengul, M.; Turan, M. Phytochemical content of some black (*Morus nigra* L.) and purple (*Morus rubra* L.) mulberry genotypes. *Food Technol. Biotechnol.* **2010**, *48*, 102–106. [CrossRef]

34. Özgen, M.; Serçe, S.; Kaya, C. Phytochemical and antioxidant properties of anthocyanin-rich *Morus nigra* and *Morus rubra* fruits. *Sci. Hortic.* **2009**, *119*, 275–279. [CrossRef]

35. Calin-Sanchez, A.; Martinez-Nicolas, J.J.; Munera-Picazo, S.; Carbonell-Barrachina, A.A.; Legua, P.; Hernandez, F. Bioactive compounds and sensory quality of black and white mulberries grown in Spain. *Plant Food Hum. Nutr.* **2013**, *68*, 370–377. [CrossRef] [PubMed]

36. Nozolino, C.; Isabelle, P.; Das, G. Seasonal changes in phenolics constituents of jack pine seedling (*Pinus banksiana*) in relation to the purpling phenomenon. *Can. J. Bot.* **1990**, *68*, 2010–2017. [CrossRef]

37. Christie, P.J.; Alfenito, M.R.; Walbot, V. Impact of low-temperature stress on general phenylpropanoid and anthocyanin pathways: enhancement of transcript abundance and anthocyanins pigmentation in maize seedlings. *Planta* **1994**, *194*, 541–549. [CrossRef]

38. Dixon, R.A.; Paiva, N.L. Stress-induced phenylpropanoid metabolism. *Plant Cell* **1995**, *7*, 1085–1097. [CrossRef]

39. Dugo, P.; Mondello, L.; Errante, G.; Zappia, G.; Dugo, G. Identification of anthocyaninsin berries by narrow-bore high-performance liquid chromatography with electrospray ionization detection. *J. Agric. Food Chem.* **2001**, *49*, 3897–3992. [CrossRef]

40. Bae, S.H.; Suh, H.J. Antioxidant activities of five different mulberry cultivars in Korea. *LWT Food Sci. Technol.* **2007**, *40*, 955–965. [CrossRef]

41. Shivashankara, K.S.; Jalikop, S.H.; Roy, T.K. Species variability for fruit antioxidant and radical scavenging ability in mulberry. *Int. J. Fruit Sci.* **2010**, *10*, 355–366. [CrossRef]
42. Seeram, N.P.; Momin, R.A.; Nair, M.G.; Bourquin, L.D. Cyclooxygenase inhibitory and antioxidant cyaniding glycosides in cherry and berries. *Phytomedicine* **2001**, *8*, 362–369. [CrossRef] [PubMed]
43. Seeram, N.P.; Zhang, Y.; Nair, M.G. Inhibition of proliferation of human cancer cell and cyclooxygenase enzymes by anthocyanidins and catechins. *Nutr. Cancer* **2003**, *46*, 101–106. [CrossRef] [PubMed]

 © 2019 by the authors. Licensee MDPI, Basel, Switzerland. This article is an open access article distributed under the terms and conditions of the Creative Commons Attribution (CC BY) license (http://creativecommons.org/licenses/by/4.0/).

 *antioxidants* 

Article

Effects of In Vitro Gastrointestinal Digestion on the Antioxidant Capacity and Anthocyanin Content of Cornelian Cherry Fruit Extract

Luminita David [1], Virgil Danciu [1], Bianca Moldovan [1,*] and Adriana Filip [2]

[1] Research Center for Advanced Chemical Analysis, Instrumentation and Chemometrics (ANALYTICA), Faculty of Chemistry and Chemical Engineering, Babeş-Bolyai University, 11 Arany Janos Street, 400028 Cluj-Napoca, Romania; muntean@chem.ubbcluj.ro (L.D.); danciuv@chem.ubbcluj.ro (V.D.)

[2] Department of Physiology, Iuliu Hatieganu University of Medicine and Pharmacy, 1-3 Clinicilor Street, 400006 Cluj-Napoca, Romania; gabriela.filip@umfcluj.ro

* Correspondence: bianca@chem.ubbcluj.ro; Tel.: +40-264-593-833

Received: 1 April 2019; Accepted: 26 April 2019; Published: 30 April 2019

Abstract: Red fruits are considered a major source of antioxidant compounds in the human diet. They usually contain anthocyanins, phenolic pigments that confer them multiple health-promoting properties. The health benefits of these bioactive phytocompounds are strongly related to their bioavailability, which has been reported to be low. The aim of the present study is to investigate the changes in antioxidant capacity and anthocyanin content of Cornelian cherry fruit extract during gastrointestinal digestion. Thus, the work was designed using a simulated in vitro digestion model. The antioxidant capacity (AA) was tested by the 2,2-azinobis (3-ethylbenzothiazolyne-6-sulphonic acid) radical cation (ABTS) method, while quantification of anthocyanins (TAC) was accomplished by the means of the pH differential method and high performance liquid chromatography (HPLC). The results showed that gastric digestion had no significant effect on the TAC of the extract, while the AA slightly increased. After duodenal digestion, only 28.33% of TAC and 56.74% of AA were maintained. Cornelian cherries' anthocyanins were stable in stomach, so they can be absorbed in order to manifest their antioxidant capacity at the cellular level. The duodenal digestion dramatically decreased the TAC and AA level in the fruit extract.

Keywords: cornelian cherry; antioxidant capacity; anthocyanins; gastrointestinal digestion

---

1. Introduction

Cornelian cherry (*Cornus mas* L.) is a plant belonging to the genus *Cornus*, growing in southeastern Europe and Asia. Its fruits possess a sour tart taste, contain a single stone, and are usually eaten raw or processed as jams, liquors, vinegars, compotes, or marmalades [1–3]. These fruits are reported to contain biologically-active compounds, such as phenolic acids, flavonoids, iridoids, triterpenoids, organic acids, and vitamin C [4–6]. The ripe fruits of *Cornus mas* have usually an attractive red color, which is due to the presence of anthocyanins, plant secondary metabolites responsible for the red, purple, or blue color of fruits, vegetables, or flowers. Anthocyanins have recently drawn much attention due to their numerous health-promoting properties, such as antioxidant, anti-inflammatory, antidiabetic, antiobesity, neuroprotective, and antiatherosclerotic effects [7]. In Cornelian cherry fruits, several anthocyanins, such as pelargonidin and cyanidin derivatives, were identified (Figure 1). Du and Francis reported the presence of five anthocyanins: cyanidin-3-*O*-galactoside, cyanidin-3-*O*-rhamnosylgalactoside, delphinidin-3-*O*-galactoside, pelargonidin-3-*O*-galactoside, and pelargonidin-3-*O*-rhamnosylgalactoside [8,9]. Later studies indicated that Cornelian cherry fruits contain normally a mixture of three to five anthocyanic pigments identified by different techniques such as high performance liquid chromatography, mass spectrometry, or nuclear magnetic resonance. The anthocyanin composition is strongly related to Cornelian cherry cultivar. Kucharska et al. confirmed the presence of five anthocyanins, delphinidin, pelargonidin, and cyanidin-3-*O*-glycosides, by investigating 26 Cornelian cherry cultivars from Poland and Ukraine [10], while Moldovan et al. reported the presence of cyanidin-3-*O*-galactoside, pelargonidin-3-*O*-glucoside, and pelargonidin-3-*O*-rutinoside in a Romanian Cornelian cherry cultivar [11].

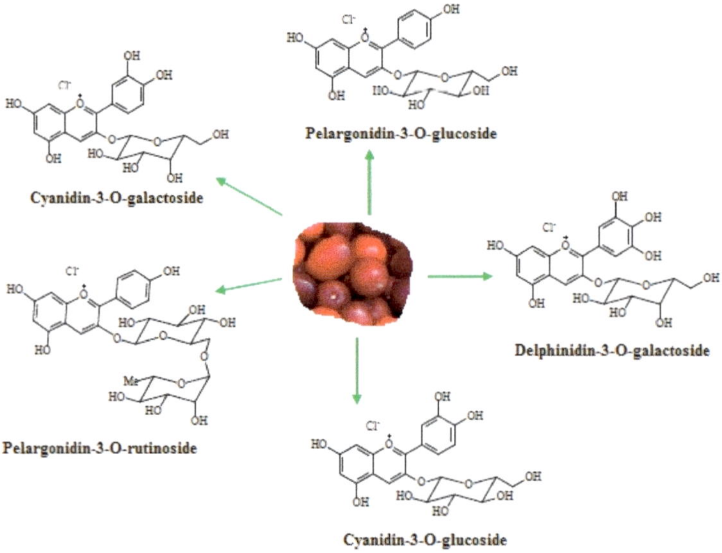

Cyanidin-3-O-galactoside

Pelargonidin-3-O-glucoside

Pelargonidin-3-O-rutinoside

Delphinidin-3-O-galactoside

Cyanidin-3-O-glucoside

Figure 1. Structure of anthocyanins from Cornelian cherry fruit.

The dietary consumption of anthocyanin-rich fruits or their derivative products may prevent numerous diseases and improve human health. Mazza et al. reported an increase in the antioxidant status of the human serum after oral intake of blueberries anthocyanins [12], although earlier studies reported that only 1% of anthocyanin compounds are present in human plasma after fruit consumption [13]. After oral administration, anthocyanins can be absorbed in stomach, as well as in intestines, either in their intact form or as metabolites [14]. After absorption, anthocyanins are transported to heart, brain, liver, kidneys, or other tissues where they can exert their antioxidant capacity and their health-promoting properties. All these results indicate that investigating anthocyanin stability in the gastrointestinal tract could deliver important information on their bioavailability and their health beneficial properties. Simulated human gastrointestinal digestion studies demonstrated that anthocyanins are rather stable in stomach, but easily degradable in small intestine mainly due to the high value of the pH [7,15,16].

The main objective of this study was to evaluate the impact of the gastrointestinal tract's biochemical conditions on the antioxidant compounds responsible for the beneficial effects on human health of Cornelian cherry fruits. Therefore, a simulated in vitro human digestion model was applied, and the stability of anthocyanins from the investigated fruits, as well as the changes in the antioxidant activity of the digested fruit extract were investigated. Although the anthocyanin profile, the total phenolic content, and the in vitro antioxidant capacity of the Cornelian cherry fruits were already elucidated by several research groups [10,11,17–21], this is the first study dealing with the impact of gastrointestinal digestion on their antioxidant activity and on the bioaccessibility of their anthocyanins.

2. Materials and Methods

2.1. Chemicals

Cyanidin-3-*O*-galactoside, pelargonidin-3-*O*-glucoside, and pelargonidin-3-*O*-rutinoside standards were purchased from Extrasynthese (Lyon, France), and pancreatin from porcine pancreas and pepsin from porcine stomach were purchased from Alfa Aesar (Karsluhe, Germany). All other chemicals and solvents were obtained from Merck (Darmstadt, Germany) and were of analytical or HPLC purity. A TYPDP1500 water distiller (Techosklo LTD., Drzkov, Czech Republic) was used to obtain the distilled water.

2.2. Preparation of Fruit Extract

Cornelian cherry (*Cornus mas* L.) fruits were manually collected at ripening stage in August 2018 in Chinteni, Cluj County, Romania. The picked fruits were selected according color, mass, and shape uniformity and stored at −18 °C until further investigation. The Cornelian cherries' anthocyanins were isolated by extraction applying an already published procedure [11]. Briefly, 90 g of homogenized fruit pulp were treated with 300 mL of food-grade acetone. The extraction was carried out at room temperature, under vigorous stirring for 60 min. The mixture was vacuum filtered through Whatman No. 1 filter paper, and the obtained solution was concentrated to a 65-mL final volume at 40 °C by a rotary evaporator to remove the acetone. The obtained concentrate was further subjected to determination of total anthocyanin content by the pH differential method, to identify anthocyanins present in the extract by HPLC analysis and in in vitro simulated gastrointestinal digestion.

2.3. Quantification and Identification of Anthocyanins

The total anthocyanin content of the Cornelian cherries' anthocyanin-rich extract was determined using the well-known spectrophotometric pH differential method [22] as previously described [23]. Briefly, samples of fruit extract were mixed with 0.025 M potassium chloride buffer solution (pH = 1) and 0.04 M sodium acetate buffer solution (pH = 4.5), respectively. The absorbance of each resulting solution was measured at 506 nm and 700 nm, respectively, against distilled water as a blank, using a UV-Vis Perkin Elmer Lambda 25 double-beam spectrophotometer (Perkin Elmer, Shelton, CT, USA). The total anthocyanin content was expressed as cyanidin-3-glucoside equivalents/liter. All measurements were performed at room temperature, in triplicate.

The identification of anthocyanins in the Cornelian cherry fruit extract was accomplished by HPLC analysis. For this purpose, an Agilent 1200 (Agilent Technologies Inc.; Santa Clara, CA, USA) equipped with a diode array detector HPLC system was used. The separation of anthocyanins from the fruit extract was performed on an Eclipse XTB-C18 (Agilent) column (150 × 4.6 mm inner diameter, particle size 5 μm), maintained at 20 °C. Samples of the extract were 2-fold diluted with 0.1% formic acid, filtered through a 0.2-μm PTFE membrane disk filter, and 20 μL were injected into the column. A mixture of two solvents, 0.1% formic acid aqueous solution (Solvent A) and 0.1% formic acid acetonitrile solution (Solvent B), was used as a mobile phase at a flow rate of 0.4 mL/min. The elution was performed by applying the following gradient profile: 0–16 min 95% A, 16–17 min 60% A, 17–20 min 5% A, and 20–21 min 95% A. The detection was performed at 506 nm. The main anthocyanins from the Cornelian cherry fruit extract were identified according to the consistency of the retention times compared to authentic anthocyanin standards. The quantification of each anthocyanin present in the extract was performed using a calibration curve of cyanidin-3-*O*-galactoside by an external standard method. All samples were injected in triplicate.

2.4. Evaluation of the Antioxidant Activity Using the ABTS Assay

The antioxidant capacity of the Cornelian cherry fruit extract was determined by the ABTS radical scavenging activity method [24] with some modifications [25]. The ABTS•+ stock solution was freshly diluted with distilled water to prepare a working solution with an absorbance between 0.6 and

0.8 recorded at 734 nm against distilled water using a spectrophotometer. The free radical scavenging capacity of the fruit extract was evaluated by adding 0.1 mL of extract sample to 6 mL diluted ABTS solution followed by incubation at room temperature, in the dark, for 15 min. After 15 min, the absorbance of the sample was monitored at 734 nm, and the antioxidant activity was calculated using a calibration curve of Trolox standard and expressed in μmol Trolox equivalents/L.

2.5. In Vitro Simulated Digestion Process

The gastrointestinal digestion was simulated according to a slightly modified protocol of Gil-Izquierdo [26]. Samples of 50 mL were subjected to an in vitro gastric digestion process by adjusting the pH of the extract to 2, by adding 1 M HCl solution. Thereafter, pepsin from porcine gastric mucosa (15.750 units EC 3.4.23.1) was added, and the samples were incubated in the absence of light, in a shaker (200 rpm), at 37 °C, for 2 h. After the simulated gastric digestion, aliquots of 5 mL were collected, rapidly cooled at −20 °C, and used further for the evaluation of anthocyanin composition and antioxidant activity of the gastric digested samples. The remaining solution from the gastric digestion step was further submitted to the in vitro simulated intestinal digestion process. The pH of the remaining solution was adjusted to 7.5 with 1 M $NaHCO_3$ solution. Then, 22.5 mL of pancreatin (2 mg/mL) from porcine pancreas (8 × USP specifications) and bile salts (25 mg/mL) solution were added, and the obtained mixture was further incubated at 37 °C for another 2 h. The obtained digested samples were properly diluted with 0.1% formic acid (3-fold for the gastric digestion samples and 1.6-fold for the intestinal digestion samples) and analyzed directly by HPLC for the determination of the anthocyanin profile, and also, the antioxidant capacity and the total anthocyanin content were evaluated. The simulation of the in vitro digestion process was performed in triplicate.

2.6. Statistical Analysis

All reported data are presented as the mean values ± the standard deviation obtained from the three replicates. Statistical analysis was conducted using XLSTAT Release 10 (Addinsoft, Paris, France) software, and the level of significance was evaluated by one-way analysis of variance (ANOVA), considering *p*-values < 0.05 to be significant.

3. Results and Discussion

Cornelian cherry fruits have been reported as a valuable source of anthocyanins, compounds with high antioxidant activities that contribute to the beneficial biological properties of these fruits. A diet rich in anthocyanins results in downregulating disease markers, improves the antioxidant function, and protects cells against the deleterious effects of oxidative stress [27]. The antioxidant activity of the Cornelian cherries' extract was evaluated using the ABTS assay and was found to be 553 μM Trolox. The anthocyanin content was determined by the pH differential method and was 360 mg Cy-3-glu equivalents/L. The HPLC analysis of the Cornelian cherry fruit extract revealed the presence of three anthocyanins (Figure 2). The results were in agreement with those previously reported by Moldovan et al. [11]. The three main anthocyanin peaks detected at 506 nm were identified by comparing the retention times and the UV absorption spectra with authentic anthocyanin standards and were assigned to cyanidin-3-*O*-galactoside (Peak 1), pelargonidin-3-*O*-glucoside (Peak 2), and pelargonidin-3-*O*-rutinoside (Peak 3).

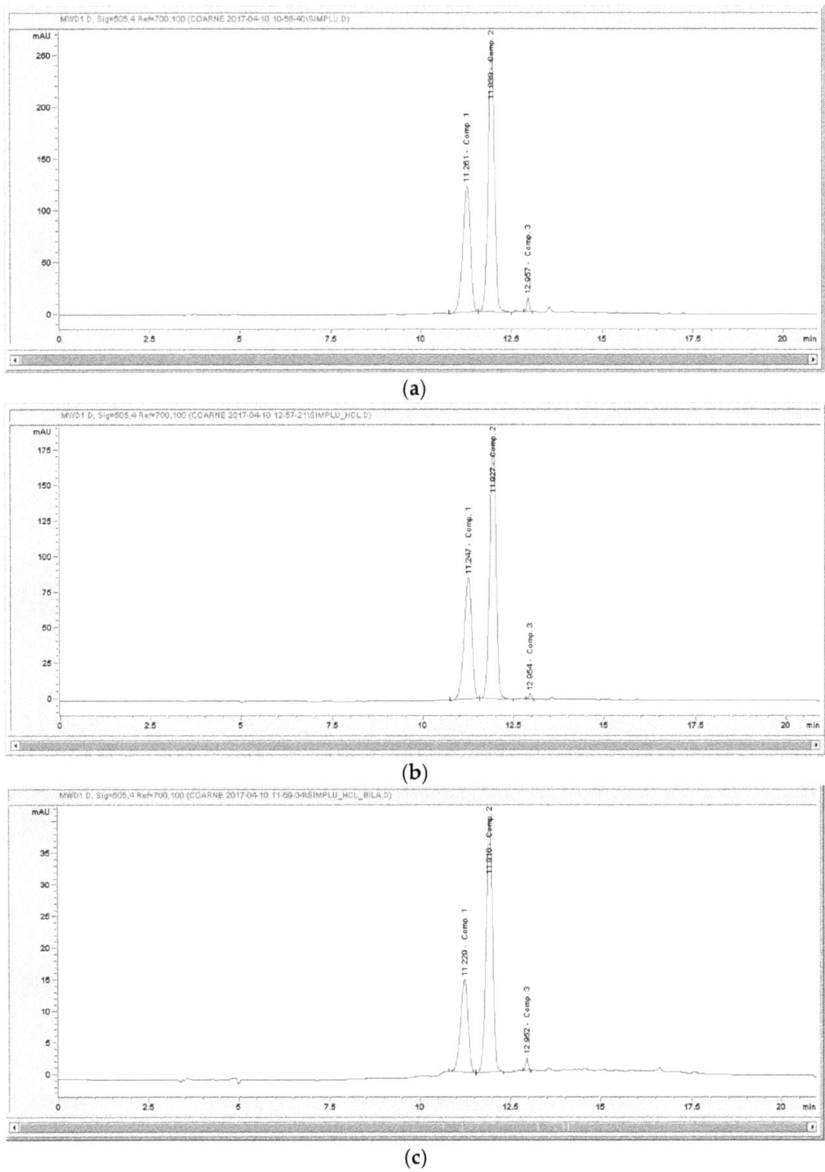

Figure 2. HPLC chromatograms (520 nm) of Cornelian cherries' anthocyanins before and after in vitro digestion: (**a**) crude extract; (**b**) gastric digestion; (**c**) intestinal digestion.

The content of the three anthocyanins is presented in Table 1. The main anthocyanin of the Cornelian cherry fruits was pelargonidin-3-*O*-glucoside (62.99% from total anthocyanin content), followed by cyanidin-3-*O*-galactoside (35.68% from total anthocyanin content). Pelargonidin-3-*O*-rutinoside has been identified in very small amounts (<2%) in the crude fruit extract.

Table 1. Anthocyanin identification and quantification (mg/L) of Cornelian cherry fruit extract samples before and after in vitro digestion.

Peak Number	Compound	Crude Extract	In Vitro Gastric Digestion	In Vitro Intestinal Digestion
1	Cyanidin-3-*O*-galactoside	128.45 ± 5.14 [a]	137.52 ± 6.05 [a]	29.9 ± 1.03 [b]
2	Pelargonidin-3-*O*-glucoside	226.78 ± 8.61 [a]	246.35 ± 10.12 [a]	70.79 ± 3.13 [b]
3	Pelargonidin-3-*O*-rutinoside	4.75 ± 0.15 [a]	2.12 ± 0.09 [b]	1.45 ± 0.06 [b]
	Recovery (%)	100	107.23	26.46

[a,b] Numbers in rows followed by different letters are significantly different ($p < 0.05$).

Anthocyanins are generally not very stable compounds. Many factors can affect the stability of these antioxidant compounds, such as pH, oxygen, chemical structure, and the presence of enzymes [28]. Thus, investigating their stability during their passage through the gastrointestinal tract is very important, in order to understand the bioavailability of these compounds. In the present study, the stability of Cornelian cherry anthocyanins was investigated using an in vitro simulated gastrointestinal digestion model. The Cornelian cherry fruit extract was incubated with pepsin at pH = 2 for the gastric digestion and, after that, with pancreatin and bile salts at pH = 7.5, to simulate small intestine digestion. After each step of the in vitro digestion, samples were analyzed in terms of their total anthocyanin content and antioxidant capacity. HPLC analysis was conducted to monitor changes in anthocyanin composition and content after each digestion step. Stomach digestion did not significantly change the qualitative or quantitative composition of the anthocyanin compounds. Figure 2b presents the HPLC chromatogram after simulated gastric digestion, and no differences between crude extract and that subjected to digestion in gastric conditions were observed. After subjecting the anthocyanin-rich extract to simulated gastric digestion, a slight increase of the total anthocyanin content to 386 mg/L was noticed. The same anthocyanins as in the original sample were recovered after stomach digestion, the recovery being 107.23%. Our findings are in good agreement with those obtained in other studies. Ryu et al. reported a slight increase after gastric digestion of Bokbunja anthocyanins [15], while Sun et al. observed the same result in the case of purple rice anthocyanins subjected to in vitro simulated digestion [29]. These findings strongly suggest that Cornelian cherry anthocyanins are highly stable when exposed to stomach conditions. The low pH value in stomach mainly contributes to the high stability of anthocyanins, which at this pH (pH = 1.5–2), occurs in the chemical structure of a stable flavylium cation [30]. The increase of the anthocyanin content could also suggest that an acidic environment and digestive enzymes improve the release of the monomeric anthocyanins from the polymeric ones, by disrupting macromolecules. Additionally, an increase in the antioxidant activity during simulated gastric digestion occurred (Figure 3). Changes in total anthocyanin content and radical scavenging capacity were not significant during the in vitro gastric digestion as compared to undigested extract ($p > 0.05$).

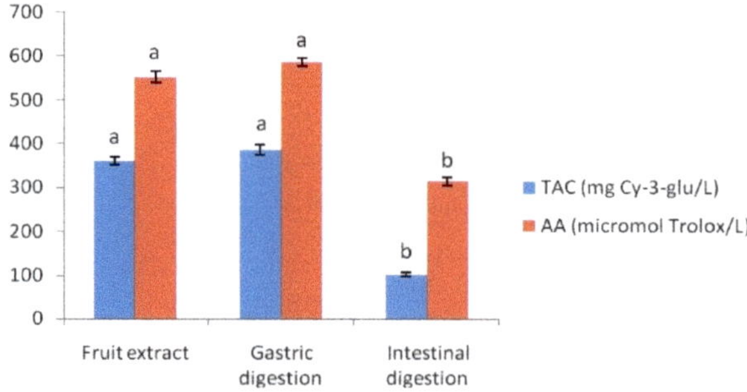

Figure 3. Changes in total anthocyanin content and antioxidant activity of Cornelian cherry fruit extract before and after in vitro gastrointestinal digestion. Means (bar value) with different letters are significantly different ($p < 0.05$).

The samples obtained after gastric digestion were further transferred to a mild alkaline intestinal environment, and pancreatic enzymes and bile salts were added in order to simulate the intestinal digestion process. After small intestine digestion, a clear decrease of the total anthocyanin content up to 26.46 mg cyanidin-3-*O*-glucoside equivalents/L was observed, reduced by more than 70% compared to the undigested sample. The low recovery of anthocyanins after intestinal digestion (26.46%) can be explained by the low stability of these compounds under alkaline conditions (pH = 7.5) attributed to the structural changes of the flavylium cation to a colorless, less stable chalcone [30]. Other studies also found that simulated intestinal digestion had a dramatic impact on the anthocyanins from different fruits. Pomegranate juice anthocyanins decreased by ~97% during this stage of digestion [31], while blueberry anthocyanins decreased more than 80% after intestinal digestion [32]. All the anthocyanin compounds present in the Cornelian fruit extract presented a significant decrease during incubation under small intestine conditions. Cyanidin-3-*O*-galactoside was the most unstable of the three investigated anthocyanins, its content being the most affected by the intestinal digestion compared to the crude extract.

The pancreatic-bile digestion significantly decreased the antioxidant capacity of the anthocyanin-rich extract of Cornelian cherries. The ABTS$\bullet^+$ scavenging activity decreased by ~43%, as compared to the undigested or gastric digested sample. The changes in the antioxidant capacity of digested Cornelian cherry fruit anthocyanins were consistent with those previously observed for red cabbage anthocyanins [33]. While the bioaccessibility of Cornelian cherry anthocyanins largely decreased during intestinal digestion, the antioxidant capacity of the extract was less affected by this phase of digestion. This fact could be explained by the formation of new antioxidant metabolites from anthocyanins degraded from the simulated intestinal fluid, which can further exert their beneficial effects on human health.

4. Conclusions

The present study evaluated for the first time the stability of anthocyanins from Cornelian cherries during their passage through the upper gastrointestinal tract by an in vitro simulation of the digestion process. The effect of the in vitro gastrointestinal digestion on the antioxidant capacity of the anthocyanin-rich extract was also investigated. The results indicated the presence of three anthocyanins, cyanidin-3-*O*-galactoside, pelargonidin-3-*O*-glucoside, and pelargonidin-3-*O*-rutinoside, in Cornelian cherry fruits, compounds that were not significantly affected by the gastric digestion. Intestinal digestion resulted in a large decrease of the anthocyanin content and antioxidant activity of the investigated fruit extract. These findings suggest that anthocyanins' stability during gastrointestinal digestion should be taken into account when estimating their bioavailability. The consumption of Cornelian cherries may be an important source of anthocyanins in the human diet, which may exert their health beneficial effects at the gastric level, while their degradation products and metabolites may act as antioxidants in small intestine.

Author Contributions: Conceptualization, methodology, writing, original draft preparation, writing, review and editing, and project administration, L.D. and B.M.; formal analysis, V.D.; investigation, L.D., V.D., B.M., and A.F.; funding acquisition, A.F.

Funding: This research was funded by the Ministry of National Education, Grant Number PN-III-P4-IC-PCE-2016-0396.

Conflicts of Interest: The authors declare no conflict of interest. The funders had no role in the design of the study; in the collection, analyses, or interpretation of data; in the writing of the manuscript; nor in the decision to publish the results.

References

1. Nouska, C.; Kazakos, S.; Mantzourani, I.; Alexopoulos, A.; Bezirtzoglou, E.; Plessas, S. Fermentation of *Cornus mas* L. juice for functional low alcoholic beverage production. *Curr. Res. Nutr. Food Sci.* **2016**, *4*, 119–124. [CrossRef]
2. Mantzourani, I.; Nouska, C.; Terpou, A.; Alexopoulos, A.; Bezirtzoglou, E.; Panayiotidis, M.; Galanis, A.; Plessas, S. Production of a novel functional fruit beverage consisting of Cornelian cherry juice and probiotic bacteria. *Antioxidants* **2018**, *7*, 163. [CrossRef] [PubMed]
3. Kawa-Rygielska, J.; Adamenko, K.; Kucharska, A.Z.; Piorecki, N. Bioactive compounds in Cornelian cherry vinegars. *Molecules* **2018**, *23*, 379. [CrossRef] [PubMed]
4. Dinda, B.; Cyriakopoulos, A.M.; Dinda, S.; Zoumpourlis, V.; Thomaidis, N.S.; Velegraki, A.; Markopoulos, C.; Dinda, M. *Cornus mas* L. (Cornelian cherry), an important European and Asian traditional food and medicine: Ethnomedicine, phytochemistry and pharmacology for its commercial utilization in drug industry. *J. Ethnopharmacol.* **2016**, *193*, 670–690. [CrossRef]
5. Moldovan, B.; David, L. Bioactive Flavonoids from *Cornus Mas* L. Fruits. *Mini-Rev. Org. Chem.* **2017**, *14*, 489–495. [CrossRef]
6. Opris, R.; Toma, V.; Olteanu, D.; Baldea, I.; Baciu, A.; Imre-Lucaci, F.; Berghian Sevastre, A.; Tatomir, C.; Moldovan, B.; Clichici, S.; et al. Effects of Silver Nanoparticles functionalized with *Cornus mas* L. extract on architecture and apoptosis in rat testicle. *Nanomedicine* **2019**, *14*, 275–299. [CrossRef] [PubMed]
7. Han, F.; Yang, P.; Wang, H.; Fernandes, Y.; Mateus, N.; Liu, Y. Digestion and absorption of red grape and wine anthocyanins through the gastrointestinal tract. *Trends Food Sci. Technol.* **2019**, *83*, 211–224. [CrossRef]
8. Du, C.-T.; Francis, F.J. Anthocyanins from *Cornus mas*. *Phytochemistry* **1973**, *12*, 2487–2489. [CrossRef]
9. Du, C.-T.; Francis, F.J. New anthocyanins from *Cornus mas*. *HortScience* **1973**, *8*, 29–30.
10. Kucharska, A.Z.; Szumny, A.; Sokol-Letowska, A.; Piorecki, N.; Klymenko, S.V. Iridoids and anthocyanins in Cornelian cherry (*Cornus mas* L.) cultivars. *J. Food Compos. Anal.* **2015**, *40*, 95–102. [CrossRef]
11. Moldovan, B.; Filip, A.; Clichici, S.; Suharovschi, R.; Bolfa, P.; David, L. Antioxidant activity of Cornelian cherry (*Cornus mas* L.) fruits extracts and the in vivo evaluation of their anti-inflammatory effects. *J. Funct. Foods* **2016**, *26*, 77–87. [CrossRef]
12. Mazza, G.; Kay, C.D.; Cottrell, T.; Holub, B.J. Absorption of anthocyanins from Blueberries and serum antioxidant status in human subjects. *J. Agric. Food Chem.* **2002**, *50*, 7731–7737. [CrossRef]
13. Manach, C.; Williamson, G.; Morand, C.; Scalbert, A.; Remesy, C. Bioavailability and bioefficacy of polyphenols in humans. I. Review of 97 bioavailability studies. *Am. J. Clinic. Nutr.* **2005**, *81*, 230S–242S. [CrossRef]
14. Fang, J. Bioavailability of anthocyanins. *Drug Metabol. Rev.* **2014**, *46*, 508–520. [CrossRef]
15. Ryu, D.; Koh, E. Stability of anthocyanins in bokbunja (*Rubus occidentalis* L.) under in vitro gastrointestinal digestion. *Food Chem.* **2018**, *267*, 157–162. [CrossRef] [PubMed]
16. McDougall, G.J.; Fyffe, S.; Dobson, P.; Stuart, D. Anthocyanins from red cabbage—Stability to simulated gastrointestinal digestion. *Phytochemistry* **2007**, *68*, 1285–1294. [CrossRef] [PubMed]
17. Moldovan, B.; Popa, A.; David, L. Effects of storage temperature on the total phenolic content of Cornelian cherry (*Cornus mas* L.) fruits extracts. *J. Appl. Bot. Food Qual.* **2016**, *89*, 208–211.
18. Hosu, A.; Cimpoiu, C.; David, L.; Moldovan, B. Study of the antioxidant property variation of Cornelian cherry fruits during storage using HPTLC and spectrophotometric assays. *J. Anal. Meth. Chem.* **2016**, *2016*, 2345375. [CrossRef] [PubMed]
19. Park, H.-M.; Hong, J.-H. Antioxidant effects of bioactive compounds isolated from pressurized steam-treated Corni fructus and their protective effects against UVB-irradiated HS68 cells. *J. Med. Food* **2018**, *21*, 1165–1172. [CrossRef]
20. DeBiaggi, M.; Donno, D.; Mellano, M.G.; Riondato, I.; Rakotoniaina, E.N.; Beccaro, G.L. *Cornus mas* (L.) fruit as a potential source of natural health promoting compounds: Physico-chemical characterization of bioactive components. *Plant Food Human Nutr.* **2018**, *73*, 89–94. [CrossRef]
21. Moldovan, B.; David, L.; Man, S.C. Impact of thermal treatment on the antioxidant activity of cornelian cherries extract. *Stud. Univ. Babes-Bolyai Chem.* **2017**, *62*, 311–317. [CrossRef]
22. Lee, J.; Durst, R.W.; Wrolstad, R.E. Determination of total monomeric anthocyanin pigment content of fruit juices, beverages, natural colorants and wines by the pH differential method: Collaborative study. *J. AOAC Int.* **2005**, *88*, 1269–1278.

23. Moldovan, B.; David, L. Influence of Temperature and Preserving Agents on the Stability of Cornelian Cherries Anthocyanins. *Molecules.* **2014**, *19*, 8177–8188. [CrossRef]

24. Re, R.; Pellegrini, N.; Proteggente, A.; Pannala, A.; Yang, M.; Rice-Evans, C. Antioxidant activity applying an improved ABTS radical cation decolorization assay. *Free Rad. Biol. Med.* **1999**, *26*, 1231–1237. [CrossRef]

25. Moldovan, B.; Ghic, O.; David, L.; Chişbora, C. The influence of storage on the total phenols content and antioxidant activity of the Cranberrybush (*Viburnum opulus* L.) fruits extract. *Rev. Chim. Bucharest.* **2012**, *63*, 463–464.

26. Gil-Izquierdo, A.; Zafrilla, P.; Tomas-Barberan, F.A. An in vitro method to simulate phenolic compound release from the food matrix in the gastrointestinal tract. *Eur. Food Res. Technol.* **2002**, *214*, 155–159. [CrossRef]

27. Holt, E.M.; Steff, E.N.; Moran, A.; Basu, S.; Steinberger, J.; Ross, J.A.; Hong, C.P.; Sinaiko, A.R. Fruit and vegetable comsumption and its relation to markers of inflammation and oxidative stress in adolescents. *J. Am. Diet. Assoc.* **2009**, *109*, 414–421. [CrossRef]

28. Yousuf, B.; Gul, K.; Wani, A.A.; Singh, P. Health benefits of anthocyanins and their encapsulation for potential use in food systems. A review. *Crit. Rev. Food Sci. Nutr.* **2016**, *56*, 2223–2230. [CrossRef]

29. Sun, D.; Huang, S.; Cai, S.; Cao, J.; Han, P. Digestion property and synergistic effect on biological activity of purple rice (*Oryza sativa* L.) anthocyanins subjected to a simulated gastrointestinal digestion in vitro. *Food Res. Int.* **2015**, *78*, 114–123. [CrossRef]

30. Castaneda-Ovando, A.; Pacheco-Hernandez, M.L.; Paez-Hernandez, M.E. Chemical studies of anthocyanins: A review. *Food Chem.* **2009**, *113*, 859–871. [CrossRef]

31. Perez-Vicente, A.; Gil-Izquierdo, A.; Garcia-Viguera, C. In vitro gastrointestinal digestion study of pomegranate juice phenolic compounds, anthocyanins and vitamin C. *J. Agric. Food Chem.* **2002**, *50*, 2308–2312. [CrossRef]

32. Correa-Betanzo, J.; Allen-Vercoe, E.; McDonald, J.; Schroeter, K.; Corredig, N.; Palliyath, G. Stability and biological activity of wild blueberry (*Vaccinium angustifolium*) polyphenols during simulated in vitro gastrointestinal digestion. *Food Chem.* **2014**, *165*, 522–531. [CrossRef] [PubMed]

33. Podsedek, A.; Redzynia, M.; Klewicka, E.; Koziolkiewicz, M. Matrix effects on the stability and antioxidant activity of red cabbage anthocyanins under simulated gastrointestinal digestion. *Biomed. Res. Int.* **2014**, *2014*, 365738. [CrossRef]

© 2019 by the authors. Licensee MDPI, Basel, Switzerland. This article is an open access article distributed under the terms and conditions of the Creative Commons Attribution (CC BY) license (http://creativecommons.org/licenses/by/4.0/).

Article

Phytochemical Composition and Antioxidant Capacity of 30 Chinese Teas

Guo-Yi Tang [1,†], Cai-Ning Zhao [1,†], Xiao-Yu Xu [1], Ren-You Gan [2,*], Shi-Yu Cao [1], Qing Liu [1], Ao Shang [1], Qian-Qian Mao [1] and Hua-Bin Li [1,*]

[1] Guangdong Provincial Key Laboratory of Food, Nutrition and Health, Department of Nutrition, School of Public Health, Sun Yat-Sen University, Guangzhou 510080, China; tanggy5@mail2.sysu.edu.cn (G.-Y.T.); zhaocn@mail2.sysu.edu.cn (C.-N.Z.); xuxy53@mail2.sysu.edu.cn (X.-Y.X.); caoshy3@mail2.sysu.edu.cn (S.-Y.C.); liuq248@mail2.sysu.edu.cn (Q.L.); shangao@mail2.sysu.edu.cn (A.S.); maoqq@mail2.sysu.edu.cn (Q.-Q.M.)

[2] Department of Food Science & Technology, School of Agriculture and Biology, Shanghai Jiao Tong University, Shanghai 200240, China

* Correspondence: renyougan@sjtu.edu.cn (R.-Y.G.); lihuabin@mail.sysu.edu.cn (H.-B.L.); Tel.: +86-21-3420-8533 (R.-Y.G.); +86-20-8733-2391 (H.-B.L.)

† These authors contributed equally to this work.

Received: 31 May 2019; Accepted: 14 June 2019; Published: 18 June 2019

Abstract: Tea has been reported to prevent and manage many chronic diseases, such as cancer, diabetes, obesity, and cardiovascular diseases, and the antioxidant capacity of tea may be responsible for these health benefits. In this study, the antioxidant capacities of fat-soluble, water-soluble, and bound-insoluble fractions of 30 Chinese teas belonging to six categories, namely green, black, oolong, dark, white, and yellow teas, were systematically evaluated, applying ferric-reducing antioxidant power and Trolox equivalent antioxidant capacity assays. In addition, total phenolic contents of teas were determined by Folin–Ciocalteu method, and the contents of 18 main phytochemical compounds in teas were measured by high-performance liquid chromatography (HPLC). The results found that several teas possessed very strong antioxidant capacity, and caffeine, theaflavine, gallic acid, chlorogenic acid, ellagic acid, and kaempferol-3-*O*-glucoside, as well as eight catechins, were the main antioxidant compounds in them. Thus, these teas could be good natural sources of dietary antioxidants, and their extracts might be developed as food additives, nutraceuticals, cosmetics, and pharmaceuticals.

Keywords: tea; *Camellia sinensis*; antioxidant activity; polyphenol; catechin; caffeine; theaflavine

---

1. Introduction

Tea is generally made from the leaves of *Camellia sinensis*, and is a very popular soft drink all over the world. Based on the fermentation degrees in an increasing order, tea can be classified into six categories, including green (unfermented), yellow (slight-fermented), white (mild-fermented), oolong (semi-fermented), black (deep-fermented), and dark (post-fermented) teas [1]. Tea has been widely associated with various health functions, such as the cardiovascular protective, anticancer, antidiabetic, antiobesity, neuroprotective, and hepatoprotective effects [2–10]. These beneficial effects can be mainly attributed to the natural antioxidant phytochemicals in tea, especially polyphenols, which may undergo big differences in different teas with diverse genotypes, maturity, producing areas, or fermentation degrees [11–15].

In the present study, the antioxidant capacity of 30 tea samples, which are the best-selling and most commonly consumed teas in China, were systematically evaluated. In addition, their total phenolic contents were determined, and main antioxidant phytochemicals were identified and quantified by

HPLC. The results should be helpful for the development of tea-based products, such as food additives, cosmetics, nutraceuticals, and pharmaceuticals.

2. Materials and Methods

2.1. Chemicals

The 2,2′-azinobis(3-ethylbenothiazoline-6-sulphonic acid) diammonium salt (ABTS), 6-hydroxy-2,5,7,8-tetramethylchromane-2-carboxylic acid (Trolox), 2,4,6-tri(2-pyridyl)-s-triazine (TPTZ), and Folin–Ciocalteu's phenol reagent were purchased from Sigma-Aldrich (St. Louis, MO, USA). The 18 standard compounds for high-performance liquid chromatography (HPLC) analysis were obtained from Derick Biotechnology Co., Ltd. (Chengdu, China). Tetrahydrofuran, methanol, formic acid, diethyl ether, and ethyl acetate were acquired from Kermel Chemical Factory (Tianjin, China). Acetic acid, sodium acetate, sodium hydroxide, hydrochloric acid, ethylenediaminetetraacetic acid, ascorbic acid, iron(III) chloride hexahydrate, iron(II) sulphate heptahydrate, potassium persulphate, sodium carbonate, ethanol, and n-hexane were bought from Damao Chemical Factory (Tianjin, China). All the reagents were of analytical or HPLC grade, and deionized water was used for all experiments.

2.2. Sample Preparation

The 30 common Chinese teas are widely consumed and famous in China. The basic information of these teas is presented in Table 1. The fat-soluble, water-soluble, and bound-insoluble fractions of these teas were acquired using tetrahydrofuran, methanol-acetic acid-water (50:3.7:46.3, *v/v/v*), and diethyl ether-ethyl acetate (1:1, *v/v*) with alkaline digestion according to the procedures described in the literature [16–18]. All extracts were preserved at −20 °C before being subjected to relevant tests.

2.3. Ferric-Reducing Antioxidant Power (FRAP) Assay

The ferric-reducing antioxidant power (FRAP) assay was conducted according to the method established by Benzie and Strain [19]. $FeSO_4$ was used as the standard, and the value was presented as μmol Fe(II)/g dry weight (DW) of the tea.

2.4. Trolox Equivalent Antioxidant Capacity (TEAC) Assay

The Trolox equivalent antioxidant capacity (TEAC) assay was performed based on the procedure reported by Re et al. [20]. Trolox was applied as the standard, and the value was displayed as μmol Trolox/g DW of the tea.

2.5. Determination of Total Phenolic Content (TPC)

Determination of total phenolic content (TPC) was carried out as described by Singleton et al. [21]. Gallic acid was adopted as the standard, and the value was described as mg gallic acid equivalent (mg GAE)/g DW of the tea.

2.6. Detection of Phytochemicals by High-Performance Liquid Chromatography (HPLC)

Caffeine, theaflavine, and polyphenols in the extracts were detected by HPLC based on the literature reported by Cai et al. [22] with small alterations. Briefly, the testing system was comprised of a Waters (Milford, MA, USA) 1525 binary HPLC pump separation module with an auto-injector, a Waters 2996 photodiode array detector (PDAD) and an Agilent Zorbax Extend-C18 column (250 × 4.6 mm, 5 μm, Santa Clara, CA, USA). Gradient elution was performed at 35 °C with the mobile phase composed of methanol (solution A) and 0.1% formic acid solution (solution B), which were routinely delivered at a flow rate of 1.0 mL/min according to the procedure: 0 min, 5% (A); 10 min, 20% (A); 15 min, 22% (A); 20 min, 25% (A); 40 min, 40% (A); 50 min, 42% (A); 60 min, 50% (A); 70 min, 95% (A); 70.10 min, 5% (A); 75 min, 5% (A). A 20 μL of extract sample was injected for HPLC analysis. The spectra were recorded between 200 and 600 nm, and the targeted compounds were identified by retention time and UV-Vis spectra in comparison with the standards and quantified based on the peak area under the maximum absorption wavelength. The value was expressed as mg/g DW of the tea.

Table 1. Basic information about the 30 Chinese teas.

No.	Name	Category	Fermentation Degree	Production Place
1	Dianhong Congou Black Tea	Black tea	Deep-fermented	Kunming, Yunnan
2	Keemun Black Tea	Black tea	Deep-fermented	Qimen, Anhui
3	Lapsang Souchong Black Tea	Black tea	Deep-fermented	Wuyishan, Fujian
4	Yichang Congou Black Tea	Black tea	Deep-fermented	Yichang, Hubei
5	Fuzhuan Brick Tea	Dark tea	Post-fermented	Anhua, Hubei
6	Liupao Tea	Dark tea	Post-fermented	Wuzhou, Guangxi
7	Pu-erh Tea	Dark tea	Post-fermented	Pu'er, Yunnan
8	Qingzhuan Brick tea	Dark tea	Post-fermented	Chibi, Hubei
9	Tibetan Tea	Dark tea	Post-fermented	Ya'an, Sichuan
10	Dianqing Tea	Green tea	Unfermented	Kunming, Yunnan
11	Dongting Biluochun Tea	Green tea	Unfermented	Suzhou, Jiangsu
12	Duyun Maojian Tea	Green tea	Unfermented	Duyun, Guizhou
13	Enshi Yulu Tea	Green tea	Unfermented	Enshi, Hubei
14	Lu'an Guapian Tea	Green tea	Unfermented	Lu'an, Anhui
15	Lushan Yunwu Tea	Green tea	Unfermented	Jiujiang, Jiangxi
16	Taiping Houkui Tea	Green tea	Unfermented	Huangshan, Anhui
17	Xihu Longjing Tea	Green tea	Unfermented	Hangzhou, Zhejiang
18	Yongxi Huoqing Tea	Green tea	Unfermented	Jingxian, Anhui
19	Fenghuang Shuixian Tea	Oolong tea	Semi-fermented	Chao'an, Guangdong
20	Luohan Chenxiang Tea	Oolong tea	Semi-fermented	Mengdingshan, Sichuan
21	Tieguanyin Tea	Oolong tea	Semi-fermented	Anxi, Fujian
22	Wuyi Rock Tea	Oolong tea	Semi-fermented	Wuyishan, Fujian
23	Gongmei White Tea	White tea	Mild-fermented	Nanping, Fujian
24	Shoumei White Tea	White tea	Mild-fermented	Nanping, Fujian
25	White Peony Tea	White tea	Mild-fermented	Nanping, Fujian
26	Huoshan Large Yellow Tea	Yellow tea	Light-fermented	Lu'an, Anhui
27	Junshan Yinzhen Tea	Yellow tea	Light-fermented	Yueyang, Hunan
28	Mengding Huangya Tea	Yellow tea	Light-fermented	Mengdingshan, Sichuan
29	Weishan Maojian Tea	Yellow tea	Light-fermented	Ningxiang, Hunan
30	Yuan'an Luyuan Tea	Yellow tea	Light-fermented	Yichang, Hubei

2.7. Data Analysis

All tests were performed in triplicate, and the values were expressed as mean ± standard deviation (SD). SPSS 22 (International Business Machines Corp, Armonk, NY, USA) and Excel 2007 (Microsoft Corporation, Redmond, WA, USA) were applied for data analysis. One-way ANOVA and post hoc Tukey test were performed to compare means of more than two samples, and p value less than 0.05 was defined as statistical significance.

3. Results

3.1. Ferric-Reducing Antioxidant Power (FRAP) Values of the Tested Teas

FRAP value was used as an important indicator for the antioxidant capacity with regard to reducing ferric ions to ferrous ions, and FRAP results of the 30 teas are displayed in Table 2. The total FRAP values ranged from 611.18 ± 5.09 to 5375.18 ± 228.43 μmol Fe (II)/g DW with a 9-fold difference. Dianqing Tea, Xihu Longjing Tea, Dongting Biluochun Tea, Yongxi Huoqing Tea, and Duyun Maojian Tea exerted the top five reducing capacities, namely 5375.18 ± 228.43, 3926.32 ± 56.00, 3845.21 ± 44.17, 3752.52 ± 96.75, and 3664.97 ± 53.33, μmol Fe(II)/g DW, respectively. Tibetan Tea exhibited the lowest reducing ability of 611.18 ± 5.09 μmol Fe(II)/g DW. In addition, according to the statistical description and non-parametric tests (Table 3), the FRAP values for the three fractions met the following order: water-soluble > bound-insoluble > fat-soluble.

Table 2. Ferric-reducing antioxidant power (FRAP) values of the 30 common Chinese teas.

No.	Name	Category	FRAP Value (μmol Fe(II)/g DW)				Mean ± SD of Categories
			Fat-Soluble Fraction	Water-Soluble Fraction	Bound-Insoluble Fraction	Total	
1	Dianhong Congou Black Tea	Black tea	11.53 ± 0.27	1142.49 ± 5.55	115.29 ± 0.54	1269.31 ± 5.88	
2	Keemun Black Tea	Black tea	5.20 ± 0.17	1034.93 ± 36.25	88.41 ± 1.93	1128.53 ± 34.52	1141.58 ± 13.92 [a]
3	Lapsang Souchong Black Tea	Black tea	15.53 ± 0.58	585.60 ± 7.42	78.32 ± 2.79	679.45 ± 8.29	
4	Yichang Congou Black Tea	Black tea	4.56 ± 0.43	1371.38 ± 5.05	113.07 ± 1.53	1489.01 ± 6.98	
5	Fuzhuan Brick Tea	Dark tea	65.53 ± 1.89	2318.76 ± 77.12	100.66 ± 3.01	2484.94 ± 80.20	
6	Liupao Tea	Dark tea	5.01 ± 0.29	922.49 ± 12.03	62.13 ± 0.70	989.62 ± 11.40	
7	Pu-erh Tea	Dark tea	5.10 ± 0.11	725.60 ± 18.04	72.77 ± 1.16	803.46 ± 16.99	1124.96 ± 23.87 [a]
8	Qingzhuan Brick tea	Dark tea	38.20 ± 0.73	648.71 ± 6.16	48.68 ± 0.96	735.59 ± 5.64	
9	Tibetan Tea	Dark tea	31.31 ± 0.69	534.04 ± 4.68	45.82 ± 0.59	611.18 ± 5.09	
10	Dianqing Tea	Green tea	250.73 ± 8.97	4937.24 ± 236.20	187.20 ± 0.88	5375.18 ± 228.43	
11	Dongting Biluochun Tea	Green tea	230.51 ± 10.31	3515.73 ± 37.33	98.96 ± 1.56	3845.21 ± 44.17	
12	Duyun Maojian Tea	Green tea	71.31 ± 1.58	3442.84 ± 50.69	150.81 ± 4.30	3664.97 ± 53.33	
13	Enshi Yulu Tea	Green tea	44.26 ± 1.55	3037.51 ± 13.42	179.64 ± 2.28	3261.41 ± 14.50	
14	Lu'an Guapian Tea	Green tea	67.92 ± 1.01	1413.60 ± 43.90	131.76 ± 3.17	1613.28 ± 43.98	3621.75 ± 81.44 [b]
15	Lushan Yunwu Tea	Green tea	93.42 ± 2.68	3396.62 ± 83.02	128.98 ± 3.08	3619.02 ± 83.57	
16	Taiping Houkui Tea	Green tea	139.26 ± 4.23	3272.18 ± 117.50	126.42 ± 2.34	3537.86 ± 112.24	
17	Xihu Longjing Tea	Green tea	196.62 ± 4.99	3650.84 ± 53.42	78.85 ± 5.71	3926.32 ± 56.00	
18	Yongxi Huoqing Tea	Green tea	90.20 ± 1.69	3512.18 ± 96.79	150.14 ± 1.39	3752.52 ± 96.75	
19	Fenghuang Shuixian Tea	Oolong tea	43.40 ± 4.37	2069.87 ± 55.49	144.26 ± 1.51	2257.52 ± 50.42	
20	Luohan Chenxiang Tea	Oolong tea	26.03 ± 0.29	1827.20 ± 12.22	79.77 ± 0.30	1933.00 ± 12.52	2013.37 ± 26.17 [a]
21	Tieguanyin Tea	Oolong tea	84.73 ± 5.69	1911.64 ± 10.10	59.82 ± 0.39	2056.20 ± 4.96	
22	Wuyi Rock Tea	Oolong tea	20.62 ± 1.40	1696.53 ± 34.87	89.60 ± 1.81	1806.75 ± 36.77	
23	Gongmei White Tea	White tea	31.03 ± 1.69	1149.16 ± 16.67	81.63 ± 1.86	1261.82 ± 18.31	
24	Shoumei White Tea	White tea	28.09 ± 1.25	934.49 ± 28.98	82.85 ± 1.39	1045.43 ± 29.22	1093.64 ± 16.82 [a]
25	White Peony Tea	White tea	32.37 ± 0.76	849.16 ± 2.04	92.16 ± 1.47	973.68 ± 2.92	
26	Huoshan Large Yellow Tea	Yellow tea	77.64 ± 2.28	2539.20 ± 14.11	89.24 ± 0.50	2706.08 ± 15.16	
27	Junshan Yinzhen Tea	Yellow tea	144.81 ± 3.73	3137.42 ± 26.71	84.27 ± 1.97	3366.50 ± 21.49	
28	Mengding Huangya Tea	Yellow tea	112.03 ± 5.95	3173.87 ± 16.22	68.02 ± 1.80	3353.92 ± 10.48	3182.34 ± 31.31 [b]
29	Weishan Maojian Tea	Yellow tea	82.53 ± 4.64	2862.76 ± 80.24	98.46 ± 1.17	3043.75 ± 78.43	
30	Yuan'an Luyuan Tea	Yellow tea	93.31 ± 3.08	3254.40 ± 29.70	93.74 ± 0.19	3441.45 ± 30.99	

Abbreviations: FRAP, ferric-reducing antioxidant power; DW, dry weight; SD, standard deviation. Different superscript lowercase letters ([a,b]) indicated statistical significance ($p < 0.05$).

Table 3. Comparison among FRAP, TEAC, and TPC of different tea fractions.

Index	Fraction	Statistical Description						p Value by Non-Parametric Test			
		MIN	Q_L	M	Q_U	MAX	Item	M-W	Moses	K-S	W-W
FRAP	1	4.56	24.68	54.89	93.34	250.73	a	<0.001	<0.001	<0.001	<0.001
	2	534.04	1009.82	1990.76	3258.84	4937.24	b	<0.001	<0.001	<0.001	<0.001
	3	45.82	78.72	90.88	127.06	187.20	c	=0.003	=1.000	=0.001	=0.025
TEAC	1	9.25	24.17	45.53	66.21	137.97	a	<0.001	<0.001	<0.001	<0.001
	2	277.41	568.63	1295.08	1881.65	2754.98	b	<0.001	<0.001	<0.001	<0.001
	3	34.34	51.52	61.42	77.67	111.45	c	=0.005	=1.000	=0.016	0.181
TPC	1	1.56	3.47	4.75	6.50	11.26	a	<0.001	<0.001	<0.001	<0.001
	2	32.26	81.37	149.58	190.77	236.50	b	<0.001	<0.001	<0.001	<0.001
	3	3.08	5.36	6.43	8.45	11.04	c	=0.006	=1.000	=0.016	=0.347

Note: 1, fat-soluble fraction; 2, water-soluble fraction; 3, bound-insoluble fraction; a, non-parametric test between 1 and 2; b, non-parametric test between 2 and 3; c, non-parametric test between 1 and 3; FRAP, ferric-reducing antioxidant power; K-S, Kolmogorov-Smirnov Test; M, median; MAX, the maximum value; MIN, the minimum value; Moses, Moses Test; M-W, Mann-Whitney test; Q_L, the lower quartile; Q_U, the upper quartile; TEAC, Trolox equivalent antioxidant capacity; TPC, total phenolic content; W-W, Wald-Wolfowitz Test.

3.2. Trolox Equivalent Antioxidant Capacity (TEAC) Values of the Tested Teas

TEAC value was an important index of free radical-scavenging capacity, and the results of the 30 Chinese teas are presented in Table 4. The total TEAC values varied from 326.32 ± 0.48 to 3004.40 ± 112.89 μmol Trolox/g DW with a 9-fold difference. Dianqing Tea, Junshan Yinzhen Tea, Mengding Huangya Tea, Weishan Maojian Tea, and Xihu Longjing Tea showed the top five free radical-scavenging capacities, namely, 3004.40 ± 112.89, 2418.71 ± 26.70, 2303.72 ± 53.67, 2250.40 ± 37.95 and 2125.92 ± 44.43 μmol Trolox/g DW, respectively. Tibetan Tea had the lowest free radical-scavenging capacity of 326.32 ± 0.48 μmol Trolox/g DW. Moreover, according to the statistical description and non-parametric tests (Table 3), TEAC values for the three fractions met the following order: water-soluble > bound-insoluble > fat-soluble bound.

Table 4. Trolox equivalent antioxidant capacity (TEAC) values of the 30 common Chinese teas.

No.	Name	Category	TEAC Value (μmol Trolox/g DW)				Mean ± SD of Categories
			Fat-Soluble Fraction	Water-Soluble Fraction	Bound-Insoluble Fraction	Total	
1	Dianhong Congou Black Tea	Black tea	44.84 ± 1.38	627.92 ± 6.80	78.68 ± 1.19	751.44 ± 7.20	
2	Keemun Black Tea	Black tea	26.43 ± 0.70	570.24 ± 17.10	64.24 ± 1.54	660.91 ± 15.26	724.28 ± 12.63 [a]
3	Lapsang Souchong Black Tea	Black tea	53.14 ± 2.44	355.48 ± 8.80	56.04 ± 1.92	464.67 ± 8.89	
4	Yichang Congou Black Tea	Black tea	24.12 ± 0.77	918.66 ± 18.82	77.34 ± 1.26	1020.11 ± 19.17	
5	Fuzhuan Brick Tea	Dark tea	40.58 ± 0.66	1172.97 ± 21.52	66.19 ± 1.19	1279.74 ± 21.17	
6	Liupao Tea	Dark tea	22.90 ± 1.29	460.32 ± 9.97	48.23 ± 1.04	531.45 ± 9.97	
7	Pu-erh Tea	Dark tea	19.45 ± 1.11	338.44 ± 5.76	48.13 ± 1.11	406.02 ± 6.98	589.43 ± 8.35 [a]
8	Qingzhuan Brick tea	Dark tea	17.85 ± 1.19	351.43 ± 3.59	34.34 ± 0.28	403.62 ± 3.15	
9	Tibetan Tea	Dark tea	9.25 ± 0.40	277.41 ± 0.00	39.66 ± 0.45	326.32 ± 0.48	
10	Dianqing Tea	Green tea	137.97 ± 3.13	2754.98 ± 113.83	111.45 ± 2.50	3004.40 ± 112.89	
11	Dongting Biluochun Tea	Green tea	127.03 ± 3.43	1883.90 ± 11.31	64.02 ± 0.88	2074.94 ± 9.06	
12	Duyun Maojian Tea	Green tea	41.05 ± 1.48	1853.93 ± 27.46	102.89 ± 1.82	1997.86 ± 25.04	
13	Enshi Yulu Tea	Green tea	29.42 ± 0.49	1639.68 ± 27.34	109.31 ± 4.09	1778.42 ± 30.71	
14	Lu'an Guapian Tea	Green tea	42.57 ± 0.31	635.75 ± 11.53	75.13 ± 0.34	753.45 ± 11.73	1964.50 ± 38.29 [b]
15	Lushan Yunwu Tea	Green tea	47.95 ± 0.54	1820.97 ± 44.12	79.77 ± 1.11	1948.69 ± 45.05	
16	Taiping Houkui Tea	Green tea	74.28 ± 0.43	1808.98 ± 26.34	77.21 ± 0.77	1960.47 ± 26.50	
17	Xihu Longjing Tea	Green tea	117.50 ± 4.53	1955.81 ± 40.78	52.61 ± 0.44	2125.92 ± 44.43	
18	Yongxi Huoqing Tea	Green tea	48.85 ± 1.35	1880.90 ± 38.23	106.61 ± 2.24	2036.36 ± 39.18	
19	Fenghuang Shuixian Tea	Oolong tea	38.72 ± 2.22	1541.83 ± 16.35	80.07 ± 0.70	1660.62 ± 18.36	
20	Luohan Chenxiang Tea	Oolong tea	73.86 ± 2.98	1376.32 ± 22.44	47.53 ± 4.07	1497.72 ± 23.35	1460.46 ± 22.23 [b]
21	Tieguanyin Tea	Oolong tea	61.30 ± 1.59	1311.56 ± 35.34	40.75 ± 1.41	1413.61 ± 33.12	
22	Wuyi Rock Tea	Oolong tea	66.52 ± 2.19	1141.73 ± 17.45	61.64 ± 1.19	1269.89 ± 14.09	
23	Gongmei White Tea	White tea	21.85 ± 0.50	628.51 ± 21.09	57.27 ± 0.21	707.63 ± 20.78	
24	Shoumei White Tea	White tea	20.03 ± 1.33	563.80 ± 7.43	57.18 ± 1.13	641.01 ± 7.33	629.61 ± 10.92 [a]
25	White Peony Tea	White tea	24.18 ± 0.23	452.93 ± 4.48	63.08 ± 0.65	540.19 ± 4.64	
26	Huoshan Large Yellow Tea	Yellow tea	46.21 ± 1.06	1278.60 ± 27.90	54.92 ± 2.72	1379.72 ± 29.46	
27	Junshan Yinzhen Tea	Yellow tea	80.05 ± 0.63	2278.99 ± 27.87	59.67 ± 1.68	2418.71 ± 26.70	
28	Mengding Huangya Tea	Yellow tea	66.10 ± 1.29	2192.70 ± 53.25	44.92 ± 1.04	2303.72 ± 53.67	2087.81 ± 42.89 [b]
29	Weishan Maojian Tea	Yellow tea	50.07 ± 1.43	2139.15 ± 38.99	61.19 ± 2.41	2250.40 ± 37.95	
30	Yuan'an Luyuan Tea	Yellow tea	56.30 ± 0.50	1971.03 ± 64.83	59.16 ± 1.42	2086.49 ± 66.69	

Abbreviations: DW, dry weight; SD, standard deviation; TEAC, Trolox equivalent antioxidant capacity. Different superscript lowercase letters ([a,b]) indicated statistical significance ($p < 0.05$).

3.3. Total Phenolic Content (TPC) of the Tested Teas

TPC was adopted to measure the total contents of phenolic compounds in the 30 Chinese teas, and the results are shown in Table 5. Briefly, the range of total TPC values was 37.25 ± 0.16 to 254.29 ± 15.51 mg GAE/g DW with a 7-fold difference. Dianqing Tea, Xihu Longjing Tea, Junshan Yinzhen Tea, Dongting Biluochun Tea, and Yuan'an Luyuan Tea possessed the top five total phenolic contents, namely 254.29 ± 15.51, 215.39 ± 11.87, 214.72 ± 3.22, 211.20 ± 2.52, and 210.05 ± 7.84 mg GAE/g DW, respectively. Tibetan Tea was observed with the lowest TPC of 37.25 ± 0.16 mg GAE/g DW. In addition, based on the statistical description and non-parametric tests (Table 3), the TPC values for the three fractions met the following order: water-soluble > bound-insoluble > fat-soluble.

3.4. Correlations among Ferric-Reducing Antioxidant Power (FRAP), Trolox Equivalent Antioxidant Capacity (TEAC), and Total Phenolic Content (TPC) Values

The correlations among FRAP, TEAC, and TPC values (based on the total values of three fractions) were determined by the simple linear regression model, and the results are presented in Figure 1. Both FRAP and TEAC values were significantly and positively correlated with TPC ($R^2 = 0.883$, $p < 0.001$ and $R^2 = 0.941$, $p < 0.001$, respectively). These results suggested that the phenolic compounds could be the main components contributing to the antioxidant activities of tea. In addition, FRAP values were positively and remarkably correlated with TEAC values ($R^2 = 0.928$, $p < 0.001$). Therefore, the antioxidants in tea could possess multiple functions regarding reducing oxidants (like Fe(III)) and scavenging free radicals (like ABTS$\bullet^+$).

Table 5. Total phenolic content (TPC) values of the 30 common Chinese teas.

No.	Name	Category	TPC Value (mg GAE/g DW)				Mean ± SD of Categories
			Fat-Soluble Fraction	Water-Soluble Fraction	Bound-Insoluble Fraction	Total	
1	Dianhong Congou Black Tea	Black tea	6.43 ± 0.55	106.68 ± 1.26	8.86 ± 0.06	121.97 ± 1.31	
2	Keemun Black Tea	Black tea	4.72 ± 0.08	96.26 ± 3.42	6.47 ± 0.03	107.44 ± 3.43	107.54 ± 1.80 [a]
3	Lapsang Souchong Black Tea	Black tea	2.31 ± 0.05	62.51 ± 1.65	6.29 ± 0.10	71.11 ± 1.63	
4	Yichang Congou Black Tea	Black tea	3.86 ± 0.23	117.54 ± 0.84	8.25 ± 0.01	129.64 ± 0.85	
5	Fuzhuan Brick Tea	Dark tea	4.21 ± 0.18	141.06 ± 3.39	6.89 ± 0.07	152.17 ± 3.57	
6	Liupao Tea	Dark tea	4.31 ± 0.13	74.28 ± 1.38	5.74 ± 0.06	84.33 ± 1.37	
7	Pu-erh Tea	Dark tea	4.10 ± 0.10	59.52 ± 0.72	4.79 ± 0.05	68.40 ± 0.81	78.16 ± 1.33 [a]
8	Qingzhuan Brick tea	Dark tea	2.37 ± 0.06	42.95 ± 0.77	3.34 ± 0.05	48.66 ± 0.75	
9	Tibetan Tea	Dark tea	1.91 ± 0.07	32.26 ± 0.20	3.08 ± 0.12	37.25 ± 0.16	
10	Dianqing Tea	Green tea	7.23 ± 0.22	236.50 ± 15.52	10.56 ± 0.21	254.29 ± 15.51	
11	Dongting Biluochun Tea	Green tea	6.66 ± 0.20	198.52 ± 2.45	6.02 ± 0.11	211.20 ± 2.52	
12	Duyun Maojian Tea	Green tea	4.12 ± 0.04	191.19 ± 2.53	9.71 ± 0.23	205.02 ± 2.74	
13	Enshi Yulu Tea	Green tea	2.34 ± 0.16	162.53 ± 3.00	11.04 ± 0.87	175.91 ± 2.01	
14	Lu'an Guapian Tea	Green tea	4.03 ± 0.05	81.52 ± 2.34	8.41 ± 0.18	93.96 ± 2.47	195.79 ± 5.45 [b]
15	Lushan Yunwu Tea	Green tea	5.46 ± 0.11	186.77 ± 4.90	8.56 ± 0.19	200.80 ± 4.76	
16	Taiping Houkui Tea	Green tea	8.65 ± 0.06	179.88 ± 1.80	8.21 ± 0.18	196.74 ± 1.82	
17	Xihu Longjing Tea	Green tea	5.90 ± 0.29	204.88 ± 11.38	4.61 ± 0.20	215.39 ± 11.87	
18	Yongxi Huoqing Tea	Green tea	5.40 ± 0.02	193.67 ± 5.26	9.75 ± 0.11	208.83 ± 5.32	
19	Fenghuang Shuixian Tea	Oolong tea	3.84 ± 0.31	188.80 ± 2.08	8.73 ± 0.28	201.36 ± 1.74	
20	Luohan Chenxiang Tea	Oolong tea	11.26 ± 0.54	150.49 ± 1.77	5.19 ± 0.10	166.94 ± 1.64	
21	Tieguanyin Tea	Oolong tea	6.44 ± 0.51	148.66 ± 1.82	3.81 ± 0.08	158.91 ± 2.17	172.11 ± 2.09 [b]
22	Wuyi Rock Tea	Oolong tea	10.67 ± 0.20	144.46 ± 2.89	6.10 ± 0.05	161.23 ± 2.83	
23	Gongmei White Tea	White tea	1.57 ± 0.04	90.11 ± 0.57	6.59 ± 0.37	98.28 ± 0.92	
24	Shoumei White Tea	White tea	1.56 ± 0.08	80.90 ± 0.81	6.40 ± 0.07	88.87 ± 0.93	84.39 ± 0.79 [a]
25	White Peony Tea	White tea	1.73 ± 0.04	57.25 ± 0.45	7.06 ± 0.07	66.04 ± 0.51	
26	Huoshan Large Yellow Tea	Yellow tea	4.79 ± 0.17	156.15 ± 1.96	5.42 ± 0.10	166.35 ± 1.93	
27	Junshan Yinzhen Tea	Yellow tea	9.35 ± 0.16	199.41 ± 3.27	5.97 ± 0.04	214.72 ± 3.22	
28	Mengding Huangya Tea	Yellow tea	6.82 ± 0.15	190.63 ± 3.98	4.22 ± 0.15	201.67 ± 3.82	198.44 ± 5.39 [b]
29	Weishan Maojian Tea	Yellow tea	4.83 ± 0.28	187.72 ± 10.14	6.87 ± 0.01	199.42 ± 10.13	
30	Yuan'an Luyuan Tea	Yellow tea	5.56 ± 0.10	198.52 ± 7.76	5.97 ± 0.08	210.05 ± 7.84	

Abbreviations: DW, dry weight; GAE, gallic acid equivalent; SD, standard deviation; TPC, total phenolic content. Different superscript lowercase letters ([a,b]) indicated statistical significance ($p < 0.05$).

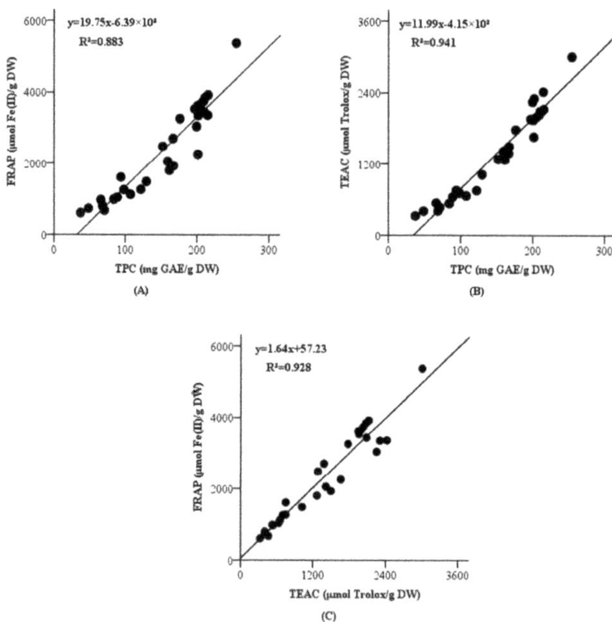

Figure 1. Correlations among FRAP and TPC values (**A**), TEAC and TPC values (**B**), FRAP and TEAC values (**C**). Abbreviations: FRAP, ferric-reducing antioxidant power; TEAC, Trolox equivalent antioxidant capacity; TPC, total phenolic content.

3.5. Systematic Cluster of the Tested Teas

Based on the FRAP, TEAC, and TPC values, a systematic cluster analysis for the 30 teas was conducted with cluster numbers from 2 to 6, and the results are summarized in Figure 2. After that, the outcomes of cluster number = 4 were further analyzed using Online Analytical Processing (OLAP) accompanied with variance analysis (ANOVA), and the results are presented in Table 6. In detail, cluster 1 contained 12 teas, which were 4 black teas, 4 dark teas, 3 white teas, and 1 green tea, with the lowest values for FRAP, TEAC, and TPC (1050.03 ± 317.40 µmol Fe(II)/g DW, 600.57 ± 194.85 µmol Trolox/g DW, and 84.66 ± 27.90 mg GAE/g DW, respectively). In addition, cluster 2 comprised all the 4 oolong teas, 1 dark tea, and 1 yellow tea, with relatively low values for FRAP, TEAC, and TPC (2207.42 ± 342.61 µmol Fe(II)/g DW, 1416.88 ± 146.80 µmol Trolox/g DW and 167.83 ± 17.30 mg GAE/g DW, respectively). Moreover, cluster 3 consisted of 7 green teas and 4 yellow teas, with apparently high values for FRAP, TEAC, and TPC (3528.45 ± 265.76 µmol Fe(II)/g DW, 2089.27 ± 180.60 µmol Trolox/g DW, and 203.61 ± 11.11 mg GAE/g DW, respectively). Furthermore, cluster 4 included only 1 tea (green tea), with FRAP, TEAC, and TPC values the highest (5375.18 ± 228.43 µmol Fe(II)/g DW, 3004.40 ± 112.89 µmol Trolox/g DW, and 254.29 ± 15.51 mg GAE/g DW, respectively). Based on the result of ANOVA, all of the differences among the 4 clusters regarding FRAP, TEAC, and TPC values were significant (all $p < 0.001$).

Table 6. Online Analytical Processing (OLAP) Cube based on systematic cluster analysis for 30 Chinese teas (cluster number = 4).

Average Linkage		FRAP	TEAC	TPC
	SUM	12600.37	7206.81	1015.94
	N	12	12	12
1	Mean	1050.03	600.57	84.66
	SD	317.40	194.85	27.90
	SUM/SUMT (%)	18.0%	17.3%	22.5%
	N/NT (%)	40.0%	40.0%	40.0%
	SUM	13244.50	8501.30	1006.96
	N	6	6	6
2	Mean	2207.42	1416.88	167.83
	SD	342.61	146.80	17.30
	SUM/SUMT (%)	18.9%	20.4%	22.3%
	N/NT (%)	20.0%	20.0%	20.0%
	SUM	38812.92	2298200	2239.75
	N	11	11	11
3	Mean	3528.45	2089.27	203.61
	SD	265.76	180.60	11.11
	SUM/SUMT (%)	55.4%	55.1%	49.6%
	N/NT (%)	36.7%	36.7%	36.7%
	SUM	5375.18	3004.40	254.29
	N	1	1	1
4	Mean	5375.18	3004.40	254.29
	SD	NA	NA	NA
	SUM/SUMT (%)	7.7%	7.2%	5.6%
	N/NT (%)	3.3%	3.3%	3.3%
	SUM	70032.96	41694.52	4516.94
	N	30	30	30
Total	Mean	2334.43	1389.82	150.57
	SD	1276.08	750.16	60.72

Abbreviations: FRAP, ferric-reducing antioxidant power; TEAC, Trolox equivalent antioxidant capacity; TPC, total phenolic content.

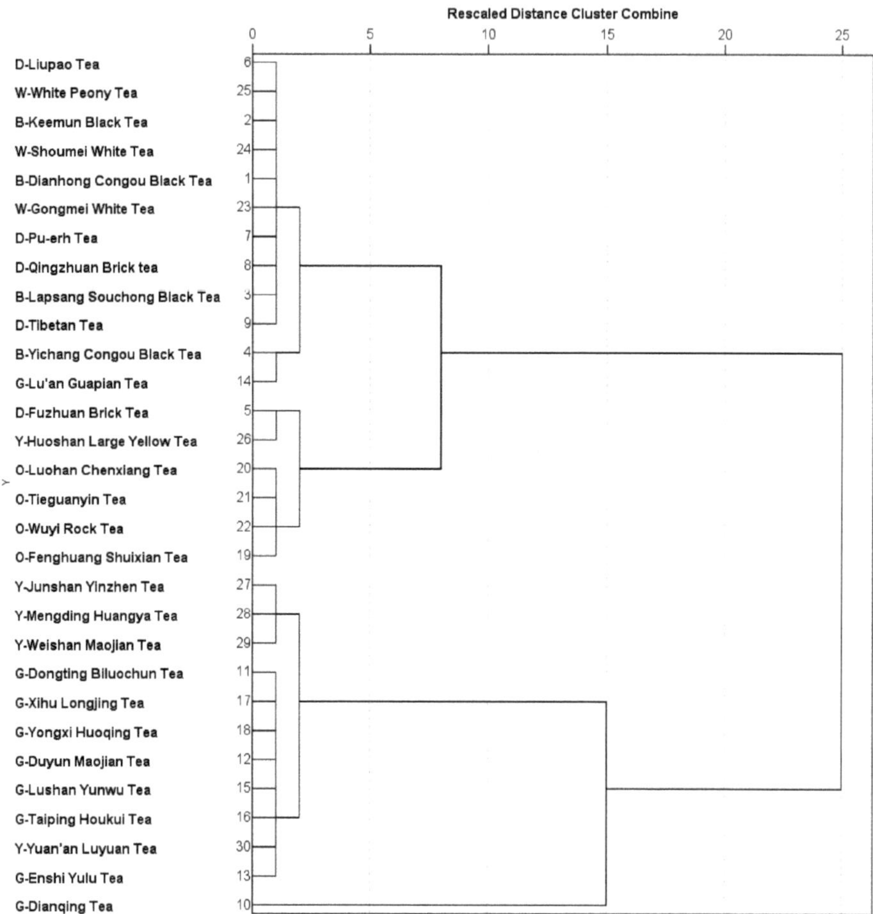

Figure 2. Dendrogram using average linkage (between groups) from systematic cluster analysis for 30 Chinese teas. B, black tea; D, dark tea; G, green tea; O, oolong tea; W, white tea; Y, yellow tea.

3.6. Contents of Phytochemical Compounds in Teas

Main phytochemicals in 30 Chinese teas, including main catechins (Table 7), caffeine, theaflavine, and other polyphenols (Table 8), were determined by HPLC-PDAD. The chromatograms of the mixed standards and the samples of Dianqing Tea and Tibetan Tea under 254 nm are shown in Figure 3. Eight catechins, including catechin, epicatechin, gallocatechin, epigallocatechin, catechin gallate, epicatechin gallate, gallocatechin gallate, epigallocatechin gallate, four other phenolic compounds, including gallic acid, chlorogenic acid, ellagic acid, and kaempferol-3-*O*-glucoside, caffeine, and theaflavine, were identified in the 30 teas, with epigallocatechin gallate, gallic acid, and caffeine detected and quantified in all Chinese tea samples.

Table 7. The contents (mg/g DW) of catechins in 30 Chinese teas.

No.	Name	Category	Catechin	Epicatechin	Gallocatechin	Epigallocatechin	Catechin Gallate	Epicatechin Gallate	Gallocatechin Gallate	Epigallocatechin Gallate
1	Dianhong Congou Black Tea	Black Tea	-	0.796 ± 0.047	1.098 ± 0.052	8.479 ± 0.500	-	2.583 ± 0.077	-	0.539 ± 0.013
2	Keemun Black Tea	Black Tea	-	0.477 ± 0.030	-	-	-	1.499 ± 0.033	-	2.164 ± 0.102
3	Lapsang Souchong Black Tea	Black Tea	-	-	-	-	-	-	-	0.761 ± 0.031
4	Yichang Congou Black Tea	Black Tea	-	0.740 ± 0.033	-	-	-	3.511 ± 0.070	0.510 ± 0.013	3.795 ± 0.089
5	Fuzhuan Brick Tea	Dark Tea	4.930 ± 0.240	10.357 ± 0.268	5.535 ± 0.128	23.430 ± 0.375	-	10.881 ± 0.105	0.933 ± 0.063	10.885 ± 0.259
6	Liupao Tea	Dark Tea	1.667 ± 0.063	3.886 ± 0.112	2.120 ± 0.150	5.440 ± 0.171	-	0.455 ± 0.037	-	0.647 ± 0.015
7	Pu-erh Tea	Dark Tea	-	1.574 ± 0.086	-	-	-	0.515 ± 0.040	-	0.584 ± 0.044
8	Qingzhuan Brick tea	Dark Tea	-	0.977 ± 0.056	1.062 ± 0.049	5.200 ± 0.140	-	0.542 ± 0.015	-	1.479 ± 0.083
9	Tibetan Tea	Dark Tea	-	-	-	2.288 ± 0.050	-	-	-	0.886 ± 0.027
10	Dianqing Tea	Green Tea	1.315 ± 0.084	5.970 ± 0.210	1.864 ± 0.080	13.094 ± 0.256	-	35.395 ± 0.568	-	59.354 ± 1.131
11	Dongting Biluochun Tea	Green Tea	0.988 ± 0.039	6.310 ± 0.272	1.824 ± 0.051	24.522 ± 0.060	-	27.893 ± 0.426	0.630 ± 0.026	43.070 ± 0.209
12	Duyun Maojian Tea	Green Tea	-	8.700 ± 0.429	2.814 ± 0.167	42.063 ± 0.126	-	18.443 ± 0.537	1.137 ± 0.062	43.056 ± 0.455
13	Enshi Yulu Tea	Green Tea	-	6.443 ± 0.166	2.135 ± 0.140	29.070 ± 0.484	-	16.774 ± 0.090	-	33.102 ± 0.594
14	Lu'an Guapian Tea	Green Tea	-	7.352 ± 0.147	3.015 ± 0.121	100.684 ± 0.561	-	7.599 ± 0.119	0.842 ± 0.044	40.161 ± 0.887
15	Lushan Yunwu Tea	Green Tea	-	6.377 ± 0.150	3.277 ± 0.150	53.447 ± 0.326	-	15.130 ± 0.431	-	48.272 ± 0.363
16	Taiping Houkui Tea	Green Tea	-	8.580 ± 0.211	3.121 ± 0.092	74.212 ± 0.226	-	11.264 ± 0.097	0.640 ± 0.020	45.016 ± 0.222
17	Xihu Longjing Tea	Green Tea	-	5.380 ± 0.216	4.002 ± 0.112	24.494 ± 0.467	0.645 ± 0.036	22.364 ± 0.869	5.844 ± 0.173	51.734 ± 0.240
18	Yongxi Huoqing Tea	Green Tea	-	6.260 ± 0.303	2.630 ± 0.165	38.486 ± 0.994	-	18.064 ± 0.181	0.601 ± 0.019	50.947 ± 0.396
19	Fenghuang Shuixian Tea	Oolong Tea	-	1.579 ± 0.089	2.509 ± 0.121	31.253 ± 0.206	-	8.435 ± 0.270	-	36.704 ± 0.362
20	Luohan Chenxiang Tea	Oolong Tea	-	7.531 ± 0.017	8.088 ± 0.092	125.439 ± 0.678	-	3.683 ± 0.102	-	22.396 ± 0.505
21	Tieguanyin Tea	Oolong Tea	0.775 ± 0.052	13.723 ± 0.216	3.938 ± 0.146	139.854 ± 1.075	-	6.471 ± 0.235	0.562 ± 0.022	23.663 ± 0.308
22	Wuyi Rock Tea	Oolong Tea	-	4.337 ± 0.223	11.528 ± 0.079	36.826 ± 0.668	0.981 ± 0.121	5.083 ± 0.122	2.261 ± 0.111	20.211 ± 0.223
23	Gongmei White Tea	White Tea	-	-	-	8.419 ± 0.143	-	3.144 ± 0.123	-	6.010 ± 0.083
24	Shoumei White Tea	White Tea	-	-	-	-	-	2.270 ± 0.062	-	3.537 ± 0.072
25	White Peony Tea	White Tea	-	1.311 ± 0.033	-	-	-	3.841 ± 0.125	-	8.539 ± 0.169
26	Huoshan Large Yellow Tea	Yellow Tea	2.040 ± 0.054	2.956 ± 0.115	11.858 ± 0.039	14.340 ± 0.135	1.608 ± 0.026	5.549 ± 0.059	10.787 ± 0.108	17.209 ± 0.177
27	Junshan Yinzhen Tea	Yellow Tea	1.366 ± 0.043	6.196 ± 0.178	2.736 ± 0.102	13.661 ± 0.196	0.351 ± 0.014	30.491 ± 0.101	1.447 ± 0.066	50.777 ± 0.224
28	Mengding Huangya Tea	Yellow Tea	-	0.968 ± 0.056	4.844 ± 0.064	22.950 ± 0.102	-	23.805 ± 0.075	2.361 ± 0.128	39.125 ± 0.082
29	Weishan Maojian Tea	Yellow Tea	-	10.062 ± 0.040	2.818 ± 0.072	45.484 ± 0.057	-	-	24.710 ± 0.247	32.856 ± 0.060
30	Yuan'an Luyuan Tea	Yellow Tea	-	5.959 ± 0.147	3.918 ± 0.051	19.877 ± 0.176	-	21.373 ± 0.027	1.388 ± 0.043	57.230 ± 0.253

DW, dry weight, "-" means not detected.

Table 8. The contents (mg/g DW) of other main phytochemicals besides catechins in 30 Chinese teas.

No.	Name	Category	Gallic Acid	Chlorogenic Acid	Caffeine	Ellagic Acid	Kaempferol-3-O-Glucoside	Theaflavine
1	Dianhong Congou Black Tea	Black Tea	2.693 ± 0.161	0.187 ± 0.005	35.283 ± 0.340	3.572 ± 0.087	1.588 ± 0.046	0.526 ± 0.019
2	Keemun Black Tea	Black Tea	2.706 ± 0.117	0.176 ± 0.005	31.452 ± 0.140	2.214 ± 0.070	-	0.542 ± 0.010
3	Lapsang Souchong Black Tea	Black Tea	1.748 ± 0.050	-	23.759 ± 0.150	-	0.385 ± 0.027	0.488 ± 0.012
4	Yichang Congou Black Tea	Black Tea	3.546 ± 0.050	0.188 ± 0.005	41.631 ± 0.312	2.614 ± 0.078	1.454 ± 0.099	0.559 ± 0.018
5	Fuzhuan Brick Tea	Dark Tea	3.097 ± 0.122	0.284 ± 0.016	27.075 ± 0.166	2.213 ± 0.067	1.002 ± 0.040	0.480 ± 0.008
6	Liupao Tea	Dark Tea	2.003 ± 0.018	-	30.565 ± 0.162	2.108 ± 0.022	0.519 ± 0.016	-
7	Pu-erh Tea	Dark Tea	1.644 ± 0.081	-	31.320 ± 0.310	2.165 ± 0.072	0.550 ± 0.043	-
8	Qingzhuan Brick tea	Dark Tea	1.507 ± 0.031	-	12.273 ± 0.040	-	-	-
9	Tibetan Tea	Dark Tea	2.203 ± 0.062	-	16.930 ± 0.101	1.553 ± 0.003	-	-
10	Dianqing Tea	Green Tea	1.430 ± 0.086	0.374 ± 0.016	39.764 ± 0.382	2.135 ± 0.037	1.605 ± 0.067	-
11	Dongting Biluochun Tea	Green Tea	0.708 ± 0.046	0.190 ± 0.006	31.993 ± 0.551	-	0.434 ± 0.030	-
12	Duyun Maojian Tea	Green Tea	1.129 ± 0.071	-	36.230 ± 0.563	1.875 ± 0.017	1.737 ± 0.090	-
13	Enshi Yulu Tea	Green Tea	1.392 ± 0.079	-	34.706 ± 0.383	1.756 ± 0.025	-	-
14	Lu'an Guapian Tea	Green Tea	0.533 ± 0.032	0.218 ± 0.009	29.232 ± 0.438	-	-	-
15	Lushan Yunwu Tea	Green Tea	0.847 ± 0.057	0.276 ± 0.013	37.778 ± 0.481	-	-	-
16	Taiping Houkui Tea	Green Tea	0.761 ± 0.043	0.241 ± 0.013	29.493 ± 0.346	-	0.347 ± 0.007	-
17	Xihu Longjing Tea	Green Tea	0.931 ± 0.043	-	38.508 ± 0.117	2.069 ± 0.097	-	-
18	Yongxi Huoqing Tea	Green Tea	1.060 ± 0.037	0.262 ± 0.011	30.783 ± 0.482	1.799 ± 0.026	-	-
19	Fenghuang Shuixian Tea	Oolong Tea	3.284 ± 0.141	-	34.770 ± 0.138	1.880 ± 0.062	1.185 ± 0.079	-
20	Luohan Chenxiang Tea	Oolong Tea	0.696 ± 0.061	0.232 ± 0.006	30.083 ± 0.287	-	0.572 ± 0.047	-
21	Tieguanyin Tea	Oolong Tea	0.294 ± 0.021	0.176 ± 0.015	14.842 ± 0.167	-	-	-
22	Wuyi Rock Tea	Oolong Tea	2.383 ± 0.142	-	25.881 ± 0.335	-	-	0.545 ± 0.011
23	Gongmei White Tea	White Tea	2.179 ± 0.038	-	27.466 ± 0.059	-	0.493 ± 0.015	-
24	Shoumei White Tea	White Tea	2.022 ± 0.026	-	25.303 ± 0.035	-	0.357 ± 0.021	-
25	White Peony Tea	White Tea	2.486 ± 0.026	-	28.758 ± 0.033	-	-	-
26	Huoshan Large Yellow Tea	Yellow Tea	3.822 ± 0.111	0.241 ± 0.009	34.201 ± 0.036	3.326 ± 0.037	0.562 ± 0.031	-
27	Junshan Yinzhen Tea	Yellow Tea	0.940 ± 0.019	-	41.457 ± 0.322	1.882 ± 0.052	1.051 ± 0.045	-
28	Mengding Huangya Tea	Yellow Tea	1.495 ± 0.073	0.313 ± 0.007	36.022 ± 0.166	3.357 ± 0.065	0.499 ± 0.030	-
29	Weishan Maojian Tea	Yellow Tea	0.752 ± 0.038	0.249 ± 0.009	37.348 ± 0.220	1.849 ± 0.039	0.688 ± 0.035	-
30	Yuan'an Luyuan Tea	Yellow Tea	0.929 ± 0.031	0.273 ± 0.007	40.737 ± 0.116	2.190 ± 0.023	1.075 ± 0.036	-

DW, dry weight, "-" means not detected.

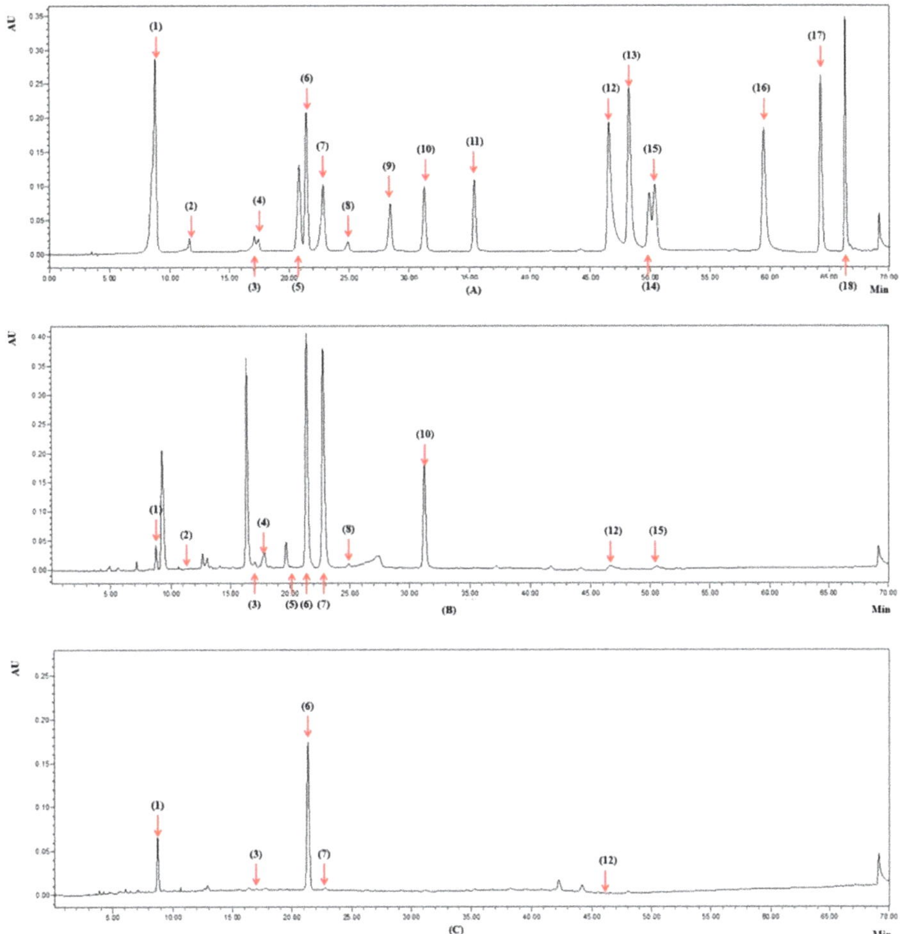

Figure 3. High-performance liquid chromatography (HPLC) Chromatograms under 254 nm of the standard compounds (**A**); Dianqing Tea (**B**); Tibetan Tea (**C**). The numbers in brackets referred to the compounds: gallic acid (**1**); gallocatechin (**2**); epigallocatechin (**3**); catechin (**4**); chlorogenic acid (**5**); caffeine (**6**); epigallocatechin gallate (**7**); epicatechin (**8**); gallocatechin gallate (**9**); epicatechin gallate (**10**); catechin gallate (**11**); ellagic acid (**12**); myricetin (**13**); quercitrin (**14**); kaempferol-3-*O*-glucoside (**15**); quercetin (**16**); theaflavine (**17**); kaempferol (**18**).

Catechins are the most abundant bioactive compounds in teas. In this study, it was found that epigallocatechin gallate was rich in the tested teas, with a range of 0.539 ± 0.013 to 59.354 ± 1.131 mg/g DW, but the difference was apparently large (up to a 110-fold difference). Green, yellow, and oolong teas were comprised of abundant epigallocatechin, but dark, black, and white teas were not. Dianqing Tea, Yuan'an Luyuan Tea, Xihu Longjing Tea, Yongxi Huoqing Tea, and Junshan Yinzhen Tea contained the top-five contents of epigallocatechin gallate, showing 59.354 ± 1.131, 57.230 ± 0.253, 51.734 ± 0.240, 50.947 ± 0.396, and 50.777 ± 0.224 mg/g DW, respectively. Dianhong Tea, with 0.539 ± 0.013 mg/g DW of epigallocatechin gallate was the lowest one. Additionally, these tea samples, especially green, yellow, and oolong teas, were also detected with remarkably high contents of epigallocatechin (2.288 ± 0.050 to 139.854 ± 1.075 mg/g DW), epicatechin (0.477 ± 0.030 to 13.723 ± 0.216 mg/g DW), and epigallocatechin gallate (0.455 ± 0.037 to 35.395 ± 0.568 mg/g DW). Tieguanyin Tea (139.854 ± 1.075 mg/g DW), Luohan

Chenxiang Tea (125.439 ± 0.678 mg/g DW), Lu'an Guapian Tea (100.684 ± 0.561 mg/g DW), Taiping Houkui Tea (74.212 ± 0226 mg/g DW), and Lushan Yunwu Tea (53.447 ± 0.326 mg/g DW) possessed the top-five contents of epigallocatechin. Tieguanyin Tea (13.723 ± 0.216 mg/g DW), Fuzhuan Brick Tea (10.357 ± 0.268 mg/g DW), Weishan Maojian Tea (10.062 ± 0.040 mg/g DW), Duyun Maojian Tea (8.700 ± 0.429 mg/g DW), and Taiping Houkui Tea (8.580 ± 0.211 mg/g DW) contained the top-five contents of epicatechin. Dianqing Tea (35.395 ± 0.568 mg/g DW), Juanshan Yinzhen Tea (30.491 ± 0.101 mg/g DW), Dongting Biluochun Tea (27.893 ± 0.426 mg/g DW), Weishan Maojian Tea (24.710 ± 0.247 mg/g DW), and Mengding Huangya Tea (23.805 ± 0.075 mg/g DW) were shown with the top-five contents of epicatechin gallate.

In addition, for other phytochemical compounds besides catechins in teas, the content of gallic acid was low in all tea samples, ranging from 0.294 ± 0.021 to 3.822 ± 0.111 mg/g DW with a 13-fold difference. Huoshan Large Yellow Tea, Yichang Congou Black Tea, Fenghuang Shuixian Tea, Fuzhuan Brick Tea, and Keemun Black Tea possessed the top-five contents of gallic acid, which were 3.822 ± 0.111, 3.546 ± 0.050, 3.284 ± 0.141, 3.097 ± 0.122, and 2.706 ± 0.117 mg/g DW, respectively. Tieguanyin Tea was shown to have the lowest content of gallic acid, which was 0.294 ± 0.021 mg/g DW. Similarly, the contents of chlorogenic acid, ellagic acid, and kaempferol-3-*O*-glucoside were also relatively low in the tested teas.

As polyphenols were suggested as the main antioxidants in teas (Figure 1), we next analyzed the relationships of different polyphenols and antioxidant activities of teas. It was found that the content of catechins had moderate positive correlations (Figure 4A,B) with FRAP values ($R^2 = 0.476$, $p < 0.001$) and TEAC values ($R^2 = 0.515$, $p < 0.001$), while the content of noncatechin polyphenols had no evident linear correlations (Figure 4C,D) with FRAP values ($R^2 = 0.001$, $p = 0.867$) and TEAC values ($R^2 = 0.002$, $p = 0.819$). These results indicate that catechins can be one of the main antioxidants in tea, but noncatechin polyphenols should not be the main contributors.

In addition to this, each tea contained relatively high caffeine content, which varied from 12.273 ± 0.040 to 41.631 ± 0.312 mg/g DW with a small difference (only a 3-fold difference). Yichang Congou Black Tea, Junshan Yinzhen Tea, Yuan'an Luyuan Tea, Dianqing Tea, and Xihu Longjing Tea comprised the top-five content of caffeine, namely 41.631 ± 0.312, 41.457 ± 0.322, 40.737 ± 0.116, 39.764 ± 0.382, and 38.508 ± 0.117 mg/g DW, respectively. The 12.273 ± 0.040 mg/g DW of caffeine in Qingzhuan Brick Tea was the lowest.

Furthermore, Wuyi Rock Tea (oolong tea) and Fuzhuan Brick Tea (dark tea) as well as all of the 4 black teas (Yichang Congou Black Tea, Keemun Black Tea, Dianhong Congou Black Tea, and Lapsang Souchong Black Tea) were found with a spot of theaflavine, and the contents were 0.545 ± 0.011, 0.480 ± 0.008, 0.559 ± 0.018, 0.542 ± 0.010, 0.526 ± 0.019, and 0.488 ± 0.012 mg/g DW, respectively.

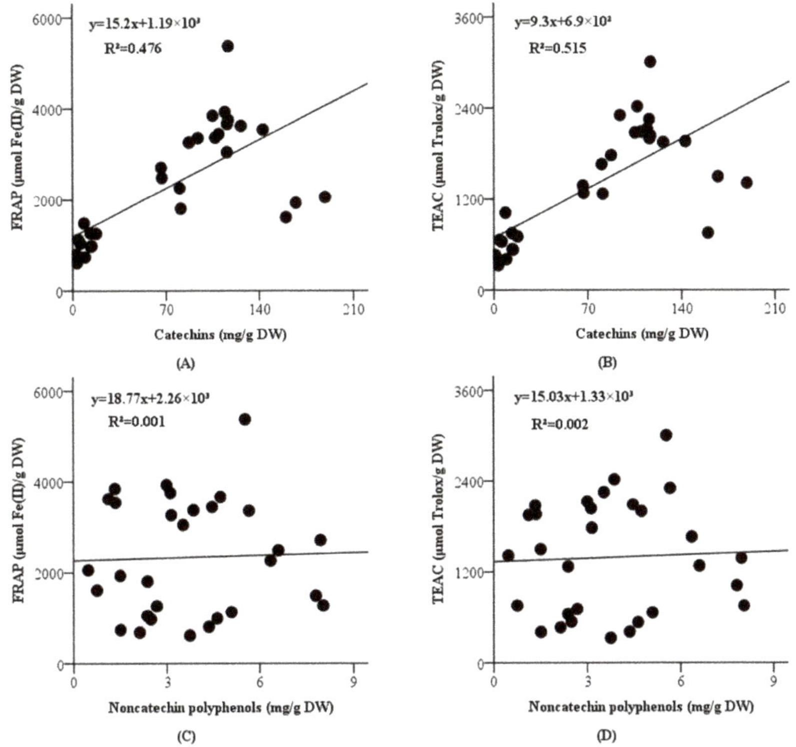

Figure 4. Correlations between FRAP and catechins (**A**), TEAC and catechins (**B**), FRAP and noncatechin polyphenols (**C**) and TEAC and noncatechin polyphenols (**D**). Abbreviations: FRAP, ferric-reducing antioxidant power; TEAC, Trolox equivalent antioxidant capacity.

4. Discussion

4.1. Antioxidant Capacities of the Tested Chinese Teas

Many natural products, such as vegetables, fruits, and medicinal plants, possess rich phytochemicals, some of which have been recognized as strong antioxidants [23–29]. These natural antioxidants are often multifunctional, and their antioxidant capacities can be generally influenced by various factors, e.g., the extraction solvents, extraction conditions, and measurement methods, resulting in the difficulty to completely illustrate the antioxidant capacities only applying a single method [30,31]. In order to maximize the extraction yields of antioxidants from tea, 3 solvents × 2 repeated extraction were adopted in this study [32]. In addition, a reliable antioxidant assessing system demands comprehensive indices, which comprise different experiments to evaluate the antioxidant capacity with diverse mechanisms of action. The FRAP assay was set up based on the power of antioxidants to reduce ferric ions to ferrous ions [19], while the TEAC assay was established on the basis of the capacity of antioxidants to scavenge the ABTS•$^{+}$ free radicals [20]. These two assays are simple, fast, repeatable, and widely used for the evaluation of antioxidant capacity [33–35]. In this study, FRAP and TEAC assays were simultaneously used to assess the antioxidant capacities of the 30 Chinese teas.

The FRAP and TEAC values of the tested teas were extremely high compared with other natural products (Table 9). That is, the antioxidant capacities of tea were higher than those of most medicinal plants, edible macro-fungi, vegetables, fruits, fruit wastes (peels and seeds), and wild fruits, as well as edible and wild flowers [33–35]. This may be explained by the apparently higher content of phenolic compounds in tea, as revealed in Table 8. Therefore, teas rich in antioxidants may be important natural sources of dietary antioxidants, and their extracts can be used to produce food additives, cosmetics, nutraceuticals, and pharmaceuticals.

Table 9. Comparison among natural products regarding FRAP, TEAC, and TPC values.

Index	Natural Product	MIN	Q_L	M	Q_U	MAX	Reference
FRAP (μmol Fe(II)/g)	30 Chinese Teas (dry)	611.2	1107.8	2156.9	3465.6	5375.2	This study
	223 Medicinal Plants (dry)	0.1	19.6	65.3	158.4	1844.9	[26]
	34 Fruit Seeds (fresh)	0.3	5.5	11.3	16.1	181.4	[36]
	48 Fruit Peels (fresh)	0.0	6.1	14.2	27.3	155.7	[36]
	56 Wild Fruits (fresh)	1.3	12.9	40.3	135.8	502.0	[37]
	49 Macro-fungi (dry)	7.9	15.1	22.1	34.5	204.7	[38]
	51 Flowers (fresh)	0.2	16.2	27.6	70.0	660.2	[39]
	10 Grape Seeds (fresh)	312.4	357.0	497.3	671.7	858.1	[18]
	30 Grape Peels (fresh)	18.3	59.9	99.9	131.9	253.0	[18]
	30 Grape Pulps (fresh)	1.3	2.9	4.9	6.7	11.8	[40]
	62 Fruits (fresh)	0.1	3.9	6.7	10.2	72.1	[24]
	56 Vegetables (fresh)	2.7	6.9	10.1	13.7	60.9	[23]
TEAC (μmol Trolox/g)	30 Chinese Teas (dry)	326.3	655.9	1396.7	2046.0	3004.4	This study
	223 Medicinal Plants (dry)	1.0	23.5	55.8	116.6	1544.4	[26]
	34 Fruit Seeds (fresh)	2.5	7.6	12.5	19.1	92.6	[36]
	48 Fruit Peels (fresh)	0.0	6.7	14.2	28.6	93.1	[36]
	56 Wild Fruits (fresh)	3.4	17.6	32.7	114.8	1140.0	[37]
	49 Macro-fungi (dry)	4.7	8.6	9.8	20.3	85.7	[38]
	51 Flowers (fresh)	0.2	8.0	12.8	35.1	191.8	[39]
	10 Grape Seeds (fresh)	207.8	227.6	274.5	345.8	473.5	[18]
	30 Grape Peels (fresh)	5.2	27.8	50.4	64.2	123.7	[18]
	30 Grape Pulps (fresh)	0.3	1.1	1.9	2.6	4.8	[40]
	62 Fruits (fresh)	0.8	2.4	3.6	5.0	80.7	[24]
	56 Vegetables (fresh)	6.9	10.3	12.8	15.3	33.6	[23]
TPC (mg GAE/g)	30 Chinese Teas (dry)	37.3	92.7	163.8	202.5	254.3	This study
	223 Medicinal Plants (dry)	0.2	3.8	8.1	14.4	98.9	[26]
	34 Fruit Seeds (fresh)	0.3	2.8	3.7	4.8	23.0	[36]
	48 Fruit Peels (fresh)	0.4	3.5	4.2	6.5	23.0	[36]
	56 Wild Fruits (fresh)	0.5	1.9	6.1	15.8	54.8	[37]
	49 Macro-fungi (dry)	2.4	4.0	4.9	6.5	44.8	[38]
	51 Flowers (fresh)	0.6	3.4	4.9	8.0	36.7	[39]
	10 Grape Seeds (fresh)	34.6	37.3	47.2	59.4	71.2	[18]
	30 Grape Peels (fresh)	1.6	6.3	10.6	13.2	25.7	[18]
	30 Grape Pulps (fresh)	0.3	0.6	0.8	1.0	1.4	[40]
	62 Fruits (fresh)	0.1	0.3	0.6	0.8	5.9	[24]
	56 Vegetables (fresh)	5.0	6.7	7.8	9.4	23.3	[23]

Abbreviations: FRAP, ferric-reducing antioxidant power; M, median; MAX, the maximum value; MIN, the minimum value; Q_L, the lower quartile; Q_U, the upper quartile; TEAC, Trolox equivalent antioxidant capacity; TPC, total phenolic content.

Moreover, the FRAP and TEAC values of water-soluble fractions were remarkably higher than those of bound-insoluble fractions, which were mildly higher than those of fat-soluble fractions. These results suggested that the components responsible for the ferric-reducing power and ABTS free radical-scavenging capacity of teas were most water-soluble compounds (approximately 87–93%) with some bound-insoluble (about 5–8%) and fat-soluble (roughly 2–5%) ones.

4.2. Antioxidant Phytochemical Components of the Tested Chinese Teas

As demonstrated previously, there were significant and remarkable correlations among FRAP, TEAC, and TPC values. These results suggested that the phenolic compounds could be the major components contributing to the antioxidant capacities of tea, which possessed multiple effects to reduce oxidants and scavenge free radicals. The outcomes demonstrated above were consistent with several previous studies, which have reported that phenolic components were the main contributors responsible for the antioxidant capacities of vegetables, macro-fungi, wild fruits, and flowers [23,37–39]. Moreover, many polyphenols have been detected in these natural products, e.g., gallic acid, chlorogenic acid, ferulic acid, anthocyanins, quercetin, rutin, myricetin, and kaempferol glycosides, which exhibited potent antioxidant capacities both in vitro and in vivo [23,37–39]. Antioxidant action can be one of the most important mechanisms of the health benefits of these natural products [41–43]. As for tea, eight catechins, caffeine, theaflavine, gallic acid, chlorogenic acid, ellagic acid, and kaempferol-3-*O*-glucoside, could be detected. Among them, epicatechin, epigallocatechin, epicatechin gallate, and epigallocatechin gallate were the most abundant polyphenols in tea, especially in the green, yellow, and oolong teas, which generally undergo a low degree of fermentation. Though tea and other natural products contain several common antioxidants, their contents in tea are generally higher.

Tea polyphenols may exert antioxidant capacities through the following mechanisms: (1) straightly reducing oxidants; (2) chelating metal ions; (3) transferring hydrogen; (4) scavenging free radicals; (5) improving activities of antioxidant enzymes; (6) increasing contents of endogenous antioxidants; and (7) regulating antioxidant-related genes [4,44–51]. All of these actions lead to the health functions of tea, such as anticancer, cardiovascular protective, neuroprotective, hepatoprotective, and renoprotective effects [6,52–56]. Thus, several teas rich in antioxidants can be developed into functional foods or nutraceuticals to prevent and treat certain oxidative stress-related chronic diseases.

4.3. Comparison of Antioxidant Phytochemicals among Different Chinese Teas

In the light of the outcomes from systematic cluster analysis accompanied by OLAP and ANOVA for cluster number = 4, green tea and yellow tea possessed remarkably high antioxidant capacities and phenolic contents, but Lu'an Guapian Tea (green tea) and Huoshan Large Yellow Tea (yellow tea) were the exceptions. In addition, oolong tea was in the middle position. Meanwhile, white tea, black tea, and dark tea exerted relatively low antioxidant capacities and phenolic contents. Thus, fermentation degree can be a crucial factor that influences the antioxidant capacity and phenolic content of tea. Tea undergoing higher fermentation degree might have lower antioxidant capacity and phenolic content, since tea polyphenols, especially catechins, may oxidize and polymerize during fermentation, generating complicated tea pigments like theaflavins, thearubigins, and theabrownins [57–59]. Moreover, the maturity of tea leaves should also be taken into consideration, because the antioxidant capacity and phenolic content would decrease accompanied with the increase of tea leaf maturity [60], which may partially explain why white tea (made of old tea leaves) exhibited relatively low antioxidant capacity and phenolic content, although it has a low fermentation degree. On the other hand, it was reported that the bioavailability of fermented tea using microbes, such as bacteria, yeasts, and fungi, could be significantly higher compared to unfermented tea [61,62]. For example, green and black teas have been observed to improve endothelial function with equal effectiveness, although green tea possesses higher antioxidant activity and phenolic content, it has a lower bioavailability [63].

5. Conclusions

In conclusion, teas here studied possessed remarkably high antioxidant capacities regarding ferric-reducing and free radical-scavenging capacities. In addition, eight catechins, caffeine, theaflavine, and several other phenolic compounds, including gallic acid, chlorogenic acid, ellagic acid, and kaempferol-3-*O*-glucoside, were detected in these Chinese teas. Compared with dark, black, and white teas, green, yellow, and oolong teas exerted stronger antioxidant capacity and contained more polyphenols, especially catechins like epicatechin, epigallocatechin, epicatechin gallate, and epigallocatechin gallate. Overall, tea is a good natural source of dietary antioxidant phytochemicals, and can be used to produce food additives, functional foods, nutraceuticals, and cosmetics.

Author Contributions: Conceptualization, R.-Y.G. and H.-B.L.; Data curation, C.-N.Z.; Formal analysis, G.-Y.T.; Funding acquisition, R.-Y.G. and H.-B.L.; Investigation, G.-Y.T., C.-N.Z., X.-Y.X., S.-Y.C., Q.L., A.S. and Q.-Q.M.;

Methodology, G.-Y.T., C.-N.Z. and H.-B.L.; Project administration, R.-Y.G. and H.-B.L.; Resources, G.-Y.T.; Software, G.-Y.T.; Supervision, R.-Y.G. and H.-B.L.; Validation, G.-Y.T., C.-N.Z., R.-Y.G. and H.-B.L.; Visualization, C.-N.Z.; Writing—original draft, G.-Y.T. and C.-N.Z.; Writing—review & editing, R.-Y.G. and H.-B.L.

Funding: This work was supported by the National Key R&D Program of China (2018YFC1604400); Shanghai Basic and Key Program (18JC1410800); Shanghai Pujiang Talent Plan (18PJ1404600).

Conflicts of Interest: The authors declare no conflict of interest.

References

1. Hashimoto, T.; Goto, M.; Sakakibara, H.; Oi, N.; Okamoto, M.; Kanazawa, K. Yellow tea is more potent than other types of tea in suppressing liver toxicity induced by carbon tetrachloride in rats. *Phytother. Res.* **2007**, *21*, 668–670. [CrossRef] [PubMed]

2. Braud, L.; Battault, S.; Meyer, G.; Nascimento, A.; Gaillard, S.; de Sousa, G.; Rahmani, R.; Riva, C.; Armand, M.; Maixent, J.M.; et al. Antioxidant properties of tea blunt ROS-dependent lipogenesis: Beneficial effect on hepatic steatosis in a high fat-high sucrose diet NAFLD obese rat model. *J. Nutr. Biochem.* **2017**, *40*, 95–104. [CrossRef]

3. Di Lorenzo, A.; Nabavi, S.F.; Sureda, A.; Moghaddam, A.H.; Khanjani, S.; Arcidiaco, P.; Nabavi, S.M.; Daglia, M. Antidepressive-like effects and antioxidant activity of green tea and GABA green tea in a mouse model of post-stroke depression. *Mol. Nutr. Food Res.* **2016**, *60*, 566–579. [CrossRef] [PubMed]

4. Guo, Y.J.; Sun, L.Q.; Yu, B.Y.; Qi, J. An integrated antioxidant activity fingerprint for commercial teas based on their capacities to scavenge reactive oxygen species. *Food Chem.* **2017**, *237*, 645–653. [CrossRef] [PubMed]

5. Kim, Y.H.; Won, Y.S.; Yang, X.; Kumazoe, M.; Yamashita, S.; Hara, A.; Takagaki, A.; Goto, K.; Nanjo, F.; Tachibana, H. Green tea catechin metabolites exert immunoregulatory effects on CD4$^+$ T cell and natural killer cell activities. *J. Agr. Food Chem.* **2016**, *64*, 3591–3597. [CrossRef] [PubMed]

6. Leung, F.P.; Yung, L.M.; Ngai, C.Y.; Cheang, W.S.; Tian, X.Y.; Lau, C.W.; Zhang, Y.; Liu, J.; Chen, Z.Y.; Bian, Z.X.; et al. Chronic black tea extract consumption improves endothelial function in ovariectomized rats. *Eur. J. Nutr.* **2016**, *55*, 1963–1972. [CrossRef] [PubMed]

7. Liu, K.; Zhou, R.; Wang, B.; Chen, K.; Shi, L.Y.; Zhu, T.D.; Mi, M.T. Effect of green tea on glucose control and insulin sensitivity: A meta-analysis of 17 randomized controlled trials. *Am. J. Clin. Nutr.* **2013**, *98*, 340–348. [CrossRef] [PubMed]

8. Nam, M.; Choi, M.S.; Choi, J.Y.; Kim, N.; Kim, M.S.; Jung, S.; Kim, J.; Ryu, D.H.; Hwang, G.S. Effect of green tea on hepatic lipid metabolism in mice fed a high-fat diet. *J. Nutr. Biochem.* **2018**, *51*, 1–7. [CrossRef]

9. Scoparo, C.T.; de Souza, L.M.; Rattmann, Y.D.; Kiatkoski, E.C.; Dartora, N.; Iacomini, M. The protective effect of green and black teas (*Camellia sinensis*) and their identified compounds against murine sepsis. *Food Res. Int.* **2016**, *83*, 102–111. [CrossRef]

10. Torello, C.O.; Shiraishi, R.N.; Della Via, F.I.; de Castro, T.; Longhini, A.L.; Santos, I.; Bombeiro, A.L.; Silva, C.; Queiroz, M.; Rego, E.M.; et al. Reactive oxygen species production triggers green tea-induced anti-leukaemic effects on acute promyelocytic leukaemia model. *Cancer Lett.* **2018**, *414*, 116–126. [CrossRef]

11. Bi, W.; He, C.N.; Ma, Y.Y.; Shen, J.; Zhang, L.H.; Peng, Y.; Xiao, P.G. Investigation of free amino acid, total phenolics, antioxidant activity and purine alkaloids to assess the health properties of non-Camellia tea. *ACTA Pharm. Sin. B* **2016**, *6*, 170–181. [CrossRef] [PubMed]

12. Pan, H.B.; Wang, F.; Rankin, G.O.; Rojanasakul, Y.; Tu, Y.Y.; Chen, Y.C. Inhibitory effect of black tea pigments, theaflavin-3/3′-gallate against cisplatin-resistant ovarian cancer cells by inducing apoptosis and G1 cell cycle arrest. *Int. J. Oncol.* **2017**, *51*, 1508–1520. [CrossRef]

13. Sun, L.J.; Warren, F.J.; Gidley, M.J. Soluble polysaccharides reduce binding and inhibitory activity of tea polyphenols against porcine pancreatic α-amylase. *Food Hydrocolloid.* **2018**, *79*, 63–70. [CrossRef]

14. Tao, L.; Park, J.Y.; Lambert, J.D. Differential prooxidative effects of the green tea polyphenol, (−)-epigallocatechin-3-gallate, in normal and oral cancer cells are related to differences in sirtuin 3 signaling. *Mol. Nutr. Food Res.* **2015**, *59*, 203–211. [CrossRef] [PubMed]

15. Wang, B.; Tu, Y.; Zhao, S.P.; Hao, Y.H.; Liu, J.X.; Liu, F.H.; Xiong, B.H.; Jiang, L.S. Effect of tea saponins on milk performance, milk fatty acids, and immune function in dairy cow. *J. Dairy Sci.* **2017**, *100*, 8043–8052. [CrossRef] [PubMed]

16. Li, L.; Chen, C.Y.O.; Chun, H.; Cho, S.; Park, K.; Lee-Kim, Y.C.; Blumberg, J.B.; Russell, R.M.; Yeum, K. A fluorometric assay to determine antioxidant activity of both hydrophilic and lipophilic components in plant foods. *J. Nutr. Biochem.* **2009**, *20*, 219–226. [CrossRef] [PubMed]

17. Nardini, M.; Cirillo, E.; Natella, F.; Mencarelli, D.; Comisso, A.; Scaccini, C. Detection of bound phenolic acids: Prevention by ascorbic acid and ethylenediaminetetraacetic acid of degradation of phenolic acids during alkaline hydrolysis. *Food Chem.* **2002**, *79*, 119–124. [CrossRef]

18. Tang, G.Y.; Zhao, C.N.; Liu, Q.; Feng, X.L.; Xu, X.Y.; Cao, S.Y.; Meng, X.; Li, S.; Gan, R.Y.; Li, H.B. Potential of grape wastes as a natural source of bioactive compounds. *Molecules* **2018**, *23*, 2598. [CrossRef] [PubMed]

19. Benzie, I.F.F.; Strain, J.J. The ferric reducing ability of plasma (FRAP) as a measure of "Antioxidant power": The FRAP assay. *Anal. Biochem.* **1996**, *239*, 70–76. [CrossRef] [PubMed]

20. Re, R.; Pellegrini, N.; Proteggente, A.; Pannala, A.; Yang, M.; Rice-Evans, C. Antioxidant activity applying an improved ABTS radical cation decolorization assay. *Free Radi. Biol. Med.* **1999**, *26*, 1231–1237. [CrossRef]

21. Singleton, V.L.; Orthofer, R.; Lamuela-Raventos, R.M. Analysis of total phenols and other oxidation substrates and antioxidants by means of Folin–Ciocalteu reagent. In *Oxidants and Antioxidants*; Packer, L., Ed.; Elsevier Academic Press Inc.: San Diego, CA, USA, 1999; Volume 299, pp. 152–178.

22. Cai, Y.Z.; Luo, Q.; Sun, M.; Corke, H. Antioxidant activity and phenolic compounds of 112 traditional Chinese medicinal plants associated with anticancer. *Life Sci.* **2004**, *74*, 2157–2184. [CrossRef] [PubMed]

23. Deng, G.F.; Lin, X.; Xu, X.R.; Gao, L.L.; Xie, J.F.; Li, H.B. Antioxidant capacities and total phenolic contents of 56 vegetables. *J. Funct. Foods.* **2013**, *5*, 260–266. [CrossRef]

24. Fu, L.; Xu, B.T.; Xu, X.R.; Gan, R.Y.; Zhang, Y.; Xia, E.Q.; Li, H.B. Antioxidant capacities and total phenolic contents of 62 fruits. *Food Chem.* **2011**, *129*, 345–350. [CrossRef] [PubMed]

25. Gan, R.Y.; Li, H.B.; Gunaratne, A.; Sui, Z.Q.; Corke, H. Effects of fermented edible seeds and their products on human health: Bioactive components and bioactivities. *Compr. Rev. Food Sci. Food Saf.* **2017**, *16*, 489–531. [CrossRef]

26. Li, S.; Li, S.K.; Gan, R.Y.; Song, F.L.; Kuang, L.; Li, H.B. Antioxidant capacities and total phenolic contents of infusions from 223 medicinal plants. *Ind. Crop. Prod.* **2013**, *51*, 289–298. [CrossRef]

27. Meng, X.; Li, Y.; Li, S.; Gan, R.Y.; Li, H.B. Natural products for prevention and treatment of chemical-induced liver injuries. *Compr. Rev. Food Sci. Food Saf.* **2018**, *17*, 472–495. [CrossRef]

28. Tang, G.Y.; Meng, X.; Li, Y.; Zhao, C.N.; Liu, Q.; Li, H.B. Effects of vegetables on cardiovascular diseases and related mechanisms. *Nutrients* **2017**, *9*, 857. [CrossRef]

29. Zhao, C.N.; Meng, X.; Li, Y.; Li, S.; Liu, Q.; Tang, G.Y.; Li, H.B. Fruits for prevention and treatment of cardiovascular diseases. *Nutrients* **2017**, *9*, 598. [CrossRef]

30. Nilsson, J.; Pillai, D.; Onning, G.; Persson, C.; Nilsson, A.; Akesson, B. Comparison of the 2,2'-azinobis-3-ethylbenzotiazoline-6-sulfonic acid (ABTS) and ferric reducing antioxidant power (FRAP) methods to asses the total antioxidant capacity in extracts of fruit and vegetables. *Mol. Nutr. Food Res.* **2005**, *49*, 239–246. [CrossRef]

31. Raudonis, R.; Raudone, L.; Jakstas, V.; Janulis, V. Comparative evaluation of post-column free radical scavenging and ferric reducing antioxidant power assays for screening of antioxidants in strawberries. *J. Chromatogr. A.* **2012**, *1233*, 8–15. [CrossRef]

32. Pastoriza, S.; Perez-Burillo, S.; Rufian-Henares, J.A. How brewing parameters affect the healthy profile of tea. *Curr. Opin. Food Sci.* **2017**, *14*, 7–12. [CrossRef]

33. Hayes, W.A.; Mills, D.S.; Neville, R.F.; Kiddie, J.; Collins, L.M. Determination of the molar extinction coefficient for the ferric reducing/antioxidant power assay. *Anal. Biochem.* **2011**, *416*, 202–205. [CrossRef]

34. Pohanka, M.; Bandouchova, H.; Sobotka, J.; Sedlackova, J.; Soukupova, I.; Pikula, J. Ferric reducing antioxidant power and square wave voltammetry for assay of low molecular weight antioxidants in blood plasma: Performance and comparison of methods. *Sensors* **2009**, *9*, 9094–9103. [CrossRef] [PubMed]

35. Van den Berg, R.; Haenen, G.; van den Berg, H.; van der Vijgh, W.; Bast, A. The predictive value of the antioxidant capacity of structurally related flavonoids using the Trolox equivalent antioxidant capacity (TEAC) assay. *Food Chem.* **2000**, *70*, 391–395. [CrossRef]

36. Deng, G.F.; Shen, C.; Xu, X.R.; Kuang, R.D.; Guo, Y.J.; Zeng, L.S.; Gao, L.L.; Lin, X.; Xie, J.F.; Xia, E.Q.; et al. Potential of fruit wastes as natural resources of bioactive compounds. *Int. J. Mol. Sci.* **2012**, *13*, 8308–8323. [CrossRef] [PubMed]

37. Fu, L.; Xu, B.T.; Xu, X.R.; Qin, X.S.; Gan, R.Y.; Li, H.B. Antioxidant capacities and total phenolic contents of 56 wild fruits from South China. *Molecules* **2010**, *15*, 8602–8617. [CrossRef] [PubMed]

38. Guo, Y.J.; Deng, G.F.; Xu, X.R.; Wu, S.; Li, S.; Xia, E.Q.; Li, F.; Chen, F.; Ling, W.H.; Li, H.B. Antioxidant capacities, phenolic compounds and polysaccharide contents of 49 edible macro-fungi. *Food Funct.* **2012**, *3*, 1195–1205. [CrossRef] [PubMed]

39. Li, A.N.; Li, S.; Li, H.B.; Xu, D.P.; Xu, X.R.; Chen, F. Total phenolic contents and antioxidant capacities of 51 edible and wild flowers. *J. Funct. Foods.* **2014**, *6*, 319–330. [CrossRef]

40. Liu, Q.; Tang, G.Y.; Zhao, C.N.; Feng, X.L.; Xu, X.Y.; Cao, S.Y.; Meng, X.; Li, S.; Gan, R.Y.; Li, H.B. Comparison of antioxidant activities of different grape varieties. *Molecules* **2018**, *23*, 2432. [CrossRef]

41. Li, Y.; Li, S.; Meng, X.; Gan, R.Y.; Zhang, J.J.; Li, H.B. Dietary natural products for prevention and treatment of breast cancer. *Nutrients* **2017**, *9*, 728. [CrossRef]

42. Xu, D.P.; Li, Y.; Meng, X.; Zhou, T.; Zhou, Y.; Zheng, J.; Zhang, J.J.; Li, H.B. Natural antioxidants in foods and medicinal plants: Extraction, assessment and resources. *Int. J. Mol. Sci.* **2017**, *18*, 18. [CrossRef] [PubMed]

43. Zheng, J.; Zhou, Y.; Li, S.; Zhang, P.; Zhou, T.; Xu, D.P.; Li, H.B. Effects and mechanisms of fruit and vegetable juices on cardiovascular diseases. *Int. J. Mol. Sci.* **2017**, *18*, 555. [CrossRef] [PubMed]

44. Bartikova, H.; Skalova, L.; Valentova, K.; Matouskova, P.; Szotakova, B.; Martin, J.; Kvita, V.; Bousova, I. Effect of oral administration of green tea extract in various dosage schemes on oxidative stress status of mice *in vivo*. *Acta Pharm.* **2015**, *65*, 65–73. [CrossRef] [PubMed]

45. Chu, K.O.; Chan, K.P.; Yang, Y.P.; Qin, Y.J.; Li, W.Y.; Chan, S.O.; Wang, C.C.; Pang, C.P. Effects of EGCG content in green tea extract on pharmacokinetics, oxidative status and expression of inflammatory and apoptotic genes in the rat ocular tissues. *J. Nutr. Biochem.* **2015**, *26*, 1357–1367. [CrossRef] [PubMed]

46. Cyboran, S.; Strugala, P.; Wloch, A.; Oszmianski, J.; Kleszczynska, H. Concentrated green tea supplement: Biological activity and molecular mechanisms. *Life Sci.* **2015**, *126*, 1–9. [CrossRef] [PubMed]

47. Fei, T.Y.; Fei, J.; Huang, F.; Xie, T.P.; Xu, J.F.; Zhou, Y.; Yang, P. The anti-aging and anti-oxidation effects of tea water extract in *Caenorhabditis elegans*. *Exp. Gerontol.* **2017**, *97*, 89–96. [CrossRef] [PubMed]

48. Liu, S.M.; Huang, H.H. Assessments of antioxidant effect of black tea extract and its rationals by erythrocyte haemolysis assay, plasma oxidation assay and cellular antioxidant activity (CAA) assay. *J. Funct. Foods.* **2015**, *18*, 1095–1105. [CrossRef]

49. Peluso, I.; Manafikhi, H.; Raguzzini, A.; Longhitano, Y.; Reggi, R.; Zanza, C.; Palmery, M. The peroxidation of leukocytes index ratio reveals the prooxidant effect of green tea extract. *Oxid. Med. Cell. Longev.* **2016**. [CrossRef]

50. Xie, H.; Li, X.C.; Ren, Z.X.; Qiu, W.M.; Chen, J.L.; Jiang, Q.; Chen, B.; Chen, D.F. Antioxidant and cytoprotective effects of Tibetan tea and its phenolic components. *Molecules* **2018**, *23*, 179. [CrossRef]

51. Zeng, L.; Luo, L.Y.; Li, H.J.; Liu, R.H. Phytochemical profiles and antioxidant activity of 27 cultivars of tea. *Int. J. Food Sci. Nutr.* **2017**, *68*, 525–537. [CrossRef]

52. Adami, G.R.; Tangney, C.C.; Tang, J.L.; Zhou, Y.; Ghaffari, S.; Naqib, A.; Sinha, S.; Green, S.J.; Schwartz, J.L. Effects of green tea on miRNA and microbiome of oral epithelium. *Sci. Rep.* **2018**, *8*, 5873. [CrossRef] [PubMed]

53. Garcia, M.L.; Pontes, R.B.; Nishi, E.E.; Ibuki, F.K.; Oliveira, V.; Sawaya, A.; Carvalho, P.O.; Nogueira, F.N.; Franco, M.D.; Campos, R.R.; et al. The antioxidant effects of green tea reduces blood pressure and sympathoexcitation in an experimental model of hypertension. *J. Hypertens.* **2017**, *35*, 348–354. [CrossRef] [PubMed]

54. Reddyvari, H.; Govatati, S.; Matha, S.K.; Korla, S.V.; Malempati, S.; Pasupuleti, S.R.; Bhanoori, M.; Nallanchakravarthula, V. Therapeutic effect of green tea extract on alcohol induced hepatic mitochondrial DNA damage in albino wistar rats. *J. Adv. Res.* **2017**, *8*, 289–295. [CrossRef] [PubMed]

55. Schimidt, H.L.; Garcia, A.; Martins, A.; Mello-Carpes, P.B.; Carpes, F.P. Green tea supplementation produces better neuroprotective effects than red and black tea in Alzheimer-like rat model. *Food Res. Int.* **2017**, *100*, 442–448. [CrossRef] [PubMed]

56. Xie, X.; Yi, W.J.; Zhang, P.W.; Wu, N.N.; Yan, Q.Q.; Yang, H.; Tian, C.; Xiang, S.Y.; Du, M.Y.; Assefa, E.G.; et al. Green tea polyphenols, mimicking the effects of dietary restriction, ameliorate high-fat diet-induced kidney injury via regulating autophagy flux. *Nutrients* **2017**, *9*, 497. [CrossRef] [PubMed]

57. Gupta, S.; Chaudhuri, T.; Seth, P.; Ganguly, D.K.; Giri, A.K. Antimutagenic effects of black tea (world blend) and its two active polyphenols theaflavins and thearubigins in Salmonella assays. *Phytother. Res.* **2002**, *16*, 655–661. [CrossRef]

58. Pereira-Caro, G.; Moreno-Rojas, J.M.; Brindani, N.; Del Rio, D.; Lean, M.; Hara, Y.; Crozier, A. Bioavailability of black tea theaflavins: Absorption, metabolism, and colonic catabolism. *J. Agr. Food Chem.* **2017**, *65*, 5365–5374. [CrossRef]

59. Wang, Q.P.; Belscak-Cvitanovic, A.; Durgo, K.; Chisti, Y.; Gong, J.S.; Sirisansaneeyakul, S.; Komes, D. Physicochemical properties and biological activities of a high-theabrownins instant Pu-erh tea produced using *Aspergillus tubingensis*. *LWT-Food Sci. Technol.* **2018**, *90*, 598–605. [CrossRef]

60. Baptista, J.; Lima, E.; Paiva, L.; Andrade, A.L.; Alves, M.G. Comparison of Azorean tea theanine to teas from other origins by HPLC/DAD/FD. Effect of fermentation, drying temperature, drying time and shoot maturity. *Food Chem.* **2012**, *132*, 2181–2187. [CrossRef]

61. Jayabalan, R.; Malbasa, R.V.; Loncar, E.S.; Vitas, J.S.; Sathishkumar, M. A review on Kombucha tea microbiology, composition, fermentation, beneficial effects, toxicity, and tea fungus. *Compr. Rev. Food Sci. Food Saf.* **2014**, *13*, 538–550. [CrossRef]

62. Zhao, D.Y.; Shah, N.P. Concomitant ingestion of lactic acid bacteria and black tea synergistically enhances flavonoid bioavailability and attenuates *D*-galactose-induced oxidative stress in mice via modulating glutathione antioxidant system. *J. Nutr. Biochem.* **2016**, *38*, 116–124. [CrossRef] [PubMed]

63. Jochmann, N.; Lorenz, M.; von Krosigk, A.; Martus, P.; Bohm, V.; Baumann, G.; Stangl, K.; Stangl, V. The efficacy of black tea in ameliorating endothelial function is equivalent to that of green tea. *Brit. J. Nutr.* **2008**, *99*, 863–868. [CrossRef] [PubMed]

© 2019 by the authors. Licensee MDPI, Basel, Switzerland. This article is an open access article distributed under the terms and conditions of the Creative Commons Attribution (CC BY) license (http://creativecommons.org/licenses/by/4.0/).

Article

Physico-Chemical Parameters, Phenolic Profile, In Vitro Antioxidant Activity and Volatile Compounds of Ladastacho (*Lavandula stoechas*) from the Region of Saidona

Ioannis K. Karabagias *, Vassilios K. Karabagias and Kyriakos A. Riganakos

Laboratory of Food Chemistry, Department of Chemistry, University of Ioannina, 45110 Ioannina, Greece; vkarambagias@gmail.com (V.K.K.); kriganak@uoi.gr (K.A.R.)
* Correspondence: ikaraba@cc.uoi.gr; Tel.: +30-697-828-6866

Received: 1 February 2019; Accepted: 25 March 2019; Published: 28 March 2019

Abstract: The aim of the present study was to characterize *Lavandula stoechas* (Ladastacho) from the region of Saidona by means of physico-chemical parameters, phenolic profile, in vitro antioxidant activity and volatile compounds. Physico-chemical parameters (pH, acidity, salinity, total dissolved solids, electrical conductivity and liquid resistivity) were determined using conventional methods. The phenolic profile was determined using high-performance liquid chromatography electrospray ionization mass spectrometry (HPLC/ESI-MS), whereas a quantitative determination was also accomplished using the total phenolics assay. In vitro antioxidant activity was determined using the 2,2-diphenyl-1-picryl-hydrazyl assay. Finally, volatile compounds were determined using headspace solid phase microextraction coupled to gas chromatography mass spectrometry (HS-SPME/GC-MS). The results showed that *Lavandula stoechas* aqueous extract had a slightly acidic pH, low salinity content and considerable electrochemical properties (electrical conductivity and liquid resistivity along with electric potential). In addition, aqueous fractions showed a significantly ($p < 0.05$) higher phenolic content and in vitro antioxidant activity, whereas phenolic compounds, such as caffeic acid, quercetin-*O*-glucoside, lutelin-*O*-glucuronide and rosmarinic acid, were identified. Finally, numerous volatile compounds were found to dominate the volatile pattern of this flowering plant, producing a strong, penetrating, cool and menthol-like odour.

Keywords: Ladastacho; characterisation; properties; HPLC/ESI-MS; HS-SPME/GC-MS; beneficial use

---

1. Introduction

The systematic cultivation and exploitation of the health benefits and beneficial applications of historical flowering plants/herbs is of great value for the modern world as new challenges arise day-by-day. The most important challenge, however, is the maintenance of the welfare and healthcare of humans in lieu of the use of synthetic drugs that may cause adverse effects.

Indeed, since ancient times, there has been all over the world a systematic use of herbs for the treatment of several health disorders. For example, in Greece, such herbs as oregano, marjoram, dill, fennel, mint, rosemary, mountain tea, sage, chamomile, thyme, parsley and basil have been used not only to flavour the cuisine or traditional dishes but also for medicinal purposes. Another important flowering plant that has a long history is Lavandula or, more commonly, lavender.

Approximately 47 species of the Lavandula genus are known, belonging to the mint family Lamiaceae. These flowering plants have a long history and have been found in numerous regions of the world, including Europe, the Mediterranean zone, Africa, Asia and India [1]. The temperate climates favor the cultivation of *Lavandula* spp. as ornamental plants or culinary herbs, and also at a commercial level for the extraction of essential oils [2]. Lavandula is grown in well-drained soils and in light places with good air circulation. It cannot grow in the shade [3,4]. The suitable soil pH may vary from acid to neutral and basic, whereas lavender can favorably grow in very alkaline soils. The optimum pH for the growth of Lavandula is between 6 and 8 [5]. In most cases, lavender is harvested by hand, depending on the usage [5].

In the Greek and Mediterranean zones, the species that grow originally are *Lavandula stoechas* [6]. Furthermore, the species of *Lavandula stoechas*, *Lavandula pedunculata* and *Lavandula dentata* were reported in Roman times [7]. The most commonly cultivated plant is the English lavender *Lavandula angustifolia* (formerly named *L. officinalis*). Some other commonly grown ornamental species are *Lavandula stoechas*, *Lavandula dentata* and *Lavandula multifida* (Egyptian lavender).

The lavender flower has been reported to have plenty of uses in practices of herbalism. The German scientific committee enhanced its use in alternative medicine, including its use for restlessness or insomnia, Roehmheld's syndrome, intestinal discomfort and cardiovascular diseases, among others [8]. The National Institutes of Health (NIH) in Maryland, U.S., stated that lavender is considered "likely" safe in food amounts and "possibly" safe in medicinal amounts. However, the NIH did not recommend the use of lavender in pregnant or breast-feeding women because of a lack of knowledge of its effects. Lavender oil should not be used in young boys due to the possible hormonal effects leading to gynecomastia [9,10]. In addition, the NIH stated that lavender may cause skin irritation and could be poisonous if consumed orally [11]. In 2005, a review on lavender essential oil reported that lavender is traditionally regarded as a 'safe' oil, whereas skin problems (contact dermatitis) may occur at only a very low frequency [12]. In a study dealing with the relationship between various fragrances and photosensitivity carried out in 2007, it was reported that, even though lavender is known to favour cutaneous photo-toxic reactions, it did not induce photohaemolysis [13].

Saidona is a beautiful village in the Messiniaki Mani. Organic olive oil, table olives, sage, mountain tea, *Lavaldula stoechas*, and other flowering plants flourish in considerable amounts in Saidona. There is a long historical tradition of the local citizens using *Lavandula stoechas*, commonly referred as "ladastacho", for the treatment of hair loss by preparing a gently boiled aqueous solution of its flowers.

Based on the above, the aim of the present work was to investigate the aqueous and methanolic extracts of *Lavandula stoechas* in terms of phenolic profile and in vitro antioxidant activity along with volatile compounds; some physico-chemical parameters were also considered. To the best of our knowledge, this is the first report in the literature that reports a combination of different parameter analyses for the characterisation of such a flowering plant grown in the Greek zone.

2. Materials and Methods

2.1. Collection of Lavandula stoechas and Preparation of Samples

Two kilograms (kg) of *Lavandula stoechas* was collected from the region of Saidona, Mani, (Messinia, Greece) (36°52′55″N 22°17′01″E) during June of 2017. *Lavandula stoechas* is shelf-grown in this region without the use of any cultivation system/procedure. The flowers were left alone to dry in a dark place and were then gently removed from the body of the plant. Finally, these were stored in aluminum foil at 4 ± 1 °C until further analysis.

2.2. Reagents and Solutions

Gallic acid (3,4,5-trihydrobenzoic acid) anhydrous for synthesis was purchased from Merck (Darmstadt, Germany). Methanol for analysis, Folin–Ciocalteu phenol reagent, sodium chloride (NaCl), sodium hydroxide (NaOH) and sodium carbonate (Na_2CO_3) were purchased from Merck. Finally, 2,2-Diphenyl-1-picrylhydrazyl (DPPH) was purchased from Sigma-Aldrich (Darmstadt, Germany).

2.3. Physicochemical Parameter Analysis

2.3.1. Determination of Acidity

For the determination of acidity, 10 g of *Lavandula stoechas* was transferred to a conical flask and 100 mL of distilled water was added. The obtained solution was titrated with a standard solution of sodium hydroxide (0.1 N) after the addition of a few drops of phenolphthalein, serving as an indicator of equivalent point. The results reported are the average values ± standard deviation values of three replicates (n = 3) and are expressed as g NaOH per 100 g of dried *Lavandula stoechas* [14].

2.3.2. Determination of pH

pH values of *Lavandula stoechas* were measured using a Delta OHM, model HD 3456.2, pH-meter (Delta OHM, Padova, Italy) (precision of ± 0.002). Samples (10 g) were diluted with 100 mL of distilled water and the homogenate was used for the pH determination. Reported results are the average values ± standard deviation values of three replicates (n = 3). All measurements were carried out at 15 ± 1 °C after immersion of the electrode (that was firstly cleaned with distilled water) in the herb aqueous solution, until constant values were reached.

2.3.3. Determination of Electrical Conductivity, Salinity and Total Dissolved Solids

Electrical conductivity, liquid resistivity, salinity and total dissolved solids of 10% ($w\ v^{-1}$) aqueous herb solutions in distilled water were measured using a Delta OHM, model HD 3456.2, conductimeter (Delta OHM, Padova, Italy) with four-ring and two-ring conductivity/temperature probes. Temperature was measured by four-wire Pt100 and two-wire Pt1000 sensors by immersion. The probe was calibrated automatically using a 1413 μS cm^{-1} conductivity standard solution (Hannah Instruments, Inc., Woonsocket, RI, USA). Results are expressed as mS cm^{-1}, Ohm (Ω), g L^{-1} and mg L^{-1}, respectively. The results reported are the average ± standard deviation values of three replicates (n = 3).

2.3.4. Extraction of Phenolic Compounds

Extraction of phenolic compounds was carried out in two independent extraction systems: methanol and water. In particular, 0.34 g of dried and gently blended *Lavandula stoechas* was placed in plastic volumetric flasks containing 30 mL of methanol or water (1.13% ($w\ v^{-1}$), (11,333.33 mg L^{-1})). The vials were then centrifuged for 2 h at 4000 rpm. The supernatant was filtered using a filtrate paper and the extracts were collected in plastic vials, having been wrapped with parafilm and aluminum foil. Finally, these were kept at −18 °C until analysis and used for the determination of phenolic compounds and in vitro antioxidant activity.

2.3.5. Analysis of *Lavandula stoechas* Phenolic Compounds Using High-Performance Liquid Chromatography Electro Spray Ionization Mass Spectrometry (HPLC/ESI-MS)

An Agilent, model 1100 series, HPLC system (Agilent, Santa Clara, California, USA) was used for the chromatographic analysis. The respected wavelength was that of 280 ± 2 nm. Water and acetonitrile (Merck, Darmstadt, Germany) were used as the mobile phase at a flow rate of 1 mL min^{-1}. The gradient elution program followed the sequence: 10% of acetonitrile then increasing to 30% for 20 min, further increasing to 40% at 30 min, to 50% at 35 min, and finally to 50% at 40 min. To avoid any memory effects, the column was eluted isocratically for 10 min before the next injection. A clear separation of the phenolic compounds was achieved using the Eclipse XDB C8 column (250 mm × 4.6 mm × 5 μm, Agilent, Santa Clara, USA) at 25 °C.

The mass spectrometer was the LC/MSD trap SL (Agilent). The MS conditions were as follows: Injection volume: 3.5 μL; Source conditions: Drying gas (nitrogen) 8 L/min at 330° C; Nebulizer pressure: 50 psi; Mass range: 100–1000; Scan mode: negative (−). Identification of phenolic compounds was achieved by comparing the mass to charge values [M−H]- of individual peaks shown at total ion chromatograms with those identified previously in the literature. Analysis of *Lavandula stoechas* samples was run in duplicate ($n = 2$).

2.3.6. Determination of in Vitro Antioxidant Activity

Preparation of DPPH free radical standard solution
A standard solution of [DPPH•] equal to ca. 0.08 mM (mmol L^{-1}) was prepared by dissolving 0.031 g of the free radical in 100 mL of methanol.

2.3.7. Preparation of the DPPH Free Radical Calibration Curve

A calibration curve of concentration versus absorbance of [DPPH•] was prepared as follows: The aforementioned standard solution of DPPH was diluted with the addition of methanol. The resulting solutions (range of 0–31 mg L^{-1}) were vortexed, left in the dark (until measurements were made), and the absorbance was measured in a UV/VIS Spectrometer (Lambda 25, PerkinElmer, Waltham, USA) at λmax of 517 nm. The calibration curve of absorbance (y) versus concentration (x) of [DPPH•] was expressed by the following equation:

$$y = 0.0243x - 0.0001; R^2 = 0.998 \tag{1}$$

2.3.8. Determination of in Vitro Antioxidant Activity of *Lavandula stoechas* Aqueous and Methanolic Extracts

The antioxidant activity of *Lavandula stoechas* aqueous and methanolic extracts was estimated in vitro using the [DPPH•] assay. For the determination of antioxidant activity, a volume of 3.0 mL of [DPPH•] solution (0.08 mM) was placed in the cuvette (final volume of 3 mL) and the absorbance of the [DPPH•] radical was measured at $t = 0$ ($A_0 = 0.7565$). Subsequently, 0.20 mL of *Lavandula stoechas* aqueous and methanolic extracts was placed in the respective cuvettes plus 2.8 mL of the [DPPH•] solution. The absorbance was measured every 15 min (regular time periods) until the value reached a plateau (steady state, A_t). The reaction between the free radical and *Lavandula stoechas* water soluble and methanolic antioxidants was accomplished at 30 min. The absorbance of the reaction mixture was measured at 517 nm.

The [DPPH•] antioxidant activity (%AA) with respect to aqueous and methanolic extract of *Lavandula stoechas* was calculated using the following equation:

$$\%AA = ((A_0 - A_t)/A_0) \times 100 \tag{2}$$

where A_0 is the initial absorbance of the [DPPH•] free radical standard solution and A_t is the absorbance of the remaining [DPPH•] free radical after reaction with *Lavandula stoechas* antioxidants, at steady state (plateau). For the estimation of *Lavandula stoechas* effective concentration EC_{50}, termed as the concentration of antioxidants that could cause a decrease in the DPPH free radical by 50%, further dilutions from the initial extracts (11,333.33 mg L^{-1}) were prepared for methanolic and aqueous extract from the initial mother solution in the range 500–11,333.33 mg L^{-1} and the value of EC_{50} was obtained from graphs of [DPPH•] remaining versus concentration of extracts. For this antioxidant test, methanol was used as the blank. Prior to absorbance measurements, the extracts were filtered using Whatman filters (Whatman plc, Buckinghamshire, UK) with a pore size of 0.45 μm. Each sample was run in triplicate ($n = 3$).

2.3.9. Determination of Total Phenolic Content

The total phenolic content of *Lavandula stoechas* was determined in its aqueous and methanolic extracts using the Folin-Ciocalteu assay according to Singleton et al. [15]. In particular, in a 5 mL volumetric flask, 0.2 mL of herb extract, followed by 2.5 mL of distilled water and 0.25 mL of Folin-Ciocalteu reagent, were added. After 3 min, 0.5 mL of saturated Na_2CO_3 (30%, $w\,v^{-1}$) was also added and the final volume of 5 mL was reached using distilled water. The flask was then stored in a dark place and the absorbance was measured after 2 h at 760 nm in a UV/VIS Spectrophotometer (SHIMADZU, UV-1280, Kyoto, Japan). Before any absorbance measurements, the solutions were filtered using Whatman filters (Whatman plc, Buckinghamshire, UK) with a pore size of 0.45 μm. For the quantification analysis, a calibration curve of standard gallic acid was constructed in wide range of concentrations: 0–6240 mg L^{-1}. The equation obtained was of the following form:

$$y = 0.0004x + 0.1479, R^2 = 0.9895 \tag{3}$$

Total phenolic content was expressed as mg of gallic acid equivalents per mL (or g L^{-1}) of *Lavandula stoechas* extract. Each sample was analysed in triplicate ($n = 3$).

3. Headspace Solid Phase Microextraction Coupled to Gas Chromatography/Mass Spectrometry (HS-SPME/GC-MS)

3.1. Headspace Extraction of Volatile Compounds

The extraction of volatile compounds of *Lavandula stoechas* was carried out using a divinyl benzene/carboxen/polydimethylsiloxane (DVB/CAR/PDMS) fiber of 50/30 μm purchased from Supelco (Bellefonte, PA, USA). Before analysis of samples, the fiber was conditioned according to the manufacturer's recommendations and was cleaned daily using the method of "clean" program to avoid any source of contamination. In particular, the injector and MS-transfer line were maintained at 260 °C and 270 °C, respectively, whereas during the "cleaning" of the fiber, the oven temperature was held at 80 °C for 0 min, and then was increased to 270 °C at 10 °C per min (2 min hold). A split ratio of 10:1 was used.

For the sample analysis, the following conditions were followed: 15 min equilibration time, 30 min sampling time, and 45 °C water bath temperature. Approximately, 0.05 g of *Lavandula stoechas* was then placed in a 15 mL vial equipped with a polytetrafluoroethylene PTFE/silicone septa screw-cap, along with a magnetic stirrer. The vials were introduced into a water bath of 45 °C under continuous stirring (600 rpm) during the headspace extraction [16].

3.2. Gas Chromatography-Mass Spectrometry Unit and Analysis Conditions

The GC unit used in the study for the gas chromatography/mass spectrometry analysis of *Lavandula stoechas* samples was an Agilent 7890A model coupled to a MS detector (Agilent 5975, Agilent, Santa Clara, USA). The DB-5MS capillary column (Agilent, Santa Clara, USA) was used in the analysis with the following characteristics: cross-linked 5% PH ME siloxane, of 60 m × 320 μm i.d., × 1 μm film thickness. Helium of excellent purity (99.999%) was the carrier gas at a flow rate of 1.5 mL min^{-1}. The injector and MS-transfer line were maintained at 250 °C and 270 °C, respectively, whereas, during the analysis, the oven temperature was maintained at 40 °C for 3 min, and then was increased to 260 °C at a rate of 8 °C min^{-1} (6 min hold). The electron impact mass spectra were recorded at a mass range of 50–550, and the ionization energy was 70 eV, whereas a split ratio of 1:2 was used to introduce the appropriate amount of sample onto the column [16].

3.3. Identification of Lavandula stoechas *Volatile Compounds*

The identification of volatile compounds was achieved using the Wiley 7, NIST 2005 mass spectral library (John Wiley & Sons, New York, USA). The Kovats indices were determined by the use of a mixture of n-alkanes (C8–C20) supplied by Supelco (Bellefonte, PA, USA). Prior to gas chromatography-mass spectrometry (GC/MS) analysis, the mixture was dissolved in n-hexane and the retention time of standards was determined according to the aforementioned temperature-programmed run. Only volatile compounds having ≥80% similarity with the Wiley library were tentatively identified using the GC-MS spectra. The method of identification was based on the combination of MS data found in the Wiley 7 NIST 2005 mass spectral library and data of Kovats index values that were determined for each volatile compound and then compared with those included in the Wiley MS library [16]. Data were expressed as a percentage (contribution of the area of each volatile to the total peak area multiplied by 100). Volatile compounds identified only in replicated samples were used in the study.

3.4. Graph Preparation and Statistical Analysis

Graphs and average ± standard deviation values of physico-chemical parameters, in vitro antioxidant activity, total phenolic content and volatile compounds were prepared using the SPPS statistics software (v.20, 2013, IBM, New York, USA) and Microsoft Office Excel spread sheets for Windows 2007 (Microsoft, Redmond, USA). A t-test and Pearson's bivariate statistics were carried out using the SPPS statistics software. The level of significance was considered to be that of $p \leq 0.05$.

4. Results and Discussion

4.1. Physico-Chemical Parameter Analysis

The pH of *Lavandula stoechas*, showing the active acidity of the sample, was slightly acidic in the range of 5.74 ± 0.03. The electric potential (E0), which in our case could show how the electric potential energy of any charged bio-molecule(s) at any location in the solution is divided by the charge of that/those molecule(s), was 73.4 ± 0.2 (mV). In addition, low salinity content was observed (0.29 ± 0.00 g L^{-1}) and the total dissolved solids content was in the range of 280 ± 2 mg L^{-1}. The electrical conductivity, a measure that defines the charged molecules in an aqueous medium, was 579 ± 1 µS cm^{-1} and the liquid resistivity, a measure that defines how strongly a material opposes the flow of electric current, was 1780 ± 20 Ω (Ohm). Finally, free acidity was in the range of 33 ± 3 g NaOH per 100 g of dried *Lavandula stoechas*. To our knowledge, the data are scarce on the physicochemical parameters of *Lavandula stoechas* aqueous extracts. Therefore, comparisons with literature data cannot be provided. In the work of Murthy et al. [14], it was reported that *Lavandula bipinnata* seed oil showed a much lower acidity: ca. 5.76 mg NaOH per g.

4.2. Phenolic Profile: Qualitative and Quantitative Determinations

4.2.1. Qualitative Analysis

The HPLC-ESI/MS analysis enabled the determination of 12 phytochemicals as shown in Table 1. Figure 1 shows a typical total ion and UV chromatogram of the aqueous *Lavandula stoechas* extract pointing out with numbers the identified phytochemicals according to retention time.

Table 1. Phytochemicals identified in methanolic and aqueous extracts of Ladastacho.

Peak Number	RT (min)	[M−H]$^-$ (*m/z*)	Phytochemicals
1	3.0	375(100), 573(46), 751(38)	5-nonadecylresorcinol
2	4.0	387(100), 607(30)	Unknown
3	8.6	179(100), 135(25)	Caffeic acid
4	11.1	473(100)	6''-O-Acetylgenistin
5	15.8	463(100), 485(20)	Quercetin 3-O-glucoside
6	16.1	461(100), 285(3)	Luteolin-O-glucuronide
7	16.6	447(100)	Luteolin-O-glucoside
8	17.7	359(100), 719(86)	Rosmarinic acid isomer
9	19.2	445(100), 269(5)	Apigenin-O-glucuronide
10	19.4	359(100), 719(30)	Rosmarinic acid
11	20.7	549(100), 507(94)	Salvianolic acid derivative
12	22.1	717(100), 519(43)	Salvianolic acid B (Lithospermic acid B)

RT, retention time (minutes).

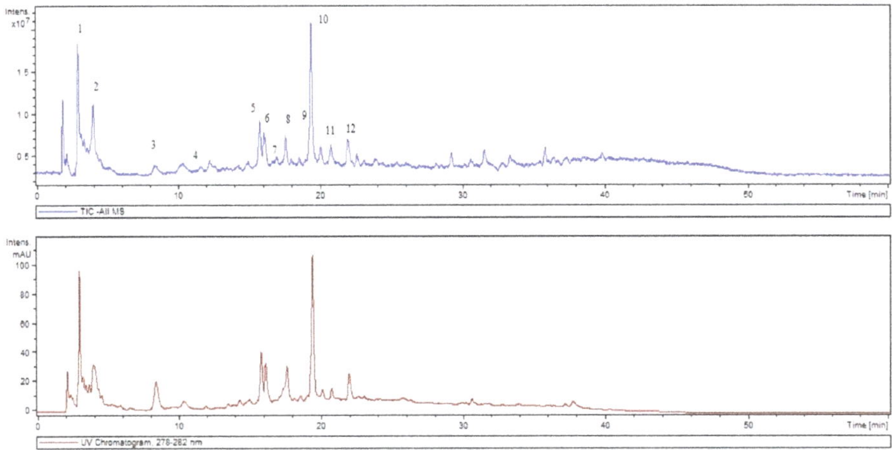

Figure 1. Total ion and UV chromatograms of *Lavadula stoechas* aqueous extract. Phytochemicals are numbered according to the retention time shown in Table 1.

Caffeic acid has been previously reported to be the main natural phenol in argan oil [17]. It has also been found in the freshwater fern *Salvinia molesta* [18], in the mushroom *Phellinus linteus* [19], in the bark of *Eucalyptus globulus* [20] and in barley grain [21]. Its antioxidant potential has been reported in vitro and also in vivo [22]. In addition, caffeic acid has been reported to possess an inhibitory effect on cancer cell proliferation in human HT-1080 fibro-sarcoma cells [23]. Despite its antioxidant potential, there are some studies that report adverse results regarding the carcinogenicity or anti-carcinogenicity of caffeic acid [24].

Rosmarinic acid is mainly found in culinary herbs, such as *Ocimum basilicum* (basil), *Ocimum tenuiflorum* (holy basil), *Melissa officinalis* (lemon balm), *Rosmarinus officinalis* (rosemary), *Origanum majorana* (marjoram) and *Salvia officinalis* (sage), thyme and peppermint or in plants with medicinal properties, such as common self-heal (*Prunella vulgaris*) or in species of the genus Stachys [25]. It is also found in other Lamiales, such as *Heliotropium foertherianum*, a plant in the family Boraginaceae. Rosmarinic acid accumulation is favored in hornworts in the fern family Blechnaceae and in species of several orders of mono- and dicotyledonous angiosperms [26], which may be found also as a derivative in Anthocerotophyta (*Anthoceros agrestis*) in the form of rosmarinic acid 3′-*O*-β-D-glucoside [27].

Regarding some potential health effects, rosmarinic acid is a potential anxiolytic as it acts as a gamma-aminobutyric acid (GABA) transaminase inhibitor, more specifically on 4-aminobutyrate transaminase [28]. Rosmarinic acid also inhibits the expression of indoleamine 2,3-dioxygenase via its cyclooxygenase-inhibiting properties [29], whereas it was found to be effective in a mouse model of Japanese encephalitis [30].

Salvianolic acid B (Sal B) is a phenolic acid derived from the dried root and rhizome of *Salvia miltiorrhiza* (Labiatae), which has been used traditionally in most Asian countries for the clinical therapy of various vascular diseases for hundreds of years. Salvia has been also used for various diseases related to blood stasis syndrome in China for thousands of years, and now it is widely used for cardiovascular diseases (CVDs) [31]. In addition, Sal B protects diverse kinds of cells from damage caused by a variety of toxic stimuli [32].

Furthermore, flavonoids may also have numerous beneficial effects in human health. For instance, apigenin and the substituted derivatives (i.e., -apigenin-7-*O*-glucuronide) isolated from the *Marrubium deserti* de Noé plant have been reported to possess antioxidant and antigenotoxic properties [33]. The same holds for Luteolin 7-glucoside, often called cynaroside, which is a flavone. It has been identified in *Campanula persicifolia* and *Campanulla rotundifolia* [34], *Teucrium gnaphalodes* [35], *Ferula varia* and *Ferula foetida* [36], the bamboo *Phyllostachys nigra* [37], dandelion coffee, *Cynara scolymus* (artichoke) [38] and *Phoenix hanceana* var. formosana [39]. What is remarkable is that antioxidant and neural cell protective effects have also been reported for the latter ethanolic extract, in which the cynaroside was identified [39].

4.2.2. Quantitative Analysis: Total Phenolic Content

The total phenolic content of the aqueous extract of *Lavandula stoechas* recorded a much higher value (ca. 4289 mg L^{-1}) compared to the methanolic one (ca. 217 mg L^{-1}), indicating that the aqueous medium (higher polarity) favoured the effective release/isolation of phytochemicals, especially phenolic acids and flavonoids, among other reducing agents (i.e., ascorbic acid). In a recent work [40], the preparation of *Lavandula angustifolia* beverages with either the method of infusion or decoction resulted in a beverage with a higher phenolic content when water, among other solvents, was used for the extraction of phytochemicals, in agreement with the present results.

The alcoholic and water extracts (1:1, *v:v*) of Romanian *Lavandula angustifolia* resulted in a significantly lower total phenolic content expressed as gallic acid (GA) equivalents (50.6 $\pm$ 3.2 mg GA g^{-1}) compared to the results of the present study [41].

4.2.3. In Vitro Antioxidant Activity

Both methanolic and aqueous extracts of *Lavandula stoechas* showed in vitro antioxidant activity during the DPPH assay. However, the aqueous extract (of higher polarity) recorded a significantly ($t = -4.196$, $df = 4$, $p = 0.014$) higher antioxidant activity. In addition, higher antioxidant activity was recorded for the higher concentration of the extracts (Figure 2). The EC$_{50}$ values were 7.05 mg mL^{-1} and 1.78 mg mL^{-1} for the methanolic and aqueous extracts, respectively.

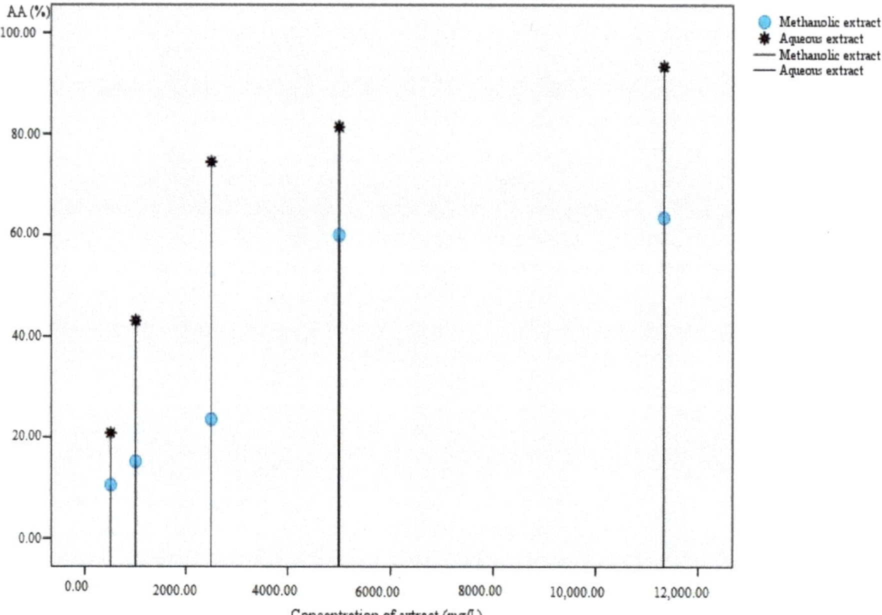

Figure 2. Antioxidant activity (AA%) of *Lavavdula stoechas* with respect to extract concentration (mg L^{-1}).

What is also of great importance is that there was a perfect correlation between the in vitro antioxidant activity and the total phenolic content of the methanolic and aqueous extracts using Pearson's bivariate statistics ($r = 1$, $p = 0.01$). The present results are in agreement with those of Hu et al. [37], who reported that solvent-extracted bamboo leaf extract (BLE) containing chlorogenic acid, caffeic acid and luteolin 7-glucoside showed in vitro antioxidant activity using the DPPH assay (among others). The authors also reported that BLE exhibited a concentration-dependent scavenging activity of the [DPPH•], in agreement with the present results.

Lavandula angustifolia extracts from the region of Southeast Romania showed a considerable in vitro antioxidant activity against the DPPH free radical, as reported by Spiridon et al. [41].

In a more recent work, *Lavandula angustifolia* (from the region of Polykarpi Pozar in Greece) aqueous extracts prepared with infusion or decoction also showed a high in vitro antioxidant activity against the DPPH free radical (>90%), in agreement with the present results [40].

4.2.4. Volatile Compounds of *Lavandula stoechas*

Fifty volatile compounds were tentatively identified using HS-SPME/GC-MS, belonging to alcohols, aldehydes, ketones, norisoprenoids and numerous terpenoids. In Table 2 is given (among other data) the percentage contribution of each volatile compound to the total ones. Figure 3 shows a typical gas chromatogram of *Lavandula stoechas*, pointing with numbers to some indicative volatile compounds. The most abundant volatiles were a-thujone (32.14%), 1,3,3-trimethyl-2-oxabicyclo[2.2.2] octane (9.77%), (+)-myrtenyl acetate (7.75%), camphor (6.07%), 1-octen-3-yl-acetate (3.94%) and fenchyl acetate (3.91%), followed by minor proportions of numerous others. A considerable number of these compounds have been reported in the essential oil of plants with edible modified stems, roots and bulbs in the Solanaceae, Tropacolaceae, Typhaceae and Zingiberaceae families, which are grown all over the world [42].

The biosynthesis of thujone in plants starts with the formation of geranyl diphosphate (GPP) from dimethylallyl pyrophosphate (DMAPP) and isopentenyl diphosphate (IPP), catalyzed by the enzyme geranyl diphosphate synthase [43]. In addition, Umlauf and Zapp [44], using carbon nuclear magnetic spectroscopy (^{13}C-NMR), reported that the isoprene units used to form thujone in plants are derived from an alternative pathway, the methylerythritol phosphate pathway (MEP).

Different opinions occur regarding the use of thujone as a food or drink supplement. In particular, the maximum thujone levels in the E.U. are related to the Artemisia species used along with the matrix (food or beverage). In that sense, the respected range may be within 0.5–35 mg kg^{-1} [45,46].

On the other hand, in the U.S., the addition of pure thujone to foods is not permitted [47]. Foods or beverages that contain Artemisia species, white cedar, oak moss, tansy or yarrow must be free of thujone [48], which practically means that these contain less than 10 mg L^{-1} of thujone [49]. Sage and sage oil (which may contain up to 50% thujone) are on the Food and Drug Administration's list of generally recognized as safe (GRAS) substances [50]. The average lethal dose value, or LD$_{50}$, of alpha-thujone, the more active of the two isomers (alpha- and beta-) found in plant-derived products, has been reported to be ca. 45 mg kg^{-1} in mice, with a 0% mortality rate at 30 mg kg^{-1} and a 100% mortality rate at 60 mg kg^{-1}, respectively [51].

The biochemical pathway of camphor production involves the presence of geranyl pyrophosphate, which, via the cyclisation of linaloyl pyrophosphate, turns to bornyl pyrophosphate, followed by hydrolysis to borneol and then oxidation to camphor. Camphor has been used in traditional medicine since ancient times in countries where it was native. Its characteristic odour and its decongestant effect on the treatment of sprains, swellings and inflammation have probably led to its use in medicine [52,53]. Camphor was also used for centuries in Chinese medicine for a variety of health issues. The lethal doses in adults are in the range of 50–500 mg kg^{-1} after oral exposure. In particular, 2 g may cause serious toxicity and 4 g is potentially lethal [54]. To the best of our knowledge, this is the first report in the literature reporting the volatile profile of Greek *Lavandula stoechas*.

Table 2. Volatile compounds identified in *Lavandula stoechas*

RT (min)	KI	RT_{lit}	Contribution (%)	Volatile compounds	Qualification	MOI
6.51	<800	<800	0.06	3-Buten-2-one (Vinyl methyl ketone)	91	MS/KI
7.14	<800	<800	0.08	3-Buten-2-ol, 2-methyl	90	MS/KI
12.31	<800	<800	0.08	Hexanal	95	MS/KI
14.62	885	891	0.04	2-Heptanone	94	MS/KI
15.31	913	915	0.28	3,3-Dimethylallyl acetate	81	MS/KI
15.78	932	-	0.04	2,4,4-Trimethylcyclopentanone	94	MS/KI
16.06	944	937	0.44	(1S)-2,6,6-Trimethylbicyclo[3.1.1]hept-2-ene (apha-Pinene)	96	MS/KI
16.48	962	958	0.48	(−)-7,7-Dimethyl-2-methylenebicyclo[2.2.1]heptane (alpha-Fenchene)	97	MS/KI
16.56	965	964	0.89	2,2-Dimethyl-3-methylidenebicyclo[2.2.1]heptane (Camphene)	97	MS/KI
16.96	982	985	0.16	3-Octanone	96	MS/KI
17.22	993	980	0.07	6,6-dimethyl-2-methylene-Bicyclo[3.1.1]heptane	95	MS/KI
18.14	1034	1023	2.36	1-methyl-4-(1-methylethyl)-Benzene	97	MS/KI
18.27	1039	1035	1.19	1-methyl-4-(prop-1-en-2-yl)Cyclohex-1-ene (dl-Limonene)	99	MS/KI
18.45	1048	1044	9.77	1,3,3-trimethyl-2-oxabicyclo[2.2.2]Octane	98	MS/KI
18.85	1066	1068	0.19	1-methyl-4-(1-methylethyl)-1,4-Cyclohexadiene	97	MS/KI
19.20	1082	1080	1.05	2-Furanmethanol, 5-ethenyltetrahydro-.alpha.,.alpha.,5-trimethyl-, cis- (cis-Linalool oxide)	91	MS/KI
19.31	1086	1096	0.19	[(1S,4S)-4-methyl-1-propan-2-yl-4-bicyclo[3.1.0]hexanyl] acetate (trans-Sabinene hydrate)	96	MS/KI
19.44	1092	-	0.28	3,4,4-trimethyl-2-Cyclohexen-1-one	83	MS/KI
19.51	1096	1099	0.10	1-methyl-4-(1-methylethylidene)-Cyclohexene	98	MS/KI
19.59	1099	1110	3.94	1-Octen-3-yl acetate	90	MS/KI
19.86	1112	1117	32.14	α: (1S,4R,5R)-4-methyl-1-(propan-2-yl)bicyclo[3.1.0]Hexan-3-one (alpha-Thujone)	95	MS/KI
20.40	1139	1121	2.80	(1R,3S,4S)-2,2,4-trimethylbicyclo[2.2.1]heptan-3-ol (alpha- Fenchyl alcohol)	97	MS/KI
21.12	1174	1146	6.07	1,7,7-Trimethylbicyclo[2.2.1]Heptan-2-one (Camphor)	97	MS/KI
21.34	1185	1164	0.66	6,6-dimethyl-4-methylidenebicyclo[3.1.1]Heptan-3-one (Pinocarvone)	81	MS/KI
21.55	1195	1164	0.40	endo-1,7,7-Trimethyl- bicyclo[2.2.1]Heptan-2-ol (Borneol)	93	MS/KI
21.67	1201	1183	0.27	1-(3-methylphenyl)-Ethanone	95	MS/KI
21.83	1209	1195	0.35	.alpha.,.alpha.4-trimethyl-3-Cyclohexene-1-methanol (alpha-Terpineol)	91	MS/KI
22.01	1219	1195	0.73	6,6-Dimethylbicyclo[3.1.1]hept-2-ene-2-carboxaldehyde (Myrtenal)	97	MS/KI
22.26	1232	1223	3.91	(2,2,4-trimethyl-3-bicyclo[2.2.1]Heptanyl) acetate (Fenchyl acetate)	96	MS/KI
23.26	1284	1280	0.11	4-Isopropenyl-1-methyl-7-oxabicyclo[4.1.0]heptan-2-one (trans-Carvone oxide)	88	MS/KI
23.55	1300	1285	2.04	Bicyclo[2.2.1]heptan-2-ol, 1,7,7-trimethyl-, acetate, (1S-endo)-	98	MS/KI

Table 2. *Cont.*

RT (min)	KI	RT_lit	Contribution (%)	Volatile compounds	Qualification	MOI
23.67	1306	-	1.61	1-methyl-4-(1-methylethylidene)-Cyclohexanol (γ-Terpineol)	91	MS/KI
23.91	1320	-	0.15	1-methylene-4-(1-methylethenyl)-Cyclohexane (Pseudolimonene)	85	MS/KI
23.96	1322	-	0.32	2,2-dimethyl-3-methylene- Bicyclo[2.2.1]heptane [Camphene, (1R,4S)-(+)-]	91	MS/KI
24.20	1336	1332	7.75	(6,6-Dimethylbicyclo[3.1.1]hept-2-en-2-yl)methyl acetate [(-)-Myrtenyl acetate]	87	MS/KI
24.75	1367	1352	0.06	1H-Cyclopenta[1,3]cyclopropa[1,2]benzene, 3a,3b,4,5,6,7-hexahydro-3,7-dimethyl-4-(1-methylethyl)-, [3aS-(3aα,3bβ,4β,7α,7aS*)-(-)-] (alpha-Cubebene)	99	MS/KI
24.79	1369	1369	0.09	2,6-Octadien-1-ol, 3,7-dimethyl-, acetate, (Z)- [cis-Geranyl acetate)	91	MS/KI
25.44	1405	1378	0.61	1,2,4-Metheno-1H-indene, octahydro-1,7a-dimethyl-5-(1-methylethyl)-, [1S-(1.alpha.,2.alpha.,3a.beta.,4.alpha.,5.alpha.,7a.beta.,8S*)]-[(+)-Cyclosativene]	94	MS/KI
25.84	1429	1396	0.07	1,4-Methano-1H-indene, octahydro-4-methyl-8-methylene-7-(1-methylethyl)-, [1S-(1.alpha.,3a.beta.,4.alpha.,7.alpha.,7a.beta.)]- [(+)-Sativene]	99	MS/KI
27.25	1515		0.08	2,3-dihydro-1,3,3-trimethyl-2-methylene-1H-Indole	80	MS/KI
27.39	1523	1505	0.06	Naphthalene, 1,2,4a,5,6,8a-hexahydro-4,7-dimethyl-1-(1-methylethyl)-, (1S,4aS,8aR)- (alpha-Muurolene)	98	MS/KI
27.49	1530	1509	0.62	Naphthalene, decahydro-4a-methyl-1-methylene-7-(1-methylethenyl)-, [4aR-(4a.alpha.,7.alpha.,8a.beta.)]-(beta-Selinene)	99	MS/KI
27.55	1534	1505	0.05	alpha-Eudesma-3,11-diene (alpha-Selinene)	99	MS/KI
27.68	1542	1540	0.12	Naphthalene, 1,2,3,5,6,8a-hexahydro-4,7-dimethyl-1-(1-methylethyl)-, (1S-cis)-	99	MS/KI
28.30	1581	1525	0.13	Naphthalene, 1,2,3,4,4a,7-hexahydro-1,6-dimethyl-4-(1-methylethyl)-	87	MS/KI
28.85	1617	-	0.08	4aH-Cycloprop[e]azulen-4a-ol, decahydro-1,1,4,7-tetramethyl-, [1aR-(1a.alpha.,4.beta.,4a.beta.,7.alpha.,7a.beta.,7b.alpha.)]- (Palustrol)	96	MS/KI
29.07	1631	1613	0.21	(4,6)]dodecane,,12-trimethyl-9-methylene-, [1R-(1R*,4R*,6R*,10S*)]- (CAS); (-)-.beta.-Caryophyllene epoxide	91	MS/KI
29.24	1643	1620	0.39	1H-Cycloprop[e]azulen-4-ol, decahydro-1,1,4,7-tetramethyl-, [1ar-(1a.alpha.,4.beta.,4a.beta.,7.alpha.,7a.beta.,7b.alpha.)]- (Viridiflorol)	99	MS/KI
29.40	1653	-	0.11	1H-Cycloprop[e]azulene, decahydro-1,1,7-trimethyl-4-methylene-, [1aR-(1a.alpha.,4a.alpha.,7.alpha.,7a.beta.,7b.alpha.)]- [(+)-Aromadendrene]	94	MS/KI

RT: retention time (minutes), KI: experimental retention indices values based on the calculations of Kovats index values (KI) using the standard mixture of alkanes. RI_lit: Retention indices of the identified compounds according to literature data included in Wiley 7 NIST MS library. Qualification: Percentage accuracy of volatile compounds identified using Wiley 7 NIST MS library data. Results reported are the average $\pm$ standard deviations values of two independent replicates ($n = 2$).

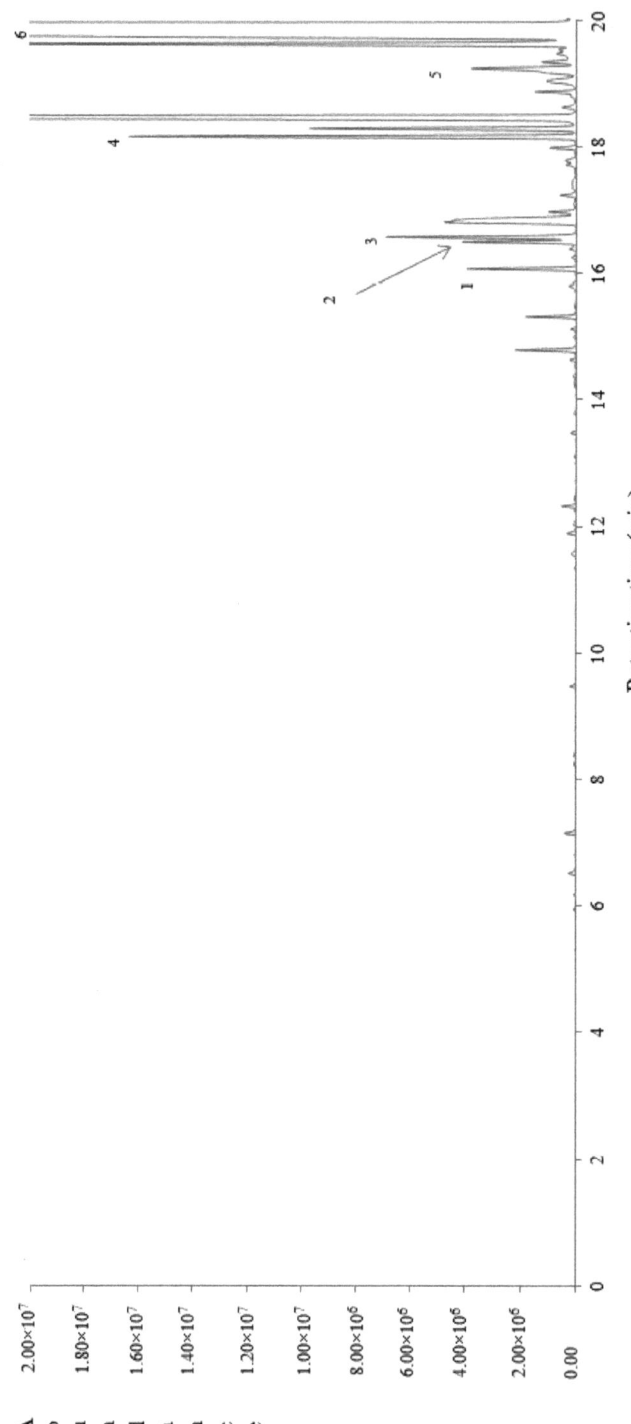

Figure 3. *Cont.*

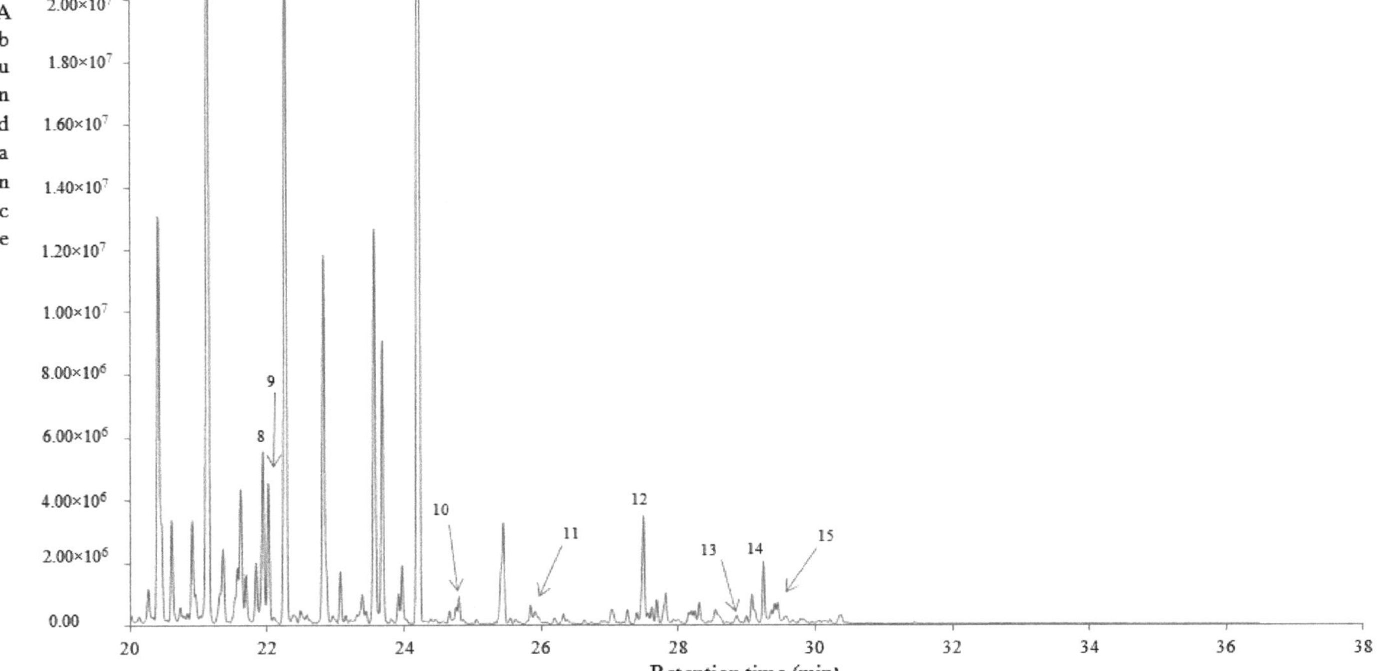

Figure 3. A typical gas chromatogram of *Lavandula stoechas*. Indicative volatile compounds (markers) are numbered according to the retention time shown in Table 2. 1: alpha-Pinene, 2: alpha-Fenchene, 3: Camphene, 4: para-Cymene, 5: cis-Linalool oxide, 6: alpha-Thujone, 7: Camphor, 8: alpha-Terpineol, 9: Myrtenal, 10: cis-Geranyl acetate, 11: (+)-Sativene, 12: beta-Selinene, 13: Viridiflorol, 14: (+)-Aromadendrene.

5. Conclusions

Lavandula stoechas from the region of Saidona proved to be a good source of phytochemicals with a high in vitro antioxidant activity, especially in its aqueous form. In addition, numerous terpenoids were tentatively identified, producing a strong, penetrating and menthol-like odour. The mode of mechanistic action by the plant may involve its cultivation and exploitation under the prism of natural "medicinal agents" that may favor the healthcare of humans. In parallel, the significance of its use as a supplement by the food industry or related industries in specific amounts may favor the preparation and distribution of health-promoting foods and beverages (HPFB) of a distinct flavour. Future research will be focused on the determination of the cytotoxicity of *Lavandula stoechas* dried proportions or aqueous extracts used in the present work (ca. 1%, $w\,v^{-1}$), based on in vivo measurements, to approve or not its prospective use in different forms in foods or beverages.

Author Contributions: Conceptualization, I.K.K.; methodology, I.K.K.; validation, I.K.K. and K.A.R.; formal analysis, I.K.K. and V.K.K.; investigation, I.K.K.; resources, K.A.R. and I.K.K.; data curation, I.K.K. and V.K.K.; writing—original draft preparation, I.K.K.; writing—review and editing, I.K.K.; visualization, I.K.K.; supervision, I.K.K.

Funding: This research received no external funding.

Acknowledgments: The authors would like thank the Mass Spectrometry Unit at the Chemistry Department of the University of Ioannina and M.G. Kontominas, who provided access to the GC/MS unit at the laboratory of Food Chemistry. The authors also acknowledge Vassiliki G. Kontogianni for her technical assistance during the HPLC/ESI-MS analysis. The APC of the article were covered through the FOODS TRAVEL AWARDS 2018 scholarship, nominated to Ioannis K. Karabagias, for his contribution to science.

Conflicts of Interest: The authors declare no conflict of interest.

References

1. Outdoor Flowering Plants-Mona Lavender. Available online: www.hgtv.com (accessed on 19 October 2018).
2. Plant Finder-Plectranthus Mona Lavender. Available online: www.missouribotanicalgarden.org (accessed on 19 October 2018).
3. Grieve, M.A. *Modern Herbal*; Dover Publications, Inc.: New York, NY, USA, 1971; Volume II, ISBN 0-486-22799-5.
4. Kathleen, N.B. *The Sunset Western Garden Book*, 7th ed.; Lane Magazine & Book Company: Los Angeles, CA, USA, 1975.
5. Matt, E. *Lavender*; University of Kentucky Center for Crop Diversification: Lexington, KY, USA, 2017.
6. Upson, T.; Andrews, S. *The Genus Lavandula. Royal Botanic Gardens*; Kew 2004; CRC Press: Boca Raton, FL, USA, 2004; ISBN 9780881926422.
7. Lis-Balchin, M. *Lavender: The Genus Lavandula*; Taylor and Francis: Abingdon, UK, 2002; ISBN 9780203216521.
8. Integrative Medicine Communications, Germany, American Botanical Council. Expanded Commission E Monograph: "Lavender Flower". 2000. Available online: cms.herbalgram.org (accessed on 18 October 2018).
9. British Broadcasting Corporation. Oils 'Make Male Breasts Develop'. February 2007. Retrieved 2018-03-17. Available online: https://www.bbc.com/news/health-43429933 (accessed on 29 December 2018).
10. British Broadcasting Corporation. More Evidence Essential Oils 'Make Male Breasts Develop. 2018. Available online: https://www.realclearscience.com/2018/03/19/more_evidence_essential_oils_make_male_breasts_develop_280504.html (accessed on 19 March 2018).
11. Lavender: Science and Safety, National Center for Complementary and Integrative Health. 2007. Available online: https://nccih.nih.gov/health/lavender/ataglance.htm (accessed on 5 November 2013).
12. Cavanagh, H.M.A.; Wilkinson, J.M. "*Lavender Essential Oil: A Review, Australian Infection Control*; CSIRO Publishing: Clayton, Australia, 2005.
13. Placzek, M.; Frömel, W.; Eberlein, B.; Gilbertz, K.P.; Przybilla, B. Evaluation of phototoxic properties of fragrances. *Acta Derm. Venereol.* **2007**, *87*, 312–316. [CrossRef] [PubMed]
14. Murthy, H.N.; Manohar, S.H.; Naik, P.M.; Lee, E.J.; Paek, K.Y. Physicochemical characteristics and antioxidant activity of *Lavandula bipinnata* seed oil. *Int. Food Res. J.* **2014**, *21*, 1473–1476.

15. Singleton, V.L.; Orthofer, R.; Lamuela-Raventós, R.M. Analysis of total phenols and other oxidation substrates and antioxidants by means of Folin-Ciocalteu reagent. *Meth. Enzymol.* **1999**, *299*, 152–178.

16. Karabagias, I.K.; Nikolaou, C.; Karabagias, V.K. Volatile fingerprints of common and rare honeys produced nn Greece: In search of PHVMs with implementation of the honey code. *Eur. Food Res. Technol.* **2018**, in press. [CrossRef]

17. Charrouf, Z.; Guillaume, D. Phenols and Polyphenols from Argania spinosa. *Am. J. Food Technol.* **2007**, *2*, 679–683. [CrossRef]

18. Choudhary, M.I.; Naheed, N.; Abbaskhan, A.; Musharraf, S.G.; Hina, S.; Rahman, A.U. Phenolic and other constituents of fresh water fern Salvinia molesta. *Phytochemistry* **2008**, *69*, 1018–1023. [CrossRef] [PubMed]

19. Lee, Y.S.; Kang, Y.S.; Jung, J.Y.; Lee, S.; Ohuchi, K.; Shin, K.H.; Kang, I.J.; Park, J.H.Y.; Shin, H.K.; Soon, S. Protein glycation inhibitors from the fruiting body of *Phellinus linteus*. *Biol. Pharm. Bull.* **2008**, *31*, 1968–1972. [CrossRef]

20. Santos, S.O.; Freire, C.R.S.; Domingues, M.R.M.A.; Silvestre, A.J.D.; Pascoal, N.C. Characterization of phenolic components in polar extracts of Eucalyptus globulus labill. bark by high-performance liquid chromatography–mass spectrometry. *J. Agric. Food Chem.* **2011**, *59*, 9386–9393. [CrossRef]

21. Quinde-Axtell, Z.; Baik, B.K. Phenolic compounds of barley grain and their implication in food product discoloration. *J. Agric. Food Chem.* **2006**, *54*, 9978–9984. [CrossRef]

22. Olthof, M.R.; Hollman, P.C.; Katan, M.B. Chlorogenic acid and caffeic acid are absorbed in humans. *J. Nutr.* **2001**, *131*, 66–71. [CrossRef]

23. Rajendra, N.P.; Karthikeyan, A.; Karthikeyan, S.; Reddy, B.V. Inhibitory effect of caffeic acid on cancer cell proliferation by oxidative mechanism in human HT-1080 fibrosarcoma cell line. *Mol. Cell Biochem.* **2011**, *349*, 11–19. [CrossRef]

24. Hirose, M.; Takesada, Y.; Tanaka, H.; Tamano, S.; Kato, T.; Shirai, T. Carcinogenicity of antioxidants BHA, caffeic acid, sesamol, 4-methoxyphenol and catechol at low doses, either alone or in combination, and modulation of their effects in a rat medium-term multi-organ carcinogenesis model. *Carcinogenesis* **1998**, *9*, 207–212. [CrossRef]

25. Clifford, M.N. Chlorogenic acids and other cinnamates. Nature, occurrence and dietary burden. *J. Sci. Food Agric.* **1999**, *79*, 362–372. [CrossRef]

26. Petersen, M.; Abdullah, Y.; Benner, J.; Eberle, D.; Gehlen, K.; Hücherig, S.; Janiak, V.; Kim, K.H.; Sander, M.; Weitzel, C.; et al. Evolution of rosmarinic acid biosynthesis. *Phytochemistry* **2009**, *70*, 1663–1679. [CrossRef]

27. Vogelsang, K.; Schneider, B.; Petersen, M. Production of rosmarinic acid and a new rosmarinic acid 3′-O-β-D-glucoside in suspension cultures of the hornwort *Anthoceros agrestis* Paton. *Planta* **2006**, *223*, 369–373. [CrossRef] [PubMed]

28. Awad, R.; Muhammad, A.; Durst, T.; Trudeau, V.L.; Arnason, J.T. Bioassay-guided fractionation of lemon balm (*Melissa officinalis* L.) using an in vitro measure of GABA transaminase activity. *Phytother. Res.* **2009**, *23*, 1075–1081. [CrossRef]

29. Lee, H.J.; Jeong, Y.I.; Lee, T.H.; Jung, I.D.; Lee, J.S.; Lee, C.M.; Kim, J.I.; Joo, H.; Lee, J.D.; Park, Y.M. Rosmarinic acid inhibits indoleamine 2,3-dioxygenase expression in murine dendritic cells. *Biochem. Pharmacol.* **2007**, *73*, 1412–1421. [CrossRef] [PubMed]

30. Swarup, V.; Ghosh, J.; Ghosh, S.; Saxena, A.; Basu, A. Antiviral and anti-inflammatory effects of rosmarinic acid in an experimental murine model of Japanese encephalitis. *Antimicrob. Agents Chemother.* **2007**, *51*, 3367–3370. [CrossRef]

31. Wang, J.; Xiong, X.; Feng, B. Cardiovascular effects of salvianolic acid B. *Evid. Based Complement. Altern. Med.* **2013**, *2013*, 16. [CrossRef] [PubMed]

32. Lin, Y.H.; Liu, A.H.; Wu, H.L.; Westenbroek, C.; Song, Q.L.; Yu, H.M.; Gert, J.; Horst, T.; Li, X.J. Salvianolic acid B, an antioxidant from *Salvia miltiorrhiza* prevents Ab25–35-induced reduction in BPRP in PC12 cells. *Biochem. Biophys. Res. Commun.* **2006**, *348*, 593–599. [CrossRef]

33. Zaabat, N.; Hay, A.E.; Michalet, S.; Darbour, N.; Bayet, C.; Skandrani, I.; Chekir-Ghedira, L.; Akkal, S.; Dijoux-Franca, M.G. Antioxidant and antigenotoxic properties of compounds isolated from Marrubium deserti de Noé. *Food Chem. Toxicol.* **2011**, *49*, 3328–3335. [CrossRef] [PubMed]

34. Teslov, L.S.; Teslov, S.V. Cynaroside and luteolin from Campanula persicifolia and C. *Rotundifolia*. *Chem. Nat. Comput.* **1972**, *8*, 117. [CrossRef]

35. Barberán, F.A.T.; Gil, M.I.; Tomás, F.; Ferreres, F.; Arques, A. Flavonoid aglycones and glycosides from Teucrium gnaphalodes. *J. Nat. Prod.* **1985**, *48*, 859–860. [CrossRef]

36. Yuldashev, M.P. Cynaroside content of the plants Ferula varia and F. *Foetida. Chem. Nat. Comput.* **1997**, *33*, 597–598. [CrossRef]

37. Hu, C.; Chun, Z.; Zhang, Y.; Kitts, D.D. Evaluation of antioxidant and prooxidant activities of bamboo Phyllostachys nigra Var. *Henonis leaf extract in vitro. J. Agric. Food Chem.* **2000**, *48*, 3170–3176. [CrossRef]

38. Nüßlein, B.; Kreis, W. Purification and characterization of a cynaroside 7-O-β-D-glucosidase from *Cynarae scolymi folium. Acta Hortic.* **2005**, *681*, 413–420. [CrossRef]

39. Lin, Y.P.; Chen, T.Y.; Tseng, H.W.; Lee, M.H.; Chen, S.T. Neural cell protective compounds isolated from Phoenix hanceana var. *Formosana. Phytochemistry* **2009**, *70*, 1173–1181. [CrossRef]

40. Deligiannidou, G.E.; Kontogiorgis, C.; Hadjipavlou-Litina Dimitra, D.; Lazari, D.; Konstantinidis, T.; Papadopoulos, A. Antioxidant contribution of lavender (*Lavandula angustifolia*), sage (*Salvia officinalis*), tilia (*Tilia tomentosa*) and sideritis (*Sideritis perfoliata*) beverages prepared at home. *SDRP J. Food Sci. Technol.* **2018**, *3*, 360–377.

41. Spiridon, I.; Colceru, S.; Anghel, N.; Teaca, C.A.; Bodirlau, R.; Armatu, A. Antioxidant capacity and total phenolic contents of oregano (*Origanum vulgare*), lavender (*Lavandula angustifolia*) and lemon balm (*Melissa officinalis*) from Romania. *Nat. Prod. Res.* **2011**, *25*, 1657–1661. [CrossRef]

42. Lim, T.K. *Edible Medicinal and Non-Medicinal Plants, Volume 12 Modified Stems, Roots, Bulbs*; Springer International Publishing: New York, NY, USA; Zürich, Switzerland, 2016.

43. Dewick, P.M. *Medicinal Natural Products: A Biosynthetic Approach*, 3rd ed.; John Wiley & Sons Ltd.: Hoboken, NJ, USA, 2009; pp. 195–197. ISBN 978-0-470-74167-2.

44. Umlauf, D.; Josef, J. Biosynthesis of the irregular monoterpene artemisia ketone, the sesquiterpene germacrene D and other isoprenoids in *Tanacetum vulgare* L. (Asteraceae). *Phytochemistry* **2004**, *65*, 2463–2470. [CrossRef]

45. European Commission. Regulation (EC) No 1334/2008 of the European Parliament and Council of 16 December 2008. Available online: https://eur-lex.europa.eu/legal-content/EN/TXT/?uri=CELEX%3A32008R1334 (accessed on 26 March 2019).

46. Opinion of the Scientific Committee on Food on Thujone Scientific Committee on Food. 2003. Available online: https://ec.europa.eu/food/sites/food/files/safety/docs/fs_food-improvement-agents_flavourings-out162.pdf (accessed on 28 October 2006).

47. Laurie, C.D.; Matulka, R.A.; Burdock, G.A. Naturally Occurring Food Toxins. *Toxins* **2009**, *2*, 2289–2332. [CrossRef]

48. Food and Drug Administration. FDA Regulation 21 CFR 172.510—Food Additives Permitted for Direct Addition to Food for Human Consumption. 2003. Available online: https://www.accessdata.fda.gov/scripts/cdrh/cfdocs/cfcfr/CFRSearch.cfm?CFRPart=172 (accessed on 28 October 2006).

49. Department of the Treasury Alcohol and Tobacco Tax and Trade Bureau Industry Circular (2007-5) 17 October 2007, Retrieved May 5 2009. Available online: https://www.ttb.gov (accessed on 29 December 2018).

50. Food and Drug Administration. Substances Generally Recognized as Safe. Archived 2005-11-30 at the Wayback Machine; 2003. Available online: https://www.fda.gov/food/ingredientspackaginglabeling/gras/ (accessed on 28 October 2006).

51. Höld, K.M.; Sirisoma, N.S.; Iked, T.; Narahashi, T.; Casida, J.E. Alpha-thujone (the active component of absinthe): Gamma-aminobutyric acid type A receptor modulation and metabolic detoxification. *Proc. Natl. Acad. Sci. USA* **2000**, *97*, 3826–3831. [CrossRef] [PubMed]

52. *Goodman and Gilman, Pharmacological Basis of Therapeutics*; Macmillan: Stuttgart, Germany, 1965; pp. 982–983.

53. Marsden, W. *The History of Sumatra Containing an Account. of the Government, Laws, Customs and Manners of the Native Inhabitants*; Cambridge University Press: Cambridge, UK, 2005.

54. Wickstrom, E. *Poisons Information Monograph: Camphor*; International Programme on Chemical Safety National Poison Center: Oslo, Norway, 1988.

© 2019 by the authors. Licensee MDPI, Basel, Switzerland. This article is an open access article distributed under the terms and conditions of the Creative Commons Attribution (CC BY) license (http://creativecommons.org/licenses/by/4.0/).

 *antioxidants*

Article

Effect of Different Cooking Methods on Polyphenols, Carotenoids and Antioxidant Activities of Selected Edible Leaves

K. D. Prasanna P. Gunathilake [1,2,*], K. K. D. Somathilaka Ranaweera [2] and H. P. Vasantha Rupasinghe [3]

[1]   Department of Food Science & Technology, Faculty of Livestock, Fisheries & Nutrition, Wayamba University of Sri Lanka, Makandura, Gonawila, Sri Lanka
[2]   Department of Food Science and Technology, Faculty of Applied Sciences, University of Sri Jayewardenepura, Gangodawila, Nugegoda, Sri Lanka; sranaweera@sjp.ac.lk
[3]   Department of Plant, Food, and Environmental Sciences, Faculty of Agriculture, Dalhousie University, Truro, NS B2N 5E3, Canada; vrupasinghe@dal.ca
*   Correspondence: gunathilakep@dal.ca; Tel.: +0094-776-594-198; Fax: +0094-229-9870

Received: 29 July 2018; Accepted: 28 August 2018; Published: 30 August 2018

Abstract: This study aimed to evaluate the effect of cooking (boiling, steaming, and frying) on polyphenols, flavonoids, carotenoids and antioxidant activity of six edible leaves. The total antioxidant capacity of the fresh and cooked leaves was determined using 2,2-diphenyl-1-picrylhydrazyl (DPPH) radical scavenging and singlet oxygen scavenging assays. The results revealed that frying caused a reduction in major bioactives and antioxidant activities in all leafy vegetables tested. However, steamed and boiled leaves of *C. auriculata* and *C. asiatica* have shown greater levels of polyphenols, flavonoids, and antioxidant capacity compared with fresh leaves. Polyphenol and flavonoid contents of boiled *S. grandiflora* and *G. lactiferum* were higher than that of their fresh form. Boiled and steamed *O. zeylanica* and *S. grandiflora* have shown higher carotenoids. Boiled and steamed leaves of *P. edulis* have shown higher antioxidant activity. The impact of cooking on the changes in bioactive concentrations and antioxidant capacities are dependent on the species and the method of cooking.

Keywords: green leafy vegetables; effect of cooking; polyphenols; antioxidant activity

---

1. Introduction

The presence and diversity of phytochemicals such as polyphenols, flavonoids, and carotenoids in vegetables are important factors for human health. Many epidemiological studies have shown that the diet rich in antioxidants play an essential role in disease prevention [1] and free radicals are known to be a significant contributor to many degenerative diseases such as cancer, cardiovascular diseases, and diabetes. Dietary antioxidants protect against free radicals such as reactive oxygen species in the human body. Provision of dietary sources of antioxidants that could function to quench or neutralize the spectrum of oxidant sources in the body is important in the prevention of oxidative damage [2]. The green leafy vegetables represent essential nutritional constituents in any balanced diet, and they contain a range of health-related phytochemicals [1]. *Centella asiatica*, *Cassia auriculata*, *Gymnema lactiferum*, *Olax zeylanica*, *Sesbania granadiflora* and *Passiflora edulis* are some of the edible leaves widely consumed as leafy vegetables in Sri Lanka and other tropical countries. These leafy vegetables possess strong antioxidative properties [1,3]. The potential exists for the discovery of synergies between foods such that there would be more than additive effects of consuming the foods in the same meal [2]. Most leafy vegetables can be consumed in their fresh form, and some of them are consumed after being cooked. However, cooking, such as boiling, steaming and frying may cause deterioration of bioactive constituents in kale, cabbage and shallot [4].

Moreover, cooking treatments soften the cell walls and facilitate the extraction of bioactives such as polyphenols [5] and carotenoids [6]. There are published investigations on the influence of cooking methods on the antioxidant activity of some European vegetables [7], African leafy vegetables [8] and brassica vegetables [9], and the different effects reported are dependent on the vegetable types. However, there are limited studies on the effect of domestic cooking on many leafy vegetables available in South Asian origin. Thus, it is important to understand the impact of domestic cooking on antioxidant activity or free radical capacity of leafy vegetables. In the present study, the effect of boiling, steaming, and frying on phytochemical contents (polyphenols, flavonoids and carotenoids) and antioxidant properties (total antioxidant capacity, 2,2-diphenyl-1-picrylhydrazyl (DPPH) radical scavenging ability and singlet oxygen scavenging ability) of six leafy vegetables (*C. auriculata*, *C. asiatica*, *O. zeylanica*, *S. grandiflora*, *G. lactiferum*, and *P. edulis*), were evaluated.

2. Materials and methods

2.1. Chemicals

All chemicals and solvents used were of analytical grade. Rutin, gallic acid, Folin–Ciocalteu reagent, methanol, *N-N*-dimethyl ρ-nitrosoamine, histidine, naphthylethylenediaminedihydrochloride, sodium nitroprusside, 2 deoxy D-ribose were obtained from Sigma Aldrich, St. Louis, MO, USA through Analytical Instrument Pvt Ltd., Colombo, Sri Lanka.

2.2. Leaf Samples

Fresh green leafy vegetable samples, *Cassia auriculata* L. ('Ranawara'), *Olax Zeylanica* ('mella'), *Centella asiatica* ('gotukola'), *Gymnema lactiferum* ('kuringan'-Ceylon cow tree), *Sesbania grandiflora* ('kathurumurunga') and *Passiflora edulis* ('passion fruit') were collected locally from the Negombo and Makandura areas of Sri Lanka, and they were cleaned and subjected to cooking treatments. Voucher plant specimens from each leafy vegetable collected were deposited in a herbarium.

2.3. Cooking Treatments

Three common cooking methods were assessed, such as boiling, steaming, and frying. The cleaned and washed leaves were cut into small pieces, and the samples (400 g) were divided into four parts (100 g each), keeping one portion as control (uncooked, stored at 4 °C in the refrigerator until use for within 24 h), and the rest was subjected to different cooking treatments, such as boiling, steaming and frying, as shown below.

2.4. Boiling

Boiling of leaves was conducted as in Jimenez-Monreal et al. [7] with some modifications. Briefly, leaf samples (100 g) were added to boiling tap water (150 mL) in a covered stainless-steel pot and cooked on a moderate flame for 5 min. The samples were drained off and cooled rapidly on plenty of ice.

2.5. Steaming

Steaming of leaves was performed as in Miglio et al. [10], with some modifications. Briefly, leafy vegetable samples were placed on a perforated tray in a stainless steel steamer covered over boiling water for 5 min under atmospheric pressure. The samples rapidly cooled on ice.

2.6. Frying

Frying of leaves was done as in Jimenez-Monreal et al. [7], with some modifications. Briefly, leafy vegetable samples were added to 500 mL of white coconut oil (Brand name-"N Joy") in a stainless steel pan at 170 °C and stirred until the sample became crisp-tender. At the end of each trial, samples were drained off and dabbed with blotting paper to allow for the absorption of exceeding oil. After cooking, the leafy vegetable samples were homogenized and stored at −18 °C.

2.7. Dry Matter Determination

All the bioactives and antioxidant activities were calculated according to the dry matter basis. Therefore, moisture contents of the cooked samples were determined according to the method described by Turkmen et al. [4].

2.8. Preparation of Methanolic Extracts

Methanolic extracts of leaves were prepared according to the method described in Gunathilake et al. [11]. One gram of cooked leafy vegetable samples were weighed and mixed with 8 mL of 70% methanol and vortexed at high speed for thirty minutes and then centrifuged (Hettich, EBA 20, Tuttlingen, Germany) for 10 min at 792× g. The extracts were subsequently filtered through a filter paper (Whatman No. 42; Whatman Paper Ltd., Maidstone, UK). The crude extracts were desolventized in a rotary evaporator (HAHNVAPOR, Model HS-2005 V, HAHNSHIN Scientific, Kyonggi-do, Korea) at 40 °C. The concentrated extracts prepared were oven dried at 40 °C for 12 h and were stored at −18 °C in air-tight screw-capped glass vials until assayed within one week. Extracts were dissolved in methanol to obtain a concentration of 3 mg/mL for each assay.

2.9. Total Polyphenol Content

The total polyphenol content of the methanolic extracts was estimated by the Folin–Ciocalteu method described by Singleton et al. [12] and with some modifications as described in Gunathilake and Rupasinghe [13]. The concentration of total phenols was expressed as mmol gallic acid equivalents (GAE) per g dry weight (DW) of cooked leaves.

2.10. Determination of Total Flavonoid Content

Total flavonoid content was measured according to a colorimetric assay method described in Zhishen et al. [14]. Total flavonoid content in the extracts of green leafy vegetables was expressed as mmol rutin equivalents (RE) per 1 g dry weight of cooked leaf sample.

2.11. Determination of Carotenoid Content

The carotenoid content was analyzed according to Türlerinde et al. [15], and carotenoid contents were reported as mg/g DW of the cooked sample.

2.12. Total Antioxidant Capacity Assay

The total antioxidant capacity of leaf extracts was analyzed according to Prieto et al. [16]. The antioxidant capacity was expressed as ascorbic acid equivalents (AAE)/g cooked leaves.

2.13. Determination of DPPH Radical Scavenging Assay

The capacity of prepared extracts to scavenge the 'stable' free radical DPPH was monitored according to Hatano et al. [17]. The percentage inhibition of the radicals due to the antioxidant activity of leaf extracts was calculated.

2.14. Singlet Oxygen Scavenging Assay

Singlet oxygen assay was performed according to the method described in Maldonado [18]. The results were calculated as mmol gallic acid equivalent per 1 g of dry weight cooked leaves.

2.15. Statistical Analysis

All samples were analyzed in triplicate, and one-way analysis of variance (ANOVA) was performed using MINITAB 15 software (State College, PA, USA). When there were significant differences ($p > 0.05$), multiple mean comparisons were carried out using the least significant difference (LSD) method.

3. Results & Discussion

3.1. The Effect of Cooking on Total Polyphenol Content

The results of the evaluation of total polyphenols in raw and cooked leafy vegetables are shown in Figure 1. Leaves of *O. zeylanica*, *S. grandiflora*, and *P. edulis* showed the significantly higher ($p < 0.05$) total polyphenol content in their raw samples compared with their cooked samples. This indicates the breaking down of polyphenols in these leafy types during cooking. However, *C. auriculata* leaves showed more than two times higher total polyphenol content in boiled and steamed leaves, compared with its raw form. In previous studies, it was reported that the leaves of *C. auriculata* are rich in polyphenols [1]. *C. asiatica* also showed approximately 200% and 139% higher total polyphenol content in steamed and boiled leaves respectively, compared with its uncooked leaves. *G. lactiferum* showed 167% higher polyphenol content in boiled leaves than in uncooked leaves. This increase in total polyphenols agrees with Turkmen et al. [4], to the extent that the cooking or wet heating could increase the phenol content in some green vegetables such as green beans, broccoli, and pepper. Further, it was reported that the basis of the increase in polyphenol content during cooking could not be categorically stated. However, it could be attributed to the possible breakdown of the complex polyphenolic compounds such as tannins present in the vegetables during heat processing to simple polyphenols [19]. Ferracane and co-authors [5] reported that the increase in total polyphenols during thermal processing might have been due to the liberation of polyphenols from the intracellular protein complexes, changes in plant cell structure, matrix modifications, or the inactivation of the polyphenol oxidases. Moreover, the heat treatments could inactivate polyphenol oxidases, preventing oxidation and polymerization of polyphenols [20]. However, the steamed leaves of *G. lactiferum* showed significantly lower ($p < 0.05$) polyphenol content, only 59% compared with its raw form. Similarly, after the boiling treatment, the total polyphenol content in *P. edulis*, *S. grandiflora*, and *O. zeylanica* have reduced concerning their original content by 55.9%, 83.5%, and 47.6%, respectively. Further polyphenol content in steamed leaves of *P. edulis*, *G. lactiferum*, *S. grandiflora* and *O. zeylanica* also have reduced by 88.5%, 59.1%, 39.9% and 63.8%, respectively with reference to their raw forms. In a previous study, it was reported that blanching of spinach, swamp cabbage, kale, shallots and cabbage for 1 min in boiling water reduced (12–26%) the total polyphenols in these vegetables [4]. However, in a study of Kao et al. [21], a significant increase in total polyphenols were also reported in Thai basil leaf and sweet potato leaf, at the initial stage of boiling for 1 min and 5 min, respectively and subsequently decreased the polyphenol content was also reported as the boiling time increased. By contrast, water cooking has a detrimental effect on polyphenols in vegetables, resulting in a complete loss of polyphenols, likely due to diffusion into the boiling water [10]. In general, frying reduced the polyphenol content in all leafy types studied. Polyphenol content was reduced by 80.7%, 18.9%, 75.2%, 75.8%, 76.3% and 20.6% in *O. zeylanica*, *C. auriculata*, *S. grandiflora*, *G. lactiferum*, *P. edulis* and *C. asiatica*, respectively. In a study of cooked vegetables, it was found that about 60% losses of polyphenols in carrots and broccoli after frying and the losses were higher than during steaming [10]. During deep-fat frying, oils undergo physicochemical changes. Moreover, the food dehydrates, and fat penetrates the food during frying, and therefore food fried in oil may contain higher levels of thermo-oxidized and polymerized products.

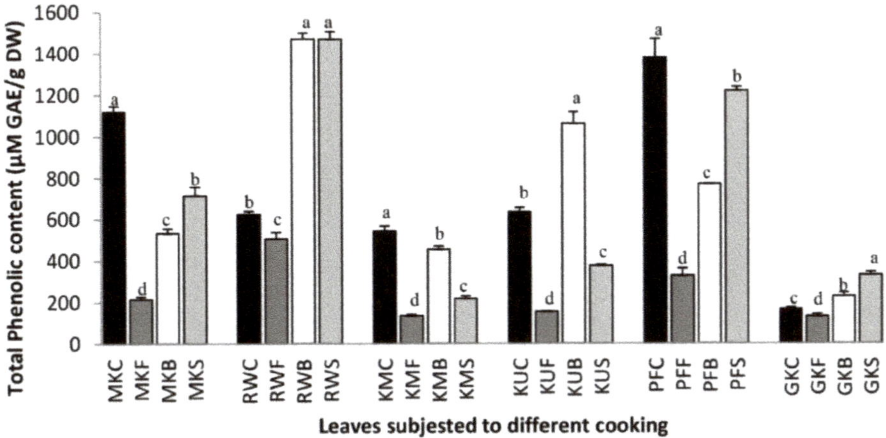

Figure 1. The total phenolic content of raw and cooked extracts of some green leafy vegetables. Values represent the means of triplicate readings. MK, *O. zeylanica*; RW, *C. auriculata*; KM, *S. grandiflora*; KU, *G. lactiferum*; PF, *P. edulis*; GK, *C. asiatica*. C—fresh leaves; F—fried; B—boiled; S—steamed. Data are means ± standard deviations of three replicate determinations. Columns with different letters for each vegetable are significantly different ($p < 0.05$). GAE: gallic acid equivalence; DW: dry weight.

3.2. The Effect of Cooking on Total Flavonoid Content

Flavonoids, widespread in most common edible fruits, vegetables, and seeds, are heat sensitive polyphenolic compounds. Therefore, the heat exposure during cooking could greatly influence their content in vegetables [22]. Total flavonoid content of raw and cooked leafy vegetables is shown in Figure 2. Total flavonoid content increased significantly ($p < 0.05$) compared with all raw forms after the boiling except *P. edulis*. The percentage increases of flavonoid content in boiled leaves of *O. zeylanica*, *C. auriculata*, *S. grandiflora*, *G. lactiferum* and *C. asiatica* were 163.3%, 140.0%, 160.2%, 197.2% and 120.1% respectively when compared with their fresh leaves. Steamed leaves contain 150.7%, 190.2%, 109.9%, 95.0% and 203.2% flavonoids in *O. zeylanica*, *C. auriculata*, *S. grandiflora*, *G. lactiferum* and *C. asiatica* respectively compared with their raw forms. These increase in flavonoid content after subsequent boiling or steaming may be related to an enhanced availability for extraction, and to a more efficient release of polyphenol or flavonoid compounds from intracellular proteins and altered cell wall structures [9]. Similarly, increased yield of flavonoids in boiled broccoli and spinach was reported previously by Mazzeo et al. [23]. However, in our study, *P edulis* have shown lower flavonoid content after the boiling (83.5%) and steaming (71.2%) with respect to the flavonoid content in its fresh leaves. As in the polyphenols, flavonoid contents in all fried leaves were significantly ($p < 0.05$) lower compared with their fresh contents. However, due to less water solubility of flavonoids in the form of flavonoid glycosides and acylated derivatives, less of these flavonoids are extracted from the plant tissues by the cooking process compared with high soluble glucuronides derivatives [24]. Therefore, most of the flavonoid glycosides and their acylated forms are retained in the tissue during the cooking process [24]. Accordingly, variation in losses and gains of flavonoids due to cooking treatments in the leafy types studied could be due to the types of cooking, the nature of leaves and the forms of the flavonoids present in the plant matrices.

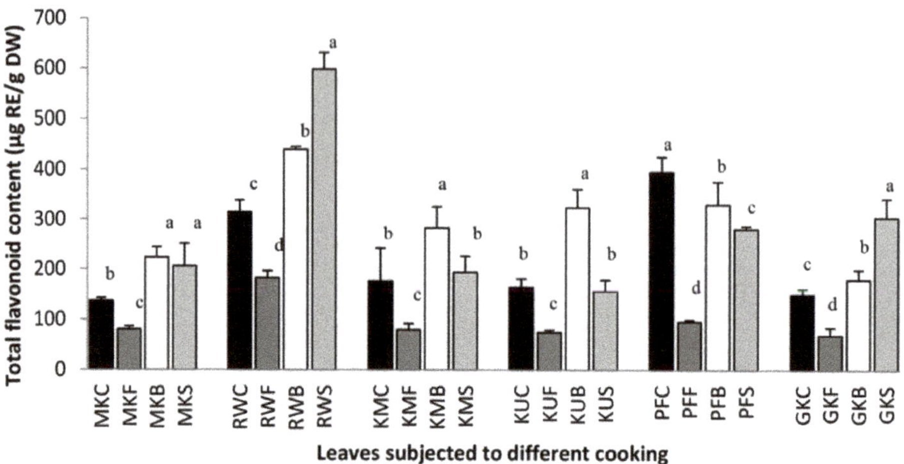

Figure 2. Total flavonoid content of raw and cooked extracts of some green leafy vegetables. Values represent the means of triplicate readings. MK, *O. zeylanica*; RW, *C. auriculata*; KM, *S. grandiflora*; KU, *G. lactiferum*; PF, *P. edulis*; GK, *C. asiatica*. C—fresh leaves; F—fried; B—boiled; S—steamed. Data are means ± standard deviations of three replicate determinations. Columns with different letters for each vegetable are significantly different ($p < 0.05$). RE: rutin equivalance.

3.3. The Effect of Cooking on Total Carotenoid Content

Carotenoid content in raw and cooked leaf samples are shown in Figure 3, and the results showed that the carotenoid contents in all cooked leaves of *C. auriculata*, *G. lactiferum* and *P. edulis* were significantly ($p < 0.05$) lower when compared with their fresh leaves. Steaming of leaves of *O. zeylanica*, *C. auriculata*, *S. grandiflora*, *G. lactiferum* and *P edulis* decreased the carotenoid content by 59.1%, 20.4%, 35.9%, 25.8% and 41.3% respectively in terms of their content in fresh leaves. In frying, carotenoid content of the leaves of *O. zeylanica*, *C. auriculata*, *S. grandiflora*, *G. lactiferum*, *Pedulis* and *C. asiatica* was significantly ($p < 0.05$) reduced by 61.9%, 66.8%, 66.5%, 68.0%, 43.0% and 55.0% respectively, compared to that of their fresh leaves. In some previous studies, a progressive reduction of carotenoids with increasing boiling time in sweet potato was also reported [25]. According to Chen and coauthors [26], the long chains of conjugated carbon-carbon double bonds present in carotenoids are susceptible to light, oxygen, heat and acid degradation. Cooking can also lead to an isomerization from the native all-*trans*-form to its *cis*-isomers [27]. However, from the nutritional standpoint, *cis*-isomers are more bioavailable than all-*trans*-carotenoids in crossing the intestinal wall as they are readily solubilized in micelles [28]. On the other hand, boiling of *O. zeylanica*, *S. grandiflora*, and *C. asiatica* leaves have shown a 19.6%, 11.9% and 79.1% increase in carotenoid content respectively, when compared to their raw contents. Carotenoid content in steamed *C. asiatica* leaves also increased by 37.4%. Some research findings also reported an increase in carotenoids in some green vegetables [29], pumpkin [30] and artichoke [5] after the boiling treatment. According to Khachik and co-authors [31], thermal processing affects the breakdown of the cellulose structure of the plant cells and the denaturation of carotenoid-protein complexes, which allows a more effective and complete extraction of carotenoids. High frying temperatures could, in fact, cause the oil to produce hydroperoxide free radicals and accelerate the degradation of carotenoids [32]. Moreover, frying decreases the initial carotenoid concentration in leafy vegetables, probably because of leaching into oil and to a higher processing temperature. Miglio et al. [10] reported that the carotenoids such as phytoene and phytofluene were completely lost during the frying process. Overall, the results of this study indicating the stability of carotenoids in foods such as leafy vegetables are highly variable.

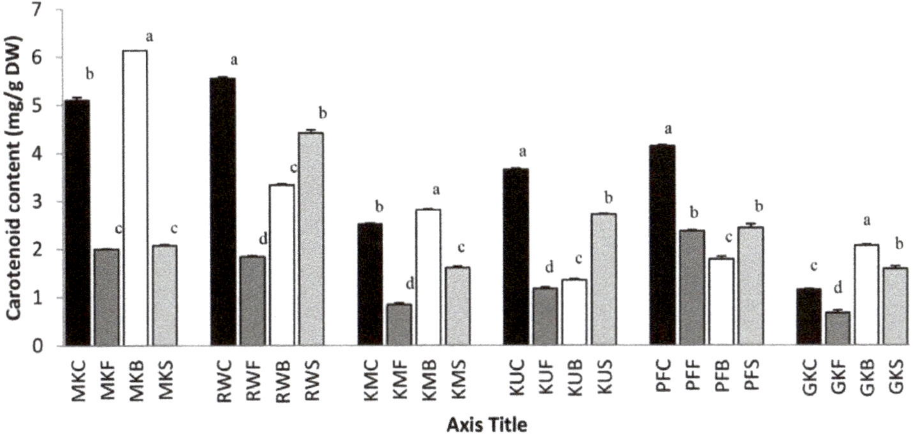

Figure 3. Total carotenoid content of raw and cooked extracts of some green leafy vegetables. Values represent the means of triplicate readings. MK, *O. zeylanica*; RW, *C. auriculata*; KM, *S. grandiflora*; KU, *G. lactiferum*; PF, *P. edulis*; GK, *C. asiatica*. C—fresh leaves; F—fried; B—boiled; S—steamed. Data are means ± standard deviations of three replicate determinations. Columns with different letters for each vegetable are significantly different ($p < 0.05$).

3.4. The Effect of Cooking on Antioxidant Activity

Three different antioxidant activity assays were used to evaluate the effect of cooking on the antioxidant capacity of leafy vegetables. The analytical parameters: linearity, detection limit, precision, and accuracy assay show that the different methods studied are useful for measuring the total antioxidant capacity in foods. This fact is important in order to quantify the changes in the antioxidant capacity of foods during processing/preservation treatment and subsequent storage [33]. Figure 4 displays the total antioxidant capacity of raw and cooked leaf samples. Results clearly show that the total antioxidant capacity of steamed leaves of *O. zeylanica* had significantly ($p < 0.05$) reduced when compared with its raw leaves. However, frying and boiling had not affected the antioxidant capacity of *O. zeylanica* leaves. Although significantly higher ($p < 0.05$) antioxidant capacity was observed in the steamed and boiled leaves of *C. auriculata* compared to that of its fresh leaves, the fried leaves showed lower antioxidant capacity. Significantly lower ($p < 0.05$) antioxidant capacities were observed in all cooked leaves of *S. grandiflora*, and *G. lactiferum* when compared with the antioxidant capacities of fresh leaves. Steamed leaves of *P. edulis* showed 10% higher antioxidant capacity than that of fresh leaves whereas fried and boiled leaves showed nearly 42% lower antioxidant capacities. Boiled *C. asiatica* leaves showed significantly higher ($p < 0.05$) total antioxidant capacity compared to its fresh forms though fried and steamed leaves showed insignificant changes.

The radical scavenging activity of leafy vegetables subjected to different cooking treatments was tested using DPPH assay. In this process, polyphenols have the ability to quench DPPH radicals, by providing hydrogen atoms or by electron donation, and to convert them to a colorless product [1]. Hence, the higher the percentage of inhibition of free radical activity, the more potent the antioxidant activity of the extract in terms of hydrogen atom-donating capacity [1]. DPPH radical scavenging ability of raw and cooked leaf samples are shown in Figure 5. In comparison to the scavenging ability of the raw form of studied leafy types, steamed leaves of *C. auriculata*, *G. lactiferum* and *C. asiatica* have shown 15.6%, 1.7% and 29.7% respectively higher scavenging ability, whereas the steamed leaves of *O. zeylanica*, *S. grandiflora*, and *P. edulis* showed 36.7%, 1.7% and 10.5% lower scavenging activity respectively. However, fried and boiled leaves of all leafy types studied showed significantly ($p < 0.05$) lower free radicals scavenging ability when compared with their raw forms. Frying reduced the scavenging ability of *O. zeylanica*, *C. auriculata*, *S. grandiflora*, *G. lactiferum*, *P edulis* and *C. asiatica* by 73.1%, 15.5%, 3.8%, 8.4% 42.5% and 4.8% while boiling of leaves have reduced the free radical scavenging ability by 70.6%, 16.3%, 4.2%, 77.3%, 21.5% and 22.5% respectively. According to our previous findings [1], it was found that the DPPH assay (r^2 = 0.903) and total antioxidant capacity assay (r^2 = 0.890) are correlated with the total phenolic content of leafy vegetable extracts.

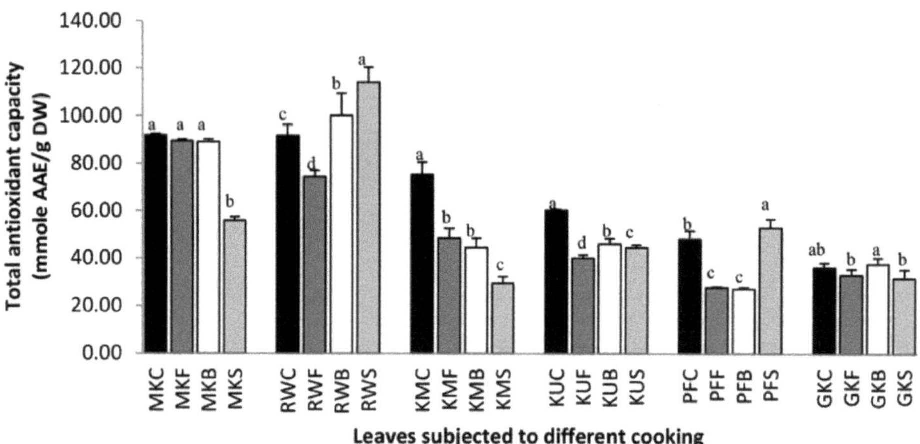

Figure 4. Total antioxidant capacity of raw and cooked extracts of some green leafy vegetables. Values represent the means of triplicate readings. MK, *O. zeylanica*; RW, *C. auriculata*; KM, *S. grandiflora*; KU, *G. lactiferum*; PF, *P. edulis*; GK, *C. asiatica*. C—fresh leaves; F—fried; B—boiled; S—steamed. Data are means ± standard deviations of three replicate determinations. Columns with different letters for each vegetable are significantly different ($p < 0.05$). AAE: ascorbic acid equivalance.

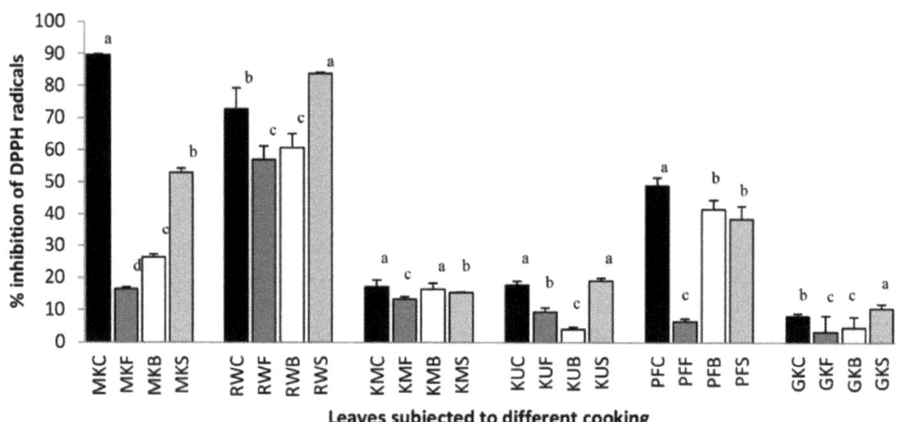

Figure 5. 2,2-diphenyl-1-picrylhydrazyl (DPPH) free radical scavenging ability of raw and cooked extracts of some green leafy vegetables. Values represent the means of triplicate readings. MK, *O. zeylanica*; RW, *C. auriculata*; KM, *S. grandiflora*; KU, *G. lactiferum*; PF, *P. edulis*; GK, *C. asiatica*. C—fresh leaves; F—fried; B—boiled; S—steamed. Data are means ± standard deviations of three replicate determinations. Columns with different letters for each vegetable are significantly different ($p < 0.05$).

Singlet oxygen is a highly energetic reactive oxygen species which induces a unique oxidation process by directly reacting with electron-rich double bonds without forming free radical intermediates in foods and biological systems [3]. Figure 6 shows the singlet oxygen scavenging ability of raw and cooked leafy types. Results showed that singlet oxygen scavenging properties had reduced during frying in all leaves compared with their fresh leaves and the percentage reduction of *O. zeylanica*, *C. auriculata*, *S. grandiflora*, *G. lactiferum*, *P. edulis* and *C. asiatica* leaves during frying was 64.9%, 65.0%, 43.0%, 72.3%, 70.4% and 71.1% respectively. Boiling of leaves increased the singlet oxygen scavenging ability by 56.0%, 31.4% and 5.4% in *S. grandiflora*, *P. edulis*, and *C. asiatica* respectively. However, boiling has reduced the scavenging activity in *C. auriculata* and *G. lactiferum* by 4.6% and 10.9% respectively and has no significant ($p > 0.05$) effect on *O. zeylanica*. Steaming of leaves has increased the singlet oxygen scavenging activity of *O. zeylanica*, *S. Grandiflora* and *C. asiatica* by 29.1%, 47.9%, and 24.7%, while it has reduced the scavenging ability of *C. auriculata*, *G. lactiferum* and *P. edulis* by 35.1%, 42.3%, and 7.9% respectively, compared with that of their fresh leaves. The decreasing singlet oxygen scavenging activity of leafy vegetables may be due to a reduction in carotenoid content in leaves as carotenoids are effective single scavenging molecules [1]. According to Gunathilake et al. [34], singlet oxygen scavenging ability of leaf extracts correlates with its carotenoid content.

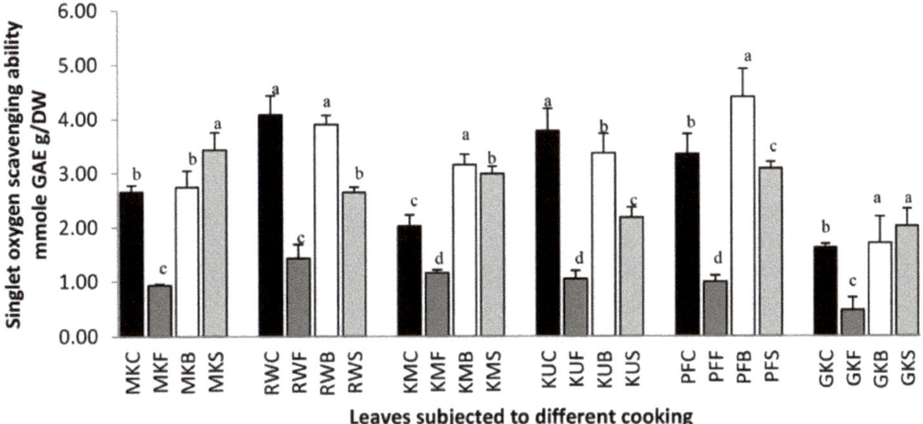

Figure 6. Singlet oxygen scavenging ability of raw and cooked extracts of some green leafy vegetables. Values represent the means of triplicate readings. MK, *O. zeylanica*; RW, *C. auriculata*; KM, *S. grandiflora*; KU, *G. lactiferum*; PF, *P. edulis*; GK, *C. asiatica*. C—fresh leaves; F—fried; B—boiled; S—steamed. Data are means ± standard deviations of three replicate determinations. Columns with different letters for each vegetable are significantly different ($p < 0.05$).

Results showed that some leaves had increased antioxidant activity due to cooking and some leafy types reduced their antioxidant activity after subsequent cooking treatment. Previous studies have shown that the thermal processing of celery [7] increased their antioxidant capacity, after the cooking treatments. It was found that boiling of several vegetables was attributed to the suppression of oxidation by antioxidants due to thermal inactivation of oxidative enzymes [20]. Furthermore, it has been suggested that processing can promote the oxidation of polyphenols to an intermediate oxidation state, which can exhibit a higher radical scavenging efficiency than the non-oxidized polyphenols [35]. Moreover, the boiling process may destroy the cell wall and subcellular compartments, and thus release potent radical-scavenging antioxidants [30]. On the other hand, some leafy vegetables have shown a reduction in their antioxidant activity after the cooking treatments. Some recent studies have also shown a significant reduction in the antioxidant activities after the thermal processing. In another study to determine the effect of thermal processing on the flavonol, rutin, and quercetin, it was shown that the radical scavenging activity decreases with decreasing amount of flavonol in the system [36]. Similarly, Puupponen-Pimia et al. [37] measured the antioxidant capacity by the DPPH assay in blanched cauliflower and reported a reduction of 23%. Overall, the antioxidant activities in various green leafy vegetables were affected differently by thermal processing. The reason could be the differing leafy type's composition of constituents determining the antioxidant capacity. While some general trends can be observed with previous findings, the effects of processing differ not only according to treatment intensity but also according to the food matrix, suggesting that each matrix should be studied separately [38]. However, it is evident that "antioxidant activity" involves complex interactions amongst intrinsic and extrinsic factors related to the food matrix and organism, which cannot be predicted using simple chemical reactions in a test tube alone and need further analysis [39].

4. Conclusions

The findings from the present study indicate that the total polyphenols, carotenoids and antioxidant capacity of selected green leafy vegetables are significantly altered during common cooking practices such as boiling, steaming, and frying. Among the cooking methods evaluated, frying reduces the polyphenols, flavonoids, carotenoids, and antioxidant activities in all leafy vegetables studied, whereas boiling and steaming have shown varying effects on polyphenols, carotenoids, and antioxidant properties, depending on the leafy type. The results of the study can be used for making recommendations on the food processing methods to be chosen for preserving the health benefits of green leafy vegetables.

Author Contributions: K.D.P.P.G. performed all the experiments, analyzed the data, and wrote the manuscript. All the authors contributed to the designing of the experiments and proofreading the manuscript.

Funding: We greatly acknowledge the financial support provided by the National Science Foundation of Sri Lanka (NSF) (Research Grant # RG/AG/2014/04).

Acknowledgments: We thank H.P.S. Jayaweera of the Wayamba University of Sri Lanka for their technical support.

Conflicts of Interest: The authors declare no conflict of interest.

References

1. Gunathilake, K.D.P.P.; Ranaweera, K.K.D.S. Antioxidative properties of 34 green leafy vegetables. *J. Funct. Foods* **2016**, *26*, 176–186. [CrossRef]
2. Prior, R.L. Oxygen radical absorbance capacity (ORAC): New horizons in relating dietary antioxidants/bioactives and health benefits. *J. Funct. Foods* **2015**, *18*, 797–810. [CrossRef]
3. Gunathilake, K.D.P.P.; Ranaweera, K.K.D.S.; Rupasinghe, H.P.V. Change of phenolics, carotenoids, and antioxidant capacity following simulated gastrointestinal digestion and dialysis of selected edible green leaves. *Food Chem.* **2018**, *245*, 371–379. [CrossRef] [PubMed]
4. Turkmen, N.; Sari, F.; Velioglu, Y.S. The effect of cooking methods on total phenolics and antioxidant activity of selected green vegetables. *Food Chem.* **2005**, *93*, 713–718. [CrossRef]
5. Ferracane, R.; Pellegrini, N.; Visconti, A.; Graziani, G.; Chiavaro, E.; Miglio, C.; Fogliano, V. Effects of different cooking methods on the antioxidant profile, antioxidant capacity, and physical characteristics of artichoke. *J. Agric. Food Chem.* **2018**, *56*, 860–8608. [CrossRef] [PubMed]
6. Chandrika, U.G.; Basnayake, B.M.L.B.; Athukorala, I.; Colombagama, P.W.N.M.; Goonetilleke, A. Carotenoid content and in vitro bioaccessibility of lutein in some leafy vegetables popular in Sri Lanka. *J. Nutr. Sci. Vitaminol.* **2010**, *56*, 203–207. [CrossRef] [PubMed]
7. Jiménez-Monreal, A.M.; García-Diz, L.; Martínez-Tomé, M.; Mariscal, M.M.M.A.; Murcia, M.A. Influence of cooking methods on antioxidant activity of vegetables. *J. Food Sci.* **2009**, *74*, H97–H103. [CrossRef] [PubMed]
8. Oboh, G. Effect of blanching on the antioxidant properties of some tropical green leafy vegetables. *LWT-Food Sci. Technol.* **2005**, *38*, 513–517. [CrossRef]
9. Wachtel-Galor, S.; Wong, K.W.; Benzie, I.F. The effect of cooking on Brassica vegetables. *Food Chem.* **2008**, *110*, 706–710. [CrossRef]

10. Miglio, C.; Chiavaro, E.; Visconti, A.; Fogliano, V.; Pellegrini, N. Effects of different cooking methods on nutritional and physicochemical characteristics of selected vegetables. *J. Agric. Food Chem.* **2008**, *56*, 139–147. [CrossRef] [PubMed]

11. Gunathilake, K.D.P.P.; Ranaweera, K.K.D.S.; Rupasinghe, H.P.V. Influence of boiling, steaming and frying of selected leafy vegetables on the in vitro anti-inflammation associated biological activities. *Plants* **2018**, *7*, 22. [CrossRef] [PubMed]

12. Singleton, V.L.; Orthofer, R.; Lamuela-Raventos, R. Analysis of total phenols and other oxidation substrates and antioxidants by means of FC reagent. *Methods Enzymol.* **1999**, *29*, 152–178.

13. Gunathilake, K.D.P.P.; Yu, L.J.; Rupasinghe, H.P.V. Reverse osmosis as a potential technique to improve antioxidant properties of fruit juices used for functional beverages. *Food Chem.* **2014**, *148*, 335–341. [CrossRef] [PubMed]

14. Jia, Z.; Tang, M.; Wu, J. The determination of flavonoid contents in mulberry and their scavenging effects on superoxide radicals. *Food Chem.* **1999**, *64*, 555–559.

15. Türlerinde, F.Ç.K.B.A.; Klorofil, A.B.; Saptanması, I.S. Spectrophotometric determination of chlorophyll-A, B and total carotenoid contents of some algae species using different solvents. *Turk. J. Bot.* **1998**, *22*, 13–17.

16. Prieto, P.; Pineda, M.; Aguilar, M. Spectrophotometric quantitation of antioxidant capacity through the formation of a phosphomolybdenum complex: Specific application to the determination of vitamin E. *Anal. Biochem.* **1999**, *269*, 337–341. [CrossRef] [PubMed]

17. Hatano, T.; Kagawa, H.; Yasuhara, T.; Okuda, T. Two new flavonoids and other constituents in liquorice root: Their relative astringency and radical scavenging effects. *Chem. Pharm. Bull.* **1988**, *36*, 1090–2097. [CrossRef]

18. Maldonado, P.D.; Rivero-Cruz, I.; Mata, R.; Pedraza-Chaverr, J. Antioxidant activity of A-typeproanthocyanidins from *Gerenium niveum* (Geraniaceae). *J. Agric. Food Chem.* **2005**, *53*, 196–200. [CrossRef] [PubMed]

19. Gunathilake, K.D.P.P.; Rupasinghe, H.P.V. Optimization of Water Based-extraction Methods for the Preparation of Bioactive-rich Ginger Extract Using Response Surface Methodology. *Eur. J. Med. Plants* **2014**, *4*, 893. [CrossRef]

20. Yamaguchi, T.; Katsuda, M.; Oda, Y.; Terao, J.; Kanazawa, K.; Oshima, S.; Inakuma, T.; Ishiguro, Y.; Takamura, H.; Matoba, T. Influence of polyphenol and ascorbateoxidases during cooking process on the radical-scavenging activity of vegetables. *Food Sci. Technol. Res.* **2003**, *9*, 79–83. [CrossRef]

21. Kao, F.J.; Chiu, Y.S.; Chiang, W.D. Effect of water cooking on the antioxidant capacity of carotenoid-rich vegetables in Taiwan. *J. Food Drug Anal.* **2014**, *22*, 202–209. [CrossRef]

22. Zhang, J.J.; Ji, R.; Hu, Y.Q.; Chen, J.C.; Ye, X.Q. Effect of three cooking methods on nutrient components and antioxidant capacities of bamboo shoot (*Phyllostachys praecox* CD Chu et CS Chao). *J. Zhejiang Univ. Sci. B* **2011**, *12*, 752–759. [CrossRef] [PubMed]

23. Mazzeo, T.; N'Dri, D.; Chiavaro, E.; Visconti, A.; Fogliano, V.; Pellegrini, N. Effect of two cooking procedures on phytochemical compounds, total antioxidant capacity, and color of selected frozen vegetables. *Food Chem.* **2011**, *128*, 627–633. [CrossRef]

24. Hiemori, M.; Koh, E.; Mitchell, A.E.I. Influence of cooking on anthocyanins in black rice (*Oryza sativa* L. *japonica* var. SBR). *J. Agric. Food Chem.* **2009**, *57*, 1908–1914. [CrossRef] [PubMed]

25. VanJaarsveld, P.J.; Harmse, E.; Nestel, P.; Rodriguez-Amaya, D.B. Retention of β-carotene in boiled, mashed orange-fleshed sweet potato. *J. Food Compos. Anal.* **2006**, *19*, 321–329. [CrossRef]

26. Chen, B.H.; Chen, T.M.; Chien, J.T. Kinetic model for studying the isomerization of alpha- and beta-carotene during heating and illumination. *J. Agric. Food Chem.* **1994**, *42*, 2391–2397. [CrossRef]

27. Burmeister, A.; Bondiek, S.; Apel, L.; Kühne, C.; Hillebrand, S.; Fleischmann, P. Comparison of carotenoid and anthocyanin profiles of raw and boiled *Solanum tuberosum* and *Solanum phureja* tubers. *J. Food Compos. Anal.* **2011**, *24*, 865–872. [CrossRef]

28. Ross, A.B.; Vuong, L.T.; Ruckle, J.; Synal, H.A.; Schulze-Konig, T.; Wertz, K.; Rümbeli, R.; Liberman, R.G.; Skipper, P.L.; Tannenbaum, S.R.; et al. Lycopene bioavailability and metabolism in humans: An accelerator mass spectrometry study. *Am. J. Clin. Nutr.* **2011**, *93*, 1263–1273. [CrossRef] [PubMed]

29. De Sá, M.C.; Rodriguez-Amaya, D.B. Optimization of HPLC quantification of carotenoids in cooked green vegetables—Comparison of analytical and calculated data. *J. Food Compos. Anal.* **2004**, *17*, 37–51. [CrossRef]

30. Azizah, A.H.; Wee, K.C.; Azizah, O.; Azizah, M. Effect of boiling and stir frying on total phenolics, carotenoids and radical scavenging activity of pumpkin (*Cucurbita moschato*). *Int. Food Res. J.* **2009**, *16*, 45–51.

31. Khachik, F.; Beecher, G.R.; Goli, M.B.; Lusby, W.R.; Smith, J.C. Separation, and identification of carotenoids and their oxidation products in the extracts of human plasma. *Anal. Chem.* **1992**, *64*, 2111–2122. [CrossRef] [PubMed]
32. Barba, F.J.; Esteve, M.J.; Tedeschi, P.; Brandolini, V.; Frígola, A. A comparative study of the analysis of antioxidant activities of liquid foods employing spectrophotometric, fluorometric, and chemiluminescent methods. *Food Anal. Methods* **2013**, *6*, 317–327. [CrossRef]
33. Gunathilake, K.D.P.P.; Ranaweera, K.K.D.S.; Rupasinghe, H.P.V. Analysis of rutin, β-carotene, and lutein content and evaluation of antioxidant activities of six edible leaves on free radicals and reactive oxygen species. *J. Food Biochem.* **2018**. [CrossRef]
34. Mayeaux, M.; Xu, Z.; King, J.M.; Prinyawiwatkul, W. Effects of cooking conditions on the lycopene content in tomatoes. *J. Food Sci.* **2006**, *71*, 461–464. [CrossRef]
35. Nicoli, M.C.; Anese, M.; Parpinel, M. Influence of processing on the antioxidant properties of fruit and vegetables. *Trends Food Sci. Technol.* **1999**, *10*, 94–100. [CrossRef]
36. Buchner, N.; Krumbein, A.; Rohn, S.; Kroh, L.W. Effect of thermal processing on the flavonols rutin and quercetin. *Rapid Commun. Mass Spectrom.* **2006**, *20*, 3229–3235. [CrossRef] [PubMed]
37. Puupponen-Pimia, R.; Hakkinen, S.T.; Aarni, M.; Suortti, T.; Lampi, A.M.; Eurola, M.; Piironen, V.; Nuutila, A.M.; Oksman-Caldentey, K.M. Blanching and long-term freezing affect various bioactive compounds of vegetables in different ways. *J. Sci. Food Agric.* **2003**, *83*, 1389–1402. [CrossRef]
38. Barba, F.J.; Esteve, M.J.; Frígola, A. High pressure treatment effect on physicochemical and nutritional properties of fluid foods during storage: A review. *Compr. Rev. Food Sci. Food Saf.* **2012**, *11*, 307–322. [CrossRef]
39. Granato, D.; Shahidi, F.; Wrolstad, R.; Kilmartin, P.; Melton, L.D.; Hidalgo, F.J.; Miyashita, K.; van Camp, J.; Alasalvar, C.; Ismail, A.B.; et al. Antioxidant activity, total phenolics and flavonoids contents: Should we ban in vitro screening methods? *Food Chem.* **2018**, *264*, 471–475. [CrossRef] [PubMed]

© 2018 by the authors. Licensee MDPI, Basel, Switzerland. This article is an open access article distributed under the terms and conditions of the Creative Commons Attribution (CC BY) license (http://creativecommons.org/licenses/by/4.0/).

 *antioxidants*

Article

Antioxidant Activities of *Dialium indum* L. Fruit and Gas Chromatography-Mass Spectrometry (GC-MS) of the Active Fractions

Muhamad Faris Osman [1], Norazian Mohd Hassan [1,*], Alfi Khatib [1] and Siti Marponga Tolos [2]

[1] Department of Pharmaceutical Chemistry, Kulliyyah of Pharmacy, International Islamic University Malaysia, Kuantan 25200, Pahang, Malaysia; farisosman@iium.edu.my (M.F.O.); alfikhatib@iium.edu.my (A.K.)

[2] Department of Computational and Theoretical Sciences, Kulliyyah of Science, International Islamic University Malaysia, Kuantan 25200, Pahang, Malaysia; smtolos@iium.edu.my

* Correspondence: norazianmh@iium.edu.my; Tel.: +60-9-570-4937

Received: 30 September 2018; Accepted: 29 October 2018; Published: 1 November 2018

Abstract: The fruit of *Dialium indum* L. (Fabaceae) is one of the edible wild fruits native to Southeast Asia. The mesocarp is consumed as sweets while the exocarp and seed are regarded as waste. This study aimed to evaluate the antioxidant activities of the fruit by using four assays, which measure its capabilities in reducing phosphomolybdic-phosphotungstic acid reagents, neocuproine, 2,2-diphenyl-picrylhydrazyl (DPPH), and inhibiting linoleic acid peroxidation. The active fractions were then analyzed by gas chromatography-mass spectrometry (GC-MS). The results showed that the seed methanol fraction (SMF) exhibited the strongest antioxidant activity with significantly higher ($p < 0.05$) gallic acid equivalence (GAE), total antioxidant capacity (TAC), and DPPH radical scavenging activity (IC_{50} 31.71; 0.88 µg/mL) than the other fractions. The exocarp dichloromethane fraction (EDF) was the discriminating fraction by having remarkable linoleic acid peroxidation inhibition (IC_{50} 121.43; 2.97 µg/mL). A total of thirty-eight metabolites were detected in derivatized EDF and SMF with distinctive classes of phenolics and amino acids, respectively. Bioautography-guided fractionation of EDF afforded five antioxidant-enriched subfractions with four other detected phenolics. The results revealed the antioxidant properties of *D. indum* fruit, which has potential benefits in pharmaceutical, nutraceutical, and cosmeceutical applications.

Keywords: *Dialium indum*; exocarp; seed; antioxidant; phenolic acids; amino acids; GC-MS analysis

1. Introduction

Antioxidants of natural origins have been shown to be beneficial in health maintenance and the reduction in the risks of chronic diseases by preventing or removing oxidative damage caused by free radicals [1]. Fruits are among good sources of antioxidants that can be an excellent alternative for the improvement of population health [2]. Wild edible fruits have gained increasing attention worldwide for their good potential to be utilized as functional foods and nutraceuticals owing to their rich content of natural antioxidants and nutrients [3,4]. However, the health promoting benefits of various wild species of edible fruits especially in Asia are still lacking systematic investigation and exploitation when compared with those in Europe and America [5]. Hence, these fruits remain underutilized as natural sources of antioxidants.

Generally, fruit consists of three main parts, namely the mesocarp (pulp), exocarp (skin or peel), and seed. Most studies have been focused on the edible mesocarp, which contains vitamins, minerals, and antioxidant metabolites. Nonetheless, certain fruits accumulate higher levels of antioxidants in their exocarp and seed when compared with the mesocarp. Polyphenols, particularly phenolic acids, and amino acids are among the main antioxidants in the exocarp and seed of fruits [6–9]. Phenolic acids are one of the most common plant phenolic antioxidants, which possess radical scavenging activities due to their phenolic hydroxyls that enable the compounds to act as reducing agents, hydrogen donors, metal ion chelators, antioxidant enzymes activators, and oxidases inhibitors [10]. On the other hand, amino acids are regarded as synergistic antioxidants, which apart from scavenging radicals also enhance the effects of primary antioxidants through the pro-oxidative chelation of metal traces and regeneration of oxidized primary antioxidants [11].

Located in the Paleotropical Kingdom, Malaysia is endowed with complex tropical rainforest ecosystems, giving rise to about 520 species of plants (trees and non-trees) that produce edible fruits or seeds [12]. *Dialium indum* L. (family Fabaceae), locally known in Malaysia as "keranji", is a wild tree bearing edible fruit. Synonyms of the plant are *Dialium laurinum* Baker and *Dialium patens* Baker [13]. *D. indum* is native to Southeast Asia as it grows specifically in the forests of Malaysia, Southern Thailand, and Indonesia. The fruit of the plant is also known as black velvet tamarind, derived from its black velvety exocarp and the flavor of the mesocarp that resembles the flesh of the tamarind fruit [14,15].

In Malaysia, the fruit has gained popularity where the name "keranji" is mentioned in a Malay folk poem. To date, only the mesocarp is consumed by locals as sweets, while the seed and the exocarp are discarded as waste. Previous studies on the antioxidant activity of *D. indum* fruit have been conducted specifically on the mesocarp, which showed appreciable in vitro chelation of ferrous ions and scavenging activity against hydroxyls, hydrogen peroxide, 2,2-diphenyl-picrylhydrazyl (DPPH), and nitric oxide radicals. These activities are suggested to be associated with the mesocarp's total ascorbic acid, β-carotene, lycopene, phenolics, and flavonoids content [16,17].

Studies on the potential health benefits in relation to antioxidant metabolites in this underutilized fruit in the literature have been very scarce. In light of this, a study on the antioxidant content and activities of the fruit is deemed necessary to fill in the research gap. Hence, the objectives of this study were: (i) to determine the antioxidant activities of *D. indum* fruit comprising various extracts and fractions of the exocarp, mesocarp, and seed; (ii) identify the metabolites present in the antioxidant active fractions and subfractions; and (iii) evaluate the correlation of the antioxidant content with the in vitro antioxidant activities.

2. Materials and Methods

2.1. Chemicals

All reagents were purchased from Merck (Darmstadt, Germany), Sigma–Aldrich (Steinheim, Germany), and Acros Organics (Thermo Fisher Scientific, Reel, Belgium), while all solvents used were of analytical grade.

2.2. Plant Materials

Sun-dried *D. indum* fruits were obtained from Bukit Ibam, Muadzam Shah, Pahang, Malaysia and identified by Shamsul Khamis, a botanist from Universiti Kebangsaan Malaysia Herbarium (UKMB). A voucher specimen (PIIUM 0257) was deposited in the Herbarium of Kulliyyah of Pharmacy, International Islamic University of Malaysia, Kuantan. Healthy and uninfected fruits were carefully selected and the exocarp (skin), mesocarp (pulp), and seed of the fruits were separated manually. Each fruit part was further dried in the drying oven (Memmert GmbH + Co. KG, Büchenbach, Germany) at 40 °C for three days, then immediately ground into powder form and kept in plastic containers at room temperature until further use.

2.3. Preparation of Extracts and Fractions

One crude extract and three fractions were prepared from each part of the *D. indum* fruit. For the crude extract, 25 g of each fruit part was macerated in 100% methanol (500 mL) at room temperature for 48 h. The extraction procedure was repeated three times. Next, the solvent was removed under reduced pressure using a rotary evaporator (RV 10, IKA®, Staufen im Breisgau, Germany) at 40 °C. Using a similar extraction procedure, 75 g of each fruit part was extracted successively with 1.5 L *n*-hexane, dichloromethane (DCM), and methanol to obtain hexane, DCM, and methanol fractions [18]. All extracts and fractions were stored at 4 °C until further use.

2.4. Reduction of Phosphomolybdic-Phosphotungstic Acid Reagents

Capability of the extracts and fractions to reduce phosphomolybdic-phosphotungstic acid reagents, otherwise known as Folin–Ciocalteu reagent, was determined according to a previous report with slight modifications [3]. Briefly, 90 μL diluted Folin–Ciocalteu reagent in deionized water (20% *v/v*) was placed in each well of a 96-flat-bottomed-well microplate. Then, an 18 μL sample solution in methanol (1000 μg/mL) was added and incubated at room temperature for 5 min, followed by the addition of 90 μL sodium carbonate in deionized water (75 g/L). The mixture was incubated for 2 h at room temperature. The absorbance was then read at 725 nm using a microplate reader (Tecan, Infinite M200 Nanoquant, Männedorf, Switzerland). Gallic acid equivalence (GAE) was determined using a gallic acid calibration curve. All samples were tested in triplicate. Results were expressed as μmol GAE per gram dry weight crude extract or fraction ± standard error of the mean (*SEM*).

2.5. Reduction of Neocuproine

The extent of the conversion of light blue-colored bis(neocuproine)copper(II) chelate to yellow-orange-colored bis(neocuproine)copper(I) chelate by antioxidants was measured using the method described by Apak, Güçlü, Özyürek, Bektaşoğlu, and Bener [19]. An aliquot of 48.8 μL copper(II) chloride in deionized water (10 mM), 48.8 μL neocuproine in 96% ethanol (7.5 mM), and 48.8 μL ammonium acetate buffer (1 M, pH 7.0) were added into wells of a 96-flat-bottomed-well microplate. Then, a 24.4 μL sample solution in 96% ethanol (1000 μg/mL) and 29.3 μL deionized water were added to make up the final volume of 200 μL in each well. The absorbance was measured at 450 nm after 30 min incubation at room temperature. The total antioxidant capacity (TAC) was determined using the standard calibration curve of trolox. All samples were measured in triplicate. Results were expressed as μmol trolox equivalence (TE) per gram dry weight crude extract or fraction ± standard error of the mean (*SEM*).

2.6. Scavenging of DPPH Radical

Serial dilution from stock solution of the samples and positive standard (quercetin) in 100% methanol (1000 μg/mL) was done in 30 mL glass vials using a micropipette, then 100 μL of each concentration was transferred to each well of the microplate. A total of 100 μL DPPH in methanol (80 μg/mL) was then added. The mixture was incubated at room temperature for 30 min and the absorbance of the test mixture was then read at 517 nm against the blanks: $A_{blank\,sample}$ (100 μL of 7.8125–1000 μg/mL sample without DPPH and 100 μL methanol) and $A_{blank\,methanol}$ (200 μL of 100% methanol). The percentage of inhibition of the DPPH radical was calculated using the equation:

$$\text{DPPH radical scavenging activity (\%)} = [1 - (A_{control} - A_{sample})/A_{control}] \times 100$$

where $A_{control}$ is the absorbance of DPPH without the sample after subtraction with $A_{blank\,methanol}$ while A_{sample} is the absorbance of samples with DPPH after subtraction with $A_{blank\,methanol}$ and $A_{blank\,sample}$. $A_{blank\,sample}$ was included in the equation to minimize the effect of varying visible colors of the different extracts to the absorbance readings. For each sample, the assay was conducted in triplicate. The concentration of extracts required to inhibit the DPPH radicals by 50% (IC_{50}) was calculated using either the logarithmic or exponential or linear regression model equation that best fitted the data for each sample ($r > 0.9$) [20].

2.7. Inhibition of Linoleic Acid Peroxidation

Briefly, 62.5 µL of varying concentrations of the samples (7.8125–1000 µg/mL) and standard (quercetin) were prepared in 1.5 mL Eppendorf tubes. The linoleic acid emulsion was freshly prepared by the emulsification of 100 µL of linoleic acid with 200 µL Tween 20 and 19.7 mL deionized water [21]. Next, 62.5 µL of varying concentrations of the *D. indum* extracts, fractions and standard were mixed with 62.5 µL of the linoleic acid emulsion, 62.5 µL of phosphate buffer (100 µM, pH 7.4), and 12.5 µL of ferrous sulfate solution (4 mM in deionized water). Linoleic acid peroxidation was started by the addition of 12.5 µL ascorbic acid (2 mM in deionized water, freshly prepared for each extract), incubated for 30 min at 37 °C, and terminated by the addition of 187.6 µL of trichloroacetic acid (10% in deionized water). The mixture was then added to 100 µL of thiobarbituric acid solution (1% in 50 mM NaOH), followed by heating in a 100 °C water bath for 10 min. The mixtures were centrifuged at $3500\times g$ for 10 min. Then, 100 µL of the supernatant was transferred into the well of a 96-flat-bottomed-well microplate and the absorbance of thiobarbituric acid-reacting substances (TBARS) in the supernatant was measured at 532 nm. For each sample concentration, the assay was conducted in triplicate [22]. The percentage of linoleic acid peroxidation inhibition and IC_{50} were calculated using a similar equation and method described in Section 2.6.

2.8. Bioautography-Guided Fractionation

The exocarp dichloromethane fraction (EDF) that exhibited strong antioxidant activities for most assays was further fractionated by column chromatography. A glass column (32 × 1000 mm) packed with silica gel 60 (70–230 mesh, 63–200 µm) was loaded with the fraction at a ratio of 130:1. The column was eluted with binary solvent systems of increasing polarity starting with 200 mL of DCM/hexane (8:2, *v*/*v*) up to 100% DCM at 5% increments to afford five subfractions (Di-1 to Di-5), then DCM/ethyl acetate (EA) (99:1, *v*/*v*) up to the volume ratio of 7:3 at 1%, 2% and 5% increments of EA successively to yield fifteen subfractions (Di-6 to Di-20). Next, DCM/acetone (95:5, *v*/*v* up to 8:2, *v*/*v*) at 5% increments of acetone, yielding two subfractions (Di-21 and Di-22). Finally, DCM/methanol (99:1, *v*/*v*) at 1% and 2% increments of methanol up to 8:2 and 9:1 ratios, respectively, giving four subfractions (Di-23 to Di-26). Antioxidant subfractions were monitored by TLC bioautographic screening using 0.4 mM DPPH in methanol as the spraying reagent and the DPPH radical scavenging activity was evaluated.

2.9. Sample Derivatization

Sample derivatization was carried out according to a method described by the manufacturer [23] with modifications. One hundred (100) µL pyridine was added to 1.5 mg of extract in a 1.5 mL Eppendorf tube and the mixture was sonicated at 30 °C for 10 min. Then, 100 µL of silylating agent *N*-methyl-*N*-(trimethylsilyl)trifluoroacetamide (MSTFA) was added, followed by incubation in a water bath at 60 °C for 15 min. Finally, the derivatized sample was syringed out using a 1 mL tuberculin syringe with needle, filtered using a 0.45 µm filter, and injected directly into the gas chromatography-mass spectrometry (GC-MS) instrument.

2.10. Metabolite Identification by GC-MS

The derivatized sample was analyzed with an Agilent 6890N Network Gas Chromatography system (Agilent, CA, USA), coupled to an Agilent 5973 Network Mass Selective Detector (Agilent, CA, USA), equipped with an Agilent 7683 Series Injector (Agilent, CA, USA). One µL aliquot of the extract was injected in splitless mode into a DB5-MS + DG column (Agilent, CA, USA) (J&W 122-5532G, 30 m × 250 µm internal diameter low-bleed fused silica capillary column coated with 5% phenyl-95% dimethylarylene siloxane of 0.25 µm thickness, with a built-in 10 m Duraguard column). Helium was used as the carrier gas and the pressure was programmed so that the helium flow was kept constant at a flow rate of 1.2 mL per min. The injector temperature was set at 250 °C.

A series of trials to optimize the column initial and final temperatures, temperature holding times, and temperature increment rates was carried out to obtain the best separation of the chromatogram peaks for EDF (Table S1). The optimized GC parameters were as follows: initial column temperature was isothermal at 100 °C for 12 min, then raised to 140 °C at a rate of 5 °C/min, held for 10 min, and increased to 250 °C at a rate of 5 °C/min. The temperature was then held at 250 °C for 10 min, making a total running time of 62 min. Chromatogram peaks were integrated by using a Chemstation integrator and the mass spectra for each of the peaks were compared with the National Institute of Standards and Technology (NIST) 2014 database library. Since no internal standard was included, only compounds with similarity indices of 90 and above were reported.

2.11. Statistical Analysis

Values are mean $\pm$ standard error of the mean (*SEM*) of triplicate analysis. Calculation and regression analysis were performed using Microsoft Excel 2016 (Microsoft Corporation, Washington, D.C., USA) and Graph 4.4.2 whereas statistical analysis was carried out using Statistical Package for Social Sciences (SPSS, version 22.0, International Business Machines Corporation, New York, NY, USA). The strength and direction of the association between two ranked variables were measured using the Spearman correlation. The Shapiro–Wilk test was used for assessing the normality of the data distribution while the homogeneity of variances was tested using Levene's test. Since the assumption of the homogeneity of variances had been violated, multiple comparisons between the various extracts were performed using one-way analysis of variance (ANOVA) of unequal variance (Welch's ANOVA) with Games Howell post hoc test. Significant difference was accepted at $p < 0.050$ or $p < 0.010$.

3. Results

3.1. Gallic Acid Equivalence (GAE)

The gallic acid equivalence for the various crude extracts and fractions is shown in Table 1. GAE values varied in the range of 92.97 $\pm$ 0.99 (for mesocarp methanol fraction, MMF) to 1405.41 $\pm$ 17.96 (for seed methanol fraction, SMF) μmol GAE/g dry extract, using a calibration curve of gallic acid ($r^2 = 0.9997$). The GAE values were further grouped as the following: low GAE (0–500 μmol GAE/g dry extract), moderate (500–1000 μmol GAE/g dry extract), and high (1000–1500 μmol GAE/g dry extract). All tested extracts and fractions were in the category of low GAE, except for SMF.

Table 1. Gallic acid equivalence (GAE) and total antioxidant capacity (TAC) of crude extracts and fractions of *D. indum* fruit.

Extract/Fraction	GAE (μmol GAE/g Dry Extract)	TAC (μmol TE/g Dry Extract)
Exocarp (E)		
EHF	95.50 $\pm$ 1.57 *	177.00 $\pm$ 13.51 *
EDF	439.44 $\pm$ 5.73 *	451.48 $\pm$ 37.83 *
EMF	258.05 $\pm$ 7.85 *	280.77 $\pm$ 2.27 *
ECM	316.77 $\pm$ 8.35 *	439.39 $\pm$ 6.26 *
Mesocarp (M)		
MHF	104.06 $\pm$ 5.48 *	185.38 $\pm$ 9.16 *
MDF	380.54 $\pm$ 1.99 *	549.52 $\pm$ 27.76 *
MMF	92.97 $\pm$ 0.99 *	104.52 $\pm$ 1.64 *
MCM	101.56 $\pm$ 1.22 *	114.63 $\pm$ 1.20 *
Seed (S)		
SHF	113.09 $\pm$ 1.77 *	259.84 $\pm$ 18.63 *
SDF	181.68 $\pm$ 1.97 *	336.20 $\pm$ 19.93 *
SMF	1405.41 $\pm$ 17.96 **	1515.79 $\pm$ 75.86 **
SCM	169.38 $\pm$ 4.05 *	222.72 $\pm$ 16.03 *

HF: hexane fraction; DF: dichloromethane fraction; MF: methanol fraction; CM: crude methanol extract. * Indicates the values in the same column are significantly different ($p < 0.050$) in comparison with SMF (marked **) as measured by one-way analysis of variance (ANOVA) of unequal variance (Welch's ANOVA) with Games Howell post hoc test.

3.2. Total Antioxidant Capacity (TAC)

Neocuproine reagent confers higher stability when compared with other chromogenic reagents such as 2,2′-azino-bis(3-ethylbenzthiazoline-6-sulfonic acid) (ABTS) and DPPH [20]. TAC for the various *D. indum* fruit extracts (Table 1) varied in the range of 104.52 ± 1.64 (for MMF) to 1515.79 ± 75.86 (for SMF) µmol TE/g dry extract using a calibration curve of trolox ($r^2 = 0.9995$). Spearman correlation was run to determine the direction and magnitude of the relationship between the GAE and TAC of all extracts. Interestingly, there was a very strong, positive correlation between GAE and TAC ($r_s = 0.929$, $n = 16$, $p < 0.010$).

3.3. DPPH Radical Scavenging Activity

Table 2 shows that only six from the twelve crude extracts and fractions of *D. indum* fruit parts had IC_{50} values within the tested concentration range (3.91–500.00 µg/mL). None of the mesocarp crude extracts and fractions reached 50% inhibition of DPPH radicals despite there being concentration-dependent DPPH scavenging activity as shown by mesocarp dichloromethane fraction (MDF) and mesocarp hexane fraction (MHF). The IC_{50} values of various crude extracts and fractions varied in the range of 31.71 ± 0.88 (for SMF) to 497.97 ± 6.43 (for exocarp hexane fraction, EHF) µg/mL and maximum percentage of inhibition varied in the range of $27.27 \pm 1.08\%$ (for MHF) to $93.11 \pm 0.22\%$ (for SMF). Spearman correlation analysis showed that there was a negative correlation between the IC_{50} values of DPPH radical scavenging assay and GAE values of the tested extracts ($r_s = -0.587$, $n = 6$, $p < 0.050$).

Table 2. 2,2-Diphenyl-picrylhydrazyl (DPPH) radical scavenging and linoleic acid peroxidation inhibition activities of *D. indum* fruit.

Extract/Fraction/Standard	DPPH Radical Scavenging		Linoleic Acid Inhibition	
	% at 500 µg/mL	IC_{50} (µg/mL)	% at 125 µg/mL	IC_{50} (µg/mL)
Exocarp (E)				
EHF	50.13 ± 0.61 *	497.97 ± 6.43 *	51.46 ± 0.62 *	103.26 ± 2.75 *
EDF	74.75 ± 0.70 *	260.82 ± 1.31 *	51.08 ± 0.84 *	121.43 ± 2.97 *
EMF	60.52 ± 0.34 *	415.78 ± 4.48 *	18.83 ± 2.12 *	NA
ECM	92.30 ± 0.08 *	127.63 ± 2.48 *	14.46 ± 0.33 *	NA
Mesocarp (M)				
MHF	27.27 ± 1.29 *	NA	33.66 ± 1.19 *	NA
MDF	37.98 ± 0.75 *	NA	11.78 ± 2.34 *	NA
MMF	NA	NA	NA	NA
MCM	NA	NA	NA	NA
Seed (S)				
SHF	NA	NA	NA	NA
SDF	NA	NA	17.80 ± 1.78 *	NA
SMF	93.11 ± 0.22	31.71 ± 0.88 *	23.42 ± 1.01 *	NA
SCM	90.99 ± 0.03 *	99.95 ± 0.98 *	20.79 ± 1.43 *	NA
Standard				
QUE	94.70 ± 0.02 **	2.40 ± 0.03 **	69.58 ± 0.03 **	44.69 ± 0.17 **

NA: not active; HF: hexane fraction; DF: DCM fraction; MF: methanol fraction; CM: crude methanol extract; QUE: quercetin. * Indicates that the values in the same column are significantly different ($p < 0.050$) in comparison with QUE (marked **) as measured by Welch's ANOVA with Games Howell post hoc test.

A total of twenty-six (26) subfractions were obtained from the fractionation of exocarp dichloromethane fraction (EDF) using column chromatography. Subfractions Di-6, Di-9, Di-11, Di-17, Di-21, Di-22, Di-23, Di-24, Di-25, and Di-26 were observed to possess clearly different TLC antioxidant bioautograms after visualization with 0.4 mM DPPH in methanol (Figure S1). Hence, DPPH radical scavenging activity of these ten EDF subfractions were determined where all selected subfractions presented concentration-dependent DPPH radical scavenging activities. However, only five subfractions, labelled as Di-21, Di-22, Di-24, Di-25, and Di-26 exhibited appreciable antioxidant activity with IC_{50} values of 83.73 ± 0.92, 87.96 ± 1.02, 53.42 ± 1.61, 37.66 ± 0.71, and 69.82 ± 1.28 µg/mL, respectively. Comparison with the IC_{50} value of EDF (260.82 ± 1.31 µg/mL) clearly signified that the five subfractions possessed stronger DPPH radical scavenging activities and thus were considered as antioxidant-enriched EDF subfractions.

3.4. Linoleic Acid Peroxidation Inhibition

Table 2 demonstrates that despite the percentage of linoleic acid peroxidation inhibition was significantly lower ($p < 0.050$) than quercetin, both EDF and EHF exhibited the highest percentage of inhibition (51.08 ± 0.84 and 51.46 ± 0.62 µg/mL, respectively) between all samples for the tested concentration range (0.98–125.00 µg/mL). Spearman correlation indicated a very weak but not statistically significant negative correlation ($r_s = -0.197$, $n = 8$, $p > 0.050$) between the maximum percentage of linoleic acid peroxidation inhibition and maximum percentage of DPPH radical scavenging of the various *D. indum* crude extracts and fractions.

3.5. GC-MS of SMF, EDF and EDF Subfractions

Based on the assays employed, SMF and EDF were considered as fractions with prominent antioxidant activities. The difference in the metabolites of both fractions were investigated using GC-MS after derivatization using MSTFA. The total ion current (TIC) chromatograms of SMF and EDF are shown in Figure 1 and the metabolites identified in both fractions are listed in Tables 3 and 4. The mass spectra data of the metabolites identified in SMF and EDF are listed in Table S2 [24–36]. Phenolics, fatty acids, and dicarboxylic acids were detected in both fractions. The total area percentage of phenolics in the TIC chromatogram of EDF was 53 times more than that of SMF. On the other hand, SMF contained amino acids, saccharides, polyol, and sesquiterpene, which were not detected in EDF.

Table 3. Metabolites identified in the *D. indum* seed methanol fraction (SMF) through Gas Chromatography-Mass Spectrometry (GC-MS) analysis.

Peak No.	RT (min)	Tentative Metabolite	Similarity Index	M^+	Molecular Formula	Area (%)
1	16.49	Proline	93	115.06	$C_5H_9NO_2$	0.09
2	19.14	Serine	91	105.04	$C_3H_7NO_3$	0.08
3	20.01	Threonine	91	119.06	$C_4H_9NO_3$	0.06
4	24.47	Malic acid	99	134.02	$C_4H_6O_5$	0.17
5	25.80	Pyroglutamic acid	95	129.04	$C_5H_7NO_3$	0.27
6	32.57	Phenylalanine	91	165.08	$C_9H_{11}NO_2$	0.16
7	32.86	Glutamic acid	98	147.05	$C_5H_9NO_4$	0.19
8	33.93	Tartaric acid	94	150.02	$C_4H_6O_6$	0.07
9	40.39	β-D-Galactofuranose	91	180.06	$C_6H_{12}O_6$	2.43
10	42.16	β-D-Glucopyranose	94	180.06	$C_6H_{12}O_6$	3.89
11	42.44	D-glucose	91	180.06	$C_6H_{12}O_6$	3.96
12	43.14	α-Cyperone	95	218.17	$C_{15}H_{22}O$	0.55
13	43.82	*myo*-Inositol	95	180.06	$C_6H_{12}O_6$	0.55
14	45.61	Palmitic acid	99	256.24	$C_{16}H_{32}O_2$	0.25
15	48.81	Linoelaidic acid	99	280.24	$C_{18}H_{32}O_2$	0.15
16	48.93	Oleic acid	99	282.26	$C_{18}H_{34}O_2$	0.38
17	49.36	Sinapic acid	99	224.07	$C_{11}H_{12}O_5$	0.07
18	49.44	Stearic acid	99	284.27	$C_{18}H_{36}O_2$	0.12
19	52.25	δ-Tocopherol	99	402.35	$C_{27}H_{46}O_2$	0.14
20	57.55	Sucrose	95	342.12	$C_{12}H_{22}O_{11}$	18.86
					Total	32.44

RT = Retention Time, C = Carbon, H = Hydrogen, O = Oxygen, N = Nitrogen, M^+ = molecular ion, *m/z*.

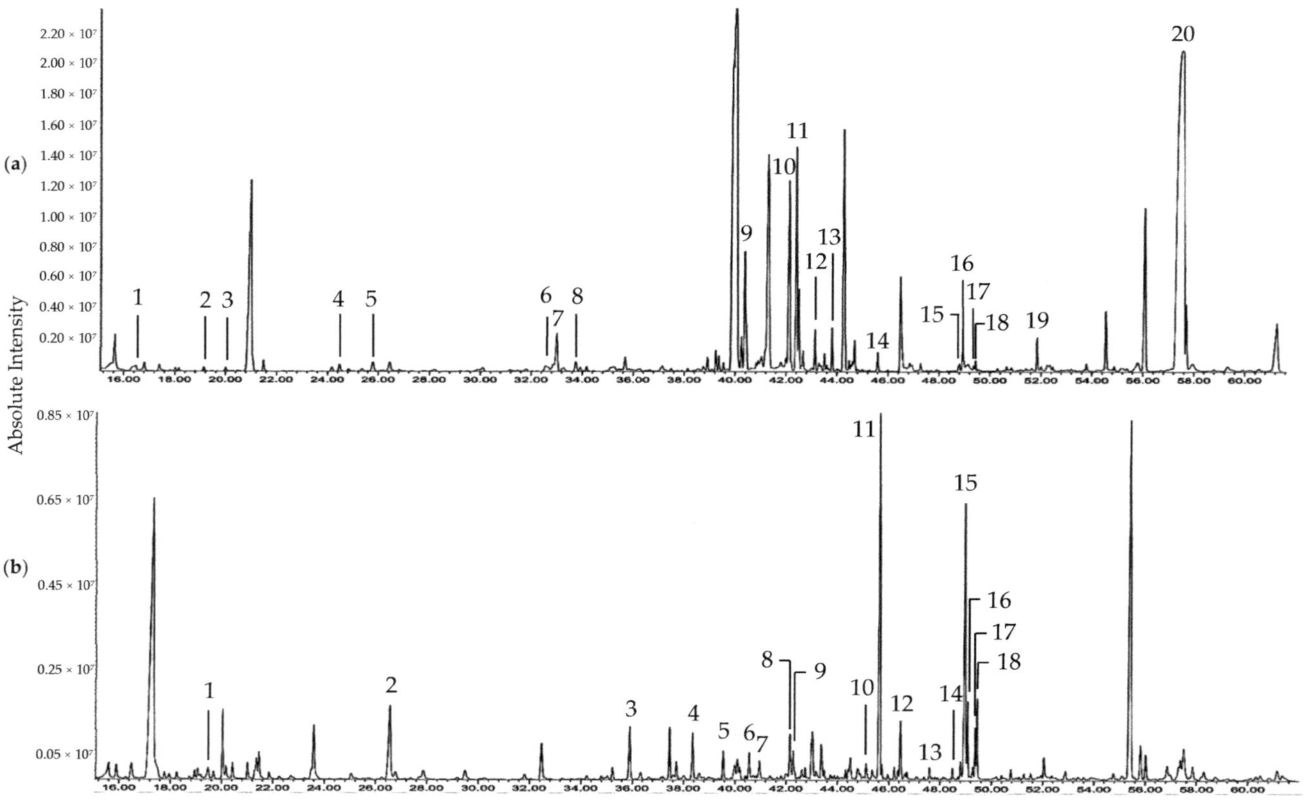

Figure 1. Total ion current (TIC) chromatograms of the trimethylsilyl (TMS)-derivatized *D. indum* seed methanol fraction (SMF) (**a**) and exocarp dichloromethane fraction (EDF) (**b**). Peak numbers refer to metabolites listed in Tables 3 and 4.

Table 4. Metabolites identified in the *D. indum* exocarp dichloromethane (DCM) fraction (EDF) through GC-MS analysis.

Peak No.	RT (min)	Tentative Metabolite	Similarity Index	M⁺	Molecular Formula	Area (%)
1	19.46	*p*-Hydroxybenzaldehyde	97	122.04	$C_7H_6O_2$	0.45
2	26.57	Vanillin	98	152.05	$C_8H_8O_3$	3.48
3	35.91	Syringic aldehyde	96	182.06	$C_9H_{10}O_4$	1.61
4	38.36	Vanillic acid	96	168.04	$C_9H_{10}O_4$	1.26
5	39.55	Azelaic acid	90	188.11	$C_9H_{16}O_4$	0.67
6	40.57	Coniferyl aldehyde	94	178.06	$C_{10}H_{10}O_3$	0.67
7	40.97	Myristic acid	99	228.21	$C_{14}H_{28}O_2$	0.50
8	42.16	Syringic acid	99	198.05	$C_9H_{10}O_5$	1.24
9	42.62	Ferulic acid	96	194.06	$C_{10}H_{10}O_4$	0.30
10	45.13	Palmitelaidic acid	99	254.23	$C_{16}H_{30}O_2$	0.45
11	45.66	Palmitic acid	99	256.24	$C_{16}H_{32}O_2$	9.71
12	46.46	Isoferulic acid	95	196.06	$C_{10}H_{10}O_4$	1.50
13	47.59	Margaric acid	98	270.26	$C_{17}H_{34}O_2$	0.24
14	48.81	Linoelaidic acid	99	280.24	$C_{18}H_{32}O_2$	0.37
15	48.98	Oleic acid	99	282.26	$C_{18}H_{34}O_2$	7.17
16	49.08	*cis*-Vaccenic acid	99	282.26	$C_{18}H_{34}O_2$	1.41
17	49.37	Sinapic acid	99	224.07	$C_{11}H_{12}O_5$	0.97
18	49.45	Stearic acid	99	284.27	$C_{18}H_{36}O_2$	1.50
					Total	33.50

RT = Retention Time, C = Carbon, H = Hydrogen, O = Oxygen, N = Nitrogen, M⁺ = molecular ion, *m/z*.

A total of nine phenolics have been identified in EDF, namely vanillic acid, syringic acid, ferulic acid, isoferulic acid, sinapic acid, vanillin, syringic aldehyde, *p*-hydroxybenzaldehyde, and coniferyl aldehyde. Numerous studies correlated the in vitro antioxidant activities of plant extracts with their phenolic contents. GC-MS of five antioxidant-enriched subfractions of EDF added another four phenolics: *p*-hydroxybenzoic acid, homovanillic acid, *p*-coumaric acid, and sinapic aldehyde, to the list of phenolics detected in EDF. The distribution of phenolic antioxidants in subfractions Di-21, Di-22, Di-24, Di-25, Di-26, and EDF is shown in Table 5.

Table 5. Area percentage (%) of phenolics in exocarp DCM fraction (EDF) and subfractions through GC-MS analysis.

Phenolics	Exocarp DCM Fraction (EDF)	EDF Subfractions				
		Di-21	Di-22	Di-24	Di-25	Di-26
Phenolic aldehydes						
p-Hydroxybenzaldehyde	0.45	ND	ND	0.02	ND	ND
Vanillin	3.48	ND	0.46	0.23	0.16	ND
Syringic aldehyde	1.61	ND	0.26	0.31	0.3	ND
Coniferyl aldehyde	0.67	ND	ND	0.25	0.11	ND
Sinapic aldehyde *	ND	ND	0.22	0.10	ND	ND
Phenolic acids						
Vanillic acid	1.26	2.07	1.30	0.73	0.39	0.45
Syringic acid	1.24	5.14	0.69	0.64	ND	ND
p-Hydroxybenzoic acid *	ND	ND	0.05	ND	ND	ND
Homovanillic acid *	ND	2.15	0.23	ND	0.01	ND
Ferulic acid	0.30	ND	ND	ND	1.20	0.44
Isoferulic acid	1.50	ND	ND	ND	ND	ND
Sinapic acid	0.97	0.98	0.17	0.89	3.48	2.15
p-Coumaric acid *	ND	ND	ND	ND	0.11	ND
Total	11.48	10.34	3.38	3.17	5.76	3.04

ND: not detected. * Indicates the phenolic was detected only in subfractions.

The total area percentage of phenolic antioxidants in the subfractions were lower than EDF, in the following decreasing order: EDF > Di-21 > Di-25 > Di-22 > Di-24 > Di-26. EDF contained the highest number of phenolic antioxidants (9 phenolic antioxidants), followed by Di-22 (8), Di-24 (8), Di-25 (8), Di-21 (4), and Di-26 (3). Vanillin was the major phenolic in EDF. Syringic acid and vanillic acid presented as the major phenolic in subfractions Di-21 and Di-22, respectively, while sinapic acid was the major phenolic in subfractions Di-24, Di-25, and Di-26.

4. Discussion

The antioxidant potential of *D. indum* fruit was evaluated by measuring the capabilities of various extracts and fractions of the different fruit parts in reducing phosphomolybdic-phosphotungstic acid reagents, neocuproine, 2,2-diphenyl-picrylhydrazyl (DPPH), and inhibiting linoleic acid peroxidation. The gallic acid equivalence (GAE) for reducing phosphomolybdic-phosphotungstic acid reagents was the highest in the DCM fractions for the exocarp (EDF) and mesocarp (MDF), while for the seed, the highest GAE was measured in the methanol fraction (SMF). The GAE of SMF was found to be the highest when compared with other local underutilized fruits of Malaysia, namely *Baccaurea angulata* [3], *Canarium odontophyllum* [37], and *Sandoricum macropodum* [38] in other studies. It can be suggested from this result that most antioxidants in the exocarp and mesocarp of *D. indum* fruit are semipolar in nature while in the seed, most of the antioxidants are polar. Low GAE values in hexane fractions might be due to the higher concentration of non-polar constituents such as fatty acids. This study also showed a strong, positive correlation (r_s = 0.929, n = 16, p < 0.010) between GAE and TAC values, which is in line with a previous study that proved the good correlation (r = 0.966) in herbal teas [39]. Both the GAE and TAC values were determined using different assays of a similar underlying antioxidant mechanism, which measure the reduction of the oxidation number of the transition metal ions by antioxidants achieved via electron transfer [40].

The strong neocuproine reducing activity, which was interpreted as the high total antioxidant capacity (TAC) of SMF, can be attributed to the presence of non-phenolic metabolites, particularly the amino acids that also showed neocuproine reducing activity in another study [41]. The following order was ranked for the DPPH radical scavenging activity of the exocarp and seed: SMF > SCM > ECM > EDF > EMF > EHF. However, this sequence was not exactly replicated in their GAE by which the following order was found: SMF > EDF > ECM > EMF > SCM > EHF. The difference in GAE and DPPH radical scavenging activities can be attributed to the difference in the accessibility of various phenolic antioxidants in the different extracts to the unpaired electron of the divalent nitrogen atom of the DPPH radical. Steric accessibility is a major determinant for redox reactions with DPPH, where small antioxidants that have better access to the radical site than their larger counterparts tend to have higher DPPH-reducing power [42]. This finding corroborates previous studies that reported a negative relationship between the IC_{50} values of DPPH radical scavenging and GAE [43–45].

Lipid peroxidation involves the formation and propagation of hydroperoxide radicals, which decompose to form byproducts such as malondialdehyde (MDA), ketones, alcohols, and hydrocarbons that can interact with sulfhydryl and amine groups in proteins, causing damage to vital proteins. The distinctive feature of this lipid peroxidation assay lies in the medium used, which is a linoleic acid emulsion, whereas in the GAE, TAC, and DPPH assays, polar media such as water, methanol, and ethanol were used. Moreover, the assay conforms to the antioxidant polar paradox hypothesis, which states that lipid-soluble antioxidants are most effective in emulsions and membranes [46]. It is interesting to note that SMF, which exhibited excellent positive results for all previous assays showed weak linoleic acid peroxidation inhibitory activity with the maximum percentage of inhibition (at 250 µg/mL) below 50%. This finding may be related to the dominance of hydrophilic antioxidants in SMF. The antioxidant activity of compounds in emulsions is attributed to the hydrophilic–lipophilic balance of their polar and hydrocarbon moieties where more lipophilic antioxidants are present in the lipid phase and at the oil-water interface at which interactions between hydroperoxides and prooxidants occur [47].

The strong DPPH radical scavenging activity of SMF can be attributed to the presence of the amino acids detected, namely proline, serine, threonine, pyroglutamic acid, phenylalanine, and glutamic acid. Amino acids were found to exert antioxidant activities by donating protons to electron-deficient radicals [48]. The percentage of amino acids in SMF was four times more than that of the detected phenolics and was higher in the seed than in the exocarp. *D. indum* is a plant in the family Fabaceae, which is a well-known plant family that produces seeds with a high content of proteins and saccharides and are consumed as nutritive food worldwide [49]. The difference in metabolite compositions between EDF and SMF is largely due to the higher percentage of phenolics in EDF (11.48%) than SMF (0.21%). The nine phenolics in EDF can be categorized into two main groups: phenolic acids and phenolic aldehydes. Previous studies have shown that the more the number of hydroxyl groups at the benzene ring, the stronger the DPPH radical scavenging activities of the phenolics and the conjugated carbon skeleton plays an important role in the antioxidant activities of the phenolics. The strength of the in vitro DPPH radical scavenging activities of the phenolics can be ranked in the following decreasing order: hydroxycinnamic acid derivatives > hydroxybenzoic acid derivatives > phenolic aldehydes [50].

Varying percentages of different phenolics have been found in the selected EDF subfractions. Vanillic acid and sinapic acid were consistently detected in EDF and all the selected subfractions, where sub-fraction Di-21 had the highest percentage of vanillic acid (2.07%) while the highest percentage of sinapic acid (3.48%) was found in sub-fraction Di-25. Higher percentages of vanillic acid in EDF, subfractions Di-21 and Di-22 could be related to their weaker DPPH radical scavenging activities than the subfractions with stronger activity, namely subfractions Di-24, Di-25, and Di-26. This finding was parallel to a study that observed an antagonistic effect of vanillic acid when combined with gallic acid, protocatechuic acid, and chlorogenic acid [51]. The stronger DPPH radical scavenging activities of the more polar EDF subfractions viz. Di-24, Di-25, and Di-26 can be related to their higher percentage of hydroxycinnamic acid derivatives (ferulic acid, sinapic acid, and *p*-coumaric acid). Hydroxycinnamic acid derivatives are more potent antioxidants than hydroxybenzoic acid derivatives due to the presence of the additional carbon–carbon double bond next to the benzene ring, thus extending the conjugated π orbital system. This configuration, in turn, stabilizes the resulting phenoxy radicals by resonance and enhances their antioxidant activities [50,52].

5. Conclusions

In summary, this study revealed the antioxidant potential of *D. indum* fruit exerted by various extracts and fractions from the different fruit parts. The fruit contains phenolics, amino acids, saccharides, fatty acids, sesquiterpene, polyols, and dicarboxylic acids that collectively contribute to its good antioxidant properties. A combination of fractionation by column chromatography, TLC bioautography, and GC-MS facilitated the identification of thirteen phenolic antioxidants in the exocarp of *D. indum* fruit for the first time. In a nutshell, *D. indum* fruit is a good natural antioxidant source that has great potential for further research and development for pharmaceutical, nutraceutical, and cosmeceutical applications.

Supplementary Materials: The following are available online at http://www.mdpi.com/2076-3921/7/11/154/s1, Table S1: Condition parameters for GC optimization, Table S2: Mass spectra data of metabolites identified in SMF and EDF, Figure S1: TLC antioxidant bioautograms of EDF subfractions (Di-1–Di-26) sprayed with 0.4 mM DPPH in methanol.

Author Contributions: M.F.O., N.M.H., and A.K. designed the experiment. S.M.T. revised the statistical analyses performed. M.F.O. carried out the experiments, analyzed the data, and prepared the manuscript. N.M.H. reviewed and corrected the manuscript. All authors approved the final manuscript.

Funding: International Islamic University Malaysia (IIUM) is acknowledged for the grant (RIGS16-123-0287) awarded to support the publication of this work.

Acknowledgments: Special thanks go to the Kulliyyah of Pharmacy, Kulliyyah of Science, and the Integrated Centre for Research Animal Care and Use, International Islamic University Malaysia Kuantan, for the provision of facilities used throughout this study.

Conflicts of Interest: The authors declare no conflict of interest.

References

1. Halliwell, B. Biochemistry of oxidative stress. *Biochem. Soc. Trans.* **2007**, *35*, 1147–1150. [CrossRef] [PubMed]
2. Kongkachuichai, R.; Charoensiri, R.; Yakoh, K.; Kringkasemsee, A.; Insung, P. Nutrients value and antioxidant content of indigenous vegetables from Southern Thailand. *Food Chem.* **2015**, *173*, 838–846. [CrossRef] [PubMed]
3. Ahmed, I.A.; Mikail, M.A.; Ibrahim, M.; Hazali, N.; Rasad, M.S.B.A.; Ghani, R.A.; Wahab, R.A.; Arief, S.J.; Yahya, M.N.A. Antioxidant activity and phenolic profile of various morphological parts of underutilised *Baccaurea angulata* fruit. *Food Chem.* **2015**, *172*, 778–787. [CrossRef] [PubMed]
4. Donno, D.; Cerutti, A.K.; Mellano, M.G.; Prgomet, Z.; Beccaro, G.L. Serviceberry, a berry fruit with growing interest of industry: Physicochemical and quali-quantitative health-related compound characterisation. *J. Funct. Food.* **2016**, *26*, 157–166. [CrossRef]
5. Wang, X.; Zhang, C.; Peng, Y.; Zhang, H.; Wang, Z.; Gao, Y.; Liu, Y.; Zhang, H. Chemical constituents, antioxidant and gastrointestinal transit accelerating activities of dried fruit of *Crataegus dahurica*. *Food Chem.* **2018**, *246*, 41–47. [CrossRef] [PubMed]
6. Kee, M.E.; Khoo, H.E.; Sia, C.M.; Yim, H.S. Fractionation of potent antioxidative components from langsat (*Lansium domesticum*) peel. *Pertanika J. Trop. Agric. Sci.* **2015**, *38*, 103–112.
7. Zefang, L.; Zhao, Z.; Hongmei, W.; Zhiqin, Z.; Jie, Y. Phenolic composition and antioxidant capacities of Chinese local pummelo cultivars' peel. *Hort. Plant J.* **2016**, *2*, 133–140. [CrossRef]
8. Wongnarat, C.; Srihanam, P. Phytochemical and antioxidant activity in seeds and pulp of grape cultivated in Thailand. *Orient. J. Chem.* **2017**, *33*, 113–121. [CrossRef]
9. Tounkara, F.; Bashari, M.; Le, G.-W.; Shi, Y.-H. Antioxidant activities of roselle (*Hibiscus sabdariffa* L.) seed protein hydrolysate and its derived peptide fractions. *Int. J. Food Prop.* **2014**, *17*, 1998–2011. [CrossRef]
10. Procházková, D.; Boušová, I.; Wilhelmová, N. Antioxidant and prooxidant properties of flavonoids. *Fitoterapia* **2011**, *82*, 513–523. [CrossRef] [PubMed]
11. Marcuse, R. Antioxidative effect of amino-acids. *Nature* **1960**, *186*, 886–887. [CrossRef] [PubMed]
12. Milow, P.; Malek, S.B.; Edo, J.; Ong, H.C. Malaysian species of plants with edible fruits or seeds and their valuation. *Int. J. Fruit Sci.* **2014**, *14*, 1–27. [CrossRef]
13. Kamarudin, M.S.; Latiff, A.; Turner, I.M. Taxonomic realignment of Malaysian vascular plants in Burkill's monumental dictionary. *Malayan Nat. J.* **2013**, *65*, 171–229.
14. Janick, J.; Paull, R.E. Fabaceae/Leguminosae. In *The Encyclopedia of Fruit and Nuts*; CAB International: Oxfordshire, UK, 2008; p. 391. ISBN 9780851996387.
15. Lasekan, O.; See, N.S. Key volatile aroma compounds of three black velvet tamarind (*Dialium*) fruit species. *Food Chem.* **2015**, *168*, 561–565. [CrossRef] [PubMed]
16. Tanjung, E.; Thalib, I.; Suhartono, E. Evaluation of antioxidant activity of some selected tropical fruits in South Kalimantan, Indonesia. *J. Trop. Life Sci.* **2014**, *4*, 210–215. [CrossRef]
17. Bamikole, A.O.; Ibidun, O.O.; Ibitayo, O.A.; Bolaji, A.O.; Idowu, O.I.; Damilola, B.B.; Abimbola, F.; Olabisi, O.T.; Joseph, A.O.; Funmilayo, A. Evaluation of antioxidant potentials of different solvent-fractions of *Dialium indum* (African black velvet tamarind) fruit pulp—In vitro. *Potravinarstvo Slovak J. Food Sci.* **2018**, *12*, 70–78. [CrossRef]
18. Ismail, M.; Bagalkotkar, G.; Iqbal, S.; Adamu, H.A. Anticancer properties and phenolic contents of sequentially prepared extracts from different parts of selected medicinal plants indigenous to Malaysia. *Molecules* **2012**, *17*, 5745–5756. [CrossRef] [PubMed]
19. Apak, R.; Güçlü, K.; Özyürek, M.; Bektaşoğlu, B.; Bener, M. Cupric ion reducing antioxidant capacity assay for food antioxidants: Vitamins, polyphenolics and flavonoids in food extracts. In *Advanced Protocols in Oxidative Stress I. Methods in Molecular Biology*; Armstrong, D., Ed.; Humana Press: New York, NY, USA, 2008; Volume 477, pp. 163–193. ISBN 9781603272186.
20. Li, W.J.; Cheng, X.L.; Liu, J.; Lin, R.C.; Wang, G.L.; Du, S.S.; Liu, Z.L. Phenolic compounds and antioxidant activities of *Liriope muscari*. *Molecules* **2012**, *17*, 1797–1808. [CrossRef] [PubMed]
21. Kapila, S.; Vibha, P.R.S. Antioxidative and hypocholesterolemic effect of *Lactobacillus casei* ssp *casei* (biodefensive properties of lactobacilli). *Indian J. Med. Sci.* **2006**, *60*, 361–370. [CrossRef] [PubMed]

22. Alimi, H.; Hfaiedh, N.; Bouoni, Z.; Sakly, M.; Rhouma, K.B. Evaluation of antioxidant and antiulcerogenic activities of *Opuntia ficus indica* f. *inermis* flowers extract in rats. *Environ. Toxicol. Pharmacol.* **2011**, *32*, 406–416. [CrossRef] [PubMed]

23. Derivatization Reagents for Selective Response and Detection in Complex Matrices. Available online: https://www.sigmaaldrich.com/content/dam/sigmaaldrich/migrationresource4/Derivatization%20Rgts%20brochure.pdf (accessed on 3 March 2018).

24. Khallouki, F.; Haubner, R.; Erben, G.; Ulrich, C.M.; Owen, R.W. Phytochemical composition and antioxidant capacity of various botanical parts of the fruits of *Prunus* × *domestica* L. from the Lorraine region of Europe. *Food Chem.* **2012**, *133*, 697–706. [CrossRef]

25. Esmaeili, N.; Ebrahimzadeh, H.; Abdi, K.; Safarian, S. Determination of some phenolic compounds in *Crocus sativus* L. corms and its antioxidant activities study. *Pharmacogn. Mag.* **2011**, *7*, 74–80. [CrossRef] [PubMed]

26. Martin, J.G.P.; Porto, E.; Corrêa, C.B.; De Alencar, S.M.; Da Gloria, E.M.; Cabral, I.S.R.; De Aquino, L.M. Antimicrobial potential and chemical composition of agro-industrial wastes. *J. Nat. Prod.* **2012**, *5*, 27–36.

27. Katona, Z.; Sass, P.; Molnár-Perl, I. Simultaneous determination of sugars, sugar alcohols, acids and amino acids in apricots by gas chromatography–mass spectrometry. *J. Chromatogr. A* **1999**, *847*, 91–102. [CrossRef]

28. Füzfai, Z.; Katona, Z.F.; Kovács, E.; Molnár-Perl, I. Simultaneous identification and quantification of the sugar, sugar alcohol and carboxylic acid contents of sour cherry, apple and ber fruits, as their trimethylsilyl derivatives, by gas chromatography–mass spectrometry. *J. Agric. Food Chem.* **2004**, *52*, 7444–7452. [CrossRef] [PubMed]

29. Roessner, U.; Wagner, C.; Kopka, J.; Trethewey, R.N.; Willmitzer, L. Simultaneous analysis of metabolites in potato tuber by gas chromatography-mass spectrometry. *Plant J.* **2000**, *23*, 131–142. [CrossRef] [PubMed]

30. Plessi, M.; Bertelli, D.; Miglietta, F. Extraction and identification by GC-MS of phenolic acids in traditional balsamic vinegar from Modena. *J. Food Compost. Anal.* **2006**, *19*, 49–54. [CrossRef]

31. Ng, L.-K.; Lafontaine, P.; Harnois, J. Gas chromatographic–mass spectrometric analysis of acids and phenols in distilled alcohol beverages. *J. Chromatogr. A* **2000**, *873*, 29–38. [CrossRef]

32. Guo, J.; Shi, Y.; Xu, C.; Zhong, R.; Zhang, F.; Zhang, T.; Niu, B.; Wang, J. Quantification of plasma myo-inositol using gas chromatography–mass spectrometry. *Clin. Chim. Acta* **2016**, *460*, 88–92. [CrossRef] [PubMed]

33. Kilani, S.; Ledauphin, J.; Bouhlel, I.; Sghaier, M.B.; Boubaker, J.; Skandrani, I.; Mosrati, R.; Ghedira, K.; Barillier, D.; Chekir-Ghedira, L. Comparative study of *Cyperus rotundus* essential oil by a modified GC/MS analysis method. Evaluation of its antioxidant, cytotoxic and apoptotic effects. *Chem. Biodivers.* **2008**, *5*, 729–742. [CrossRef] [PubMed]

34. NIST Chemistry WebBook. Available online: https://webbook.nist.gov/ (accessed on 22 September 2018).

35. Zhang, K.; Zuo, Y. GC-MS determination of flavonoids and phenolic and benzoic acids in human plasma after consumption of cranberry juice. *J. Agric. Food Chem.* **2004**, *52*, 222–227. [CrossRef] [PubMed]

36. Lytovchenko, A.; Beleggia, R.; Schauer, N.; Isaacson, T.; Leuendorf, J.E.; Hellmann, H.; Rose, J.K.C.; Fernie, A.R. Application of GC-MS for the detection of lipophilic compounds in diverse plant tissues. *Plant Methods* **2009**, *5*, 4. [CrossRef] [PubMed]

37. Shakirin, F.H.; Prasad, K.N.; Ismail, A.; Yuon, L.C.; Azlan, A. Antioxidant capacity of underutilized Malaysian *Canarium odontophyllum* (dabai) Miq. fruit. *J. Food Compost. Anal.* **2010**, *23*, 777–781. [CrossRef]

38. Ikram, E.H.K.; Eng, K.H.; Jalil, A.M.M.; Ismail, A.; Idris, S.; Azlan, A.; Nazri, H.S.M.; Diton, N.A.M.; Mokhtar, R.A.M. Antioxidant capacity and total phenolic content of Malaysian underutilized fruits. *J. Food Compost. Anal.* **2009**, *22*, 388–393. [CrossRef]

39. Apak, R.; Güçlü, K.; Ozyürek, M.; Karademir, S.E.; Erçağ, E. The cupric ion reducing antioxidant capacity and polyphenolic content of some herbal teas. *Int. J. Food Sci. Nutr.* **2006**, *57*, 292–304. [CrossRef] [PubMed]

40. Huang, D.; Boxin, O.U.; Prior, R.L. The chemistry behind antioxidant capacity assays. *J. Agric. Food Chem.* **2005**, *53*, 1841–1856. [CrossRef] [PubMed]

41. Imer, F.; Aldemir, E.; Kiliç, H.; Sonmezoğlu, I.; Apak, R. The protective effect of amino acids on the copper(II)-catalyzed autoxidation of ascorbic acid. *J. Food Drug Anal.* **2008**, *16*, 46–53.

42. Pyrzynska, K.; Pękal, A. Application of free radical diphenylpicrylhydrazyl (DPPH) to estimate the antioxidant capacity of food samples. *Anal. Methods* **2013**, *5*, 4288–4295. [CrossRef]

43. Clarke, G.; Ting, K.; Wiart, C.; Fry, J. High correlation of 2,2-diphenyl-1-picrylhydrazyl (DPPH) radical scavenging, ferric reducing activity potential and total phenolics content indicates redundancy in use of all three assays to screen for antioxidant activity of extracts of plants from the Malaysian rainforest. *Antioxidants* **2013**, *2*, 1–10. [CrossRef] [PubMed]

44. Javadi, N.; Abas, F.; Mediani, A.; Hamid, A.A.; Khatib, A.; Simoh, S.; Shaari, K. Effect of storage time on metabolite profile and alpha-glucosidase inhibitory activity of *Cosmos caudatus* leaves—GCMS based metabolomics approach. *J. Food Drug Anal.* **2015**, *23*, 433–441. [CrossRef] [PubMed]

45. Lai, H.; Lim, Y. Evaluation of antioxidant activities of the methanolic extracts of selected ferns in Malaysia. *Int. J. Environ. Sci. Dev.* **2011**, *2*, 442–447. [CrossRef]

46. Barden, L.; Barouh, N.; Villeneuve, P.; Decker, E. Impact of hydrophobicity on antioxidant efficacy in low-moisture food. *J. Agric. Food Chem.* **2015**, *63*, 5821–5827. [CrossRef] [PubMed]

47. Yehye, W.A.; Rahman, N.A.; Alhadi, A.A.; Khaledi, H.; Ng, S.W.; Ariffin, A. Butylated hydroxytoluene analogs: Synthesis and evaluation of their multipotent antioxidant activities. *Molecules* **2012**, *17*, 7645–7665. [CrossRef] [PubMed]

48. Wang, J.; Hu, S.; Nie, S.; Yu, Q.; Xie, M. Reviews on mechanisms of in vitro antioxidant activity of polysaccharides. *Oxid. Med. Cell. Longev.* **2016**, *2016*, 1–13. [CrossRef]

49. Gulewicz, P.; Martinez-Villaluenga, C.; Kasprowicz-Potocka, M.; Frias, J. Non-nutritive compounds in Fabaceae family seeds and the improvement of their nutritional quality by traditional processing—A review. *Pol. J. Food Nutr. Sci.* **2014**, *64*, 75–89. [CrossRef]

50. Mathew, S.; Abraham, T.E.; Zakaria, Z.A. Reactivity of phenolic compounds towards free radicals under in vitro conditions. *J. Food Sci. Technol.* **2015**, *52*, 5790–5798. [CrossRef] [PubMed]

51. Palafox-Carlos, H.; Gil-Chávez, J.; Sotelo-Mundo, R.R.; Namiesnik, J.; Gorinstein, S.; González-Aguilar, G.A. Antioxidant interactions between major phenolic compounds found in "Ataulfo" mango pulp: Chlorogenic, gallic, protocatechuic and vanillic acids. *Molecules* **2012**, *17*, 12657–12664. [CrossRef] [PubMed]

52. Yamagami, C.; Akamatsu, M.; Motohashi, N.; Hamada, S.; Tanahashi, T. Quantitative structure-activity relationship studies for antioxidant hydroxybenzalacetones by quantum chemical- and 3-D-QSAR (CoMFA) analyses. *Bioorg. Med. Chem. Lett.* **2005**, *15*, 2845–2850. [CrossRef] [PubMed]

© 2018 by the authors. Licensee MDPI, Basel, Switzerland. This article is an open access article distributed under the terms and conditions of the Creative Commons Attribution (CC BY) license (http://creativecommons.org/licenses/by/4.0/).

Article

Antiradical and Xanthine Oxidase Inhibitory Activity Evaluations of *Averrhoa bilimbi* L. Leaves and Tentative Identification of Bioactive Constituents through LC-QTOF-MS/MS and Molecular Docking Approach

Qamar Uddin Ahmed [1,*], Alhassan Muhammad Alhassan [1,2,*], Alfi Khatib [1], Syed Adnan Ali Shah [3,4], Muhammad Mahmudul Hasan [1] and Murni Nazira Sarian [1]

[1] Department of Pharmaceutical Chemistry, Kulliyyah of Pharmacy, International Islamic University Malaysia, Pahang DM, Kuantan 25200, Malaysia; alfikhatib@iium.edu.my (A.K.); mhasan222@gmail.com (M.M.H.); murninazira88@gmail.com (M.N.S.)
[2] Department of Pharmaceutical and Medicinal Chemistry, Faculty of Pharmaceutical Sciences, Usmanu Danfodiyo University, Sokoto P.M.B. 2346, Nigeria
[3] Atta-ur-Rahman Institute for Natural Products Discovery (AuRIns), Universiti Teknologi MARA, Bandar Puncak Alam, Shah Alam 40450, Malaysia; syedadnan@salam.uitm.edu.my
[4] Faculty of Pharmacy, Puncak Alam Campus, Universiti Teknologi MARA, Bandar Puncak Alam 42300, Malaysia
*   Correspondence: quahmed@iium.edu.my (Q.U.A.); alhassan.alhassan@udusok.edu.ng (A.M.A.); Tel.: +60-09-571-6400-4932 (Q.U.A.); +2348139766462 (A.M.A.)

Received: 3 July 2018; Accepted: 25 September 2018; Published: 8 October 2018

Abstract: The objective of the present study was to investigate the antiradical and xanthine oxidase inhibitory effects of *Averrhoa bilimbi* leaves. Hence, crude methanolic leaves extract and its resultant fractions, namely hexane, chloroform, and n-butanol were evaluated for 2,2-diphenyl-1-picrylhydrazyl (DPPH) radical scavenging effect and xanthine oxidase inhibitory activity. The active constituents were tentatively identified through LC-QTOF-MS/MS and molecular docking approaches. The n-butanol fraction of *A. bilimbi* crude methanolic leaves extract displayed significant DPPH radical scavenging effect with IC_{50} (4.14 $\pm$ 0.21 µg/mL) ($p < 0.05$), as well as xanthine oxidase inhibitory activity with IC_{50} (64.84 $\pm$ 3.93 µg/mL) ($p < 0.05$). Afzelechin 3-*O*-alpha-L-rhamnopyranoside and cucumerin A were tentatively identified as possible metabolites that contribute to the antioxidant activity of the n-butanol fraction.

Keywords: *Averrhoa bilimbi*; Oxalidaceae; DPPH; xanthine oxidase; LC-QTOF-MS/MS; molecular docking; tentative bioactive constituents

---

1. Introduction

Metabolic activity in the human body produces several reactive chemical species which are known as free radicals. Though some of these free radicals like superoxide (O^{2-}), hydroxyl radical (OH), nitric oxide radical (NO), and hydrogen peroxide (H_2O_2) play beneficial roles in the maintenance of homeostasis, their presence in excessive amounts is often considered detrimental to the body system [1,2]. High levels of free radicals in the body are associated with increased risk of developing metabolic disorders like diabetes mellitus, and cardiovascular diseases [3–5]. Free radicals are frequently involved in the activation of signal transduction systems such as mitogen-activated protein kinases-dependent pathways which interfere with different gene expressions, homeostasis, and the development of diseases [5]. Similarly, increased activity of xanthine oxidase has been linked to the development of oxidative stress and metabolic diseases. There is significant correlation between increased xanthine oxidase activity and some metabolic syndrome markers viz. superoxide dismutase, abdominal circumference, high-sensitivity C-reactive protein content, and body mass indices [6]. Increased xanthine oxidase activity has also been implicated in the development of obesity which is considered a key factor in the pathogenesis of diabetes and cardiovascular diseases [7].

Plants have the inherent ability to biosynthesize several compounds with antioxidant properties which could be explored in the search for a better antioxidant agent for the enhancement of health and prevention of diseases [8,9]. *Averrhoa bilimbi* (family Oxalidaceae) is an Asian medicinal plant commonly known in the Malay language as "belimbing buluh". *A. bilimbi* has been reported to possess significant antioxidant and antidiabetic properties [10]. Topical administration of *A. bilimbi* crude leaves extract has been reported to impede ultraviolet light-induced oxidative damage in albino mice [11]. Oral administration of the methanolic extract (250 and 500 mg/kg,) has been reported to prevent CCl_4-induced hepatic damage in experimental animal model [12]. Crude ethanol and water extracts, as well as the semi-purified butanol fraction of *A. bilimbi* leaves have been reported to display significant ($p < 0.05$) antihyperglycemic activity, i.e., comparable to metformin in streptozotocin-induced diabetic rats [13–15]. However, there is a paucity of information on the phytoconstituents of this plant that could be responsible for its antioxidant properties. Free radicals and xanthine oxidase are believed to be associated with the development of oxidative stress and related metabolic disorders [3–7]. Hence, this study investigates the in vitro antiradical and xanthine oxidase inhibitory effects of *A. bilimbi* leaves. An attempt has also been made to identify the active constituents of *A. bilimbi* leaves through LC-QTOF-MS/MS analysis and the molecular docking approach.

2. Materials and Methods

2.1. Materials

Allopurinol, xanthine, xanthine oxidase, potassium di-hydrogen phosphate (KH_2PO_4), di-potassium hydrogen phosphate (K_2HPO_4), DPPH, and ascorbic acid were purchased from Sigma-Aldrich Chemicals (St. Louis, MO, USA). Dimethylsulphoxide (DMSO), hydrochloric acid (HCl), absolute ethanol, methanol, chloroform, n-butanol, hexane, and other reagents of analytical grade were obtained from Merck (Darmstadt, Germany).

2.2. Plant Collection and Processing

Fresh leaves (1 kg) of *A. bilimbi* were obtained in July 2016 from the garden at Indera Makhota in the neighbourhood of International Islamic University Malaysia (IIUM), Kuantan Campus. The plant was identified and authenticated by the taxonomist at Taman Pertanian, Indera Makhota, 25200 Kuantan, Pahang Darul Makmur, Malaysia. Subsequently, the sample was deposited in the herbarium of Kulliyyah of Pharmacy, IIUM, Kuantan, Pahang DM, Malaysia (voucher no: NMPC-QAAB-12). The fresh plant material was initially dried in a PROTECH laboratory air dryer (FDD-720-Malaysia) at 45 °C. Thereafter, the dried plant material was pulverised into a powdered form using Fritsch Universal Cutting Mill-Pulverisette 19-Germany. The powdered material was kept in airtight polythene bags at 4 °C prior to further use.

2.3. Extraction and Fractionation

Total leaf extract was prepared by soaking powdered leaves (350 g) in methanol (2 L) at room temperature for 48 h, followed by filtration and concentration of the filtrate under reduced pressure using a rotary evaporator (Buchi, Switzerland). The entire procedure was repeated 3 times to yield 40.5 g (11.6%) of crude methanolic leaves extract. Afterwards, the dried crude methanolic leaves extract was subjected to liquid-liquid fractionation. The extract was first suspended in 10% aqueous ethanol, then transferred into the separating funnel and sequentially extracted with hexane, chloroform, n-butanol, and double distilled water to afford hexane fraction (11.67 g), chloroform fraction (8.26 g), n-butanol (8.30 g), and aqueous fraction (12.14 g), respectively.

2.4. Preliminary Phytochemical Screening

The crude methanolic leaves extract of *A. bilimbi* was initially subjected to various phytochemical tests to detect the occurrence of a different class of secondary metabolites which are associated with the existence of flavonoids (Shinoda test) [16], terpenoids (Salkowski test), saponins (frothing test) [17], alkaloids (Dragendorff's and Mayer's tests) and anthraquinones (free and combined) [18].

2.5. LC-QTOF-MS and LC-QTOF-MS/MS Analyses of Bioactive Constituents of n-Butanol Fraction

The LC-QTOF-MS and LC-QTOF-MS/MS analyses of the active n-butanol fraction, which showed significant antiradical (DPPH radical scavenging activity) and xanthine oxidase inhibitory effects, were carried out using an Agilent 1290 Infinity Ultra high performance liquid chromatographic system (LC) (Agilent, Santa Clara, CA, USA) coupled to an Agilent 6520 Accurate-Mass Q-TOF mass spectrometer (Agilent, Santa Clara, CA, USA) equipped with dual electrospray ionization (ESI) source. The n-butanol fraction was prepared at a concentration of 0.01 mg/mL in methanol, and was first filtered using a 0.45 μm nylon syringe filter before an aliquot was collected (3.0 μL) and injected into the LC system. The sample was separated on a reverse phase column [Agilent Zorbax Eclipse XDB-C18, Narrow-Bore 2.1 × 150 mm, 3.5 micron (P/N: 930990-902)] (Agilent, Santa Clara, CA, USA). The mobile phase included A–0.1% formic acid in water, and B–0.1% formic acid in acetonitrile. The column was eluted using a gradient mode under the following conditions. Auto sampler temperature: 6 °C; column temperature: 25 °C; flow rate: 0.5 mL/min; linear gradient from 5% B to 100% B from 0 min to 25 min. Full scan mass spectrometry was carried out using ESI positive ionization mode resulting in full mass spectrum with mass range (m/z) of 100–3200. The system workstation was equipped with a spectra library of several compounds which was used in the tentative identification of major, possibly bioactive constituents based on their accurate mass and elemental composition.

2.6. DPPH Free Radical Scavenging Activity

The DPPH (2,2-diphenyl-1-picrylhydrazyl) free radical scavenging activity of *A. bilimbi* crude methanolic leaves extract, and its fractions viz. hexane, chloroform, and n-butanol fractions, was evaluated according to the protocol described by Nickavar et al. and Barontini et al. [19,20] with some modifications. Briefly, 1 mL of different concentrations of test samples in methanol (4.8 μg/mL–1000 μg/mL) were prepared and respectively treated with 2 mL of 0.1 mM of freshly prepared DPPH solution in methanol, and finally, diluted using 1 mL of deionized water. The mixture was allowed to stay in an incubator at 30 °C for 30 min and absorbance at 517 nm was recorded using a microplate reader (Infinite M200 Nanoquant Tecan, Männedorf, Switzerland). Methanol was used as the blank, whereas DPPH, methanol, and water (2:1:1) served as controls. Ascorbic acid was used as positive standard and IC_{50} values in μg/mL were determined. The IC_{50} values were calculated by the linear regression of plots. The percentage DPPH free stable radical scavenging activity was computed using the following equation:

$$\% \text{ radical scavenging activity} = (Abs_{control} - Abs_{sample}/Abs_{control}) \times 100.$$

where, $Abs_{control}$ = absorbance of control, Abs_{sample} = absorbance of extract, fraction or ascorbic acid.

2.7. Xanthine Oxidase Inhibitory Activity Assay

The xanthine oxidase inhibitory activity of *A. bilimbi* crude methanolic leaves extract and its fractions viz. hexane, chloroform and n-butanol fractions was determined by following the protocol described by Thiombiano et al. [21]. The assay was carried out in 96 well plates. The reaction mixture was prepared in such a way that it contains 50 μL of test solution together with 30 μL of freshly xanthine oxidase solution (0.2 units/mL). The assay mixture was pre-incubated at 37 °C for 15 min. Thereafter, the reaction was initiated by the addition of 60 µL of substrate solution (0.15 mM of xanthine). The mixture was placed in the incubator set at 37 °C and allowed to stay for 30 min. The reaction was terminated by the addition of 50 µL of 0.5 M HCl. A known potent xanthine oxidase inhibitor, viz. allopurinol, was used as positive control. The absorbance at 295 nm was measured using microplate reader (Infinite M200 Nanoquant Tecan, Tecan Group Ltd., Männedorf, Switzerland). The percentage of xanthine oxidase inhibition was calculated using the following equation:

$$\% \text{ xanthine oxidase inhibition} = (Abs_{control} - Abs_{sample}/Abs_{control}) \times 100.$$

where, $Abs_{control}$ = activity of enzyme without extract/fraction, and Abs_{sample} = enzyme activity in presence of extract/fraction or allopurinol.

2.8. Molecular Docking

The phenolic compounds identified through analysis in the n-butanol fraction of *A. bilimbi* crude leaves methanolic extract were considered as possible molecules that contribute to the antioxidant properties. Hence, they were docked into the active site of the crystal structure of xanthine oxidase. The 3D structure for bovine xanthine oxidase, co-crystallized with standard inhibitor, quercetin (PDB ID: 3NVY), was obtained from a protein data bank (PDB) (http://www.rcsb.org/) [22]. Software including Molecular Graphics Library MGLTools 1.5.6 (The Scripps Research Institute, San Diego, CA, USA), AutoDock Tools 4.2 (www.scripps.edu), (The Scripps Research Institute, San Diego, CA, USA) [23], and Discovery Studio Visualizer 4.0 (www.accelrys.com) Bovia, San Diego, CA, USA) [24] were used for the molecular docking experiments. Crystallographic waters and co-crystallized ligand were removed, and polar hydrogens were added to a macromolecule using AutoDock 4.2, after which the structure was saved in the PDBQT file format that contains a protein structure with hydrogen in all polar residues. The 2D structures of ligands were drawn through ChemDraw® (PerkinElmer, Waltham, MA, USA) and converted to 3D format using ChemBio3D Ultra® 14 Suit (PerkinElmer, Waltham, MA, USA), which were further subjected to energy minimization using molecular mechanics 2 (MM2) force field and saved in PDB file format. The ligands were then prepared for docking by computing the charges, and the structures were saved in PDBQT file format via AutoDock 4.2. Molecular docking experiment of ligands was performed following the protocol described by Zhang et al. [25]. The size and center of the coordinates of the grid box was first validated by re-docking the co-crystallized ligand into the active sites of the receptor. The rotational bonds of the ligands were considered to be flexible, while those of the receptor were kept rigid. A grid box was centered on binding site of the co-crystallized ligand. The box size was set to 50, 50 and 50 Å^3 (x, y and z, respectively) and the grid spacing to 0.375 Å. The grid maps for atoms were calculated and genetic algorithm (GA) was used for searching; the population size was set to 150, 100 runs, and 5 million energy evaluations. The ligands (identified compounds) were then docked into the active site of the enzyme using the grid box parameter obtained from the re-docking of co-crystallized ligand as reference.

2.9. Statistical Analysis

All measurements were carried out in triplicate. The 50% inhibitory concentration (IC$_{50}$) was calculated by nonlinear regression analysis and expressed as mean ± standard deviation (SD). Means were compared for statistically significant differences by one-way analysis of variance (ANOVA) using GraphPad Prism 7 (La Jolla, San Diego, CA, USA) with a 95% confidence interval.

3. Results and Discussion

The results of the various phytochemical tests carried out on *A. bilimbi* crude methanolic leaves extract are shown in Table 1. The results indicate the presence of alkaloids, flavonoids, terpenoids, and saponins (which have positive inference), while the extract showed negative results for the presence of free and combined anthraquinones.

Table 1. Phytochemical screening for chemical class identification of *Averrhoa bilimbi* crude methanolic leaves extract.

Bioactive	Test/Procedure	Observation	Inference
Alkaloid	Dragendorff's reagent Mayer's reagent	Orange red ppt. Cream ppt.	+ +
Saponin	Frothing test	Frothing	++
Terpenoid	Chloroform + H_2SO_4	Salkowski test: Brownish red ring at the junction	+++
Flavonoid	Flavones: Mg filling Flavonols: Zn pellet	Shinoda test: Reddish brown Amber	++ ++
Free anthraquinone	Chloroform + NH_3 (10%)	No change from dirty green	−
Combined anthraquinone	10% HCl + Chloroform	No change from colourless	−

+++: Strong intensity reaction, ++: Medium intensity reaction, +: Weak intensity reaction, −: Undetected.

3.1. DPPH Radical Scavenging Activity

The DPPH radical exhibits absorption maxima at $\lambda = 517$ nm in its stable state. This absorption decreases when the stable radical undergoes reduction to a hydrazine derivative in the presence of antioxidant agent; consequently, the purple colour of the DPPH reagent is changed to pale yellow [26]. The reduction is achieved either through the donation of a hydrogen atom to DPPH, or the removal of oxygen. The results of the DPPH radical scavenging activity which demonstrated the antiradical effect for crude methanolic extract and its subsequent fractions, as well as reference standard (ascorbic acid), are shown in Table 2.

Table 2. Free radical scavenging (DPPH) activity of *Averrhoa bilimbi* crude methanolic leaves extract and its fractions.

Sample	IC_{50} (µg/mL)
Crude methanolic extract	10.53 ± 0.72 *
Hexane fraction	>1000
Chloroform fraction	13.44 ± 1.00 *
n-Butanol fraction	4.14 ± 0.21 *
Ascorbic acid (positive control)	5.52 ± 0.29

Values are mean of triplicate experiments ± standard deviation. The results were analysed using one way Analysis of variance (ANOVA). The superscript (*) indicates significant activity ($p < 0.05$). IC_{50} represents the concentration required to scavenge 50% of available radicals. The IC_{50} values were calculated by linear regression of plots.

The results of the DPPH assay showed that the n-butanol fraction displays the highest activity, followed by the crude methanolic leaves extract and the chloroform fraction, while the hexane fraction did not show any significant radical scavenging activity. The radical scavenging activity displayed by the n-butanol fraction was comparable to that of the ascorbic acid, which was used as reference standard. The substantial antioxidant activity of the n-butanol fraction could be due to the presence of phenolic constituents which are frequently soluble in polar solvents and are known to display significant radical scavenging activity [8,27]. The IC_{50} value for the chloroform (13.44 ± 1.00) fraction also signifies a significant radical scavenging effect, which suggests the presence of some active constituents.

3.2. Xanthine Oxidase Inhibitory Activity

Xanthine oxidase as the name implies is an enzyme that is responsible for the conversion of xanthine and hypoxanthine to uric acid through oxidation reaction. Increased activity of xanthine oxidase leads to over production of uric acid and resultant hyperuricemia [28]. Increased levels of uric acid in the blood are implicated in the development of gouty arthritis, as well as increased reactive oxygen species, which are involved in various pathological processes [29]. Table 3 shows the result for the xanthine oxidase inhibitory activity of *A. bilimbi* crude methanolic leaves extract and its resultant fractions. The n-butanol fraction displayed significant ($p < 0.05$) activity with 50% effective concentration (IC_{50}) of 64.84 ± 3.93 µg/mL, while the hexane and chloroform fractions displayed very weak inhibitory activities. The xanthine oxidase inhibitory activity observed with this fraction suggests that it contains substantial amount of phytoconstituents with xanthine oxidase inhibitory potential.

Table 3. IC_{50} for xanthine oxidase inhibitory activity of *Averrhoa bilimbi* crude methanolic leaves extract and its fractions.

Sample	IC_{50} µg/mL
Crude methanolic leaves extract	>1000
Hexane fraction	>1000
Chloroform fraction	>1000
n-Butanol fraction	64.84 ± 3.93 *
Allopurinol (positive control)	16.21 ± 0.91

Values are mean of triplicate experimental results with standard deviation. The results were analysed using one way ANOVA. The symbol asterisk (*) indicates significant activity ($p < 0.05$).

Xanthine oxidase is present in large quantities in the liver, where it plays vital role in purine nucleotide metabolism. The activity of xanthine oxidase has been linked to the generation of reactive oxygen species such as superoxide (O^{2-}), which is known to play vital roles in the development and progression of important pathological conditions like diabetes mellitus and related complications [30]. This result suggests that inhibition of xanthine oxidase could be part of the mechanism of *A. bilimbi* activity against oxidative tissue damage.

3.3. LC-QTOF-MS Analysis of n-Butanol Fraction of Averrhoa bilimbi Crude Methanolic Leaves Extract

The results of the LC-QTOF-MS analysis of active n-butanol fraction of *A. bilimbi* crude methanolic leaves extract are shown in Figure 1 and Table 4. Several compounds which were detected in the active n-butanol fraction have been identified from the LC-MS compound library by matching their accurate mass. There were also some unknown compounds that could not be identified because they were not present in the compound database. The list of identified compounds include two phenolic constituents, namely, 5,7,4′-trihydroxy-6-(1-ethyl-4-hydroxyphenyl) flavone-8-C-glucoside (cucumerin A) (Figure 2) and afzelechin 3-O-alpha-L-rhamnopyranoside (Figure 3) which could partly be responsible for the antiradical and xanthine oxidase inhibitory activity of the n-butanol fraction of the *A. bilimbi* crude methanolic leaves extract.

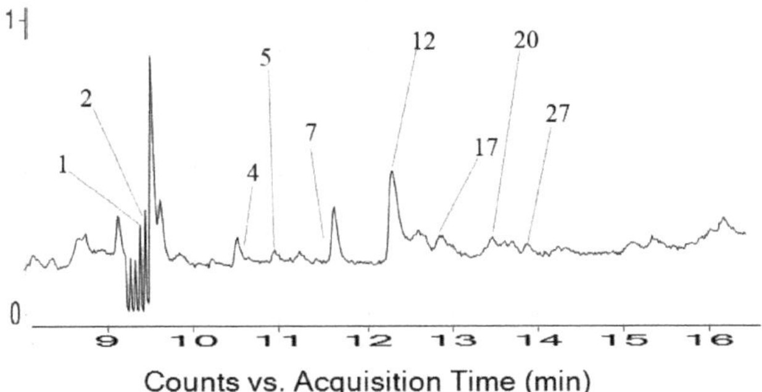

Figure 1. LC-MS spectrum of n-butanol fraction of *Averrhoa bilimbi* crude methanolic leaves extract (For compounds and retention time, please refer to Table 4).

Table 4. Results of LC-QTOF-MS analysis of n-butanol fraction.

Number	Tentative Compounds	Retention Time	Molecular Formula/Molecular Weight (M⁺)
1	5,7,4′-Trihydroxy-6-(1-ethyl-4-hydroxyphenyl)flavone0-8-C-glucoside (Cucumerin A)	9.305	$C_{29}H_{28}O_{11}$/552.185
2	Afzelechin 3-*O*-alpha-L-rhamnopyranoside	9.305	$C_{21}H_{24}O_9$/420.142
3	Ethyl 3-(*N*-butylacetamido)propionate	9.997	$C_{11}H_{21}NO_3$/215.152
4	Elaeokanine C	10.749	$C_{12}H_{21}NO_2$/211.156
5	2-Ethyl-dodecanoic acid	10.958	$C_{14}H_{28}O_2$/228.208
6	Isoavocadienofuran	11.063	$C_{17}H_{26}O$/246.198
7	(5alpha,8beta,9beta)-5,9-Epoxy-3,6-megastigmadien-8-ol	11.451	$C_{13}H_{20}O_2$/208.144
8	Diglycidyl resorcinol ether	12.120	$C_{12}H_{14}O_4$/222.090
9	19-Hydroxycinnzeylanol 19-glucoside	12.212	$C_{26}H_{42}O_{13}$/562.264
10	Xestoaminol C	12.236	$C_{14}H_{31}NO$/229.240
11	Phytosphingosine	12.266	$C_{18}H_{39}NO_3$/317.293
12	2-Hydroxyhexadecanoic acid	12.334	$C_{16}H_{32}O_3$/272.236
13	Pentadecanal	12.408	$C_{15}H_{30}O$/226.230
14	Anapheline	12.707	$C_{13}H_{24}N_2O$/224.188
15	Palmitic amide	12.713	$C_{16}H_{33}NO$/255.256
16	Tetradecylamine	12.769	$C_{14}H_{31}N$/213.246
17	Pentadecanoyl-EA	12.825	$C_{17}H_{35}NO_2$/285.266
18	Codonopsine	12.879	$C_{14}H_{21}NO_4$/267.147
19	Enigmol	13.365	$C_{18}H_{39}NO_2$/301.297
20	7-Hexadecen-1-ol	13.430	$C_{16}H_{32}O$/240.245
21	(Z)-2-Amino-1-hydroxyoctadec-4-en-3-one	13.498	$C_{18}H_{35}NO_2$/297.266
22	Dihydroceramide C2	13.500	$C_{20}H_{41}NO_3$/343.308
23	6-Hydroxysphingosine	13.502	$C_{18}H_{37}NO_3$/315.277
24	Methyl 8-[2-(2-formyl-vinyl)-3-hydroxy-5-oxo-cyclopentyl]-octanoate	13.585	$C_{17}H_{26}O_5$/310.179
25	Dehydrophytosphingosine	13.753	$C_{18}H_{37}NO_3$/315.277
26	14-Methyl-8-hexadecen-1-ol	13.795	$C_{17}H_{34}O$/254.261
27	Nonadecanal	13.800	$C_{19}H_{38}O$/282.291
28	Oleoyl Ethanolamide	15.294	$C_{20}H_{39}NO_2$/325.296
29	Linoleamide	15.308	$C_{18}H_{33}NO$/279.256

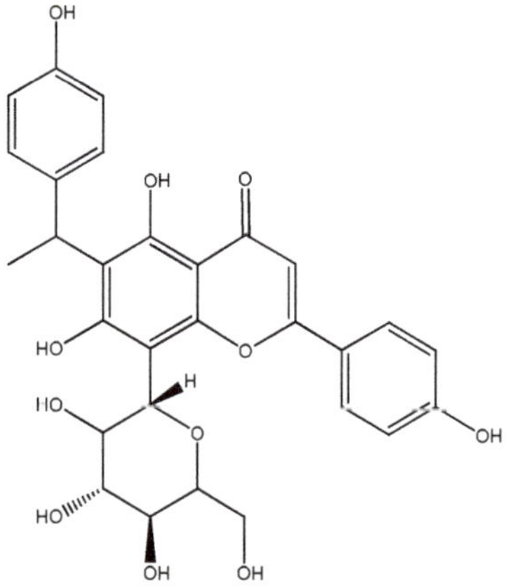

Figure 2. Tentative structure of 5,7,4′-trihydroxy-6-(1-ethyl-4-hydroxyphenyl)flavone-8-C-glucoside (cucumerin A).

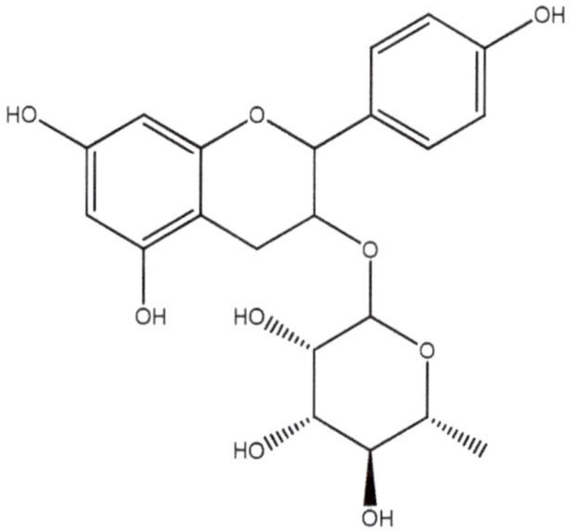

Figure 3. Tentative structure of afzelechin 3-*O*-alpha-L-rhamnopyranoside.

The two compounds, viz. cucumerin A and afzelechin 3-*O*-alpha-L-rhamnopyranoside, are flavonoid glycosides. The antioxidant activities of flavonoid glycosides have been well studied. Based on our literature search, we discovered that there is no available literature focusing specifically on the radical scavenging effect or xanthine oxidase inhibitory activity of afzelechin 3-*O*-alpha-L-rhamnopyranoside and cucumerin A. However, previous research findings have shown that related flavonoid glycosides displayed significant antioxidant potential through DPPH radical scavenging effect and xanthine oxidase inhibitory activity [31,32]. Thus, it can be rightly construed that the antioxidant activity of the n-butanol fraction of *A. bilimbi* crude methanolic leaves extract could partly be due to the presence of phytoconstituents which were identified through the LC-QTOF-MS analysis.

3.4. Identification of Compounds Structures through Fragmentation Analysis (LC-QTOF-MS/MS)

A Q-TOF LC-MS system was used to analyze samples with some modifications [33]. LC-QTOF-MS/MS analysis was carried out to study the fragmentation pattern of 5,7,4′-trihydroxy-6-(1-ethyl-4-hydroxyphenyl)flavone-8-glucoside (cucumerin A) (Figures 4 and 5) and afzelechin 3-O-alpha-L-rhamnopyranoside (Figures 6 and 7), in order to further characterize the structure of these compounds. The MS/MS spectrum and fragmentation pattern of cucumerin A are shown in Figures 4 and 5, respectively. The spectrum shows the presence of fragment ions that are the characteristic of

C-glycosides. The ion peaks at m/z 403 and m/z of 421 representing [M + H-150]$^+$ and [M + H-132]$^+$ are typical of flavonoids with C-glycosides [34]. The peak at m/z 421 (base peak) represents the most stable ion fragment. In addition, a neutral loss of water molecules was observed, which produced fragment ions at [M + H-18]$^+$, [M + H-36]$^+$ and [M + H-54]$^+$. A peak at m/z 341 [M + H-132]$^+$, which indicates the loss of the B ring of the flavone skeleton and 1-ethyl-4-hydroxyphenyl ring attached to C-6 of the compound, further confirmed its structure. Based on this fragmentation pattern, we resolve the structure of this compound as 5,7,4′-trihydroxy-6-(1-ethyl-4-hydroxyphenyl) flavone-8-glucoside (cucumerin A). Figures 6 and 7 show the MS/MS spectrum and fragmentation pattern of afzelechin 3-*O*-alpha-L-rhamnopyranoside, respectively. The peak at m/z 275 represents the aglycone afzelechin ion, which was produced from the loss of the glycone moiety (O-rhamnosyl). This form of fragmentation pattern is characteristic of flavone-O-glycosides [35]. The aglycone ion underwent further fragmentation to yield an ion peak at m/z 107, which is characteristic of the flavan. The spectrum also revealed the loss of water molecules with peaks at m/z 403 and 367 representing [M + H-18]$^+$ and [M + H-54]$^+$. Based on this information, we tentatively identified this compound as afzelechin 3-*O*-alpha-L-rhamnopyranoside.

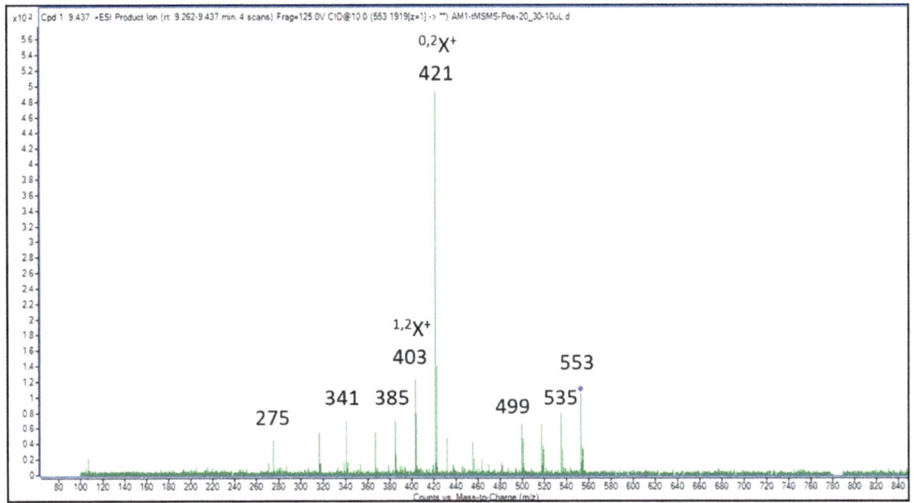

Figure 4. MS/MS spectrum of cucumerin A.

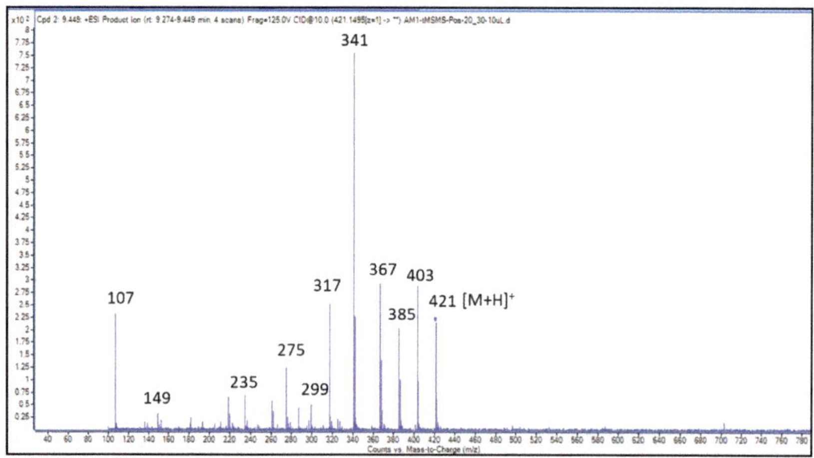

Figure 5. Fragmentation pattern of cucumerin A.

Figure 6. MS/MS spectrum of afzelechin 3-*O*-alpha-L-rhamnopyranoside.

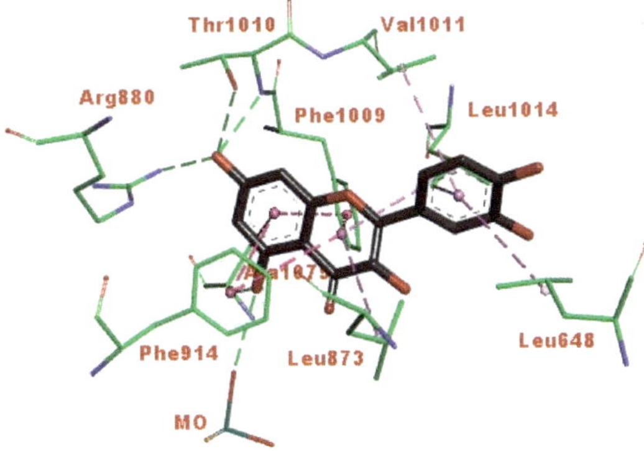

Figure 7. Fragmentation pattern of afzelechin 3-*O*-alpha-L-rhamnopyranoside.

3.5. Molecular Docking

The structure of vertebrate xanthine oxidase exists as a homodimer with each of its functional unit having four redox sites, viz. an active-site molybdenum center, a pair of spinach ferredoxin-like [2Fe-2S] clusters, and a redox cofactor as flavin adenine dinucleotide (FAD) [36]. Figure 8 shows the active site of bovine xanthine oxidase (PDB code: 3NVY) with quercetin as the co-crystallized ligand. The amino acid residues in the active site include Arg880, Thr1010, Glu808, Phe1009, Ala1079, Phe914, Val1011, Leu1014, Leu873, and Leu648. The 7-OH of quercetin formed hydrogen-bonding interactions with Arg880 and Thr1010. The benzopyran ring displayed pi-pi stacking interactions with Phe914 and Phe1009, while 5-OH formed hydrogen-bonding interaction with the Mo. These interactions are considered as important requirements for the inhibitory activity of flavonoids against xanthine oxidase [36].

Figure 8. Binding interaction of quercetin (co-crystallized ligand) with the active site residues of xanthine oxidase (protein data bank (PDB) code: 3NVY). Green and purple dashes represent hydrogen-bonding and hydrophobic interactions, respectively.

Figure 9 shows the binding pose of the afzelechin 3-*O*-alpha-L-rhamnopyranoside in the active site of xanthine oxidase. This compound displayed a binding pose similar to quercetin. The 7-OH formed hydrogen-bonding interactions with Arg880 and Thr1010, while the benzopyran ring exhibited pi-pi stacking interaction with Phe914. These interactions suggest that this compound may exert inhibitory activity against the enzyme. A pi-hydrophobic interaction was observed between the aromatic ring B of the flavan and Leu648. The glycone moiety displayed extensive hydrogen-bonding interactions with Lys771, Glu802, and Asn768, which may also play a role in the xanthine oxidase inhibitory activity of this molecule.

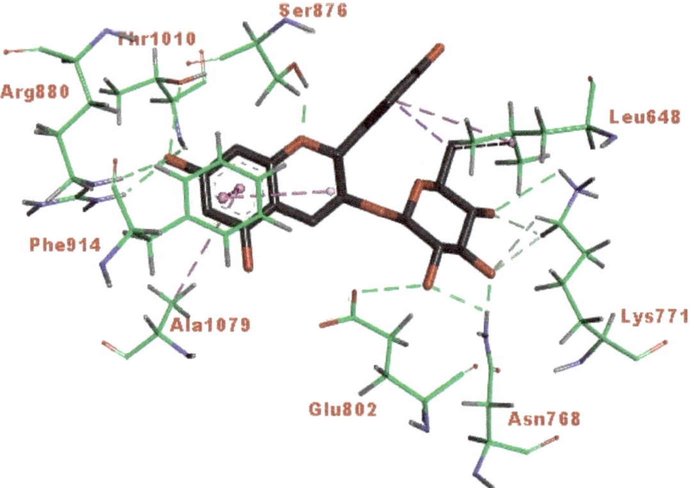

Figure 9. Binding interaction of afzelechin 3-*O*-alpha-L-rhamnoside with the active site residues of xanthine oxidase (PDB code: 3NVY). Green and purple dashes represent hydrogen-bonding and hydrophobic interactions, respectively.

The binding pose of cucumerin A in the active site of xanthine oxidase is shown in Figure 10. Cucumerin A displayed similar interaction with quercetin and afzelechin. The 4-OH of the phenyl ethyl group binds into the catalytic pocket, exhibiting hydrogen bonding with Arg880 and Thr1010, as well as pi-pi stacking with Phe914 and Phe1010. The 7-OH of the flavonoid ring formed hydrogen bonding with Ser879, while the glycone moiety interacted with Asp872 as well as Ser876 through hydrogen bonding. These interactions show that cucumerin A could also contribute to the xanthine oxidase inhibitory effect of the butanol fraction of the *A. bilimbi* crude methanolic leaves extract. Figure 11 shows the binding pose of allopurinol, a known inhibitor of xanthine oxidase. The C=O group displayed hydrogen-bonding interaction with Arg 880, while the aromatic ring was stacked by pi-pi hydrophobic interactions with Phe914 and Phe1009. Additional hydrophobic interactions were observed with Ala 1078 and Ala1089. These interactions were similar to what was observed in the case of quercetin, cucumerin A, and afzelechin 3-*O*-alpha-L-rhamnopyranoside. However, it did not exhibit a hydrogen-bonding interaction with Thr1010, as observed with the aforementioned three flavonoids.

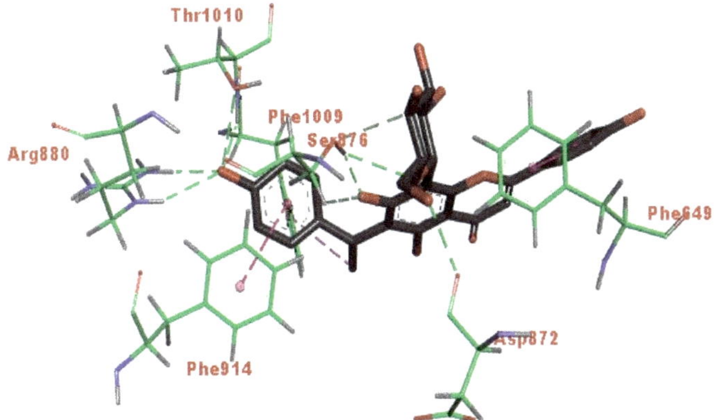

Figure 10. Binding interaction of cucumerin A with the active site of xanthine oxidase (PDB code: 3NVY). Green and purple dashes represent hydrogen-bonding and hydrophobic interactions, respectively.

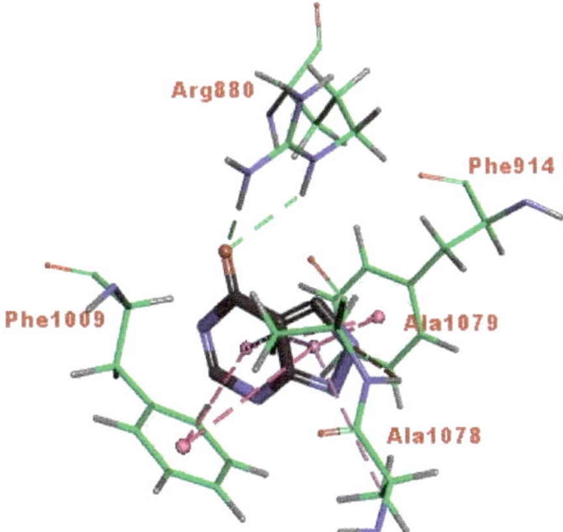

Figure 11. Binding interactions of allopurinol with the active site residues of xanthine oxidase (PDB code: 3NVY). Green and purple dashes represent hydrogen-bonding and hydrophobic interactions, respectively.

Table 5 shows the docking score of the two phenolic compounds, the co-crystallized ligand and allopurinol. The lower the free binding energy, the greater the interactions of the compound with the enzyme active site residues, which may consequently lead to better inhibitory effect. Cucumerin A and afzelechin 3-*O*-alpha-L-rhamnopyranoside have lower free binding energies than both quercetin and allopurinol. This further suggests that these two compounds could be responsible for the significant xanthine oxidase inhibitory effect of *A. bilimbi* leaves.

Table 5. Docking scores.

Ligand	Free Binding Energy (Kcal/mol)	Estimated Ki
Afzelechin 3-*O*-alpha-L-rhamnopyranoside	−11.43	4.18 nM
Cucumerin A	−12.0	1.59 nM
Quercetin	−10.18	34.57 nM
Allopurinol	−5.66	70.74 uM

4. Conclusions

A. bilimbi is an important medicinal plant widely used in traditional medicine. The findings of this research study demonstrate that the leaves of *A. bilimbi* possess strong in-vitro antioxidant capacity, which is partly displayed through scavenging of free radicals and inhibition of xanthine oxidase activity. This antioxidant property was found to be significant in the n-butanol fraction, as revealed by the IC_{50} value. LC-MS-QTOF analysis, as well as molecular docking studies, led to the tentative identification of 5,7,4′-trihydroxy-6-(1-ethyl-4-hydroxyphenyl) flavone-8-glucoside (cucumerin A) and afzelechin 3-*O*-alpha-L-rhamnopyranoside as the compounds that are likely to be responsible for the antioxidant effect of this plant. However, other phytoconstituents which were not present in the library of the LC-MS system could not be identified from their accurate mass, and may also play some roles. Further research is needed to obtain such bioactive compounds in pure form for complete pharmacological evaluations.

Author Contributions: Q.U.A. conceived and designed the experiments, Q.U.A. and A.M.A. performed the experiments and prepared the article, A.K., S.A.A.S., M.M.H. and M.N.S. discussed and analysed some of the data obtained during the experiments. All authors revised the paper thoroughly and approved the final manuscript prior to submission.

Funding: Corresponding author (Q.U.A.) is grateful to the Research Management Centre, IIUM for providing financial assistance through Research Initiative Grant Schemes (RIGS 16-294-0458) to accomplish this work.

Acknowledgments: LCMS Laboratory, Jeffrey Cheah School of Medicine and Health Sciences, Monash University Malaysia is acknowledged for LCMS analysis.

Conflicts of Interest: The authors declare no conflict of interest.

References

1. Rahal, A.; Kumar, A.; Singh, V.; Yadav, B.; Tiwari, R.; Chakraborty, S.; Dhama, K. Oxidative stress, prooxidants, and antioxidants: The interplay. *BioMed Res. Int.* **2014**, *2014*, 761264. [CrossRef] [PubMed]

2. Phaniendra, A.; Jestadi, D.B.; Periyasamy, L. Free radicals: Properties, sources, targets, and their implication in various diseases. *Indian J. Clin. Biochem.* **2015**, *30*, 11–26. [CrossRef] [PubMed]

3. Lobo, V.; Patil, A.; Phatak, A.; Chandra, N. Free radicals, antioxidants and functional foods: Impact on human health. *Pharmacogn. Rev.* **2010**, *4*, 118–126. [CrossRef] [PubMed]

4. Matough, F.A.; Budin, S.B.; Hamid, Z.A.; Alwahaibi, N.; Mohamed, J. The role of oxidative stress and antioxidants in diabetic complications. *Sult. Qaboos Univ. Med. J.* **2012**, *12*, 5–18. [CrossRef]

5. Asmat, U.; Abad, K.; Ismail, K. Diabetes mellitus and oxidative stress-a concise review. *Saudi Pharm. J.* **2016**, *24*, 547–553. [CrossRef] [PubMed]

6. Feoli, A.M.P.; Macagnan, F.E.; Piovesan, C.H.; Bodanese, L.C.; Siqueira, I.R. Xanthine oxidase activity is associated with risk factors for cardiovascular disease and inflammatory and oxidative status markers in metabolic syndrome: Effects of a single exercise session. *Oxid. Med. Cell. Longev.* **2014**, *2014*, 587083. [CrossRef] [PubMed]

7. Tam, H.K.; Kelly, A.S.; Metzig, A.M.; Steinberger, J.; Johnson, L.A. Xanthine oxidase and cardiovascular risk in obese children. *Child. Obes.* **2014**, *10*, 175–180. [CrossRef] [PubMed]

8. Sarian, M.N.; Ahmed, Q.U.; Siti Zaiton, M.S.; Alhassan, M.A.; Suganya, M.; Vikneswari, P.; Sharifah, N.A.S.M.; Alfi, K.; Latip, J. Antioxidant and antidiabetic effects of flavonoids: A structure-activity relationship based study. *BioMed Res. Int.* **2017**, *2017*, 8386065. [CrossRef] [PubMed]

9. Hasan, M.M.; Ahmed, Q.U.; Soad, S.Z.M.; Tunna, T.S. Animal models and natural products to investigate in vivo and in vitro antidiabetic activity. *Biomed. Pharmacother.* **2018**, *101*, 833–841. [CrossRef] [PubMed]

10. Alhassan, A.M.; Ahmed, Q.U. *Averrhoa bilimbi* L.: A review of its ethnomedicinal uses, phytochemistry, and pharmacology. *J. Pharm. Bioallied Sci.* **2016**, *8*, 265–271. [PubMed]

11. Precious, L.A.; Coren, J.P.; Celestine, L.A.; Mary, R.M.; Janina, C.E.; Rheinmark, L.S. Topical administration of *Averrhoa bilimbi* Linn. leaves crude extract prevents UVB-induced oxidative damage in albino mice. *The STETH* **2012**, *6*, 29–41.

12. Nagmoti, D.M.; Yeshwante, S.B.; Wankhede, S.S.; Juvekar, A.R. Hepatoprotective effect of *Averrhoa bilimbi* Linn. against carbon tetrachloride induced hepatic damage in rats. *Pharmacologyonline* **2010**, *3*, 1–6.

13. Pushparaj, P.; Tan, C.H.; Tan, B.K.H. Effects of *Averrhoa bilimbi* leaf extract on blood glucose and lipids in streptozotocin-diabetic rats. *J. Ethnopharmacol.* **2000**, *72*, 69–76. [CrossRef]

14. Pushparaj, P.N.; Tan, B.K.H.; Tan, C.H. The mechanism of hypoglycemic action of the semi-purified fractions of *Averrhoa bilimbi* in streptozotocin-diabetic rats. *Life Sci.* **2001**, *70*, 535–547. [CrossRef]

15. Tan, B.K.H.; Tan, C.H.; Pushparaj, P.N. Anti-diabetic activity of the semi-purified fractions of *Averrhoa bilimbi* in high fat diet fed-streptozotocin-induced diabetic rats. *Life Sci.* **2005**, *76*, 2827–2839. [CrossRef] [PubMed]

16. Firdouse, S.; Alam, P. Phytochemical investigation of extract of *Amorphophallus campanulatus* tubers. *Int. J. Phytomed.* **2011**, *3*, 32–35.

17. Ayoola, G.A.; Coker, H.A.; Adesegun, S.A.; Adepoju-Bello, A.A.; Obaweya, K.; Ezennia, E.C.; Atangbayila, T.O. Phytochemical screening and antioxidant activities of some selected medicinal plants used for malaria therapy in Southwestern Nigeria. *Trop. J. Pharm. Res.* **2008**, *7*, 1019–1024.

18. Iqbal, E.; Salim, K.A.; Lim, L.B. Phytochemical screening, total phenolics and antioxidant activities of bark and leaf extracts of *Goniothalamus velutinus* (Airy Shaw) from Brunei Darussalam. *J. King Saud Univ. Sci.* **2015**, *27*, 224–232. [CrossRef]

19. Nickavar, B.; Kamalinejad, M.; Izadpanah, H. In vitro free radical scavenging activity of five *Salvia* species. *Pak. J. Pharm. Sci.* **2007**, *20*, 291–294. [PubMed]

20. Barontini, M.; Bernini, R.; Carastro, I.; Gentili, P.; Romani, A. Synthesis and DPPH radical scavenging activity of novel compounds obtained from tyrosol and cinnamic acid derivatives. *New J. Chem.* **2014**, *38*, 809–816. [CrossRef]

21. Thiombiano, A.M.E.; Adama, H.; Jean, B.M.; Bayala, B.; Nabèrè, O.; Samson, G.; Roland, M.N.T.; Moussa, C.; Martin, K.; Millogo, F.; et al. In vitro antioxidant, lipoxygenase and xanthine oxidase inhibitory activity of fractions and macerate from *Pandiaka angustifolia* (Vahl) Hepper. *J. Appl. Pharm. Sci.* **2014**, *4*, 9–13.

22. Protein Data Bank. Available online: http://www.rcsb.org/ (accessed on 3 July 2018).

23. Molecular Graphics Laboratory, The Scripps Research Institute, San Diego, USA. Available online: www.scripps.edu (accessed on 3 July 2018).

24. Discovery Studio Visualizer 4.0, Bovia, San Diego, USA. Available online: www.accelrys.com (accessed on 3 July 2018).

25. Zhang, H.J.; Hu, Y.J.; Xu, P.; Liang, W.Q.; Zhou, J.; Liu, P.G.; Cheng, L.; Pu, J.B. Screening of potential xanthine oxidase inhibitors in *Gnaphalium hypoleucum* DC. by immobilized metal affinity chromatography and ultrafiltration-ultra performance liquid chromatography-Mass spectrometry. *Molecules* **2016**, *21*, 1242. [CrossRef] [PubMed]

26. Mishra, K.; Ojha, H.; Chaudhury, N.K. Estimation of antiradical properties of antioxidants using DPPH assay: A critical review and results. *Food Chem.* **2012**, *130*, 1036–1043. [CrossRef]

27. Złotek, U.; Mikulska, S.; Nagajek, M.; Świeca, M. The effect of different solvents and number of extraction steps on the polyphenol content and antioxidant capacity of basil leaves (*Ocimum basilicum* L.) extracts. *Saudi J. Biol. Sci.* **2016**, *23*, 628–633. [CrossRef] [PubMed]

28. Maiuolo, J.; Oppedisano, F.; Gratteri, S.; Muscoli, C.; Mollace, V. Regulation of uric acid metabolism and excretion. *Int. J. Cardiol.* **2015**, *213*, 8–14. [CrossRef] [PubMed]

29. Dalbeth, N.; Lauterio, T.J.; Wolfe, H.R. Mechanism of action of colchicine in the treatment of gout. *Clin. Ther.* **2014**, *36*, 1465–1479. [CrossRef] [PubMed]

30. Hao, S.; Zhang, C.; Song, H. Natural products improving hyperuricemia with hepatorenal dual effects. *Evid. Based Complement. Altern. Med.* **2016**, *2016*, 7390504. [CrossRef] [PubMed]

31. Tatsimo, S.; Tamokou, J.; Havyarimana, L.; Csupor, D.; Forgo, P.; Hohmann, J.; Kuiate, J.R.; Tane, P. Antimicrobial and antioxidant activity of kaempferol rhamnoside derivatives from *Bryophyllum pinnatum*. *BMC Res. Notes* **2012**, *5*, 158. [CrossRef] [PubMed]

32. Spanou, C.; Veskoukis, A.S.; Kerasioti, T.; Kontou, M.; Angelis, A.; Aligiannis, N.; Skaltsounis, A.L.; Kouretas, D. Flavonoid glycosides isolated from unique legume plant extracts as novel inhibitors of xanthine oxidase. *PLoS ONE* **2012**, *7*, e32214. [CrossRef] [PubMed]

33. Lawal, U.; Sze, W.L.; Khozirah, S.; Intan, S.I.; Alfi, K.; Faridah, A. α-Glucosidase inhibitory and antioxidant activities of different *Ipomoea aquatica* cultivars and LC–MS/MS profiling of the active cultivar. *J. Food Biochem.* **2016**, *41*, e12303. [CrossRef]

34. Colombo, R.; Yariwake, J.H.; McCullagh, M. Study of C-and O-glycosylflavones in sugarcane extracts using liquid chromatography: Exact mass measurement mass spectrometry. *J. Braz. Chem. Soc.* **2008**, *19*, 483–490. [CrossRef]

35. Martucci, M.E.P.; De Vos, R.C.; Carollo, C.A.; Gobbo-Neto, L. Metabolomics as a potential chemotaxonomical tool: Application in the genus *Vernonia* Schreb. *PLoS ONE* **2014**, *9*, e93149. [CrossRef] [PubMed]

36. Cao, H.; Pauff, J.M.; Hille, R. X-ray crystal structure of a xanthine oxidase complex with the flavonoid inhibitor quercetin. *J. Nat. Prod.* **2014**, *77*, 1693–1699. [CrossRef] [PubMed]

© 2018 by the authors. Licensee MDPI, Basel, Switzerland. This article is an open access article distributed under the terms and conditions of the Creative Commons Attribution (CC BY) license (http://creativecommons.org/licenses/by/4.0/).

Article

Development and Validation of an Analytical Method for Carnosol, Carnosic Acid and Rosmarinic Acid in Food Matrices and Evaluation of the Antioxidant Activity of Rosemary Extract as a Food Additive

Seung-Hyun Choi [1,†], Gill-Woong Jang [1,†], Sun-Il Choi [1], Tae-Dong Jung [1], Bong-Yeon Cho [1], Wan-Sup Sim [1], Xionggao Han [1], Jin-Sol Lee [1], Do-Yeon Kim [1], Dan-Bi Kim [2] and Ok-Hwan Lee [1,*]

[1] Department of Food Science and Biotechnology, Kangwon National University, Chuncheon 24341, Korea; zzaoszz@naver.com (S.-H.C.); jkw5235@naver.com (G.-W.J.); docgotack89@hanmail.net (S.-I.C.); lgtjtd@naver.com (T.-D.J.); bongyeon.cho92@gmail.com (B.-Y.C.); simws9197@naver.com (W.-S.S.); xionggao414@hotmail.com (X.H.); amylbl32@naver.com (J.-S.L.); doy0918@naver.com (D.-Y.K.)

[2] Korea Food Research Institute, Wanju-gun 55365, Korea; dbkim1022@kfri.re.kr

* Correspondence: loh99@kangwon.ac.kr; Tel.: +82-33-250-6454; Fax: +82-33-259-5565

† These authors contributed equally to this work.

Received: 11 March 2019; Accepted: 23 March 2019; Published: 26 March 2019

Abstract: Antioxidants are used to prevent the oxidation of foods. When used for food additive purposes, the dosage should be regulated and the functionality evaluated to ensure stability. In this study, we performed a method validation for the quantitative analysis of rosemary extract residues and evaluated the antioxidant activity of rosemary extract in food matrices. The validated method was able to determine rosemary extract under the optimized high-performance liquid chromatography-photodiode array (HPLC-PDA) conditions. Furthermore, the antioxidant activity was evaluated by peroxide value, acid value, and in terms of the residual antioxidant levels in lard oil. For HPLC-PDA analysis, the limit of detection and quantification of rosemary extracts was ranged from 0.22 to 1.73 µg/mL, 0.66 to 5.23 µg/mL and the recoveries of the rosemary extracts ranged from 70.6 to 114.0%, with relative standard deviations of between 0.2% and 3.8%. In terms of antioxidant activity, carnosic acid performed better than carnosol. Furthermore, by evaluation of the residual antioxidant level using HPLC, we found that carnosic acid is more stable in lard oil than carnosol. These results indicate that rosemary extract can be used as an antioxidant and that the analytical method is suitable for the determination of rosemary extract in various food samples.

Keywords: rosemary extracts; method validation; antioxidant activity; HPLC; food matrices

1. Introduction

Lipid oxidation is a major cause of quality degradation in fat-containing foods. All foods, especially those with high fats, are vulnerable to lipid peroxidation and free radical attack, leading to food decay and rotten odors. The formation of free radicals in food causes undesirable changes, destroying vitamins and other substances, reducing the nutritional value, and affecting the health, safety, color, flavor, and texture of the food [1,2]. Autoxidation of lipids generally involves a free radical chain reaction, which is initiated by exposure of the lipids to heat, ionizing radiation, light, metalloprotein catalysts, or metal ions. Thus, the inhibition of free radicals is of practical importance in protecting lipids from oxidation [3].

Antioxidants are widely used in foods to delay or prevent oxidation. Being easily oxidized, they exchange electrons with peroxides and free radicals, preventing these compounds from undergoing reactions with other substances in the food matrix that would lead to damage. The addition of antioxidants to foods prevents the formation of various flavors and rotten odors, thereby extending the shelf life [4–6].

Rosmarinus officinalis L. (Lamiaceae), commonly referred to as rosemary, is a plant used worldwide in cooking as an herb. Rosemary contains a large number of phenolinc compounds including phenolic diterpenes such as carnosol, carnosic acid, epi- and iso-rosemanols, rosmanol, and the phenolic ester, rosmarinic acid [7,8]. Rosemary extract also contains a large number of phenolic compounds, including carnosol, carnosic acid and rosmarinic acid, and has been widely used in the food industry due to its inherent high antioxidant activity [9]. In some countries, rosemary extract is designated as an antioxidant food additive, with defined acceptance criteria.

It is designated as "Rosemary Extract" (China) or "Extracts of rosemary" (European Union) in countries where it has been approved as a food additive. In China, the addition of rosemary extract is permitted to a maximum concentration of ~300.0–700.0 µg/mL, depending on food type. In the European Union, the maximum residue level of rosemary extract is defined by the sum of carnosol and carnosic acid, since these are the main active ingredients of rosemary extract: Depending on the type of food, ~30.0–250.0 µg/mL is permitted [10,11]. However, in South Korea, the addition of rosemary extract to food is not permitted. Consequently, there is a need for a validated analytical method for detecting the addition of rosemary extract as an unauthorized antioxidant, in foods commonly consumed in South Korea. There have been no studies on the analysis of rosemary extract to evaluate its content as a food additive in permitted food types. Previous studies on the analysis of rosemary have been reported, such as the analysis of active ingredients including carnosol and carnosic acid in extracts, and the analysis of active ingredients in rosemary [8,12]. If rosemary extract is used as a food additive, to ensure food safety its concentration should be monitored to determine whether it exceeds the allowable level. Therefore, an analysis method that is capable of accurately analyzing rosemary extract in foods is required. However, to date, no research has been conducted into the development of an analytical method to determine rosemary extract in food matrices.

The addition of an antioxidant to food prevents the food from oxidizing during storage; over time, the antioxidant loses its activity and its concentration decreases. The residual antioxidant level is one of the important factors in the role of antioxidants in food. However, most previous studies have evaluated the antioxidative activity of the major active substances of rosemary extract, and, to date, no studies have investigated the antioxidant activity with residual levels of rosemary extract in food [13,14].

In this study, we developed and validated a high-performance liquid chromatography-photodiode array (HPLC-PDA) method for the determination of rosemary extract in various primary domestic and imported food product matrices, including edible oils, processed meat products, and dressings. Also, to demonstrate the effective application of the established method on real samples, various edible oil, processed meat product, and dressing samples were collected from grocery markets in South Korea and their rosemary extract contents were determined. In addition, to evaluate its function as an antioxidant, rosemary extract was added to lard oil and the storage stability was evaluated by measuring the residual antioxidant level, peroxide value and acid value.

2. Materials and Methods

2.1. Materials

Analytical standards, such as rosmarinic acid, carnosol, and carnosic acid, were obtained from Sigma-Aldrich (St. Louis, MO, USA). All solvents were suitable for analysis and were purchased from J.T. Baker (Phillipsburg, NJ, USA).

2.2. Food Materials

Ninety food samples including thirty edible oils, thirty processed meat products, and thirty dressings, were purchased in local markets in South Korea. The shelf life of the samples was sufficient for the investigation. To validate the procedure, edible oils, processed meat products and dressings that were found to be free of rosemary extract, were selected. Prior to analysis, all samples were stored under the storage conditions indicated on the product label.

2.3. Optimization of HPLC Conditions

Rosemary extracts were tested using HPLC, and the optimum analytical conditions were determined. We first evaluated the comparison of the different analytical methods, as it achieved the best results. Then, analytical parameters such as column type and column temperature were evaluated to provide optimum separation conditions for rosemary extract in edible oils, processed meat products and dressings. The sensitivity of the analytical method was evaluated based on the maximum allowable levels of rosemary extract in foreign countries (within the European Union and China).

2.4. Optimization of Extraction Method

Using the optimized HPLC-PDA analysis method, the optimum preparation method for removing rosemary extract from three matrices (edible oil, processed meat products, and dressing) was determined. The recovery was used as an index for establishing the optimal sample preparation method. Sample preparation for rosemary extract analysis was optimized using previously reported methods [15,16].

2.5. Sample Preparation

The samples were prepared according to the modified method of Kim et al. (2016) [15]. The samples were weighed (5 g each) into a 50 mL conical tubes, rinsed with 15 mL of n-hexane and then transferred to a separatory funnel containing 5 mL of n-hexane, followed by extraction with 150 mL portions of n-hexane-saturated acetonitrile. For preparing the processed meat products and the dressing samples, the organic solvent layer was transferred to another separatory funnel to remove the residues which interfered with the separations. The acetonitrile phase was collected in a concentrate flask, and then evaporated to a volume of 3 to 4 mL using a water bath ($\leq$40 °C, EYELA, SB-1200, Tokyo, Japan) with a vacuum rotary evaporator (EYELA, N-1200A, Tokyo, Japan). The flask was rinsed with small portions of solvent (acetonitrile:iso-propanol, 1:1, *v/v*) which were then transferred to a 10 mL volumetric flask. The rinsing step was repeated until exactly 10 mL was collected in volumetric flask. The samples were filtered through a 0.45 μm syringe filter (Millex-HV, Millipore, Bedford, MA, USA).

2.6. HPLC Instrument Conditions

The HPLC apparatus was a Waters 2695 separation module HPLC system (Waters Co., Milford, MA, USA) equipped with a pump, an autosampler, a column oven, and a Waters 996 photodiode array detector. The analytical column was a Shiseido Capcell Pak C_{18} UG120 (Shiseido, 4.6 mm $\times$ 250 mm, 5.0 μm, Tokyo, Japan). The column temperature was maintained at 30 °C. The mobile phase was composed of A (1% acetic acid in water) and B (methanol) with a gradient elution as follows: 0–20 min, linear from 10 to 65% A; 20–40 min, linear from 65 to 100% A; 40–45 min, maintained at 100% A; 45–47 min, linear from 100 to 10% A; and then finally, holding for 3 min. The mobile phase was filtered through a 0.45 μm membrane filter (Whatman, Amersham, UK) and degassed under vacuum. The flow rate was set 1.0 mL/min, and the injection volume was 20 μL. The antioxidants were determined at 284 nm. Data acquisition and remote control of the HPLC system were performed using Empower software (Waters Co., Milford, MA, USA).

2.7. Method Validation

The HPLC method for the determination of rosemary extract in three food matrices (edible oils, processed meat products, and dressings) was validated in terms of linearity, trueness, precision, limit of detection (LOD) and quantification (LOQ), according to the guidelines of the International Conference on Harmonization (2005) [17]. The matrix-matched calibration curves were prepared by spiking prepared extracts of blank edible oil, processed meat product, and dressing in seven concentrations from 1.56 to 100.0 µg/mL for rosmarinic acid, and from 6.25 to 400.0 µg/mL for carnosic acid and carnosol. During analytical method development and validation, matrix-matched calibration curves from each assessed analytical method were evaluated to ensure that the sensitivity and linearity were consistent with the experimental observations, and that there were no matrix peak interferences. The linearity evaluation was performed by calculating the linearity of the calibration curve determined using a range of concentrations of each compound. The selectivity was assessed by examining the chromatogram to confirm that there were no ingredients that could interfere with the target analyte of rosemary extract in edible oils, processed meat products, and dressings. The LOD and LOQ were calculated for the analytes in edible oils, processed meat products, and dressings. The trueness and precision were determined in three food matrices at three different concentration levels (rosmarinic acid: 5.0, 10.0, and 20.0 µg/mL; carnosic acid and carnosol 25.0, 50.0, and 100.0 µg/mL). Both inter-day (three repeats on three different days) and intra-day (three repeats during a single day) method validation experiments were performed.

2.8. Function Evaluation as an Antioxidant

2.8.1. Experimental Condition

Lard oil was used to evaluate the functionality of the rosemary extract as an antioxidant. In other countries, rosemary extracts defined the sum of carnosol and carnosic acid, is permitted at 50 µg/mL in lard oil. Therefore, these two compounds, except rosmarinic acid, were used for the functional evaluation. Two antioxidant components of rosemary extract (carnosic acid and carnosol) were added to lard oil to provide a combined concentration of 50 µg/mL. In addition, to compare the effects of synthetic antioxidants (carnosol and carnosic acid standards) and natural antioxidants, rosemary leaf extract was prepared and analyzed by HPLC to calculate its carnosol and carnosic acid concentrations. This leaf extract was then added to lard oil to provide a combined carnosol and carnosic acid concentration of 50 µg/mL. The mixture of carnosol and carnosic acid standards was also added to lard oil to provide a combined concentration of 50 µg/mL. To provide a positive control, butylated hydroxyanisole (BHA) was added to lard oil at a concentration of 50 µg/mL. The prepared samples were stored in a dry oven at 50 °C for 42 days. The peroxide value, acid value, and residual antioxidant level of each sample were evaluated every 7 days.

2.8.2. Measurement of Antioxidant Activity

Antioxidant activity was evaluated using the peroxide value and acid value. The peroxide value and acid value of each sample were analyzed by modifying the Official Methods and Recommended Practices of the American Oil Chemists' Society (AOCS, 1993) Cd 8-53 and Te 1a-64, respectively [18]. The peroxide value was proceeded as follows. First, 2 g of sample was dissolved in 10 mL of acetic acid and chloroform mixture (3:2, *v/v*) in a 100 mL Erlenmeyer flask and 0.4 mL of KI saturated solution was added and kept in a dark place for 10 minutes. After that, 12 mL of distilled water and 0.4 mL of 1% starch solution were added in succession, and the mixture was shaken. The resultant was titrated with 0.01 N $Na_2S_2O_3$ solution. Acid value was measured by the following method. 5 g of sample is precisely weighed, placed in a stoppered Erlenmeyer flask, and dissolved in 100 mL of a neutral ethanol/ether mixture (1:2). Using phenolphthalein solution as an indicator, titrate with 0.1 N potassium hydroxide ethanolic standard solution until pale red color persists for 30 s. The peroxide value and the acid value were measured every seven days. The peroxide value and the acid value results were respectively expressed as milli-equivalents of hydroperoxides per kg of lipids (meq/kg) and in terms of the number of mg of KOH required to neutralize the free fatty acids contained in 1 g of lipid (mg KOH/g).

2.8.3. Measurement of Residual Antioxidant Level

The residual amount of antioxidant compound was evaluated using HPLC. Sample preparation was carried out as described in Section 2.5. The residual amount of antioxidant in each experimental group was expressed as the rate of change of antioxidant over time. The results of each test group measured on the first day are displayed as 100%, while the later results are displayed as a percentage of the first day's result.

2.9. Statistical Analysis

The peroxide value, acid value, and residual antioxidant level results were analyzed by ANOVA and Duncan's multiple range tests using SAS 9.4 (SAS Institute Inc., Cary, NC, USA). Significance was indicated at a p-value of < 0.05.

3. Results and Discussion

3.1. Optimization of HPLC Conditions

The chromatographic conditions for analyzing the rosemary extract were investigated and optimized. The sensitivity of the analytical method was determined based on the residual concentration of rosemary extracts permitted in foreign countries (China and within the European Union) [10,11]. The analytical method for the target analyte separation was based on gradient elution with acetonitrile, methanol or acetic acid solution as the mobile phase [19–21].

In the gradient elution method, using acetonitrile and acetic acid solution, ghost peak detection and tailing of the carnosic acid peak were problematic. The gradient elution method using methanol, acetonitrile, and acetic acid solution resulted in tailing of the carnosol and carnosic acid peaks at high concentration (data not shown). The gradient elution method using methanol and acetic acid solution provided better peak separation and peak shapes than those of the other methods. Accordingly, we selected the latter gradient elution method for determination of rosemary extract in food matrices.

To compare the separation efficiency of the C_{18} columns from a range of manufacturers, we analyzed the standard solution (at low, medium, high concentration; rosmarinic acid: 5.0, 10.0, and 20.0 µg/mL; carnosic acid and carnosol 25.0, 50.0, and 100.0 µg/mL) and the sample blank. The column types used in this comparison were Shiseido Capcell Pak C_{18} UG 120 (4.6 × 250 mm, 5.0 µm), and Waters Sunfire C_{18} (4.6 × 250 mm, 5.0 µm). The retention times varied depending on the manufacturer of the C18 column. The Shiseido Capcell Pak C_{18} UG 120 showed better separation and the retention times of the analytes were faster than found with other C_{18} columns (data not shown). Therefore, we selected the Shiseido Capcell Pak C_{18} UG 120 column for further study.

The column temperature was adjusted to the separation efficiency of the analytes and the pressure inside the analytical device. Since the column temperature is an important factor that affects chromatographic separation, changes in column temperature can be useful for efficient separation [22–24]. As the temperature increases, the retention time of the peak decreases, without affecting component separation. Nevertheless, a column temperature of 30 °C was selected since the highest overall peak area was observed at this temperature (data not shown).

The analytical method for rosemary extract determination was optimized by comparing the results with different mobile phases, column manufacturers, and column temperatures. Finally, the optimal HPLC analytical conditions of rosemary extracts was described in Section 2.6.

3.2. Optimization of Extraction Method

The sample preparation is important in chromatographic applications. One of the main goals of this step is to concentrate the analyte to remove the interfering matrix components and particulates;

therefore, increasing the sensitivity [25]. First, we assessed whether the method of Suh et al. (2011) could be applied to rosemary extracts [16]. This method reports the pretreatment of erythorbic acid, an antioxidant, in processed meat products. The antioxidant compound was extracted by liquid-liquid extraction using 2% metaphosphoric acid solution. The recovery rate results from this method showed that rosmarinic acid had a high recovery rate, but that carnosol and carnosic acid were not detected.

We also assessed the Kim et al. (2016) method, which analyzed seven synthetic antioxidants in edible oils [15]. The synthetic antioxidants were extracted by liquid-liquid extraction method using n-hexane-saturated acetonitrile as the extraction solvent. With this method, the recoveries of rosmarinic acid, carnosol, and carnosic acid were all excellent, and there was minimal use of organic solvents. However, when applied to the processed meat products and dressings, precipitates formed in the separating funnel; therefore, it was difficult to separate the organic layers efficiently. Consequently, we modified the pretreatment process to establish the optimal pretreatment conditions. For samples of the three food matrices, the rosemary extract recoveries ranged from 73.1 to 116.5% with the modified method. Therefore, we selected a sample preparation method modified from Kim et al. (2016), since this provided the highest rosemary extract recovery from the three matrices (Figure 1). The modified method is presented in Section 2.5.

3.3. Method Validation

3.3.1. Selectivity, Linearity, LOD, and LOQ

Selectivity in chromatography refers to the extent to which an analytical method can determine the analyte in a matrix without interference from other components that are expected to be present in the food matrix [26]. The chromatograms of rosemary extract standards obtained using the HPLC-PDA method are shown in Figure 1. No interferences or co-eluting peaks are not observed and the analytes are separated efficiently in the chromatograms in edible oils, processed meat products, and dressings using the Capcell Pak C_{18} column. The linearity of the established method was evaluated in the range 1.56–100 µg/mL for rosmarinic acid, and 6.25–400 µg/mL for carnosol and carnosic acid. The correlation coefficients (R^2) of the rosemary extracts were in the range of 0.9987–1.0000. For HPLC-PDA, the LOD and LOQ are shown in Table 1. The LOD lay in the range of 0.38–0.78 µg/mL in edible oils, 0.38–1.50 µg/mL in processed meat products, and 0.22–1.73 µg/mL in dressings. The LOQ lay in the range of 1.14–2.38 µg/mL in edible oils, 1.15–4.55 µg/mL in processed meat products, and 0.66–5.23 µg/mL in dressings. The sensitivity of the HPLC-PDA method is suitable for the quantitative analysis of rosemary extracts below the maximum acceptable levels in foreign countries (China and within the European Union).

Table 1. Correlation coefficients of the calibration curves, and limit of detection (LOD) and quantification (LOQ).

Matrix	Analytes	Range (µg/mL)	Slope	Intercept	Correlation Coefficient (R^2) [1]	LOD (µg/mL)	LOQ (µg/mL)
Oil	Rosmarinic acid	1.56–100	34,243.3	−27,953.9	0.9994	0.38	1.14
	Carnosol	6.25–400	7926.7	8343.5	1.0000	0.78	2.38
	Carnosic acid	6.25–400	5585.2	−13,120.1	0.9995	0.65	1.96
Processed meat products	Rosmarinic acid	1.56–100	34,954.9	−33,771.5	0.9995	0.38	1.15
	Carnosol	6.25–400	8165.6	−2906.3	1.0000	1.05	3.17
	Carnosic acid	6.25–400	5657.9	−25,961.3	0.9987	1.50	4.55
Dressing	Rosmarinic acid	1.56–100	33,781.5	−45,267.7	0.9992	0.22	0.66
	Carnosol	6.25–400	7941.4	−17,908.1	0.9997	1.39	4.21
	Carnosic acid	6.25–400	5526.4	−9656.3	0.9997	1.73	5.23

[1] Correlation coefficient was calculated using the average of the results measured three times.

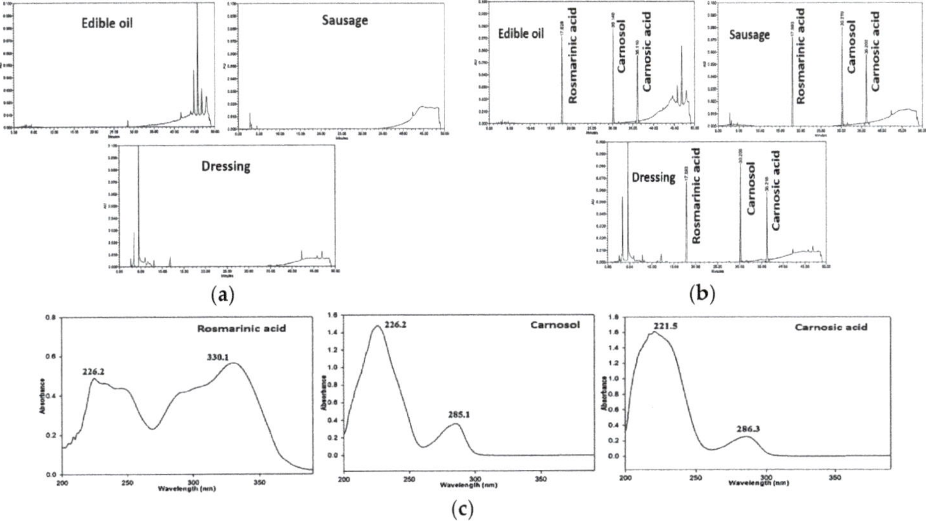

Figure 1. Chromatograms of rosemary extracts obtained using HPLC-PDA. (**a**) Blank samples; (**b**) Spiked samples; (**c**) Analyte spectra.

3.3.2. Trueness and Precision

The recovery results for the rosemary extracts are presented in Table 2. The trueness of the developed analytical method was determined from the recovery rates. The spiking levels for the recovery study were 5.0, 10.0, and 20.0 µg/mL for rosmarinic acid, and 25.0, 50.0, and 100.0 µg/mL for carnosol and carnosic acid in the edible oil, processed meat product, and dressing matrices. The average recoveries in edible oil were 70.6–90.8% for rosmarinic acid, 82.6–93.8% for carnosol, and 85.1–92.5% for carnosic acid. In processed meat products, the average recoveries were 72.8–106.9% for rosmarinic acid, 99.1–108.2% for carnosol, and 99.4–115.2% for carnosic acid. In dressings, the average recoveries were 80.4–116.3% for rosmarinic acid, 101.5–114.0% for carnosol, and 99.2–116.5% for carnosic acid. In Codex Alimentarius Commission Guidelines (CAC/GL 40), the average recovery criteria applicable to concentrations of residues at the levels ≥ 10 µg/mL ≤ 100 µg/mL and > 1 µg/mL ≤ 10 µg/mL are 70–120% and 60–120%, respectively [27]. The results of recovery satisfied this guideline.

Table 2. Recoveries of spiked rosemary extracts (three different concentrations) from edible oil, processed meat products, and dressing (n = 3).

Matrix	Analytes		Concentration (µg/mL)	Mean ± SD (µg/mL)	RSD [1] (%)	Recovery (%)
			5.0	4.5 ± 0.1	1.5	90.8
		Intra-day	10.0	7.2 ± 0.1	1.8	72.2
	Rosmarinic		20.0	14.1 ± 0.0	0.2	70.6
	acid		5.0	4.9 ± 0.2	3.6	89.7
		Inter-day	10.0	7.1 ± 0.2	2.5	71.0
Oil			20.0	14.3 ± 0.4	2.4	71.3
			25.0	21.4 ± 0.1	0.5	85.6
		Intra-day	50.0	42.4 ± 0.3	0.6	84.8
	Carnosol		100.0	93.5 ± 0.2	0.2	93.5
			25.0	20.6 ± 0.6	3.0	82.6
		Inter-day	50.0	4.19 ± 1.6	3.8	83.8
			100.0	93.8 ± 2.8	3.0	93.8

Table 2. *Cont.*

Matrix	Analytes		Concentration (µg/mL)	Mean ± SD (µg/mL)	RSD [1] (%)	Recovery (%)
Processed meat products	Carnosic acid	Intra-day	25.0	23.1 ± 0.1	0.6	92.5
			50.0	43.6 ± 0.1	0.3	87.1
			100.0	88.6 ± 0.3	0.4	88.9
		Inter-day	25.0	22.4 ± 0.7	3.1	89.6
			50.0	42.6 ± 1.3	3.1	85.1
			100.0	88.1 ± 2.3	2.6	88.1
	Rosmarinic acid	Intra-day	5.0	5.4 ± 0.0	0.8	106.9
			10.0	8.7 ± 0.0	0.8	86.7
			20.0	14.7 ± 0.0	0.4	73.5
		Inter-day	3.13	5.1 ± 0.1	1.5	102.1
			12.5	8.4 ± 0.1	0.8	84.2
			25.0	14.6 ± 0.2	1.2	72.8
	Carnosol	Intra-day	25.0	27.0 ± 0.2	0.8	108.2
			50.0	53.4 ± 0.2	0.4	106.8
			100.0	99.3 ± 0.6	0.6	99.3
		Inter-day	25.0	25.4 ± 0.2	0.6	101.8
			50.0	52.2 ± 0.5	1.0	104.4
			100.0	99.1 ± 0.9	0.9	99.1
	Carnosic acid	Intra-day	25.0	28.8 ± 0.2	0.6	115.2
			50.0	55.5 ± 0.3	0.6	111.0
			100.0	101.4 ± 1.4	1.4	101.4
		Inter-day	25.0	26.8 ± 0.2	0.8	107.2
			50.0	54.5 ± 1.2	2.1	108.9
			100.0	99.4 ± 1.5	1.5	99.4
Dressing	Rosmarinic acid	Intra-day	5.0	5.8 ± 0.0	0.5	116.3
			10.0	9.5 ± 0.0	0.3	95.3
			20.0	16.1 ± 0.1	0.4	80.5
		Inter-day	5.0	5.5 ± 0.0	0.2	109.3
			10.0	9.2 ± 0.1	0.7	92.0
			20.0	16.1 ± 0.2	1.3	80.4
	Carnosol	Intra-day	25.0	28.5 ± 0.1	0.4	114.0
			50.0	52.3 ± 0.2	0.4	104.5
			100.0	101.5 ± 0.2	0.2	101.5
		Inter-day	25.0	28.4 ± 0.1	0.4	113.5
			50.0	52.1 ± 0.2	0.3	104.1
			100.0	102.8 ± 1.6	1.5	102.8
	Carnosic acid	Intra-day	25.0	29.1 ± 0.1	0.2	116.5
			50.0	51.3 ± 0.2	0.4	102.5
			100.0	99.2 ± 0.3	0.3	99.2
		Inter-day	25.0	27.7 ± 0.3	1.1	110.8
			50.0	51.0 ± 1.1	2.1	101.9
			100.0	99.3 ± 0.9	0.9	99.3

[1] RSD is the relative standard deviation.

The precision of the rosemary extract analysis was evaluated using the intra-day and inter-day RSDs. For all samples, the intra-day and inter-day precisions ranged respectively as follows: From 0.2 to 1.8% and from 0.2 to 3.6% for rosmarinic acid; from 0.2 to 0.8% and from 0.3 to 3.8% for carnosol; and, from 0.2 to 1.4% and from 0.8 to 3.1% for carnosic acid. The relative standard deviation (RSD) analysis results are in accordance with the CAC/GL 40.

3.4. Sample Collection and Monitoring of Residual Rosemary Extract Levels

The method developed and validated in this study was used to monitor the presence of residual rosemary extracts in edible oils, processed meat products, and dressings in domestic markets. A total of ninety samples of edible oils, processed meat products, and dressings were collected from grocery markets in South Korea. Using the HPLC-PDA analysis method, chromatographic peaks in each sample were identified by comparison of the retention time and the PDA spectrum of the rosemary extract standard. Across the ninety analyzed samples, no rosemary extracts were found in any of the samples.

3.5. Assessment of Functionality as Antioxidant

3.5.1. Antioxidant Activity

To evaluate the antioxidative activity of rosemary extract as a food additive, carnosol and carnosic acid were added to lard oil and the peroxide and acid values were evaluated.

The antioxidant activity results for the rosemary extracts are shown in Figure 2. The peroxide value of the lard oil increased across the storage period, from 27.71 to 454.95 meq/kg in the control group (with no antioxidant added to the lard oil). The peroxide value of the lard oil with added antioxidants increased as follows: From 26.61 to 433.77 meq/kg in the carnosic acid + carnosol group; from 27.22 to 429.34 meq/kg in the carnosol group; from 26.84 to 416.71 meq/kg in the carnosic acid group; from 27.09 to 414.69 meq/kg in the rosemary leaf extract group; and, from 27.12 to 107.52 meq/kg in the BHA group. The acid value of the lard oil increased across the storage period, from 0.56 to 8.56 mg KOH/g in the control group (with no antioxidant added to the lard oil). The acid value of the lard oil with added antioxidants increased as follows: From 0.56 to 4.83 mg KOH/g in the carnosic acid + carnosol group; from 0.56 to 4.84 mg KOH/g in the carnosol group; from 0.56 to 4.65 mg KOH/g in the carnosic acid group; from 0.56 to 3.53 mg KOH/g in the rosemary leaf extract group; and, from 0.56 to 1.67 mg KOH/g in the BHA group. Acid value analysis showed that there were no significant differences between the experimental groups until day 14. However, a significant difference was observed between the experimental groups after day 21. The acid value was highest in the control group, followed by carnosol + carnosic acid group, carnosol group, carnosic acid group, rosemary leaf extract group, and butylated hydroxyanisole (BHA) group, showing a similar tendency to that observed for the peroxide value. In terms of the peroxide value and the acid value, between the two components of rosemary extracts used as acceptance criteria, carnosic acid was found to inhibit oxidation more effectively than carnosol. These results are in good agreement with reports by Frankle et al. (1996) and Hopia et al. (1996) who demonstrated that carnosic acid has superior antioxidant ability compared with that of carnosol in oil [13,14]. Rosemary leaf extract showed better antioxidant activity than those of the other compounds, with the exception of BHA, which is a synthetic phenolic antioxidant and positive control. This is because rosemary leaf extract contains a range of antioxidant compounds, in addition to carnosol and carnosic acid.

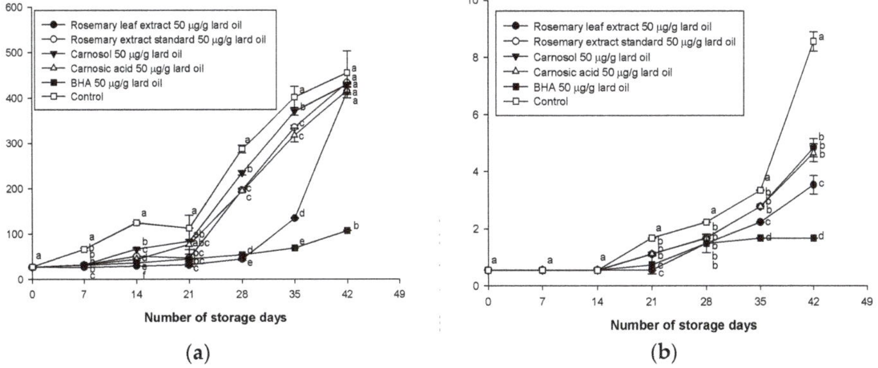

Figure 2. Antioxidant activity of rosemary extracts in lard oil during storage. (**a**) Peroxide value; (**b**) Acid value. [a–d] Different letters indicate statistically significant different among the groups at *p*-value < 0.05.

3.5.2. Oxidation Stability

The residual antioxidant level in each experimental group was expressed as the percentage change of the residual level as a function of the storage period, by setting the HPLC result measured at day 0 to 100% (Figure 3). Across the storage period, the residual antioxidant level decreased significantly for all experimental groups, with the exception of BHA.

In particular, carnosol was mostly lost during the first seven days of storage, and the residual amount of carnosol was close to 0%. After seven days, only the residual level of BHA continued to decrease, since the other antioxidants were mostly lost by the seventh day, and thereafter the residual levels remained constant. The results of the change in residual antioxidant levels in lard oil showed a similar tendency to that of the acid value and the peroxide value results.

Thorsen & Hildebrandt (2003) reported that when carnosol and carnosic acid were dissolved in solvents, carnosol degraded more rapidly and was more unstable than carnosic acid [8]. Our results are consistent with this result.

This experiment was performed by thermal oxidation to shorten the duration of the experiment.

As a result of the residual amount measurement, it was confirmed that the ingredients of the rosemary extract were less thermostable than the BHA, which is a synthetic antioxidant. Despite the fact that the thermal stability was low and the residual amount was decreased, it was confirmed that components of rosemary extracts showed antioxidant ability in the results of peroxide value and the acid value. In addition, considering that the food groups using rosemary extract, such as oil and sausage, are stored and distributed at room temperature or low temperature, it is considered that they will actually have an larger antioxidant effect for a longer period of time. Therefore, the rosemary extract can be used as an antioxidant.

We have determined that the oxidation rate is effectively reduced at the legal maximum of 50 μg/mL, based on the peroxide value, the acid value, and the residual antioxidant level. Since carnosic acid has a higher antioxidant capacity than carnosol, rosemary extracts with a higher carnosic acid to carnosol ratio will be a more effective antioxidant.

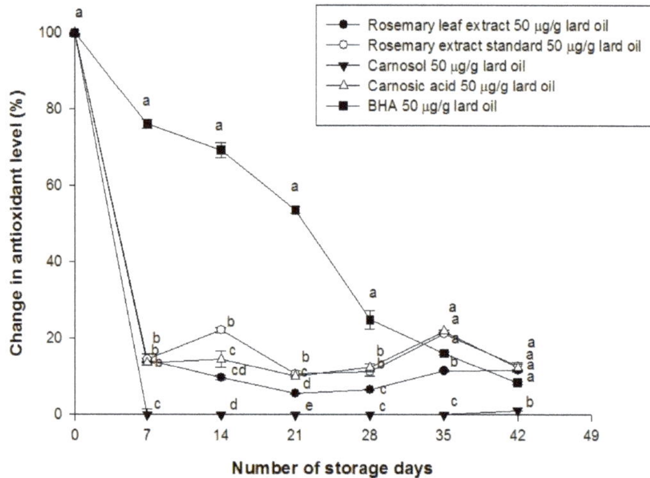

Figure 3. Change of residual antioxidant levels in lard oil during storage. Different letters indicate statistically significant different among the groups at *p*-value <0.05.

4. Conclusions

A quantitative method for the identification of rosemary extract in edible oils, processed meat products, and dressings was developed and validated using HPLC-PDA. The analytical method has been verified in accordance to the ICH guidelines. The validated HPLC-PDA analysis method is suitable for the determination of rosemary extract in edible oils, processed meat products and dressings. The validated method was verified through real sample analysis; none of the tested samples showed the presence of rosemary extracts. In addition, a functional evaluation of rosemary extract as an antioxidant was carried out. We confirmed that carnosic acid has superior antioxidative and oxidative stability compared to carnosol. These results demonstrate that the validated method is suitable for identification and quantification of rosemary extract and can be used to verify the safety of edible oils, processed meat products, and dressings that containing rosemary extracts residues. In addition, functional evaluation of the rosemary extract as an antioxidant has provided an effective use for rosemary extracts.

Author Contributions: Methodology, S.-H.C., G.-W.J., B.-Y.C., J.-S.L., D.-Y.K. and O.-H.L.; validation, S.-H.C., G.-W.J. and S.-I.C.; resources, D.-B.K. and X.H.; data curation, T.-D.J. and W.-S.S.; writing—original draft preparation, S.-H.C. and G.-W.J.; writing—review and editing, S.-H.C., G.-W.J. and O.-H.L.; supervision, O.-H.L.; project administration, O.-H.L.; funding acquisition, O.-H.L.

Funding: This research was funded Basic Science Research Program through the National Research Foundation of Korea (NRF) funded by the Ministry of Education, grant number NRF-2017R1D1A3B06028469.

Acknowledgments: This work was supported by the Basic Science Research Program through the National Research Foundation of Korea (NRF) funded by the Ministry of Education (NRF-2017R1D1A3B06028469) and has been worked with the support of a research grant of Kangwon National University in 2017.

Conflicts of Interest: The authors declare no conflict of interest.

References

1. Škrinjar, M.; Kolar, M.H.; Jelšek, N.; Hraš, A.R.; Bezjak, M.; Knez, Ž. Application of HPLC with electrochemical detection for the determination of low levels of antioxidants. *J. Food Compos. Anal.* **2007**, *20*, 539–545. [CrossRef]

2. Shahidi, F.; Zhong, Y. Lipid oxidation and improving the oxidative stability. *Chem. Soc. Rev.* **2010**, *39*, 4067–4079. [CrossRef] [PubMed]

3. Bandonienė, D.; Venskutonis, P.R.; Gruzdienė, D.; Murkovic, M. Antioxidative activity of sage (*Salvia officinalis* L.), savory (*Satureja hortensis* L.) and borage (*Borago officinalis* L.) extracts in rapeseed oil. *Eur. J. Lipid Sci. Technol.* **2002**, *104*, 286–292. [CrossRef]

4. Acworth, I.N.; Bailey, B. *The Handbook of Oxidative Metabolism*; ESA Inc.: Chelmsford, MD, USA, 1997; pp. 1–4.

5. Chen, Q.; Shi, H.; Ho, C.T. Effects of rosemary extracts and major constituents on lipid oxidation and soybean lipoxygenase activity. *J. Am. Oil Chem. Soc.* **1992**, *69*, 999. [CrossRef]

6. De Abreu, D.P.; Losada, P.P.; Maroto, J.; Cruz, J.M. Evaluation of the effectiveness of a new active packaging film containing natural antioxidants (from barley husks) that retard lipid damage in frozen Atlantic salmon (*Salmo salar* L.). *Food Res. Int.* **2010**, *43*, 1277–1282. [CrossRef]

7. Couto, R.O.; Conceição, E.C.; Chaul, L.T.; Oliveira, E.M.; Martins, F.S.; Bara, M.T.F.; Rezende, K.R.; Alves, S.F.; Paula, J.R. Spray-dried rosemary extracts: Physicochemical and antioxidant properties. *Food Chem.* **2012**, *131*, 99–105. [CrossRef]

8. Thorsen, M.A.; Hildebrandt, K.S. Quantitative determination of phenolic diterpenes in rosemary extracts: Aspects of accurate quantification. *J. Chromatogr. A* **2003**, *995*, 119–125. [CrossRef]

9. Yesil-Celiktas, O.; Sevimli, C.; Bedir, E.; Vardar-Sukan, F. Inhibitory effects of rosemary extracts, carnosic acid and rosmarinic acid on the growth of various human cancer cell lines. *Plant Foods Hum. Nutr.* **2010**, *65*, 158–163. [CrossRef] [PubMed]

10. National Health and Family Planning Commission (NHFPC). *National Standard of The People's Republic of China: National Food Safety Standard for Food Additive Use (GB2760-2014)*; National Health and Family Planning Commission of the People's Republic of China: Beijing, China, 2014.

11. EC. Commission regulation (EU) No 1129/2011 of 11 November 2011 amending Annex II to regulation (EC) No 1333/2008 of the european parliament and of the council by establishing a union list of food additives. *Off. J. Eur. Union L* **2011**, *295*, 12.

12. Oliveira, A.S.; Ribeiro-Santos, R.; Ramos, F.; Castilho, M.C.; Sanches-Silva, A. UHPLC-DAD Multi-Method for Determination of Phenolics in Aromatic Plants. *Food Anal. Methods* **2018**, *11*, 440–450. [CrossRef]

13. Hopia, A.I.; Huang, S.W.; Schwarz, K.; German, J.B.; Frankel, E.N. Effect of different lipid systems on antioxidant activity of rosemary constituents carnosol and carnosic acid with and without α-tocopherol. *J. Agric. Food Chem.* **1996**, *44*, 2030–2036. [CrossRef]

14. Frankel, E.N.; Huang, S.W.; Aeschbach, R.; Prior, E. Antioxidant activity of a rosemary extract and its constituents, carnosic acid, carnosol, and rosmarinic acid, in bulk oil and oil-in-water emulsion. *J. Agric. Food Chem.* **1996**, *44*, 131–135. [CrossRef]

15. Kim, J.M.; Choi, S.H.; Shin, G.H.; Lee, J.H.; Kang, S.R.; Lee, K.Y.; Lim, H.S.; Kang, T.S.; Lee, O.H. Method validation and measurement uncertainty for the simultaneous determination of synthetic phenolic antioxidants in edible oils commonly consumed in Korea. *Food Chem.* **2016**, *213*, 19–25. [CrossRef] [PubMed]

16. Suh, H.J.; Choi, S.H. Assessment of Estimated Daily Intakes of Propyl Gallate, EDTA (ethylenediamine tetra acetate), and Erythorbic Acid in Korea. *Korean J. Food Sci. Technol.* **2011**, *43*, 766–772. [CrossRef]

17. International Conference on Harmonization (ICH). International Conference on Harmonization of Technical Requirements for Registration of Pharmaceuticals for Human Use. ICH Harmonised Tripartite Guidline. Validation of Analytical Procedures: Text and Methodology. Available online: https://www.ich.org/fileadmin/Public_Web_Site/ICH_Products/Guidelines/Quality/Q2_R1/Step4/Q2_R1__Guideline.pdf (accessed on 11 March 2019).

18. AOCS. *Official Methods and Recommended Practices of the American Oil Chemists' Society*; AOCS: Washington, DC, USA, 1993.

19. Teruel, M.R.; Garrido, M.D.; Espinosa, M.C.; Linares, M.B. Effect of different format-solvent rosemary extracts (*Rosmarinus officinalis*) on frozen chicken nuggets quality. *Food Chem.* **2015**, *172*, 40–46. [CrossRef]

20. Puangsombat, K.; Smith, J.S. Inhibition of heterocyclic amine formation in beef patties by ethanolic extracts of rosemary. *J. Food Sci.* **2010**, *75*, 2. [CrossRef]

21. Smith, J.S.; Ameri, F.; Gadgil, P. Effect of marinades on the formation of heterocyclic amines in grilled beef steaks. *J. Food Sci.* **2008**, *73*, 6. [CrossRef]

22. Stuppner, H.; Sturm, S.; Konwalinka, G. HPLC analysis of the main oxindole alkaloids from Uncaria tomentosa. *Chromatographia* **1992**, *34*, 597–600. [CrossRef]

23. Pan, L.; LoBrutto, R.; Kazakevich, Y.V.; Thompson, R. Influence of inorganic mobile phase additives on the retention, efficiency and peak symmetry of protonated basic compounds in reversed-phase liquid chromatography. *J. Chromatogr. A* **2004**, *1049*, 63–73. [CrossRef]

24. Dolan, J.W. Temperature selectivity in reversed-phase high performance liquid chromatography. *J. Chromatogr. A* **2002**, *965*, 195–205. [CrossRef]

25. Saad, B.; Sing, Y.Y.; Nawi, M.A.; Hashim, N.; Mohamed Ali, A.S.; Saleh, M.I.; Sulaiman, S.F.; Talib, K.M.; Ahmad, K. Determination of synthetic phenolic antioxidants in food items using reversed-phase HPLC. *Food Chem.* **2007**, *105*, 389–394. [CrossRef]

26. Marın, A.; Garcıa, E.; Garcıa, A.; Barbas, C. Validation of a HPLC quantification of acetaminophen, phenylephrine and chlorpheniramine in pharmaceutical formulations: Capsules and sachets. *J. Pharm. Biomed. Anal.* **2002**, *29*, 701–714. [CrossRef]

27. Codex Alimentarius Commission. Guidelines on Good Laboratory Practice in Residue Analysis, CAC/GL 40-1993. Available online: http://www.fao.org/fao-who-codexalimentarius/codex-texts/guidelines/en/ (accessed on 10 March 2019).

© 2019 by the authors. Licensee MDPI, Basel, Switzerland. This article is an open access article distributed under the terms and conditions of the Creative Commons Attribution (CC BY) license (http://creativecommons.org/licenses/by/4.0/).

 *antioxidants*

Article

Antioxidant Activities, Phenolic Profiles, and Organic Acid Contents of Fruit Vinegars

Qing Liu [1], Guo-Yi Tang [1], Cai-Ning Zhao [1], Ren-You Gan [2,*] and Hua-Bin Li [1,*]

[1]   Guangdong Provincial Key Laboratory of Food, Nutrition and Health, Department of Nutrition, School of Public Health, Sun Yat-sen University, Guangzhou 510080, China; liuq248@mail2.sysu.edu.cn (Q.L.); tanggy5@mail2.sysu.edu.cn (G.-Y.T.); zhaocn@mail2.sysu.edu.cn (C.-N.Z.)

[2]   Department of Food Science & Technology, School of Agriculture and Biology, Shanghai Jiao Tong University, Shanghai 200240, China

*   Correspondence: renyougan@sjtu.edu.cn (R.-Y.G.); lihuabin@mail.sysu.edu.cn (H.-B.L.); Tel.: +86-21-3420-8517 (R.-Y.G.); +86-20-873-323-91 (H.-B.L.)

Received: 6 March 2019; Accepted: 22 March 2019; Published: 27 March 2019

Abstract: Fruit vinegars are popular condiments worldwide. Antioxidants and organic acids are two important components of the flavors and health benefits of fruit vinegars. This study aimed to test the antioxidant activities, phenolic profiles, and organic acid contents of 23 fruit vinegars. The results found that the 23 fruit vinegars varied in ferric-reducing antioxidant power (FRAP, 0.15–23.52 μmol Fe(II)/mL), Trolox equivalent antioxidant capacity (TEAC, 0.03–7.30 μmol Trolox/mL), total phenolic content (TPC, 29.64–3216.60 mg gallic acid equivalent/L), and total flavonoid content (TFC, 2.22–753.19 mg quercetin equivalent/L) values. Among the 23 fruit vinegars, the highest antioxidant activities were found in balsamic vinegar from Modena (Galletti), Aceto Balsamico di Modena (Monari Federzoni), red wine vinegar (Kühne), and red wine vinegar (Galletti). In addition, polyphenols and organic acids might be responsible for the antioxidant activities of fruit vinegars. The most widely detected phenolic compounds in fruit vinegars were gallic acid, protocatechuic acid, chlorogenic acid, caffeic acid, and *p*-coumaric acid, with tartaric acid, malic acid, lactic acid, citric acid, and succinic acid the most widely distributed organic acids. Overall, fruit vinegars are rich in polyphenols and organic acids and can be a good dietary source of antioxidants.

Keywords: fruit vinegar; antioxidant capacity; phenolics; flavonoid; organic acid

---

1. Introduction

Traditional vinegar is made from cereals and has been consumed for a long time. Another type of vinegar, fruit vinegar, made from fruit or fruit juices, has become increasingly popular in recent years because consumers are paying more attention to the functional properties of food products.

Oxidative stress is one of the main causes of certain chronic diseases such as liver, neurodegenerative, and cardiovascular diseases [1,2], which can be prevented by antioxidants [3]. Fruits are rich in antioxidants and are widely consumed by humans. Fruit vinegars can retain a number of antioxidants from fruit or fruit juices [4] and possess relatively high antioxidant capacities compared to wine and fruit juices [5]. Furthermore, fruit vinegars can increase antioxidant capacities of diets [6] as the fermentation process can produce functional components such as organic acids [7] which are not, or are only rarely, present in raw fruit materials. Fruit vinegars have also been reported to possess several health benefits, such as suppressing obesity-induced oxidative stress [8], regulating lipid metabolism, and decreasing liver damage [9], which can be at least partly due to the antioxidant activity of fruit vinegars [10]. Hence, it is valuable to determine and compare the antioxidant capacities of different fruit vinegars.

In addition, raw fruit materials are the main sources of phenolic compounds in fruit vinegars [11], and phenolics play key roles in the organoleptic properties and health effects of fruit vinegars. However, scientific research has reported that different fruit vinegars vary in their phenolic composition and contents [12], due to differences among the raw materials and manufacturing processes [7]. Only one study revealed the differences in phenolic profiles among three fruit vinegars [13]. Therefore, the phenolic profile is another important factor in measuring the value of fruit vinegars.

Fermentation is a key process in the production of fruit vinegars, during which most organic acids are produced through chemical and microbial actions [7]. Organic acids can contribute to the organoleptic qualities of fruit vinegars [14]. Furthermore, organic acids demonstrate antimicrobial activities [15] and can control blood glucose levels and regulate lipid abnormalities [16]. The organic acids in fruit vinegars have been found to be different from those in traditional cereal vinegars [13]. Therefore, it is valuable to understand the organic acid profile of fruit vinegars.

This research, therefore, was conducted to determine the antioxidant activities, total phenolic contents (TPC), and total flavonoid contents (TFC) of 23 commonly-consumed fruit vinegars. Moreover, the main phenolic compounds and organic acids were also identified and quantified in the 23 fruit vinegars. This study provides a good reference for the public as to consuming fruit vinegars rich in antioxidant phenolics and organic acids.

2. Materials and Methods

2.1. Chemicals and Materials

The chemicals for the determination of ferric-reducing antioxidant power (FRAP), Trolox equivalent antioxidant capacity (TEAC), TPC, TFC and phenol analysis were bought according to the paper [17] we published previously. Eighteen standard phenolic compounds, including gallic acid, protocatechuic acid, gallo catechin, chlorogenic acid, cyanidin-3-glucoside, caffeic acid, epicatechin, catechin gallate, p-coumaric acid, ferulaic acid, melatonin, 2-hydroxycinnamic acid, rutin, resveratrol, daidzein, equol, quercetin, and genistein were purchased from Sigma-Aldrich (St. Louis, MO, USA). Standard organic acids, including ascorbic acid, lactic acid, citric acid, and succinic acid, were obtained from Sigma-Aldrich, and oxalic acid, tartaric acid, and malic acid were bought from National Institutes for Food and Drug Control (Beijing, China). Phosphoric acid and potassium phosphate monobasic used for organic acid analysis were of analytical grade and bought from Damao Chemical Factory (Tianjin, China) and Yongda Chemical Reagent Company (Tianjin, China), respectively. Double-distilled water was used in all the experiments. The 23 fruit vinegars (Table 1) were bought from online shopping platforms and local markets in Guangzhou, China, and were stored at 4 °C before use.

2.2. Determination of FRAP, TEAC, TPC, and TFC Values

The FRAP, TEAC, TPC, and TFC values were evaluated based on the methods published previously [17], and were expressed as μmol Fe(II)/mL, μmol Trolox/mL, mg gallic acid equivalent (mg GAE)/L, and mg of quercetin equivalent (mg QE)/L, respectively.

2.3. Phenolic Composition Analysis

The phenolic components in 23 fruit vinegars were analyzed by High Performance Liquid Chromatography coupled with Photometric Diode Array detector (HPLC-PDA) (Waters, Milford, MA, USA) based on the literature [17]. Separation was conducted using an Agilent Zorbax Extend-C18 column (250 × 4.6 mm, 5 μm) (CA, USA) at 40 °C. Mobile phase A was formic acid solution (0.1%, *v/v*), and B was methanol. The procedure of gradient elution was set as: 0 min, 5% (B); 15 min, 20% (B); 20 min, 30% (B); 25 min, 37% (B); 40 min, 40% (B); 60 min, 50% (B); 65 min, 50% (B); 65.1 min, 5% (B); and 70 min, 5% (B). The spectra were scanned between 200 and 600 nm. Peak area was used to quantify phenolic compounds and the results were expressed as μg/mL.

2.4. Organic Acid Analysis

HPLC-PDA was used to analyze the organic acids in 23 fruit vinegars based on the literature [13] with slight modifications. Separation was conducted using an Agilent TC-C18(2) column (250 × 4.6 mm, 5 μm) at 35 °C with a mobile phase of 0.01 mol/L monopotassium phosphate buffer solution (pH = 2.5). The injection volume was 20 μL and the flow rate was 1 mL/min. The spectra were recorded at 210 nm. Peak area was used to quantify organic acids and the results were expressed as μg/mL.

2.5. Data Analysis

Each test was conducted in triplicate, and the results are shown as mean ± standard deviation (SD). SPSS 22.0 (IBM, Somers, NY, USA) and Excel 2007 (Redmond, WA, USA) were used to analyze the statistical differences. One-way analysis of variance (ANOVA) and the *post hoc* Tukey test were conducted to compare the differences among the means in more than two samples. The Pearson test was used for the correlation analysis. Statistical significance was defined as $p < 0.05$.

3. Results and Discussion

3.1. Antioxidant Activities of Fruit Vinegars

The antioxidant activities of phenolic compounds and other phytochemicals in natural foods are often multifunctional and more than one method is required to measure the antioxidant capacities of fruit vinegars because methods measure different aspects of antioxidant capacities [18]. In this study, FRAP and TEAC assays were both used to evaluate the antioxidant activities of fruit vinegars. The FRAP assay measures the ability to reduce a ferric tripyridyltriazine complex to the ferrous complex [19], while the TEAC method determines the ability to scavenge ABTS$^{•+}$ free radicals [20].

The FRAP values of the 23 fruit vinegars ranged from 0.15 to 23.52 μmol Fe(II)/mL, while the TEAC values ranged from 0.03 to 7.30 μmol Trolox/mL (Table 1). These results were consistent with a previous study in which the FRAP value of a wine vinegar was 9.50 mmol Fe(II)/L, while the TEAC value was 3.12 mmol Trolox/L [5]. The five highest FRAP values, in decreasing order, were found in balsamic vinegar of Modena (Galletti) (23.52 μmol Fe(II)/mL), Aceto Balsamico di Modena (Monari Federzoni) (13.39 μmol Fe(II)/mL), red wine vinegar (Galletti) (8.04 μmol Fe(II)/mL), red wine vinegar (Kühne) (5.35 μmol Fe(II)/mL), and apple vinegar beverage (Long He Kuan) (3.99 μmol Fe(II)/mL). Similarly, results for the fruit vinegars with the highest TEAC values were generally consistent with the results for the FRAP values (Table 1).

3.2. TPC and TFC Values

The TPC values of fruit vinegars ranged from 29.64 to 3216.60 mg GAE/L (Table 1). The fruit vinegars with the highest TPC values were balsamic vinegar of Modena (Galletti) (3216.60 mg GAE/L), followed by Aceto Balsamico di Modena (Monari Federzoni) (1901.92 mg GAE/L), red wine vinegar (Galletti) (993.51 mg GAE/L), red wine vinegar (Kühne) (654.95 mg GAE/L), and apple vinegar (Heng Shun) (495.52 mg GAE/L). The TPC values of the tested fruit vinegars were in accordance with the findings of Ren et al. [13], in which the TPC values of fruit vinegars ranged from 274.08 to 754.50 mg GAE/L, but only three fruit vinegars were tested in that study. On the other hand, the TFC values ranged from 2.22 to 753.19 mg QE/L (Table 1). The fruit vinegars with the highest TFC values were Aceto Balsamico di Modena (Monari Federzoni) (753.19 mg QE/L), followed by balsamic vinegar of Modena (Galletti) (699.67 mg QE/L), red wine vinegar (Kühne) (51.47 mg QE/L), red wine vinegar (Galletti) (50.34 mg QE/L), and apple vinegar (Heng Shun) (31.39 mg QE/L).

Table 1. The ferric-reducing antioxidant power (FRAP), Trolox equivalent antioxidant capacity (TEAC), total phenolic contents (TPC), and total flavonoid contents (TFC) values of 23 fruit vinegars.

No.	Product	Producing Place	FRAP value (μmol Fe(II)/mL)	TEAC value (μmol Trolox/mL)	TPC value (mg GAE/L)	TFC value (mg QE/L)
1	Apple vinegar (Hai Tian)	Foshan, China	0.84 ± 0.01 [k]	0.29 ± 0.01 [j]	149.77 ± 3.94 [h]	7.85 ± 0.35 [d]
2	Apple vinegar (Ba Zhen)	Dongguan, China	1.43 ± 0.04 [i,j]	0.43 ± 0.01 [i,j]	249.18 ± 2.42 [g]	20.77 ± 0.16 [d]
3	Apple vinegar (Guang Wei Yuan)	Guangzhou, China	0.80 ± 0.02 [k]	0.23 ± 0.01 [j]	123.67 ± 1.51 [h,i]	12.88 ± 0.08 [d]
4	Apple vinegar (Heng Shun)	Zhenjiang, China	2.96 ± 0.01 [f]	1.01 ± 0.02 [g]	495.52 ± 20.59 [e]	31.39 ± 0.43 [c,d]
5	Apple vinegar (Zi Lin)	Taiyuan, China	1.83 ± 0.01 [h]	0.69 ± 0.02 [h]	398.17 ± 8.25 [f]	12.78 ± 0.29 [d]
6	Apple vinegar (Cu Bo Shi)	Xinxiang, China	0.23 ± 0.01 [l]	0.03 ± 0.00 [k]	66.90 = 0.87 [h,i]	5.86 ± 0.10 [d]
7	Apple vinegar (Sempio)	Seoul, South Korea	0.22 ± 0.01 [l]	0.04 ± 0.00 [k]	43.75 ± 0.34 [i]	2.22 ± 0.16 [d]
8	Apple vinegar (Galletti)	Gremona, Italy	1.28 ± 0.02 [i,j]	0.49 ± 0.02 [i]	256.13 ± 1.86 [g]	12.32 ± 0.16 [d]
9	Apple vinegar (Kühne)	Hamburg, Germany	1.37 ± 0.01 [i,j]	0.53 ± 0.02 [i]	198.14 ± 0.33 [g,h]	11.14 ± 0.28 [d]
10	Apple cider vinegar (Heinz)	Pittsburgh, America	1.05 ± 0.04 [j,k]	0.31 ± 0.01 [j]	163.28 ± 0.15 [h]	13.11 ± 0.14 [d]
11	Apple cider vinegar (Xin He)	Jinan, China	2.40 ± 0.05 [g]	0.90 ± 0.03 [g]	426.72 ± 14.52 [e,f]	28.04 ± 0.51 [c,d]
12	Apple vinegar beverage (Hua Sheng Tang)	Zhongshan, China	0.19 ± 0.01 [l]	0.06 ± 0.00 [k]	62.19 ± 0.40 [h,i]	5.41 ± 0.22 [d]
13	Apple vinegar beverage (Tian Di Yi Hao)	Jiangmen, China	0.15 ± 0.00 [l]	0.04 ± 0.00 [k]	29.64 ± 0.16 [i]	5.79 ± 0.14 [d]
14	Apple vinegar beverage (Long He Kuan)	Beijing, China	3.99 ± 0.04 [e]	1.69 ± 0.03 [e]	469.10 ± 8.79 [e,f]	4.52 ± 0.14 [d]
15	Red wine vinegar (Galletti)	Gremona, Italy	8.04 ± 0.11 [c]	3.17 ± 0.06 [c]	993.51 ± 23.19 [c]	50.34 ± 2.43 [c]
16	Red wine vinegar (Kühne)	Hamburg, Germany	5.35 ± 0.09 [d]	2.09 ± 0.03 [d]	654.95 ± 39.52 [d]	51.47 ± 0.99 [c]
17	Italian red wine vinegar (Ponti)	Ghemme, Italy	3.71 ± 0.02 [e]	1.36 ± 0.01 [f]	396.40 ± 2.68 [f]	19.75 ± 0.49 [d]
18	White wine vinegar (Galletti)	Gremona, Italy	1.53 ± 0.03 [i]	0.52 ± 0.01 [i]	229.60 ± 2.14 [g,h]	5.58 ± 0.30 [d]
19	White wine vinegar (Kühne)	Hamburg, Germany	1.28 ± 0.03 [i,j]	0.46 ± 0.02 [i]	153.90 ± 2.13 [h]	6.54 ± 0.16 [d]
20	Italian white wine vinegar (Ponti)	Ghemme, Italy	1.21 ± 0.03 [j]	0.30 ± 0.00 [j]	117.00 ± 2.82 [h,i]	5.09 ± 0.14 [d]
21	Balsamic vinegar of Modena (Galletti)	Modena, Italy	23.52 ± 0.33 [a]	7.30 ± 0.16 [a]	3216.60 ± 132.67 [a]	699.67 ± 24.08 [b]
22	Aceto Balsamico di Modena (Monari Federzoni)	Bomporto, Italy	13.39 ± 0.25 [b]	4.49 ± 0.09 [b]	1901.92 ±16.06 [b]	753.19 ± 36.85 [a]
23	Fruit vinegar (Wan Jia Xiang)	Taiwan, China	0.89 ± 0.03 [k]	0.26 ± 0.01 [j]	194. 65 ± 0.32 [g,h]	23.54 ± 0.99 [c,d]

[a,b,c,d,e,f,g,h,i,j,k,l] different letters within a parameter indicate statistical significance at $p < 0.05$.

Taking the FRAP, TEAC, TPC, and TFC values together, balsamic vinegar of Modena (Galletti), Aceto Balsamico di Modena (Monari Federzoni), red wine vinegar (Kühne), and red wine vinegar (Galletti) showed the highest antioxidant capacities and phenolic contents among the 23 fruit vinegars tested. In general, fruit vinegars made from red grapes, especially balsamic vinegar (balsamic vinegar of Modena (Galletti) and Aceto Balsamico di Modena (Monari Federzoni)), possessed stronger antioxidant activities. This finding is consistent with a previous study in which balsamic vinegars made from red grapes displayed significantly higher TPC values, radical scavenging, and oxidant reducing activities compared to fruit vinegars made from red grapes, white grapes, and apples [21]. In addition, antioxidant activities were found to be higher in red grape balsamic vinegars than in red wine vinegars, probably due to the phenolic contents in different fruit vinegars being affected by the raw materials, such as red grapes, white grapes and apples, and manufacturing processes [22,23].

3.3. Correlations Among FRAP, TEAC, TPC, and TFC Values

The FRAP values of the 23 fruit vinegars were highly correlated with the TEAC values ($R^2 = 0.989$) (Table 2), indicating that the components responsible for reducing oxidants were consistent with those scavenging free radicals in fruit vinegars. In addition, a moderate correlation ($R^2 = 0.832$) was found between the TPC and TFC values (Table 2), indicating that flavonoids were not the only phenolic compounds in fruit vinegars. In addition, the FRAP and TEAC values both showed high positive correlations with TPC values ($R^2 = 0.990$ and 0.971, respectively) (Table 2), suggesting that phenolic components contribute to both the oxidant-reducing and radical scavenging activities of fruit vinegars. In a previous study conducted by Dávalos et al. [12], the antioxidant activities and TPC values of wine vinegars exerted a positive correlation ($p < 0.01$), consistent with our finding. On the other hand, the FRAP and TEAC values both showed moderate correlations with TFC values ($R^2 = 0.804$ and 0.767, respectively) as shown in Table 2. Although the four values were correlated with each other, further studies are still needed to evaluate the specific compounds that contribute to each value in fruit vinegars, as most of the methods were based on the same reaction mechanism.

Table 2. Correlation analysis among FRAP, TEAC, TPC and TFC values.

Correlation Coefficient (R^2)	FRAP Value	TEAC Value	TPC Value	TFC Value
FRAP value	1	0.989 *	0.990 *	0.804 *
TEAC value	-	1	0.971 *	0.767 *
TPC value	-	-	1	0.832 *
TFC value	-	-	-	1

* indicates statistical significance at $p < 0.01$.

3.4. Polyphenols and Organic Acids in Fruit Vinegars

Some studies have suggested that fruit vinegars possess the ability to improve oxidative stress-related disorders, such as obesity [8], liver damage [9], and diabetes [24]. Our results indicated that polyphenols in fruit vinegars were the major ingredients contributing to the antioxidant activities. We therefore further investigated the main phenolic compounds in fruit vinegars (Table 3). Retention time and UV spectra were used to recognize the phenolic compounds, and the peak areas were used to quantify these phenolic compounds (Figure 1). It was found that gallic acid, protocatechuic acid, chlorogenic acid, caffeic acid, and *p*-coumaric acid were the most widely detected phenolic compounds in 23 fruit vinegars. In addition, the highest concentrations of gallic acid, protocatechuic acid, chlorogenic acid, caffeic acid, and *p*-coumaric acid were found in Balsamic vinegar of Modena (Galletti) (12.56 µg/mL), balsamic vinegar of Modena (Galletti) (3.29 µg/mL), apple vinegar (Zi Lin) (10.91 µg/mL), balsamic vinegar of Modena (Galletti) (3.58 µg/mL), and balsamic vinegar of Modena (Galletti) (1.97 µg/mL), respectively. The data we obtained were in accordance with the data in Phenol-Explorer (a database on polyphenol content in foods), in which the contents of gallic acid, protocatechuic acid caffeic acid, and *p*-coumaric acid in vinegars were 2.59 ±, 0.81, 0.28, 0.29 mg/100 mL, respectively [25]. Balsamic vinegar of Modena (both Galletti and Monari Federzoni) contained high gallic acid and *p*-coumaric acid contents which might be responsible for the strong antioxidant capacities and high phenolic contents and needs further study. Many polyphenols, such as gallic acid, protocatechuic acid, chlorogenic acid, caffeic acid, and *p*-coumaric acid found in fruit vinegars, have been reported to suppress oxidative stress-related damages [26–29].

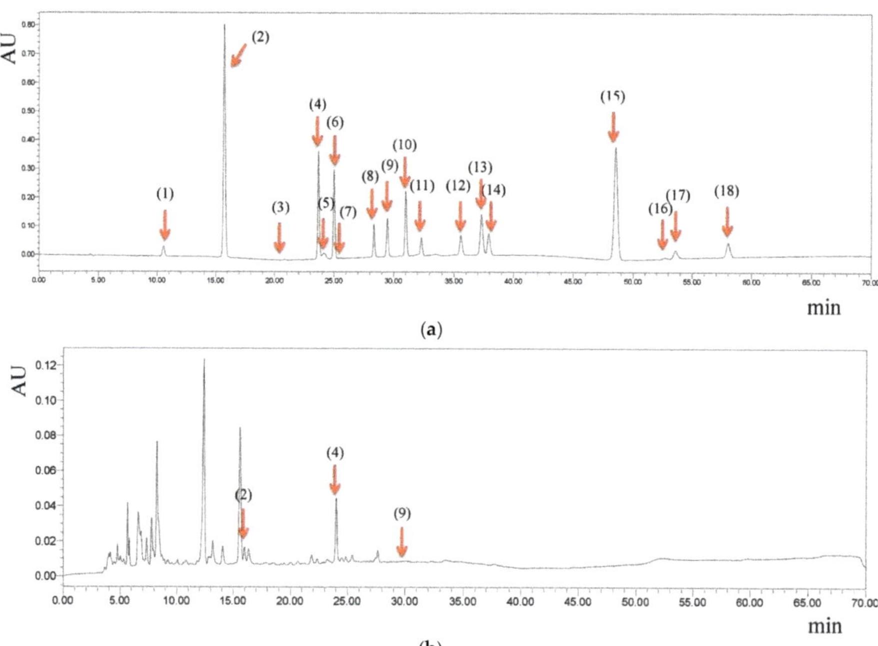

Figure 1. The chromatograms of phenolic compound standards (**a**) and apple vinegar (Zi Lin) (**b**) under 254 nm. Peak identification, retention time and maximum absorption: (1) gallic acid, 10.543 min, 271.3 nm; (2) protocatechuic acid, 15.723 min, 259.4 nm; (3) gallo catechin, 20.821 min, 270.1 nm; (4) chlorogenic acid, 23.661 min, 326.0 nm; (5) cyanidin-3-glucoside, 24.145 min, 279.6 nm; (6) caffeic acid, 24.987 min, 323.6 nm; (7) epicatechin, 25.580 min, 278.4 nm; (8) catechin gallate, 28.334 min, 277.2 nm; (9) *p*-coumaric acid, 29.454 min, 309.3 nm; (10) ferulaic acid, 31.018 min, 323.6 nm; (11) melatonin, 32.325 min, 221.7 nm; (12) 2-hydroxycinnamic acid, 35.562 min, 276.0 nm; (13) rutin, 37.296 min, 255.9 nm; (14) resveratrol, 37.908 min, 304.6 nm; (15) daidzein, 48.543 min, 248.8 nm; (16) equol, 52.706 min, 280.8 nm; (17) quercetin, 53.591 min, 254.7 nm; (18) genistein, 58.002 min, 259.4 nm.

Organic acids are another important component of fruit vinegars. The main organic acids in fruit vinegars and their contents are shown in Table 4. Retention times were used to identify organic acids, and peak areas were used to quantify the contents (Figure 2). In the 23 fruit vinegars tested, tartaric acid, malic acid, lactic acid, citric acid, and succinic acid were the most widely detected organic acids, with the highest content found in white wine vinegar (Kühne) (1566.48 µg/mL), apple vinegar (Guang Wei Yuan) (7691.98 µg/mL), apple vinegar (Guang Wei Yuan) (2541.64 µg/mL), apple vinegar (Guang Wei Yuan) (6485.24 µg/mL), and apple vinegar (Cu Bo Shi) (1775.77 µg/mL), respectively.

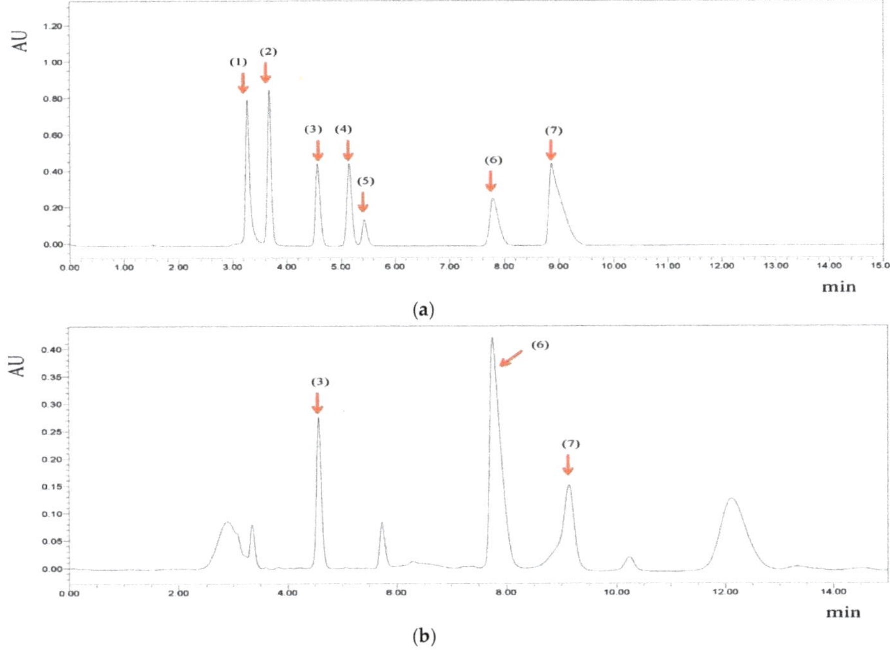

Figure 2. The chromatograms of the organic acid standards (**a**) and apple vinegar (Cu Bo Shi) (**b**) under 210 nm. Peak identification and retention time: (1) oxalic acid, 3.268 min; (2) tartaric acid, 3.677 min; (3) malic acid, 4.565 min; (4) ascorbic acid, 5.150 min; (5) lactic acid, 5.429 min; (6) citric acid, 7.795 min; (7) succinic acid, 8.869 min.

Fruit vinegars are popular all over the world due to their good flavor and health benefits. Phenolic compounds and organic acids are the main components that contribute to the sensory qualities and health benefits of fruit vinegars. Some research has indicated that different grape varieties possess different phenolic contents and composition [30], and similarly apple varieties [31], depending on factors like cultivars, growing environments, and ripeness stage [17]. Factors, such as yeast strains [32], acetic acid bacteria, and production technology [33] within the fermentation processes, could also affect the phenolic profiles of fruit vinegars [4,34]. In addition, phenolic compounds have been widely explored for the abilities in preventing chronic diseases, including anti-cancer, anti-obesity, anti-aging and anti-diabetes activities [35]. According to our results, a usual serving of fruit vinegar (10 mL) contains approximately 0.30–32.67 mg GAE of polyphenols, and the FRAP and TEAC values of 10 mL fruit vinegar are approximately 1.50–235.20 μmol Fe(II) and 0.30–73.00 μmol Trolox, respectively. As the phenolic contents and composition in fruit vinegars varied, further studies are needed to explore the bioavailability and health benefits of fruit vinegars in vivo.

Organic acids in fruit vinegars are produced through hydrolysis, biochemical metabolism and microbial actions in the fermentation process. In fruit vinegars, the contents and types of organic acids affect their sensory qualities and also their health functions. Our results indicated that fruit vinegars possessed abundant organic acids with different and complex compositions. Among all the tested organic acids, tartaric acid, malic acid, lactic acid, citric acid, and succinic acid were most widely distributed in the 23 fruit vinegars tested, and the results were consistent with another study [36]. Organic acids exert some health benefits such as antimicrobial activities [15], controlling blood glucose levels, and regulating lipid abnormalities [16]. Fruit vinegars were found to contain more complex compositions of organic acids than cereal vinegars [13], indicating fruit vinegars possess a richer taste and fruit flavor compared with conventional cereal vinegars.

Table 3. Main phenolic compounds and their contents in 23 fruit vinegars.

No.	Products	Gallic Acid	Protocatechuic Acid	Chlorogenic Acid	Caffeic Acid	*p*-Coumaric Acid	Ferulic Acid
1	Apple vinegar (Hai Tian)	-	-	2.79 ± 0.16 [c]	-	-	-
2	Apple vinegar (Ba Zhen)	-	-	0.32 ± 0.01 [d]	-	-	-
3	Apple vinegar (Guang Wei Yuan)	-	-	-	-	-	-
4	Apple vinegar (Heng Shun)	-	1.54 ± 0.05 [b]	2.99 ± 0.21 [c]	-	-	-
5	Apple vinegar (Zi Lin)	-	0.82 ± 0.04 [e]	10.91 ± 0.80 [a]	-	0.17 ± 0.01 [e]	-
6	Apple vinegar (Cu Bo Shi)	-	-	-	-	-	-
7	Apple vinegar (Sempio)	-	0.08 ± 0.00 [h]	0.11 ± 0.00 [d]	-	-	-
8	Apple vinegar (Galletti)	-	0.37 ± 0.01 [g]	3.13 ± 0.11 [c]	-	-	-
9	Apple vinegar (Kühne)	-	1.09 ± 0.04 [d]	5.30 ± 0.29 [b]	-	0.10 ± 0.00 [e]	-
10	Apple cider vinegar (Heinz)	-	1.00 ± 0.06 [d]	0.23 ± 0.00 [d]	-	-	-
11	Apple cider vinegar (Xin He)	-	-	4.67 ± 0.21 [b]	-	-	-
12	Apple vinegar beverage (Hua Sheng Tang)	-	-	-	-	-	-
13	Apple vinegar beverage (Tian Di Yi Hao)	-	0.19 ± 0.01 [h]	0.59 ± 0.02 [d]	-	-	-
14	Apple vinegar beverage (Long He Kuan)	-	-	-	-	-	-
15	Red wine vinegar (Galletti)	4.10 ± 0.18 [d]	0.47 ± 0.04 [f,g]	-	1.48 ± 0.10 [c]	1.13 ± 0.05 [c]	-
16	Red wine vinegar (Kühne)	9.99 ± 0.58 [b]	1.38 ± 0.05 [c]	-	-	1.39 ± 0.01 [b]	-
17	Italian red wine vinegar (Ponti)	4.36 ± 0.33 [d]	0.49 ± 0.01 [f]	-	1.73 ± 0.01 [b]	0.81 ± 0.02 [d]	-
18	White wine vinegar (Galletti)	-	-	-	-	0.18 ± 0.00 [e]	-
19	White wine vinegar (Kühne)	-	0.32 ± 0.01 [g]	-	0.32 ± 0.01 [d]	0.15 ± 0.01 [e]	0.31 ± 0.01
20	Italian white wine vinegar (Ponti)	-	0.16 ± 0.00 [h]	-	-	-	-
21	Balsamic vinegar of Modena (Galletti)	12.56 ± 0.86 [a]	3.29 ± 0.05 [a]	-	3.58 ± 0.14 [a]	1.97 ± 0.05 [a]	-
22	Aceto Balsamico di Modena (Monari Federzoni)	7.50 ± 0.60 [c]	-	-	-	1.17 ± 0.06 [c]	-
23	Fruit vinegar (Wan Jia Xiang)	-	-	-	-	-	-

[a,b,c,d,e,f,g] Different uppercase letters within a column indicate statistical significance at *p* < 0.05.

Table 4. Main organic acids and their contents in 23 fruit vinegars.

No.	Products	Tartaric Acid (μg/mL)	Malic Acid (μg/mL)	Lactic Acid (μg/mL)	Citric Acid (μg/mL)	Succinic Acid (μg/mL)
1	Apple vinegar (Hai Tian)	-	480.72 ± 5.12 [d]	-	-	-
2	Apple vinegar (Ba Zhen)	14.71 ± 0.27 [e]	372.61 ± 7.95 [d]	-	-	-
3	Apple vinegar (Guang Wei Yuan)	-	7691.98 ± 435.24 [a]	2541.64 ± 107.29 [a]	6485.24 ± 389.42 [a]	-
4	Apple vinegar (Heng Shun)	-	1771.35 ± 117.77 [c]	-	-	-
5	Apple vinegar (Zi Lin)	-	613.05 ± 41.96 [d]	-	-	1637.36 ± 67.61 [a]
6	Apple vinegar (Cu Bo Shi)	-	1707.76 ± 105.42 [c]	-	5263.43 ± 215.07 [b]	1775.77 ± 113.27 [a]
7	Apple vinegar (Sempio)	-	588.03 ± 11.69 [d]	-	-	-
8	Apple vinegar (Galletti)	-	402.11 ± 23.26 [d]	-	-	-
9	Apple vinegar (Kühne)	-	236.38 ± 3.01 [d]	-	-	-
10	Apple cider vinegar (Heinz)	-	-	-	-	-
11	Apple cider vinegar (Xin He)	-	4102.22 ± 253.95 [b]	-	-	991.08 ± 13.89 [b]
12	Apple vinegar beverage (Hua Sheng Tang)	-	573.18 ± 18.66 [d]	-	-	-
13	Apple vinegar beverage (Tian Di Yi Hao)	-	594.91 ± 42.89 [d]	-	-	-
14	Apple vinegar beverage (Long He Kuan)	-	4386.73 ± 326.85 [b]	50.20 ± 1.06 [b]	787.96 ± 24.81 [c]	-
15	Red wine vinegar (Galletti)	383.29 ± 5.74 [d]	-	-	-	252.60 ± 6.50 [c]
16	Red wine vinegar (Kühne)	948.01 ± 54.86 [b]	-	-	-	-
17	Italian red wine vinegar (Ponti)	1030.86 ± 75.33 [b]	187.36 ± 4.47 [d]	-	-	-
18	White wine vinegar (Galletti)	760.55 ± 8.73 [c]	-	-	-	346.04 ± 6.56 [c]
19	White wine vinegar (Kühne)	1566.48 ± 94.47 [a]	-	-	-	-
20	Italian white wine vinegar (Ponti)	1066.64 ± 65.49 [b]	-	-	-	-
21	Balsamic vinegar of Modena (Galletti)	96.15 ± 1.15 [e]	-	-	-	-
22	Aceto Balsamico di Modena (Monari Federzoni)	393.77 ± 13.37 [d]	603.48 ± 15.38 [d]	-	-	-
23	Fruit vinegar (Wan Jia Xiang)	-	-	-	-	-

[a,b,c,d] Different uppercase letters within a column indicate statistical significance at $p < 0.05$.

4. Conclusions

The antioxidant activities, TPC, and TFC of 23 fruit vinegars were studied. The FRAP values of the fruit vinegars were in the range of 0.15–23.52 μmol Fe(II)/mL, and the TEAC values ranged from 0.03 to 7.30 μmol Trolox/mL. The TPC and TFC values ranged from 29.64 to 3216.60 mg GAE/L and from 2.22 to 753.19 mg QE/L, respectively. Balsamic vinegar of Modena (Galletti), Aceto Balsamico di Modena (Monari Federzoni), red wine vinegar (Kühne), and red wine vinegar (Galletti) exhibited the highest antioxidant activities among 23 fruit vinegars tested. The high correlation of FRAP and TEAC with TPC values indicated that the abilities of fruit vinegars to reduce oxidants and to scavenge free radicals were mainly attributed to polyphenols. Several phenolic compounds, including, gallic acid, protocatechuic acid, chlorogenic acid, caffeic acid, and *p*-coumaric acid, and organic acids including tartaric acid, malic acid, lactic acid, citric acid, and succinic acid, were mainly detected in fruit vinegars. Some of the phenolic compounds (such as gallic acid and *p*-coumaric acid) might be responsible for the high antioxidant content and strong antioxidant activities of fruit vinegars and need to be further explored. The polyphenols and organic acids of fruit vinegars might contribute to their antioxidant activities, flavors, and health effects. Overall, fruit vinegars can be good natural sources of dietary antioxidant polyphenols and organic acids, which should be of interest to food scientists, nutritionists, the public, and food producers.

Author Contributions: Conceptualization, Q.L., R.-Y.G and H.-B.L.; Formal analysis, Q.L. and G.-Y.T.; Funding acquisition, R.-Y.G.; Investigation, Q.L., G.-Y.T. and C.-N.Z.; Project administration, H.-B.L.; Resources, C.-N.Z.; Software, Q.L.; Supervision, R.-Y.G. and H.-B.L.; Validation, Q.L. and H.-B.L.; Writing – original draft, Q.L.; Writing – review & editing, R.-Y.G. and H.-B.L.

Funding: This study was funded by the Shanghai Basic and Key Program (18JC1410800), Shanghai Pujiang Talent Plan (No. 18PJ1404600), the Agri-X Interdisciplinary Fund of Shanghai Jiao Tong University (No. Agri-X2017004), and the Shanghai Agricultural Science and Technology Key Program (18391900600).

Conflicts of Interest: The authors declare no conflict of interest.

References

1. Liguori, I.; Russo, G.; Curcio, F.; Bulli, G.; Aran, L.; Della-Morte, D.; Gargiulo, G.; Testa, G.; Cacciatore, F.; Bonaduce, D. Oxidative stress, aging, and diseases. *Clin. Interv. Aging* **2018**, *13*, 757–772. [CrossRef]

2. Ezhilarasan, D. Oxidative stress is bane in chronic liver diseases: Clinical and experimental perspective. *Arab. J. Gastroenterol.* **2018**, *19*, 56–64. [CrossRef] [PubMed]

3. Tan, B.L.; Norhaizan, M.E.; Liew, W.; Rahman, H.S. Antioxidant and oxidative stress: A mutual interplay in age-related diseases. *Front. Pharmacol.* **2018**, *9*, 1162. [CrossRef] [PubMed]

4. Bakir, S.; Toydemir, G.; Boyacioglu, D.; Beekwilder, J.; Capanoglu, E. Fruit antioxidants during vinegar processing: Changes in content and *in vitro* bio-accessibility. *Int. J. Mol. Sci.* **2016**, *17*, 1658. [CrossRef]

5. Pellegrini, N.; Serafini, M.; Colombi, B.; Del Rio, D.; Salvatore, S.; Bianchi, M.; Brighenti, F. Total antioxidant capacity of plant foods, beverages and oils consumed in Italy assessed by three different in vitro assays. *J. Nutr.* **2003**, *133*, 2812–2819. [CrossRef]

6. Ninfali, P.; Mea, G.; Giorgini, S.; Rocchi, M.; Bacchiocca, M. Antioxidant capacity of vegetables, spices and dressings relevant to nutrition. *Br. J. Nutr.* **2005**, *93*, 257–266. [CrossRef] [PubMed]

7. Chen, H.Y.; Chen, T.; Giudici, P.; Chen, F.S. Vinegar functions on health: Constituents, sources, and formation mechanisms. *Compr. Rev. Food. Sci. Food Saf.* **2016**, *15*, 1124–1138. [CrossRef]

8. Halima, B.; Sonia, G.; Sarra, K.; Houda, B.; Fethi, B.; Abdallah, A. Apple cider vinegar attenuates oxidative stress and reduces the risk of obesity in high-fat-fed male wistar rats. *J. Med. Food* **2018**, *21*, 70–80. [CrossRef] [PubMed]

9. Bouazza, A.; Bitam, A.; Amiali, M.; Bounihi, A.; Yargui, L.; Koceir, E.A. Effect of fruit vinegars on liver damage and oxidative stress in high-fat-fed rats. *Pharm. Biol.* **2016**, *54*, 260–265. [CrossRef] [PubMed]

10. Coelho, E.; Genisheva, Z.; Oliveira, J.M.; Teixeira, J.A.; Domingues, L. Vinegar production from fruit concentrates: Effect on volatile composition and antioxidant activity. *J. Food Sci. Technol.-Mysore* **2017**, *54*, 4112–4122. [CrossRef]

11. Solieri, L.; Giudici, P. Vinegars of the world. In *Vinegars of the World*; Solieri, L., Giudici, P., Eds.; Springer: New York, NY, USA, 2009; pp. 1–16.

12. Ho, C.W.; Lazim, A.M.; Fazry, S.; Zaki, U.; Lim, S.J. Varieties, production, composition and health benefits of vinegars: A review. *Food Chem.* **2017**, *221*, 1621–1630. [CrossRef] [PubMed]

13. Zhang, J.; Tian, Z.; Wang, J.; Wang, A. Advances in antimicrobial molecular mechanism of organic acids. *Acta Vet. Zootechn. Sin.* **2011**, *42*, 323–328.

14. Petsiou, E.I.; Mitrou, P.I.; Raptis, S.A.; Dimitriadis, G.D. Effect and mechanisms of action of vinegar on glucose metabolism, lipid profile, and body weight. *Nutr. Rev.* **2014**, *72*, 651–661. [CrossRef] [PubMed]

15. Liu, Q.; Tang, G.Y.; Zhao, C.N.; Feng, X.L.; Xu, X.Y.; Cao, S.Y.; Meng, X.; Li, S.; Gan, R.Y.; Li, H.B. Comparison of antioxidant activities of different grape varieties. *Molecules* **2018**, *23*, 2432. [CrossRef] [PubMed]

16. Ren, M.M.; Wang, X.Y.; Tian, C.R.; Li, X.J.; Zhang, B.S.; Song, X.Z.; Zhang, J. Characterization of organic acids and phenolic compounds of cereal vinegars and fruit vinegars in China. *J. Food Process Preserv.* **2017**, *41*, e12937. [CrossRef]

17. Deng, G.F.; Lin, X.; Xu, X.R.; Gao, L.L.; Xie, J.F.; Li, H.B. Antioxidant capacities and total phenolic contents of 56 vegetables. *J. Funct. Food.* **2013**, *5*, 260–266. [CrossRef]

18. Moon, J.K.; Shibamoto, T. Antioxidant assays for plant and food components. *J. Agric. Food Chem.* **2009**, *57*, 1655–1666. [CrossRef]

19. Frankel, E.N.; Meyer, A.S. The problems of using one-dimensional methods to evaluate multifunctional food and biological antioxidants. *J. Sci. Food Agric.* **2000**, *80*, 1925–1941. [CrossRef]

20. Sinanoglou, V.J.; Zoumpoulakis, P.; Fotakis, C.; Kalogeropoulos, N.; Sakellari, A.; Karavoltsos, S.; Strati, I.F. On the characterization and correlation of compositional, antioxidant and colour profile of common and balsamic vinegars. *Antioxidants* **2018**, *7*, 139. [CrossRef]

21. Giudici, P.; Gullo, M.; Solieri, L.; Falcone, P.M. Technological and microbiological aspects of traditional balsamic vinegar and their influence on quality and sensorial properties. In *Advances in Food and Nutrition Research*; Taylor, S.L., Ed.; Elsevier Academic Press Inc.: San Diego, CA, USA, 2009; Volume 58, pp. 137–182.

22. Tesfaye, W.; Morales, M.L.; Garcia-Parrilla, M.C.; Troncoso, A.M. Wine vinegar: Technology, authenticity and quality evaluation. *Trends Food Sci. Technol.* **2002**, *13*, 12–21. [CrossRef]

23. Davalos, A.; Bartolome, B.; Gomez-Cordoves, C. Antioxidant properties of commercial grape juices and vinegars. *Food Chem.* **2005**, *93*, 325–330. [CrossRef]

24. Hmad, H.B.; Khlifi, S.; Jemaa, H.B.; Jemmousi, H.; Ben Slama, F.; Abdallah, A. Hypoglycemic and hypolipidemic effects of apple cider vinegar in Tunisian type 2 diabetic patients. *Acta Physiol.* **2017**, *221*, 250.

25. Yigitturk, G.; Acara, A.C.; Erbas, O.; Oltulu, F.; Yavasoglu, N.; Uysal, A.; Yavasoglu, A. The antioxidant role of agomelatine and gallic acid on oxidative stress in STZ induced type I diabetic rat testes. *Biomed. Pharmacother.* **2017**, *87*, 240–246. [CrossRef] [PubMed]

26. Phenol-Explorer. Available online: http://phenol-explorer.eu/contents/food/140 (accessed on 6 March 2019).

27. Safaeian, L.; Emami, R.; Hajhashemi, V.; Haghighatian, Z. Antihypertensive and antioxidant effects of protocatechuic acid in deoxycorticosterone acetate-salt hypertensive rats. *Biomed. Pharmacother.* **2018**, *100*, 147–155. [CrossRef] [PubMed]

28. Wang, X.M.; Xi, Y.; Zeng, X.Q.; Zhao, H.D.; Cao, J.K.; Jiang, W.B. Effects of chlorogenic acid against aluminium neurotoxicity in ICR mice through chelation and antioxidant actions. *J. Funct. Food* **2018**, *40*, 365–376. [CrossRef]

29. Agunloye, O.M.; Oboh, G.; Ademiluyi, A.O.; Ademosun, A.O.; Akindahunsi, A.A.; Oyagbemi, A.A.; Omobowale, T.O.; Ajibade, T.O.; Adedapo, A.A. Cardio-protective and antioxidant properties of caffeic acid and chlorogenic acid: Mechanistic role of angiotensin converting enzyme, cholinesterase and arginase activities in cyclosporine induced hypertensive rats. *Biomed. Pharmacother.* **2019**, *109*, 450–458. [CrossRef] [PubMed]

30. Fu, L.; Xu, B.T.; Xu, X.R.; Gan, R.Y.; Zhang, Y.; Xia, E.Q.; Li, H.B. Antioxidant capacities and total phenolic contents of 62 fruits. *Food Chem.* **2011**, *129*, 345–350. [CrossRef]

31. Alarcon-Flores, M.I.; Romero-Gonzalez, R.; Vidal, J.; Frenich, A.G. Evaluation of the presence of phenolic compounds in different varieties of apple by Ultra-High-Performance liquid chromatography coupled to tandem mass spectrometry. *Food Anal. Meth.* **2015**, *8*, 696–709. [CrossRef]

32. Valles, B.S.; Bedrinana, R.P.; Tascon, N.F.; Garcia, A.G.; Madrera, R.R. Analytical differentiation of cider inoculated with yeast (*Saccharomyces cerevisiae*) isolated from Asturian (Spain) apple juice. *LWT-Food Sci. Technol.* **2005**, *38*, 455–461. [CrossRef]

33. Raspor, P.; Goranovic, D. Biotechnological applications of acetic acid bacteria. *Crit. Rev. Biotechnol.* **2008**, *28*, 101–124. [CrossRef] [PubMed]

34. Xia, T.; Yao, J.H.; Zhang, J.; Duan, W.H.; Zhang, B.; Xie, X.L.; Xia, M.L.; Song, J.; Zheng, Y.; Wang, M. Evaluation of nutritional compositions, bioactive compounds, and antioxidant activities of Shanxi aged vinegars during the aging process. *J. Food Sci.* **2018**, *83*, 2638–2644. [CrossRef] [PubMed]

35. Zhang, Y.J.; Gan, R.Y.; Li, S.; Zhou, Y.; Li, A.N.; Xu, D.P.; Li, H.B. Antioxidant phytochemicals for the prevention and treatment of chronic diseases. *Molecules* **2015**, *20*, 21138–21156. [CrossRef] [PubMed]

36. Lin, J.T.; Liu, S.C.; Shen, Y.C.; Yang, D.J. Comparison of various preparation methods for determination of organic acids in fruit vinegars with a simple Ion-Exclusion liquid chromatography. *Food Anal. Meth.* **2011**, *4*, 531–539. [CrossRef]

© 2019 by the authors. Licensee MDPI, Basel, Switzerland. This article is an open access article distributed under the terms and conditions of the Creative Commons Attribution (CC BY) license (http://creativecommons.org/licenses/by/4.0/).

 *antioxidants*

Article

Antimicrobial and Antibiofilm Activity against *Staphylococcus aureus* of *Opuntia ficus-indica* (L.) Mill. Cladode Polyphenolic Extracts

Federica Blando [1], Rossella Russo [2], Carmine Negro [3], Luigi De Bellis [3] and Stefania Frassinetti [2,*]

[1] Institute of Sciences of Food Production (ISPA), National Research Council (CNR), Research Unit of Lecce, Via Prov. le Lecce-Monteroni, 73100 Lecce, Italy; federica.blando@ispa.cnr.it

[2] Institute of Biology and Agricultural Biotechnology (IBBA), National Research Council (CNR), Research Unit of Pisa, Via Moruzzi 1, 56124 Pisa, Italy; rossella.wilbur@gmail.com

[3] Department of Biological and Environmental Sciences and Technologies (DiSTeBA), Salento University, 73100 Lecce, Italy; carmine.negro@unisalento.it (C.N.); luigi.debellis@unisalento.it (L.D.B.)

* Correspondence: frassinetti@ibba.cnr.it; Tel.: +39-050-315-2704

Received: 26 March 2019; Accepted: 28 April 2019; Published: 2 May 2019

Abstract: Plant extracts are a rich source of natural compounds with antimicrobial properties, which are able to prevent, at some extent, the growth of foodborne pathogens. The aim of this study was to investigate the potential of polyphenolic extracts from cladodes of *Opuntia ficus-indica* (L.) Mill. to inhibit the growth of some enterobacteria and the biofilm formation by *Staphylococcus aureus*. *Opuntia ficus-indica* cladodes at two stages of development were analysed for total phenolic content and antioxidant activity by Oxygen Radical Absorbance Capacity (ORAC) and Trolox equivalent antioxidant capacity (TEAC) (in vitro assays) and by cellular antioxidant activity in red blood cells (CAA-RBC) (ex vivo assay). The Liquid Chromatography Time-of-Flight Mass Spectrometry (LC/MS–TOF) analysis of the polyphenolic extracts revealed high levels of piscidic acid, eucomic acid, isorhamnetin derivatives and rutin, particularly in the immature cladode extracts. *Opuntia* cladodes extracts showed a remarkable antioxidant activity (in vitro and ex vivo), a selective inhibition of the growth of Gram-positive bacteria, and an inhibition of *Staphylococcus aureus* biofilm formation. Our results suggest and confirm that *Opuntia ficus-indica* cladode extracts could be employed as functional food, due to the high polyphenolic content and antioxidant capacity, and used as natural additive for food process control and food safety.

Keywords: *Opuntia ficus-indica* cladode; flavonoid; antioxidant; antimicrobial activity; antibiofilm activity

---

1. Introduction

In recent years, there has been an increased interest in natural antimicrobials, especially those obtained from plants. Some plant species are a rich source of natural compounds with antimicrobial properties, which are able to prevent, at some extent, the growth of foodborne pathogens, and extend the shelf life of the food [1].

Pathogens involved in foodborne diseases or food processing plant contamination are often capable to adhere and form biofilms. These structures are organized communities of bacterial cells enclosed in a self-produced polymeric matrix, composed of polysaccharides, proteins and other organic components, adhering to inert or living surfaces [2]. It is known that bacteria within biofilms are more resistant to antibiotics and other chemical agents than planktonic cells in suspension and their increased tolerance towards antimicrobial agents reduces the effectiveness to the treatment of biofilm-related infections [3,4]. Bacterial biofilms spread widely and play important roles in many industrial activities.

In the dairy industry or other food processing industries or food-contact surfaces, biofilm formation is a potential source of contamination and can lead to serious hygiene problems and economic losses [4].

Staphylococcus aureus is a well-known pathogen living as biofilm in a wide variety of environments and represents a severe risk of food contamination. It is has been found frequently on surfaces of food processing plants and it is responsible for infections related to consumption of fresh and processed foods [5].

Phenolic compounds occurring in vegetable foods and medicinal plants have been extensively investigated against a wide range of microorganisms. Several studies demonstrated the antimicrobial activity of dietary polyphenols, their activity on bacterial growth being mainly related to the strain, the polyphenol structure, and the dosage assayed [6–8]. Plant extracts, rich in polyphenols, has been reported to inhibit the biofilm formation by *Staphylococcus aureus*, including methicillin-resistant *Staphylococcus aureus* MRSA, *Escherichia coli* and *Pseudomonas aeruginosa* [9,10].

Opuntia ficus-indica (L.) Mill. (cactus pear) is a succulent species native from America, afterwards domesticated in other countries, occupying arid and semiarid zones. The first economic importance of *Opuntia ficus-indica* relies on the production of edible fruits, consumed fresh or transformed. Cladodes (succulent stem), known as pads or nopals, are also consumed at young stage in Mexico and United States in several different food preparation, or at older stage as forage, when there is shortage of fresh forage due to droughts [11,12].

The aims of this study were to investigate the potential of polyphenolic extracts from cladodes of *Opuntia ficus-indica* to inhibit the growth of some enterobacteria and the biofilm formation by *Staphylococcus aureus*. Moreover, cladode extracts were chemically and biochemically characterized, as high performance liquid chromatography time-of-flight mass spectrometry (HPLC/MS-TOF) profile and antioxidant capacity on different in vitro and ex vivo systems.

2. Materials and Methods

2.1. Chemicals and Reagents

Reagents were purchased from various suppliers as follows: authentic standards of rutin, isoquercitrin, isorhamnetin, isorhanmetin 3-*O*-glucoside, narcissin, kaempferol (Extrasynthèse, Genay, France); FCR (Folin-Ciocalteu's reagent), gallic acid, *p*-hydroxybenzoic acid, sodium carbonate, sodium hydroxide, 6-hydroxy-2,5,7,8-tetramethylchroman-2-carboxylic acid (Trolox), 2,2′-Azino-bis(3-ethylbenzothiazoline-6-sulfonic acid) diammonium salt (ABTS), 2,2′-Azobis (2-methylpropionamidine) dihydrochloride (AAPH), fluorescein sodium salt and 2′,7′-dichlorofluorescein diacetate (DCFH-DA), as well as acetonitrile (HPLC grade), ethanol, methanol, formic acid (Sigma-Aldrich, St. Louis, MO, USA). In all experiments, Milli-Q (Merck Millipore, Darmstadt, Germany) water was used.

2.2. Plant Material and Extraction

Opuntia ficus-indica cladodes were collected from plants producing purple fruits ('*Rossa*' variety), in the Salento countryside (Apulia Region, Italy, N 40°21′18″; E 17°59′47″).

Cladodes were collected from the same plants at two developmental stages: immature (from spring shoots, fully developed cladodes at 3 weeks of development) and mature (from fall shoots, old cladodes at around 24 weeks).

Cladodes were cleaned from spines and glochids, cut in small pieces, processed in a Waring blender with liquid nitrogen, freeze-dried using a Freezone® 2.5 model 76530 lyophilizer (Labconco Corp., Kansas City, MO, USA) for 48 h and stored at −20 °C. This product was defined as the dry weight (DW) of cladode. Extraction was done in triplicate from 500 mg (DW) macerated with 25 mL aqueous methanol (80%) overnight at 4 °C. After centrifugation (4000× *g*), the supernatant was recovered and evaporated in vacuo at 32 °C using a R-205 Büchi rotavapor (Büchi Labortechnik AG, Essen, Germany), then re-suspended in distilled water to a concentration of 50 mg/mL. Extracts were then filtered on a 0.45-μm CA syringe filter (Filtres Fioroni, Ingré, France) and portioned in 1-mL aliquot, stored at −20 °C until analysis.

2.3. Identification and Quantification of Phenolic Compounds

The identification and quantification of phenolic compounds in cladode extracts was performed in triplicate (an analysis for each of the three extractions/replicas) using an Agilent 1200 Liquid Chromatography system (Agilent Technologies, Palo Alto, CA, USA) and the chromatographic conditions and column were the same already reported by Sabella et al. [13]. The identification of phenolic compounds was confirmed by a TOF LC/MS system (Agilent 6320, Agilent Technologies, Palo Alto, CA, USA), equipped with a dual ESI interface operating in negative ion mode [13].

The identified phenolic compounds were quantified by the external standard method using a six-point calibration curve of *p*-hydroxybenzoic acid (0.5–100 mg/L), rutin (0.5–50 mg/L), isorhamnetin (0.5–50 mg/L), and narcissin (0.5–50 mg/L).

2.4. In Vitro Antioxidant Activity

The Folin–Ciocalteu reducing capacity assay, the antioxidant activity by Oxygen Radical Absorbance Capacity (ORAC) and Trolox equivalent antioxidant capacity (TEAC) were evaluated in cladode extracts, as described in [14]. A rapid microplate methodology, using a microplate reader (Victor^{TM} X3, Perkin Elmer, Waltham, MA, USA) and 96-well plates (Costar, 96-well black round bottom plate, Corning) were used. All experiments were performed in triplicate, and two independent assays were performed for each sample.

2.5. Ex Vivo Antioxidant Activity

2.5.1. Cellular Antioxidant Activity (CAA-RBC) Assay in Red Blood Cells

The antioxidant activity of *Opuntia* cladode extracts was evaluated in an ex vivo erythrocytes system as described in Frassinetti et al. [15] as well as erythrocytes preparation. Trolox was used as a standard and the fluorescence was read at 485 nm excitation and 535 nm emission by using a Victor^{TM} X3 microplate reader. Human erythrocytes from random subjects were exposed to a peroxyl radical generator, the AAPH, following one hour pre-treatment with 500 and 1000 μg/mL of cladode extracts. Each value was express according to the formula:

$$\text{CAA unit} = 100 - \left(\int_{SA} \Big/ \int_{CA} \right) \times 100$$

where $\int_{SA}$ is the integrated area of the sample curve and $\int_{CA}$ is the integrated area of the control curve [16].

2.5.2. Erythrocytes Oxidative Hemolysis

Hemolysis of human erythrocytes was induced by thermal decomposition of AAPH as previously described [15]. Briefly, erythrocytes were incubated with Trolox 500 μM (antioxidant standard) and cladode extracts at 37 °C for 1 h, followed by incubation with 50 mM AAPH at 37 °C for 4 h. The erythrocytes hemolysis was evaluated by spectrophotometric reading (λ = 540 nm) of the hemoglobin released in the supernatant, and expressed as percentage compared to the control (AAPH-treated erythrocytes).

2.6. Antimicrobial Activity

2.6.1. Bacterial Media

Nutrient Broth (NB), Nutrient Agar (NA), Mueller Hinton Broth (MHB), Mueller Hinton Agar (MHA), Mc Farland standard 0.5, Tripticase Soy Broth (TSB), were purchased from Oxoid (Basingstone, UK).

2.6.2. Bacterial Strains and Growth Conditions

The pathogenic bacterial strains were obtained from the American Type Culture Collection (ATCC). Gram-negative bacteria [*Escherichia coli* (ATCC 25922), *Salmonella enterica* ser. *Typhimurium* (ATCC 14028), and *Enterobacter aerogenes* (ATCC 13048)], and Gram-positive bacteria [*Enterococcus faecalis* (ATCC 29212) and *Staphylococcus aureus* (ATCC 25923)] were employed to test the antimicrobial activities of cladode extracts. All bacteria strains were grown on NB and MHB and incubated overnight at 37 °C under aerobic conditions.

2.6.3. Inhibition Assay–MINIMUM Inhibitory Concentration (MIC)

The minimal inhibitory concentration (MIC) against selected bacteria was determined according to [7] with some modifications. Cladode extracts were diluted in sterile water to the concentration of 2000 μg/mL; then, further dilutions were made up to the concentration of 50 μg/mL.

Tested pathogenic microorganisms were cultured in MHB at 37 °C for 16 h. Initially, the cultures were diluted to match the turbidity of 0.5 McFarland standard; thereafter, further dilutions with sterile MHB made it possible to obtain a suspension of about $1–5 \times 10^5$ CFU/mL. Aliquot of bacterial suspensions (50 μL) were added to a sterile 96-well plate containing 100 μL of MHB and 100 μL of cladode extract dilutions. A positive control (without cladode extract) was included on each microplate.

The plates were incubated in aerobic conditions at 37 °C for 24 h. A microplate reader (Eti-System Fast Reader Sorin Biomedica, Modena, Italy) was used to record the optical density (OD) at 600 nm. The MIC was defined as the lowest concentration of cladode extract able to inhibit the microorganism's growth.

2.7. Biofilm Production and Inhibition Assay

The biofilm production was determined using the method described by [2] with some modifications. The assay was performed using two *Staphylococcus* strains: the biofilm producer *Staphylococcus aureus* (ATCC 35556) and *Staphylococcus epidermidis* (ATCC 12228), used as a negative control, since it does not produce biofilm.

The strains were activated by culturing in 5 mL of TSB at 37 °C for 24 h. After 24 h, the optical density (OD) was measured at 600 nm and appropriate dilutions were made in TSB + 1% sucrose, to obtain an optical density of 0.1 corresponding to about 10^6 cells/mL. The assay was performed in sterile 96-well polystyrene plate (Greiner Bio-One Gmbh, Austria). Briefly, 100 μL *Staphylococcus aureus* cells were inoculated and cultured with or without 100 μL of cladodes extract (at concentrations ranging from 50 to 1500 μg/mL), without shaking at 37 °C. After 24 h incubation, non-adherent cells were removed by dipping each sample three times in sterile PBS. Samples were fixed at 60 °C for 1 h and the biofilms were stained with 0.1% solution of crystal violet in water, according to [17]. After staining, samples were washed thrice with distilled water. The quantitative analysis of biofilm production was performed by adding 125 μL of 30% acetic acid to de-stain the samples. Afterwards, the OD at 492 nm was detected using the microplate reader. The percentage of biofilm inhibition was determined by the formula:

$$Biofilm\ reduction\ \% = \frac{OD\ control\ -\ OD\ sample}{OD\ control} \times 100\%$$

2.8. Statistical Analysis

Statistical analysis has been carried out using GraphPad Prism, version 6.00 for Windows (GraphPad software, San Diego, CA, USA). Results were expressed as mean values ± standard deviation (SD) of analysis of data deriving from extractions and assays made in triplicate. Differences between samples were analyzed by one-way analysis of variance (ANOVA) with Bonferroni's multiple comparison test; p-value lower than 0.05 was considered statistically significant.

3. Results and Discussion

3.1. Polyphenolic Composition of Opuntia ficus-indica Cladode Extracts

The polyphenolic extracts from immature and mature cladodes revealed a predominant presence of piscidic and eucomic acids. These phenolic acids were identified upon their retention time, UV-Visible spectra and fragmentation pattern; also their mass spectral characteristics (by TOF LC/MS) were compared with the available literature [18,19] (Figure 1).

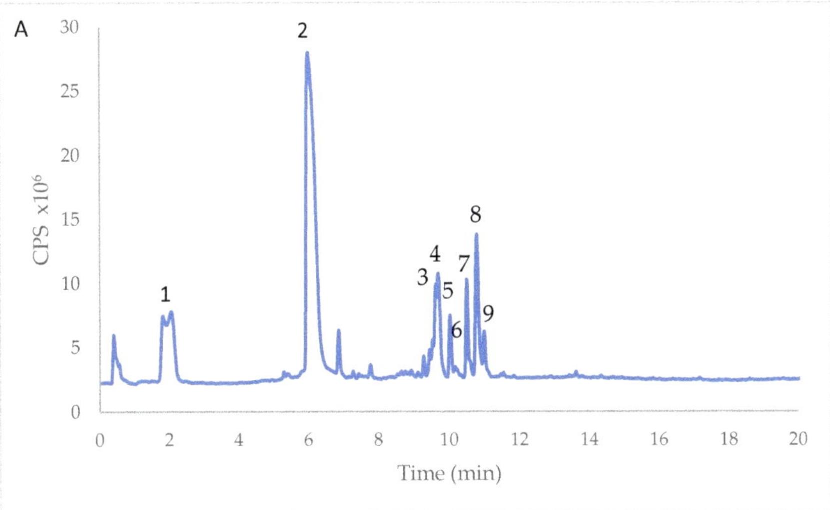

Figure 1. High performance liquid chromatography time-of-flight mass spectrometry (HPLC/MS-TOF) chromatogram of *Opuntia ficus-indica* extract from immature cladodes (*imm*). The chromatographic profile of mature cladode (*m*) extract was similar, except for isorhamnetin glucoside (peak 9), only present in *imm*. The numbers indicating peaks refer to the identified compounds reported in Table 1.

The presence of piscidic acid and eucomic acid in *Opuntia* cladode has been reported [18,20]; others reported the occurrence of hydroxybenzoic acids derivatives (2- or 4- or 3,4-dihydroxy benzoic acids) [21].

In the mature cladodes, we found a higher content of both piscidic and eucomic acids in respect to the value reported (on a similar development stage) by [18], threefold and tenfold, respectively (Table 2).

Table 1. Identification of phenolic acids and flavonols in *Opuntia ficus-indica* extracts (from *imm* and *m*) by means of High performance liquid chromatography time-of-flight mass spectrometry HPLC/TOF $(M - H)^{-}$.

Peak	RT	Formula $[M - H]^{-}$	m/z Exp	m/z Calc	Δ (ppm)	Compound Name	Reference
1	1.988	$C_{11}H_{11}O_7$	255.0512	255.0510	−0.67	Piscidic acid	Ginestra et al., 2009 [18]
2	6.170	$C_{11}H_{11}O_6$	239.0567	239.0561	−2.52	Eucomic acid	Ginestra et al., 2009 [18]
3	9.628	$C_{34}H_{41}O_{20}$	769.6770	769.6773	−0.39	Isorhamnetin rhamnosyl rutinoside	Santos-Zea et al., 2011 [19]
4	9.695	$C_{33}H_{39}O_{20}$	755.6512	755.6508	0.53	Isorhamnetin glucosyl rhamnosyl pentoside	Santos-Zea et al., 2011 [19]
5	10.022	$C_{27}H_{29}O_{16}$	609.1459	609.1461	0.34	Rutin	Ginestra et al., 2009 [18]
6	10.034	$C_{27}H_{29}O_{16}$	609,1470	609,1461	−1.43	Isorhamnetin sambubioside	Santos-Zea et al., 2011 [19]
7	10.503	$C_{27}H_{29}O_{15}$	593.1517	593.1512	−0.88	Kampherol glucosyl rhamnoside	Santos-Zea et al., 2011 [19]
8	10.785	$C_{28}H_{31}O_{16}$	623.1645	623.1618	−4.37	Narcissin (isorhamnetin rutinoside)	Santos-Zea et al., 2011 [19]
9	10.987	$C_{22}H_{21}O_{12}$	477.1045	477.1038	−1.42	Isorhamnetin glucoside	Ginestra et al., 2009 [18]

Table 2. Phenolic acids and flavonols content of *Opuntia ficus-indica* extracts, from immature and mature cladodes. Phenolic acids were quantified using the calibration curve of *p*-hydroxybenzoic acid. Flavonols were determined using calibration curves with the closest appropriate standard (rutin, isorhamnetin glucoside or isorhamnetin rutinoside). Mean values ($n = 3$) ± standard deviation (SD) are expressed as dry weight (DW) or fresh weight (FW). The same letters (*a–h*) in the column indicate that mean values are not significantly different ($p < 0.05$).

Immature Stage	mg/g DW	mg/100g FW
(1) Piscidic acid	1.984 ± 0.019 *c*	14.681 ± 0.146 *c*
(2) Eucomic acid	13.506 ± 0.143 *a*	99.944 ± 1.062 *a*
(4) Isorhamnetin rhamnosyl rutinoside	0.411 ± 0.003 *d*	3.041 ± 0.027 *e*
(5) Isorhamnetin glucosyl rhamnosyl pentoside	0.296 ± 0.004 *e*	2.192 ± 0.033 *d*
(6) Rutin	2.030 ± 0.023 *c*	15.022 ± 0.171 *c*
(8) Narcissin (isorhamnetin rutinoside)	3.188 ± 0.042 *b*	23.598 ± 0.302 *b*
(9) Isorhamnetin glucoside	0.465 ± 0.021 *d*	3.444 ± 0.161 *e*
Mature Stage	**mg/g DW**	**mg/100g FW**
(1) Piscidic acid	3.281 ± 0.032 *b*	28.257 ± 0.279 *b*
(2) Eucomic acid	1.616 ± 0.02 *f*	13.917 ± 0.174 *f*
(4) Isorhamnetin rhamnosyl rutinoside	0.187 ± 0.012 *e*	1.618 ± 0.104 *d*
(5) Isorhamnetin glucosyl rhamnosyl pentoside	0.266 ± 0.007 *e*	2.293 ± 0.067 *d*
(6) Rutin	0.752 ± 0.004 *g*	6.474 ± 0.034 *g*
(8) Narcissin (isorhamnetin rutinoside)	1.160 ± 0.029 *f*	9.985 ± 0.25 *h*
(9) Isorhamnetin glucoside	ND	ND

In the immature cladodes, we found a much higher value, as sum of piscidic and eucomic acids, even if the ratio is inverted: in immature cladode the predominant form is eucomic acid, instead in mature cladode piscidic acid is predominant (Table 2). Anyway, the amount of piscidic and eucomic acids found in *O. ficus-indica* cladode, particularly at immature stage, is exceptionally high.

Flavonol compounds (mainly isorhamnetin derivatives and rutin) were also identified (Figure 1 and Table 1). Isorhamnetin derivatives and rutin have been already described to be present in cladode extracts of different *Opuntia* species [18–23]. In those papers, isorhamnetin was reported as predominant core aglycone for flavonoids present in *Opuntia* sp., with low occurrence of quercetin and kaempferol. We confirmed these findings (Figure 1 and Table 2).

The content of isorhamnetin glucoside in our immature sample was similar to that one reported by Guevara-Figueroa and co-workers [21], in a wild variety (Amarillo) of *Opuntia ficus-indica*, instead our reported content of narcissin was much higher, as well as the rutin content (Table 2). In a mature cladode sample, the content of isorhamnetin derivatives (totaling 1.76 mg/g DW) was similar to that one reported in the above study (1.69 mg/g DW) [21]. The total flavonols content in our immature samples (6.78 mg/g DW) was in agreement with the content (8.82 mg/g DW) reported in young cladodes of *Opuntia ficus-indica* [23]. In mature sample, the total flavonols content (totalling 2.51 mg/g DW) was again in agreement with [21] who found ~3.5 mg /g DW.

Isorhamnetin derivatives, piscidic and eucomic acids present in such a high amount in cladode of *Opuntia ficus-indica* indicate that nopal can be considered a promising plant for the development of polyphenol-based commercial products. The effect of these phenolic compounds has been evaluated in vitro for contrasting hypercholesterolemia [24], and against UVA-induced oxidative stress on human keratinocytes [25], suggesting a possible pharmaceutical use of cladode extracts.

3.2. In Vitro Antioxidant Activity

Cladode extracts analysis revealed a much higher total phenol content, as well as antioxidant capacity, at immature stage of development, than mature cladode (Table 3). Therefore, the habit of Mexican population to eat 'nopalitos' (that is young prickly pads of 3–4 weeks of age) relays on a scientific basis of higher health benefit.

Table 3. Total phenols and antioxidant capacity, expressed as Oxygen Radical Absorbance Capacity (ORAC) and Trolox equivalent antioxidant capacity (TEAC), of *Opuntia ficus-indica* extracts, from immature and mature cladodes. Total phenols are expressed as gallic acid equivalents (GAE), TEAC and ORAC assays are expressed as Trolox Equivalent (TE). Mean values ($n = 11$) ± standard deviation (SD) are expressed on a dry weight (DW) or fresh weight (FW) basis. The same letters (a,b) in the same column indicate that mean values are not significantly different ($p < 0.05$).

	Total Phenols		TEAC		ORAC	
	mg GAE/g DW	mg GAE/100 g FW	µmol TE/g DW	µmol TE/100 g FW	mmol TE/g DW	mmol TE/100 g FW
Immature stage	54.02 ± 1.62 *a*	399.61 ± 11.98 *a*	12.55 ± 1.14 *a*	92.87 ± 8.51 *a*	0.88 ± 0.08 *a*	6.52 ± 0.56 *a*
Mature stage	18.63 ± 0.94 *b*	160.14 ± 8.09 *b*	8.23 ± 0.72 *b*	70.85 ± 6.22 *b*	0.29 ± 0.04 *b*	2.47 ± 0.36 *b*

Total phenol content (Table 3) was higher than that one reported in some studies [21,26] but in agreement with [23] (for immature cladode), and similar to that one reported in mature cactus pads from Tenerife [27] (for mature cladode); in fact, the phenol level results higher in immature cladodes if compared with mature, e.g., about 4 g/kg in our immature cladode samples from Apulia, and 2.6 g/kg in mature cladodes as reported by Rocchetti et al. for Sicilian *Opuntia ficus-indica* [26].

The antioxidant capacity measured by ABTS assay resulted higher for immature cladode than mature (12.5 vs 8.2 µmol TE/g DW, respectively) (Table 3). Only one study dealt with TEAC of cladode extracts reporting a several-fold higher value [28]. The ORAC values for our immature and mature cladode (Table 3) agreed with the literature: ~770 µmol TE/g DW [28]; a range from 260–380 µmol TE/g

DW, in young pads [19]; 657 µmol TE/g DW [29]. The ORAC value for immature cladode extracts was comparable to the value reported for blueberry; instead, that one for mature cladode was comparable to strawberry/raspberry values [30].

Therefore, from the experiments assessing the antioxidant capacity of cladodes extracts, we can conclude that both immature and mature cladodes are a good source of antioxidant functional compounds.

3.3. Ex Vivo Antioxidant Activity

3.3.1. Cellular Antioxidant Activity Assay in Red Blood Cells (CAA-RBC)

As shown in Figure 2A, cladodes pre-treated erythrocytes exhibited a significantly higher cellular antioxidant activity (CAA unit = 30 ± 1) compared to untreated cells (control, CAA = 0; $p < 0.01$), but lower than Trolox (CAA unit = 45 ± 0.79) used as standard. The cellular antioxidant activity of mature and immature cladodes was similar for both samples (30 ± 1 and 29 ± 0.9, respectively).

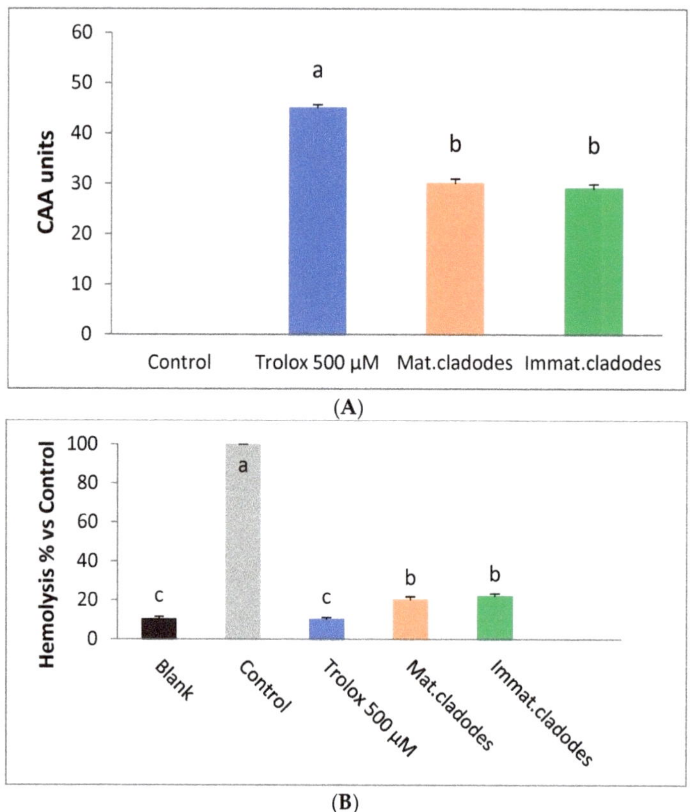

Figure 2. (**A**) Effects of *Opuntia ficus-indica* cladode extracts on the cellular antioxidant activity (CAA) in human erythrocytes. Trolox was used as reference standard. (**B**) Effects of *Opuntia ficus-indica* cladodes extracts on 2,2′-Azobis (2-methylpropionamidine) dihydrochloride (AAPH)-induced hemolysis in human erythrocytes. Trolox was used as reference standard. Assays were carried out in triplicate and the results were expressed as mean values ± SD. Different letters indicate significant differences ($p \leq 0.05$). One-way ANOVA with Bonferroni's multiple comparison test.

3.3.2. Haemolysis Assay

Since erythrocyte is a unique cell model with simple metabolism and sensitivity to oxidation, oxidative haemolysis is widely used to evaluate antioxidant activity. Therefore, cladode extracts were tested on human erythrocytes to counteract the oxidative hemolysis induced by peroxyl radicals produced by AAPH thermal decomposition. The anti-hemolytic activity of cladode extracts was compared with that from erythrocytes exposed to AAPH alone (control). As shown in Figure 2B both *Opuntia* cladodes extracts exhibited a strong anti-hemolytic effect (about 80%). This effect was statistically significant ($p \leq 0.001$) compared to the control (that induced 100% of hemolysis) but lower than the Trolox (about 90% hemolysis inhibition). No statistically significant differences were found between mature and immature cladodes anti-hemolytic effects.

3.4. Antimicrobial Activity

The antimicrobial activity against selected enteric bacterial strains was measured evaluating the O.D. at 600 nm in the presence of increasing doses of cladode extract. Different dilutions of methanol did not affect the bacterial growth (data not shown).

Gram-negative microorganisms (*Escherichia coli* ATCC 25922, *Salmonella typhimurium* ATCC 14028 and *Enterobacter aerogenes* ATCC 13048) were inhibited at a concentration of 2000 µg/mL of mature cladode extract, whereas immature cladode extract was more effective inhibiting at a concentration of 1500 µg/mL (Table 4). The two Gram-positive bacteria (*Staphylococcus. aureus* ATCC 25923 and *Enterococcus. faecalis* ATCC 29212) showed MIC values of 1500 µg/mL for mature and 1000 for immature cladode extract (Table 4). The MIC against planktonic cells of the biofilm producer *Staphylococcus aureus* ATCC 35556 was 1000 and 700 µg/mL for mature and immature extracts, respectively (Table 4).

Table 4. Minimal inhibitory concentrations (MIC) of *Opuntia ficus-indica* mature (*m*) and immature cladode extracts (*imm*) against selected bacterial strains.

Strain		MIC (*m*) µg/mL	MIC (*imm*) µg/mL
Escherichia coli	ATCC 25922	2000	1500
Salmonella typhimurium	ATCC 14028	2000	1500
Enterobacter aerogenes	ATCC 13048	2000	1500
Enterococcus faecalis	ATCC 29212	1500	1000
Staphylococcus aureus	ATCC 25923	1500	1000
Staphylococcus aureus	ATCC 35556	1000	700

Our results agree with several other studies showing that the inhibitory effect of phenolic compounds from natural extracts are more effective to Gram-positive than Gram-negative bacteria [7]. The susceptibility of bacteria to drugs depends on the characteristics of the drug (hydrophobicity or hydrosolubility) and on the microbial membrane composition [31].

The antimicrobial activity of plant phenolics has been extensively investigated against many different microorganisms [6]. We previously reported the antimicrobial activity of bergamot whole-fruit powder against potentially pathogenic bacteria [32]. In general, phenolic compounds are involved in membrane damage, protein and cell wall binding and enzyme inactivation; they can act as pro-oxidants, leading to damage of DNA, lipids and other biological molecules [33].

Few authors have reported the antimicrobial activity of *Opuntia ficus-indica*. Ginestra and co-workers [18] reported that different phytochemical fractions of *Opuntia ficus-indica* did not show antimicrobial activity against the tested bacterial strains. On the other hand, it has been reported the antimicrobial activity of alcoholic and aqueous extracts of *Opuntia* cladodes against *Vibrio cholerae* and *Proteus mirabilis* [34]. Moreover, other authors [35] described the antimicrobial activity of *Opuntia* cladodes against *Escherichia coli* and *Staphylococcus aureus*, with a minimum bactericidal concentration (MBC) of 4 mg/mL and 1 mg/mL, respectively.

The antimicrobial activity of *Opuntia* cladodes extracts may be related to its high content of polyphenols, especially isorhamnetin that has been already reported exerting antimicrobial activity [36].

3.5. Biofilm Inhibition

The effects of cladode extracts on biofilm formation by *Staphylococcus aureus* ATCC 35556, a strong biofilm producer, was also investigated.

As shown in Figure 3, both cladode extracts (mature and immature) at concentration of 500 µg/mL did not inhibited the biofilm formation. At 1000 µg/mL, only the immature cladode extract inhibited significantly the biofilm formation with an inhibition rate of 80%. At 1500 µg/mL concentration, both extracts significantly inhibited the biofilm formation, with an inhibition rate of 71% for the mature cladode and 85% for the immature cladode.

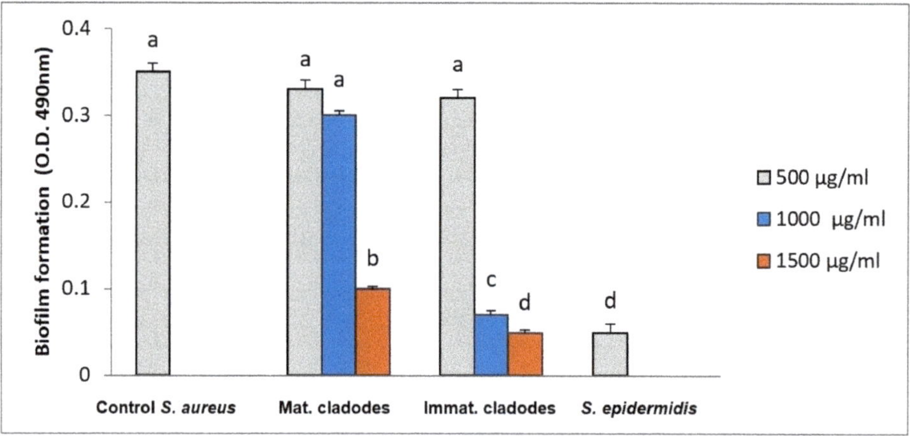

Figure 3. Effects of different concentrations of *Opuntia ficus-indica* cladode extracts on biofilm formation by *Staphylococcus aureus* ATCC 35556. *Staphylococcus epidermidis* ATCC 12228 was used as negative control. Assays were carried out in triplicate and the results were expressed as mean values ± SD. Different letters indicate significant differences ($p \leq 0.05$). One-way ANOVA with Bonferroni's multiple comparison test.

The inhibition of biofilm development at concentration of *Opuntia* cladode extracts higher than MIC demonstrated that the bacterial cells in a biofilm are more resistant to antimicrobial agents compared to the planktonic cells, which is a well-known feature. Our results concerning the inhibition of *Staphylococcus aureus* MRSA biofilm formation agree with previous works [10,37]. In our study, the inhibition of biofilm could be explained by the presence of flavonoids in cladodes extracts. In fact, flavonoids such as quercetin, apigenin, luteolin and rutin were found to be effective in the inhibition of *Staphylococcus aureus* biofilm [38,39].

4. Conclusions

The results obtained in this work indicate that cladodes of *Opuntia ficus-indica* are rich in phenolic compounds, particularly *p*-hydroxybenzoic acid derivatives, and possess in vitro antioxidant activity which results higher in the immature cladodes. The extracts exhibited in vivo antioxidant capacity with a strong anti-hemolytic effect. A selective inhibition against potentially pathogenic bacteria and a significant inhibition of the biofilm formation were shown by the cladodes at both developmental stages. To our knowledge, this is the first report on the characterization of *Opuntia ficus-indica* cladodes extracts for their antimicrobial and anti-biofilm properties. In conclusion, our results indicate that *Opuntia ficus-indica* cladode extracts, especially the immature ones, could be used for food purpose as functional food (being comparable to the antioxidant value of blueberries), as well as for food process control and food safety as recently proposed by Rocchetti et al. [26].

Author Contributions: Conceptualization, methodology, investigation, writing—original draft preparation: F.B. and S.F.; methodology, investigation, formal analysis: R.R., C.N.; writing—review and editing: L.D.B.

Funding: This research received no external funding

Conflicts of Interest: The authors declare no conflicts of interest.

References

1. Gyawali, R.; Hayek, S.A.; Ibrahim, S.A. Plant extracts as antimicrobials. In *Handbook of Natural Antimicrobials for Food Safety and Quality*; Taylor, T.M., Ed.; Elsevier: Cambridge, UK, 2015; pp. 31–47.
2. Di Ciccio, P.; Vergara, A.; Festino, A.R.; Paludi, D.; Zanardi, E.; Ghidini, S.; Ianieri, A. Biofilm formation by *Staphilococcus aureus* on food contact surfaces: Relationship with temperature and cell surface hydrophobicity. *Food Control* **2015**, *50*, 930–936. [CrossRef]
3. Stewart, P.S.; Costerton, J.W. Antibiotic resistance of bacteria in biofilms. *Lancet* **2001**, *358*, 135–138. [CrossRef]
4. Satpathy, S.; Sen, S.K.; Pattanaik, S.; Raut, S. Review on bacterial biofilm: An universal cause of contamination. *Biocatal. Agric. Biotechnol.* **2016**, *7*, 56–66. [CrossRef]
5. Doulgeraki, A.I.; Di Ciccio, P.; Ianieri, A.; Nychas, G.E. Methicillin-resistant food-related *Staphylococcus aureus*: A review of current knowledge and biofilm formation for future studies and applications. *Res. Microbiol.* **2017**, *168*, 1–15. [CrossRef]
6. Daglia, M. Polyphenols as antimicrobial agents. *Curr. Opin. Biotechnol.* **2012**, *23*, 174–181. [CrossRef] [PubMed]
7. Delgado Adámez, J.; Gamero Samino, E.; Valdés Sánchez, E.; González-Gómez, D. In vitro estimation of the antibacterial activity and antioxidant capacity of aqueous extracts from grape seeds (*Vitis vinifera* L.). *Food Control* **2012**, *24*, 136–141. [CrossRef]
8. Chibane, L.B.; Degraeve, P.; Ferhout, H.; Bouajila, J.; Oulahal, N. Plant antimicrobial polyphenols as potential natural food preservatives. *J. Sci. Food Agric.* **2018**, *99*, 1457–1474. [CrossRef] [PubMed]
9. Nostro, A.; Guerrini, A.; Marino, A.; Tacchini, M.; Di Giulio, M.; Grandini, A.; Saraçoğlu, H.T. In vitro activity of plant extracts against biofilm-producing food-related bacteria. *Int. J. Food Microbiol.* **2016**, *238*, 33–39. [CrossRef]
10. Brambilla, L.Z.S.; Endo, E.H.; Cortez, D.A.G.; Dias Filho, B.P. Anti-biofilm activity against *Staphylococcus aureus* MRSA and MSSA of neolignans and extract of *Piper regnellii*. *Rev. Bras. Farmacognosia* **2017**, *27*, 112–117. [CrossRef]
11. Stintzing, F.C.; Schieber, A.; Carle, R. Phytochemical and nutritional significance of cactus pear. *Eur. Food Res. Technol.* **2001**, *212*, 396–407. [CrossRef]
12. Hernández-Urbiola, M.I.; Pérez-Torrero, E.; Rodríguez-Garcia, M.E. Chemical analysis of nutritional content of prickly pads (*Opuntia ficus-indica* L.) at varied ages in an organic harvest. *Int. J. Environ. Res. Public Health* **2011**, *8*, 1287–1295.
13. Sabella, E.; Luvisi, A.; Aprile, A.; Negro, C.; Vergine, M.; Nicolì, F.; Miceli, A.; De Bellis, L. *Xylella fastidiosa* induces differential expression of lignification related-genes and lignin accumulation in tolerant olive trees cv. Leccino. *J. Plant Physiol.* **2018**, *220*, 60–68. [CrossRef]
14. Albano, C.; Negro, C.; Tommasi, N.; Gerardi, C.; Mita, G.; Miceli, A.; De Bellis, L.; Blando, F. Betalains, phenols and antioxidant capacity in cactus pear [*Opuntia ficus-indica* (L.) Mill.] fruits from Apulia (South Italy) genotypes. *Antioxidants* **2015**, *4*, 269–280. [CrossRef]
15. Frassinetti, S.; Gabriele, M.; Caltavuturo, L.; Longo, V.; Pucci, L. Antimutagenic and antioxidant activity of a selected lectin-free common bean (*Phaseolus vulgaris* L.) in two cell-based models. *Plant Foods Hum. Nutr.* **2015**, *70*, 35–41. [CrossRef] [PubMed]
16. Wolfe, K.L.; Liu, R.H. Cellular antioxidant activity (CAA) assay for assessing antioxidants, foods, and dietary supplements. *J. Agric. Food Chem.* **2007**, *55*, 8896–8907. [CrossRef] [PubMed]
17. O'Toole, G.A. Microtiter dish biofilm formation assay. *J. Vis. Exp.* **2011**, *47*. [CrossRef]

18. Ginestra, G.; Parker, M.L.; Bennet, R.N.; Robertson, J.; Mandalari, G.; Narbad, A.; Lo Curto, R.B.; Bisignano, G.; Faulds, C.B.; Waldron, K.W. Anatomical, Chemical, and Biochemical Characterization of Cladodes from Prickly Pear [*Opuntia ficus-indica* (L.) Mill.]. *J. Agric. Food Chem.* **2009**, *57*, 10323–10330. [CrossRef]

19. Santos-Zea, L.; Guitiérrez-Uribe, J.A.; Serna-Saldivar, O. Comparative analysis of total phenols, antioxidant activity, and flavonol glycoside profile of cladode flours from different varieties of *Opuntia* spp. *J. Agric. Food Chem.* **2011**, *59*, 7054–7061. [CrossRef] [PubMed]

20. Di Lorenzo, F.; Silipo, A.; Molinaro, A.; Parrilli, M.; Schiraldi, C.; D'Agostino, A.; Izzo, E.; Rizza, L.; Bonina, A.; Bonina, F.; Lanzetta, R. The polysaccharide and low molecular weight components of *Opuntia ficus indica* cladodes: Structure and skin repairing properties. *Carbohydr. Polym.* **2017**, *157*, 128–136. [CrossRef]

21. Guevara-Figueroa, T.; Jiménez-Islas, H.; Reyes-Escogido, M.L.; Mortensen, A.G.; Laursen, B.B.; Lin, L.-W.; Barba de la Rosa, A.P. Proximate composition, phenolic acids, and flavonoids characterization of commercial and wild nopal (*Opuntia* spp.). *J. Food Compos. Anal.* **2010**, *23*, 525–532. [CrossRef]

22. Ressaissi, A.; Attia, N.; Falé, P.L.; Pacheco, R.; Teixeira, V.H.; Machuqueiro, M.; Borges, C.; Serralheiro, M.L.M. Aqueous extracts from nopal (*Opuntia ficus-indica*): Antiacetylcholinesterase and antioxidant activity from phenolic bioactive compounds. *Int. J. Green Herb. Chem.* **2016**, *5*, 337–348.

23. Medina-Torres, L.; Vernon-Carter, E.J.; Gallegos-Infante, J.A.; Rocha-Guzman, N.E.; Herrera-Valencia, E.E.; Calderas, F.; Jiménez-Alvarado, R. Study of the antioxidant properties of extracts obtained from nopal cactus (*Opuntia ficus-indica* L.) cladodes after convective drying. *J. Sci. Food Agric.* **2011**, *91*, 1001–1005. [CrossRef] [PubMed]

24. Ressaissi, A.; Attia, N.; Falé, P.L.; Pacheco, R.; Victor, B.L.; Machuqueiro, M.; Borges, C.; Serralheiro, M.L.M. Isorhamnetin derivatives and piscidic acid for hypercholesterolemia: Cholesterol permeability, HMG-CoA reductase inhibition, and docking studies. *Arch. Pharmacal Res.* **2017**, *40*, 1278–1286. [CrossRef]

25. Petruk, G.; Di Lorenzo, F.; Imbimbo, P.; Silipo, A.; Bonina, A.; Rizza, L.; Piccoli, R.; Monti, D.M.; Lanzetta, R. Protective effect of *Opuntia ficus-indica* L. cladodes against UVA-induced oxidative stress in normal human keratinocytes. *Bioorg. Med. Chem. Lett.* **2017**, *27*, 5485–5489. [CrossRef]

26. Rocchetti, G.; Pellizzoni, M.; Montesano, D.; Lucini, L. Italian *Opuntia ficus-indica* Cladodes as Rich Source of Bioactive Compounds with Health-Promoting Properties. *Foods* **2018**, *7*, 24. [CrossRef]

27. Pérez Méndez, L.; Tejera Flores, F.; Darias Martín, J.; Rodríguez Rodríguez, E.M.; Díaz Romero, C. Physicochemical characterization of cactus pads from *Opuntia dillenii* and *Opuntia ficus indica*. *Food Chem.* **2015**, *188*, 393–398. [CrossRef] [PubMed]

28. Corral-Aguayo, R.D.; Yahia, E.M.; Carrillo-Lopez, A.; González-Aguilar, G. Correlation between some nutritional components and the total antioxidant capacity measured with six different assays in eight horticultural crops. *J. Agric. Food Chem.* **2008**, *56*, 10498–10504. [CrossRef]

29. Avila-Nava, A.; Calderón-Oliver, M.; Medina-Campos, O.N.; Zou, T.; Gu, L.; Torres, N.; Pedraza-Chaverri, J. Extract of cactus (*Opuntia ficus indica*) cladodes scavenge reactive oxygen species in vitro and enhances plasma antioxidant capacity in humans. *J. Funct. Foods* **2014**, *10*, 13–24. [CrossRef]

30. Wu, X.; Beecher, G.R.; Holden, J.M.; Haytowitz, D.B.; Gebhardt, S.E.; Prior, R.L. Lipophilic and hydrophilic antioxidant capacities of common foods in the United States. *J. Agric. Food Chem.* **2004**, *52*, 4026–4037. [CrossRef] [PubMed]

31. de Almeida, C.G.; Garbois, G.D.; Amaral, L.M.; Diniz, C.C.; Le Hyaric, M. Relationship between structure and antibacterial activity of lipophilic N-acyldiamines. *Biomed. Pharmacother.* **2010**, *64*, 287–290. [CrossRef] [PubMed]

32. Gabriele, M.; Frassinetti, S.; Caltavuturo, L.; Montero, L.; Dinelli, G.; Longo, V.; Pucci, L. *Citrus bergamia* powder: Antioxidant, antimicrobial and anti-inflammatory properties. *J. Funct. Foods* **2017**, *31*, 255–265. [CrossRef]

33. Duda-Chodak, A.; Tarko, T.; Satora, P.; Sroka, P. Interaction of dietary compounds, especially polyphenols, with the intestinal microbiota: A review. *Eur. J. Nutr.* **2015**, *54*, 325–341. [CrossRef]

34. El-Mostafa, K.; El Karrassi, Y.; Badreddine, A.; Andreoletti, P.; Vamecq, J.; El Kebbai, M.S.; Latruffe, N.; Lizard, G.; Nasser, B.; Cherkaoui-Malki, M. Nopal cactus (*Opuntia ficus-indica*) as a source of bioactive compounds for nutrition, health and disease. *Molecules* **2014**, *19*, 14879–14901. [CrossRef] [PubMed]

35. Sánchez, E.; Rivas Morales, C.; Castillo, S.; Leos-Rivas, C.; García-Becerra, L.; Ortiz Martínez, D.M. Antibacterial and Antibiofilm Activity of Methanolic Plant Extracts against Nosocomial Microorganisms. *Evid. Based Complement. Altern. Med.* **2016**, *2016*, 1–8. [CrossRef]

36. Bhattacharya, D.; Koley, H. Antibacterial activity of polyphenolic fraction of kombucha against enteric bacterial pathogens. *Curr. Microbiol.* **2016**, *73*, 885–896. [CrossRef]
37. Bazargani, M.; Rohloff, J. Antibiofilm activity of essential oils and plant extracts against *Staphylococcus aureus* and *Escherichia coli* biofilms. *Food Control* **2016**, *61*, 156–164. [CrossRef]
38. Slobodníková, L.; Fialová, S.; Rendeková, K.; Kováč, J.; & Mučaji, P. Antibiofilm activity of plant polyphenols. *Molecules* **2016**, *21*, 1717. [CrossRef]
39. Al-Shabib, N.A.; Husain, F.M.; Ahmad, I.; Khan, M.S.; Khan, R.A.; Khan, J.M. Rutin inhibits mono and multi-species biofilm formation by foodborne drug resistant *Escherichia coli* and *Staphylococcus aureus*. *Food Control* **2017**, *79*, 325–332. [CrossRef]

© 2019 by the authors. Licensee MDPI, Basel, Switzerland. This article is an open access article distributed under the terms and conditions of the Creative Commons Attribution (CC BY) license (http://creativecommons.org/licenses/by/4.0/).

Article

Optimization of Ultrasound-Assisted Extraction of Phenolic Compounds from Black Locust (*Robiniae Pseudoacaciae*) Flowers and Comparison with Conventional Methods

Ivana Savic Gajic [1,*], Ivan Savic [1], Ivana Boskov [1], Stanko Žerajić [1], Ivana Markovic [2] and Dragoljub Gajic [3]

[1] Faculty of Technology, University of Nis, Bulevar oslobodjenja 124, 16000 Leskovac, Serbia
[2] Technical Faculty, University of Belgrade, Vojske Jugoslavije 12, 19210 Bor, Serbia
[3] Gajic Associates, Doktora Pantica 77, 14000 Valjevo, Serbia
[*] Correspondence: savicivana@tf.ni.ac.rs; Tel.: +381-16-247-203

Received: 30 June 2019; Accepted: 25 July 2019; Published: 27 July 2019

Abstract: The aim of this study was to optimize the ultrasound-assisted extraction of phenolic compounds from black locust (*Robiniae pseudoacaciae*) flowers using central composite design. The ethanol concentration (33–67%), extraction temperature (33–67 °C), and extraction time (17–33 min) were analyzed as the factors that impact the total phenolic content. The liquid-to-solid ratio of $10 \ cm^3 \ g^{-1}$ was the same during extractions. The optimal conditions were found to be 59 °C, 60% (v/v) ethanol, and extraction time of 30 min. The total phenolic content (TPC = 3.12 g_{GAE} 100 g^{-1} dry plant material) and antioxidant activity (IC_{50} = 120.5 $\mu g \ cm^{-3}$) of the extract obtained by ultrasound-assisted extraction were compared with those obtained by maceration (TPC = 2.54 g_{GAE} 100 g^{-1} dry plant material, IC_{50} = 150.6 $\mu g \ cm^{-3}$) and Soxhlet extraction (TPC = 3.22 g_{GAE} 100 g^{-1} dry plant material, IC_{50} = 204.2 $\mu g \ cm^{-3}$). The ultrasound-assisted extraction gave higher total phenolic content and better antioxidant activity for shorter extraction time so that it represents the technique of choice for the extraction of phenolic compounds. The obtained extract, as the source of antioxidants, can be applied in the pharmaceutical and food industries.

Keywords: extraction; phenolic compounds; antioxidants; central composite design; optimization

---

1. Introduction

Black locust (*Robinia pseudoacacia*) is a deciduous tree originating from the Southeastern region of North America. Today, it can be found in all regions of the world with a moderate climate [1]. In Serbia, it is located in Vojvodina where it was used for bonding the "living" sand, as well as in the plains and valleys of northern Serbia. Black locust grows up to 20 m in height and 90 cm in tree diameter. Initially, its plantation was for ornamental purposes, but there are studies that mention the importance of a mixed plantation of this species, because it is a nitrogen fixing species that promotes a growing facilitation effect of other trees [2]. It has a smooth bark, thorny branches, and a rare crochet. The flowers are a cluster with characteristically white color, pleasant and intense aroma, and sweet taste.

Black locust flowers are a source of vitamin C (40 mg 100 g^{-1}), resin, tannins [3], essential oil, monoterpenes (cis-β-ocimene ~26.6%, (E)-α-bergamotene ~8.9%, pinens, β-pinens, limonene), diterpene, triterpene, terpene alcohols (linalool ~33.1%), and small amount of poisonous robinin (kaempferol 3,7-di-O-glucoside) that completely disappears after thermal treatment and drying [4–7]. The content of unsaturated fatty acids in the flowers is about seven times higher than the content that of saturated fatty acids [8]. Secundiflorol, mucronulatol, isomucronulatol, and isovestitol were identified in ethanol extracts of the whole plant [9]. The presence of flavonoid glycosides, such as flavonol

3,7-di-O-glycosides, were confirmed in the black locust flower [6]. Stefova et al. [10] reported that the quercetin content is low while the kaempferol content is high in the flower extract. Sarikurkcu et al. [8] extracted the bioactive compounds from black locust flowers by the Soxhlet extractor using ethyl acetate, acetone, and methanol, and prepared the aqueous extract by maceration. The total phenolic content (TPC) and total flavonoid content (TFC) were higher in the acetone and methanol extracts compared to the aqueous extract. In order to investigate the impact of drying methods on the antioxidant properties and TPC, the ethanol extracts were prepared from the flowers previously treated by sun drying, hot-air drying, freeze-drying, and microwave-vacuum drying [11]. The highest TPC was determined in the samples treated using the freeze-drying method.

Medical use of black locust is limited only to the flower, because the other parts, especially bark, are toxic due to the high concentration of robinin [6]. It has long been used as a folk drug for alleviating the colds and coughs, stomach cramps, rheumatic pains, migraine, fever, and skin diseases [11,12]. These flowers reduce the level of cholesterol, and have diuretic and laxative effects. Phenolic compounds from the black locust flower have antioxidant [8,11,13], antimicrobial [13–15], and anticancer properties [10]. The techniques used for the extraction of phenolic compounds from black locust flowers are summarized in Table 1.

Table 1. The experimental conditions for extraction of phenolic compounds from black locust flowers previously reported in the literature.

Extraction Technique	Solvent	Temperature (°C)	Extraction Time (min)	Optimization of Factors	Ref.
Magnetic stirrer	Ethanol	Room	720	No	[11]
Reflux	Acetone	-	30	No	[10]
Maceration	Methanol	Room	-	No	[6]
Maceration	Water	100	15	No	[8]
Soxhlet	Ethyl acetate, acetone, methanol	-	300	No	[8]
Soxhlet	Ethanol	-	-	No	[14]
Microwave	Ethanol	-	2	Box–Behnken design	[16]

Because of the relatively long extraction time, large solvent consumption, higher cost of equipment, and degradation of compounds, the reflux, maceration, and Soxhlet extractions tend to be replaced by more environment-friendly extraction techniques. Ultrasound-assisted extraction (UAE) has become a good alternative method for extracting phenolic compounds since it can offer a high reproducibility for shorter extraction times and reduced consumption of the solvent [17,18]. The UAE is commonly carried out at lower temperatures so that the thermal degradation of bioactive compounds in the extract can be prevented.

TPC is related to the antioxidants, mainly phenolics in this case, in the extracts. The extracts' phenolic composition is conditioned by liquid–solid ratio, extraction time, extraction temperature, particle size, and any other extraction variables [19]. The eco-friendly and nontoxic organic solvents are recommended by the US Food and Drug Administration for the extraction of bioactive compounds from plant materials [20]. Water and ethanol were proved as the extraction solvents of choice to reach good yields of phenolic compounds from plant materials [21,22] due to their nontoxicity. Having that in mind, the ethanolic solutions were used to extract the phenolic compounds from black locust flowers.

The one-variable-at-a-time (OVAT) approach analyzes the impact of one factor on the defined response, while all other factors are constant. In this way, the great number of experimental runs is necessary for optimizing the observed process. The interactions between the factors cannot be analyzed using the OVAT approach. The application of OVAT enables to obtain the local optimal extraction conditions, which do not correspond to the real (global) optimal conditions. In order to overcome these problems, the response surface methodology (RSM) is commonly used. This methodology predicts the system response, analyzes the interactions between the factors, defines the relationship between the response and factors, and optimizes the extraction conditions with a limited number of experiments [23,24]. Yang et al. [25] optimized the extraction parameters (extraction time, solvent concentration, and liquid-to-solid ratio) of UAE in terms of common bean (*Phaseolus vulgaris* L.) antioxidants using the two-level factorial design. Izadiyan and Hemmateenejad [26] performed multiresponse optimization of the factors (extraction temperature, solvent concentration, and extraction time) affecting the UAE of Iranian *Ocimum basilicum* using a central composite design (CCD). Yin et al. [27] applied the CCD to optimize the factors (extraction temperature, extraction time, and powder dosage) of the UAE of natural anthocyanin from purple sweet potato for silk fabric dyeing. Živković et al. [28] determined the optimal conditions for the UAE of phenolic compounds from pomegranate peel using CCD, and investigated the relationship between extraction time, solvent concentration, liquid-to-solid ratio, and extraction temperature.

Since the use of RSM has not been reported yet for modeling the UAE of phenolic compounds from black locust flowers, the aim of this study was to generate a polynomial equation that describes the extraction process. The impact of ethanol concentration, extraction temperature, and extraction time were estimated using the CCD. The optimal extraction conditions were optimized using a numerical optimization method. The identification and quantification of phenolic compounds were carried out using high-performance liquid chromatography (HPLC). The TPC and antioxidant activity of extract obtained using UAE were compared to maceration and Soxhlet extraction. The structural changes of plant materials after using these extraction techniques were observed by scanning electron microscopy (SEM).

2. Materials and Methods

2.1. Plant Materials

Black locust flowers (*Robiniae pseudoacaciae flos*) were purchased from Dr Josif Pancic (Belgrade, Serbia). The moisture content (10.6%, w/w) of the plant material was determined by measuring the weight before and after the drying at 105 °C in a hot air oven to a constant weight. The dried flowers were ground using an electric grinder to the particle size that passed through a 0.5 mm sieve.

2.2. Reagents

Ethanol (96%, v/v) was purchased from Zorka Pharma (Sabac, Serbia). Folin–Ciocalteu reagent, gallic acid (97%), and quercetin were purchased from Merck (Darmstadt, Germany). Rutin trihydrate (purity 97%) was purchased from Alfa Aesar (A Johnson Matthey Company, Heysham, United Kingdom). Epigallocatechin, and ferulic acid, 2,2-diphenyl-1-picrylhydrazyl (DPPH), formic acid (HPLC grade), and methanol (HPLC grade) were purchased from Sigma Aldrich (St. Louis, MO, USA).

2.3. Extraction of Phenolic Compounds from Black Locust Flowers

2.3.1. Ultrasound-Assisted Extraction

The extractions of phenolic compounds from black locust flowers were performed in an ultrasonic cleaning bath (Sonic, Nis, Serbia) with dimensions: $30 \times 15 \times 20$ cm. The bath was filled with distilled water up to one third of its total volume (about 3.0 dm^3). The operating frequency was 40 kHz, while the total power was 3×50 W. The powder of black locust flowers (2 g) was transferred into a flask of 100 cm^3 and extracted with 20 cm^3 of the ethanol solution at the defined temperature. The liquid-to-solid ratio of 10 cm^3 g^{-1} was the same during extractions, because the evaporation of the used solvents was prevented using the reflux condenser. After extraction, the flasks were immediately cooled to room temperature using chilled water. The extracts were separated from the solid matrix by vacuum filtration and subjected to further analysis. In order to determine the concentration of extracts, 1 cm^3 of the aliquot was dried at 105 °C in a laboratory oven.

2.3.2. Maceration

The powder of black locust flowers (2.0 g) was extracted with 20 cm^3 of 60% (v/v) ethanol at 25 °C for 24 h. The extract was further treated and analyzed as in the case of UAE.

2.3.3. Soxhlet Extraction

The extraction was carried out using 5.0 g of black locust flowers powder and 500 cm^3 of 60% (v/v) ethanol at boiling temperature for 6 h. The extract was subjected to the treatment and analysis as in the previous cases.

2.4. Total Phenolic Compounds

The TPC was determined according to the Folin–Ciocalteu colorimetric method [29]. Briefly, 0.1 cm^3 of the extract was mixed with 1 cm^3 of the Folin–Ciocalteu reagent previously diluted tenfold with distilled water and 1 cm^3 of sodium carbonate (7%, w/v). The TPC was expressed as gram of gallic acid equivalents (GAE) per 100 g dry plant material (d.p.m.). A series of methanolic solutions of gallic acid (0.005–0.300 mg cm^{-3}) was prepared by diluting the stock solution (1 mg cm^{-3}) to construct the calibration curve. Instead of sodium carbonate, an equivalent amount of distilled water was added to the blank solution. Absorbance of the samples was measured at 760 nm and room temperature after incubation of 90 min in relation to distilled water on the Varian Cary 100 spectrophotometer (Mulgrave, Victoria, Australia) in the quartz cuvettes (1 × 1 cm).

2.5. DPPH Assay

The methanolic solution of DPPH radicals (1 cm^3) prepared to the concentration of 3 × 10^{-4} mol dm^{-3} was added to 2.5 cm^3 of the analyzed extracts [30]. Furthermore, the blank solution composited of 1 cm^3 of methanol and 2.5 cm^3 of the extracts, as well as the control solution composited of 1 cm^3 of DPPH radicals and 2.5 cm^3 of methanol were prepared. Absorbance was measured at 517 nm in relation to methanol after incubation for 30 min at room temperature. The inhibition of DPPH radicals (%) was calculated as follows (Equation (1)):

$$I_{DPPH}(\%) = \frac{A_c - (A_s - A_B)}{A_c} \times 100 \tag{1}$$

where, I_{DPPH} is the inhibition of DPPH radicals expressed in %, A_s is the absorbance of the samples treated with DPPH solution, A_B is the absorbance of the blank, and A_C is the absorbance of control.

2.6. HPLC Analysis

The identification and quantification of the extract's phenolic compounds under optimal conditions were carried out using the previously described HPLC method [31]. The separation was achieved using a Zorbax Eclipse XDB-C$_{18}$ column (4.6 × 250 mm, 5 μm) (Agilent Technologies, Santa Clara, California, USA). Based on the standards of rutin, quercetin, gallic acid, epigallocatechin, and ferulic acid, the extracts' phenolic compounds were identified. The phenolic contents were expressed as mg 100 g^{-1} dry plant material.

2.7. Morphological Analysis of Black Locust Flowers

The morphology of black locust flowers before and after the UAE, maceration, and Soxhlet extraction was analyzed using the Vega-3 LMU Scanning Electron Microscope (Tescan, Brno, Czech Republic) under high vacuum conditions. The SEM was used at an accelerating voltage at 20 kV and a magnification of 2000 × (10 μm). The plant material was mounted on a metal grid with double-sided adhesive tape, and then the bead surfaces were coated with chrome under vacuum.

2.8. Experimental Design

The CCD with three factors was deployed to determine the optimal extraction conditions for phenolic compounds. The ethanol concentration (%, X_1), extraction temperature (°C, X_2) and extraction time (min, X_3) were investigated at the five levels (-1.68, -1, 0, $+1$, $+1.68$). The levels of coded and actual factors are shown in Table 2. The UAE is commonly performed at the lower temperatures (20–70 °C) compared to conventional extraction procedures [32]. This method is desirable for the extraction of thermosensitive phenolic compounds from various plant species. For this reason, the temperature was observed in the range of 33–67 °C. The experimental design included the data set that belongs to the points of factorial design (2^3), axial points (2×3), and central point (4). The central point was repeated four times to determine the statistical parameters of the proposed model. Due to statistical calculations, the factors X_i were coded as x_i according to Equation (2):

$$x_i = \frac{X_i - X_0}{\delta X} \tag{2}$$

where X_0 is the value of X_i at the central point and δX is the step change. The Design Expert 12.0.0 (Stat Ease, Minneapolis, MN, USA) software was used to obtain the analysis of variance (ANOVA), regression coefficients, and regression equation. The data of response were fitted using a second-order polynomial equation (Equation (3)):

$$Y = \beta_0 + \beta_1 x_1 + \beta_2 x_2 + \beta_3 x_3 + \beta_{11} x_1^2 + \beta_{22} x_2^2 + \beta_{33} x_3^2 + \beta_{12} x_1 x_2 + \beta_{13} x_1 x_3 + \beta_{23} x_2 x_3 + \varepsilon \tag{3}$$

where Y is the predicted response; β_0 is the intercept; β_1, β_2, and β_3 are the linear coefficients of x_1, x_2, x_3, respectively; β_{11}, β_{22}, and β_{33} are the squared coefficients of x_1, x_2, and x_3, respectively; β_{12}, β_{13}, β_{23} are the coefficients of interaction between x_1 and x_2, x_1 and x_3, x_2 and x_3, respectively; ε is the residual.

Table 2. Experimental design space for the extraction of phenolic compounds from black locust flowers.

Factors	Coded	Levels				
		-1.68	-1	0	$+1$	$+1.68$
ethanol concentration [%]	x_1	33	40	50	60	67
extraction temperature [°C]	x_2	33	40	50	60	67
extraction time [min]	x_3	17	20	25	30	33

2.8.1. Statistical Analysis of the Regression Model

ANOVA with 95% confidence level was carried out to analyze the significance of the model and equation terms. The sum of squares (SS), degree of freedoms (df), mean squares (MS), F- and p-values were used as the statistical parameters. The statistical significance of the terms was analyzed based on p-value (Prob > F). The model terms are statistically significant, if the p-value is less than 0.0500. The coefficient of determination (R^2), adjusted correlation coefficient (Adj-R^2), and predicted correlation coefficient (Pred-R^2) were used to express the quality of the regression model. The model's significance was checked using an F-test.

2.8.2. Optimization of Phenolic Compounds' Extraction

The extraction was optimized using a numerical optimization method in order to maximize the yield of phenolic compounds. Before optimization, the weighted factor was assigned to 1. The weight is important to define the form of response desirability function. It is desirable to have the value in the range of 1–10. The higher value of the weight indicates the greater importance of the response. The importance of goal was adjusted at the default value of 3. This parameter can have a value between 1 (least important) and 5 (most important).

3. Results and Discussion

3.1. Modeling of Phenolic Compounds' Extraction Using CCD

Modeling of the extraction of phenolic compounds from black locust flowers was carried out according to the matrix of CCD with 18 experimental runs. The combinations of different factor levels and TPC are given in Table 3. The coded factors are also presented in parenthesis. TPC in the extracts was in the range 2.33–3.15 g_{GAE} 100 g^{-1} d.p.m.

Table 3. Matrix of central composite design for three factors with total phenolic content.

Standard Order	Ru Order	X_1, Ethanol Concentration [%]	X_2, Extraction Temperature [°C]	X_3, Extraction Time [min]	Y, Total Phenolic Content [g_{GAE} 100 g^{-1} d.p.m.] Experimental	Predicted
11	1	50 (0)	33 (−1.68)	25 (0)	2.71	2.72
8	2	60 (+1)	60 (+1)	30 (+1)	3.15	3.19
9	3	33 (−1.68)	50 (0)	25 (0)	2.65	2.73
10	4	67 (+1.68)	50 (0)	25 (0)	3.02	3.02
5	5	40 (−1)	40 (−1)	30 (+1)	2.58	2.55
17	6	50 (0)	50 (0)	25 (0)	2.85	2.88
16	7	50 (0)	50 (0)	25 (0)	2.91	2.88
3	8	40 (−1)	60 (+1)	20 (−1)	2.71	2.69
12	9	50 (0)	67 (+1.68)	25 (0)	3.12	3.04
7	10	40 (−1)	60 (+1)	30 (+1)	3.11	3.16
18	11	50 (0)	50 (0)	25 (0)	2.92	2.88
13	12	50 (0)	50 (0)	17 (−1.68)	2.33	2.35
6	13	60 (+1)	40 (−1)	30 (+1)	3.00	3.04
2	14	60 (+1)	40 (−1)	20 (−1)	2.81	2.78
15	15	50 (0)	50 (0)	25 (0)	2.91	2.88
4	16	60 (+1)	60 (+1)	20 (−1)	2.48	2.54
14	17	50 (0)	50 (0)	33 (+1.68)	3.02	2.97
1	18	40 (−1)	40 (−1)	20 (−1)	2.48	2.46

The ANOVA results at 95% confidence level are depicted in Table 4. The significance of terms in the second order polynomial equation was estimated using ANOVA. The model's F-value of 34.64 was higher than the critical value of 3.39 so that the model can be considered as statistically significant. Since the lack-of-fit F-value of 4.3 was lower than the critical value of 9.01, the lack-of-fit can be considered as not statistically significant relative to the pure error (0.0031). There is a 12.96% chance that the lack-of-fit F-value this large could occur due to noise. The interaction between ethanol concentration and extraction temperature, as well as the quadratic terms of ethanol concentration and extraction temperature were not statistically significant terms. The statistically significant model F-value and not significant lack-of-fit F-value indicate the adequacy of the proposed model [33].

R^2 of 0.984 implies that 98.4% of the variation in the yield of phenolic compounds could be explained by the regression model (Table 4). Pred-R^2 of 0.821 was in reasonable agreement with the Adj-R^2 of 0.947, while Adj-R^2 was close to R^2. The coefficient of variation of 1.99% (C.V. < 10%) indicates the low deviation between the experimental and predicted values of the response, and the high degree of precision and reliability. Adequate precision of 19.8 indicates an adequate signal so that this model can be used to navigate the design space. This parameter is a measure of the signal to noise ratio, and it is desirable to have the value higher than 4 [34].

The polynomial model that describes the extraction process and represents the interaction between factors and response is presented in Table 5.

Table 4. Analysis of variance (ANOVA) of the quadratic response surface model.

Source	SS	df	MS	F-Value	p-Value Prob > F
Model	0.9795	9	0.1088	34.64	<0.0001 *
X_1	0.1023	1	0.1023	32.57	0.0005 *
X_2	0.1180	1	0.1180	37.56	0.0003 *
X_3	0.4652	1	0.4652	148.04	<0.0001 *
X_1X_2	0.1105	1	0.1105	35.15	0.0004 *
X_1X_3	0.0162	1	0.0162	5.16	0.0528 **
X_2X_3	0.0761	1	0.0761	24.20	0.0012 *
X_1^2	0.0078	1	0.0078	2.50	0.1529 **
X_2^2	0.0001	1	0.0001	0.05	0.8351 **
X_3^2	0.0839	1	0.0839	26.72	0.0009 *
Residual	0.0251	8	0.0031		
Lack-of-fit	0.0221	5	0.0044	4.30	0.1296 **
Pure error	0.0031	3	0.0010		
Corrected total	1.0046	17			
Std. Dev.	0.06	R^2	0.975		
Mean	2.82	Adj-R^2	0.947		
C.V. %	1.99	Pred-R^2	0.821		
PRESS	0.18	Adequate precision	19.8		

* statistically significant; ** not statistically significant.

Table 5. Estimated regression coefficients in the polynomial equation.

Coefficient	Parameter Estimate	Standard Error
Intercept	2.898	0.028
X_1	0.087	0.015
X_2	0.093	0.015
X_3	0.185	0.015
X_1X_2	−0.118	0.020
X_1X_3	0.045	0.020
X_2X_3	0.098	0.020
X_1^2	−0.025	0.016
X_2^2	0.003	0.016
X_3^2	−0.082	0.016

The not statistically significant terms could be excluded from the second order polynomial equation in order to improve the prediction ability of the proposed model. The regression coefficients indicate that the linear effects have a positive impact on the response. The quadratic effects of ethanol concentration and extraction time, as well as the interaction between ethanol concentration and extraction temperature have a negative impact on the response. The extraction time of the linear effects had the highest impact on TPC, followed by extraction temperature and ethanol concentration.

The adequacy of the model was also evaluated by the residuals, which represent the difference between the observed and predicted values of the response [35]. The residuals are thought of as the elements of variation unexplained by the regression model. The obtained residuals are plotted against the expected values in the normal probability plot (Figure 1). The obtained plots of the model after excluding nonstatistically significant terms indicate that the residuals are normally distributed. The slight deviation of points from the straight line in the reduced model indicates a better prediction of the regression model.

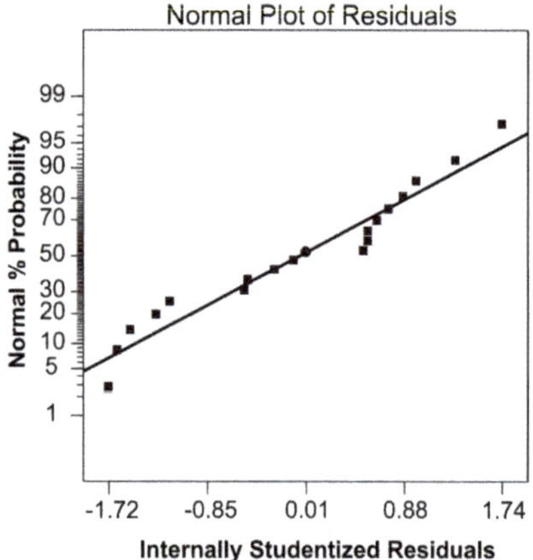

Figure 1. Normal probability plot of studentized residuals for the reduced polynomial model.

Cook's distances for the reduced polynomial model is depicted in Figure 2. Based on these values, the regression changes can be estimated when the case is deleted. Cook's distances were less than the limit of 1.0 so that there were no outliers in the given dataset.

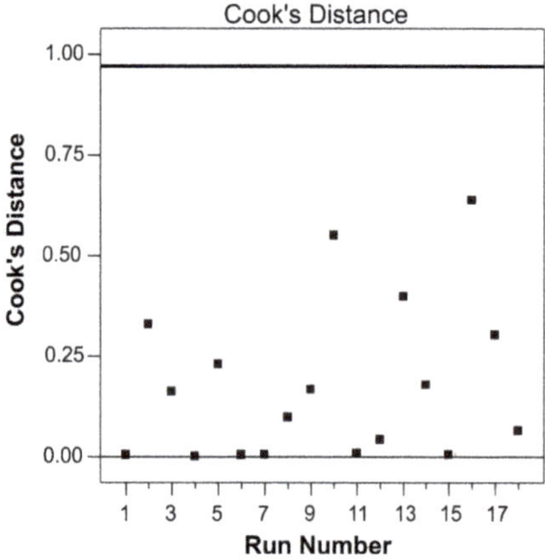

Figure 2. Cook's distance for the reduced polynomial model.

3.2. The Impacts of Factors on the Response Surface

Figure 3a illustrates the interaction between ethanol concentration and extraction temperature for extraction time of 25 min. The yield of phenolic compounds increased with increasing ethanol concentration [36]. This impact on TPC is more significant for shorter extraction times. By means of analyzing the response shape, it can be concluded that there is a strong interaction between these factors. The increase of extraction temperature leads to increased TPC [36], but only using lower ethanol concentrations. The interaction between ethanol concentration and extraction time at 50 °C is depicted in Figure 3b. The impact of ethanol concentration on the response is significant at longer extraction times, while the effect of ethanol concentration at the shorter extraction times is almost negligible. The increase of extraction time is also significant at higher ethanol concentration levels and has a positive impact on the TPC [33]. Saturation in the response is achieved after extraction time of 30 min. The impacts of extraction temperature and time using 50% (v/v) ethanol are presented in Figure 3c. The extraction time has a more pronounced impact at higher extraction temperatures.

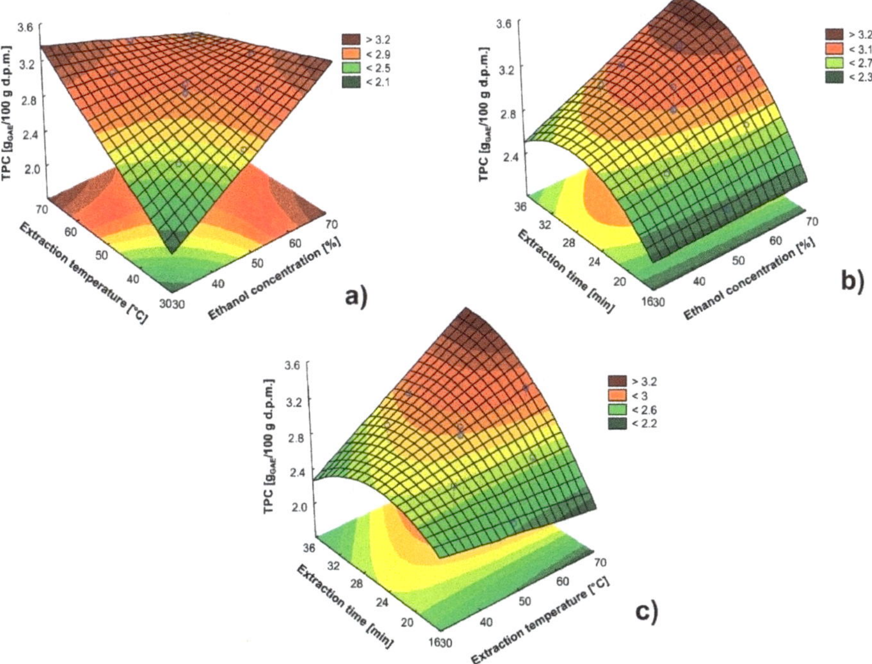

Figure 3. The impacts of: (**a**) ethanol concentration and extraction temperature for 25 min; (**b**) ethanol concentration and extraction time at 50 °C; and (**c**) extraction temperature and extraction time using 50% (v/v) ethanol on the TPC.

3.3. Optimization of the Extraction

The yield of phenolic compounds was maximized to obtain the optimal conditions for the extraction of these bioactive compounds [35]. Prior to applying the optimization method, the factor levels were ranged between −1 and +1. The optimal conditions were achieved for 60% (v/v) ethanol, 59 °C, and 30 min at the liquid-to-solid ratio of 10 cm^3 g^{-1}. The predicted TPC under these conditions was 3.17 g$_{GAE}$ 100 g^{-1} d.p.m., while the TPC was found to be 3.12 g$_{GAE}$ 100 g^{-1} d.p.m. Based on the good agreement between obtained and predicted TPCs, it can be concluded that the proposed model is adequate.

Sarikurkcu et al. [8] determined the TPC of 56.74 mg$_{GAE}$ g^{-1} acetone extract, 36.42 mg$_{GAE}$ g^{-1} methanol extract, and 27.17 mg$_{GAE}$ g^{-1} aqueous extract of black locust flowers. Ji et al. [11] found the highest TPCs of 47.30 mg$_{GAE}$ g^{-1} d.p.m. for the freeze-drying method, and the lowest TPC of 29.15 mg$_{GAE}$ g^{-1} d.p.m. for sun drying method. The results of TPC obtained in this paper are in accordance with available data for different extraction techniques. Unlike previous studies, the extraction time for UAE is shorter so that the energy-efficient procedure was developed. The reduction in energy consumption was achieved by applying the ultrasound and advanced mathematical approach compared to other available procedures.

3.4. HPLC Analysis

The identification and quantification of phenolic compounds in the extract of black locust flowers were carried out based on the retention times and UV spectra of the standards using the reversed-phase high-performance liquid chromatographic method with ultraviolet detection (RP-HPLC–UV) The contents of rutin (56.9 mg 100 g^{-1} d.p.m., R_t = 32.478 min, λ_{max} = 250 nm), epigallocatechin (10.10 mg 100 g^{-1} d.p.m., R_t = 18.180, λ_{max} = 250 nm), ferulic acid (6.76 mg 100 g^{-1} d.p.m., R_t = 30.789 min, λ_{max} = 320 nm), and quercetin (2.44 mg 100 g^{-1} d.p.m., R_t = 50.096, λ_{max} = 250 nm) were quantified. The lowest content of identified phenolic compounds was in the case of quercetin [10]. Veitch et al. [6] identified flavonol 3,7-di-O-glycosides, flavonoid robinin, glucosyl analogue of robinin, kaempferol, and isorhamnetin in methanolic extracts obtained by maceration of black locust flowers. Truchado et al. [4] also found robinin in nectar collected from black locust flowers (Bologna, Italy) using the HPLC–MS method. In addition to these studies, there are no available data related to the chromatographic analysis of given plant material.

3.5. Comparison of Ultrasound-Assisted Extraction with Maceration and Soxhlet Extraction

The TPC and half maximal inhibitory concentration (IC_{50}) were determined for the extracts obtained by UAE, maceration, and Soxhlet extraction to compare the efficiency of extraction techniques (Table 6). The results, which refer to the TPC, indicate that the yields of phenolic compounds are almost the same for the UAE and Soxhlet extractions. The UAE has been proven to be a more efficient and profitable extraction technique of phenolic compounds compared to the Soxhlet extraction, since the extraction time was significantly shorter [25,37]. This fact is the result of cavitation, which causes the enhancement of mass transfer of bioactive compounds through the destroyed cell walls.

Table 6. The comparison of UAE with maceration and Soxhlet extraction.

Extraction Technique	Extraction Temperature [°C]	Ethanol Concentration [%]	Extraction Time [h]	TPC [g_{GAE} 100 g^{-1} d.p.m.]	IC_{50} [µg cm^{-3}]
UAE	59	60	0.5	3.12	120.5
Maceration	25	60	24	2.54	150.6
Soxhlet Extraction	90	60	6	3.22	204.2

The extract obtained by UAE gave better antioxidant activity than those obtained by Soxhlet extraction and maceration. The higher temperatures used in the Soxhlet extraction can probably cause the degradation of thermolabile phenolic compounds, leading to weak antioxidant activity. In the literature, there are data indicating that the methanolic extract of black locust flowers has the highest antioxidant activity (471.75 mg Trolox g^{-1} extract) compared to the extracts obtained using ethyl acetate, acetone, and water [8]. Ji et al. [11] determined that the ethanolic extracts of black locust flowers have the highest antioxidant activity previously dried by lyophilization.

The results of this study are hard to compare with available data since the extracts were obtained using different solvents and assay for the determination of antioxidant activity. The obtained black locust extracts were represented as the main source of phenolic compounds with expressed antioxidant activity that can be used in the pharmaceutical and cosmetic industries.

3.6. Morphological Analysis

SEM analysis was used to reveal the change in the plant material after application of different extraction techniques. The untreated flowers showed closed cells and rough surfaces (Figure 4a).

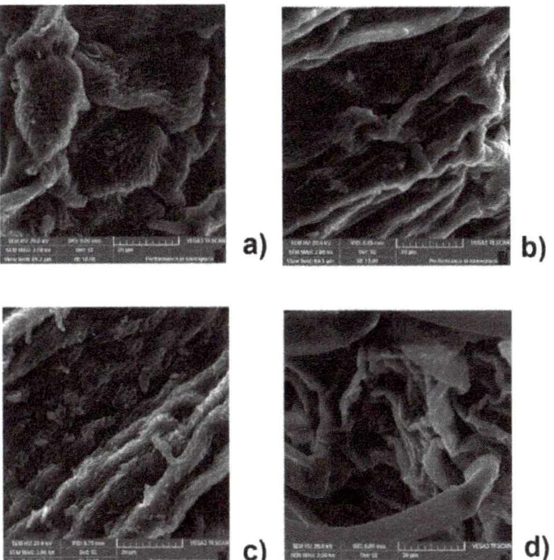

Figure 4. SEM (scanning electron microscopy) micrographs of black locust flowers: (**a**) without treatment; (**b**) after maceration; (**c**) after Soxhlet extraction, and (**d**) after the UAE (Ultrasound-assisted extraction).

In the samples subjected to the extraction, the physical modification of cell walls with different degradation degrees can be noticed. The SEM micrographs of the sample after maceration (Figure 4b) were not considerably different from those of untreated samples, but only puny damage was observed on the external surface of the flower. In Soxhlet extraction (Figure 4c), the profuse rupture of the cellular matrix and small broken pieces were dispersed due to the treatment with high temperature. The surface of the flower was greatly destroyed after treatment with the UAE (Figure 4d). This indicates that ultrasound treatment affects the structure of the cell due to the high, localized pressures induced by cavities [38,39], which should result in the instantaneous release of the soluble compounds into the solvent. Unlike this mechanism of UAE, the maceration depends on the permeation and solubilization in order to bring the compound out of the cell [31,40].

4. Conclusions

The second-order polynomial model provided adequate mathematical description of the UAE of phenolic compounds from black locust flowers. ANOVA analysis confirmed that extraction time has the highest impact on the TPC, followed by the extraction temperature and ethanol concentration. The optimal extraction conditions were 60% (v/v) ethanol, 59 °C, and 30 min at the liquid-to-solid ratio of $10 \text{ cm}^3 \text{ g}^{-1}$. Under these conditions, the TPC was found to be 3.12 g_{GAE} 100 g^{-1} d.p.m. Compared to the maceration and Soxhlet extraction, the UAE has a higher yield of phenolic compounds for shorter extraction time and better antioxidant activity. Therefore, the UAE represents an effective, reliable, and feasible technique for preparing extracts enriched with phenolic compounds, such as rutin quantified by HPLC analysis. The structural changes in the flowers obtained by the different techniques were evaluated using SEM analysis. The micrographs indicate that the different action mechanisms are characteristic for the used extraction techniques. Further studies will be directed to more detailed identification and quantification of other phenolic compounds available in the extracts obtained under optimal conditions and to investigate their biological activity.

Author Contributions: I.S. and I.S.G., conceived, designed, and performed the experiments. D.G., S.Ž., I.B., and I.M., participated in the data analysis and contributed intellectually to the manuscript.

Funding: This work was supported by the Ministry of Education, Science and Technological Development of the Republic of Serbia under the project TR-34012. Ivana Boskov is the recipient of a scholarship from the Ministry of Education, Science and Technological Development of the Republic of Serbia.

Conflicts of Interest: The authors declare no conflict of interest.

References

1. Lavin, M.; Sousa, M. Phylogenetic systematics and biogeography of the tribe *Robinieae* (Leguminosae). *Syst. Bot. Monogr.* **1995**, *45*, 1–165. [CrossRef]

2. Katiki, L.M.; Ferreira, J.F.; Gonzalez, J.M.; Zajac, A.M.; Lindsay, D.S.; Chagas, A.C.S.; Amarante, A.F. Anthelmintic effect of plant extracts containing condensed and hydrolyzable tannins on *Caenorhabditis elegans*, and their antioxidant capacity. *Vet. Parasitol.* **2013**, *192*, 218–227. [CrossRef] [PubMed]

3. Buzhdygan, O.Y.; Rudenko, S.S.; Kazanci, C.; Patten, B.C. Effect of invasive black locust (*Robinia pseudoacacia* L.) on nitrogen cycle in floodplain ecosystem. *Ecol. Model.* **2016**, *319*, 170–177. [CrossRef]

4. Truchado, P.; Ferreres, F.; Bortolotti, L.; Sabatini, A.G.; Tomaas-Barberaan, F.A. Nectar flavonol rhamnosides are floral markers of acacia (*Robinia pseudacacia*) honey. *J. Agric. Food Chem.* **2008**, *56*, 8815–8824. [CrossRef] [PubMed]

5. Wang, X.Y.; Tang, L.; Zhao, L.; Luan, Y.N.; Zhang, Z.Y. Determination of polyphenols in flowers of *R. pseudoacacia* L. by Folin-Ciocaileu method. *Food Drug* **2010**, *12*, 332–334.

6. Veitch, N.C.; Elliott, P.C.; Kite, G.C.; Lewis, G.P. Flavonoid glycosides of the black locust tree, *Robinia pseudoacacia* (Leguminosae). *Phytochemistry* **2010**, *71*, 479–486. [CrossRef] [PubMed]

7. Xie, J.; Sun, B.; Yu, M. Constituents of top fragrance from fresh flowers of *Robinia pseudoacacia* L. occurring in China. *Flavour Fragr. J.* **2006**, *21*, 798–800. [CrossRef]

8. Sarikurkcu, C.; Kocak, M.S.; Tepe, B.; Uren, M.C. An alternative antioxidative and enzyme inhibitory agent from Turkey: *Robinia pseudoacacia* L. *Ind. Crops Prod.* **2015**, *78*, 110–115. [CrossRef]

9. Tian, F.; McLaughlin, J.L. Bioactive flavonoids from the black locust tree, *Robinia pseudoacacia*. *Pharm. Biol.* **2000**, *38*, 229–234. [CrossRef]

10. Stefova, M.; Kulevanova, S.; Stafilov, T. Assay of flavonols and quantification of quercetin in medicinal plants by HPLC with UV-diode array detection. *J. Liq. Chromatogr. Relat. Technol.* **2001**, *24*, 2283–2292. [CrossRef]

11. Ji, H.F.; Du, A.L.; Zhang, L.W.; Xu, C.Y.; Yang, M.D.; Li, F.F. Effects of drying methods on antioxidant properties in *Robinia pseudoacacia* L. flowers. *J. Med. Plants Res.* **2012**, *6*, 3233–3239.

12. Funk, D.T.; Roach, B.A. *Black Locust, a Bibliography*; US Deparment of Agriculture, Forest Service, Central States Forest Experiment Station: Columbus, OH, USA, 1961.

13. Marinas, I.C.; Oprea, E.; Geana, E.I.; Chifiriuc, C.; Lazar, V. Antimicrobial and antioxidant activity of the vegetative and reproductive organs of *Robinia pseudoacacia*. *J. Serb. Chem. Soc.* **2014**, *79*, 1363–1378. [CrossRef]

14. Rosu, A.F.; Bita, A.; Calina, D.; Rosu, L.; Zlatian, O.; Calina, V. Synergic antifungal and antibacterial activity of alcoholic extract of the species *Robinia pseudoacacia* L. (Fabaceae). *Eur. J. Hosp. Pharm. Sci. Pract.* **2012**, *19*, 216. [CrossRef]

15. Bhalla, P.; Bajpai, V.K. Chemical composition and antibacterial action of *Robinia pseudoacacia* L. flower essential oil on membrane permeability of foodborne pathogens. *J. Essent. Oil Bear. Plants* **2017**, *20*, 632–645. [CrossRef]

16. Xiang, C.G.; Xiang, N.; Huang, C.L.; Sun, H.B.; Li, W.F. Optimization of microwave extraction of flavonoids from *Robinia pseudoacacia* L. flowers using response surface methodology. *Food Sci.* **2011**, *22*, 006.

17. He, B.; Zhang, L.L.; Yue, X.Y.; Liang, J.; Jiang, J.; Gao, X.L.; Yue, P.X. Optimization of ultrasound-assisted extraction of phenolic compounds and anthocyanins from blueberry (*Vaccinium ashei*) wine pomace. *Food Chem.* **2016**, *204*, 70–76. [CrossRef] [PubMed]

18. Martinez-Guerra, E.; Gude, V.G.; Mondala, A.; Holmes, W.; Hernandez, R. Microwave and ultrasound enhanced extractive-transesterification of algal lipids. *Appl. Energy* **2014**, *129*, 354–363. [CrossRef]

19. Mokrani, A.; Madani, K. Effect of solvent, time and temperature on the extraction of phenolic compounds and antioxidant capacity of peach (*Prunus persica* L.) fruit. *Sep. Purif. Technol.* **2016**, *162*, 68–76. [CrossRef]

20. Bartnik, D.D.; Mohler, C.M.; Houlihan, M. Methods for the Production of Food Grade Extracts. U.S. Patent 20060088627, 27 April 2006.
21. d'Alessandro, L.G.; Kriaa, K.; Nikov, I.; Dimitrov, K. Ultrasound assisted extraction of polyphenols from black chokeberry. *Sep. Purif. Technol.* **2012**, *93*, 42–47. [CrossRef]
22. Drosou, C.; Kyriakopoulou, K.; Bimpilas, A.; Tsimogiannis, D.; Krokida, M. A comparative study on different extraction techniques to recover red grape pomace polyphenols from vinification byproducts. *Ind. Crop. Prod.* **2015**, *75*, 141–149. [CrossRef]
23. Quiroz, J.Q.; Torres, A.C.; Ramirez, L.M.; Garcia, M.S.; Gomez, G.C.; Rojas, J. Optimization of the microwave-assisted extraction process of bioactive compounds from annatto seeds (*Bixa orellana* L.). *Antioxidants* **2019**, *8*, 37. [CrossRef] [PubMed]
24. Le, B.; Golokhvast, K.S.; Yang, S.H.; Sun, S. Optimization of microwave-assisted extraction of polysaccharides from *Ulva pertusa* and evaluation of their antioxidant activity. *Antioxidants* **2019**, *8*, 129. [CrossRef] [PubMed]
25. Yang, Q.Q.; Gan, R.Y.; Ge, Y.Y.; Zhang, D. Ultrasonic treatment increases extraction rate of common bean (*Phaseolus vulgaris* L.) antioxidants. *Antioxidants* **2019**, *8*, 83. [CrossRef] [PubMed]
26. Izadiyan, P.; Hemmateenejad, B. Multi-response optimization of factors affecting ultrasonic assisted extraction from Iranian basil using central composite design. *Food Chem.* **2016**, *190*, 864–870. [CrossRef] [PubMed]
27. Yin, Y.; Jia, J.; Wang, T.; Wang, C. Optimization of natural anthocyanin efficient extracting from purple sweet potato for silk fabric dyeing. *J. Clean. Prod.* **2017**, *149*, 673–679. [CrossRef]
28. Živković, J.; Šavikin, K.; Janković, T.; Ćujić, N.; Menković, N. Optimization of ultrasound-assisted extraction of polyphenolic compounds from pomegranate peel using response surface methodology. *Sep. Purif. Technol.* **2018**, *194*, 40–47. [CrossRef]
29. Singleton, V.L.; Orthofer, R.; Lamuela-Raventós, R.M. Analysis of total phenols and other oxidation substrates and antioxidants by means of Folin-Ciocalteu reagent. *Methods Enzymol.* **1999**, *299*, 152–178.
30. Molyneux, P. The use of the stable free radical diphenylpicrylhydrazyl (DPPH) for estimating antioxidant activity. *Songklanakarin J. Sci. Technol.* **2004**, *26*, 211–219.
31. Xu, D.P.; Zheng, J.; Zhou, Y.; Li, Y.; Li, S.; Li, H.B. Ultrasound-assisted extraction of natural antioxidants from the flower of Limonium sinuatum: Optimization and comparison with conventional methods. *Food Chem.* **2017**, *217*, 552–559. [CrossRef]
32. Chemat, F.; Rombaut, N.; Sicaire, A.G.; Meullemiestre, A.; Fabiano-Tixier, A.S.; Abert-Vian, M. Ultrasound assisted extraction of food and natural products. Mechanisms, techniques, combinations, protocols and applications. A review. *Ultrason. Sonochem.* **2017**, *34*, 540–560. [CrossRef]
33. Chavan, Y.; Singhal, R.S. Ultrasound-assisted extraction (UAE) of bioactives from arecanut (*Areca catechu* L.) and optimization study using response surface methodology. *Innov. Food Sci. Emerg. Technol.* **2013**, *17*, 106–113. [CrossRef]
34. Asfaram, A.; Ghaedi, M.; Yousefi, F.; Dastkhoon, M. Experimental design and modeling of ultrasound assisted simultaneous adsorption of cationic dyes onto ZnS: Mn-NPs-AC from binary mixture. *Ultrason. Sonochem.* **2016**, *33*, 77–89. [CrossRef] [PubMed]
35. Maran, J.P.; Manikandan, S.; Nivetha, C.V.; Dinesh, R. Ultrasound assisted extraction of bioactive compounds from *Nephelium lappaceum* L. fruit peel using central composite face centered response surface design. *Arab. J. Chem.* **2017**, *10*, S1145–S1157. [CrossRef]
36. Khan, M.K.; Abert-Vian, M.; Fabiano-Tixier, A.S.; Dangles, O.; Chemat, F. Ultrasound-assisted extraction of polyphenols (flavanone glycosides) from orange (*Citrus sinensis* L.) peel. *Food Chem.* **2010**, *119*, 851–858. [CrossRef]
37. Zoumpoulakis, P.; Sinanoglou, V.; Siapi, E.; Heropoulos, G.; Proestos, C. Evaluating modern techniques for the extraction and characterisation of sunflower (*Hellianthus annus* L.) seeds phenolics. *Antioxidants* **2017**, *6*, 46. [CrossRef] [PubMed]
38. Vajic, U.J.; Grujic-Milanovic, J.; Zivkovic, J.; Savikin, K.; Godevac, D.; Miloradovic, Z.; Bugarski, B.; Mihajlovic-Stanojevic, N. Optimization of extraction of stinging nettle leaf phenolic compounds using response surface methodology. *Ind. Crop. Prod.* **2015**, *74*, 912–917. [CrossRef]

39. Ali, A.; Lim, X.Y.; Chong, C.H.; Mah, S.H.; Chua, B.L. Optimization of ultrasound-assisted extraction of natural antioxidants from Piper betle using response surface methodology. *LWT* **2018**, *89*, 681–688. [CrossRef]
40. Jovanović, A.A.; Đorđević, V.B.; Zdunić, G.M.; Pljevljakušić, D.S.; Šavikin, K.P.; Gođevac, D.M.; Bugarski, B.M. Optimization of the extraction process of polyphenols from *Thymus serpyllum* L. herb using maceration, heat-and ultrasound-assisted techniques. *Sep. Purif. Technol.* **2017**, *179*, 369–380. [CrossRef]

© 2019 by the authors. Licensee MDPI, Basel, Switzerland. This article is an open access article distributed under the terms and conditions of the Creative Commons Attribution (CC BY) license (http://creativecommons.org/licenses/by/4.0/).

Article

Optimization of Ultrasound-Assisted Extraction of Polyphenols from *Myrtus communis* L. Pericarp

Nadia Bouaoudia-Madi [1], Lila Boulekbache-Makhlouf [1,*], Khodir Madani [1], Artur M.S. Silva [2], Sofiane Dairi [1,3], Sonia Oukhmanou–Bensidhoum [1] and Susana M. Cardoso [2,*]

[1] Laboratoire de Biomathématiques, Biophysique, Biochimie, et Scientométrie (L3BS),
    Faculté des Sciences de la Natureet de la Vie, Université de Bejaia, 06000 Bejaia, Algeria
[2] QOPNA & LAQV-REQUIMTE, Department of Chemistry, University of Aveiro, 3810-193 Aveiro, Portugal
[3] Faculté des Sciences de la Nature et de la Vie, Université de Jijel, 18000 Jijel, Algeria
*   Correspondence: lilaboulekbachemakhlouf@yahoo.fr (L.B.-M.); susanacardoso@ua.pt (S.M.C.)

Received: 17 May 2019; Accepted: 26 June 2019; Published: 2 July 2019

Abstract: Response surface methodology (RSM) was used to optimize the extraction of phenolics from pericap of *Myrtus communis* using ultrasound-assisted extraction (UAE). The results were compared with those obtained by microwave-assisted extraction (MAE) and conventional solvent extraction (CSE) methods. The individual compounds of the optimized extract obtained by UAE were identified by ultra-high-performance liquid chromatography coupled with diode array detection and electrospray ionization mass spectrometry (UHPLC-DAD-ESI-MS^n). The yield of total phenolic compounds (TPC) was affected more significantly by ethanol concentration, irradiation time, liquid solvent-to-solid ratio ($p < 0.0001$) and amplitude ($p = 0.0421$) and optimal parameters conditions set by the RSM model were 70% (v/v), 7.5 min and 30%, respectively. The experimental yield of TPC (241.66 ± 12.77 mg gallic acid equivalent/g dry weight) confirmed the predicted value (235.52 ± 9.9 mg gallic acid equivalent/g dry weight), allowing also to confirm the model validity. Under optimized conditions, UAE was more efficient than MAE and CSE in extracting antioxidants, which comprised mostly myricetin glycosides. Globally, the present work demonstrated that, compared to MAE and CSE, UAE is an efficient method for phenolic extraction from *M. communis* pericarp, enabling to reduce the working time and the solvent consumption.

Keywords: myrtle; ultrasound-assisted extraction; phenolic compounds; antioxidant capacity; liquid chromatography analysis; mass spectrometry

1. Introduction

The *Myrtus* genus, belonging to the Myrtaceae family, comprises about 50 species that are native of the Mediterranean basin. Among those, *M. communis* is an aromatic evergreen perennial sub-shrub (high: 1–3 m) with white flowers (blossoming time: June to July) and dark blue ripe berries [1]. It is native to Southern Europe, North Africa and West Asia and widespread in the Mediterranean region. Its fruits are consumed either raw or processed in diverse products such as canned fruits, yogurts, beverages, jams and jellies. In addition, there has been a growing interest in the use of berry extracts as ingredients in functional foods and dietary [2]. Volatile oils, tannins, anthocyanins, fatty acids, sugars, and organic acids such as citric and malic acids are important components of these fruits [3]. In general, myrtle berries are accepted as being rich in phenolic compounds, which in turn are associated to the fruits claimed health effects, including the prevention of degenerative diseases, such as cancer and cardiovascular diseases [4]. In fact, phenolic compounds are accepted as potent antioxidants due to their double bonds and hydroxyl groups, being capable of preventing the oxidation of free radicals that may damage physiological molecules cells, such as lipid proteins and DNA [5]. Many studies have shown a positive relationship between the phenolics content and the antioxidant capacity of fruits and vegetables [4–6]. Moreover, the regular consumption of fruits and vegetables is believed to prevent oxidative stress events and oxidative-stress related diseases [7–9].

Due to countless beneficial characteristics of phenolic compounds in human health, research has been intensified, aiming to find fruits, vegetables, plants, agricultural and agroindustrial residues as sources of these bioactive components. Obtaining such compounds often requires many long and costly steps, such as extraction, isolation and identification [5], and often result in thermal degradation of various bioactive constituents [6,7]. In this context, the development of new extraction methods is one of the major challenges in technological innovation towards the direction of "Green chemistry" [8,9]. Among them, ultrasound-assisted extraction (UAE) and microwave-assisted extraction (MAE) are particularly attractive because of their simplicity, low cost of equipment, efficiency in extracting analytes from different matrices and the requirement of low energy, reduced quantity of solvent and/or time consumption [10], compared to conventional extraction methods, which have several disadvantages, such as the use of volatile and hazardous solvents, the long extraction time and more recovery energy [11]. The enhancement of the extraction process by ultrasounds is attributed to the disruption of the cell walls, reduction of the particle size and the increased mass transfer of the cell content to the solvent, caused by the collapse of the bubbles produced by acoustic cavitation [8]. The processing parameters optimization and interpretation of experiments compared to others has been previously done through response surface methodology (RSM) [12]. This latter has been shown to be a powerful tool in optimizing experimental conditions (factors) to maximize the response. With the experimental results of a response surface design, a mathematical polynomial model, describing the relation between a response (dependent variable) and the considered factors (independent variable), is built. The mathematical model, usually a second-order polynomial model, can be visualized graphically by drawing 2D contour plots or 3D response surface plots [13]. The model allows determining the optimum value of the independent variables (X_i), as well as those of the dependent ones (Y).

Previous studies focusing on phenolic compounds and/or the antioxidant abilities of myrtle pericarp have been performed with extracts obtained by conventional methods [14–16], while, to our knowledge, there is no available information on the optimization of ultrasonic procedure for the extraction of phenolic compounds from this matrix, using a safer solvent such as ethanol, which is an organic solvent used in the food and pharmaceutical industries [17]. Therefore, the present study aimed at the optimization of UAE process parameters using RSM, including ethanol concentration, extraction time, irradiation amplitude and liquid-to-solid ratio, to maximize the content of the extracted phenolics. Levels of phenolic compounds and the antioxidant activity of pericarp *M. communis* extract obtained under the optimum setting parameters (UAE-OPT extract) were compared with those of extracts obtained by microwave-assisted extraction (MAE) and conventional solvent extraction (CSE) methods, using previously established conditions [18]. Then, the individual phenolic compounds present in the optimized extract obtained by UAE were identified by UHPLC-DAD-ESI-MS[n].

2. Materials and Methods

2.1. Plant Material

The fruits of *M. communis* were harvested from spontaneous plants in Adakar, Bejaia, located in the northeast of Algeria. The collected samples were identified by the Vegetable Ecological Laboratory of the Algiers University, Algeria and a voucher specimen was deposited at the Herbarium of Natural History Museum of Aix-en-Provence, France, under the voucher number D-PH-2013-37-12. Berries were washed and then dried in a static oven at 40 °C for one week. Pericarps were separated manually from seeds and further grounded in an electrical grinder (A11Basic, IKA, Retsch, Germany), which was then sieved to obtain a fine powder (<250 μm).

2.2. Extraction of Phenolic Compounds

2.2.1. Ultrasound Extraction

UAE was performed in an ultrasonic apparatus (Vibra cell, VCX 75115 PB, SERIAL No. 2012010971 MODEL CV 334, SONICS, Newtown, Connecticut, USA) with a working frequency fixed at 20 kHz. For extraction, 1 g of the pericarp powder was placed in a 250 mL amber glass bottle containing ethanol. The suspension was exposed to acoustic waves under distinct setting parameters (solvent concentration, irradiation time, ultrasound amplitude and solvent-to-solid ratio). The temperature was maintained constant by circulating external cold water and checking the temperature using a T-type thermocouple [5]. Indeed, ultrasound is considered a non-thermal technology, since it increases only the local temperature without affecting the surrounding environment [19]. After extraction, the solution was filtered through a sintered glass filter of porosity 2.

To determine the effect of ethanol concentration, irradiation time, ultrasound amplitude and solvent-to-solid ratio on the extraction yield of phenolic compounds from myrtle pericarp, RSM was applied with a Box–Behnken Design (BBD) [5]. This design resulted in the testing of four factors in a single block of 30 sets of test conditions (Table 1). The constant values for irradiation time, liquid-to-solid ratio and ethanol concentration in the UAE trials were 10 min, 50 mL/g and 50% (*v/v*), respectively.

$$Y = \beta_0 + \sum\nolimits_{i=0}^{k} \beta_i X_i + \sum\nolimits_{i=1}^{k} \beta_{ii} X^2 + \sum\nolimits_{i>1}^{k} \beta_{ij} X_i X_j + E \tag{1}$$

where $X_i, X_j, \ldots, X_k$ are the independent variables affecting the responses Y (the yield of total phenolic compounds); β_i, β_{ii} and β_{ij} are the regression coefficients for linear, quadratic and interaction terms, respectively; and k is the number of variables.

The factor levels were coded as −1 (low), 0 (central point or middle) and 1 (high), respectively, according to Equation (2):

$$X_i = (X_i - X_0)/\Delta X \ldots \tag{2}$$

where X_i is the coded value of the variable X_i; X_0 is the value of X at the center point; and ΔX is the step change.

2.2.2. Microwave-Assisted Extraction

Phenolic extracts were obtained using a domestic microwave oven (Samsung MW813ST, Kuala Lumpur, Malaysia) adapted by adding of a condenser [18]. The apparatus operated at a frequency of 2450 MHz and a maximum output power of 1000 W with a 100 W increment. The size of the heating cavity was 37.5 cm (L) × 22.5 cm (W) × 38.6 cm (D). The applied extraction conditions corresponded to those previously optimized [18]. A volume of 32 mL of 42% ethanol concentration was added to 1 g of pericarp *Myrtus* powder in a flat-bottomed flask. The mixture was irradiated at 500 W for 62 s. The resultant extract was then filtered through a sintered glass filter of porosity 2 and was stored at 4 °C until further analysis.

2.2.3. Conventional Solvent Extraction

Conventional solvent extract followed the procedure established by Dahmoune et al. [5]. One gram of myrtle powder was placed in a conical flask, and 50 mL of 50% (*v/v*) ethanol were added. After stirring for 2 h, the mixture was filtered through a sintered glass filter of porosity 2 and the extract was stored at 4 °C until further use.

2.3. Analytical Determinations

2.3.1. Total Phenolic and Flavonoid Contents

The total phenolic content (TPC) of the UAE, MAE and CSE extracts was assessed according to the method of George et al. [20] and expressed as mg of gallic acid equivalent (GAE) per gram of myrtle pericarp powder on dry weight (DW) basis (mg GAE g^{-1} DW). The total flavonoid content was estimated by the aluminum trichloride method according to Quettier-Deleu et al. [21] and the results were expressed as mg of quercetin equivalent per g of myrtle pericarp powder, on a DW basis.

2.3.2. Total Monomeric Anthocyanins and Condensed Tannin Contents

Total monomeric anthocyanin content was determined by the pH-differential method [22], and the results were expressed as mg cyanidin-3-*O*-glucoside equivalents per g of myrtle pericarp powder on a DW basis. The condensed tannin content was determined by the HCl–vanillin method as described by Aidi Wannes et al. [23] and the results were expressed as mg catechin equivalents per g of myrtle pericarp powder on DW basis.

2.3.3. Antioxidant Activity

The antioxidant activity of all samples was tested by using two different tests, namely 1,1-diphenyl-2-picrylhydrazyl radical (DPPH$^{\bullet}$) scavenging activity and reducing power methods [24]. DPPH$^{\bullet}$ solution (60 µM) was prepared in absolute methanol and reaction was performed by the adding of 3 mL of this solution to 1 mL of the extracts, during 20 min at 37 °C in the dark. Thereafter, the absorbance was measured at 515 nm. The inhibition rate of the extracts was calculated according Equation (3).

$$\% \, Scavenging = \frac{(A_{control} - A_{extract})}{A_{control}} \times 100 \tag{3}$$

where $A_{control}$ is the absorbance of DPPH$^{\bullet}$ and distilled water A_{sample} is the absorbance of DPPH$^{\bullet}$ and sample extract. α-tocopherol and BHA (250 µg/mL) were used as positive controls.

For reducing power assay, 1 mL of desired dilution was mixed with 2.5 mL of sodium phosphate buffer (0.2 M, pH 6.6) and 2.5 mL of 1% (*m/v*) potassium ferricyanide $K_3[Fe(CN)_6)]$, followed by incubation in a water bath at 50 °C for 20 min and the addition of 2.5 mL of 10% (*m/v*) trichloroacetic acid. At last, an aliquot of the resulting solution (1 mL) was added to 5 mL of distilled water and 1 mL of 0.1% (*m/v*) of FeCl3·6H2O. Note that this method estimates the ability to reduce Fe^{3+} to Fe^{2+}. Antioxidant compounds present in the samples form a colored complex with potassium ferricyanide, trichloroacetic acid and ferric chloride, which is measured at 700 nm.

2.4. Identification of Phenolic Compounds by UHPLC-DAD-ESI-MS^n

The phenolic compounds of the UAE-OPT extract were characterized by UHPLC-DAD-ESI-MS^n (DAD: diode array detector; ESI: electrospray ionization) analysis on an Ultimate 3000 (Dionex Co., USA) apparatus equipped with an ultimate 3000 Diode Array Detector (Dionex Co., San Jose, CA, USA) and coupled to a mass spectrometer. Analysis was run on a Hypersil Gold (Thermo Scientific, San Jose, CA, USA) C18 column (100 mm length; 2.1 mm i.d.; 1.9 µm particle diameter, end-capped) and its temperature was maintained at 30 °C. The mobile phase was composed of (A) 0.1% of formic acid (*v/v*) and acetonitrile (B). The solvent gradient started with 5% of Solvent B, reaching 40% at 14 min and 100% at 16 min, followed by the return to the initial conditions. The flow rate was 0.1 mL min^{-1} and UV–Vis spectral data for all peaks were accumulated in the range 200–700 nm while the chromatographic profiles were recorded at 280, 340 and 530 nm.

The mass spectrometer consisted of a Thermo LTQ XL (Thermo Scientific, San Jose, CA, USA) ion trap MS (mass spectrometer) apparatus equipped with an ESI source, operating in negative and positive modes, under the pre-established conditions [25].

2.5. Statistical Analysis

Each extraction trial and all the analyses were carried out in three independent analysis performed in triplicate. The influence of individual factors on the TPC yield (single-factor experiment) was estimated by Analysis of Variance (ANOVA) and Tukey's post hoc test with a 95% confidence level, while data obtained from the BBD and Central Composite Design (CCRD) trials were analyzed through ANOVA for the response variable to evaluate the model significance and suitability. Significant and highly significant levels were set for $p < 0.05$ and $p < 0.01$, respectively. The John's MacIntosh Product (Version 7.0, SAS, Cary, NC, USA) and Design-Expert (Trial version 10.0, SAS, Cary, NC, USA) software packages were used to construct the BBD and CCRD and to analyze all the results. Principal Component Analysis (PCA) was applied to detect the relationships between contents of phenolic compounds, flavonoids, anthocyanins, tannins, as well as antioxidant activity and their extraction methodologies i.e., UAE, MAE and CSE. All tests were done in triplicate.

3. Results and Discussion

3.1. Optimization of UAE Conditions

3.1.1. Modeling and Fitting the Model Using RSM

The experimental design and subsequent response allied to TPC are summarized in Table 1, with results from TPC recovery varying in the range of 79–235 mg GAE/g DW.

Table 1. Central composite design with the observed responses and predicted values for yield of total phenolic compounds of *Myrtus communis* pericarp using the UAE method.

Run	X_1-Ethanol (%, v/v)	X_2-Irradiation Time (min)	X_3-Amplitude (%)	X_4-Solvent-to Solid Ratio (mL/g)	TPC Recovery (mg GAE/g DW)
1	50	2.5	50	25	134.93 ± 11.37
2	70	10	70	30	195.24 ± 0.99
3	30	5	30	20	105.29 ± 11.72
4	70	5	30	30	221.73 ± 3.64
5	50	7.5	50	25	200.50 ± 12.82
6	50	7.5	50	15	78.90 ± 10.45
7	50	7.5	50	25	200.90 ± 28.02
8	50	7.5	50	25	200.10 ± 23.11
9	50	7.5	90	25	200.02 ± 13.23
10	70	10	30	30	214.07 ± 14.66
11	50	7.5	10	25	210.72 ± 2.70
12	50	7.5	50	25	210.21 ± 15.39
13	10	7.5	50	25	170.43 ± 9.38
14	30	10	70	20	159.56 ± 10.02
15	30	10	70	30	228.39 ± 12.96
16	50	7.5	50	25	203.34 ± 16.40
17	70	10	70	20	200.42 ± 14.47
18	70	10	30	20	185.51 ± 13.34
19	50	7.5	50	35	195.94 ± 12.80
20	50	12.5	50	25	190.43 ± 15.47
21	70	5	70	30	142.96 ± 9.60
22	30	10	30	20	118.92 ± 12.08
23	70	5	70	20	115.45 ± 16.12
24	30	5	30	30	219.12 ± 18.72
25	30	10	30	30	180.77 ± 9.38
26	70	5	30	20	161.68 ±9.42
27	30	5	70	30	179.22 ± 9.70
28	90	7.5	50	25	235.21 ±17.36
29	50	7.5	50	25	210.21 ± 16.60
30	30	5	70	20	111.38 ± 7.63

TPC results are expressed as means ± standard deviation; GAE, gallic acid equivalent; UAE, ultrasound-assisted extraction; DW, dry weight.

The least square technique was used to calculate the regression coefficients of the intercept, linear, quadratic, and interaction terms [26] (Table 2). Notably, the linear parameters, namely ethanol concentration, irradiation time and liquid–solid ratio ($p < 0.0001$), followed by amplitude ($p = 0.0421$) significantly affected the extraction content of phenolic compounds. The quadratic terms X_2^2 and X_4^2 were highly significant at the level $p < 0.001$, while the X_1^2 and X_3^2 terms were insignificant

($p > 0.05$). Regarding TPC yield, the interaction of ethanol concentration with amplitude of ultrasound (X_1–X_3) and with liquid to solid ratio (X_1–X_4), and that of irradiation time amplitude of ultrasound (X_2–X_3) were highly significant ($p < 0.0001$), followed by irradiation time with liquid-to-solid ratio ($p = 0.0054$), amplitude of ultrasound with liquid-to-solid ratio ($p < 0.0094$) and ethanol concentration with irradiation time ($p = 0.0367$). Those significant terms played a dominant role in myrtle pericarp extraction by ultrasound. Indeed, those significant terms played a dominant role in myrtle pericarp extraction by ultrasound. This is justified by the analyses of variance, as represented in Table 2, with significant p-values ($p < 0.0001$) for the linear parameters, namely ethanol concentration (X_1), irradiation time (X_2) and liquid–solid ratio (X_4), the quadratic terms X_2^2 and X_4^2, and the interaction terms (X_1–X_3), (X_1–X_4) and (X_2–X_3). However, insignificant p-values ($p > 0.0001$) were obtained for the third linear parameter, amplitude (X_3), the quadratic terms X_1^2 and X_3^2 and the interaction terms (X_1–X_2), (X_2–X_4) and (X_3–X_4).

Table 2. Estimated regression coefficients for the quadratic polynomial model and analyzes of variance (ANOVA) for the experimental results.

Parameters	Estimated Coefficients	Standard Error	DF [a]	Sum of Squares	F Ratio [b]	Prob > F
Model			14	46971.996	43.8356	<0.0001
Intercept	205.032	3.912523			52.40	<0.0001
Linear						
X_1-Ethanol	10.99875	1.736676	1	2903.340	37.9327	<0.0001
X_2-Time	14.04375	1.736676	1	4733.446	61.8434	<0.0001
X_3-Amplitude	−3.994583	1.736676	1	382.961	5.0035	0.0421
X_4-Ratio	27.390417	1.736676	1	18005.638	235.2474	<0.0001
Quadratic						
X_1^2	−1.526438	1.717541	1	60.454	0.7898	0.3892
X_2^2	−11.56144	1.717541	1	3468.113	45.3116	<0.0001
X_3^2	−0.888938	1.717541	1	20.503	0.2679	0.6128
X_4^2	−17.87644	1.717541	1	8291.469	108.3279	<0.0001
Interaction						
X_1–X_2	5.049375	2.187167	1	407.939	5.3298	0.0367
X_1–X_3	−11.46062	2.187167	1	2101.535	27.4570	<0.0001
X_1–X_4	−12.58812	2.187167	1	2535.37	33.1252	<0.0001
X_2–X_3	15.196875	2.187167	1	3695.120	48.2775	<0.0001
X_2–X_4	−7.198125	2.187167	1	829.008	10.8312	0.0054
X_3–X_4	−6.580625	2.187167	1	692.874	9.0525	0.0094
Lack of fit			10	94.0987	4.550	0.0911
Pure error			4			
R^2					0.9776	
Adjusted R^2					0.9553	
C.V. %	3.71%.					
RMSE	8.7186					
CorTotal [c]			28	48043.545		

[a] Degree of freedom; [b] the model mean square to error mean square ratio; [c] corrected total. DF, degree of freedom; F Ratio, freedom ratio; Prob, probability; C.V., coefficient of variation.

Based on the significant terms, the regression equation for the UAE efficiency was obtained as follows:

$$Y = 205.032 + 10.998X_1 + 14.043X_2 - 3.994X_3 + 27.390X_4 + 5.049X_1X_2 - 11.460X_1X_3 - 10.588X_1X_4 + 15.196X_2X_3 - 7.198X_2X_4 - 6.580X_3X_4 - 1.526X_1^2 - 11.561X_2^2 - 0.888X_3^2 - 17.876X_4^2 \qquad (4)$$

Note that the p-value can be employed to check the interaction strength between independent factors. From this analysis, p-value <0.0001 indicated that the response surface quadratic model was significant, which means that the model represented the data satisfactorily. The adjusted coefficient of determination (R^2adj) and the coefficient of determination (R^2) were 0.9553 and 0.9776, respectively, which implied that the sample variations of 97.76% for the UAE efficiency of myrtle pericarp phenols were attributed to the independent variables, and only 2.24% of the total variations could not be explained by the model, indicating a good degree of correlation between experimental and predict values of the TPC yield. In addition, the low value of coefficient of variance (3.71%) clearly indicated that the model was reproducible and reliable [27]. All these results indicate that the model could work well for the prediction of TPC in the myrtle pericarp extracts.

3.1.2. Response Surface Analysis (RSA)

To provide a better understanding of the interaction between factors, the 3D response surface plot was constructed (Figure 1) using Equation (4). The graphs were generated by plotting the response using the z-axis against two independent variables, while keeping the other independent variable at the fixed level. Figure 1A–C shows the interactions between the ethanol concentration and each of the three other factors, namely irradiation time, amplitude and liquid-to-solid ratio, respectively, on the recovery of TPC. As shown, an increase of ethanol concentration from 20% to 80% (*v/v*), or extraction time from 5 to 10 min resulted in a rapid enhancement of TPC with a maximum of 235.21 mg GAE/g being recovered with an irradiation time of 7.5 min and ethanol concentration of 70% (*v/v*).

The high phenolic content indicates that the mixture ethanol/water at 70% (*v/v*) allowed the solubilization of phenolics from *M. communis* pericarp, thus confirming the results of the single factor experiments [28] that explained the efficiency of the ultrasonic method by the fact that sonication improved the hydration and fragmentation process and hence facilitates the mass transfer of solutes to the extraction solvent. For the extraction yield of TPC performed at fixed extraction time and liquid-to-solid ratio, with varying ethanol concentration and amplitude (Figure 1B), it was possible to conclude that maximum recovery (210.05 mg GAE/g) was achieved for 70% (*v:v*) of ethanol and an ultrasound amplitude of 35%. This fact can be explained by the larger amplitude ultrasonic wave that promotes the liquid medium to produce more cavitation bubbles, thus resulting in a stronger pressure, capable of destroying the cell wall and accelerating mass transfer [29]. Figure 1C shows an enhancement of TPC that reached a peak value of 230.15 mg GAE/g, for 70% (*v:v*) ethanol and about 30 mL/g of liquid-to-solid ratio. A higher ratio corresponds to a greater concentration difference between the exterior solvent and the interior tissues of *Myrtus* pericarp. It prominently prompted the TPC to be rapidly dissolved, which resulted in an increase in the extraction yield. The response surface plot for the significative interactive effect of irradiation time and amplitude of ultrasound on the response value at a fixed ethanol concentration and liquid-to–solid ratio is shown in Figure 1D. A higher TPC was obtained with the irradiation time at 10 min and amplitude of 30%; these results confirm those reported in the literature [30,31]. Figure 1E shows an interaction between extraction time and the liquid-to-solid ratio ($p < 0.05$). The best content (148 mg GAE/g) was found with the solid–liquid ratio of about 30 mL/g and the radiation time of 10 min. The increase of the ethanol proportion required high sonication intensity to generate the cavitation bubbles. However, a higher increase in the liquid-to-solid ratio diminished the supply of ultrasonic energy density and negatively affected the extraction yield.

The yield of TPC constantly improved with the increase of both amplitude of ultrasound and liquid-to-solid ratio, reaching a maximum when X_3 and X_4 became 32% and 20% (*v/v*), respectively (Figure 1F). Beyond this level, the yield of TPC reduced with the increase of X_1 and X_4. Hence, the interactive effect of X_3 and X_4 was remarkable. Overall, these results indicate that the TPC extraction yield was more significantly affected ($p < 0.0001$) by linear parameters, namely ethanol concentration, irradiation time and liquid-to-solid ratio.

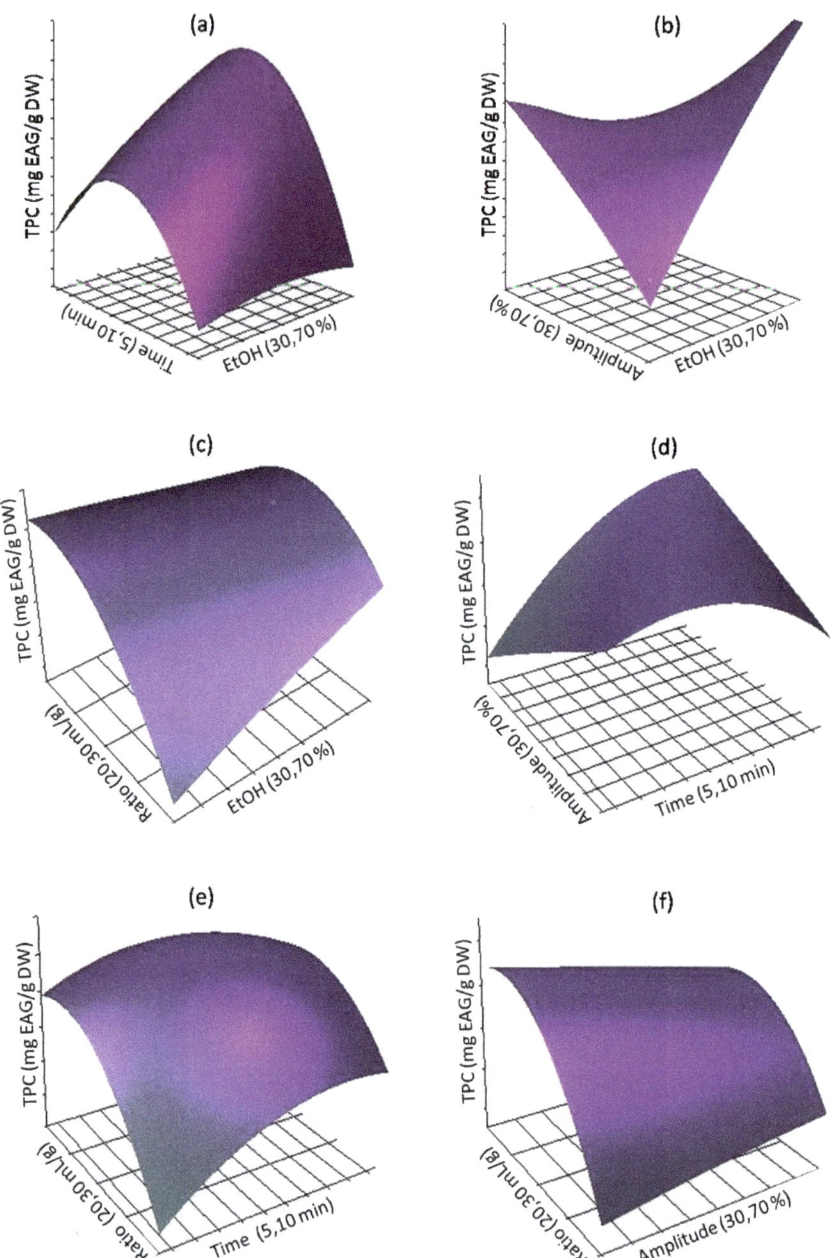

Figure 1. Response surface analysis for the Total phenolic compounds (TPC) with UAE with respect to: (**a**) ethanol concentration and irradiation time; (**b**) ethanol concentration and amplitude; (**c**) ethanol concentration and solvent-to-solid ratio; (**d**) extraction time and amplitude; (**e**) extraction time and solvent-to-solid ratio; and (**f**) amplitude and solvent-to-solid ratio.

3.1.3. Validation and Verification of the Predictive Model

According to the result of response surface and prediction by this built model, the optimal conditions were thus obtained for the following conditions: ethanol at 70% (*v/v*), 7.5 min extraction time, 30% amplitude and a liquid-to-solid ratio of 28 mL/g. To ensure that the predicted result was not biased to the practical value, experimental rechecking was performed using these deduced optimal conditions. The predicted extraction yield of TPC in UAE-OPT was 235.52 ± 9.9 mg GAE/g, that was consistent with the experimental yield of 241.66 ± 12.77 mg GAE/g DW (Table 3). The results showed no significant difference between the experimental and the predicted values. This strong correlation between experimental and the predicted values indicates that the response of regression model is adequate to reflect the expected optimization for the extraction of antioxidants from *M. communis* pericarp.

3.2. Comparison between UAE, MAE and CSE Methods

Remarkably, the highest TPC was obtained by UAE (241.60 ± 12.77 mg GAE/g). This corresponded to four and three times higher than that obtained by MAE and CSE, respectively, thus indicating that the application of UAE has a positive effect on the extraction of TPC (Table 3). The highest levels of TPC in UAE-OPT extract was reflected by its higher amounts of flavonoids, anthocyanins and tannins (18.99 ± 1.31 mg QE/g; 25.06 ± 0.36 mg/g; 35.56 ± 0.36 mg CE/g, respectively). These findings are consistent with those reported in the literature [32] and are mainly attributed to the fact that ultrasound radiation can facilitate mass transfer and accelerate the extracting process so that the extraction of bioactive compounds may be improved. Hence, according to the overall data, it is possible to conclude that the herein optimized UAE process yields higher levels of bioactive compounds in a short time and requires less solvent consumption than MAE and CSE. Note that, in this study, the operating temperature in the UAE-OPT was kept constant at room temperature, excluding any heating effect. This might positively or negatively influence the polyphenols recovery depending on the applied amplitude.

Table 3. Comparison of extraction yield of polyphenols obtained by optimized ultrasound-assisted (UAE-OPT), microwave-assisted (MAE) and conventional solvent (CSE) methods.

Method	EtOH (%)	Time (min)	US amp (w/power)	Liq:sol (mL/g)	TPC (mg GAE/g)	Flavonoids (mgQE/g)	Anthocyanin (mg/g)	Tannins (mg CE/g)	DPPH (%)	RP (Abs 700 nm)
UAE-OPT	70	7.5	30	28	241.66 ± 12.77 [a]	18.99 ± 1.31 [a]	25.06 ± 0.36 [a]	35.56 ± 0.36 [a]	90.71 ± 0.23 [a]	0.568 ± 0.002 [b]
MAE	42	62	500	32	119.59 ± 8.40 [b]	11.5 ± 0.01 [b]	5.64 ± 0.06 [c]	31.70 ± 1.00 [b]	87.16 ± 0.28 [b]	0.439 ± 0.006 [b]
CSE	50	7200		50	76.40 ± 7.27 [c]	6.95 ± 0.20 [c]	6.96 ± 0.72 [b]	30.70 ± 0.17 [c]	88.03 ± 1.04 [b]	0.429 ± 0.001 [b]
BHA									26.98 ± 0.69 [c]	1.37 ± 0.03 [a]
α-tocopherol									17.17 ± 0.4 [d]	0.53 ± 0.01 [b]

Contents of TPC, flavonoids, anthocyans and tannins are means ± standard deviation. Different letters in the same row indicate significant differences ($p < 0.05$) according to the ANOVA test. BHA, butylated hydroxyanisole; CE, catechin equivalents; DPPH, 1,1-diphenyl-2-picrylhydrazyl radical; EtOH, ethanol; GAE, gallic acid equivalents; Liq:sol, liquid-to-solid ratio; QE, quercetin; RP, ferric reducing antioxidant power; US amp, ultrasound amplitude.

The antioxidant capacity of the extracts was assessed by DPPH$^\bullet$ scavenging and ferric reducing antioxidant power assays. The results show that UAE-OPT extract presented higher DPPH$^\bullet$ scavenging ability (90.71% inhibition) when compared to CSE (88.03% inhibition) and MAE (87.16% inhibition) extracts. The same tendency was also observed for reducing power, since the absorbance at 700 nm for UAE-OPT extract was considerably higher than those obtained for MAE and CSE (0.439 ± 0.006 and 0.429 ± 0.01, respectively). This means that UAE method is more efficient for the recovery of antioxidants than the herein tested microwave and conventional solvent extraction methods, a fact that is probably due to its superior richness in phenolic components, including flavonols [33] and as evidenced in the following section. This information was also confirmed by PCA analysis. PCA was applied to the extracts (UAE-OPT, MAE and SCE) for phenolic compounds (TPC, flavonoids, anthocyanins and tannins) and antioxidant activity, where the two chosen factors justified 100.0% of total variance. The resulting plots allowed selecting the better extraction method of different compounds of myrtle pericarp, and clearly divided the samples into three groups, depending on the extraction method (Figure 2). For PC1, which explains 95.91% of the total variance, the first group showed a positive correlation with PC1, thus confirming that UAE was the best extraction method for phenolic compounds with potent antioxidant activity. The highest correlation was found between antioxidant activity (DPPH$^\bullet$ and RP essay) and anthocyanins, hence suggesting that these compounds might have a key influence on the antioxidant capacity of the extracts. The best correlation between the MAE and TPC, flavonoids and tannins were observed in the second group. PC2 explains only better 4.09% of the experimental variability, which could essentially be associated to the CSE method and anthocyanins content (the third group).

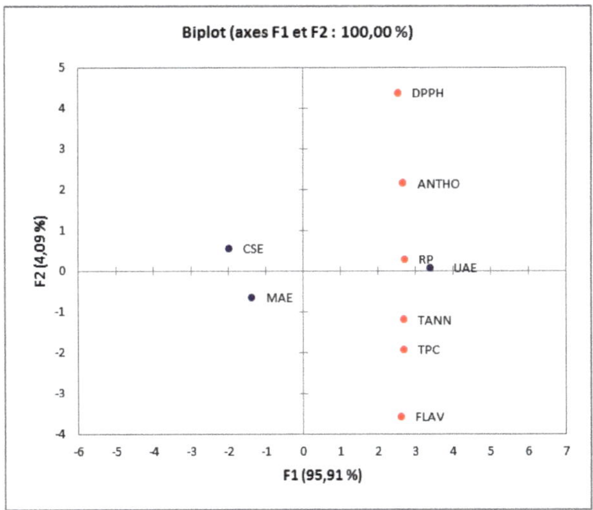

Figure 2. Principal component analysis of phenolic compounds for *M. communis* pericarp with UAE, MAE and CSE. FLAV, flavonoids; ANTHO, anthocyanins; TANN, tannins.

3.3. Identification of Phenolics by UHPLC-DAD-ESI-MS^n Analysis

The UAE-OPT extract was analyzed by UHPLC-DAD-ESI-MS^n to further elucidate its phenolic profile. The registered chromatogram at 280 nm is shown in Figure 3 and the UV-Vis and MS^n spectral data of eluted peaks are summarized in Table 4.

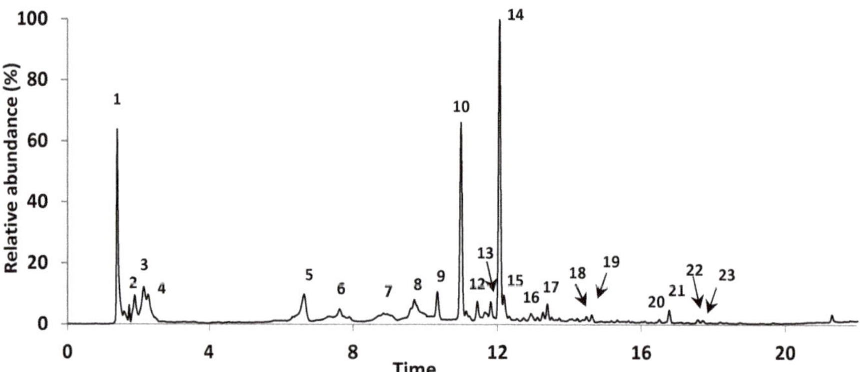

Figure 3. Chromatographic profile at 280 nm of *M. communis* pericarp extract obtained by UAE extraction at optimized conditions. Numbers in the figure correspond to the eluted UHPLC peaks for which UV and MS data are summarized in Table 4.

Among the distinct phenolic groups found in the extract, flavonols were the prevalent components. Overall, eleven flavonol glycosides were detected, being myricetin glycosides, namely myricetin-*O*-hexoside and myricetin-*O*-deoxyhexoside (eluted in Peaks 10/11 and 14/15, respectively) the major abundant ones, which probably correspond to myricetin-3-*O*-galactoside and myricetin-3-*O*-rhamnoside, since these are known to be present as main phenolic components in distinct organs of *M. communis* plant [34–38].

Besides the above compounds, four other myricetin glycosides were found in the extract. The compound eluted in Peak 9, showing a [M-H]$^-$ at m/z 631, corresponded to myricetin-*O*-galloyl-hexoside, since the main fragments in MS^2 spectrum were formed by the loss of 152 Da (equivalent to a galloyl moiety) and 332 Da (equivalent to the simultaneous loss of galloyl and hexosyl units). This could possibly correspond to myricetin 3-(6″-*O*-galloyl galactoside), which has been previously reported in leaves [32,34,36,38] and berries [16]. In addition, the compounds with [M-H]$^-$ at m/z 449 (Peak 13) and at m/z 625 (co-eluted in Peak 18) were, respectively, assigned to myricetin-*O*-pentoside and myricetin-*O*-hexosyl-deoxyhexoside, according to their fragmentation pattern, which showed the loss of a pentosyl (132 Da) and deoxyhexosyl plus hexosyl (308 Da) moieties, respectively. In turn, the compound eluted in Peak 19 at 14.6 min with a pseudomolecular ion at m/z 569 and fragment ions at m/z 485 (equivalent to galloyl ester moiety) and 317 (myricetin) was tentatively assigned to a galloylester of myricetin.

The three remaining flavonols detected in the UAE-OPT extract were assigned to quercetin and kaempferol derivatives. From those, the compound eluted in Peak 12 was characterized by a [M-H]$^-$ at m/z 615 and fragment ions at m/z 463 (−152 Da, loss of galloyl group) and 301 (−162 Da, loss of an hexosyl group), and was tentatively assigned to quercetin-*O*-hexoside-gallate on the basis of data reported in the literature [39–41]. This compound has been already reported in Myrtaceae family, namely in *Eucalyptus* species [42–44] and two other species from the same family, namely *Myrcia multiflora* extracts [45] and *Eugenia edulis* [46]. In addition, the compound eluted in Peak 17 with a deprotonated ion at m/z 447 and a base peak fragment ion at m/z 301 (−146, equivalent to the loss of a deoxyhexose unit), was identified as quercetin-*O*-deoxyhexoside according to literature data, probably corresponding to quercetin-3-*O*-rhamnoside [39,40]. This last flavonoid was previously detected in pericarp [29], berries [31,47] and leaves [30,31,35] of *M. communis*. Finally, the flavonol eluted in Peak 16 ([M-H]$^-$ at m/z 447) presented the main fragment ion at m/z 285 in the MS^2 spectrum, which in turn showed a fragmentation pattern coherent with kaempferol. Based on UV-Vis spectra (UV$_{max}$ at 265 and 353) and MS^n spectral data, this compound was assigned to kaempferol-*O*-hexoside.

Table 4. UHPLC-DAD-ESI-MS^n data for *M. communis* pericarp extract obtained under optimized UAE conditions.

No. Peak	t_R (min)	λmax (nm)	(*m/z*)	MS^n Ions (*m/z*)	Probable Compound
1	1.4	275	191 [a]	MS^2[191]: 173, 127, 111, 93	Quinic acid
			341 [a]	MS^2[341]: 179	Caffeoyl-*O*-hexoside
2	1.9	276	331	MS^2[331]: 271, 169, 241, 211, 193, 125; MS^3[271]: 211, 169	Galloyl-*O*-hexoside
3	2.1	273	343	MS^2[343]: 191, 169, 125	Galloyl quinic acid (isomer 1)
4	2.3	271	169	MS^2[169]: 125	Gallic acid
5	6.7	274	343	MS^2[343]: 191, 169, 125	Galloyl quinic acid (isomer 2)
		276, 525	463 [a,b]	MS^2[463]: 301, 300, 337, 315	Delphinidin-*O*-hexoside
6	7.6		495 [a]	MS^2[495]: 343, 325, 191, 169	Digalloyl quinic acid
			483 [a]	MS^2[483]: 271, 331, 313, 439, 193, 169; MS^3[271]: 211, 169	Digalloyl hexoside
7	8.8	274, 525	477 [b]	MS^2[477]: 315, 314	Petunidin-*O*-hexoside
8	9.7	274, 525	647 [a,b]	MS^2[647]: 495, 477	Petunidin-*O*-galloyl-hexoside derivative
			491 [a,b]	MS^2[491]: 329	Malvidin-*O*-hexoside
9	10.3	265, 356	631	MS^2[631]: 479, 299, 317	Myricetin-*O*-galloyl-hexoside
10	11.0	260, 356	479	MS^2[479]: 316, 317	Myricetin-*O*-hexoside (isomer 1)
11	11.2	260, 356	479	MS^2[479]: 316, 317	Myricetin-*O*- hexoside (isomer 2)
12	11.4	265, 356	615	MS^2[615]: 463, 301; MS^2[463]: 179, 151	Quercetin *O*-hexoside- gallate
13	11.8	263, 356	449	MS^2[449]: 316, 317	Myricetin-*O*-pentoside
14	12.1	261, 351	463	MS^2[463]: 316, 317	Myricetin-*O*-deoxyhexoside (isomer 1)
15	12.2	261, 351	463	MS^2[463]: 316, 317	Myricetin-*O*-deoxyhexoside (isomer 2)
16	12.9	265, 353	447	MS^2[447]: 285; MS^3[285]: 267, 257, 241	Kaempferol-*O*-hexoside
17	13.4	257, 350	447	MS^2[447]: 301; MS^3[301]: 179, 151	Quercetin-*O*-eoxyhexoside
18	14.5	265, 352	431 [a]	MS^2[431]: 271; MS^3[271]: 211, 169	Galloyl derivative
			625 [a]	MS^2[625]: 479, 317	Myricetin-*O*-hexosyl-deoxyhexoside
19	14.6	273, 350	569	MS^2[569]: 485, 317	Myricetin-*O*-galloyl ester
20	16.5	276	583	MS^2[583]: 271, 565, 313, 211, 331; MS^3[271]: 211, 169	Gallomyrtucommulone F
21	16.8	276	567	MS^2[567]: 271, 313, 211, 169; MS^3[271]: 211, 169	Gallomyrtucommulone C
22	17.6	276	467	MS^2[467]: 271, 313, 169, 211; MS^3[271]: 211, 169	Gallomyrtucommulone-type (isomer 1)
23	17.8	276	467	MS^2[467]: 271, 313, 169, 211; MS^3[271]: 211, 169	Gallomyrtucommulone-type (isomer 2)

Peak numbers correspond to those depicted in Figure 3; [a] co-eluted compounds in a peak fraction; [b] the respective [M]$^+$ ions were registered in the positive mode. t_R, retention time; λmax, wavelength of maximum absorbance.

Besides flavonols, other flavonoids in UAE-OPT extract corresponded to anthocyanins that were eluted from 7.6 min to 9.7 min (Peaks 6–8). Note that, in general, anthocyanins are preferred detected as $[M]^+$ in ESI in the positive mode, while typically they show $[M-2H]^-$ in the negative mode [47], as represented in Table 4. Overall, according to UV-Vis and MS^n spectral data, these compounds were assigned to delphinidin, petunin and malvidin derivatives. In more detail, the compound in Peak 6 exhibiting a $[M-2H]^-$ at m/z 463 and a base peak MS^2 fragment ion at m/z 301 (−162 Da) was assigned to delphinidin-*O*-hexoside by comparison with data reported in the literature [48–50]. In turn, petunidin-*O*-hexoside and a petunidin-*O*-hexoside derivative were eluted in Peaks 7 and 8, respectively. The first showed a $[M-2H]^-$ at m/z 477 and a main MS^2 fragment ions at m/z 315/314 [48–50] while ions corresponding to petunidin-*O*-hexoside and its hydrated form (at m/z 477 and m/z 495, respectively) were predominant in MS^2 spectrum of the latter compound. The petunidin-*O*-hexoside derivative was co-eluted with malvidin-*O*-hexoside ($[M-2H]^-$ at m/z 477→329). Note that, except for petunidin-*O*-hexoside, hexosides of delphinidin, petunin and malvidin have already been described in distinct organs of *M. communis*, including pericarp [14,30,31,51,52].

Several non-flavonoid compounds could also be observed in UEA-OPT extract, including caffeoyl hexoside, gallic acid and galloyl derivatives. The first ($[M-H]^-$ at m/z 341→179, eluted in Peak 1) was the only hydroxycinnamic acid found in the extract. Gallic acid ($[M-H]^-$ at m/z 169→125, eluted in Peak 4), has been described in the literature for extracts obtained from the pericarp [16] berries [16,30] and leaves [38,52].

Regarding galloyl derivatives (typical UV_{max} at 273–276 nm), these enclosed esters of mono- or di-galloyl groups with a hexose or quinic acid unit, or even with myrtucommulone-type groups. In detail, the compound eluted in Peak 2 with a $[M-H]^-$ at m/z 331 and corresponding fragments at m/z 271, 169, 241, 211, 193 and 125, was assigned to a galloyl hexoside [53], presumably galloyl-3-*O*-β-D-galactoside-6-*O*-gallate, since this latter has been previously reported in *M. communis* leaves [2,38]. Besides, two isomers of galloyl quinic acid ($[M-H]^-$ at m/z 343→191, 169, 125) could be found in Peaks 3 and 5, while a digalloyl hexoside ($[M-H]^-$ at m/z 483→271, 331, 313, 439, 193, 169) and digalloyl quinic acid ($[M-H]^-$ at m/z 495→343, 325, 191, 169) were detected as co-eluted compounds in Peak 6. All these galloyl derivatives have been previously detected in *M. communis* leaves [35,38,52].

Moreover, four gallomyrtucommulone-type derivatives were found in UAE-OPT extract. All these compounds showed a UV_{max} at 276 nm, and similar fragment ions in MS^n spectra, including ions at m/z 331, 313, 271 and 211, which are typically formed in galloylhexoside [43]. Indeed, the ion at m/z 331 correspond to the galloyl hexoside moiety, while ions at m/z 271 and m/z 211 result from the cross-ring fragmentation of the hexose unit in the galloyl hexoside moiety and that at m/z 313 can be formed due to the loss of water molecule from the latter. Among these compounds, those eluted in Peaks 20 and 21 ($[M-H]^-$ at m/z 583 and 567, respectively) were assigned to gallomyrtucommulone F and gallomyrtucommulone C, in accordance to previous data reported in *M. communis* leaves [36]. Besides these two compounds, the extract also contained two isomeric unidentified gallomyrtucommulone-type derivatives (MW 468 Da) that presumably vary in their acyl chain regarding those previously identified.

4. Conclusions

The response surface methodology was successfully employed to optimize total phenolic extraction yield from dried *M. communis* pericarp by non-conventional solvent extraction process, namely using UAE. As compared to MAE and CSE extractions, the proposed UAE method allowed a higher phenolic recovery yield and antioxidant activity with a short working time and a lower solvent consumption. The quantification of the amounts of phenolic compounds in the three types of extract complemented with PCA analysis also allowed concluding that the *M. communis* pericarp extract obtained under optimal UAE experimental conditions contained higher levels of flavonoids, tannins and anthocyanin than the remaining extracts and, particularly, the latter phenolic components could be correlated to its antioxidant activity. According to UHPLC-DAD-ESI-MS^n analysis, flavonols, particularly myricetin-*O*-hexoside and myricetin-*O*-deoxyhexoside, were the prevalent phenolic components of UAE-OPT.

Author Contributions: N.B.-M. performed the experiments (extraction, optimization and quantification of phenolics and antioxidant activity), analyzed the data and wrote the original draft; L.B.-M. contributed to conceptualization, supervision, data curation, and writing—review and editing; K.M. conceptualized and supervised the research; A.M.S.S. contributed resources and performed writing—review and editing; S.D. contributed to the methodology; S.O.-B. contributed to the investigation; S.M.C. contributed to the supervision, data curation and writing—review and editing.

Funding: Foundation for Science and Technology (FCT), the European Union, the National Strategic Reference Framework (QREN), the European Regional Development Fund (FEDER), and Operational Programme Competitiveness Factors (COMPETE), for funding the Organic Chemistry Research Unit (QOPNA) (FCT UID/QUI/00062/2019), through national funds and where applicable co-financed by the FEDER, within the PT2020 Partnership Agreement. Project AgroForWealth (CENTRO-01-0145-FEDER-000001), funded by Centro2020, through FEDER and PT2020, financed the research contract of Susana M. Cardoso.

Conflicts of Interest: The authors have declared no conflict of interest.

References

1. Nuvoli, F.; Spanu, D. Analisie prospettive economiche dell'utilizzazionze industriale del mirto. *Riv. Ital. EPPOS* **1996**, *12*, 231–236. [CrossRef]

2. Sumbul, S.; Ahmad, M.A.; Asif, M.; Akhtar, M.; Saud, I. Physicochemical and phytochemical standardization of berries of *Myrtus communis* Linn. *J. Pharm. Bioall. Sci.* **2012**, *4*, 322–326. [CrossRef]

3. Messaoud, C.; Boussaid, M. *Myrtus communis* berry color morphs: A comparative analysis of essential oils, fatty acids, phenolic compounds, and antioxidant activities. *Chem. Biodivers.* **2011**, *8*, 300–310. [CrossRef]

4. Liu, R.H. Health benefits of fruit and vegetables are from additive and synergistic combinations of phytochemicals. *Am. J. Clin. Nutr.* **2003**, *78*, 517S–520S. [CrossRef]

5. Dahmoune, F.; Boulekbache, L.; Moussi, K.; Aoun, O.; Spigno, G.; Madani, K. Valorization of *Citrus limon* residues for the recovery of antioxidants: Evaluation and optimization of microwave and ultrasound application to solvent extraction. *Ind. Crops Prod.* **2013**, *50*, 77–87. [CrossRef]

6. Đorđević, T.; Antov, M. Ultrasound assisted extraction in aqueous two-phase system for the integrated extraction and separation of antioxidants from wheat chaff. *Sep. Purif. Technol.* **2017**, *182*, 52–58. [CrossRef]

7. De Castro, M.D.L.; García-Ayuso, L.E. Soxhlet extraction of solid materials: An outdated technique with a promising innovative future. *Anal. Chim. Acta* **1998**, *369*, 1–10. [CrossRef]

8. Chemat, F.; Rombaut, N.; Sicaire, A.-G.; Meullemiestre, A.; Fabiano-Tixier, A.-S.; Abert-Vian, M. Ultrasound assisted extraction of food and natural products. Mechanisms, techniques, combinations, protocols and applications. A review. *Ultrason. Sonochem.* **2017**, *34*, 540–560. [CrossRef]

9. Živković, J.; Šavikin, K.; Janković, T.; Ćujić, N.; Menković, N. Optimization of ultrasound-assisted extraction of polyphenolic compounds from pomegranate peel using response surface methodology. *Sep. Purif. Technol.* **2018**, *194*, 40–47. [CrossRef]

10. Jacotet-Navarro, M.; Rombaut, N.; Deslis, S.; Fabiano-Tixier, A.-S.; Pierre, F.-X.; Bily, A.; Chemat, F. Towards a "dry" bio-refinery without solvents or added water using microwaves and ultrasound for total valorization of fruit and vegetable by-products. *Green Chem.* **2016**, *18*, 3106–3115. [CrossRef]

11. Zu, G.; Zhang, R.; Yang, L.; Ma, C.; Zu, Y.; Wang, W.; Zhao, C. Ultrasound-assisted extraction of carnosic acid and rosmarinic acid using ionic liquid solution from *Rosmarinus officinalis*. *Int. J. Mol. Sci.* **2012**, *13*, 11027–11043. [CrossRef] [PubMed]

12. Yan, Y.-L.; Yu, C.-H.; Chen, J.; Li, X.-X.; Wang, W.; Li, S.-Q. Ultrasonic-assisted extraction optimized by response surface methodology, chemical composition and antioxidant activity of polysaccharides from *Tremella mesenterica*. *Carbohydr. Polym.* **2011**, *83*, 217–224. [CrossRef]

13. Bezerra, M.A.; Santelli, R.E.; Oliveira, E.P.; Villar, L.S.; Escaleira, L.A. Response surface methodology (RSM) as a tool for optimization in analytical chemistry. *Talanta* **2008**, *76*, 965–977. [CrossRef] [PubMed]

14. Montoro, P.; Tuberoso, C.I.G.; Piacente, S.; Perrone, A.; De Feo, V.; Cabras, P.; Pizza, C. Stability and antioxidant activity of polyphenols in extracts of *Myrtus communis* L. berries used for the preparation of myrtle liqueur. *J. Pharm. Biomed.* **2006**, *41*, 1614–1619. [CrossRef] [PubMed]

15. Amensour, M.; Sendra, E.; Abrini, J.; Bouhdid, S.; Pérez-Alvarez, J.A.; Fernández-López, J. Total phenolic content and antioxidant activity of myrtle (*Myrtus communis*) extracts. *Nat. Prod. Commun.* **2009**, *4*, 819–824. [CrossRef] [PubMed]

16. Tuberoso, C.I.G.; Rosa, A.; Bifulco, E.; Melis, M.P.; Atzeri, A.; Pirisi, F.M.; Dessì, M.A. Chemical composition and antioxidant activities of *Myrtus communis* L. berries extracts. *Food Chem.* **2010**, *123*, 1242–1251. [CrossRef]

17. Jovanović, A.A.; Đorđević, V.B.; Zdunić, G.M.; Pljevljakušić, D.S.; Šavikin, K.P.; Gođevac, D.M.; Bugarski, B.M. Optimization of the extraction process of polyphenols from *Thymus serpyllum* L. herb using maceration, heat-and ultrasound-assisted techniques. *Sep. Purif. Technol.* **2017**, *179*, 369–380. [CrossRef]

18. Bouaoudia-Madi, N.; Boulekbache-Makhlouf, L.; Kadri, N.; Dahmoune, F.; Remini, H.; Dairi, S.; Oukhmanou-Bensidhoum, S.; Madani, K. Phytochemical analysis of *Myrtus communis* plant: Conventional versus microwave assisted-extraction procedures. *J. Complement. Integr. Med.* **2017**, *14*. [CrossRef]

19. Benmeziane, A.; Boulekbache-Makhlouf, L.; Mapelli-Brahm, P.; Khaled Khodja, N.; Remini, H.; Madani, K.; Meléndez-Martínez, A.J. Extraction of carotenoids from cantaloupe waste and determination of its mineral composition. *Food Res. Int.* **2018**, *111*, 391–398. [CrossRef]

20. George, S.; Brat, P.; Alter, P.; Amiot, M.J. Rapid Determination of Polyphenols and Vitamin C in Plant-Derived Products. *J. Agric. Food Chem.* **2005**, *53*, 1370–1373. [CrossRef]

21. Quettier-Deleu, C.; Gressier, B.; Vasseur, J.; Dine, T.; Brunet, C.; Luyckx, M.; Cazin, M.; Cazin, J.-C.; Bailleul, F.; Trotin, F. Phenolic compounds and antioxidant activities of buckwheat (*Fagopyrum esculentum* Moench) hulls and flour. *J. Ethnopharmacol.* **2000**, *72*, 35–42. [CrossRef]

22. Cheok, C.Y.; China, N.L.; Yusof, Y.A.; Talib, R.A.; Law, C.L. Optimization of total monomeric anthocyanin (TMA) and total phenolic content (TPC) extractions from mangosteen (*Garcinia mangostana* Linn.) hull using ultrasonic treatments. *Ind. Crops Prod.* **2013**, *50*, 1–7. [CrossRef]

23. Wannes, W.A.; Mhamdi, B.; Sriti, J.; Jemia, M.B.; Ouchikh, O.; Hamdaoui, G.; Kchouk, M.E.; Marzouk, B. Antioxidant activities of the essential oils and methanol extracts from myrtle (*Myrtus communis* var. italica L.) leaf, stem and flower. *Food Chem. Toxicol.* **2010**, *48*, 1362–1370. [CrossRef] [PubMed]

24. Zou, Y.; Lu, Y.; Wei, D. Antioxidant activity of a flavonoid-rich extract of *Hypericum perforatum* L. in vitro. *J. Agric. Food Chem.* **2004**, *52*, 5032–5039. [CrossRef]

25. Pereira, O.R.; Catarino, M.D.; Afonso, A.F.; Silva, A.M.S.; Cardoso, S.M. Salvia elegans, Salvia greggii and Salvia officinalis Decoctions: Antioxidant Activities and Inhibition of Carbohydrate and Lipid Metabolic Enzymes. *Molecules* **2018**, *23*, 3169. [CrossRef] [PubMed]

26. Jacotet-Navarro, M.; Rombaut, N.; Fabiano-Tixier, A.-S.; Danguien, M.; Bily, A.; Chemat, F. Ultrasound versus microwave as green processes for extraction of rosmarinic, carnosic and ursolic acids from rosemary. *Ultrason. Sonochem.* **2015**, *27*, 102–109. [CrossRef] [PubMed]

27. Dahmoune, F.; Moussi, K.; Remini, H.; Belbahi, A.; Aoun, O.; Spigno, G.; Madani, K. Optimization of Ultrasound-Assisted Extraction of Phenolic Compounds from *Citrus sinensis* L. Peels using Response Surface Methodology. *Chem. Eng. Trans.* **2014**, *37*, 889–894. [CrossRef]

28. Dahmoune, F.; Spigno, G.; Moussi, K.; Remini, H.; Cherbal, A.; Madani, K. *Pistacia lentiscus* leaves as a source of phenolic compounds: Microwave-assisted extraction optimized and compared with ultrasound-assisted and conventional solvent extraction. *Ind. Crops Prod.* **2014**, *61*, 31–40. [CrossRef]

29. Ghafoor, K.; Choi, Y.H.; Jeon, J.Y.; Jo, I.H. Optimization of ultrasound-assisted extraction of phenolic compounds, antioxidants, and anthocyanins from grape (*Vitis vinifera*) seeds. *J. Agric. Food Chem.* **2009**, *57*, 4988–4994. [CrossRef] [PubMed]

30. Wannes, W.A.; Marzouk, B. Differences between myrtle fruit parts (*Myrtus communis* var. italica) in phenolics and antioxidant contents. *J. Food Biochem.* **2013**, *37*, 585–594. [CrossRef]

31. Pereira, P.; Cebola, M.-J.; Oliveira, M.C.; Bernardo-Gil, M.G. Supercritical fluid extraction *vs* conventional extraction of myrtle leaves and berries: Comparison of antioxidant activity and identification of bioactive compounds. *J. Supercrit. Fluid* **2016**, *113*, 1–9. [CrossRef]

32. Romani, A.; Pinelli, P.; Mulinacci, N.; Vincieri, F.; Tattini, M. Identification and quantitation of polyphenols in leaves of *Myrtus communis* L. *Chromatographia* **1999**, *49*, 17–20. [CrossRef]

33. Boldbaatar, D.; El-Seedi, H.R.; Findakly, M.; Jabri, S.; Javzan, B.; Choidash, B.; Göransson, U.; Hellman, B. Antigenotoxic and antioxidant effects of the Mongolian medicinal plant *Leptopyrum fumarioides* (L): An in vitro study. *J. Ethnopharmacol.* **2014**, *155*, 599–606. [CrossRef] [PubMed]

34. Romani, A.; Coinu, R.; Carta, S.; Pinelli, P.; Galardi, C.; Vincieri, F.F.; Franconi, F. Evaluation of antioxidant effect of different extracts of *Myrtus communis* L. *Free Radic. Res.* **2004**, *38*, 97–103. [CrossRef] [PubMed]

35. Romani, A.; Campo, M.; Pinelli, P. HPLC/DAD/ESI-MS analyses and anti-radical activity of hydrolyzable tannins from different vegetal species. *Food Chem.* **2012**, *130*, 214–221. [CrossRef]

36. Taamalli, A.; Iswaldi, I.; Arráez-Román, D.; Segura-Carretero, A.; Fernández-Gutiérrez, A.; Zarrouk, M. UPLC–QTOF/MS for a rapid characterisation of phenolic compounds from leaves of *Myrtus communis* L. *Phytochem. Anal.* **2014**, *25*, 89–96. [CrossRef] [PubMed]

37. Hasdemir, B.; Yaşa, H.; Onar, H.Ç.; Yusufoğlu, A.S. Investigation of essential oil composition, polyphenol content, and antioxidant activity of *Myrtus communis* L. from Turkey. *JOTCSA* **2016**, *3*, 427–438. [CrossRef]

38. Yoshimura, M.; Amakura, Y.; Tokuhara, M.; Yoshida, T. Polyphenolic compounds isolated from the leaves of *Myrtus communis*. *J. Nat. Med.* **2008**, *62*, 366–368. [CrossRef]

39. Boulekbache-Makhlouf, L.; Medouni, L.; Medouni-Adrar, S.; Arkoub, L.; Madani, K. Effect of solvents extraction on phenolic content and antioxidant activity of the byproduct of eggplant. *Ind. Crops Prod.* **2013**, *49*, 668–674. [CrossRef]

40. Saldanha, L.L.; Vilegas, W.; Dokkedal, A.L. Characterization of flavonoids and phenolic acids in Myrcia bella cambess. Using FIA-ESI-IT-MSn and HPLC-PAD-ESI-IT-MS combined with NMR. *Molecules* **2013**, *18*, 8402–8416. [CrossRef]

41. Sobeh, M.; ElHawary, E.; Peixoto, H.; Labib, R.M.; Handoussa, H.; Swilam, N.; El-Khatib, A.H.; Sharapov, F.; Mohamed, T.; Krstin, S. Identification of phenolic secondary metabolites from *Schotia brachypetala Sond.* (Fabaceae) and demonstration of their antioxidant activities in *Caenorhabditis elegans*. *PeerJ* **2016**, *4*, e2404. [CrossRef] [PubMed]

42. Cadahía, E.; Conde, E.; García-Vallejo, M.; Fernández de Simón, B. High pressure liquid chromatographic analysis of polyphenols in leaves of *Eucalyptus camaldulensis*, *E. globulus* and *E. rudis*: Proanthocyanidins, ellagitannins and flavonol glycosides. *Phytochem. Anal.* **1997**, *8*, 78–83. [CrossRef]

43. Amakura, Y.; Yoshimura, M.; Sugimoto, N.; Yamazaki, T.; Yoshida, T. Marker constituents of the natural antioxidant Eucalyptus leaf extract for the evaluation of food additives. *Biosci. Biotechnol. Biochem.* **2009**, *73*, 1060–1065. [CrossRef]

44. Okamura, H.; Mimura, A.; Niwano, M.; Takahara, Y.; Yasuda, H.; Yoshida, H. Two acylated flavonol glycosides from *Eucalyptus rostrata*. *Phytochemistry* **1993**, *33*, 512–514. [CrossRef]

45. Cascaes, M.; Guilhon, G.; Andrade, E.; Zoghbi, M.; Santos, L. Constituents and pharmacological activities of Myrcia (Myrtaceae): A review of an aromatic and medicinal group of plants. *Int. J. Mol. Sci.* **2015**, *16*, 23881–23904. [CrossRef]

46. Hussein, S.A.; Hashem, A.N.; Seliem, M.A.; Lindequist, U.; Nawwar, M.A. Polyoxygenated flavonoids from *Eugenia edulis*. *Phytochemistry* **2003**, *64*, 883–889. [CrossRef]

47. Barboni, T.; Venturini, N.; Paolini, J.; Desjobert, J.-M.; Chiaramonti, N.; Costa, J. Characterisation of volatiles and polyphenols for quality assessment of alcoholic beverages prepared from Corsican *Myrtus communis* berries. *Food Chem.* **2010**, *122*, 1304–1312. [CrossRef]

48. Bochi, V.C.; Godoy, H.T.; Giusti, M.M. Anthocyanin and other phenolic compounds in Ceylon gooseberry (*Dovyalis hebecarpa*) fruits. *Food Chem.* **2015**, *176*, 234–243. [CrossRef]

49. Dias, T.; Bronze, M.R.; Houghton, P.J.; Mota-Filipe, H.; Paulo, A. The flavonoid-rich fraction of Coreopsis tinctoria promotes glucose tolerance regain through pancreatic function recovery in streptozotocin-induced glucose-intolerant rats. *J. Ethnopharmacol.* **2010**, *132*, 483–490. [CrossRef]

50. Lopes-Lutz, D.; Dettmann, J.; Nimalaratne, C.; Schieber, A. Characterization and quantification of polyphenols in Amazon grape (*Pourouma cecropiifolia Martius*). *Molecules* **2010**, *15*, 8543–8552. [CrossRef]

51. Scorrano, S.; Lazzoi, M.R.; Mergola, L.; Di Bello, M.P.; Del Sole, R.; Vasapollo, G. Anthocyanins Profile by Q-TOF LC/MS in *Myrtus communis* Berries from Salento Area. *Food Anal. Methods* **2017**, *10*, 2404–2411. [CrossRef]

52. Romani, A.; Pinelli, P.; Galardi, C.; Mulinacci, N.; Tattini, M. Identification and quantification of galloyl derivatives, flavonoid glycosides and anthocyanins in leaves of *Pistacia lentiscus* L. *Phytochem. Anal.* **2002**, *13*, 79–86. [CrossRef] [PubMed]

53. Abu-Reidah, I.M.; Ali-Shtayeh, M.S.; Jamous, R.M.; Arráez-Román, D.; Segura-Carretero, A. HPLC–DAD–ESI-MS/MS screening of bioactive components from *Rhus coriaria* L. (Sumac) fruits. *Food Chem.* **2015**, *166*, 179–191. [CrossRef] [PubMed]

© 2019 by the authors. Licensee MDPI, Basel, Switzerland. This article is an open access article distributed under the terms and conditions of the Creative Commons Attribution (CC BY) license (http://creativecommons.org/licenses/by/4.0/).

 *antioxidants*

Article

Optimization of the Microwave-Assisted Extraction Process of Bioactive Compounds from Annatto Seeds (*Bixa orellana* L.)

Julian Quintero Quiroz [1,*], Angélica Celis Torres [1], Luisa Muñoz Ramirez [1], Mariluz Silva Garcia [2], Gelmy Ciro Gomez [1] and John Rojas Camargo [1]

[1] College of Pharmaceutical and Food Sciences, University of Antioquia, Calle 67 No. 53-108, University Campus, Medellín 050010, Colombia; angelica.celis@udea.edu.co (A.C.T.); luisa.munozr@udea.edu.co (L.M.R.); gelmy.ciro@udea.edu.co (G.C.G.); jhon.rojas@udea.edu.co (J.R.C.)
[2] Institute of Food Science and Technology (INTAL), Cra. 50g #12S-91, Itagüi 055412, Colombia; marisylva87@gmail.com
* Correspondence: julian.quintero@udea.edu.co; Tel.: +57-421-965-91

Received: 6 December 2018; Accepted: 31 January 2019; Published: 6 February 2019

Abstract: This study deals with the extraction, optimization, and evaluation of the antioxidant and antimicrobial activities of bioactive compounds obtained from the seeds of annatto using microwave-assisted extraction as compared to leaching. Annatto seeds were subjected to a microwave treatment of 2450 MHz and power of 700 watts using a response surface design involving four factors: pH (4–11), solvent concentration (ethanol) (50–96%), solvent-to-seed ratio (2–10), and microwave exposure time (0–5 min). The contents of polyphenol compounds and bixin were taken as response variables. Subsequently, the antioxidant and antimicrobial activities were assessed at the optimal processing conditions predicted by the experimental design. Microwaves, solvent concentration, and the solvent-to-seed ratio showed a statistically significant effect for the extraction of polyphenol compounds and bixin. Thus, microwaves accelerated the extraction of those compounds and the slight increase in temperature caused some degradation of the polyphenol compounds. The microwave-assisted extraction increased the contents of polyphenols and bixin along with their antioxidant activity as compared to leaching extraction. However, this technique does not significantly improve the antimicrobial activity against *Bacillus cereus* and *Staphylococcus aureus*.

Keywords: antimicrobial activity; antioxidant activity; polyphenol compounds; bixin

---

1. Introduction

The reddish or orange extract obtained from the seeds of annatto (*Bixa orellana* L.) has a great coloring strength, making this extract useful in the food and cosmetic sectors. In addition, the extract possesses inflammatory, antioxidant, and antimicrobial properties [1,2]. These pharmacological activities are due to diverse bioactive compounds such as polyphenols (i.e., hypolatin and caffeic acid) and carotenoids (i.e., bixin or 6-methyl hydrogen (9Z)-6,6′-diapocarotene-6) [3]. The antioxidant activity of these compounds is attributed to their high structural conjugation and capability to couple with singlet oxygen which is responsible for denaturation of proteins belonging to cell membranes, leading to cell lysis [1,4,5].

Recent reports deal with the biological activity of annatto extract obtained by traditional techniques, which are compelling for the pharmaceutical and food companies. For instance, Zhang and collaborators assessed the ability of bixin to promote transcription of a new Ets-related factor (Nrf2) and E74-like factor 2 (ELF2) in a murine model with inflammation in the lung tissue due to SiO_2 particles. This transcription factor regulates the inducible expression of numerous genes of detoxifying enzymes and antioxidants by binding to a specific DNA sequence known as the antioxidant response element (ARE), resulting in protection against several pathologies such as cancer, liver toxicity, and inflammation [6]. Further, the activation of Nrf2 in mice potentiates the antioxidant and healing abilities and decreased lung tissue inflammation after inhalation of bixin particles [2]. On the other hand, Viuda and collaborators determined the in vitro antioxidant and antimicrobial activities of annatto seed extract. They used five different hydrogen-donating or radical-scavenging tests including the 2,2′-diphenyl-1-picrylhydrazyl (DPPH) stable radical, ferric-reducing antioxidant power (FRAP), thiobarbituric acid reactive substance (TBARS), Rancimat, and ferrous ion (Fe^{2+}) chelating activity. On the other hand, the microdilution in broth method was used to evaluate the antimicrobial activity against *Listeria innocua* (Spanish Type Culture Collection, University of Valencia, Research Building, Burjassot, Spain. (CECT) 910), *Aeromonas hydrophila* (CECT 5734), *Bacillus cereus* (American Type Culture Collection (ATCC) 11778), and *Pseudomona aeruginosa* (ATCC 9027). They found the annatto seed extract as an alternative preservative to replace the synthetic ones such as butyl-hydroxy toluene (BHT) in food matrices due to its extensive antioxidant activity (93.01 ± 1.22, 75.88 ± 0.21, 25.41 ± 3.52, 10.53 ± 0.31, and 1.34 ± 0.07 for DPPH (inhibition, %), TBARS (inhibition, %), FRAP (TEAC), FIC (chelating effect, %), and rancimat (AAI), respectively). Conversely, it had a mild antimicrobial action compared to some synthetic antimicrobials (1.024 mg/mL, 256 mg/mL, 512 mg/mL, and 256 mg/mL for *P. aeruginosa*, *B. cereus*, *L. innocua*, and *A. hydrophila*, respectively) [1].

Maceration, leaching, extraction with supercritical fluids (CO_2), dispersive liquid–liquid, sonication, enzymatic extraction, microextraction, and microwave-assisted extraction (MAE) are the most widely used methods for the extraction of bioactive molecules from plants [7]. However, the extraction technique affects the functional activities of the extract and thus, conventional approaches require long processing times, exposing the active compounds to degradation and resulting in low efficiency. On the other hand, some emerging extraction techniques such as supercritical CO_2 extraction are costly and involve high energy consumption [8]. Further, the use of mechanical methods, spouted bed, sonication, microwaves, and supercritical fluids (CO_2) as emerging extraction technologies, have been focused simply on the extraction of bixin and norbixin [7,9–11]. Most of these studies highlight the coloring capacity of the extract, and others evaluate the effect of the extraction technique on the antioxidant and antimicrobial activities separately [11,12].

The microwave-assisted extraction (MAE) involves the formation of high-energy electromagnetic waves with a frequency, oscillation period, and wavelength ranging from 300 MHz–30 GHz, 3×10^{-9} s–33×10^{-12} s, and 1–10 mm, respectively. These waves have the ability to change the molecular rotation and ionic mobility of the medium without altering the sample. This rotation generates the adsorption and dissipation of energy in the medium, causing a generalized heating, forcing the rapid migration of all the active compounds from the solid-phase to the solvent-phase [7,13–17]

For this reason, the goal of this study was to define the optimal microwave-assisted extraction (MAE) conditions to render the bioactive compounds of annatto seeds without affecting their antioxidant and antimicrobial activities. The solvent concentration, pH, treatment time and solvent-to-seed ratio were taken as independent variables employing a surface response experimental design.

2. Materials and Methods

2.1. Materials

Sodium carbonate (lot a0594092339), anhydrous sodium acetate (lot 882032), absolute ethanol (lot k4958171742), potassium persulphate (lot k40707991048), methanol (lot l823009611), acetic acid glacial (lot k47109963539), acetone (lot k43912814243), sodium hydroxide (lot b0895598319), concentrated hydrochloric acid (lot k41017317016), Folin–Ciocalteu phenol reagent (lot hc43368401), Mueller–Hinton broth (lot b1223098542), tetrahydrofuran (lot dg643), and

2,4,6-tri-(2-pyridyl)-1,3,5-triazine (lot l58155738107) were obtained from Merck (Darmstadt, Germany). Dimethyl sulfoxide (lot 190260) was obtained from PanReac AppliChem (Barcelona, Spain). 2,2′-Azino-bis(3-ethylbenzothiazoline-6-sulfonic acid), 2, 2-diphenyl-1-picrylhydrazil (067k lot 1154), and diammonium salt (lot slbp9592v) were obtained from Sigma-Aldrich (St. Louis, MO, USA). ($\pm$)-6-Hydroxy-2,5,7,8-tetramethylchromane-2-carboxylic acid (Trolox) (97% lot stbb6668) was obtained from Aldrich Chem (St. Louis, MO, USA). Ferric chloride. $6H_2O$ (lot 9n005099n) was obtained from Carlo Erba (Barcelona, Spain). 3-(4,5-Dimethylthiazol-2-yl)-2,5-diphenyltetrazolium bromide (MTT) (lot p31b064) was obtained from Alfa Aesar (Haverhill, MA, USA). Annatto seeds were purchased from a farmers market of Medellín city (Colombia), which were sun-dried until reaching a moisture content of $10.58 \pm 0.98\%$.

2.2. Experimental Design

The MAE conditions were optimized using a Box–Behnken experimental design (BBD), employing the Design Expert Software® Version 8.0.6 (Stat-Ease, MN, USA). The dependent variables for the two extraction methods are described in Table 1. From these conditions, 30 experimental runs were established and presented in a randomized pattern. The amount of total polyphenols determined by the Folin–Ciocalteu method, and the bixin (Bix) content were taken as independent variables [18].

Table 1. Factor levels of the independent variables according to the Box–Behnken experimental design (BBD).

Independent Variables	Symbol	Coded Levels		
		−1	0	+1
Treatment time (min)	X_1	0	2.5	5
pH	X_2	4	7.5	11
Solvent concentration (ethanol) (%)	X_3	50	73	96
Solvent-to-seed ratio (X_4:1)	X_4	2	6	10

The solvent pH at the concentrations determined by the BBD was adjusted with 0.1 M sodium hydroxide and acetic acid solutions. The microwave unit was operated at a frequency of 2450 MHz and power of 700 W (LG Ms-147xc, Seoul, South Korea), submitting the sample to 30-s cycles until reaching the desirable total time (Figure 1). In order to optimize the extraction conditions and investigate the effect of the independent variables the method of multiple regression of least squares was used. The coefficient of determination (r^2) and the adjusted coefficient of determination (r^2-adj model) were used as fitting parameters of the regression models. The experimental data were adjusted to the following polynomial equation:

$$Y = \beta_0 + \beta_1 X_1 + \beta_2 X_2 + \beta_3 X_3 + \beta_4 X_4 + \beta_{11} X_1 X_1 + \beta_{22} X_2 X_2 + \beta_{33} X_3 X_3 + \beta_{44} X_4 X_4 + \beta_{12} X_1 X_2 + \beta_{13} X_1 X_3 + \beta_{14} X_1 X_4 + \beta_{23} X_2 X_3 + \beta_{24} X_2 X_4 + \beta_{34} X_3 X_4. \tag{1}$$

where, Y represents the predicted response, β_0, is the intercept of the model, β_1, β_2, β_3, β_4, β_{11}, β_{22}, β_{33}, β_{44} and β_{12}, β_{13}, β_{14}, β_{23}, β_{24} and, β_{34} are linear and interaction coefficients, respectively, and X_1, X_2, X_3, and, X_4 are independent variables. The analysis of variance (ANOVA) was used to investigate the statistical significance of the independent variables from the models obtained (with a confidence level of 95%). The optimization of the extraction process was achieved by maximizing the extraction of polyphenol compounds and bixin having the same weight (weight of 1) and minimizing the solvent-to-seed ratio. The accuracy of the optimal conditions was determined with the desirability values of the dependent factors. The calculation of the relative and absolute errors was accomplished between the responses predicted by the model versus the ones obtained experimentally under optimal conditions.

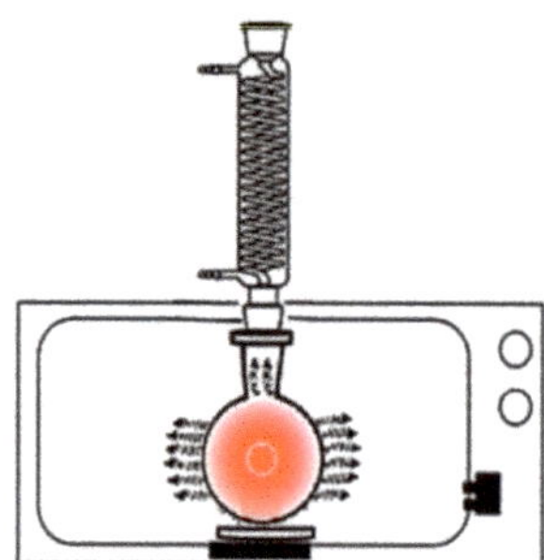

Figure 1. Schematics for microwave-assisted extraction (MAE) from annatto seeds.

2.3. Characterization of Optimal Annatto Seed Extracts

The optimal extract obtained by the MAE was compared with the one obtained by leaching. The latter was obtained with ethanol using at optimum concentration, pH, and seeds-to-solvent ratio for the MAE, with a continuous agitation for 48 h. The results are presented as means and standard deviation (SD). The analysis was performed using the Statgraphics® Centurion XVI software (Madrid, Spain).

2.3.1. Total Polyphenol Concentration

The total polyphenol concentration in the annatto seed extract was determined using the Folin–Ciocalteau method [19]. Briefly, 20 μL of sample were diluted in 1.58 μL of distilled water. Then, 100 μL of Folin–Ciocalteau reagent and 300 μL of 20% sodium carbonate were added and mixed. The absorbance of the colored complex generated was read after 1 h of storage under darkness. The maximum absorbance was read at 725 nm in a UV/VIS spectrophotometer (UV-1700, Shimadzu®, Kyoto, Japan). The experiments were performed in triplicate and the results are expressed as mg of gallic acid (GA) per grams of seeds (mg GA/g seed).

2.3.2. Quantification of Bixin

Here, 100 μL of sample were added to 2 mL of tetrahydrofuran and diluted to 10 mL with acetone to obtain an absorbance of less than 0.15 at 487 nm. The concentration of bixin in the sample was determined from a calibration curve built in a spectrophotometer (UV-1700, Shimadzu®, Kyoto Japan) using the following equation (Equation (2)) [20]:

$$Bixin\ (\%) = \frac{A \times 100 \times V}{A_{1cm}^{1\%} \times 100} \tag{2}$$

where

$A_{1cm}^{1\%}$ = 3090 (1 g/100 mL)$^{-1}$ × 1 cm^{-1} (specific absorptivity coefficient of bixin in acetone) [20];
A = Absorbance value of the sample; and
V = Dilution volume (mL) of the sample.

2.3.3. Antioxidant Activity (ABTS^{+} Method)

The ABTS assay was performed following the method described by Contreras et al. 2011 [21]. Briefly, 100 μL of sample (diluted appropriately with water) were mixed with 1 mL of ABTS^{+} solution. The degraded color was read after 30 min at 730 nm using a spectrophotometer (UV-1700, Shimadzu®, Kyoto, Japan). A Trolox calibration curve was conducted for quantification purposes and the results are expressed as Trolox equivalents (TE) or μmol TE/L.

2.3.4. Ferric Reducing Antioxidant Power (FRAP)

The FRAP was measured as previously described by Benzie and Strain (1996) with modifications [22]. Briefly, 90 µL of deionized water were mixed with 30 µL of sample (diluted appropriately with water) and added to 900 µL of the FRAP reagent (pre-warmed at 37 °C). The sample was incubated for 30 min at 37 °C. Subsequently, the sample absorbance at 593 nm was measured using a spectrophotometer (UV-1700, Shimadzu®, Kyoto, Japan). A Trolox® calibration curve was employed for quantification purposes and the results are expressed as µmol TE/L.

2.3.5. Antioxidant Activity (DPPH Method)

The DPPH assay was performed as previously described [23]. Briefly, 2 mL of 0.5 mM DPPH reagent was mixed with 2 mL of methanol and 0.2 mL of sample (diluted appropriately with water). The absorbance peak at 517 nm was recorded after 30 min of incubation under darkness (UV-1700, Shimadzu®, Kyoto, Japan). A Trolox® calibration curve was done for quantification purposes and the results are expressed as µmol TE/L.

2.3.6. Antimicrobial Activity of Annatto Seed Extracts

The antibacterial activity was determined employing the colorimetric microdilution method with broth incubated with *Bacillus cereus* (ATCC 11778) and *Staphylococcus aureus* (ATCC 6538). Bacterial strains were incubated for 24 h at 37 °C in a Mueller–Hinton broth and adjusted to a final density of 10^6 CFU/mL before inoculation. Samples were dissolved in dimethylsulfoxide (DMSO) to reach a final concentration of 4.096 mg/L. Double serial dilutions were made at a concentration ranging from 4 to 4096 mg/L. Then, 96-well microplates were prepared by adding 20 µL of sample on the respective dilution medium followed by addition of 220 µL of Mueller-Hinton broth. The last row containing only 220 µL of Mueller–Hinton broth and 10 µL of inoculum was used as a negative control. The final volume in each well was 250 µL. Once the samples were homogenized, they were incubated for 5 h at 37 °C. After incubation, 25 µL of 3-(4,5-dimethylthiazol-2-yl)-2,5-diphenyl-tetrazolium bromide (MTT) (Alfa Aesar, Germany), dissolved in DMSO (800 mg/L) were added to each well and incubated for 1 h to allow viable microorganisms to metabolize the yellow MTT dye in formazan. The minimum inhibitory concentration (MIC) was considered as the concentration of the first well that did not show any color change (from yellow to purple). The procedure was repeated three times for each microorganism and pH (4, 7, and 11) [1].

2.4. Statistical Analysis

The results are presented as means and standard deviation (SD), according to the normality of the data. The analysis was performed using the Statgraphics® centurion XVI software (Madrid, Spain).

3. Results

3.1. Experimental Design

The experimental matrix was composed of 30 experimental runs. Experiments were performed in triplicate and the tabulated results from each experimental run are described in Table 2. The maximum values achieved in the extraction process were 4.36 ± 0.04 mg GA/g seed and $0.51 \pm 0.01\%$ of bixin. These values were obtained at 2.5 min time for MAE, pH of 11, and a solvent-to-seed ratio of 6:1. The only variation between the extractions corresponded to the solvent concentration used at a 50% for the maximum extraction of polyphenol compounds and 96% for the maximum extraction of bixin.

ANOVA was used to evaluate the significance of the quadratic polynomial models. For each term in the models, a large F-value and a small p-value would imply a more significant effect on the respective response variable [24]. The ANOVA (Table 3) shows how factors such as treatment time, solvent concentration and the solvent-to-seed ratio had a statistically significant effect ($p < 0.05$) for the MAE.

Table 2. Experimental matrix and results obtained from the MAE.

Treatment Number	MAE (min)	pH	Solvent Concentration (Ethanol) (%)	Solvent-to-Seed Ratio (X_4:1)	MAE	
					Polyphenols (mg GA/g Seed) $n = 3$	Bixin (%) $n = 3$
1	2.5	7.5	73.0	10.0	2.83 ± 0.04	0.13 ± 0.00
2	2.5	4.0	50.0	6.0	1.53 ± 0.22	0.03 ± 0.00
3	2.5	7.5	50.0	10.0	2.85 ± 0.19	0.05 ± 0.00
4	5.0	4.0	96.0	6.0	1.96 ± 0.06	0.33 ± 0.04
5	2.5	7.5	73.0	6.0	1.96 ± 0.15	0.05 ± 0.01
6	5.0	11.0	73.0	6.0	2.01 ± 0.08	0.13 ± 0.02
7	5.0	11.0	73.0	6.0	1.35 ± 0.05	0.32 ± 0.01
8	2.5	11.0	96.0	6.0	3.94 ± 0.00	0.51 ± 0.01
9	5.0	7.5	50.0	6.0	3.79 ± 0.07	0.04 ± 0.01
10	2.5	7.5	50.0	2.0	0.53 ± 0.08	0.27 ± 0.03
11	2.5	4.0	73.0	6.0	4.16 ± 0.24	0.19 ± 0.00
12	2.5	4.0	73.0	2.0	0.83 ± 0.06	0.12 ± 0.01
13	2.5	7.5	73.0	6.0	1.89 ± 0.14	0.05 ± 0.01
14	2.5	7.5	96.0	6.0	1.72 ± 0.15	0.19 ± 0.00
15	2.5	11.0	73.0	10.0	3.25 ± 0.57	0.04 ± 0.01
16	2.5	11.0	50.0	6.0	4.36 ± 0.04	0.03 ± 0.00
17	2.5	7.5	96.0	6.0	2.43 ± 0.01	0.58 ± 0.00
18	2.5	7.5	96.0	2.0	0.66 ± 0.01	0.38 ± 0.03
19	0.0	7.5	73.0	6.0	1.96 ± 0.08	0.06 ± 0.01
20	5.0	7.5	96.0	10.0	2.10 ± 0.01	0.13 ± 0.00
21	2.5	7.5	73.0	2.0	0.70 ± 0.01	0.07 ± 0.02
22	0.0	7.5	73.0	6.0	1.90 ± 0.23	0.04 ± 0.00
23	2.5	11.0	73.0	2.0	0.68 ± 0.03	0.17 ± 0.01
24	0.0	7.5	73.0	10.0	2.15 ± 0.16	0.10 ± 0.01
25	2.5	4.0	73.0	6.0	2.28 ± 0.01	0.12 ± 0.03
26	0.0	7.5	73.0	6.0	2.02 ± 0.03	0.08 ± 0.00
27	0.0	7.5	50.0	6.0	3.90 ± 0.30	0.08 ± 0.00
28	2.5	7.5	73.0	2.0	1.57 ± 0.30	0.25 ± 0.01
29	0.0	4.0	73.0	10.0	1.96 ± 0.15	0.33 ± 0.01
30	5.0	7.5	73.0	6.0	1.91 ± 0.18	0.06 ± 0.02

Values are expressed as mean ± standard deviation ($n = 3$). MAE: microwave-assisted extraction, GA: gallic acid.

Table 3. ANOVA for the response variables in the MAE.

Variable	Polyphenols (mg GA/g Seed)	Bixin (%)
	p-Value	*p*-Value
Model	<0.0001	<0.0001
X_1, treatment time (min)	<0.0001	<0.0001
X_2, pH	0.532	0.112
X_3, solvent concentration (%)	<0.0001	<0.0001
X_4, solvent-to-seed ratio (X4:1)	<0.0001	0.677
X_2X_3	0.001	>0.050
X_2X_4	0.044	0.001
X_3X_4	>0.050	0.004
X_2X_2	>0.050	<0.0001
X_4X_4	<0.0001	>0.050
X_1X_1	<0.0001	<0.0001
Lack of fit	0.0405	0.137
r^2	0.931	0.914
r^2-adj	0.901	0.875

The response surface plots obtained from the polynomial equations are shown in Figure 2. All the models were subjected to an optimization process and the polynomial equations for the response variables are described as follows:

$$\text{Ln (Polyphenols)} = 0.70 + 0.03 \times X_2 - 0.26 \times X_3 + 0.53 \times X_4 + 0.26 \times X_1 - 0.42 \times X_2X_3 + 0.18 \times X_2X_4 - 0.41 \times X_4{}^2 + 0.29 \times X_1{}^2 \tag{3}$$

$$\text{Ln (Bixin)} = -0.59 - 1.12 \times X_2 - 0.04 \times X_3 + 0.76 \times X_4 - 0.15 \times X_1 + 0.04 \times X_2X_4 - 0.01 \times X_3X_4 + 0.06 \times X_2{}^2 + 0.01 \times X_4{}^2 + 0.10 \times X_1{}^2 \tag{4}$$

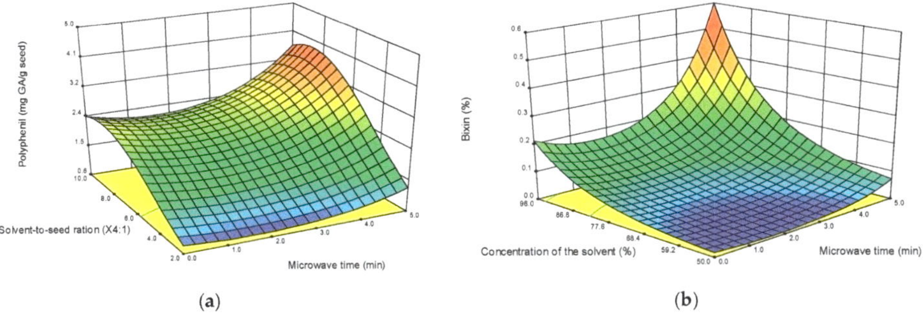

(**a**)　　　　(**b**)

Figure 2. Response surface plots of solvent concentration (ethanol) and treatment conditions for: (**a**) polyphenol compounds and (**b**) bixin.

The response surfaces plots depict how the application of microwaves accelerates the process of extraction of the bioactive compounds. When a longer treatment time was applied the yield of bioactive compounds was larger. Once both dependent variables were optimized (polyphenol compounds and bixin), the estimated conditions for the MAE along with their relative error were as listed in Table 4. The absolute bias for the MAE process was high, and thus the experimental results obtained were greater than those predicted by the models.

Table 4. Predicted local maximum for the optimization of the MAE applying a Box–Behnken experimental design.

pH	Solvent (Ethanol) (%)	Solvent-to-Seed Ratio (X_4:1)	Treatment Time (min)	Polyphenol (mg GA/g Seed)	Bixin (%)
7.00	96	5.95	5.00	2.69	0.58
	Experimental result			3.08 ± 0.01	0.58 ± 0.02
	Relative error			−0.39	0.00
	Absolute error (%)			14.41	0.55

Values are expressed as mean ± standard deviation ($n = 3$).

The absolute error compares the results predicted by the optimized model to the experimental ones obtained under optimal conditions. It indicates that the bixin extraction models using MAE are more precise than those obtained for the extraction of polyphenol compounds.

3.2. Effect of the MAE on the Antimicrobial and Antioxidant Activities of the Extract

The antioxidant and antimicrobial properties of the extracts obtained at the optimal operation conditions of the MAE were compared to those obtained by leaching. Table 5 shows larger antimicrobial and antioxidant activities for the MAE than those obtained by leaching. This is explained by the higher content of polyphenol and bixin compounds present in MAE.

Table 5. Comparison of the antioxidant and antimicrobial activities of annatto extracts obtained by MAE and leaching.

Extract		MAE	Leaching
Bixin	(%)	0.576 [b] ± 0.015	0.165 [a] ± 0.002
Polyphenols	(mg GA/g seed)	3.078 [b] ± 0.012	0.343 [a] ± 0.003
ABTS	(μM Trolox/L extract)	577 [b] ± 5	174 [a] ± 8
FRAP	(μM Trolox/L extract)	316 [b] ± 10	127 [a] ± 2
DPPH	(μM Trolox/L extract)	1043 [b] ± 50	811 [a] ± 5
Bacillus cereus	pH 11 (mg/L)	16 [a]	128 [b]
	pH 7 (mg/L)	16 [a]	128 [b]
	pH 4 (mg/L)	16 [a]	128 [b]
Staphylococcus aureus	pH 11 (mg/L)	8 [a]	32 [b]
	pH 7 (mg/L)	8 [a]	32 [b]
	pH 4 (mg/L)	8 [a]	32 [b]

Different superscript letters within row indicate significant differences ($p < 0.05$) according to LSD-Fisher; values are expressed as mean ± standard deviation ($n = 3$). ABTS: 2,2′-azino-bis(3-ethylbenzothiazoline-6-sulphonic acid), FRAP: Ferric Reducing Antioxidant Power, and DPPH: 2,2-Diphenyl-1-picrylhydrazyl

4. Discussion

The evaluation of the variables studied and optimized responses; microwave treatment time, pH, solvent concentration (%), and solvent-to-seed ratio (X_4:1), showed a significant effect ($p < 0.05$) on the extraction of the polyphenol compounds and bixin from annatto seeds. Table 2 shows the variability of the results obtained in each experimental unit, with results varying from 0.53 to 4.36 mg GA/g seeds for polyphenol compounds and from 0.03 to 0.58% for bixin. This variability is confirmed by the ANOVA (Table 3), showing that all the independent variables studied had a greater influence in the extraction process for both metabolites.

The influence of these conditions on the extraction process is mainly due to (1) the increase in the mass transfer capacity and/or the extraction power of the system, (2) the affinity of the bioactive compounds towards the solvent, and (3) the effect of the heat generated on the micro-domains of the system [25]. By increasing the solvent-to-seed ratio, the saturation point of the system is favored increasing the mobility of the seeds in the agitated system, generating a greater mass transfer of the bioactive compounds to solvent [25]. The amount of solvent in the system not only affected the extraction capability of the system, but also had a statistically significant interaction effect (X_3X_4) (Table 3), improving the affinity of the compounds towards the extractive solvent resulting in an enhanced mass transfer process. Polyphenol compounds present in annatto extract (i.e., apigenin, hypolaetin, and the caffeic acid derivative) possess a larger polarity as compared to bixin, which is a highly conjugated long chain molecule, also known for having a very low solubility. Therefore, increasing the concentration of ethanol favors the release of bixin by having a higher solvent content to form hydrogen bonds, and in the case of polyphenol compounds the extraction is favored by its polarity [3,26,27].

On the other hand, the effect of pH was not significant in its linear term for the extraction of the polyphenol compounds and bixin. This result is opposite to that of Rubio-Senent and collaborators, who demonstrated that the activity and solubility of polyphenol compounds and bixin were affected by pH. Further, bixin, is a carotenoid having a highly conjugated structure and a carboxyl end-group. As a result, its solubility at neutral and alkaline pH was limited. Conversely, the polyphenol compounds possess an aromatic ring structure which is attached to hydroxyl groups or other planar rings. These molecules are slightly acidic and their solubility and activity are maximized at neutral and slightly acidic pH [28]. The insignificance of pH in the system can be attributed to the powerful effect of other independent variables studied which overshadowed its effect.

The microwave treatment time was significant ($p < 0.05$) since it increased the extraction power for the polyphenol compounds and bixin. However, polyphenol compounds were more sensitive to this extraction process than bixin (Figure 2a) since once the maximum point of extraction was reached, the content of polyphenols steadily decreased. This is explained by the heat generated during the microwave treatment. Thus, the longer the treatment, the larger the temperature reached in the system, causing solvent volatilization along with a partial degradation of the polyphenol compounds. Some authors have reported that the bioactivity of these secondary metabolites are affected by the extraction mechanism and heat treatments [7,29]. On the other hand, the optimal extraction conditions were found when a pH of 7, solvent concentration of 96%, solvent-to-seed ratio of 6:1, and a microwave time of 5 min were in place. These conditions rendered the largest yield of polyphenol and bixin compounds. It is then worthy to mention that the antioxidant and antimicrobial activities of the optimized extract were probably due to its largest content of polyphenol and bixin compounds, which in turn were responsible for their antioxidant and antimicrobial activities. These results coincide with those reported by Vasu et al. in 2010, which extracted only bixin from *B. orellana* seeds by applying microwaves for a time of 18 min as compared to a leaching extraction for 80 min. As a result, the bixin extraction raised from 8.2% to 16.3%. They also claimed that MAE is an alternative green technology and superior to conventional extraction for the extraction of bixin [12].

On the one hand, the antioxidant activity increased for the extract obtained by MAE, with a 5-and 10-fold increase of bixin and polyphenol compounds, respectively, as compared to the extract obtained by leaching. Moreover, the extract obtained by MAE presented antimicrobial activity against *B. cereus* and *S. aureus*. There were statistically significant differences ($p < 0.05$) for both activities between treatments, where MAE presented greater in vitro activities thanks to the high contents of polyphenols and antioxidant compounds. However, a direct proportional association was not appreciated between the increase of the bioactive compounds and the increments of the antimicrobial and antioxidant activities. Thus, the antioxidant activity was the most affected variable by the extraction method. The results of the antimicrobial and antioxidant activities of the lyophilized extract are comparable with those reported by some authors [1,12,30].

5. Conclusions

The MAE technique increased the yield of polyphenol compounds and bixin extracted from the annatto seeds, reducing energy consumption and increasing the antioxidant capability of extracts obtained. Despite the fact that the active compounds had a larger yield as compared to leaching, the antimicrobial activity did not present significant differences.

Author Contributions: Conceptualization, J.Q.Q., G.C.G., and J.R.C.; Investigation, J.Q.Q., A.C.T., L.M.R., and M.S.G.; writing—Original draft preparation, J.Q.Q., A.C.T., L.M.R., and M.S.G.; Writing—Review and editing, G.C.G. and J.R.C.; Supervision, G.C.G., and J.R.C.; funding acquisition, J.Q.Q., G.C.G., and J.R.C.

Funding: This work was sponsored by Colciencias through the grant 727 of 2015 for the formation of Colombian PhDs and 715 for research and development projects in engineering-2015. The authors also thank the sustainability strategy of CODI 2018-2019.

Conflicts of Interest: The authors declare no conflict of interest.

References

1. Viuda, M.; Ciro, G.L.; Ruiz, Y.; Zapata, J.E.; Sendra, E.; Pérez-Álvarez, J.A.; Fernández-López, J. In vitro Antioxidant and Antibacterial Activities of Extracts from Annatto (*Bixa orellana* L.) Leaves and Seeds. *J. Food Saf.* **2012**, *32*, 399–406. [CrossRef]
2. Zhang, H.; Xue, L.; Li, B.; Tian, H.; Zhang, Z.; Tao, S. Therapeutic potential of bixin in PM2.5 particles-induced lung injury in an Nrf2-dependent manner. *Free Radic. Biol. Med.* **2018**, *126*, 166–176. [CrossRef] [PubMed]
3. Campos, R.; Yamashita, F.; Cesar, F.; Zerlotti, A. Simultaneous extraction and analysis by high performance liquid chromatography coupled to diode array and mass spectrometric detectors of bixin and phenolic compounds from annatto seeds. *J. Chromatogr. A* **2011**, *1218*, 57–63. [CrossRef]
4. Butnariu, M. Methods of Analysis (Extraction, Separation, Identification and Quantification) of Carotenoids from Natural Products. *J. Ecosyst. Ecography* **2016**, *6*, 1–19. [CrossRef]

5. Shahid-ul-Islam, L.J.; Mohammad, F. Phytochemistry, biological activities and potential of annatto in natural colorant production for industrial applications—A review. *J. Adv. Res.* **2016**, *7*, 499–514. [CrossRef] [PubMed]

6. Zhang, B.; Tomita, Y.; Qiu, Y.; He, J.; Morii, E.; Noguchi, S.; Aozasa, K. E74-like factor 2 regulates valosin-containing protein expression. *Biochem. Biophys. Res. Commun.* **2007**, *356*, 536–541. [CrossRef] [PubMed]

7. Yolmeh, M.; Habibi Najafi, M.B.; Farhoosh, R. Optimisation of ultrasound-assisted extraction of natural pigment from annatto seeds by response surface methodology (RSM). *Food Chem.* **2014**, *155*, 319–324. [CrossRef] [PubMed]

8. Lianfu, Z.; Zelong, L. Optimization and comparison of ultrasound/microwave assisted extraction (UMAE) and ultrasonic assisted extraction (UAE) of lycopene from tomatoes. *Ultrason. Sonochem.* **2008**, *15*, 731–737. [CrossRef]

9. Alcázar-Alay, S.C.; Osorio-Tobón, J.F.; Forster-Carneiro, T.; Meireles, M.A.A. Obtaining bixin from semi-defatted annatto seeds by a mechanical method and solvent extraction: Process integration and economic evaluation. *Food Res. Int.* **2017**, *99*, 393–402. [CrossRef]

10. Taham, T.; Silva, D.O.; Barrozo, M.A.S. Improvement of bixin extraction from annatto seeds using a screen-topped spouted bed. *Sep. Purif. Technol.* **2016**, *158*, 313–321. [CrossRef]

11. Rodrigues, L.M.; Alcázar-Alay, S.C.; Petenate, A.J.; Meireles, M.A.A. Bixin extraction from defatted annatto seeds. *C. R. Chim.* **2014**, *17*, 268–283. [CrossRef]

12. Soumya, V.; Venkatesh, P.; Kothandam, H.P. Microwave facilitated extraction of bixin from *Bixa orellana* and it's in-vitro antioxidant activity. *Der Pharm. Lett.* **2010**, *2*, 479–485.

13. Zou, T.B.; Jia, Q.; Li, H.W.; Wang, C.X.; Wu, H.F. Response surface methodology for ultrasound-assisted extraction of astaxanthin from *Haematococcus pluvialis*. *Mar. Drugs* **2013**, *11*, 1644–1655. [CrossRef] [PubMed]

14. Teng, H.; Jo, I.H.; Choi, Y.H. Optimization of ultrasonic-assisted extraction of phenolic compounds from chinese quince (*Chaenomeles sinensis*) by response surface methodology. *J. Appl. Biol. Chem.* **2010**, *53*, 618–625. [CrossRef]

15. Da Porto, C.; Porretto, E.; Decorti, D. Comparison of ultrasound-assisted extraction with conventional extraction methods of oil and polyphenols from grape (*Vitis vinifera* L.) seeds. *Ultrason. Sonochem.* **2013**, *20*, 1076–1080. [CrossRef] [PubMed]

16. Ahmed, M.; Akter, M.S.; Eun, J.B. Optimization conditions for anthocyanin and phenolic content extraction form purple sweet potato using response surface methodology. *Int. J. Food Sci. Nutr.* **2011**, *62*, 91–96. [CrossRef] [PubMed]

17. Ghafoor, K.; Choi, Y.H.; Jeon, J.Y.; Jo, I.H. Optimization of ultrasound-assisted extraction of phenolic compounds, antioxidants, and anthocyanins from grape (*Vitis vinifera*) seeds. *J. Agric. Food Chem.* **2009**, *57*, 4988–4994. [CrossRef]

18. Barbosa, M.; Borsarelli, C.; Mercadante, A. Light stability of spray-dried bixin encapsulated with different edible polysaccharide preparations. *Food Res. Int.* **2005**, *38*, 989–994. [CrossRef]

19. Swain, T.; Hillis, W.E. The phenolic constituents of *Prunus domestica*. I.—The quantitative analysis of phenolic constituents. *J. Sci. Food Agric.* **1959**, *10*, 63–68. [CrossRef]

20. FAO/WHO. *Compendium of Food Additive Specifications*; FAO: Rome, Italy, 2006; Volume 67, pp. 3–4, ISBN 9251055599.

21. Contreras-Calderón, J.; Calderón-Jaimes, L.; Guerra-Hernández, E.; García-Villanova, B. Antioxidant capacity, phenolic content and vitamin C in pulp, peel and seed from 24 exotic fruits from Colombia. *Food Res. Int.* **2011**, *44*, 2047–2053. [CrossRef]

22. Benzie, I.F.F.; Strain, J. The Ferric Reducing Ability of Plasma (FRAP) as a Measure of "Antioxidant Power": The FRAP Assay. *Anal. Biochem.* **1996**, *239*, 70–76. [CrossRef] [PubMed]

23. Nurhuda, H.H.; Maskat, M.Y.; Mamot, S.; Afiq, J.; Aminah, A. Effect of blanching on enzyme and antioxidant activities of rambutan (*Nephelium lappaceum*) peel. *Int. Food Res. J.* **2013**, *20*, 1725–1730.

24. Quanhong, L.; Caili, F. Application of response surface methodology for extraction optimization of germinant pumpkin seeds protein. *Food Chem.* **2005**, *92*, 701–706. [CrossRef]

25. Sinha, K.; Chowdhury, S.; Saha, P.D.; Datta, S. Modeling of microwave-assisted extraction of natural dye from seeds of *Bixa orellana* (Annatto) using response surface methodology (RSM) and artificial neural network (ANN). *Ind. Crops Prod.* **2013**, *41*, 165–171. [CrossRef]

26. Da Costa, C.L.S.; Chaves, M.H. Extração de pigmentos das sementes de *Bixa orellana* L.: Uma alternativa para disciplinas experimentais de química orgânica. *Quim. Nova* **2005**, *28*, 149–152. [CrossRef]
27. Novoa Vidal, A.; Motidome, M.; Mancini-filho, J.; Linares Fallarero, A.; Tanae, M.; Torres Brando, M.L.; Lapa, A.J. Actividad antioxidante y ácidos fenólicos del alga marina *Bryothamnion triquetrum* (S.G. Gmelim) Howe. *Rev. Bras. Cienc. Farm.* **2001**, *37*, 373–382.
28. Rubio-Senent, F.; Fernández-Bolaños, J.; García-Borrego, A.; Lama-Muñoz, A.; Rodríguez-Gutiérrez, G. Influence of pH on the antioxidant phenols solubilised from hydrothermally treated olive oil by-product (alperujo). *Food Chem.* **2017**, *219*, 339–345. [CrossRef] [PubMed]
29. Zhang, L.L.; Xu, M.; Wang, Y.M.; Wu, D.M.; Chen, J.H. Optimizing ultrasonic ellagic acid extraction conditions from infructescence of platycarya strobilacea using response surface methodology. *Molecules* **2010**, *15*, 7923–7932. [CrossRef]
30. Yolmeh, M.; Habibi-najafi, M.B.; Shakouri, S.; Hosseini, F.; Resource, N.; Branch, S. Comparing Antibacterial and Antioxidant Activity of Annatto Dye Extracted by Conventional and Ultrasound-Assisted Methods. *Zahedan J. Res. Med. Sci.* **2015**, *17*, 29–33. [CrossRef]

 © 2019 by the authors. Licensee MDPI, Basel, Switzerland. This article is an open access article distributed under the terms and conditions of the Creative Commons Attribution (CC BY) license (http://creativecommons.org/licenses/by/4.0/).

 *antioxidants*

Article

Far Infrared Irradiation Enhances Nutraceutical Compounds and Antioxidant Properties in *Angelica gigas* Nakai Powder

Md Obyedul Kalam Azad [1,2], Jing Pei Piao [1], Cheol Ho Park [1] and Dong Ha Cho [1,*]

1   College of Biomedical Science, Kangwon National University, Chuncheon 24341, Korea; azadokalam@gmail.com (M.O.K.A.); jinchwu@hotmail.com (J.P.P.); chpark@kangwon.ac.kr (C.H.P.)
2   Head of Research and Technology, Rentia Plant Factory, Chuncheon 24341, Korea
*   Correspondence: chodh@kangwon.ac.kr

Received: 14 November 2018; Accepted: 10 December 2018; Published: 11 December 2018

Abstract: The aim of this study was to investigate the effect of far infrared irradiation (FIR) on nutraceutical compounds, viz. total phenolic content, total flavonoids, and antioxidant capacity, of *Angelica gigas* Nakai (AGN). The FIR treatment was applied for 30 min with varied temperatures of 120, 140, 160, 180, 200, 220, and 240 °C. Results showed that FIR increased total phenolic and flavonoid content in AGN at 220 °C. The HPLC results revealed higher quantities of decursin (62.48 mg/g) and decursinol angelate (41.51 mg/g) at 220 °C compared to control (38.70 mg/g, 27.54 mg/g, respectively). The antioxidant capacity of AGN was also increased at 220 °C, as measured by 1,1-diphenyl-2-picrylhydrazyl (DPPH), ferric reducing antioxidant power (FRAP), and the phosphomolybdenum (PPMD) method. A further increase of the FIR temperature caused a reduction of compound content. In addition, the results also showed a strong correlation between phenolic content and antioxidant properties of AGN powder. These findings will help to further improve the nutraceutical profile of AGN powder by optimizing the FIR conditions.

Keywords: AGN; FIR; phenolic; flavonoid; antioxidant capacity

1. Introduction

Angelica gigas Nakai (AGN) is one of the most important traditional herbal medicinal plants in Korea. AGN possesses abundant nutraceutical properties, with strong antioxidant capacity [1]. The roots of AGN contain coumarin derivatives decursin (D) and decursinol angelate (DA) [2] which have several pharmacological properties such as anti-amnestic activity [3] anti-bacterial action [4], anti-allergic effect [5], and anti-tumor activity [6]. AGN is mainly used for the treatment of menopausal syndromes in Korea; therefore, it is called 'female ginseng' [7].

Plant bioactive compounds have a strong intermolecular covalent bond with large molecular weight. This complex molecular structure makes them less functional [8]. Many methods, including mechanical, chemical, and radiation approaches, are being employed to improve the plant antioxidant profiles through the stretching and bending of molecules. Among those, far infrared irradiation (FIR) is an efficient and convenient method to liberate the molecules from their complex crystalline structure and reduce intermolecular energy [9]. FIR is a subdivision of the electromagnetic spectrum which is used in the food processing industry to induced biological activities and prevent food quality degradation [10].

FIR has the capacity to transfer heat through molecular vibration to the center of the materials without degrading the constituent of the molecules. FIR cleaves covalent bonds and liberates antioxidants such as phenolic acids, flavonoids, tannins, and carotenoids [7,10].

It is reported that FIR increases the quantities of nutraceutical compounds, while improving antioxidant property, anti-inflammatory, and inhibitory activity in the A549 Cell line in *Chrysanthemum indicum* L. [11]. In particular, FIR enhances phenolic content and antioxidant capacity of buckwheat flour [12], sprouting bean flour [13], citrus cack [14], and rice hull [15]. However, to the best of our knowledge, there are no reports published on the effect of FIR on nutraceutical compounds in AGN. Therefore, the objective of this study was to induce nutraceutical compounds and increase antioxidant capacity through the application of FIR treatment in AGN powder.

2. Materials and Methods

2.1. Application of FIR Treatment and Preparation of AGN Extract

Angelica gigas Nakai (AGN) was purchased from Pyeongchang local market, Korea. AGN was dried in an oven at 50 °C for 24 h and powdered using a grinder. The powder was meshed using a 200 μm sieve to obtain a uniform particle size of the powder. Two grams of powder were mixed with 4 mL of water in a glass petri dish and exposed to an FIR dryer (HKD-10; Korea Energy Technology, Seoul, Korea) at 120, 140, 160, 180, 200, 220 and 240 °C for 30 min. Afterward, 1 g of the treated powder and control sample (without FIR application) was suspended in 100 mL of 80% ethanol and kept over-night in a shaker at room temperature. The extracts were filtered through Advantech 5B filter paper (Tokyo Roshi Kaisha Ltd., Saitama, Japan) and dried using a vacuum rotatory evaporator (EYLA N-1000, Tokyo, Japan) in 40 °C water bath to obtain crude extracts. The crude extracts were freeze-dried to achieve a moisture content of <3–5%. Dried crude extracts were diluted using 80% ethanol to prepare 1000 mg/L stock solution and kept at −20 °C for further analysis.

2.2. Estimation of Total Phenolic Content

Total phenolic (TP) content of FIR treated and control AGN samples were determined by the Folin Ciocalteu assay [16]. In brief, a sample aliquot of 1 mL of extract (1 mg/mL) was added to a test tube containing 0.2 mL of phenol reagent (1 N). The volume was increased by adding 1.8 mL of deionized water and the solution was vortexed and left for 3 min for reaction. Furthermore, 0.4 mL of Na_2CO_3 (10% in water, v/v) was added and the final volume (4 mL) was adjusted by adding 0.6 mL of deionized water. The absorbance was measured at 725 nm after incubation for 1 h at room temperature. The TP content was calculated from a calibration curve using gallic acid and expressed as mg of gallic acid equivalent (GAE) per g dry weight (dw).

2.3. Determination of Total Flavonoid Content

The total flavonoid content (TF) content was quantified according to Ghimeray et al. [12] with slight modifications. Shortly, a 0.5 mL aliquot of the sample (1 mg/mL) was mixed with 0.1 mL of 10% aluminum nitrate and 0.1 mL of potassium acetate (1 M) solution. To this mixture, 3.3 mL of distilled water was added to make the total volume of 4 mL. The mixture was vortexed and incubated for 40 min. The TF content was measured using a spectrophotometer (UV-1800 240 V, Shimadzu Corporation, Kyoto, Japan) at 415 nm. The TF content was expressed as mg/g coumarin equivalents on a dry weight basis.

2.4. HPLC Analysis of AGN Extract with or without FIR Treatments

An HPLC system (CBM-20A, Shimadzu Co, Ltd., Kyoto, Japan) with two gradient pump systems (LC-20AT, Shimadzu), a C18 column (Kinetex, 100 × 4.6 mm, 2.6 micron, Phenomenex), an auto-sample injector (SIL-20A, Shimadzu), a UV detector (SPD-10A, Shimadzu) and a column oven (35 °C, CTO-20A, Shimadzu) were used to analyze D and DA. Solvent A was water with 0.4% formic acid, and solvent B was acetonitrile. A gradient elution was used (0–15 min, 33–45% B; 15–30 min, 45–55% B; 30–40 min, 55–80% B; 40–45 min, 80–33% B). The flow rate was 1.0 mL/min, the injection volume was 10 μL, and the detection wavelength was 329 nm. The standard samples for the D and DA analysis were prepared at concentrations of 10, 20, 40, 60 and 80 µg/mL. As D and DA are the main active ingredients of the AGN therefore, they were separated and then quantified by HPLC.

2.5. Antioxidant Capacity Analysis

2.5.1. DPPH Free Radical Scavenging Capacity

The antioxidant capacity was determined on the basis of the scavenging activity of the stable 2, 2-diphenyl-1 picryl hydrazyl (DPPH) free radical according to methods described by Braca et al. [17]. The DPPH solution was prepared (5.914 mg of DPPH powder dissolved in 100 mL of methanol) to maintain an absorbance range of 1.1–1.3 by spectrophotometer. Briefly, 1 mL of stock solution was added to 3 mL of DPPH solution. The blank sample was prepared with 1 mL of distilled water instead of stock extract in 3 mL of DPPH solution. The mixture was shaken vigorously and left to stand at room temperature in the dark for 30 min. The absorbance was measured at 517 nm using a spectrophotometer (UV-1800 240 V, Shimadzu Corporation, Kyoto, Japan). The percent inhibition activities of the treated and control AGN samples were calculated against a blank sample using the following equation:

Inhibition (%) = [(blank sample − extract sample)/blank sample] × 100.

2.5.2. Ferric Reducing Antioxidant Power Assay (FRAP)

The reducing power of the samples was estimated according to the FRAP assay as described by Yu et al. [18]. In brief, 1 mL of stock solution (1 mg/mL) was mixed with 1 mL of 0.2 M phosphate buffer maintaining a pH of 6.6. The mixture was then incubated at 50 °C for 20 min. After incubation, 1 mL of trichloro-acetic acid (TCA) was added to the solution and centrifuged at 3000 rpm for 10 min. The collected supernatant was diluted with distilled water at 1:1 ratio. Finally, 0.25 mL of 0.1% ferric chloride was added and the absorbance was measured at 700 nm by a spectrophotometer.

2.5.3. Phosphomolybdenum Method (PPMD)

The total antioxidant capacity of AGN stock solution (1 mg/mL) was assayed according to the PPMD method described by Prieto et al. [19]. In brief, 1 mL of stock solution was added with 3 mL of 0.6 M sulfuric acid, 28 mM sodium phosphate and 4 mM ammonium molybdate solution. The reaction mixture was incubated at 95 °C for 150 min. The absorbance of the mixture was measured at 695 nm by spectrophotometer against a blank. The antioxidant capacity was expressed as the absorbance of the sample.

3. Statistical Analysis

All data were expressed as mean ± SD of triplicate measurements. The obtained results were compared among the different FIR temperature using a paired *t*-test in order to observe the significant differences at the level of 5%. The paired *t*-test between the mean values of the treated samples and control were analyzed by MINITAB (version 16.0).

4. Results and Discussion

4.1. Effect of FIR Irradiation on Total Phenolic and Flavonoid Contents of AGN Powder

Total phenolic and total flavonoid content of the FIR treated and control AGN was demonstrated in Table 1. It is shown that total phenolic (22.65 mg/g) and total flavonoid content (7.87 mg/g) was significantly increased at 220 °C of FIR treatment which is two times higher than of the control (12.75 mg/g, 2.51 mg/g, respectively). In the same way, it is shown in Table 2 and Figure 1 that the quantities of the main active ingredients decursin (62.48 mg/g) and decursinol angelate (41.51 mg/g) were about two and 1.5 times higher at 220 °C than the control (38.70 mg/g, 27.94 mg/g, respectively).

Table 1. Total phenolic (TP) and total flavonoid (TF) content of *Angelica gigas* Nakai treated by FIR irradiation.

FIR Treatment	TP (mg/g GAE dw)	TF (mg/g CUE dw)
Control	12.75 ± 0.25 e	2.51 ± 0.07 c
120 °C	13.19 ± 0.26 d	2.60 ± 0.05 c
140 °C	13.54 ± 0.54 d	2.73 ± 0.11 c
160 °C	14.03 ± 0.28 d	2.76 ± 0.08 c
180 °C	17.75 ± 0.53 c	3.94± 0.07 c
200 °C	20.63 ± 0.41 b	5.21 ± 0.15 b
220 °C	22.65 ± 0.22 a	7.87 ⊥ 0.14 a
240 °C	17.53 ± 0.35 c	5.55 ± 0.16 b

Each value is expressed as the mean ± SD (n = 3). Values labeled with different letters in a column are significantly different (p < 0.05). GAE: gallic acid equivalent; CUE: coumarin equivalent; dw: dry weight.

Table 2. HPLC quantification of decursin and decursinol angelate content of *A. gigas* Nakai treated by FIR irradiation.

FIR Treatment	Decursin (mg/g dw)	Decursinol Angelate (mg/g dw)
Control	38.70 ± 2.11 d	27.94 ± 1.28 c
120 °C	52.89 ± 1.89 c	32.05 ± 0.92 bc
140 °C	53.12 ± 1.55 b	32.14 ± 1.31 bc
160 °C	53.52 ± 1.05 b	32.28 ± 0.95 bc
180 °C	54.27 ± 1.59 b	33.51 ± 1.07 b
200 °C	55.01 ± 1.62 b	34.21 ± 0.99 b
220 °C	62.48 ± 2.38 a	41.51 ± 1.42 a
240 °C	56.32 ± 1.43 b	36.05 ± 1.32 b

Each value is expressed as the mean ± SD (n = 3). Values labeled with different letters in a column are significantly different (p < 0.05).

It is stated that FIR is magnificently applied in the drying of various food materials [20]. FIR creates internal heating via molecular vibration of the materials i.e., molecules absorb the radiation of certain wavelengths and energy, causing excited vibration [21]. FIR significantly induced plant secondary plant metabolites through the breakdown of a covalent bond of long-chain polymers [22]. Previous studies observed that FIR application increased phenolic content in soybean sprout powder [13], rice hull [15], ginseng, garlic, tomato, grapes, and onion [22]. In our study, the highest phenolic and flavonoid compounds, including D and DA, were attained at 220 °C of FIR treatment. Since D and DA have several pharmacological properties, therefore, FIR would be a suitable stimuli to enhance these compounds.

Previously, Ghimeray et al. [12] obtained the highest quantities of total phenols, total flavonoids, and quercetin content of buckwheat at 120 °C for 60 min of the FIR treatment. On the other hand, Azad et al. [13] observed the highest phenolic and isoflavonoid content in soybean sprout powder at 120 °C with an exposure time of 120 min. In our study, the bioactive compounds content decreased as the FIR temperature was increased. The FIR treatment at 220 °C found to be more adventurous in this study. However, Adak et al. [23] showed that the content of phenolic compounds and anthocyanins in strawberries was reduced when the FIR temperature was above 80 °C. Therefore, it might be assumed that the optimum temperature and exposure time of the FIR depends on the plant materials.

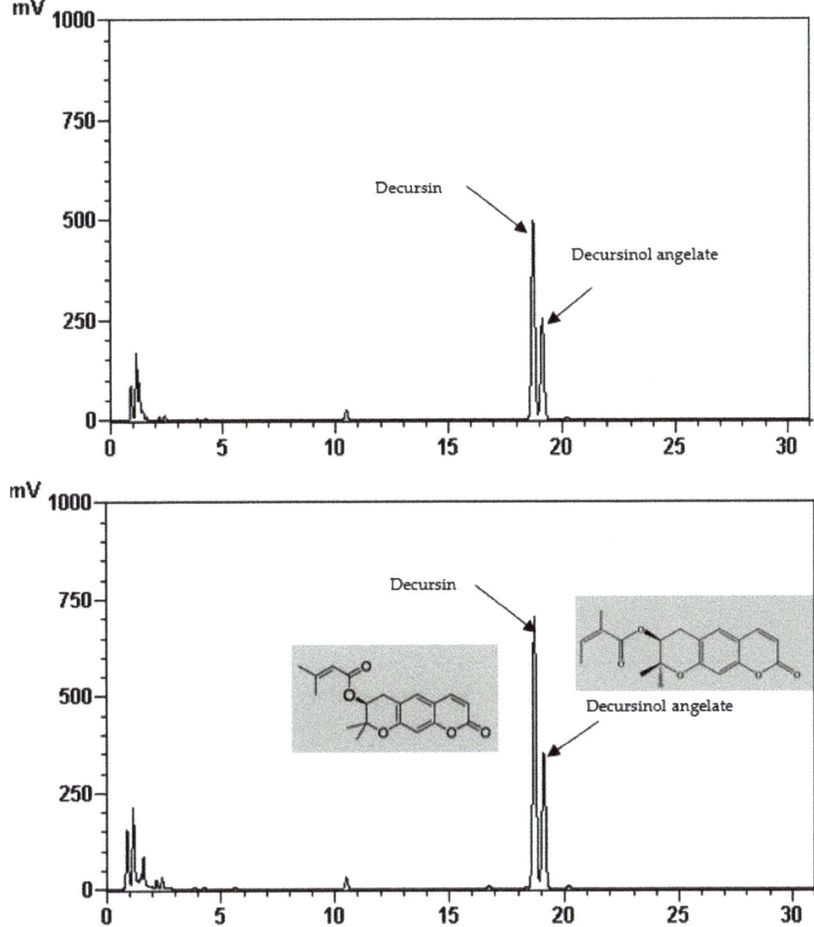

Figure 1. HPLC chromatogram of decursin and decursinol angelate of control (above) and treated AGN at 220 °C (down).

4.2. Effect of FIR Irradiation on Antioxidant Capacity of AGN Powder

Plant bioactive compounds have an enormous antioxidant capacity due to their redox properties which allow them to act as reducing agents, hydrogen donators, metal chelators and single oxygen quenchers [24].

The effect of FIR on the free radical antioxidant capacity of the AGN powder was measured by the DPPH, FRAP and PPMD method. DPPH is a stable free radical compound widely used to test the free radical scavenging ability of various materials [25]. FRAP measures the reducing potential of an antioxidant reacting with a ferric tripyridyltriazine complex, producing a colored ferrous tripyridyltriazine. PPMD is based on the reduction of Mo (VI) to Mo (V) by the sample analyte and the subsequent formation of a green phosphate/Mo(V) complex [19].

It was shown that the antioxidant capacity of the AGN was increased by FIR treatment. Antioxidant capacity was found to be two times higher at 220 °C (79%) of FIR compared to control (37%), according to the DPPH analysis. In the FRAP method, antioxidants cause the reduction of the Fe3+/ferricyanide complex to the ferrous form and activity is measured as the increase in the absorbance at 700 nm. In this assay, the yellow color of the test solution changes to various shades of green and blue depending on the reducing power of antioxidant samples [26]. In FRAP assay, the absorbance was three times higher at 220 °C of FIR treatment compared to control. On the other hand, the absorbance was four times higher in treated AGN at 220 °C compared to control measured by PPMD. The reducing power of the treated sample increased with increasing concentrations in a strongly linear manner. It is clearly shown that antioxidant properties of the AGN powder were increased at 220 °C FIR in all analytical methods i.e., DPPH, FRAP, and PPMD (Figures 2–4).

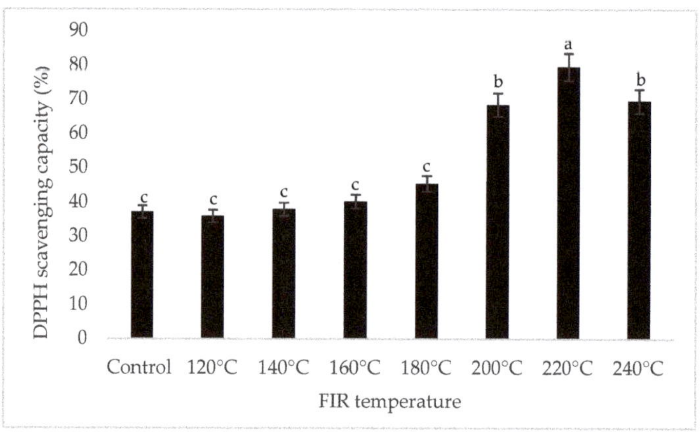

Figure 2. DPPH free radical scavenging activity of *A. gigas* Nakai treated by FIR irradiation. Each value is expressed as the mean ± SD (n = 3). Different lowercase letters within the row indicate significant differences (p <0.05) according to ANOVA. FIR: far infrared irradiation.

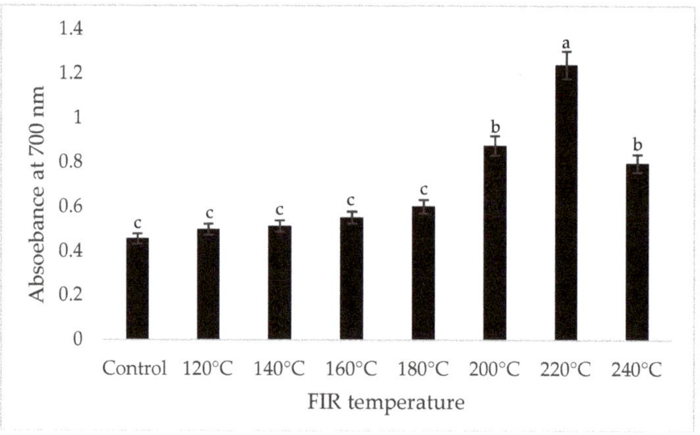

Figure 3. Reducing power of *A. gigas* Nakai treated by FIR irradiation (FRAP assay). Values are the mean ± SD (n = 3). Different lowercase letters within the row indicate significant differences (p <0.05) according to ANOVA. FIR: far infrared irradiation.

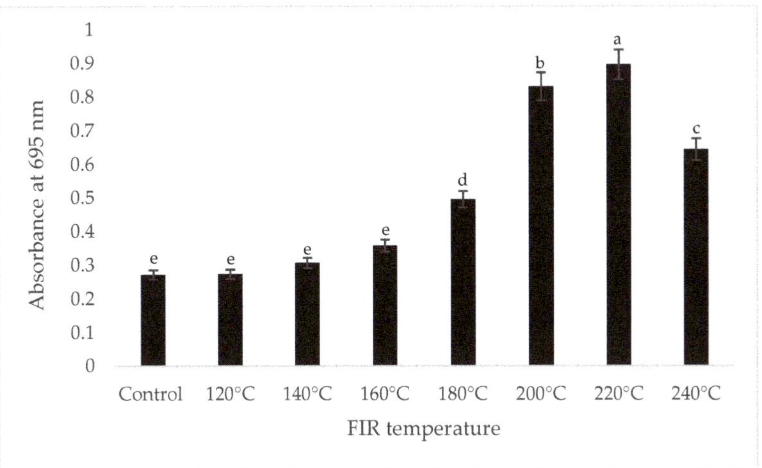

Figure 4. Total antioxidant capacity of *A. gigas* Nakai treated by FIR irradiation (PPMD assay). Values are the mean ± SD (*n* = 3). Different lowercase letters within the row indicate significant differences (*p* <0.05) according to ANOVA. FIR: far infrared irradiation.

Previous studies showed that FIR increased the antioxidant capacity of plant food materials [12,13,15]. It was also reported that total phenolic content and antioxidant capacity has a highly significant linear correlation [23]. The increases in antioxidant activity of treated samples is due to the increase of the total polyphenol and flavonoid compounds [27].

It is reported that D and DA are not directly correlated to the total antioxidant capacity in AGN; however, the increased content of D and DA improve the pharmacological profile [27]. Significant correlations between the content of total phenolic compounds and antioxidant capacity (R^2 = 0.852) were observed in this study (Figure 5), and are supported by the findings of Cai et al. [28] and Djeridane et al. [29].

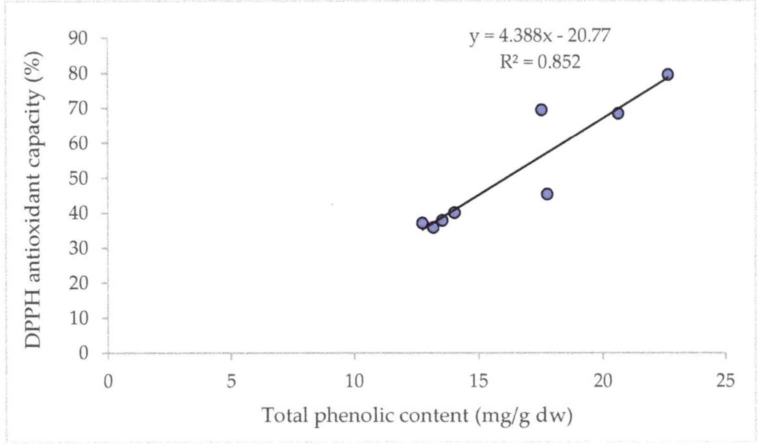

Figure 5. Linear regression between total phenolic content and antioxidant capacity (DPPH assay).

5. Conclusions

Our results have demonstrated that FIR should be considered as suitable stimuli to enhance nutraceutical compounds and preserved antioxidant properties in AGN powder. The optimum FIR condition obtained for AGN is at 220 °C for 30 min. A significant linear correlation was found between phenolic content and antioxidant properties of AGN. In the current study, only FIR temperature is optimized; however, duration of FIR treatment needs to be optimized in order to attain the highest content of nutraceutical properties from AGN powder.

Author Contributions: M.O.K.A., analyzed data and wrote the manuscript, J.P.P. performed the experiment, C.H.P. revised the manuscript and D.H.C. designed and supervised the study.

Funding: This study was supported by a research grant from Kangwon National University, 2016. (Grant No. 520160107).

Conflicts of Interest: The authors declare no conflict of interest.

References

1. Son, S.H.; Kim, M.J.; Chung, W.Y.; Son, J.A.; Kim, Y.S.; Kim, Y.C.; Kang, S.S.; Lee, S.K.; Park, K.K. Decursin and decursinol inhibit VEGF-induced angiogenesis by blocking the activation of extracellular signal-regulated kinase and c-Jun N terminal kinase. *Cancer Lett.* **2009**, *280*, 86–92. [CrossRef]

2. Ahn, K.S.; Sim, W.S.; Lee, I.K.; Seu, Y.B.; Kim, I.H. Decursinol angelate: A cytotoxic and protein kinase C activating agent from the root of *Angelica gigas*. *Planta Med.* **1997**, *63*, 360–361. [CrossRef] [PubMed]

3. Kim, D.H.; Jung, W.Y.; Park, S.J.; Kim, J.M.; Lee, S.J.; Kim, Y.C.; Ryu, J.H. Antiamnesic effect of ESP-102 on Ab1-42-induced memory impairment in mice. *Pharmacol. Biochem. Behav.* **2010**, *97*, 239–248. [CrossRef] [PubMed]

4. Lee, S.; Lee, Y.S.; Jung, S.H.; Shin, K.H.; Kim, B.K.; Kang, S.S. Anti-tumor activities of decursinol angelate and decursin from *Angelica gigas*. *Arch. Pharm. Res.* **2003**, *26*, 727–730. [CrossRef] [PubMed]

5. Joo, S.S.; Park, D.S.; Shin, S.H.; Jeon, J.H.; Kim, T.K.; Choi, Y.J.; Lee, S.H.; Kim, J.S.; Park, S.K.; Hwang, B.Y.; et al. Anti-allergic effects and mechanisms of action of the ethanolic extract of *Angelica gigas* in dinitrofluorobenzene induced inflammation models. *Environ. Toxicol. Pharmacol.* **2010**, *30*, 127–133. [CrossRef] [PubMed]

6. Ahn, Q.; Jeong, S.J.; Lee, H.J.; Kwon, H.Y.; Han, I.; Kim, H.S.; Lee, H.J.; Lee, E.O.; Ahn, K.S.; Jung, M.H.; et al. Inhibition of cyclooxygenase-2-dependent surviv in mediates decursin-induced apoptosis in human KBM-5myeloid leukemia cells. *Cancer Lett.* **2010**, *298*, 212–221. [CrossRef]

7. Sarker, S.; Naharl, D. Natural medicine: The genus Angelica. *Curr. Med. Chem.* **2004**, *11*, 1479–1500. [CrossRef] [PubMed]

8. Khoddami, A.; Wilkes, M.A.; Robert, T.H. Techniques for analysis of plant phenolic compounds. *Molecules* **2013**, *18*, 2328–2375. [CrossRef]

9. Peleg, H.; Naim, M.; Rouseff, R.L.; Zehavi, U. Distribution of bound and free phenolic acids in oranges (*Citrus sinensis*) and grapefruits (*Citrus paradise*). *J. Sci. Food Agric.* **1991**, *57*, 417–426. [CrossRef]

10. Krishnamurthy, K.; Khurana, K.; Jun, S.; Irudayaraj, J.; Demirci, A. Infrared heating in food processing: An overview. *Compr. Rev. Food Sci. Food Saf.* **2008**, *7*, 2–13. [CrossRef]

11. Kim, W.W.; Ghimeray, A.K.; Wu, J.C.; Eom, S.H.; Lee, B.G.; Kang, W.S.; Cho, D.H. Effect of far infrared drying on antioxidant property, anti-inflammatory activity, and inhibitory activity in A549 Cells of Gamguk (*Chrysanthemum indicum* L.) flower. *Food Sci. Biotechnol.* **2012**, *21*, 261–265. [CrossRef]

12. Ghimeray, A.K.; Sharma, P.; Hu, W.; Cheng, W.; Park, C.H.; Rho, H.S.; Cho, D.H. Far infrared assisted conversion of isoflavones and its effect on total phenolics and antioxidant activity in black soybean seed. *J. Med. Plants Res.* **2013**, *7*, 1129–1137.

13. Azad, M.O.K.; Kim, W.W.; Park, C.H.; Cho, D.H. Effect of artificial LED light and far infrared irradiation on phenolic compound, isoflavones and antioxidant capacity in soybean (*Glycine max* L.) sprout. *Foods.* **2018**, *7*, 174. [CrossRef] [PubMed]

14. Jeong, S.M.; Kim, S.Y.; Kim, D.R.; Jo, S.C.; Nam, K.C.; Ahn, D.U. Effect of heat treatment on antioxidant activity of citrus peels. *J. Agric. Food Chem.* **2004**, *52*, 3389–3393. [CrossRef] [PubMed]

15. Lee, S.; Kim, J.; Jeong, S.; Kim, D.; Ha, J.; Nam, K.; Ahn, D. Effect of far-infrared radiation on the antioxidant activity of rice hulls. *J. Agric. Food Chem.* **2003**, *51*, 4400–4403. [CrossRef] [PubMed]

16. Singleton, V.L.; Rossi, J.A. Colorimetry of total phenolics with phosphor molybdic-phosphotungstic acid reagents. *Am. J. Enol. Vitic.* **1965**, *16*, 144–458.

17. Braca, A.; Fico, G.; Morelli, I.; Simone, F.; Tome, F.; Tommasi, N. Antioxidant and free radical scavenging activity of flavonol glycosides from different Aconitum species. *J. Ethnopharmacol.* **2003**, *8*, 63–67. [CrossRef]

18. Yu, L.; Zhao, M.; Wang, J.S.; Cui, C.; Yang, B.; Jiang, Y.; Zhao, Q. Antioxidant, immunomodulatory and anti-breast cancer activities of phenolic extract from pine (*Pinusmassoniana Lamb*) bark. *Innov. Food Sci. Emerg. Technol.* **2008**, *9*, 122–128. [CrossRef]

19. Prieto, P.; Pineda, M.; Aguilar, M. Spectrophotometric quantitation of antioxidant capacity through the formation of a phosphomolybdenum complex: Specific application to the determination of vitamin E. *Anal. Biochem.* **1999**, *269*, 337–341. [CrossRef] [PubMed]

20. Sandu, C. Infrared radiative drying in food engineering: A process analysis. *Biotechnol. Prog.* **1986**, *2*, 109–119. [CrossRef] [PubMed]

21. Meeso, N. Far-infrared heating in paddy drying process. In *New Food Engineering Research Trends*; Urwaye, A.P., Ed.; Nova Science Publishers, Inc.: New York, NY, USA, 2008; pp. 225–256.

22. Eom, S.H.; Park, H.J.; Seo, D.W.; Kim, W.W.; Cho, D.H. Stimulating effects of far infra-red ray radiation on the release of antioxidative phenolics in grape berries. *Food Sci. Biotechnol.* **2009**, *18*, 362–366.

23. Adak, N.; Heybeli, N.; Ertekin, C. Infrared drying of strawberry. *Food Chem.* **2017**, *219*, 109–116. [CrossRef] [PubMed]

24. Piluzza, G.; Bullitta, S. Correlations between phenolic content and antioxidant properties in twenty-four plant species of traditional ethnoveterinary use in the Mediterranean area. *Pharm. Biol.* **2011**, *49*, 240–247. [CrossRef] [PubMed]

25. Sakanaka, S.; Tachibana, Y.; Okada, Y. Preparation and antioxidant properties of extracts of Japanese persimmon leaf tea (kakinohacha). *Food Chem.* **2005**, *89*, 569–575. [CrossRef]

26. Prasad, K.N.; Yang, B.; Dong, X.; Jiang, G.; Zhang, H.; Xie, H.; Jiang, Y. Flavonoid contents and antioxidant activities from *Cinnamomum* species. *Innov. Food Sci. Emerg. Technol.* **2009**, *10*, 627–632. [CrossRef]

27. Lee, K.K.; Oh, Y.C.; Cho, W.K.; Ma, J.Y. Antioxidant and anti-inflammatory activity determination of one hundred kinds of pure chemical compounds using offline and online screening HPLC assay. *Evid. Complement. Altern. Med.* **2015**, 165457. [CrossRef]

28. Cai, Y.Z.; Luo, Q.; Sun, M.; Corke, H. Antioxidant activity and phenolic compounds of 112 Chinese medicinal plants associated with anticancer. *Life Sci.* **2004**, *74*, 2157–2184. [CrossRef]

29. Djeridane, A.; Yousfi, M.; Nadjemi, B.; Boutassouna, D.; Stocker, P.; Vidal, N. Antioxidant activity of some Algerian medicinal plants extracts containing phenolic compounds. *Food Chem.* **2006**, *97*, 654–660. [CrossRef]

© 2018 by the authors. Licensee MDPI, Basel, Switzerland. This article is an open access article distributed under the terms and conditions of the Creative Commons Attribution (CC BY) license (http://creativecommons.org/licenses/by/4.0/).

MDPI

St. Alban-Anlage 66

4052 Basel

Switzerland

Tel. +41 61 683 77 34

Fax +41 61 302 89 18

www.mdpi.com

Antioxidants Editorial Office

E-mail: antioxidants@mdpi.com

www.mdpi.com/journal/antioxidants

www.ingramcontent.com/pod-product-compliance
Lightning Source LLC
LaVergne TN
LVHW070013070726
842759LV00012B/68